Ecology of Australian Freshwater Fishes

To our parents

Ralph and Kate Humphries

George and Anne Walker

Ecology of Australian Freshwater Fishes

Editors
Paul Humphries
Charles Sturt University
Keith Walker
The University of Adelaide

CSIRO PUBLISHING

National Library of Australia Cataloguing-in-Publication entry

Ecology of Australian freshwater fishes/edited by Paul Humphries and Keith Walker.

9780643097438 (hbk.)
9780643097445 (epdf)
9780643108301 (epub)

Includes bibliographical references and index.

Fishery management – Australia.
Freshwater fishes – Ecology – Australia.
Freshwater ecology – Australia.

Humphries, Paul.
Walker, K. F. (Keith Forbes)

333.9560994

Published by

CSIRO PUBLISHING
36 Gardiner Road, Clayton VIC 3168
Private Bag 10, Clayton South VIC 3169
Australia

Telephone: [+613] 9545 8555
Local call: 1300 788 000 (Australia only)
Fax: +61 3 9662 7555
Email: csiropublishing@csiro.au
Web site: www.publishing.csiro.au

Front cover: Spotted galaxias, *Galaxias truttaceus* (© Neil Armstrong)

Set in 10/13 Adobe Minion Pro and ITC Stone Sans
Edited by Adrienne de Kretser, Righting Writing
Cover and text design by James Kelly
Typeset by Desktop Concepts Pty Ltd, Melbourne
Index by Indexicana
Printed by Ingram Lightning Source

Mar26_RP_ILS

Foreword

In 30 years of field work in Australian inland waters, one of my most memorable field trips was a three-week trip with Adelaide fish ecologist, the late Jim Puckridge, sampling the Coongie Lakes of central Australia. Heavy rains some months earlier had flooded these arid-zone waterways and I recall one clear windless morning out on the boat with Jim hauling in multi-mesh nets teeming with callop (golden perch) to measure and release. The nets had been set for only a few minutes and already we were bringing them in with hundreds of fish. Pelicans bobbed around the boat, my arms ached from the exertion and my mind reeled at the immense productivity of fish in this newly filled lake in an expanse of rolling sand dunes.

At night, Jim would sit under the impossibly bright stars of the Australian desert sky and, in his quiet articulate way, describe his theories of how these fish used clusters of floods to recruit, how they survived the long dry periods, and what the likely food webs were that supported the immense densities of fish and waterbirds when the inland waters flooded. Inspired but physically drained, I would squirm into my sleeping bag and dream of silver streams of fish, rolling red dunes and sheets of blue water under a vast cloudless sky…

That was nearly 20 years ago. We have learned a lot more about the ecology of Australia's freshwater fishes since then. Tantalising theories have been proposed for how the fishes of inland waters cope with the variable conditions typical of these habitats. Meanwhile, there have been exciting finds of new species in northern tropical streams and a greater appreciation of how various threats such as river regulation and climate change are altering fish communities across the country. Until now, most of this literature has been scattered across scientific papers, consultancy reports, researchers' note books and anglers' observations.

Fortunately, Paul Humphries and Keith Walker decided to bring together our current knowledge of the ecology of Australian fishes in this timely book. These two scientists are well qualified for this undertaking because, between them, they have studied the aquatic ecology of rivers, lakes and estuaries for some sixty years, including some research on freshwater and estuarine fishes from Western Australia to the Murray-Darling Basin and Tasmania. They also have considerable depth in scientific editing (for example, Paul has edited works in historical ecology as well as fish and river ecology, and Keith was the Asia-Pacific Editor of the successful journal 'River Research and Applications' for 20 years), both are actively involved in science communication and community liaison, and both are experienced teachers.

In this book, leading fish ecologists have written chapters on topics ranging from fish evolution and biogeography through to their habitats, movement, reproduction and assemblage structure, much of it work that has only been done in the last few years. To me, the book emphasises how we can apply these findings to protect our native fish species but also reveals how much more there is yet to learn. Keen anglers will enjoy this book which will greatly enrich their future fishing experiences. Freshwater ecologists will find this a valuable trove of information, an effective synthesis of our current understanding of fish ecology, and an excellent source of ideas and hypotheses to test. General readers will savour the readable style, copious pictures and fascinating insights into the lives of fish of Australia's inland waters.

And I bet Jim would have enjoyed it too.

Andrew Boulton
Adjunct Professor, School of Environmental and Rural Science, University of New England, New South Wales

Contents

Contributors

Mark Adams Evolutionary Biology Unit, South Australian Museum, Adelaide, South Australia 5000

Angela H. Arthington Australian Rivers Institute, Griffith University, Queensland 4111

Stephen R. Balcombe Australian Rivers Institute, Griffith University, Queensland 4111

Nicholas R. Bond Australian Rivers Institute, Griffith University, Queensland 4111

Gerry P. Closs Department of Zoology, University of Otago, Dunedin 9054, New Zealand

David A. Crook Research Institute for the Environment and Livelihoods, Charles Darwin University, Northern Territory 0909

Peter C. Gehrke SMEC Australia, South Brisbane, Queensland 4101

Bronwyn M. Gillanders School of Earth and Environmental Sciences, The University of Adelaide, South Australia 5005

Michael P. Hammer Museum and Art Gallery of the Northern Territory, PO Box 4646, Darwin, Northern Territory 0801

John H. Harris Department of Environmental Management and Ecology, La Trobe University, Wodonga, Victoria 3689 (Mail: 568 Bootawa Rd, Tinonee, New South Wales 2430)

Jane M. Hughes Australian Rivers Institute, Griffith University, Queensland 4111

Paul Humphries Institute for Land, Water and Society, School of Environmental Sciences, Charles Sturt University, Albury, New South Wales 2640

Mark J. Kennard Australian Rivers Institute, Griffith University, Queensland 4111

Alison J. King Research Institute for the Environment and Livelihoods, Charles Darwin University, Northern Territory 0909

John D. Koehn Arthur Rylah Institute for Environmental Research, Department of Sustainability and Environment, Heidelberg, Victoria 3084

Mark Lintermans Institute for Applied Ecology, University of Canberra, Australian Capital Territory 2617

Nicole G. McCasker School of Environmental Sciences, Charles Sturt University, Albury, New South Wales 2640

Simon J. Nicol Oceanic Fisheries Programme, Secretariat of the Pacific Community, BP D5 – 98848 Noumea, New Caledonia

Bradley J. Pusey Australian Rivers Institute, Griffith University, Queensland 4111

Rick J. Stoffels CSIRO Land and Water, Murray-Darling Freshwater Research Centre, PO Box 991, Wodonga, Victoria 3689

Peter J. Unmack National Evolutionary Synthesis Center, Suite A200, 2024 W Main Street, Durham, North Carolina 27705-4667, USA

Keith F. Walker School of Earth and Environmental Sciences, The University of Adelaide, South Australia 5005 (Mail: PO Box 331, Yankalilla, South Australia 5203)

Qifeng Ye SARDI Aquatic Sciences, PO Box 120, Henley Beach, South Australia 5022

Preface

The science of Australian freshwater fishes is maturing quickly and is poised for rapid advances, yet the fauna is imperilled by environmental changes and alien species. We believe that now is opportune for a synthesis of many facts and ideas that are widely scattered in time and space, to take stock of what we know and expose what we do not, and to suggest future paths. There already are excellent guides to the fishes, and so many research papers and reviews of biogeography, ecology, physiology and taxonomy that there must be scope for a comprehensive text or treatise, but we have in mind the need to encourage more scientists into research, perhaps even research careers. For that reason, we imagined this book as an introduction.

The *Ecology of Australian Freshwater Fishes* is a primer, especially on ecological issues. It describes the origins, diversity and ecology of our fishes and the ways they have become adapted to some of the most challenging environments on planet Earth. It explains why we should learn from, but not completely rely upon, the perspectives of scientists from the Northern Hemisphere. What do we know of the origins, evolution and ecology of our continental fauna? How different are our species from fishes in other parts of the world? How have they responded to multiple invasions of alien species, and how might they respond to more? What is their likely fate in face of environmental changes driven by a hotter, drier climate and intensified competition for water? How can we conserve our fishes, given what we know? And again – what more do we need to know?

Our authors are 23 ecologists, all with Australian experience. Together, they are a broad cross-section of professional people. They work in universities, with natural resource management agencies or professional consultancies. Many are on governmental committees concerned with conservation and management of freshwater fishes, and on the editorial boards of journals, and all are well-known as speakers and authors. Many more managers and scientists could have contributed; we wish that there had been a way to involve them all, but it is some consolation that much of their work is acknowledged throughout.

We thank all of our authors for their contributions, commitment and patience with our editorial demands. A special thank-you to John Harris, who stepped in at a late stage, under pressure, to rally contributors and assemble missing material. The book might have been delayed but for John's dedication and professionalism.

This book could not have been realised without a host of other people involved in discussions, reading drafts, contributing data and reviewing chapters. To all, we are grateful.

Our reviewers included Mark Adams, Jim Barrett, Lee Baumgartner, Luciano Beheregaray, Christopher Burridge, Steven Cooke, John Craig, Keith Gido, Dean Jerry, Matt Jones, Bob Kearney, Hubert Keckeis, Sam Lake, Helen Larson, Bill Matthews, Tom Mommsen, Andy Moore, Dave Morgan, Peter Moyle, Tim Page, Peter Pridmore, Tom Rayner, Stuart Rowland, Zeb Tonkin and Brenton Zampatti.

Assistance with production of chapters, figures and tables came from Pheona Anderson, Danielle Armitage, Lee Baumgartner, Andrew Berghuis, Keith Bishop, Ben Broadhurst, Steven Brooks, Paul Brown, Gavin Butler, Andrew Carroll, Romain Causse, Stuart Chilcott, Rhian Clear, Keryn Cobden, Craig Copeland, Peter Davies, John Diggle, John Douglas, Cynthia Downs, Brendan Ebner, Iain Ellis, Peter Gibson, Arkellah Hall, Scott Hardie, Gabrielle Holder, Simon Hoyle, Max Humphries, Michael Hutchison, Peter Jackson, Stacey Kopf, Mark Lintermans, Rebecca Lloyd, Allan Lugg, Jarod Lyon, Jed Macdonald, David Makin, Martin Mallen-Cooper, Dale McNeil, Andy Moore, Dave Morgan, Tim Page, Bill Phillips, Janet Pritchard, Tarmo Raadik, Joe Roman, John Russell, Liesl Schiller, Günther Schmida, Bob Simpson, Danielle Stewart, Michelle Stoffels, Ivor Stuart, Ben Stewart-Koster, Charles Todd, Ian Wooden and Danny Wright.

We thank Charles Sturt University's Institute for Land, Water and Society for funding the inception workshop for this project, and for support in referencing and indexing. Greg Fry's advice and assistance with referencing was invaluable. Our thanks also to John

Manger, Tracey Millen, Deepa Travers and CSIRO Publishing for expert advice and help in production.

We are indebted to our colleague Andrew Boulton for encouragement and his gracious Foreword. We are grateful also to photographer Neil Armstrong for permission to use his image of a spotted galaxias as our cover image.

To our teachers and students, past and present, thank you for your keen interest and may this book go some way to repay you! Paul especially thanks Sam Lake for his enthusiasm, encouragement and support over many years. Keith would like to acknowledge an immense debt to WD (Bill) Williams AO (1936–2002), a pioneer in Australian limnology and a kindly mentor. Together, we wish to honour the inspirational contributions to fish biology by RM (Bob) McDowall (1939–2011).

Not least, we thank our families for their love, patience, indulgence and good humour during the gestation of this book.

Paul Humphries, Albury, New South Wales
Keith Walker, Tunkalilla, South Australia
September 2012

1 The ecology of Australian freshwater fishes: an introduction

Paul Humphries and Keith F. Walker

For those eager to know more of the fishes in the rivers, lakes and other wetlands of Australia, this book is a beginning. We expect that it will raise many questions, knowing that they could lead to discoveries and new understanding. With your curiosity aroused, we hope that you will be inspired to join the quest for discovery that unites our authors.

We begin by considering what it is about fishes that captivates people, and scientists in particular. Our focus then moves progressively to the origins of freshwater fishes and the nature of the Australian fauna, the relationships between people and fishes and the history of inquiry by explorers, naturalists, taxonomists, biologists and ecologists and, of course, fishers. With that, we will have paved the way for advice from a cross-section of the researchers and managers at the forefront of fish ecology in Australia.

MYSTIQUE

Fishes have always played a role in human economy and culture (Allen 1972; Balme 1995). Aside from their immense value as food, they feature in the history and folklore of many peoples, including indigenous Australians (Smith 1970) (Fig. 1.1). Images of fishes in art feature in a series of enlightening articles by Peter and Marilyn Moyle in *Environmental Biology of Fishes*, 1992–95.

Our fascination with fishes reflects their beauty, abundance, ubiquity and variety. Marine species include diminutive (8 mm) gobies and gargantuan (20 m) whale sharks, the grotesque blob-fish and the serenely beautiful leafy sea-dragons and wrasses. In fresh waters there is reputedly the smallest of all vertebrates, *Paedocypris progenetica* (Cyprinidae: 7.9 mm), and giants including the arapaima, Chinese paddlefish, giant Mekong catfish and the sturgeons. Other freshwater curiosities are archerfishes, desert pupfishes, electric eels, lungfishes, piranhas, salamanderfish and the upside-down catfish and vampire catfish. From desert springs to Antarctic shores, from tidal mud-flats to deep oceans and from underground rivers to waterfalls and mountain lakes, fishes occur in nearly all kinds of aquatic environments. An exceptional few venture on to land; others can take briefly to the air. Some are solitary; others are gregarious. Some move little; others cross the boundaries between rivers, lakes and the oceans.

Primal fears are aroused by fish such as the piranhas and sharks. There are literary tales such as Ernest Hemingway's *The Old Man and the Sea* (1952). There are also religious traditions such as fish in Buddhism

Figure 1.1: Arnhem Land Aboriginal rock painting depicting fish, probably barramundi (foreground and right) and fork-tailed catfish (right) (photo: B Wiedemann, Northern Territory Government Photo Slide Collection. Courtesy Northern Territory Library).

as a symbol of wakefulness – because they cannot close their eyes and appear not to sleep – and biblical stories, such as the parable of loaves and fishes (Mark 6: 30–44). Indeed, the symbol of Christianity is a fish, suggested by the Greek acronym ICHTHYS (ΙΧΘΥΣ) for 'Jesus Christ, the Son of God, our Saviour'.

Just as the oceans are a source of living and mineral wealth, fresh waters have health-giving and inspirational qualities. Consider, for example, the reverence that people have for the waters of Lourdes in France or the Ganges in India. Among countless spiritual associations, Henry David Thoreau wrote, 'Who hears the rippling of rivers will not utterly despair of anything' (*A Week on the Concord and Merrimack Rivers*, 1849). And from Mark Twain, 'The face of the water, in time, became a wonderful book – a book that was a dead language to the un-educated passenger, but which told its mind to me without reserve, delivering its most cherished secrets as clearly as if it uttered them without a voice. And it was not a book to be read once and thrown aside, for it had a new story to tell every day' (*Life on the Mississippi*, 1883).

Our reverence for fresh water reflects our dependence on it, and fishes, as denizens, are a tangible reminder. They are a source of protein, minerals and oils, and they offer culinary delights and sport for anglers. Recreational fishing is no less popular in Australia than in other parts of the world, drawing millions of dollars annually (Chs 9, 12), and anglers ensure that trout and other alien species are actively stocked in rivers, lakes and reservoirs. Ironically, millions of dollars are allocated also to protect native species and to contain the destructive effects of species such as common carp (*Cyprinus carpio*) (Ch. 11). Recreational fishing does much to promote management to improve angling, but research in pursuit of new knowledge receives much less support. Freshwater fisheries in Australia are funded poorly relative to their marine counterparts, reflecting their comparatively low economic value (Ch. 9).

FISHES IN SCIENCE

Origins

Many of the first fishes lived not in the sea, as you might suppose, but in fresh water. They were ostracoderms, jawless vertebrates with bodies encased by armour or scales; they arose in the Cambrian Period, 510 million years ago, and although they were extinct by the end of the Devonian, 350 million years ago, their descendants persist today as lampreys and hagfishes.

The first jawed fishes, the acanthodians, arose 410 million years ago and became extinct before the end of the Permian, 250 million years ago, but their legacy also persists because they share a common ancestor with modern fishes. The placoderms, another group of jawed fishes, reigned in marine and freshwater environments through the Devonian, and their descendants probably include the sharks and rays. The sharks and rays and bony fishes are structurally so different that they might have arisen independently from jawless ancestors.

The first bony fishes emerged 395 million years ago, and they too were of freshwater origin. From them arose the ray-finned fishes, which remain dominant today. Land vertebrates owe their origins not to the ray-finned fishes but to an allied group, the lobe-finned fishes, which persist as lungfishes and coelacanths (Clack 2009). In the Devonian, lobe-finned fishes crawled from the water to become amphibians, beginning another evolutionary saga.

The fishes are not a mere stepping-stone in vertebrate evolution. They have thrived for more than 500 million years and have evolved into the largest and most diverse group of vertebrates. As we will see, they are part of the fabric of virtually all aquatic

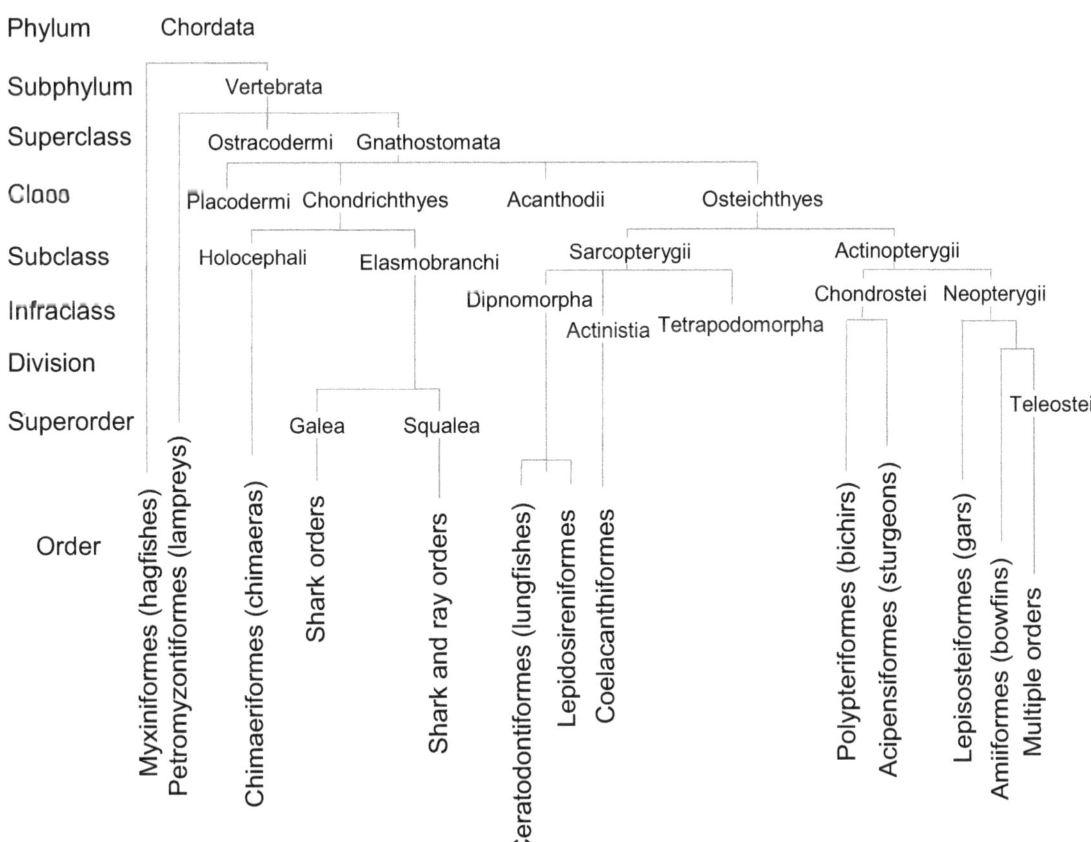

Figure 1.2: A phylogeny of fishes.

environments. To appreciate their significance, you might pause to consider what the world might be like if there were no fishes!

Modern fishes are in three taxonomic classes (Fig. 1.2): the Agnatha (a polyphyletic group including jawless lampreys and hagfishes), Chondrichthyes (cartilaginous fishes, including chimaeras, sharks, skates and rays) and Osteichthyes (bony fishes). Among the bony fishes are the Actinistia (coelacanths), Dipnomorpha (lungfishes) and Actinopterygii (ray-finned fishes). The ray-finned fishes contain the Chondrostei (bichirs, paddlefish, sturgeons) and Neopterygii, most of them among the Teleostei (the bony fishes, including herring, perch and salmon).

This is a bare outline and those seeking more detail will discover that, while there is broad agreement about major groups, there is no consensus over the names or affiliations of subordinate groups (Long 2010). For example, the Actinistia (formerly Crossopterygii) and Dipnomorpha (Dipnoi) may be grouped as Sarcopterygii. Fish phylogeny continues to evolve; ideas change and much remains for discovery.

The world's fish fauna includes more than 30 000 species, of which 40% live in fresh waters and a tiny proportion (< 1%) migrate to and from the sea. There are big differences in the numbers of freshwater species associated with the continents. Australia (with New Guinea) has about 500 species; in comparison, Europe has about 360 species, Africa and tropical Asia each have more than 3000 species, North America has more than 1000 species and South and Central America together may have more than 5000 species (Lundberg *et al.* 2000). There are greater differences still in the composition and diversity of the continental faunas (Ch. 2) and, despite many alien species introduced in the last 200 years, the Australian fauna retains much of its unique character (Ch. 11). We will return to this point shortly.

Ichthyology

The science of ichthyology (Greek *ichthys*, fish; *logos*, study) is as old as science itself. Its origins were more than two millennia ago, when Aristotle (384–322 BC), the father of natural history, described the form, function and relationships of some 115 species of fishes from the Aegean Sea (Günther 1880).

Considering the diversity, abundance and ubiquity of fishes and their cardinal roles in the structure and function of ecosystems, it is not surprising that ichthyology has featured strongly in the growth of the natural sciences. Disciplines like biogeography, physiology, behaviour and population ecology have all benefited from studies of fishes. And there have been many incidental benefits, as in the development of voltaic cells and other applications of electricity, stemming from studies of fish anatomy and physiology (Walker 1937; Kellaway 1946; Wu 1984).

Aristotle's ideas were not challenged until the 16th century, when science began to stir after centuries of torpor. The French naturalist Pierre Belon distinguished aquatic invertebrates and vertebrates and classified the latter based on size, skeletal structure, movement, number of limbs, morphology and habitat. His concept of 'cetaceans', while not unreasonable at the time, generously included whales and dolphins, amphibians, seals, hippopotamus, beavers, sharks and rays, sturgeons and marine bony fish and even mermaids. Two contemporaries, Hippolyte Salviani and Guillaume Rondelet, built on Belon's ideas, although they too were prone to inconsistencies in nomenclature and classification.

In the 17th century, Willem Piso and Georg Margrav made observations on Brazilian fishes, Giovanni Borelli described the ways that fishes swim and the function of the swim-bladder, Marcello Malpighi studied swordfish optic nerves, Jan Swammerdam investigated digestion and Joseph Du Verney studied respiration. British ichthyologists John Ray and Francis Willoughby separated fishes into those with bony skeletons and cartilaginous skeletons. They recognised that fishes possess blood, gills and a ventricle with only one chamber, and that some have scales and others do not. This meant that the whales, dolphins and porpoises clearly were not like other 'fishes' in Belon's taxonomy, although Ray and Willoughby lacked the confidence to make a complete division. Shortly afterward, Peter Artedi classified fishes as Malacopterygii, Acanthopterygii, Branchiostegi or Chondropterygii, using names that we recognise today. So sound was Artedi's work that fellow naturalist Carl Linnaeus needed only to rationalise the binomial nomenclature of fishes. Soon many more

Figure 1.3: Georges Cuvier, French anatomist, palaeontologist and administrator, and one of the world's most influential ichthyologists (National Library of Australia).

scientists, including Johann Reinhold Forster and his son Georg (see below), would stand on the shoulders of Artedi and Linnaeus.

The Renaissance was productive for science and philosophy, exemplified by Isaac Newton's contributions to physics and mathematics, the formation of scientific societies and studies of the nature of electricity and magnetism (and, incidentally, the ether and 'occult' forces). Ichthyology also moved apace. In the late 1700s, Marcus Bloch explored the classification of fishes, Bernard Germain de Lacépède produced his luminary *Histoire des Poissons*, anatomists and physiologists investigated how fishes function, and methods emerged for aquaculture of salmonids (trout and salmon). Perhaps it was Georges Cuvier, a French anatomist, palaeontologist and administrator, who contributed more than any other. He and student Achille Valenciennes produced 22 volumes of the monumental *Histoire Naturelle des Poissons* (Fig. 1.3) between 1828 and 1848 (despite Cuvier's death in 1832). This encouraged studies of comparative anatomy, evoking proto-evolutionary ideas from Étienne Geoffroy Saint-Hilaire, moving ichthyology to the frontier of science (Appel 1987) and laying foundations on which Charles Darwin and Alfred Russell Wallace later would build (Knapp 1999; Fagan 2007).

The 19th century brought more descriptions of regional faunas and seminal studies of fish biology. Pioneers included Richardson, Jenyns and Günther (Britain), Agassiz, Storer, Baird, Girard, Cope and Jordan (USA), Quoy and Gaimard, Lesson and Blanchard (France), Haeckel, Kner, Rüppell and Klunzinger (Germany), Steindachner (Austria) and others. The birth of the American Fisheries Society in 1870 and the American Society for Ichthyology and Herpetology in 1914, and a proliferation of journals devoted to fish and fisheries, were other milestones in the maturation of ichthyology as a science.

The 19th century also brought the first investigations of fishes in Australia. Here was a new-found continent where discoveries of bizarre animals and plants had already revolutionised global natural history. So different were the fishes and their environment, that 'natural history' prevailed and science faltered. It would be some years before Australian ichthyology was to earn its scientific credentials.

Natural history

The discipline of ecology, sometimes referred to as 'scientific natural history', is traceable to the works of Aristotle, Herodotus, Theophrastus and other Grecian scholars (Egerton 2001). Yet it took 2000 years to replace the Aristotelian view of nature as immutable (a view surely antithetical to ecology) with one where change was seen as an inevitable, intrinsic and necessary part of the natural world (Benson 2000). The enlightened thinkers of the Renaissance provided the stimulus, and expeditions by intrepid explorer-scientists, especially in the 19th century, were catalysts for rapid progress.

Alexander von Humboldt, the father of biogeography, opened the way for ecological ideas after an expedition to South America with Aimé Bonpland in 1799–1804 (von Humboldt 1995). He made new observations of freshwater fishes and their environment, including a treatise on electric eels (von Humboldt and Bonpland 1805). In his wake, Darwin, Wallace and Joseph Hooker developed proto-ecological perspectives, inquiring into the adaptations of animals and plants. This was a path to ideas about natural selection.

Other 19th century European scientists, including Ernst Haeckel (who coined *oecologie*, from the Greek *oikos*, for household or habitat), emphasised that the relationships of organisms with their environment were a product of evolution and subject to continuous change. Haeckel's so-called Law of Recapitulation ('ontogeny recapitulates phylogeny') (Haeckel 1876) was influential, likening the pharyngeal arches of embryonic mammals to those of fishes and suggesting a line of ascent.

In the early 20th century, more proto-ecological ideas were developed for plants by Oscar Drude and Eugenius Warming in Europe, and taken up by Frederic Clements and Henry Chandler Cowles (Benson 2000) in the USA. Clements' ideas about plant succession and climax communities were especially influential. Other contributions came from Charles Adams ('ecology as a marriage of evolution, history and taxonomy') and Victor Shelford ('ecology as a marriage of physiology and experimentation') (Ilerbaig 1999). Shelford, in particular, studied fishes in ponds and streams (Shelford 1911a, b, c). Charles Elton in the USA and later Arthur Tansley in Britain rejected Clements' 'community as an organism' concept, and Tansley advocated the now-familiar term 'ecosystem', combining biological, physical and chemical components of the natural world (Jax 1998).

Around this time, in the early 20th century, scientists rediscovered a paper that had been written by an American, Stephen Forbes, in 1887. In *The lake as a microcosm*, Forbes foreshadowed the ecosystem concept, writing of matter cycled in a system in dynamic equilibrium (Forbes 1878, 1887; Benson 2000). His ideas were taken further in Raymond Lindeman's classic paper of 1942, elaborating a concept of trophodynamics, wherein energy from the sun moves progressively through discrete trophic levels and matter is cycled (Lindeman 1942). Lindeman was a protégé of George Evelyn Hutchinson, doyen of aquatic ecologists and a scientist who helped to lay the foundations for modern, quantitative approaches to ecology.

By this time, ecology had achieved respectability and a new era was to unfold in the latter part of the 20th century. Freshwater ecology and fish ecology advanced in its company and studies of fishes provided important new ideas, as they had in Forbes' lake studies. We now turn to the history of freshwater ecology and fish and fisheries ecology, before sharpening our focus on the Australian fauna.

Freshwater ecology

After Lindeman, freshwater ecology continued to explore ecosystem perspectives. Scientists such as Howard Odum, John Teal and Ramon Margalef led the way, venturing energy budgets for lotic (flowing-water) ecosystems (Odum 1957; Teal 1957; Margalef 1960). In 1970, with publication of Noel Hynes' *The Ecology of Running Waters*, lotic ecology achieved a first synthesis, drawing together scattered fragments and anecdotes that often mark the beginnings of scientific disciplines. Hynes challenged older perceptions of streams as mere drainage channels occupied by loose assemblages of animals and plants. His work paved the way for the 'river continuum concept' and other influential ideas (see 'Fishes in freshwater ecosystems').

It was no coincidence that, around this time, new ideas emerged about the ecology of freshwater fishes. Books devoted to the ecology of tropical (Lowe-McConnell 1975; Welcomme 1979; Goulding 1981a) and temperate fishes (Carlander 1977; Matthews and Heins 1987) led the way, all of them 'required reading' for present-day researchers. In the years that followed, there were advances in understanding of trophic interactions and behaviour (Werner), the structure and dynamics of populations (Angermeier, Elliott, Grossman, Matthews, Rahel, Schlosser), competition and predator–prey relationships (Fausch, Fraser, Gilliam, Keast, Persson, Mittlebach), alien species (Moyle), life history strategies (Reznick), food webs (Carpenter, Kitchell, Penczak, Power, Winemiller) and diadromy (McDowall) and behaviour (Pitcher).

In the approach to the new millennium, there was a metaphorical explosion of authors and ideas in freshwater ecology. This defies a brief summary and, in any case, we wish to defer to the authors of following chapters. A highlight – perhaps a lowlight – of the foregoing list of leading researchers is that only two antipodean ecologists are named (R.M. McDowall, a New Zealander, and J. Allen Keast, an Australian). This does not deprecate the work of other antipodeans; rather, it shows the consequences of our historical emphasis on autecology and regional perspectives rather than on global ones. It was a motive for this book and it implies a challenge for us all.

Fish ecology

Fish ecology is where ichthyology, ecology, physiology and fisheries science intersect (Magnuson 1991). The fusion produced many formative papers and texts in the 1950s and 1960s, and it intensified in the late 20th century. This is a young science, despite its origins in natural history.

Fish ecology has benefited from an intimate association with commerce. Fisheries have existed for a millennium (Hoffmann 1996) – much longer if artisanal fisheries are included – and they have been a driver for research and management. Many early studies that we now regard as ecology, distinct from natural history or taxonomy, stemmed from fishery imperatives (Magnuson 1991). Forbes' work is an exception.

Freshwater fishes have attracted much less commercial attention than marine species, although fisheries for the lake sturgeon (*Acipenser fulvescens*: Peterson *et al.* 2007), various salmon species (Gustafson *et al.* 2006), Nile perch (*Lates niloticus*: Paterson and Chapman 2009) and even Murray cod (*Maccullochella peelii*: Rowland 1989) all have been significant in the last 150 years or so. Progress in management and research reflects the imbalance. This is less true of recreational fisheries, as brown trout (*Salmo trutta*) and rainbow trout (*Oncorhynchus mykiss*), for example, are among the most-studied of all fishes (e.g. Allen 1951).

The boundaries between freshwater and marine fishes and fisheries are arbitrary and by no means impermeable, and ecologists need to maintain a broad view. An early example of the cross-fertilisation of ideas is the work of Johan Hjort on marine fishes in northern Europe (Hjort 1914). Hjort realised that larval mortality was critical for the maintenance of populations, and raised fundamental questions about larval food resources and the influence of the environment. His work is still influential in both marine and freshwater science.

In the mid 20th century, Shelby Gerking's studies of the structure, dynamics and movements of stream fishes in the USA (e.g. Gerking 1949) were 'pure' ecology, unrelated to commercial enterprise. So too were the trophic studies of Allen Keast (e.g. Keast 1965). K. Radway Allen's comprehensive study of brown trout in the Horokiwi Stream, New Zealand (Allen 1951), combined ecology, fish and fisheries ecology. The first real 'fish ecology' text probably was *The Ecology of Fishes* by a Russian, George Nikolsky (Nikolsky 1963). Other works had applied perspectives, venturing ecological ideas from aquaculture (Dendy 1963; Magnuson 1991) and fisheries science (e.g. Rounsefell and Everhart 1953; Ricker 1954; Beverton and Holt 1957; Le Cren 1965; Elliott 1966; Cushing 1968; Bagenal 1969; Nikolskii 1969). Fisheries science has flourished since, with a proliferation of specialised journals (*Transactions of the American Fisheries Society, North American Journal of Fisheries Management, Fisheries, Fisheries Management and Ecology, Journal of Fish Biology, Fisheries Research, Canadian Journal of Fisheries and Aquatic Sciences, Reviews in Fish and Fisheries, Fisheries and Oceans, Fishery Bulletin* and others). Knowledge of environmental factors affecting the distribution, growth, disease, feeding, spawning, early life history, recruitment, movement and migration, behaviour and population dynamics of fishes has grown exponentially. Few fisheries are managed in an exemplary manner, however, probably not for lack of knowledge but because political and economic imperatives often override those of science (Chs 9, 12).

In the last decade or so, important texts have appeared, including those by Lucas *et al.* (2001), Fuiman and Werner (2002), Hart and Reynolds (2002), Sloman *et al.* (2006), Berra (2007), Dufour *et al.* (2008), Burton (2010), Helfman *et al.* (2009) and deserved classics such as those by Matthews (1998) and Wootton (1998). You should bookmark these as essential reading.

We turn now from global issues to sharpen our focus. To begin, we consider the distinctive nature of the Australian fauna and, by implication, problems that may result from viewing our fauna solely from a Northern Hemisphere perspective.

THE AUSTRALIAN FAUNA

Ghosts of Gondwana

Australia's native freshwater fishes are survivors of an evolutionary odyssey that began in the Cambrian, about 510 million years ago, when landmasses were joined as Pangaea. That union was undone during the Jurassic, 150–200 million years ago, freeing Gondwana and Laurasia as antecedents of modern landmasses. Gondwana included Antarctica, Africa, Madagascar, South America, Australia, New Guinea and New Zealand, plus Arabia and India, although the latter were destined to drift northward and join Eurasia.

Gondwana itself began to fragment in the Jurassic, about 167 million years ago, when Antarctica, Madagascar, India and Australia broke from Africa and South America. India separated from Antarctica 120 million years ago, and New Zealand separated 85–130 million years ago. Australia first began to sever its Antarctic connection 80 million years ago, in the Late Cretaceous, and continued to drift through the Eocene, 40 million years ago.

Plate tectonics is an active field of geophysical research, of course, and there are many unresolved issues. The biogeographic implications are contentious in some respects (e.g. Upchurch 2008), but the big picture is clear, and part of a paradigm that shapes our understanding of the history of the global fauna and flora.

The founding fishes were confined to an island continent that offered little security for any organisms in lakes and rivers. Amid long periods of aridity and changing sea levels, nearly all of the Gondwanan stocks perished, and there was no access for freshwater fishes that were to evolve later in other parts of the world. As with most other groups of fauna and flora, Australia's fishes evolved in relative isolation. The modern native freshwater fish fauna includes about 256 species (in Australia, excluding New Guinea: Allen *et al.* 2002), but freshwater groups that dominate other continental faunas, including the Centrarchidae, Characidae, Cichlidae, Cobitidae, Cyprinidae, Cyprinodontidae, Percidae, Poeciliidae and Salmonidae are not among them. That has changed, of course, in the last 200 years (Ch. 11).

Among the native fishes, only three species of Gondwanan origin were able to endure the challenges of geology, climate and evolution. Two are species of *Scleropages*, including the southern saratoga (*S. leichhardti*) from the Fitzroy system in northern Queensland and the northern saratoga (*S. jardinii*), from rivers flowing to the Gulf of Carpentaria and in southern Papua New Guinea. These are osteoglossids (arowanas, or bony tongues), among the Osteichthyes, with modern relatives in Africa, Asia and South America (Allen *et al.* 2002) (Fig. 1.2). The Australian and Asian lineages diverged during the Early Cretaceous, 140 million years ago (Kumazawa and Nishida 2000). The third Gondwanan refugee is the Australian lungfish (*Neoceratodus forsteri*), now confined to rivers in south-eastern Queensland (Frentiu *et al.* 2001). With allies in Africa and South America, this is one of the Dipnomorpha. Modern members of this group are united by their ability to breathe air, and by their origin in the Devonian, 380 million years ago. *Neoceratodus* today is little different from fossils more than 100 million years old (Allen *et al.* 2002).

What little is known of the history of the ancient Australian fish fauna is revealed, piecemeal, by fossils discovered at places such as Canowindra, Gogo, Grenfell, Korumburra, Riversleigh and Talbragar (Johnson 2004; Long *et al.* 2006; Young 2006). This is a fascinating story, but a diversion from a more pressing question. Aside from the three living fossils, where did the modern fishes come from?

In fact, most fish species in Australian inland waters have evolved from marine and estuarine families, but not necessarily in recent evolutionary time, as some older literature suggests. Those families may not be uniquely Australian but their freshwater species do show a high level of endemism – about 90% of our native freshwater species occur nowhere else (Allen *et al.* 2002). Not surprisingly, the distributions of families and species do not obviously reflect Gondwanan origins (McDowall 2002). Moreover, some of the species have polyphyletic origins and it is not possible to describe a simple line of ascent. The fauna includes some 55 species of galaxioids (Aplochitonidae, Galaxiidae, Retropinnidae, Lepidogalaxiidae), four species of lampreys (Geotriidae, Mordaciidae), a few taxa with tropical and subtropical Asia-Pacific affinities and perhaps 10 species of Percichthyidae (McDowall 2002). The galaxioids are dominant, although many are threatened by environmental changes and the depredations of alien salmonids, especially trout (McDowall 2006; Ch. 11). The percichthyids are another important group and surprisingly little known, although researchers are making rapid inroads using molecular markers (Moore *et al.* 2010; Chs 2, 3). Indeed, genetic studies of percichthyids such as golden perch (*Macquaria ambigua*: Faulks *et al.* 2010b; Macquarie perch (*M. australasica*: Faulks *et al.* 2010a, 2011) and other, non-percichthyid species (e.g. Jerry 2008) will surely provide new insights into the biogeography and evolution of the native fishes.

Changing environments

The exigencies of life in Australian inland waters, over many eons, have ensured that most of the modern native fishes are hardy, opportunistic, often highly mobile species with life cycles attuned to an erratic climate. They are able to contend with unpredictable flow regimes, including all magnitudes of floods and droughts, and fluctuations of oxygen, salinity, temperature and other indicators of water quality (e.g. Boulton and Brock 1999).

It is not the present climate that has shaped the fish fauna, of course, but palaeoclimates that have changed repeatedly, both gradually and abruptly, over evolutionary time. Rainfall, hence stream flow, in Australia has been highly variable for thousands of years (Ayliffe *et al.* 1998) and the habitats available to fishes have ranged from dry through moist to wet. The south-eastern part of the continent has offered most security for aquatic organisms and, not surprisingly, it harbours more than half of the present native fish fauna (McDowall 2002).

We might suppose that, as evolutionary products of a changeable environment, the native fishes are predisposed to cope well with the changes visited on them in the last 200 years. Yet these challenges are unprecedented in their nature, scale, diversity and intensity. The native fishes must now contend with alien fishes such as common carp, eastern gambusia (*Gambusia holbrooki*) and trout, alien organisms other than fish, diseases and parasites and even alien chemicals, such as pesticides. There have been changes in the physical environment, such as dams, weirs, levees and diversions, and changes in the quantity and quality of runoff through land clearing, eutrophication, salinisation and other by-products of agricultural and urban development. Above all, the pace of change may be so rapid as to defy adaptations through physiology, ecology and evolution.

In the aftermath of these changes, many species of native fishes have declined sharply in range and abundance (Cadwallader 1978; Gehrke and Harris 2001). Today, in the Murray-Darling Basin, one-quarter of all fish species are aliens and, among them, the common carp accounts for nearly 60% of the total fish biomass (Davies *et al.* 2010; Ch. 11). The environment has changed irrevocably and the consequences may be even more profound than they were for the disintegration of Gondwana.

ENCOUNTERS WITH AUSTRALIAN FISHES

Indigenous people

The Dreamtime exploits of a spiritual messenger called Ngurunderi are part of the traditional lore of Aboriginal tribes who lived along the Murray River (e.g. Ngarrindjeri-Tendi 2006). One tale recounts a time when Ngurunderi encountered Pondi, the giant Murray cod, and gave chase. In his flight, Pondi gouged the channel of the river and his thrashing tail carved billabongs and cliffs along the banks. Eventually, in Lake Alexandrina, Pondi was speared and parts of his body were thrown back to become the other fishes of the river. Pondi himself remained, but as a vestige of his former size. Another tale, from the Ngarrindjeri people of the lower Murray, regarded the bones in *thukeri* (the unpalatable bony herring, *Nematalosa erebi*) as a punishment imposed by Ngurunderi on two men who refused to share their catch. These stories have been retold many times and, while details vary, they eloquently bind the birth of the river and its fish fauna.

Fishes were a staple for tribes living near rivers, lakes and seashores throughout Australia and indigenous people would have had an immense knowledge of the various species (Humphries 2007). The writings of colonial Europeans preserve some fragments of information (e.g. Scott 2005) but much has been lost because, among those first fishers, knowledge was handed down by word of mouth rather than pen and paper and because most were displaced from their homelands and traditional lives.

Aside from early journals and letters, the names for the fishes and other knowledge passed on by Aboriginal people, there is evidence of sophisticated fish-catching technology in the form of hooks, spears, nets and fish traps and the use of poisons (e.g. Humphries 2007). There are remains of ancient stone fish traps at Brewarrina on the Barwon River, where stones were arranged to shepherd fish into enclosures where they could easily be captured. Along river banks and lake shores throughout Australia there are middens containing the shells of freshwater mussels, the gastroliths of crayfish and the otoliths and other bones of fish. Indeed, middens like these are associated with the earliest evidence of Aboriginal occupation, at places such as Lake Mungo in New South Wales (Bowler *et al.* 1970). The midden materials suggest that, over time, the pressure of exploitation, especially of Murray cod and golden perch, was sufficient to cause changes in the average sizes of individuals (Balme 1995) and there is similar evidence for mussels (Walker 1981).

European discovery

Europeans began to explore the southern oceans from the early 1600s and many parties included a naturalist or a ship's surgeon with a part-time interest in fauna and flora. William Dampier was one of the first to record animals and plants, during his travels along the coast of Western Australia (Whitley 1964, 1970; Finney 1984). Fishes often excited attention and some beautiful drawings have been preserved. Specimens returned to Europe prompted taxonomic descriptions (Paxton and McGrouther 1997), often by people who had neither seen the fish alive nor set foot in Australia. Their legacy of sometimes spurious associations and fanciful names has proven very resistant to change (see 'Common names').

Two of the earliest maritime expeditions were by James Cook, mapping part of the eastern coast of Australia and paving the way for later British colonisation. On the first expedition, plant and animal specimens were gathered by Joseph Banks and Daniel Solander, a student of Carl Linnaeus, but freshwater fishes were overlooked. Cook's second expedition, with naturalists Johann Reinhold Forster and son Georg on board, did record freshwater fishes, but from New Zealand rather than Australia (Humphries 2003b). It was Forster who first introduced Europeans to the giant kokopu (*Galaxias argenteus*), a galaxiid from the South Island.

Around the beginning of the 19th century, François Péron and Charles-Alexandre Lesueur collected two galaxiids (spotted galaxias, *G. truttaceus* and common galaxias, *G. maculatus*, from Tasmania) during travels with French explorer Nicolas Baudin in 1798–1801 (Humphries 2003b). That expedition returned to France with tens of thousands of specimens, including many marine fishes but only those two freshwater species. The spotted galaxias was described by Cuvier and Valenciennes (Fig. 1.4) but the common galaxias was to languish in obscurity for some years. Eventually, in 1842, it was described by English clergyman Leonard Jenyns, but from a specimen collected on Tierra del Fuego by Charles Darwin rather than the one collected earlier by Péron.

One reason for the comparative neglect of freshwater fishes was that few early naturalists would venture far inland, fearful of unmapped territory, hostile natives, a harsh climate and lack of water and no doubt discouraged by the reluctance of naval commanders to provide resources for collecting (Hoare 1982; Duyker 2006). In time, these impediments were overcome. Péron, René-Primavère Lesson on Louis Isadore Duperry's expedition of 1822–25, John Richardson from James Clark Ross' expedition of 1839–43 and, famously, Charles Darwin on board *The Beagle*, on Robert Fitzroy's expedition of 1831–36, all added freshwater fish specimens to the cabinets of royalty and museums in Europe.

Land exploration began, in earnest, in the early 1800s. George Evans' crossing of the Blue Mountains in 1813 was in the vanguard of forays by Europeans into outback Australia. Evans named the Macquarie and Fish rivers, and remarked on the abundance of large fish that probably were Murray cod (Mackaness 1965). Soon after, John Oxley explored the Lachlan and Macquarie rivers, where fishes again were plentiful, a mainstay for local indigenous people and welcome food for the exploring party (Oxley 1820). Hamilton Hume and William Hovell crossed the Murray River near Albury in 1824 and reported that 'the river abounds with that species of cod fish which is common in all the western rivers. In the lagoons they [indigenous people] caught a kind of bream or carp, of the weight of about two pounds, and of the finest flavour' (Andrews 1981).

Figure 1.4: Spotted galaxias, *Galaxias truttaceus*. (a) The type specimen from the Muséum National d'Histoire Naturelle, Paris (MNHM). (b) Charles-Alexandre Lesueur's pencil drawing (Muséum d'histoire naturelle, Le Havre). (c) The type specimen from Cuvier and Valenciennes' *Histoire Naturelle des Poissons*, 1836.

Surveyor Charles Sturt followed the Macquarie River to the Darling River in 1828–29 and in 1830 he travelled down the Murrumbidgee to the Murray and Lake Alexandrina, making occasional notes on fishes and their use by indigenous people. Others lavished praise on the culinary qualities of Murray cod but Sturt and his men were unimpressed, despite their stores spoiling: 'the only fish we could take was the common cod or perch; and, without sauce or butter, it is insipid enough' (Sturt 1982, p. 113).

It was not until 1838, when surveyor and naturalist Thomas Mitchell had completed three expeditions, that descriptions were published of Murray cod, silver perch (*Bidyanus bidyanus*) and freshwater catfish (*Tandanus tandanus*: Mitchell 1839). Other explorers, including Cunningham, Forrest, Giles, Gregory and Eyre, traversed the continent from the 1840s to 1880s; they noted the freshwater fishes they found but left us with anecdotes and asides rather than scientific data. One salient encounter was between Ludwig Leichhardt and his party with southern saratoga (*Scleropages leichhardti*) splashing noisily in the river near his camp (Leichhardt 1847). Appropriately enough, that species now bears his name.

Common names

The first Europeans to encounter *Terra Australis* and its fauna and flora referred to native pines, bears, cats and tigers and trout, when these bore only a fanciful resemblance to their northern namesakes. The animals and plants were different, sometimes bizarre; the indigenous names seemed incomprehensible, conveyed no sense of taxonomy and varied between regions.

For some animals, indigenous names have been recalled and promoted to common use. For others, not least the native fishes, confusion remains. The 'Murray cod' is a percichthyid and not a member of the true cods (Gadidae). The 'golden perch' also is a percichthyid and not a true perch (Percidae). The 'freshwater catfish' is unrelated to the fork-tailed catfishes (Ariidae). If we were to choose from original names, we might refer to 'pondi', 'callop' and 'tandan', respectively. We avoid this confusion by using the binomial nomenclature introduced by Linnaeus. There is no ambiguity about *Maccullochella peelii*, *Macquaria ambigua* and *Tandanus tandanus* and the latter, at least, preserves an original name, thanks in part to the nomenclator, Thomas Mitchell (thus, *Tandanus tandanus* Mitchell, 1839).

There is little chance that names so ingrained as 'Murray cod' will soon be replaced by 'pondi' or 'goodoo', despite some rationalisations (for example, the older name 'bony bream' has been overtaken by the more accurate 'bony herring'). We should not place too much literal trust in names such as 'Murray cod', 'sleepy cod' (*Oxyeleotris lineolata*) or 'trout cod' (*Maccullochella macquariensis*) and we should not be led to think that the native 'silver perch' (Terapontidae) is close kin to the alien redfin perch (*Perca fluviatilis*: Percidae).

Despite progress towards standardised common names, there is no national standard. As a result, there is scope for confusion aplenty in everyday use and no less in the older scientific literature. For reference, we have appended a list of common names that respects most recent opinions (see Appendix).

In passing, an interesting illustration of muddled common names is that the indigenous name 'barramundi' applied originally to the saratoga (*Scleropages* spp.) but was appropriated for 'marketing purposes' by the Queensland Fisheries Service (<http://www.dpi.qld.gov.au>) in the 1980s. It supplanted other, less-appealing names ('Asian sea bass', 'giant perch') and 'barramundi' now refers exclusively to that species (*Lates calcarifer*). It is popularly believed that the name 'saratoga' is a corruption of 'ceratodus', a superseded name for the Australian lungfish (*Neoceratodus forsteri*) and the two forms do bear a superficial resemblance to one another (e.g. large scales, large pectoral fins, roughly similar body shape).

Classification

The taxonomy of Australian freshwater fishes remained a domain for European naturalists until the early 1900s (Fig. 1.5). Natural history was in vogue and collecting, sometimes likened to stamp collecting, was haphazard and undisciplined (Whitley 1964, 1970; Finney 1993). Much collecting was undertaken by private collectors on behalf of naturalists looking for new species to describe. François Louis de la Porte, Comte de Castelnau and French Consul-General in Melbourne in 1864–77, fulminated 'not one of the Australian Colonies has a particular work on one single branch of its zoology, whereas every State of North America has a complete series of valuable works on each branch of that science' (Castelnau 1872, pp. 31–32).

Castelnau rightly observed that knowledge of Australia's freshwater fishes, in particular, was extremely poor, although 20–30 freshwater species had been described formally by that time. From Victoria had come a short essay by Frederick McCoy in the Intercolonial Exhibition of 1866–67, a paper by Albert Günther (British Museum) and descriptions of new species by explorers and others who had received specimens from the explorers, but there was little more. Castelnau was embittered by criticisms of his own work by Günther but he was correct in that most specimens had been sent to taxonomists like Cuvier, Valenciennes and de Lacépède in France, Jenyns, Richardson and Günther in England, Klunzinger in Berlin and Steindachner in Vienna. Among the explorer-scientists, only Thomas Mitchell and perhaps Wilhelm Blandowski had the expertise to describe new species. Blandowski was a Prussian naturalist who, with Gerard Krefft, had journeyed to the

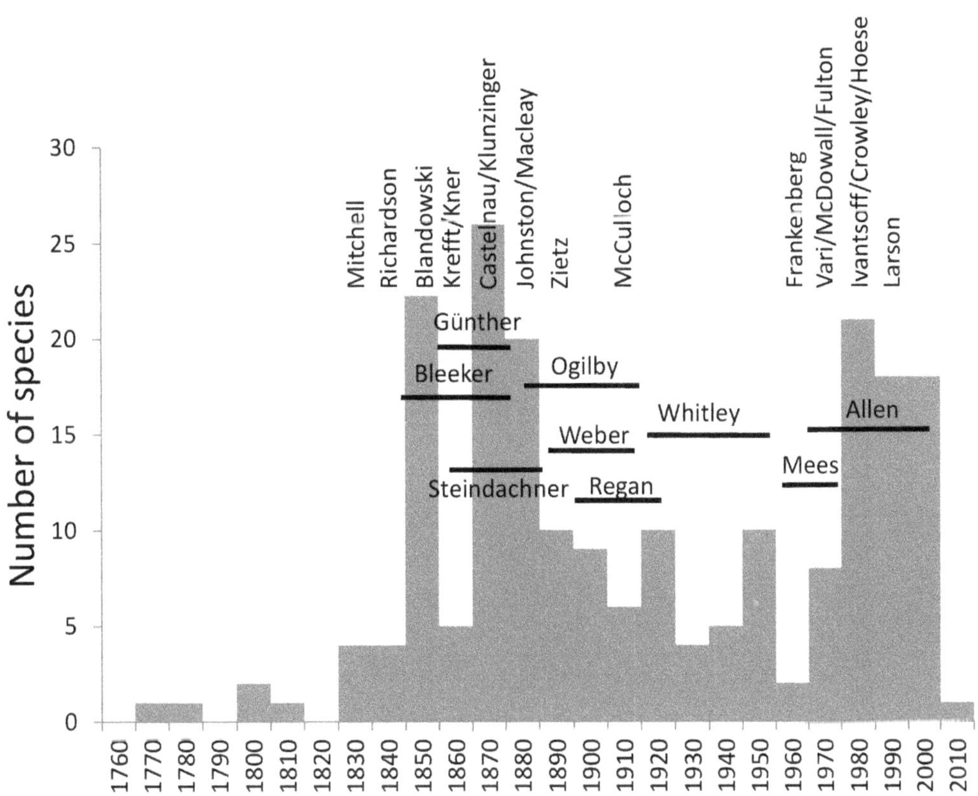

Figure 1.5: Time-line of descriptions of Australian freshwater fishes. Names are those of nomenclators who contributed two or more species. Blandowski's 'uncredited' species are included (see text).

Murray-Darling junction in 1856–57 (Humphries 2003a, 2009). His work came to naught because of acrimony with his employer, the Philosophical Institute of Victoria, leaving Murray cod, silver perch, freshwater catfish and carp gudgeons (*Hypseleotris* spp.) as the only native species described by taxonomists in Australia.

In 1861, Krefft became curator at the Australian Museum in Sydney, where he remained until 1874 when a scandal involving missing gold and pornographic pictures brought his dismissal (Nancarrow 2009). He energetically described species of fish and sent many specimens to Günther in Britain. Perhaps his most significant find was the iconic Australian lungfish, first known only from fossil teeth and thought to be an amphibian (Whitley 1927, 1929). Several other species, including the crimson-spotted rainbowfish (*Melanotaenia duboulayi*), the Murray River rainbowfish (*M. fluviatilis*) and southern purple-spotted gudgeon (*Mogurnda adspersa*), were described by Castelnau (1878) from commercial catches in the 1870s. About this time, Frederick McCoy began his impressive *Prodromus of the Zoology of Victoria*, containing illustrations and descriptions of several freshwater fish (McCoy 1885) (Fig. 1.6). Some illustrations in the Prodromus were by Arthur Bartholomew; others were by Ludwig Becker, who died with Robert O'Hara Burke and William John Wills on their infamous expedition in 1860. In Tasmania, government statistician Robert Johnston began taxonomic studies of Tasmanian species and James Ogilby, first at the Australian Museum and later at the Queensland Museum, contributed descriptions of galaxiids and gudgeons.

By the turn of the 20th century, freshwater fish taxonomy was gaining momentum from the contributions of Ogilby, David Stead and Allan McCulloch in New South Wales and Edgar Waite in South Australia.

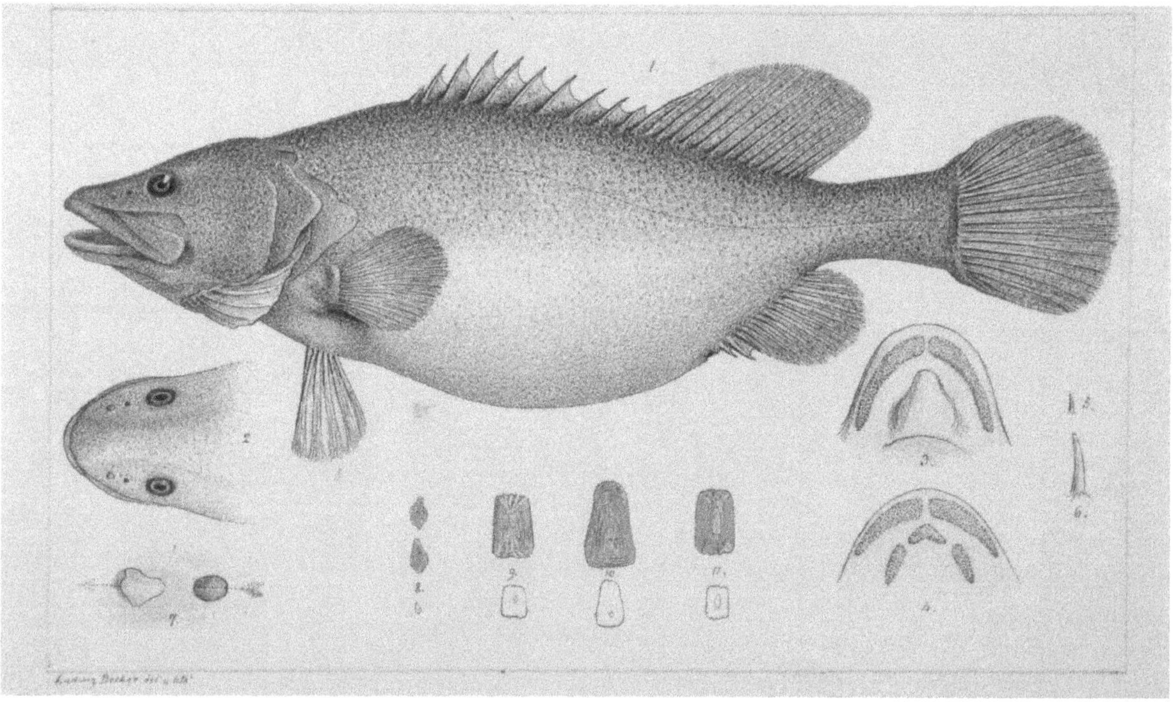

Figure 1.6: Ludwig Becker's illustration of Murray cod for Frederick McCoy's 'Prodromus of Victoria', 1878–80.

Eric Scott wrote of Tasmanian fish and Bruce Shipway of Western Australian species. At the Australian Museum, Gilbert Whitley extended and reviewed the taxonomy of freshwater species, especially the hardyheads (Atherinidae) (Fig. 1.7a).

Whitley's name especially warrants a highlight. Dozens of freshwater, estuarine and marine fishes were described by this prolific scientist and historian, and the years from the mid 1920s until the early 1960s might justly be called the 'Whitley Era' of Australian fish taxonomy. Indeed, his luminous writings on the history of Australian ichthyology are the source of much information in this chapter. Allan McCulloch (the man for whom *Maccullochella* was named) took Whitley under his wing at the Australian Museum in 1922, and after McCulloch's death by suicide in 1925 Whitley became Curator of Fishes. He was also President of the Royal Zoological Society of New South Wales (1940–41, 1959–60, 1973–74) and the Linnean Society of New South Wales (1967) (Murray and Roach 2002). The prize for best student paper at annual congresses of the Australian Society for Fish Biology is named in his honour.

Progress slowed after Whitley's retirement in 1964, although there were notable contributions by Gerlof Mees in Western Australia in the early 1960s, especially his description of the enigmatic salamanderfish (*Lepidogalaxias salamandroides*: Mees 1961). There followed a minor resurgence, with more than 15 species described in each subsequent decade. Roger Frankenberg, Wayne Fulton and Bob McDowall in the 1960s and 1970s highlighted the significance of Australia, especially Tasmania, as a biodiversity hotspot for Galaxiidae. Other taxonomists included Vari and Hutchins (Terapontidae), Crowley and Ivantsoff (Atherinidae), Hoese (Eleotridae) and Larson (Gobiidae). In the aftermath of the Whitley Era, the years from the 1970s to the present day might be called the 'Allen Era', when Gerry Allen and co-authors have named more than 20 Australian freshwater species (and many other, non-freshwater species) (Fig. 1.7b). This work benefited greatly from advances using molecular markers, and reviews of several groups are now underway using mitochondrial DNA. There are exciting prospects for biogeography, phylogeny and taxonomy (Chs 2, 3).

Figure 1.7: Australian fish biologists. (a) Gilbert Whitley (National Library of Australia). (b) Gerry Allen (photo: G Allen). (c) John Lake (National Archives of Australia).

Biology

In the tradition of natural history, the first biological studies were fragmentary, anecdotal and of historical rather than scientific interest. We can only lament the lost opportunities for careful description and analysis and for recording the knowledge of indigenous Australians, although there are some consolations in early chronicles (Blandowski 1857; Thompson 1893; Gilmore 1934). Among the most comprehensive natural histories are those by George Bennett, curator at the Australian Museum in 1835–41, in his books *Wanderings in New South Wales, Batavia, Pedir Coast, Singapore, and China* (Bennett 1834) and *Gatherings of a Naturalist in Australasia* (Bennett 1982). In the first, Bennett described freshwater fishes in New South Wales and commented on their palatability, size and diet and the ways they could be caught. He also wrote of Murray cod, noting the presence in their stomachs of caterpillars and freshwater mussels.

The first systematic study, however, was by Blandowski and Krefft, on their expedition to the junction of the Murray and Darling rivers in 1856–57 (Fig. 1.8). Blandowski collected eight new species of fish, among thousands of other specimens. Although he claimed 19 different forms (Blandowski 1857, p. 134), several had already been described and others were life stages of the same species. He respected the knowledge of the local Nyeri Nyeri people, who brought him fishes, related their habits and habitats, how big they grew, what they ate, what they tasted like and, not least, explained their cultural significance (Blandowski 1857; Krefft 1866; Humphries 2007).

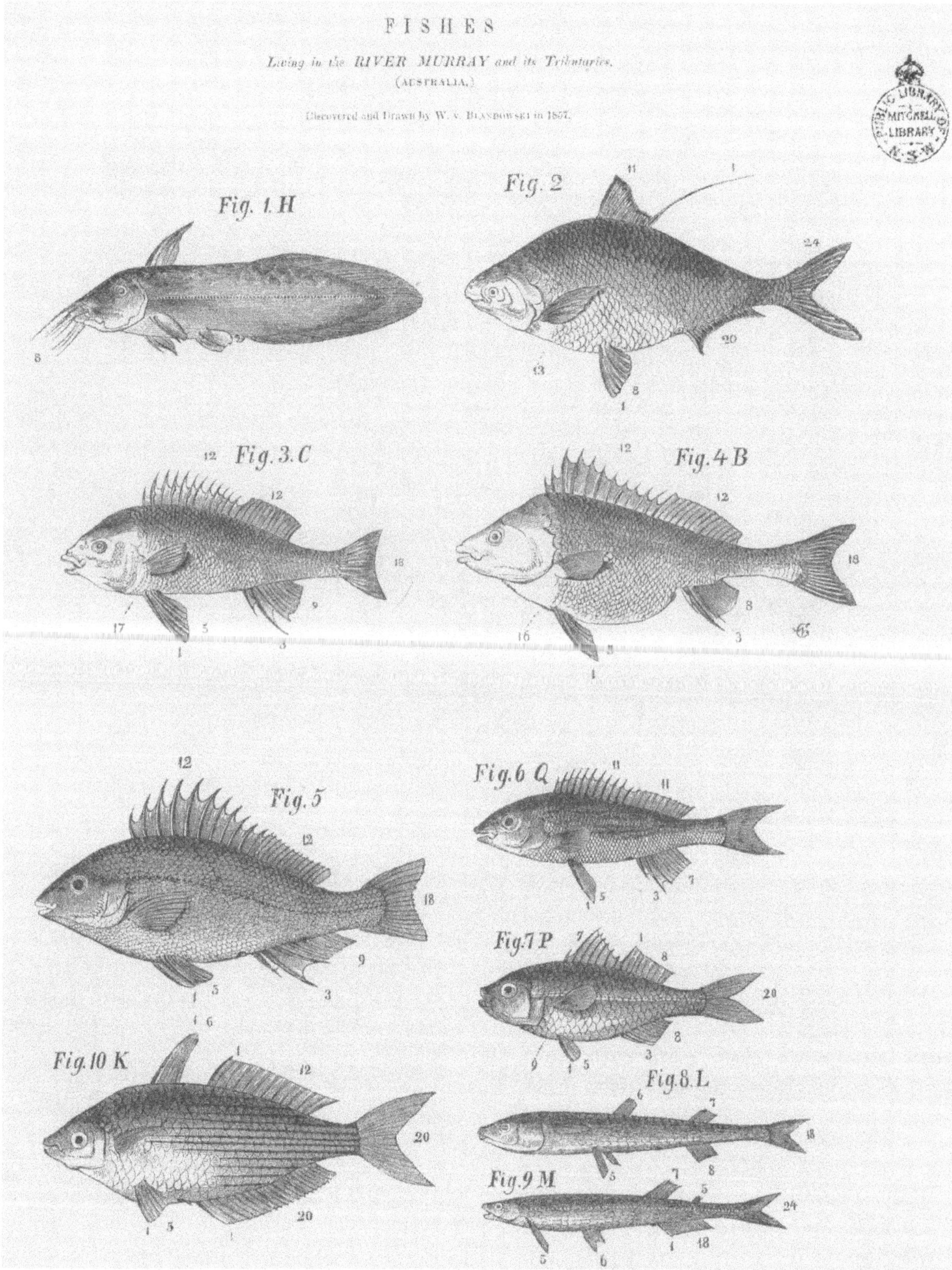

Figure 1.8: Fishes collected by Wilhelm Blandowski and Gerard Krefft at the junction of the Murray and Darling Rivers, 1856–57 (State Library of New South Wales).

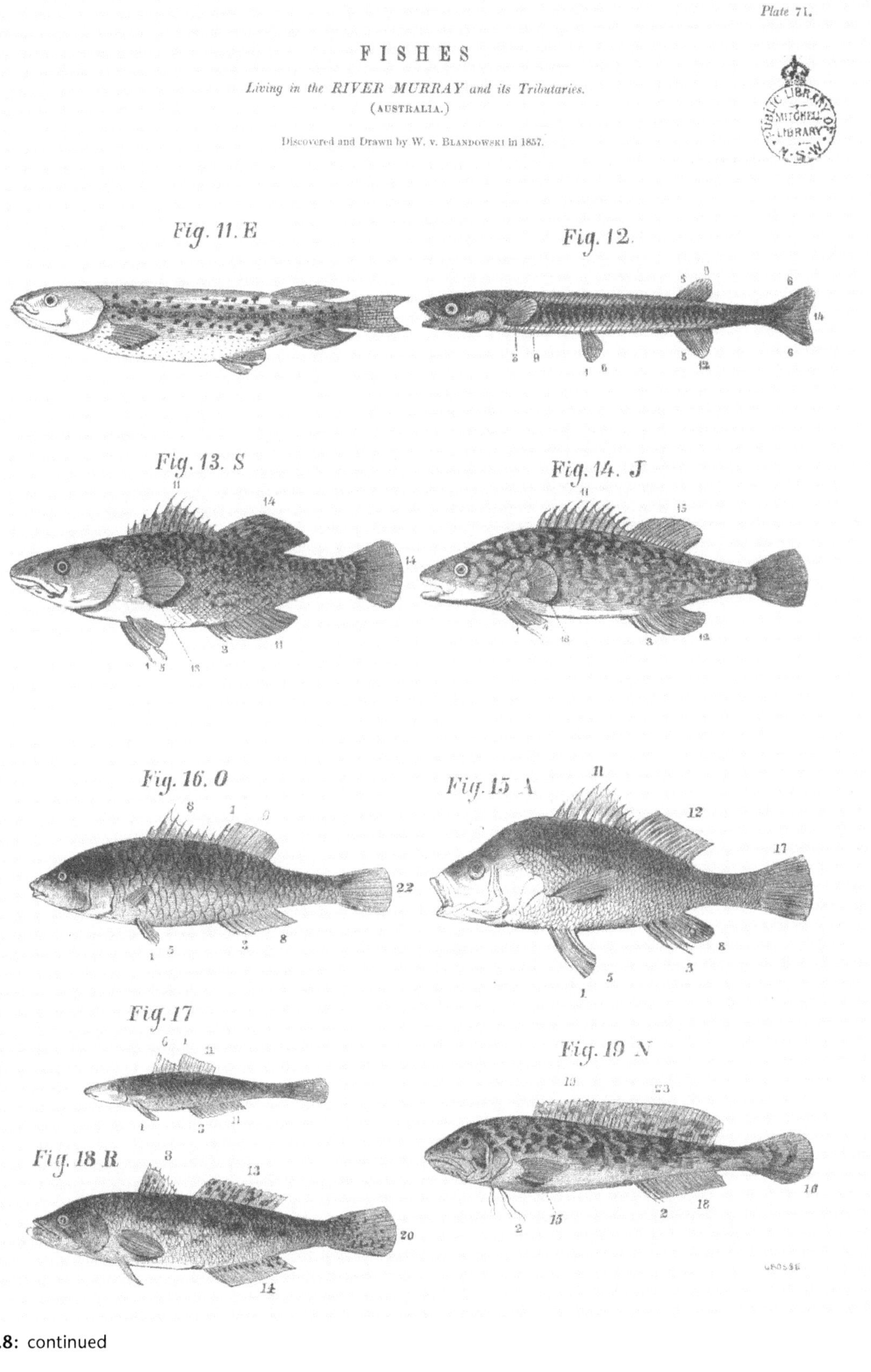

Figure 1.8: continued

Some strategic, not necessarily systematic, biological studies followed the advent of acclimatisation societies, formed to import desirable alien species and to move native species within Australia or export them to other parts of the world (Ch. 11). First among them, the Acclimatisation Society of Victoria, was formed in 1861 by Edward Wilson, editor of *The Argus* newspaper in Melbourne (Finney 1993). Shortly after, acclimatisation societies began in New South Wales, South Australia, Tasmania and Queensland, alongside other organisations with similar goals. Their activities are recorded in contemporary annual reports, newspapers and books (Wilson 1859; Nicols 1882; Seager 1888; Pasonson 2000; Low 2001).

The acclimatisation societies ushered in decades of rampant introductions, including common carp, redfin perch, goldfish (*Carassius auratus*), brown trout and rainbow trout. Indeed, the first freshwater fisheries regulations were to protect alien rather than native species (Finney 1993). As early as 1857, Edward Wilson advocated transferring Murray cod to rivers outside the Murray-Darling Basin. With inland fisheries newly established around this time, there were inevitable conflicts between those who favoured native species, particularly Murray cod, and those who favoured salmonids and other alien species. The newspapers of the day told of the concerns, priorities and conflicts.

As short-sighted as these activities later proved to be, they did promote research and inspire rudimentary ecological thinking. Concern over the sustainability of fisheries in New South Wales led to a Royal Commission led by William Macleay (1880). Marine issues dominated but there was some regard for fisheries for Murray cod, golden perch, silver perch and freshwater catfish. A Select Committee in Victoria (1892) and another Royal Commission in New South Wales (1894–96) shed more light on the biology of native species and associated fisheries. Notwithstanding their marine emphasis, these reports, alongside books (Tenison-Woods 1883; Thompson 1893; Roughley 1916), do say something of the status of freshwater fishes before there was extensive environmental degradation, albeit long after European settlement and the commencement of commercial fisheries.

Fisheries records, reports and legislation contain important data (Ch. 9) but they are of course biased towards 'target' species. For example, salmonids are targeted in Tasmania, Murray cod and golden perch in Victoria, New South Wales and South Australia and barramundi (*Lates calcarifer*) in Queensland and the Northern Territory. Western Australia's only claim to an inland scale fishery may be that for silver cobbler (*Neoarius midgleyi*) in Lake Argyle, in the Kimberley region. Inevitably, studies have favoured target species and reports such as those on Murray cod by Dannevig (1903) and Dakin and Kesteven (1938) are notable.

Ecology

Ecological studies on Australian freshwater fishes also have 19th-century roots. There were important studies by Alfred Dunbavin Butcher (Butcher 1945, 1946) and others in the early 20th century, and one of the historic milestones was J.O. Langtry's survey of the Murray River and tributaries near Albury in 1949–50 (Cadwallader 1977). Langtry's work was not published by the Victorian Fisheries and Wildlife Department until 1960, but in retrospect it was a shining example of emergent new understanding. Langtry drew on catch data, advice from local fishers and extensive knowledge of the Murray and he made many recommendations for research and conservation. His reports presage the declines of many native fish in that region and are a sober reminder that alien species were well-established more than 60 years ago.

In Langtry's time, studies were becoming more ambitious and more scientific. Researchers were ready to embrace concepts, ideas and techniques developed in other countries; they had access to useful, systematic, quantitative data and could advance from mere descriptions towards questions and answers. In fact, the post-war era fostered a surge in all forms of Australian science through the Commonwealth Scientific and Industrial Research Organisation (CSIRO), universities, museums and government agencies.

In the mid 1950s, Aubrey Nicholls of CSIRO began studies of trout in Tasmania, leading to numerous papers in 1958–73 (e.g. Nicholls 1958). Also in Tasmania, Maurice Blackburn (Blackburn 1950) described the Tasmanian whitebait (*Lovettia sealii*), a diadromous species supporting a recreational fishery. Alan Weatherley investigated the distribution, growth and behaviour of brown trout, redfin perch and tench (*Tinca tinca*) (Weatherley 1958, 1959, 1961, 1963).

There were also the beginnings of research by Jim Thomson on mullet (Mugilidae) (Thomson 1955) and Gordon Grigg on the Australian lungfish (Grigg 1965).

In 1960, an Inland Fisheries Research Station was established at Narrandera, New South Wales, with John Lake as head biologist (Fig. 1.7c). With colleagues Leyton Llewellyn and Hamar Midgley, Lake's research related the life histories of Murray-Darling fishes to their environment and raised insightful questions about adaptations, cues for spawning and early life history. His most influential work was based on experiments in ponds at Narrandera, and although he published only two related papers (Lake 1967b, c), he wrote much more widely on Murray-Darling and Australian freshwater fishes (Lake 1967b, 1971, 1978). Incidentally, it is tempting to suppose that the name for Narrandera was inspired by Ngurunderi, the Dreamtime hero who created the fishes of the Murray, but it is not so: it honours the local indigenous tribe and refers to goannas or other lizards.

In the next decade or so, ecological research flourished. Potter and colleagues investigated lampreys (Potter and Strahan 1968; Potter 1970); MacKay studied firetail gudgeon (*Hypseleotris galii*) and golden perch (MacKay 1973a, c, 1974); Pollard described the biology of common galaxias (Pollard 1971a, b, 1972, 1974); Llewellyn described the life histories of several Murray-Darling species (Llewellyn 1973, 1974, 1979); Davis studied freshwater catfish (Davis 1977a, c); Cadwallader described the diet of Victorian species (Cadwallader and Rogan 1977; Cadwallader *et al.* 1980); Glover and Sim pioneered research on the ecology of desert fishes (Glover and Sim 1978a, b; Glover 1979, 1982); Beumer investigated spangled perch (*Leiopotherapon unicolor*: Beumer 1979a, b) and eels (Anguillidae: Beumer 1979a); Jackson investigated the biology of river blackfish (*Gadopsis marmoratus*) and its interactions with brown trout (Jackson 1978, 1980; Jackson and Williams 1980) and the diet of Australian grayling (*Prototroctes maraena*: Jackson 1976); Fletcher studied the impact of salmonids on mountain galaxias (*Galaxias olidus*: Fletcher 1979); and Midgley and Merrick studied the ecology of Queensland freshwater fishes (Merrick and Midgley 1976, 1981; Midgley 1977; Merrick *et al.* 1983).

The 1960s and 1970s paved the way for a new generation of researchers and managers, symbolised by formation in 1971 of the Australian Society for Fish Biology (ASFB). The ASFB website (<http://www.asfb.org.au>) is a gateway for newcomers, providing easy access to other researchers through active working groups, regular meetings and links to cognate groups such as Native Fish Australia (<http://www.nativefish.asn.au>).

Our historical review rests at this point, as most of our present-day researchers are part of that generation and some are authors of the chapters that follow. We now turn to the ecological roles of fishes and how these have been accommodated in thinking about freshwater ecosystems.

FISHES IN FRESHWATER ECOSYSTEMS

Ecology is by definition a holistic science, although ecologists often have special interests in particular groups of animals or plants. That is not a bad thing, provided that an interest in fishes or some other taxon does not blind us to the big picture. We should think about fishes in the context of their environment and their interactions with other organisms. For some problems, it may help to assume a fish's viewpoint – to try to 'think like a fish'!

Trophic roles are central to many studies (Ch. 6): what fish eat, what eats them and who they compete with for food, and how these activities affect other organisms directly, through changes in abundance, and indirectly, through behaviour (Carpenter *et al.* 2001; Power *et al.* 2008). For example, top predators may structure animal and plant communities (Sergio *et al.* 2008) and detritivores can influence nutrient cycling (Taylor *et al.* 2006). Migratory fishes transport nutrients from freshwater to marine systems and *vice versa*, linking aquatic and terrestrial food webs (Ch. 5). Indeed, nutrients carried in the bodies of migratory salmon are important for the growth of forests in parts of North America (Allen *et al.* 2005). Fishes are also environmental modifiers, changing the physical and biological structure of freshwater systems through nest building (e.g. freshwater catfish) and foraging (e.g. common carp). Common carp in particular are ecosystem engineers, reflecting a prodigious capacity to alter their environment (Weber and Brown 2009).

Table 1.1: Some models relevant to freshwater fish ecology

Model/concept	Authors	Description	Relevance for fishes	Relevance in Australia
Trophic cascades in lakes	Carpenter and Kitchell (1993)	Fluctuations in predator populations alter nutrient cycling, algal biomass and production	Identifies structuring role via trophic interactions in lakes	Few large permanent lakes; potentially also streams
Fish zones	Shelford (1911a); Gerking (1949); Huet (1959)	River habitat zones	Regions inhabited by assemblages	Used in monitoring (Gehrke and Harris 1996)
River continuum concept	Vannote *et al.* (1980); Minshall *et al.* (1985)	Longitudinal movement of matter and energy through processing of imported material	More species with stream size	Mid-elevation wooded streams
Serial discontinuity concept[1]	Ward and Stanford (1983, 1995)	Dams reset resource gradients downstream	Dams interfere with sediment transport, food and productivity, movements	Describes downstream effects of dams; flow requirements for fishes
Flood pulse concept	Junk *et al.* (1989); Bayley (1991); Tockner *et al.* (2000)	Flood pulse drives floodplain–river systems, links communities	Floodplain as nursery and food source; channel for dispersal, spawning, refuge	Floodplain rivers only. Nature of pulse clarified by Walker *et al.* (1995), Puckridge *et al.* (1998)
Flood recruitment model[1]	Harris and Gehrke (1994)	Fish life history and abundance shaped by irregular large floods in Australian rivers	Some fishes spawn on floodplain; others rely on plankton washed to channel	For Murray-Darling Basin fishes but relevant to other boom–bust systems (e.g. Lake Eyre Basin)
Riverine productivity model	Thorp and Delong (1994, 2002)	Matter and energy from local phytoplankton, benthic algae and aquatic plants, and from riparian zone via leaves and organic carbon	Riverine littoral important for younger stages	High-gradient streams, impoundments and tailwaters
Low-flow recruitment hypothesis[1]	Humphries *et al.* (1999)	Some fishes spawn and recruit during seasonal, warm, low-flow periods in dry regions when food supplies are at peak	Explains why many species do not breed during floods	For Murray-Darling Basin fishes (Milton and Arthington 1984)

Model/concept	Authors	Description	Relevance for fishes	Relevance in Australia
Inshore retention concept[1]	Schiemer *et al.* (2001)	Water retention in large rivers encourages zooplankton and provides habitat for young fishes, but is limited by flow regulation	Emphasises water retention for recruitment in big rivers	Lowland rivers
Riverscape concept	Fausch *et al.* (2002)	Merges dynamic landscape model of population ecology and life histories of riverine fishes	Reconciles scales of management with those significant for fishes	All rivers
Riverine ecosystem synthesis	Thorp *et al.* (2006, 2008)	Rivers as hydrogeomorphic patches. Combines other models	Concerns distribution and diversity of species, community structure and ecosystem processes	All rivers

[1] Concepts, models or hypotheses as spin-offs from the parent model or concept above.

Most ecological studies of Australian fishes refer to lotic (flowing water) environments, because there are few large, permanent lakes apart from glacial lakes in central Tasmania and volcanic lakes in western Victoria. Another factor is that few native species reside wholly in lakes (Chs 4, 10). Nevertheless, the concept of trophic cascades, one of the defining ecological ideas of recent decades, arose from studies of lakes and lacustrine fishes (Chs 6, 11).

In sufficient numbers, piscivorous fishes can structure lake communities by preying on lower carnivores, such as zooplanktivorous fishes, which in turn affect the abundance of zooplankton, hence phytoplankton, primary production and nutrient cycling (Carpenter and Kitchell 1993). The presence or absence of top predators therefore can influence whether or not a lake is prone to algal blooms. The numbers of piscivores can be artificially manipulated, as a tool for restoration in eutrophic lakes (Lathrop *et al.* 2002). These trophic cascades occur in coastal and estuarine habitats (Jackson *et al.* 2001b) as well as inland waters (Lake *et al.* 2007; Power *et al.* 2008; Humphries and Winemiller 2009) and they highlight the significance of fishes as drivers of ecosystem structure.

In the later 20th century, fishes played a part in the development of the classification of river zones (Huet 1959; Hynes 1970) but those classifications soon gave way to predictive models that did not always explicitly include fishes. Some of these models and ideas are outlined in Table 1.1. The first of them, the river continuum concept (RCC), portrayed a stream as a physical and chemical continuum along which communities of organisms contributed to the progressive rendering of organic matter imported from the catchment (Vannote *et al.* 1980). The concept focused on the role of bacteria, fungi and macroinvertebrates and said little of fishes except to predict that the number of fish species should increase with stream size. While the RCC often is cited as a useful model for mid-elevation streams in forested catchments, its validity for other streams, especially lowland rivers, is less clear. A corollary to the RCC, the riverine productivity model, emphasises contributions to production from the littoral zones of large rivers (Thorp and Delong 1994; Thorp *et al.* 1998), but again the roles of fishes are understated.

An alternative model for lowland rivers, the flood pulse concept (FPC), emphasises the floodplain as a source of matter and energy, and the channel as a distributary, and highlights the role of annual flooding as an ecosystem driver (Junk *et al.* 1989; Bayley 1991, 1995; Junk and Bayley 2008). Unlike the RCC, the FPC makes explicit the role of fishes in floodplain rivers; it represents the floodplain as a nursery and acknowledges the importance of food from the floodplain for fishes in the channel.

Recently, the riverine ecosystem synthesis has been proposed as a meta-model, combining parts of the foregoing concepts (Thorp *et al.* 2006, 2008). It views rivers as downstream arrangements of large habitat patches formed by geomorphology and climate. Hypotheses ventured as part of the synthesis relate to the distribution and diversity of species and their interactions with ecosystem processes. All are pertinent for fishes.

Fausch *et al.* (2002) posed challenges for stream fish ecologists that are not yet fully addressed. Their riverscape concept, built on other ideas (e.g. Schlosser 1991, 1995; Elder and Schlosser 1995; Schlosser and Kallemeyn 2000), included these imperatives for lotic studies:

- Research and management must be conducted in a spatially explicit manner to avoid loss of information, lack of interpretability and inability to find explanations and make predictions.
- Scales for research and conservation should be matched, spatially and temporally. If, in a year or a lifetime, a fish moves hundreds of kilometres, what value is there in sampling and monitoring at the reach scale? If we do not address this issue (and most studies do not), how can we expect to learn the true nature of patterns and processes important to fishes?
- The hierarchical nature of stream ecosystems, with continuous downstream flow of matter and energy, needs to be reconciled with the up- and downstream linkages that are mediated by the movements of fishes and invertebrates.

CONSERVATION

Indigenous Australians have had significant impacts on freshwater fishes (Humphries 2007), as have aboriginal peoples elsewhere in the world (e.g. Butler 2000; Grayson 2001), but the range, abundance and diversity of the native fauna have declined catastrophically since the arrival of Europeans. Of 256 wholly freshwater fish species now recognised (Ch. 2), 58 (23%) are listed federally as threatened, eight as critically endangered and another seven as endangered (Ch. 12). Species such as the Murray hardyhead (*Craterocephalus fluviatilis*: Wedderburn *et al.* 2008) have approached extinction but, with the possible exceptions of distinct local forms of mountain galaxias (Raadik 2011) and undescribed species of cod and golden perch (Chs 2, 3), no Australian freshwater fish species is known to have become extinct. Nonetheless, there is a clear warning from North America, where 40% of freshwater fish species are considered imperilled and many are extinct in the wild (Jelks *et al.* 2008).

The reasons for the declines in native species broadly include habitat alteration, river regulation, barriers to movements, interactions with alien species and overfishing (Ch. 12). The impacts in Australia have been like those in other parts of the world, especially where post-industrial societies have moved into regions with artisanal fisheries (Griffiths 1997; Crosby 2000; Humphries and Winemiller 2009). Indeed, the patterns of degradation and disturbance in fresh waters are like those in estuarine and coastal marine systems throughout the world (Jackson *et al.* 2001b).

While native fishes have declined, alien species have prospered (Low 2001). Many alien species have been introduced to Australia, some have become pests, and most attempts to contain or eradicate them have failed (Chs 11, 12). The ranges of several alien species continue to increase; for example, common carp and eastern gambusia have reached Tasmania in recent years.

The future prospects for native freshwater fishes are complicated by the onset of global climate change caused by human activity. Australian waterways will become warmer and flows will become even more variable than now, with more frequent, more intense floods and droughts. Rising sea levels are likely to flood the lower reaches of rivers. While climate change has been a hallmark of Australia's past, the scale and pace of projected changes could exceed anything experienced before now (CSIRO 2008; Steffen 2009; Steffen *et al.* 2009). Our native fishes already are confronted by challenges that are more sweeping, more intense and more rapid than at any other time in their long evolutionary history and it seems that the challenges will redouble as we enter a new era of changes fuelled by human industry. For many native species, the prognosis is uncertain; for some, it is bleak (Ch. 13). The legacies of our modern era surely will overwhelm the ghosts of Gondwana.

CONCLUSION

Twelve chapters follow, expanding and extending issues touched on lightly above and outlining the state-of-the-art in the ecology of Australian freshwater fishes. Our subject matter is so wide-ranging and so dynamic that, even if we were to venture a comprehensive treatise, it would be out-of-date tomorrow. What follows are first steps that, depending on the inclinations and perceptions of each reader, could be a passing interest, a window on the world or a path to a career. The most telling questions here may be ones that are not asked.

2 Biogeography

Peter J. Unmack

INTRODUCTION

Biogeography is the study of patterns in the distributions of organisms and the ways those patterns have arisen. While documenting patterns is straightforward, trying to explain them often is not. Many explanations are based on historical events, and there may be multiple possible explanations which cannot be distinguished. We are limited in our ability to conduct experiments to test hypotheses, but we can use fragmentary evidence to offer explanations for observed patterns due to unobservable events in the past. Our interpretations are strongly linked to our understanding of Earth history (geology, climatology etc.) and to an assumption that any molecular or morphological markers used are appropriate and not misleading. If our ideas and assumptions are wrong, our interpretations could fail. Consider, for example, how early interpretations were overturned by emergence of the theory of continental drift. Thus, biogeography is a controversial subject and most biogeographic explanations should be viewed with healthy scepticism.

This chapter is intended to complement rather than replace previous works. It summarises key information and offers new perspectives on patterns of endemism and relationships between Australia's biogeographic provinces. Readers with broader interests should consult McDowall (1981), Allen (1982, 1989, 1991), Merrick and Schmida (1984), Bishop and Forbes (1991), Harris (1995), Unmack (2001a), Allen *et al.* (2002), Pusey *et al.* (2004), Merrick (2006) and the World Wildlife Fund's (WWF) Freshwater EcoRegions website (<http://www.feow.org/index.php>). Biogeographic overviews of the global freshwater fish fauna include those by Matthews (1998), Banarescu (1990), Lundberg *et al.* (2000), Berra (2007) and Abell *et al.* (2008), and ichthyological texts those by Moyle and Cech (2004), Barton (2007) and Helfman *et al.* (2009).

AUSTRALIAN FISHES IN A GLOBAL CONTEXT

Currently, more than 32 000 fish species are recognised globally, and each year more than 300 new species are described (Eschmeyer and Fong 2012). Over 80% of the known species belong to two groups within the Teleostei (Fig. 1.2), the Ostariophysi and Acanthopterygii.

Freshwater fishes number about 15 170 species (Eschmeyer *et al.* 2010), and about 63% are ostariophysans, including catfishes (Siluriformes), characins (Characiformes) and minnows, loaches and their relatives (Cypriniformes). There are about 16 764 marine species (Eschmeyer *et al.* 2010). Most are acanthopterygians, including groups like the billfishes, dories, flatfishes, groupers and puffer-fishes. Freshwater acanthopterygians are comparatively few; they include killifishes and livebearers (Cyprinodontiformes) and species from the cichlids, darters, perches, freshwater sunfishes, gouramies and snakeheads (Perciformes).

The freshwater fishes of Australia and New Guinea are distinctive because they are dominated by acanthopterygian fishes rather than the ostariophysans which predominate in fresh waters on other continents (except Antarctica, where there are no freshwater fishes). Indeed, the only freshwater representatives of many acanthopterygian families are Australian species. Acanthopterygians also dominate on the island of Madagascar which, like Australia, has been long isolated from other land masses (88+ million years) (Sparks and Stiassny 2003). Others occur on Pacific islands, but most are diadromous species able to disperse between islands during their marine phase (McDowall 2001). While Australia's extraordinary mammal fauna receives a lot of attention, our fish fauna is no less distinctive.

Australia's fauna reflects long geological isolation. The most recent continental connection has been with Antarctica, but the two began to separate around 95 million years ago (mya) and the process was complete around 37 mya (Veevers 1984). The northern and southern continents have not been close until recently, in geological time, and no other southern continent has been near Australia since the break-up of Gondwana. Because most ostariophysan fishes require freshwater connections for dispersal (they cannot cross marine barriers), Australia lacks this group apart from members of two families with marine representatives, the catfish families Ariidae and Plotosidae.

Recent thinking is that two to six Australian fish families (Neoceratodontidae, Osteoglossidae and perhaps Lepidogalaxiidae, Melanotaeniidae, Pseudomugilidae, Percichthyidae) evolved in fresh water. Other families evolved either from marine ancestors or came from freshwater fishes with marine life history phases (e.g. Anguillidae, Galaxiidae, Geotriidae, Gobiidae, Kuhliidae, Latidae, Mordaciidae, Retropinnidae). A common misapprehension is that, given the marine affinities of most species, most of Australia's fish fauna has evolved in the last few million years. This is clearly not the case for most families (Hills 1956; Lundberg *et al.* 2000; Unmack 2001a). There is fossil evidence for several large freshwater species (Hills 1934, 1943), including *Percalates antiquus* (Eocene), 'Terapontidae' (Eocene), *Scleropages* cf. *leichhardti* (Oligocene-Miocene) and *Maccullochella* (Miocene). Other evidence comes from phylogenetic relationships (Melanotaeniidae, >80 mya: Sparks and Smith 2004) and molecular clocks (*Craterocephalus*, 43–75 mya: Unmack and Dowling 2010). It is unfortunate that our knowledge of Australian freshwater fish fossils is scant and that the information we do have is biased towards large species. Some extensive Miocene and Pliocene fossil collections do exist, from sites around Lake Eyre, Riversleigh and Murgon, but they remain in museum collections, unstudied.

Australia has two endemic families, the Lepidogalaxiidae and Neoceratodontidae (extant species only), but the number rises to four if families shared with New Guinea (Melanotaeniidae, Pseudomugilidae) are included. Many genera (32 of 89, or 36%) and species (190 of 256, or 74%), however, are endemic to Australia. When genera and species shared exclusively with New Guinea are included, these numbers become 47 of 89 (53%) genera and 233 of 256 (91%) species (Table 2.1).

Australia and New Guinea are home to major freshwater radiations by families and genera that otherwise have widespread marine distributions. For example, Plotosidae, Terapontidae and species of *Ambassis* range from southern Africa to the western Pacific and north to Japan, and the Apogonidae and Atherinidae are even more widespread (Berra 2007). The existence of marine and freshwater representatives in these groups raises questions that still await answers. For example, it has been assumed that the ancestral species are of marine rather than freshwater origin. If so, was there a single marine ancestor, or multiple ancestors, and might some freshwater species have returned to the sea? For species of *Craterocephalus*, the one group closely examined so far, there appears to have been a single freshwater invasion from the sea, although one species, Munro's hardyhead (*C. munroi*), has returned to estuarine environments (Unmack and Dowling 2010).

Geologically, New Guinea is tied to Australia (as parts of one continent) and we cannot sensibly explore the fishes of one without reference to the other. While they are now separated by sea water, there have been many terrestrial connections in geological time, including the last glacial maximum 18 000 years ago. Essentially, all families in New

Table 2.1: Species richness and generic richness and endemism for Australia, biogeographic provinces and subprovinces

Area	Species richness	Endemics (+NG)	% endemics (+NG)	Generic richness	Endemics (+NG)	% endemics (+NG)	Species/genus ratio
Australia	**256**	**190 (233)**	**74 (91)**	**89**	**32 (47)**	**36 (53)**	**35**
Provinces							
Eastern	77	30	39	45	4	9	58
Bass	28	4	14	14	0	0	50
Southern Tasmania	20	7	35	10	0	0	50
Murray-Darling	38	12	32	22	0	0	58
South-western	14	11	79	8	4	50	57
Pilbara	15	7	47	13	1	8	87
Kimberley	44	14	32	24	2	8	55
Northern	125	36 (78)	29 (62)	56	2 (7)	4 (13)	45
Central Australian	33	18	55	18	3	17	55
Palaeo	5	0	0	5	0	0	100
Subprovinces							
North-eastern	70	11 (23)	16 (33)	37	2	5	
W Depauperate Northern	41	3	7	25	0	0	
W Speciose Northern	61	8 (10)	13 (16)	33	0	0	
E Depauperate Northern	45	2 (3)	4 (7)	31	0 (1)	0 (3)	
E Speciose Northern	62	0 (4)	0 (6)	36	1	3	
Southern Pilbara	4	0	0	3	0	0	
North West Cape	2	2	100	2	1	50	
Northern Pilbara	13	3	23	11	0	0	

In the 'endemics' columns, the numbers in parentheses represent totals with addition of species/genera limited to New Guinea (NG). The species/genus ratio represents the percentage of species within each geographic area that are from different genera.

Guinea are shared with Australia but there are many additional endemic genera and species, especially in the northern half of the island. About 51 genera and 61 freshwater species are shared and many more species have close phylogenetic relationships. The freshwater 'ecoregions' of New Guinea (geographically distinct animal and plant communities) are described by Allen (1991) and on the WWF website (<http://www.feow.org/index.php>).

New Zealand is not geologically linked to Australia but its freshwater fish fauna is related. Almost all genera in New Zealand are shared with Australia and many species have a juvenile marine phase that would facilitate dispersal across the Tasman. McDowall (1990, 2010) provides an excellent discussion on the origins and biogeography of the New Zealand freshwater fishes.

HISTORICAL SETTING

Geological background

Mary White's popular books are an excellent review of the history of Australia in terms of flora and geology (White 1994, 1998, 2000). Various state-oriented geology texts provide good technical summaries, including editions for South Australia (Drexel and Preiss 1995) and Victoria (Birch 2003). In addition, Byrne *et al.* (2008) have reviewed conditions in Central Australia over the last 20 million years.

Geological perspectives remind us of time scales and the immense changes that occur in geologically active regions (e.g. uplift rates in New Guinea of 600 m per million years: Veevers 1984). Australia probably has experienced less mountain-building and tectonic activity than any other major landmass over the last 80-odd million years. The last major mountain-building episode was in the Eastern Highlands, about 90 mya, and most other mountain ranges were formed 300–600 mya. The major sedimentary basins and river basins have been around for 50 million years (Veevers 1991). The inland seas that divided Australia as three to four islands 90–100 mya (Frakes *et al.* 1987a), and the uplift and subsequent erosion of the Eastern Highlands, would have had major effects on the distributions of freshwater fishes and would have contributed to the isolation and subsequent evolution of families and/or genera that persist today. There is scant evidence tying these events directly to the modern fauna, however, owing to the time that has passed, with many opportunities for dispersal (e.g. Bostock *et al.* 2006). Australia's mountains are old and eroded and the land surface generally has the lowest average elevation of any continent, implying that there have been few significant topographic barriers through most of the Tertiary (the last 65 million years). Erosion rates through the Mesozoic and Cenozoic are estimated as roughly 10 m per million years (Gale 1992).

Although there is some debate in the geological literature over the scale of recent (post-Eocene) tectonic activity in Australia, there has been activity in the Eastern Highlands in Victoria (Brown 2008; Holdgate *et al.* 2008, 2011; Braun *et al.* 2009; Vandenberg 2010, 2011; Norvick 2011), the Flinders Ranges in South Australia (Alley and Benbow 1995), the Eucla Basin in Western Australia (Sheard and Smith 1995), along the Murray River (Stephenson 1986; Stone 2006; McLaren *et al.* 2011) and elsewhere. Again, it is difficult to relate these events directly to modern biogeographic patterns. Yet, in a land of low topographic relief, even small tectonic movements can have significant effects on drainage patterns.

Australia may have been quiet in tectonic terms, but it did experience extensive volcanism during the Tertiary (Johnson 1989), including most of the Eastern Highlands, from northern Queensland to eastern South Australia. While most activity was more than 30 mya, south-eastern South Australia, western Victoria and north-eastern Queensland experienced lava flows until 5000–13 000 years ago. Most impacts would have been in river valleys, as lava flows downhill and follows drainage lines. While these events would be devastating for local fish populations, the affected places eventually would be recolonised from tributaries and downstream reaches. Lava flows may dam rivers, causing them to spill into an adjacent basin, although no examples are documented in Australia. Lava flows may have levelled some regions along drainage divides (e.g. parts of western Victoria, Atherton Tablelands near Mareeba in Queensland) and could have provided opportunities for fish to move between basins.

There was a major sea-level rise in the Late Cretaceous, re-forming Australia as three to four large islands (Frakes *et al.* 1987a). By 90 mya, the coastline would have resembled that of today (Veevers 1984). Smaller marine incursions occurred after this time but these would have reduced the size of basins rather than inundated entire drainage systems. Marine geological deposits are significant for the habitats of cave fishes in Western Australia (Humphreys 2001), the limestone gorge and narrowed floodplain of the Murray River in South Australia (Twidale *et al.* 1978) and karstic (water-eroded limestone) habitats in south-eastern South Australia (Sheard and Smith 1995).

Coastal areas of southern and western Australia were inundated at times during the Eocene, Oligocene, Miocene and Pliocene. One of the recent inundations was in the Eucla Basin in the Great Australian Bight during the mid Miocene, forming the Nullarbor Plain (Benbow 1990; Sheard and Smith 1995). Another region affected by marine incursions is the lower Murray-Darling Basin, which was inundated several times between 12–32 mya (Brown and Radke 1989) and again 3–5 mya (Frakes *et al.* 1987b; Stephenson and Brown 1989). In addition, the coastline from Portland, Victoria, west to the Murray mouth in South Australia has extended southward over the past 2 million years, after marine inundation followed by minor uplift and sediment deposition (Joyce *et al.* 2003). As before, these events are not easily linked to biogeographic patterns, but formation of the Nullarbor Plain did contribute to the east–west separation of aquatic biota across southern Australia. The connection and disconnection of mainland Australia and the islands of Tasmania and New Guinea were other significant events for aquatic biota, and fishes in particular.

Long-term climate change

The most dramatic changes to Australian environments have been due to long-term climate change. Today, 70% of Australia is arid and much of the western half of the continent (see 'Palaeo Province') lacks fishes other than spangled perch (*Leiopotherapon unicolor*). Yet the plant fossil record shows that, during the early Tertiary, Australia was wetter, with extensive forests (White 1994). Subsequent changes in the climate were part of a drying global climate in the Tertiary, as the Australian landmass drifted slowly northward. The details are uncertain, but fossil evidence indicates that Australia has become drier over the last 45 million years (Kershaw *et al.* 1994; Martin 1994). Other evidence comes from geology, with extensive palaeo-drainages across much of central, southern and western Australia (Van de Graaff *et al.* 1977). Areas of extreme aridity spread gradually from western and south-central Australia to eastern and northern regions. Central southern Australia experienced increasing aridity by around 15 mya and central Australia has been strongly affected over the last 4 million years (Byrne *et al.* 2008). Given the lack of topographic barriers, it seems likely that long-term climatic shifts (500 000 years or more) have had most influence on present-day biogeographic patterns. Examples may be current disjunct distributions of species (e.g. those with northern occurrences in the Waterpark Creek–Shoalwater Bay region, Queensland, that otherwise have continuous distributions in their disjunct southern ranges). Undoubtedly, extirpations of species in river basins also occurred during periods of climatic extremes.

MOVEMENT BETWEEN RIVER BASINS

Barriers to movement

Almost all freshwater fishes are obligate aquatic organisms; that is, they cannot survive out of fresh water (the exceptions are diadromous species, species capable of diapause and those able to move across land). Virtually all require freshwater connections to disperse to new places, and these connections typically are bounded by drainage divides (the boundaries between river basins) and sea water. Drainage divides determine where rain water travels after it falls on the land. For most rivers, the sea is the terminus, but there are endorheic systems, like the rivers of the Lake Eyre Basin, where the water does not reach the sea.

River basins are, in effect, a fundamental unit in aquatic biogeography, as they define the spatial and temporal scope of water bodies accessible to fishes and other biota. The boundaries often are physiographically well-defined and tend to be stable for long periods of time (potentially, millions of years). Fishes generally cannot move easily between basins unless there is a hydrological connection. Thus, we tend to treat the river basin as a default biogeographical unit, as it is the largest area that has consistent aquatic connections and tends to be isolated from other basins.

One caveat is that where a fish lives within a basin will affect how it responds to the addition or removal of barriers. For instance, upland stream fishes are more likely to be in the vicinity of drainage rearrangements (see 'Drainage changes') and lowland fishes are more likely to be influenced by sea-level changes. Understanding how the distributions of freshwater fishes change in these ways is a fundamental part of biogeography (see reviews by Banarescu 1990; Unmack 2001a).

Six factors influence the ways that fishes move between river basins, namely drainage divides, sea water, climate, the ecology of individual species, rains of fish and adventitious animal transport. These are outlined below.

Drainage changes

Bishop (1995) identified three types of drainage rearrangement: beheading, river capture and diversion (the last divided further as channel migration, divide-topping catastrophes and tectonism). His conclusion – that drainage re-alignments occur less often than suggested in the geological and biological literature – is a contentious issue, for several reasons.

One is that whenever populations of aquatic species occur in two or more non-adjacent basins (their mouths are separated by other rivers), the default explanation almost always is that it must have been due to river capture. This explanation is often invoked without regard for geomorphic evidence, and alternatives are rarely considered. The elbow of capture and barbed drainage seen on topographic maps are often cited as evidence of river capture. These are tributaries meeting the main river at angles opposite to those expected, looking as if they might once have met a river flowing in the opposite direction. This is consistent with river capture but it is weak evidence; for example, the angle of confluent streams often is determined by underlying geology. If biogeographic speculations are accepted uncritically, then river captures have happened frequently and recently. Two examples of rivers having crossed basin boundaries (but perhaps not river capture in the usual sense) are, first, Two Ocean Pass in Wyoming, where North Two Ocean Creek splits to form Atlantic Creek and Pacific Creek, each flowing to a different ocean and, second, South American portals including the Casiquiare River in Venezuela which flows from the Orinoco River to the Negro River (Winemiller *et al.* 2008; Willis *et al.* 2010). Examples of connections across drainage divides include Isa Lake in Wyoming, which sits atop the drainage divide and connects two different river basins, and the Rupununi River, which connects to the Amazon Basin through annual flooding across the Rupununi savannah via Lake Amuku in Guyana (Lowe-McConnell 1964; Watkins *et al.* 2004).

Another reason that drainage rearrangements are difficult to study is that they are caused by erosive processes; thus, much of the evidence for the event is removed as part of the process. Finally, many presumed examples of river capture in the geological literature have been overturned by later studies (Unmack 2001a). The point here is that drainage rearrangements may be less common, more irregular events than the biogeographic literature suggests. In the absence of solid evidence (Burridge *et al.* 2006), they are not a persuasive explanation for most across-drainage divide patterns.

If fishes are commonly found across drainage divides, and drainage rearrangements are not a universal explanation, how can we explain this pattern? In fact, fish do swim across drainage divides. There often are shallow divides separating basins and in some places there are seasonal or permanent connections (e.g. Casiquiare River, Lake Isa, Two Ocean Pass). More common still are dry, low-relief regions across drainage divides (see 'Murray-Darling Province'). Given intense rainfall, there could be short-term aquatic connections sufficient to allow fish to swim across land and across divides. These events may be extremely rare, but in time they are likely to occur. Indeed, there is clear evidence of the capabilities of fish to move across normally terrestrial environments after intense rainfall (see 'Rains of fishes'). Shallow drainage divides are often limited in their geographic extent but they are common globally as well as in Australia, where much of the landscape is low relief.

Sea-water barriers

Most freshwater fishes are unable to tolerate sea-water salinity (35 g L^{-1}) and those that can do not survive long under those conditions. Thus, the sea is a strong barrier to dispersal. The barrier may change, however, with changing sea levels. Sea-level changes occur at two scales. Long-term changes can be caused by variations in water quantity (water trapped in glaciers and polar ice), in the uplift and subsidence of continental margins and the volume of oceanic basins (Partridge *et al.* 1995). Short-term changes are related to climatic variations that occur every 100 000–150 000 years (see 'Climate') and the degree of change is again related to the volume of water contained in glaciers. There were big sea-level changes during the Pleistocene and the Miocene (Miller *et al.* 2005) and changes at that scale have probably occurred over much of Earth's history, although their magnitudes would have varied with the extent of glaciation.

The last glacial cycle resulted in a fall in sea level of between -120 and -135 m (Clark and Mix 2002) and similar lows occurred in earlier times (Miller *et al.* 2005). Sea levels have been higher, but only a few tens of metres higher than today. The effects of sea-level changes are correlated with the width of the continental shelf. If the shelf is narrow and steep, the coastline during low sea levels may change little; if the shelf is wide, the coastline would change substantially. Lower sea levels allow isolated drainages to join, and reconfigure coastal wetlands and landforms, influencing connectivity between river systems. In a few places, lowered sea levels have resulted in massive drainage changes, such as the formation of 'Lake Carpentaria' and the connection of northern Australian drainages with New Guinea. A similar, smaller pattern occurred in south-eastern Australia, with the formation of 'Lake Bass' and the integration of drainages from northern Tasmania and central southern Victoria (Fig. 2.1).

A potential mechanism for movements of fish across marine barriers is transportation in river flood-plumes (Wolanski and Jones 1981; Grimes and Kingsford 1996). There are records of *Percalates* found 5 km out to sea after flooding (Williams 1970), suggesting that this idea warrants more study.

Climatic barriers

The climate of Australia, like the sea level, has changed continually in response to independent short- and long-term processes. Through the Tertiary, the trend was from warm, moist conditions towards a drier climate, driven mainly by continental drift. Australia and South America have moved away from Antarctica, allowing circumpolar currents to form and preventing warm water mixing with cooler Antarctic oceans. Antarctica began to form glaciers around 33 mya; these increased gradually, then rapidly around 5 mya as South America broke free. Around 3 mya, major glaciation began in the Northern Hemisphere (Partridge *et al.* 1995).

Short-term changes have caused climatic shifts every 100 000–150 000 years (Lüthi *et al.* 2008). These are driven by Milankovitch Cycles, related to three irregularities in the Earth's orbit, namely changes in the shape of orbit, the wobble of the planetary axis and changes in the direction of the axis of rotation. These cycles affect the amount of solar radiation received at the surface of the Earth (Bennett 1990).

Early biogeographers placed great importance on the last glaciation (10 000–20 000 years ago), when massive changes occurred in biotic communities over short periods (Bennett 1990; Lomolino *et al.* 2009). Yet events of similar magnitude occurred in previous glacial cycles (Lüthi *et al.* 2008) and there is no reason to suppose that these were not significant for communities. Indeed, molecular studies have suggested a minor role for Late Pleistocene events in phylogeographic patterns within species, whereas speciation events are much older (Page *et al.* 2004; Unmack and Dowling 2010).

More problems arise in linking phylogeography with Pleistocene climates. One is that molecular clocks have wide confidence intervals, usually spanning more than one full glacial cycle, making it difficult to attribute changes to any one cycle. Another is that fast-evolving molecular markers (e.g. microsatellites) may be better suited to addressing this issue, at least for the most recent cycles. Finally, the suggestion that any one glacial cycle is responsible for an observed pattern overlooks the fact that similar climatic shifts often occurred before and after each cycle (Lüthi *et al.* 2008). This implies that climate may not have been responsible for the pattern.

Freshwater fishes are limited in their ability to avoid the effects of climate change as their capacity for movement is limited by hydrological connections. They must move or tolerate conditions or be extirpated. Recolonisation for many species may be

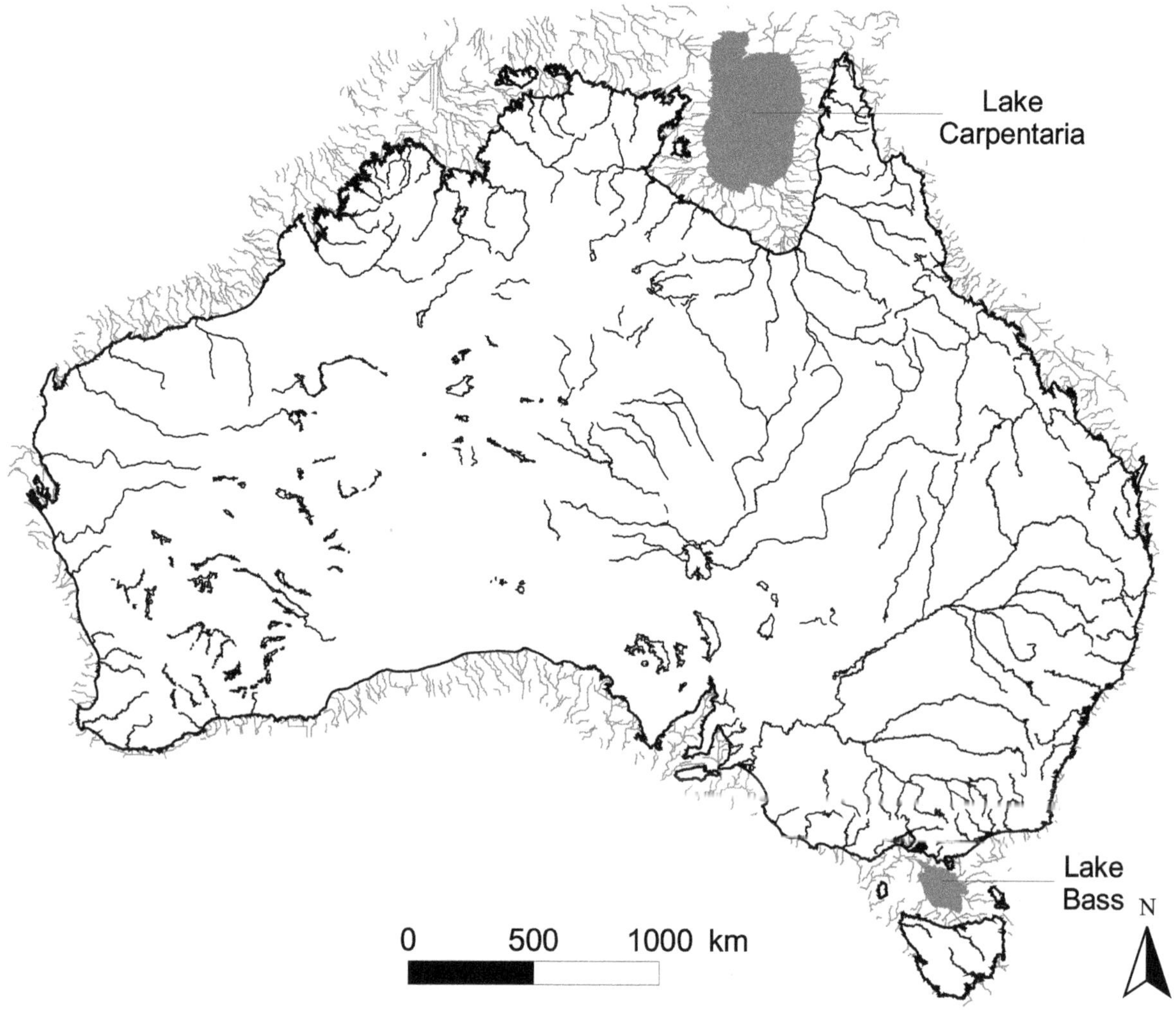

Figure 2.1: Australian low sea-level drainage patterns to 135 m below sea level, derived in ArcGIS from a bathymetric 30 arc-second (c. 1 km) dataset from the Australian Geological Survey Organisation. Major rivers are also shown. The margin of the figure represents the 135 m below sea-level contour.

problematic, given the isolation of drainage basins. The orientation, altitudinal range and size of drainage basins each have a role in allowing fishes to persist. Rivers with north–south axes are likely to cross more climatic zones than east-west rivers, and the same is true of rivers with major altitudinal changes (Moyle and Herbold 1987; Oberdorff *et al.* 1997). As climate changes, these zones shift and fish may move with them, within the basin. Fish near the southern extent of their geographic range may not be close to their low-temperature limits (because cooler temperatures would have prevailed at peak glaciation), but those in the northernmost parts of their range may approach their high-temperature limits (because warmer conditions prevail now than would have prevailed at peak glaciation). This effect should be stronger for species in east–west drainages than for those in north–south drainages (or ones that have broad latitudinal extent, such as the Murray-Darling Basin). The direction of the effect would be opposite, of course, in the Northern Hemisphere.

Other evidence for climatic effects is the pattern of upstream–downstream distributional limits in basins with broad north–south extents. For instance, southern species such as river blackfish (*Gadopsis marmoratus*) and mountain galaxias (*Galaxias olidus*) occur in upstream reaches in the northern parts of their ranges, whereas northern species such as crimson-spotted rainbowfish (*Melanotaenia duboulayi*), Pacific blue-eye (*Pseudomugil signifer*), firetail gudgeon (*Hypseleotris galii*) and flathead gudgeons (*Philypnodon* spp.) favour downstream reaches in the southern parts of their ranges. Of course, climatic constraints may not simply affect species at the limits of their physiological tolerances; other factors, such as temperature thresholds for reproduction or the length of the growing season (Cussac *et al.* 2004), may be more important.

Ecology of species

Only some fishes are likely to utilise a connection between two river basins. The opportunity for dispersal will be available only to those in the vicinity of the connection and it may be accepted only if there is a motivation to move, such as access to improved habitat or other resources. The ability of immigrants to cohabit with other species may then determine whether or not they persist, as will the presence of conspecific individuals (Waters 2011). This situation is illustrated by fishes in the Casiquiare River, a portal between the Orinoco and Negro rivers in Venezuela (Winemiller *et al.* 2008).

In the central latitudes of Australia (i.e. parts of the Pilbara, Central Australian, northern Murray-Darling and central Eastern provinces), there is a striking lack of congeneric sympatry. That is, it is rare for two or more species in the same genus to co-occur. There are exceptions, including the widespread carp gudgeon genus *Hypseleotris*, which has two to five sympatric species (except in the Northern Province). Other exceptions include grunters (*Leiopotherapon*, Pilbara Province), eels (*Anguilla*) and gudgeons (*Gobiomorphus*, Eastern Province). In contrast, the Northern Province is home to 16 (of 58) genera with sympatric species, and in south-eastern Australia nine genera commonly have sympatric species. This suggests that when two congeners come into contact, they either introgress or one is extirpated.

The home ranges of individuals also would influence a species' capacity to disperse and occupy new environments. A species that typically has a small home range, or low fecundity or vagility, is likely to have a small geographic range, to be vulnerable to local extinctions and to show genetic differences between populations. The same would apply for species with narrow environmental needs. In contrast, species with broad requirements (general diet, broad habitat preferences, extended breeding season, tolerant offspring) are likely to be more widespread, to show less propensity for genetic divergence and to be more likely to persist over time. The ecology of individual species clearly is relevant to understanding biogeographic patterns, but often is poorly understood. This would be a fruitful area for research.

Rains of fishes

Fishes, frogs and other animals occasionally fall from the sky (Whitley 1972). How significant are these rains for patterns in biogeography (Unmack 2001a)? What is the nature and strength of storms that cause rains of fishes and how often do they occur in different parts of the world? How far can fish be transported and still survive? From what height and at what velocity can fish be dropped yet still survive? These are intriguing questions, but it is difficult to believe that these events influence biogeographic patterns in ways comparable to, say, fish moving between river basins via headwater connections.

Most claims of rains of fish are mistaken. A storm passes, people find fish flopping about on the ground and naturally assume they have fallen from the sky. Yet fishes are rarely found on rooftops or in water tanks. Meteorologists have suggested that fishes could be carried high by updrafts before falling, but they would be frozen at high altitudes and in any case would hit the ground at such speed that the chances of survival would be very low. In Australia, so-called 'rains' usually refer to spangled perch, often the only species present in inland areas. Spangled perch have astounding dispersal abilities: they have been seen swimming 16 km over six hours, along flooded wheel ruts (Shipway 1947). They can occur in huge numbers in waterholes and after heavy rainfall they spread quickly via overland flow. Their sudden appearances following rain once gave rise to folklore suggesting that they were able to burrow and aestivate in dry mud (Lake 1978). As would be expected, they show very low levels of genetic divergence (Bostock *et al.* 2006); indeed, spangled perch is the most widespread of Australia's freshwater fishes.

Adventitious transport

We sometimes hear of fish appearing in farm dams and other habitats that they supposedly could not have reached by swimming (Lake 1978, p. 58). At least for places on private property, it is unlikely that people could have introduced them. Perhaps the fish arrived as eggs transported by birds?

Adventitious transport of small aquatic invertebrates and plants by birds and other animals is commonplace, but usually only very small items are transported (Figuerola *et al.* 2003, 2005). There appear to be no records of fish eggs being moved in this way, but it is conceivable for some species. As the eggs would need to be broadcast in an area with birds present, this would preclude substrate spawners. They would need to be well-attached to the bird yet readily detached in water when the bird lands, and they would need to tolerate dehydration. And a new population from eggs would require male and female individuals to hatch, survive, mate and reproduce. The combination of probabilities must be very small.

It is also theoretically possible for individual fish to be picked up by piscivorous birds such as cormorants and pelicans and then dropped. Most fish could be transported only a short distance this way, and they would have to survive trauma from capture and transport. Again, the probability would be small.

Historically, indigenous people may have moved fishes between regions but there is no clear evidence in support. When new habitats are formed, they are often colonised by fish within a few years. It is tempting to invoke extraordinary events and unobserved events, and no doubt they do occur, but people tend to underestimate the ability of fishes to disperse via overland flows. In general, the evidence for adventitious transport of freshwater fishes by other animals is not persuasive, and even if there were one or two well-documented examples they would not undermine the general rule.

Waterfalls as barriers

Many waterfalls are barriers to movements of most fishes but it is unclear how strong a barrier, especially over long periods. They may be insurmountable limits (e.g. some fishes in the Burdekin River, Queensland are limited to reaches downstream of Burdekin Falls: Pusey *et al.* 1998a), but some species occur above and below waterfalls (e.g. gudgeons (*Mogurnda*), rainbowfish (*Melanotaenia exquisita*, *M. nigrans*): Larson and Martin 1990). How is this possible?

A waterfall may be younger than the local fish assemblage, or the fish may have colonised the headwaters from another drainage, or they may have climbed the waterfall. Populations above waterfalls have been included in regional studies (e.g. Atherton Tableland, Queensland: Hurwood and Hughes 1998; McGlashan and Hughes 2000) but it seems that no-one has yet investigated the effects of waterfalls on population genetic structure. In the Northern Hemisphere, salmonids often occur above waterfalls (Northcote 2010) and studies of these and other species generally indicate small but significant genetic differences (Tatarenkov *et al.* 2010). Finally, several native Australian fishes are exceptional climbers. These include eels, lampreys, the climbing galaxias (*Galaxias brevipinnis*) and Cox's gudgeon (*Gobiomorphus coxii*) (Merrick and Schmida 1984). All are diadromous species and all occur above waterfalls in south-eastern Australia.

In most cases, waterfalls create strong upstream barriers for some species, but usually not downstream barriers (Pusey *et al.* 1998a; Hardman *et al.* 2002). That is, most of the species above a waterfall usually also occur below it. The Bloomfield River cod (*Guyu wujalwujalensis*) and Mitchell gudgeon (*Kimberleyeleotris hutchinsi*), however, are only known from above waterfalls (Pusey and Kennard 1996; Morgan *et al.* 2011a). Endemic species above waterfalls are known from North America (e.g. Shoshone Falls on the Snake River, Idaho: Smith 1978; Cumberland River Falls, Tennessee: Stearns and Etnier 1986) and South America (e.g. Río Curuá and Río Madeira, Brazil: Birindelli *et al.* 2009; Torrente-Vilara *et al.* 2011). In most cases, the endemic species above waterfalls do not occur at the waterfall itself but occupy upstream habitats. While the presence of species above and not below waterfalls is unusual, there may be other reasons, such as predation (e.g. trout below waterfalls; *Galaxias* species above: McDowall 2006).

AUSTRALIA'S FRESHWATER FISH FAUNA

Assembling a list of Australian freshwater fishes is less easy than it might seem. Some 'freshwater' species spend parts of their lives in estuaries or the sea, and some 'marine' species spend considerable time in fresh water. Species that occupy marine habitats for significant times are often excluded from freshwater biogeographic analyses because they are not constrained by sea-water barriers (Unmack 2001a).

With this in mind, Australian freshwater fishes are of five kinds.

1 Species that live in fresh water and never occur in estuarine or marine environments. These are the focus for this book.
2 Species that occur mainly in fresh water but also in tidal reaches and estuaries; for example flathead gudgeons (*Philypnodon*), empire gudgeon (*Hypseleotris compressa*), *Gobiomorphus*, nurseryfish (*Kurtus*), Australian bass (*Percalates novemaculeata*), Australian smelt (*Retropinna semoni*), Pacific blue-eye and various flounders. These generally are considered freshwater species but they may occasionally 'cheat' by crossing marine barriers.
3 Species with a diadromous life cycle, spending parts of their lives in marine and freshwater environments. These include some galaxiids, Australian grayling (*Prototroctes maraena*), barramundi (*Lates calcarifer*), flagtails (Kuhliidae), gobies and others.
4 Species that are common in estuaries but also enter fresh waters. Many gobies (Gobiidae) belong here, as do some hardyheads (Atherinidae) and estuary perch (*Percalates colonorum*).
5 Species that are mainly marine but often enter fresh water, sometimes for long periods, and may travel long distances inland. These include bull shark (*Carcharhinus leucas*), oxeye herring (*Megalops cyprinoides*) and mangrove jack (*Lutjanus argentimaculatus*) (see Allen *et al.* 2002).

As this chapter concerns species in fresh waters, the following discussion is limited to fishes in categories 1–3 above (that is, estuarine and marine species are excluded).

Another caveat here is to acknowledge undescribed species and changes in generic status, most of which are based on molecular data and are not yet formally published. Molecular work has demonstrated that for many groups the current taxonomy is inadequate and almost every study has revealed undescribed species (Crowley and Ivantsoff 1990; Musyl and Keenan 1992, 1996; Page *et al.* 2004; Wong *et al.* 2004c; Hammer *et al.* 2007). As this work progresses, some widespread species are likely to be found to contain multiple species where, despite morphological similarities, the degree of genetic divergence and the geographic pattern of genetic discontinuities are beyond those typical of single species. In other words, these groups may presently contain cryptic species (see Ch. 3; Bickford *et al.* 2007).

In some cases, molecular data have suggested that changes are warranted at a generic and even family level. For example, *Percalates colonorum* and *P. novemaculeata*, two species formerly assigned to *Macquaria*, are not monophyletic with other *Macquaria* species (Jerry *et al.* 2001). There are also differences in larval morphological features between these species and the golden perch *Macquaria ambigua*, supporting the reassignment to *Percalates* (Trnski *et al.* 2005). Indeed, recent work based on multiple nuclear genes places the two species as a separate family, remote from the Percichthyidae (Near *et al.* 2012). Another example concerns molecular data for the freshwater cobbler '*Tandanus*' *bostocki*. Recent work (Unmack, unpubl.) suggests that this is not monophyletic with *Tandanus* nor closely related to other plotosid genera, and it is likely that will be placed in a new, monotypic genus. I also consider the Bloomfield River cod *Guyu wujalwujalensis* to be part of *Macquaria*. The molecular evidence shows this as a sister species to Macquarie perch (*M. australasica*) and shows these two as sister species to golden perch (Jerry *et al.* 2001; Unmack, unpubl.).

The issue of undescribed taxa is emphasised by five deficiencies in traditional taxonomic work on

Australian freshwater fishes. These are (a) a paucity of museum vouchers and poor geographic coverage for many species (freshwater fishes are rarely included in faunal surveys), (b) few researchers outside museums retain formalin-preserved voucher specimens, (c) there is a lack of sustained taxonomic effort, (d) there is a comparative lack of taxonomic interest in freshwater fishes and (e) there has been a decline in the number of Australian fish taxonomists since the early 1990s (Leis *et al.* 2007). As a consequence, some current taxonomy is based on preliminary studies before the 1980s, often based on few specimens from limited areas.

Taxonomic research on Australian freshwater fishes (and, it must be acknowledged, other fauna and flora) has all but stalled for want of funding. Prior to 1991, seven state museums had employed a total of 11 curators or supporting scientists in this field (Paxton and McGrouther 1997; Leis *et al.* 2007) but today only three museums have a fish curator and there is only one associated scientist. Some former curators are describing fishes in retirement, but there are few employed fish taxonomists and opportunities for trainees are scarce indeed. The irony is that robust taxonomies are now more achievable than ever, not least because of the advent of molecular methods.

This will explain why undescribed taxa are highlighted above. It is easy to understand why this is a concern, made more serious by the lack of commitment to taxonomic research. For example, from the viewpoint of biodiversity conservation, some of these new taxa may be threatened and patterns of species richness and endemism based on inaccurate taxonomy could be misleading. In the following discussion, I use the current names for all 34 new cryptic species recognised herein, but a Roman numeral is added to indicate cryptic species within taxon (e.g. Hyrtl's catfish: *Neosilurus hyrtlii* I, *N. hyrtlii* II, *N. hyrtlii* III). Designations of cryptic species are conservative and generally based on combined mitochondrial and nuclear DNA sequences and allozyme electrophoresis. There are perhaps another 60 species pending confirmation, distributed over many genera (e.g. *Ambassis*, *Galaxias*, *Glossamia*, *Hypseleotris*, *Melanotaenia*, *Mogurnda*, *Philypnodon*). Tables 2.2 and 2.3 provide summations based on both the taxonomy presented here and the 60 species that are pending confirmation.

With those caveats, there currently are 256 known species in 36 families in Australian inland waters. This includes five recently described or promoted species, plus 34 undescribed taxa beyond those listed by Allen *et al.* (2002). In terms of species richness, the eight most speciose families (22% of families) account for 186 species, or 73% of all Australian species (Table 2.2). Despite the disparities in species richness across families, many of the less-speciose families are widespread, apart from the Neoceratodontidae (Australian lungfish) and Lepidogalaxiidae (salamanderfish). A similar pattern exists within genera. Of 89 genera, the 11 most speciose (with six or more species) account for 104 species, or 41% of the total (Table 2.3).

The first attempt to define freshwater biogeographic regions for fishes (and molluscs) was the 'fluvifaunula' concept of Iredale and Whitley (1938), dividing Australia into nine provinces. Lake (1971) matched fish distributions to 12 major drainage divisions and his classification appeared in many subsequent publications. Unmack (2001a) reconsidered the distributions of 156 strictly freshwater fishes (from categories 1–2 above) with regard for differences between the faunas of rivers, and recognised 31 regions (Fig. 2.2). These regions were further grouped as 11 provinces, based on analyses that maximised differences in species composition as well as highlighting concentrations of endemic taxa.

The terms 'region' and 'province' used here are broadly synonymous with 'biome', 'realm', 'biotic community', 'bioregion' and 'ecoregion'. Some boundaries between 'regions' and 'provinces' are not clear geographic demarcations (e.g. the boundary between Bass and Eastern provinces) and should be thought of as fuzzy boundaries. The scheme shown here is modified to reflect new records and discoveries, and includes a re-examination of within-province patterns (cf. Unmack 2001a). There are changes to the northern extent of the South-Western Province, given new records of western galaxias (*Galaxias occidentalis*) and western pygmy perch (*Nannoperca vittata*) (Arrowsmith River: Morgan and Gill 2001). The Pil-

Table 2.2: Australian freshwater fishes: species richness by family

Family	Species richness
Eleotridae	31 (59)
Percichthyidae	26 (28)
Terapontidae	25 (26)
Gobiidae	24 (27)
Galaxiidae	23 (38)
Melanotaeniidae	21 (25)
Plotosidae	20
Atherinidae	16
Ambassidae	8
Ariidae	7
Retropinnidae	7
Pseudomugilidae	6 (7)
Soleidae	6
Anguillidae	5
Clupeidae	4
Toxotidae	3
Kuhliidae	2
Mordaciidae	2
Osteoglossidae	2
Synbranchidae	2 (4)
Apogonidae	1 (5)
Belonidae	1
Bovichtidae	1
Cynoglossidae	1
Engraulidae	1
Geotriidae	1
Hemiramphidae	1
Kurtidae	1
Latidae	1
Lepidogalaxiidae	1
Mugilidae	1
Muraenidae	1
Neoceratodontidae	1
Odontobutidae	1
'Percichthyidae'	1
Tetrarogidae	1

The 'Percichthyidae' is represented by *Percalates* and will eventually be given a new name.
Numbers in parentheses are estimates with species of uncertain taxonomy included.

bara Province boundary is moved southward to include the Irwin River, which lacks native fishes but has the same turtle species as other Pilbara drainages (*Chelodina steindachneri*: D. Morgan, Murdoch University, pers. comm.). The Pilbara Province now has three subprovinces, following Morgan and Gill (2004) and Morgan *et al.* (2004a), and the Northern Province now has five rather than two subprovinces (Fig. 2.2).

Species richness by region and province is shown in Figure 2.3. The most striking pattern is that most western, central and southern regions have high levels of endemism, whereas those from northern Australia (VOR) around the coast to eastern Victoria (SEV) all have low endemism, with the exception of north-eastern Queensland (NEQ). Thus, the endemic line in Figure 2.4 divides Australia into regions of high and low endemism. The separation mainly reflects isolation, as the larger, high-endemism region is fragmented by aridity, drainage divides and other factors that impede dispersal of fishes (Unmack 2001a). In contrast, species across northern and eastern Australia tend to be widespread and there is a lower proportion of endemics (Table 2.1).

BIOGEOGRAPHIC PROVINCES

South-Western Province

The South-Western Province has one of the lowest levels of species richness but the highest proportions of endemic species (79%), genera (50%) and families of any province. Most of the species have near-relatives in south-eastern Australia (e.g. *Galaxias*, *Galaxiella*, *Nannoperca*), but the genera *Bostockia* and *Nannatherina* are endemic, as is the local freshwater cobbler (nominally '*Tandanus*' *bostocki*, but see 'Australia's freshwater fish fauna'). The Lepidogalaxiidae has origins in the Cretaceous (or earlier) and appears to be a sister group to the Euteleostei, containing most living fishes (Li *et al.* 2010). Another endemic is the western pygmy perch, which includes up to four species in this province (*N. vittata* I-IV; Unmack *et al.* 2011). The only species shared with other provinces have marine life history phases (pouch lamprey, *Geotria australis*; common galaxias, *Galaxias maculatus*; spotted galaxias, *G. truttaceus*) or are mainly estuarine (Swan River goby, *Pseudogobius olorum*).

These patterns are not limited to fishes, as south-western Western Australia is one of the world's top 25

Table 2.3: Australian freshwater fish species richness by genus

Genus	Species richness
Craterocephalus	16
Galaxias	13 (28)
Melanotaenia	13 (17)
Hypseleotris	11 (15)
Neosilurus	11
Nannoperca	9
Craterocephalus I	7
Glossogobius	7 (10)
Ambassis	6
Craterocephalus II	6
Macquaria	6
Mogurnda	6 (22)
Retropinna	6
Anguilla	5
Chlamydogobius	5
Gadopsis	5 (6)
Hephaestus	5
Pseudomugil	5 (6)
Rhadinocentrus	5
Scortum	5
Stiphodon	5
Galaxiella	4
Maccullochella	4
Oxyeleotris	4
Paragalaxias	4
Syncomistes	4 (5)
Craterocephalus III	3
Leiopotherapon	3
Leptachirus	3
Neoarius	3
Pingalla	3
Porochilus	3
Tandanus	3
Toxotes	3
Awaous	2
Bidyanus	2
Brachirus	2
Gobiomorphus	2
Iriatherina	2
Kimberleyeleotris	2
Kuhlia	2
Mordacia	2
Nematalosa	2
Ophisternon	2 (4)
Philypnodon	2 (10)
Sciades	2
Scleropages	2
'Oxyeleotris'	1
Amniataba	1
Anodontiglanis	1
Bostockia	1 (2)
Bostrychus	1
Bunaka	1
Cairnsichthys	1
Cinetodus	1
Clupeoides	1
Cynoglossus	1
Denariusa	1
Geotria	1
Giurus	1
Glossamia	1 (5)
Gymnothorax	1
Hannia	1
Hemiarius	1
Kurtus	1
Lates	1
Lepidogalaxias	1
Lovettia	1
Milyeringa	1
Nannatherina	1
Neoceratodus	1
Neochanna	1
Neosiluroides	1
Notesthes	1

Table 2.3: Australian freshwater fish species richness by genus (continued)

Genus	Species richness
Parambassis	1
Percalates	1
Potamalosa	1
Prototroctes	1
Pseudaphritis	1
Scaturiginichthys	1
Schismatogobius	1
Sicyopus	1
Sicyopterus	1
Smilosicyopus	1
Stenogobius	1
Strongylura	1
Synclidopus	1
'Tandanus'	1
Thryssa	1
Trachystoma	1
Variichthys	1
Zenarchopterus	1

The genera *'Oxyeleotris'* and *'Tandanus'* refer to *O. nullipora* and *T. bostocki*, respectively; these belong in other genera and await formal description (Thacker and Unmack 2005; Unmack, unpubl.). The genus *Craterocephalus* is included whole and as three species-groups ('eyresii' (I), 'stercusmuscarum' (II), 'stramineus' (III): see Unmack and Dowling 2010).
Numbers in parentheses are estimates with species of uncertain taxonomy included.

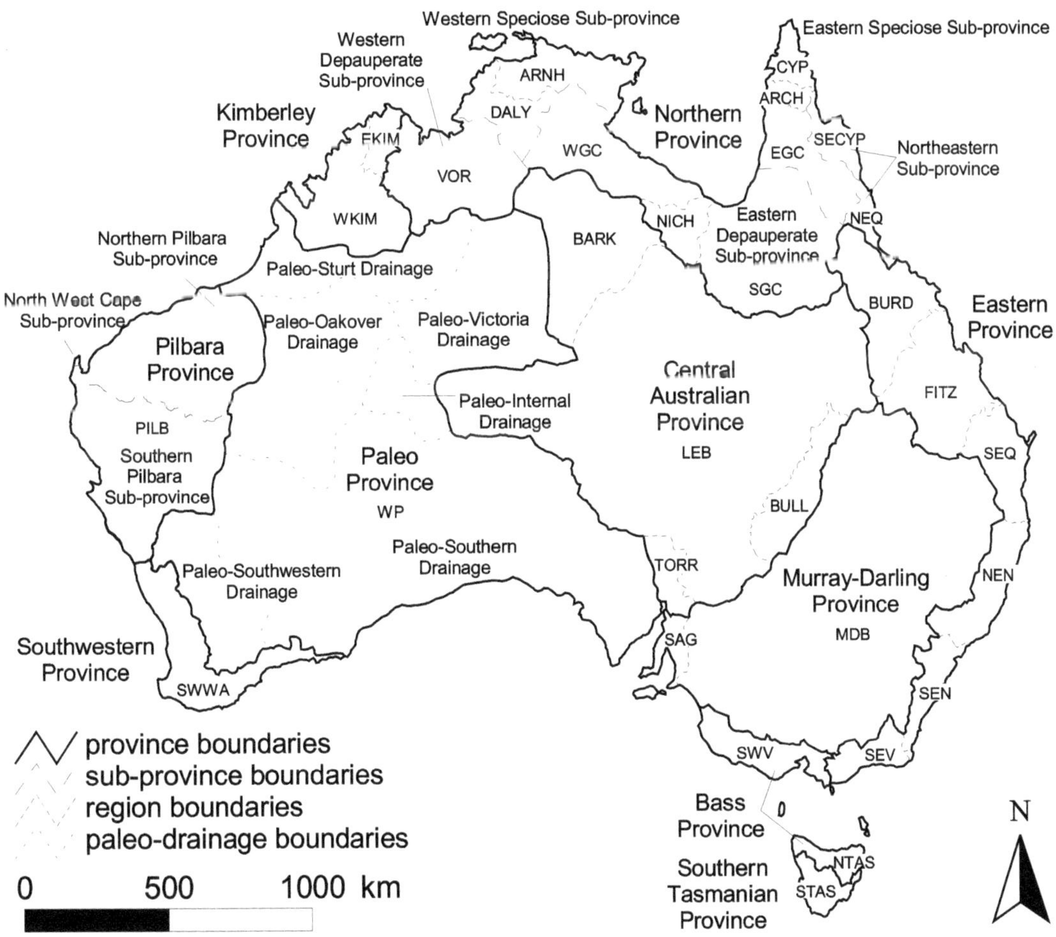

Figure 2.2: Freshwater fish biogeographic provinces, subprovinces and regions in Australia (modified from Unmack 2001a).

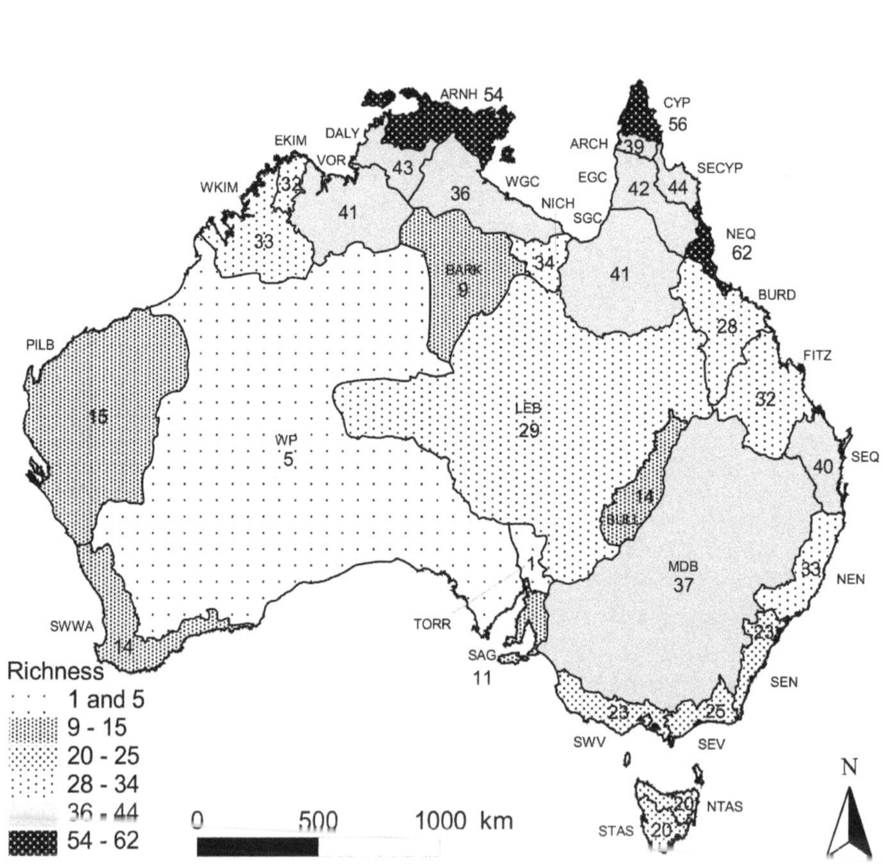

Figure 2.3: Australian freshwater regions with freshwater fish species richness indicated by number and differential shading.

hotspots for endemism (Myers *et al.* 2000). This is due partly to prolonged isolation by arid regions to the north and east. Earlier in the Tertiary, southern Australia was wetter, with aquatic habitats distributed from east to west. The central area began to dry in the Oligocene and stream flow there essentially ceased by the mid Miocene (Van de Graaff *et al.* 1977; Benbow 1990; Martin 1994). Fishes were confined to the wettest corner of Western Australia, isolated from other species and unable to migrate.

Pilbara Province

The Pilbara Province is another area isolated by aridity, but less so than for the South-Western Province because the Kimberley Province is comparatively near. Most species here either are endemic (47%) or widespread in Australia. The deep hardyhead (*Craterocephalus cuneiceps*) is unusual in that its nearest relatives are in eastern Australia (Unmack and Dowling 2010), whereas most other Pilbara species have near-relatives in the Kimberley Province. Another endemic species, the golden carp gudgeon (*Hypseleotris aurea*), is related to species in northern Australia (Thacker and Unmack 2005), as is the Fortescue grunter (*Leiopotherapon aheneus*: Bostock *et al.* 2006). Two other species, the cave gudgeon (*Milyeringa veritas*) and the blind cave eel (*Ophisternon candidum*), are endemic to a subterranean system in the North West Cape. This is one of only a few places in the world where there are two sympatric blind cave fish (Proudlove 2006). The remaining species are widespread or are related to widespread species and two may be new species, namely the bony herring *Nematalosa erebi* II and Hyrtl's catfish *Neosilurus hyrtlii* IV.

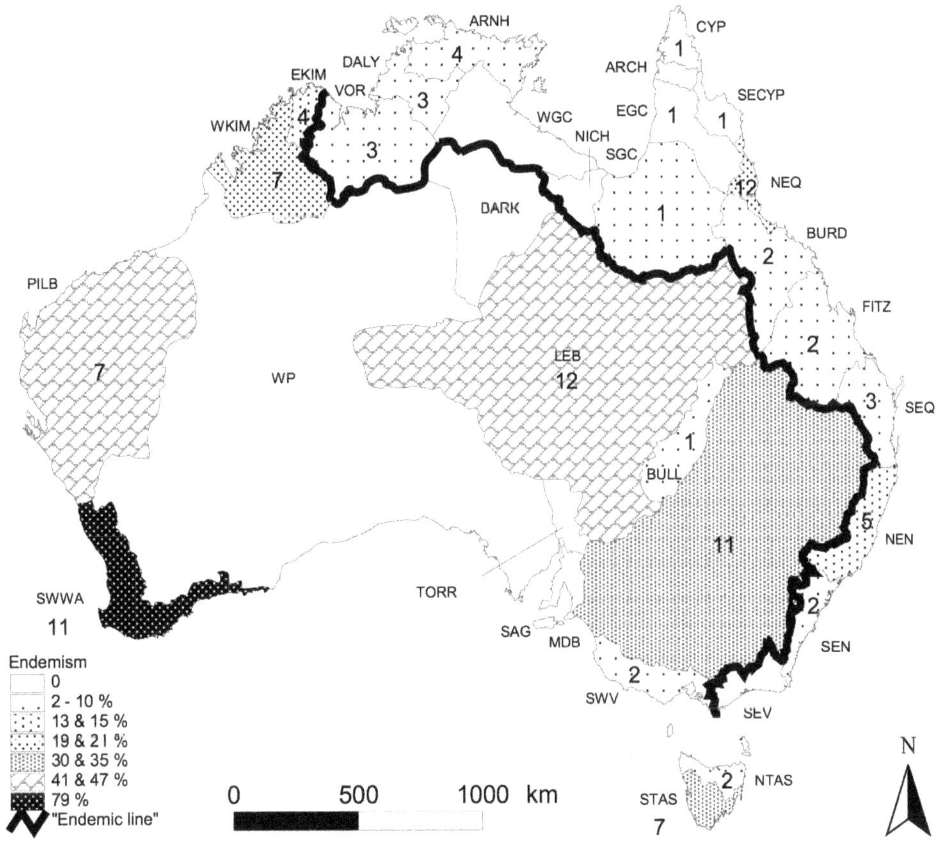

Figure 2.4: Numbers of Australian freshwater fishes endemic to each region, indicated by number and differential shading based on the percentage of endemics. The thick black 'endemic line' separates regions with low endemism to the north and east from the remainder of Australia, most of which has relatively high endemism.

In two other species that have been genetically examined, namely the spangled perch and western rainbowfish (*Melanotaenia australis*), the Pilbara populations show genetic differences that reflect their present isolation (Unmack 2005, unpubl.; Bostock *et al.* 2006). Morgan and Gill (2004) suggested that Pilbara Province should be split into three subprovinces (Southern Pilbara, Northern Pilbara, North West Cape), to highlight differences in species richness and acknowledge the subterranean fauna.

Kimberley Province

This province has widespread species, endemic species (32%) and two endemic genera (*Kimberleyeleotris, Hannia*). Most of the rivers flow through gorges that probably were critical refuges in past dry climates as they are today; they constrain the valley and force hyporheic water (water flowing through the river bed) to the surface (Unmack 2001b). The gorges provide some of the most persistent aquatic habitats in an otherwise arid region, as well as providing unique habitats that may promote local species richness.

Several families have endemic species here as well as high species richness; this is true especially of Eleotridae (five species, nine total), Terapontidae (three, nine), Plotosidae (two, five) and Melanotaeniidae (two, four). A notable absentee is the Pseudomugilidae, a family present in all other northern and eastern coastal provinces. Kimberley Province is the westernmost limit for a number of species, due to aridity and the scarcity of rivers to the south. The resident fishes generally have their nearest relatives in the Northern Province, reflecting a lack of geographic barriers. A review of the fish fauna of this region and parts of the adjacent Northern Province is provided by Morgan *et al.* (2011a).

Northern Province

This province represents a vast northern area with high species richness (125 species) but comparatively few endemics (36 species, or 29%); indeed, nearly half of the species (62) are shared with New Guinea. Most of the endemic species are confined to small areas, especially in the Wet Tropics region of Queensland.

The outstanding endemic here is a percichthyid, the Bloomfield River cod, with a tiny range and far from its closest relative, Macquarie perch, whose northernmost limit is 2000 km away in central New South Wales (Jerry *et al.* 2001; Pusey *et al.* 2001). Another highlight is the endemic genus *Cairnsichthys*, which may be ancestral among Melanotaeniidae (Sparks and Smith 2004). The Terapontidae has seven endemic species and the Atherinidae and Melanotaeniidae each have six. Eight families account for 94 species (78%), with 19 species of Gobiidae, 16 species of Terapontidae, 14 species of Eleotridae and 12 species of Melanotaeniidae. Another highlight in Northern Province is the presence of an amphidromous fauna in the Wet Tropics region. Ten amphidromous species (mostly gobies) have recently been recorded for the first time in Australia (Thuesen *et al.* 2011a). Undoubtedly this number will increase; many streams remain unsurveyed, most populations are small and the species with access to these streams may change over time. This fauna is found primarily in small, steep catchments which lack species restricted to fresh water. Most of the species otherwise are found on oceanic islands.

My original scheme recognised three subprovinces (Speciose Northern, Depauperate Northern, North-Eastern: Unmack 2001a) but, as the first two are not contiguous and some faunal differences exist, it is better to distinguish five subprovinces: Western Depauperate, Western Speciose, Eastern Depauperate, Eastern Speciose and North-Eastern (Fig. 2.2). The differences between the speciose and depauperate subprovinces reflect north–south differences in climate (the latter are more arid and have lower species richness: Table 2.1). The easternmost area, the North-Eastern Subprovince, has a number of endemic species and includes the Wet Tropics, an area with steep stream gradients and substantially more rainfall than anywhere else in Australia.

Several factors contribute to the wide distributions of many species in Northern Province. Some drainage divides are low relief, providing easier opportunities for fish to move overland between basins. For instance, several fishes are shared between the headwaters of the Daly, Roper and Alligator rivers, all rising on a relatively flat plateau. Few differences exist in the fauna on the eastern and western sides of Cape York, where there is little topographic relief, high rainfall and potential for flooding, increasing the chances for fishes to disperse.

Perhaps the main influence is low sea level, when the entire region between Australia and New Guinea was dry land and the rivers were joined (Fig. 2.1). The situation is complex, as during the most recent low sea levels the region was more arid than today (Webster and Streten 1972; Williams 1984) and probably unsuitable for some fishes. In the present Gulf of Carpentaria there is a large depression, 'Lake Carpentaria', which at times has variously contained fresh or brackish water (Reeves *et al.* 2008). The lake could have precluded species mixing between rivers and it may have restricted eastward movement from streams flowing north from Arnhem Land in the Northern Territory. Some species are widespread in northern Australia or southern New Guinea but are not shared (Unmack 2001a). The ecological differences between rivers in the two regions could prevent colonisation by some species. Genetic studies (on fishes and aquatic invertebrates) have demonstrated connectivity within the past two to three glacial cycles, based on shared or similar haplotypes (de Bruyn *et al.* 2004; Baker *et al.* 2008; Huey *et al.* 2008, 2010; Masci *et al.* 2008), and older connections, perhaps in the Early Pleistocene (Unmack 2005; Cook and Hughes 2010; Unmack and Dowling 2010).

Eastern Province

Eastern Province is an enigma, covering a vast range from north to south along the eastern coast but lacking strong faunal differences between any two adjacent rivers. Yet the southernmost and northernmost rivers share no species. The province appears as a classic biogeographic 'filter' (Lomolino *et al.* 2009), where species are gradually lost and gained along a geographic gradient. The filter agent appears to be the climatic gradient, for at least some species. It suggests the ranges of these species are controlled by physiological factors such as tolerance to temperature (e.g. effects of temperature on reproduction). Some species may be limited by rainfall, as patterns of stream runoff vary considerably along the north–south axis. Biotic interactions could also be important.

Despite the great extent of Eastern Province it has a modest number of endemic species (30 species, 39%), mainly because there have been frequent connections with surrounding provinces. The outstanding endemic here (indeed, in all of Australia) is the Australian lungfish (*Neoceratodus forsteri*), the sole extant member of the Neoceratodontidae. This species is restricted to the Mary and Burnett rivers in south-eastern Queensland although it also occurs in the Brisbane River, where it may have been introduced (Kemp 1996; Frentiu *et al.* 2001; Pusey *et al.* 2004). Historically, lungfishes were widespread in Australia and there are at least 13 described fossil species (Unmack 2001a), but all but one were lost as the climate dried through the Miocene and Pliocene.

Several rivers, including the Burdekin and Fitzroy, have multiple endemic species and the latter is home to the southern saratoga (*Scleropages leichhardti*), whose nearest relatives are in northern Australia and southern New Guinea. Molecular work suggests that the Burnett, Mary and Brisbane rivers may harbour other endemic *Hypseleotris* and *Philypnodon* species (Thacker *et al.* 2007; Hammer 2008).

Other unusual endemic species include the non-parasitic lamprey (*Mordacia praecox*) from the southern Eastern Province, and Oxleyan pygmy perch (*Nannoperca oxleyana*), a relative of the Yarra pygmy perch (*N. obscura*) and the southern pygmy perch (*N. australis*), both from south-eastern Australia (Unmack *et al.* 2011). Finally, two species of *Maccullochella* (Mary River cod, *M. mariensis* and Clarence River cod, *M. ikei*) occur in the province, although other *Maccullochella* of unknown affinities were present historically in the Brisbane and Richmond rivers (Rowland 1993; Nock *et al.* 2010).

Southern Tasmanian Province

Nearly all the fishes in this province have marine life history stages and the four endemic *Galaxias* species may be landlocked populations derived from diadromous species (Andrews 1976; Burridge *et al.* 2012). The most significant endemics are in the genus *Paragalaxias*, with three of four species occurring only in this province. Despite abundant aquatic environments, the province appears to have been long-isolated from most of Australia and emigration opportunities for resident fishes would have been limited as they could move only to the north. The province may have experienced more extreme cold conditions than the rest of Australia, limiting spawning opportunities for some species. In addition, southern Tasmania has a narrow continental shelf that would have hindered movements between rivers during low sea levels.

Bass Province

Bass Province illustrates the role that sea-level changes may have in shaping fish distributions. The fish faunas of northern Tasmania and the westernmost two-thirds of Victoria are similar, despite their current marine separation (of 28 species, 17 are shared). The absence from Tasmania of mountain galaxias and Yarra pygmy perch, both widespread in southern Victoria, is remarkable. Four species only are endemic to Bass Province and all have narrow ranges; they are the saddled galaxias (*Galaxias tanycephalus*, a landlocked relative of the spotted galaxias), a variant of the dwarf galaxias (*Galaxiella pusilla* II), Arthurs galaxias (*Paragalaxias mesotes*) and variegated pygmy perch (*Nannoperca variegata*).

Murray-Darling Province

Murray-Darling Province is flanked by more basins than any other (there are 20 east of the Murray River mouth, a few immediately west of the mouth, and Bulloo River and Cooper Creek in Central Australian Province). Thus, the provincial fish fauna is a blend of north-eastern, north-western, southern and central eastern influences. It includes 12 endemic species, most of them with near-relatives in adjacent drainages.

Connections and faunal exchanges between Eastern Province and Murray-Darling Province have occurred and it is now clear which coastal river basins have been involved. The most species shared with Murray-Darling Province are in the Brisbane River (Eastern Province) but the molecular evidence indicates that many of the recent connections have been via the Burnett River, with some links to the Fitzroy, Clarence, Hunter, Hawkesbury and Glenelg rivers. Murray-Darling genera with close relatives in the Burnett River include *Ambassis*, *Craterocephalus*, *Hypseleotris*, *Melanotaenia*, *Mogurnda*, *Philypnodon*, *Porochilus* and *Tandanus* (Unmack 2005; Thacker *et al.* 2007, 2008; Jerry 2008; Unmack and Dowling 2010). The connection may have occurred across a plain that separates the upper Boyne River (a Burnett tributary) and the upper reaches of Burra Creek (tributary to the Condamine River, hence the Murray-Darling system). Shared similar and identical haplotypes in both provinces show that movements in both directions have occurred.

A second key question is whether Murray-Darling Province fishes evolved in the basin and then moved into coastal Eastern Province drainages, or originated in Eastern Province and then invaded Murray-Darling Province, or whether some species had independent origins (i.e. their nearest relatives are in other provinces). Some researchers have argued that Murray-Darling Province populations of *Tandanus* and *Macquaria* evolved from coastal ancestors, based on molecular phylogenetic trees showing their derived position relative to coastal populations (Faulks *et al.* 2008, 2010a; Jerry 2008). A similar pattern exists for *Hypseleotris* sp. 'Midgley' (Thacker *et al.* 2007). In other cases (e.g. western carp gudgeon, *Hypseleotris klunzingeri*: Thacker *et al.* 2007; southern purple-spotted gudgeon, *Mogurnda adspersa*: Faulks *et al.* 2008), the data are ambiguous as a reciprocally monophyletic (single) split separating coastal and inland populations offers no insight into origins. Golden perch also may show a single split.

Determining directionality and geographic origins from phylogenetic data is fraught, for several reasons. In most cases, the results are based solely on mitochondrial DNA (mtDNA), which may reveal patterns unlike those identified by nuclear markers (Ballard and Whitlock 2004; Ch. 3). Second, multiple appropriate outgroups (i.e. all sister species plus other congeners) need to be included in analyses. Third, there are dubious assumptions about ancestral ranges (e.g. it is commonly assumed that the distribution of lineages is the same today as in the past, when this is unlikely to have been true of all taxa). Fourth, single-gene phylogenies are often discordant with the phylogeny of populations or species under consideration (Funk and Omland 2003), even without the above-mentioned limitations. A fifth factor is that within-taxon hybridisation, lineage replacement (i.e. overwriting old genetic signatures by new immigrants) and competitive exclusion (Waters 2011) are likely to have featured in some species. This would disguise the fact that populations existed prior to the latest immigration of a new lineage and would affect interpretation of origins. In some cases, the extinction of older lineages may be due to an influx of novel (invading) mtDNA lineages.

To illustrate the point that origins are complex and dynamic, and cannot be determined unambiguously from single-gene gene trees, consider the mtDNA genealogy for fly-specked hardyhead *Craterocephalus stercusmuscarum* IV (*C. s. fulvus*). In this taxon, the southernmost Eastern Province population is sister to all other populations, representing both Eastern and Murray-Darling province lineages (Unmack and Dowling 2010), suggesting an Eastern Province origin. The next split is between an exclusively Eastern Province clade and a Murray-Darling Province and Burnett River (Eastern Province) clade. This is consistent with an invasion from coastal basins into Murray-Darling Province, followed by an invasion back to the Burnett River and Eastern Province. A more parsimonious, alternative explanation is for a single invasion from Eastern Province (Burnett) into Murray-Darling Province, but this implies long-term persistence of a unique lineage in the Burnett River without mixing or replacement by other Eastern Province lineages during periods of low sea level (which seems unlikely). While based only on mtDNA data, these ambiguities highlight how biogeographic interpretations would change if certain populations had become extinct (i.e. the southernmost Eastern Province population). If the current mtDNA lineage in the Burnett River were to 'invade' and replace those in surrounding Eastern Province basins, this would remove all mtDNA evidence of past relationships.

Replacement of a pre-existing mtDNA lineage with another does not mean, of course, that the genetic signature of the pre-existing population is erased. Thus, no study based on a single gene can ever provide more than a cursory snapshot of reality. Chapter 3 shows that the elucidation of phylogenetic and phylogeographic patterns requires a number of informative and independent genes for genealogical concordance. This is especially so for Murray-Darling Province fishes, where the potential for multiple movements in either direction makes statements about origins very speculative.

High levels of recent genetic mixing are indicated by molecular evidence for some species (golden perch, flathead gudgeon: Thacker *et al.* 2008; Faulks *et al.* 2010b, c) and would be promoted by extensive migrations (e.g. golden perch: Reynolds 1983). Some species now separated by big distances have similar haplotypes across the basin, implying long-term mixing (e.g. Australian smelt; Murray River rainbowfish, *Melanotaenia fluviatilis*; fly-specked hardyhead; Murray hardyhead, *C. fluviatilis*; carp gudgeons, *Hypseleotris* spp.; Murray cod, *Maccullochella peelii*; Macquarie perch; southern purple-spotted gudgeon; dwarf flathead gudgeon, *Philypnodon macrostomus*) (Unmack 2005; Hammer *et al.* 2007; Thacker *et al.* 2007, 2008; Faulks *et al.* 2008; Nock *et al.* 2010; Rourke *et al.* 2010; Unmack and Dowling 2010; Woods *et al.* 2010). A few others have higher levels of structuring, suggesting less-frequent mixing of populations (e.g. Darling hardyhead, *Craterocephalus amniculus*: Adams *et al.* 2011).

Some species are widespread across the southern half of the Murray-Darling Basin but with outlying populations in the Macquarie River, the most southerly Darling tributary. These include flathead galaxias (*Galaxias rostratus*), trout cod (*Maccullochella macquariensis*), Macquarie perch, flathead gudgeon and dwarf flathead gudgeon (also in the Condamine River). These presumably were all more widespread in the northern part of the province, along with species like southern pygmy perch, currently restricted to the southern part.

Central Australian Province

Central Australian Province is similar to Murray-Darling Province, being surrounded by other basins and having a corresponding admixture of species. There are strong north-eastern, northern and south-eastern influences and 18 endemic species, eight of them restricted to desert springs. The phylogenetic relationships with surrounding areas are not always clear, but recent connections have existed to Gulf of Carpentaria drainages (Northern Province), enabling exchanges of glassfish (*Ambassis*, apparently in two invasions from Gulf drainages: Huey *et al.* 2011a), bony herring (Masci *et al.* 2008) and eastern rainbowfish (*Melanotaenia splendida*: Unmack 2005). Connections have existed also with the Burdekin River (Eastern Province), facilitating transfers of fly-specked hardyhead (Unmack and Dowling 2010), eastern rainbowfish (Unmack 2005), Midgley's carp gudgeon (Thacker *et al.* 2007) and glassfish (Unmack and M. Adams, unpubl.). Some species are related to those in the Murray-Darling Province, including bony herring (Hughes and Hillyer 2006), eastern rainbowfish (Unmack 2005), golden perch (Faulks *et al.* 2010b) and western carp gudgeon (Thacker *et al.* 2007), and Australian smelt represents an older exchange (Hughes and Hillyer 2006; Hammer *et al.* 2007). The locations and times of these connections are unknown.

The outstanding endemic here is the enigmatic Cooper Creek catfish (*Neosiluroides cooperensis*), known only from the Cooper Creek drainage. This is the sole member of a genus that is a distinct genetic lineage among the Plotosidae (Unmack, unpubl.). It has the best eyesight of any Australian plotosid, yet lives in water consistently so turbid that vision is impossible. It probably has the lowest fecundity per body weight of any Central Australian fish but nothing more is known of its reproduction or ecology (Unmack 1996).

Other endemics occur in mound springs. Dalhousie Springs, South Australia, has four endemic species of which one, the Dalhousie hardyhead (*Craterocephalus dalhousiensis*), voluntarily moves into water at 41.8°C, the hottest fish habitat recorded in Australia and one of the hottest records in the world (Wager and Unmack 2000). Most other species in Dalhousie Springs occur in water up to 39–40°C.

The goby genus *Chlamydogobius* has diversified in the province, with four species in mound-spring complexes and in the Finke River (Larson 1995). The redfin blue-eye (*Scaturiginichthys vermeilipinnis*) is endemic to tiny springs at Edgbaston, Queensland, where it co-occurs with Edgbaston goby (*Chlamydogobius squamigenus*). Owing to the shallowness and low discharge typical of mound springs, these species confront extraordinary physiological challenges, with temperature swings of 21°C or more over several hours daily (Wager and Unmack 2004). One other species, while not endemic, has a remarkable disjunct distribution. The fly-specked hardyhead *Craterocephalus stercusmuscarum* III is widespread in north-eastern Australia but in Central Australian Province it occurs only in Cooper Creek (Wager and Unmack 2000; Unmack and Dowling 2010). One population occurs in a single, spring-fed pool on Myross Station (near Edgbaston) and another is 1000 km away in McDonnell Creek, tributary to Lake Blanche.

Only five species have been recorded from Lake Eyre itself, namely Lake Eyre hardyhead (*Craterocephalus eyresii*), spangled perch, golden perch, bony herring and Australian smelt. The role that the lake plays in isolating the faunas of tributary rivers is being tested using molecular data (Huey *et al.* 2008, 2011a; Masci *et al.* 2008; Faulks *et al.* 2010a, c). It has been assumed that the lake is a major barrier even when full (Glover 1982, 1990; Wager and Unmack 2000). Two issues confronting any species trying to disperse via the lake are that the rivers need to flow into the lake at the appropriate time and the connections must remain for sufficient time for species to migrate upstream. Salinity in the lake tends to rise rapidly, limiting the time that most species could persist. Of those species recorded from the lake, Australian smelt is found only in Cooper Creek and all the others occur in most tributaries (Wager and Unmack 2000). Masci *et al.* (2008) and Faulks *et al.* (2010b, c) found no strong evidence for contemporary gene flow among bony herring or golden perch in the tributaries. A similar result is reported for *Ambassis* and Hyrtl's catfish (Huey *et al.* 2008, 2011a). Thus, despite the presence of some of these species in Lake Eyre when full, at least four seem not to have dispersed between the tributaries in some thousands of years.

As part of Central Australian Province, the Bulloo River clearly has faunal relationships with Cooper Creek and all but one species in the Bulloo occurs also in the Cooper. The exception is a plotosid catfish known from two specimens collected in 1974 (Wager and Unmack 2000) and related to the softspine catfish (*Neosilurus mollespiculum*), endemic to the Burdekin River in Queensland. Other species have been recorded from the Bulloo River historically, including Welch's grunter (*Bidyanus welchi*), Barcoo grunter (*Scortum barcoo*) and Flinders Ranges purple-spotted gudgeon (*Mogurnda clivicola*), but none has been reported in recent years. A survey is needed to clarify the status of the Bulloo fish fauna.

Palaeo Province

By default, the rest of Australia is Palaeo Province. There are several subprovinces, reflecting likely past drainage connections (Palaeo–Victoria, Palaeo–Sturt, Palaeo–Internal, Palaeo–Oakover, Palaeo–Southwestern, Palaeo–Southern). The few provincial records are almost all spangled perch. Some populations may have been introduced (e.g. Lander and Hanson rivers, Northern Territory: Duguid *et al.* 2005) and undiscovered populations may occur in the more remote areas of South Australia and Western Australia. One site with other species is Sturt Creek, near the border of Western Australia and the Northern Territory, where spangled perch cohabits with Hyrtl's catfish and western rainbowfish, both related to populations immediately north of Sturt Creek (Unmack 2005, unpubl.).

EXTINCTION

Extinction is not often discussed in biogeographic terms because we usually have scant information from fossils or other lines of evidence. From what we do know of patterns of distribution and phylogeography, and changes in Australian environments through the Tertiary, extinctions have been common. The disjunct distributions of many fish species imply frequent extirpations that have influenced many, indeed all, of the provincial faunas. While fossil data for Australian fishes are limited, most demonstrate that the genera present in the fossil record persist today but that some local extirpation has occurred (Hills 1956; Unmack 2001a).

Aridity will have played a major role. The dry, fishless expanse of central southern Australia must once have harboured populations of pygmy perches (*Nannoperca*) and other southern genera (e.g. *Gal-*

axias, *Galaxiella*) now confined to south-western and south-eastern drainages. Atherinids in the *Craterocephalus* 'eyresii' and 'stramineus' groups are geographically disjunct (Unmack and Dowling 2010), implying that populations have been extirpated from intermediate drainages. In some cases this would have been due to aridity; in others it could be related to habitat changes and/or interactions with other species. In the southern Gulf of Carpentaria, disappearances of northern species are almost certainly due to aridity at the borders between specious and depauperate subprovinces in the Northern Province (Unmack 2001a). Other examples are cited in the descriptions of provinces, above.

Negative biotic interactions also appear to have had a major role. The *Melanotaenia* 'australis' group (McGuigan *et al.* 2000) and *Craterocephalus* 'disjunct' group (Unmack and Dowling 2010) each have isolated species on the Atherton Tableland, 1200 km from their nearest sister species in north-western Australia, and intervening drainages are now inhabited by other, less closely related lineages that have invaded and replaced them. A similar situation exists in eastern Australia, with a 900 km gap between sister species Utchee rainbowfish (*M. utcheensis*) and crimson-spotted rainbowfish now occupied by eastern rainbowfish (McGuigan *et al.* 2000). Some disjunctions, including the allopatry of blackbanded rainbowfish (*M. nigrans*), exquisite rainbowfish (*M. exquisita*) and slender rainbowfish (*M. gracilis*) in all but one drainage, may be due to negative biotic interactions. In other cases (e.g. McCulloch's rainbowfish, *M. maccullochi*, banded rainbowfish, *M. trifasciata*), the disjunction may reflect a lack of suitable habitat.

CONCLUSION

Australia is an old, flat, stable continent that probably has been inhabited by most of the extant fish groups throughout the Tertiary, if not longer. The flat landscape means low barriers between drainage basins, providing fishes with opportunities to expand their ranges. The broad, shallow continental shelf to the north would have facilitated genetic mixing between east and west, and with New Guinea to the north. Increasing aridity, especially since the Miocene, would have caused extirpations of groups from the limits of their ranges and extinctions of some species (including Miocene lungfishes, described earlier). Recurrent climatic shifts throughout the Pleistocene would have had major effects (Ehlers and Gibbard 2007). Species that survived the earlier climatic changes probably persisted through the more recent changes. Local extirpations would have occurred in each climatic cycle and populations would have expanded and contracted within and between river basins, but most speciation and many phylogeographic patterns date from the Early Pleistocene or before.

Systematic work on Australian fishes has been limited and molecular analyses suggest that the current inventory of described species is far short of the true number of species (e.g. Allen *et al.* (2002) listed 206 species; this chapter lists 256 species, or 316 including all undescribed species). Recent outstanding discoveries include redfin blue-eye from springs in 1990 (Ivantsoff *et al.* 1991); Bloomfield River cod from a single stream reach in 1993 (Pusey and Kennard 2001); *Synclidopus hogani*, a sole (Soleidae) from the tidal freshwater reaches of the Daintree River in 2002 (Johnson and Randall 2008); and a relative of western pygmy perch only from a 1.2 km stream reach in the Mitchell River Basin, Western Australia, in 2010 (D. Morgan, Murdoch University, pers. comm.). Many parts of remote northern Australia, especially Arnhem Land and the Kimberley, are poorly represented in fish records (Morgan *et al.* 2011a). Even sampled areas such as the Wet Tropics have yielded 10 new species records for Australia, from sampling in different habitats using alternative techniques such as snorkelling (Thuesen *et al.* 2011a). In other cases, small, geographically distinct populations of widespread species have been identified, including hardyheads all formerly referred to as Lake Eyre hardyhead (Crowley and Ivantsoff 1990), *Chlamydogobius* (Larson 1995), species of *Mogurnda* (Allen and Jenkins 1999) and Utchee rainbowfish (McGuigan 2001). The mountain galaxias may contain up to 15 species (Raadik 2011)! Undoubtedly, there are many more species to be discovered.

There is scope for expanded biogeographic studies, too. Phylogenetic relationships in many groups are a mystery, or are limited to knowledge of a few genes or have limited taxonomic coverage. Opportunities abound for fine-scale research in population genetics. For example, there are hybrid zones between species of *Melanotaenia*, *Mogurnda* and perhaps *Craterocephalus*, especially in the Atherton Tableland. There are also ecological biogeographic questions pending research (Matthews 1998), concerning species richness, factors limiting the ranges of species, physiological tolerances, the ecology of individual species, niche modelling and the effects of climate change, both in terms of palaeoclimates and future global warming scenarios. Many more novel questions could be asked, limited only by your creativity!

3 Evolutionary processes and biodiversity

Michael P. Hammer, Mark Adams and Jane M. Hughes

INTRODUCTION

Freshwater fishes are the world's most speciose vertebrates and their variety staggers the imagination. There are species with suction-cap oral disks, saw-like rostrums, flamboyant barbels, poisonous spines and even lungs. Some lack eyes and scales; some have fins better adapted to climbing than swimming. Others guard their eggs or young; some keep them in their mouths, in a pouch, or even attached to their heads! Flashes of colour are a feature of many species, but even dull or camouflaged species can have spectacular nuptial displays. Movements can be local sorties, regular forays between fresh water and sea water or opportunistic journeys over thousands of kilometres during floods, and might include burrowing into sediments to survive dry times. Within and between populations there are likely to be variations in such biological traits as appearance, habitat use, diet, spawning, disease resistance, movements and behaviour. The remarkable variability of fishes contributes to ecologically diverse aquatic ecosystems and to diversity in populations, families, genera, species and individuals, collectively 'biological diversity' or simply 'biodiversity' (Fig. 3.1).

This chapter is a primer on genetics for fish ecologists, recognising that advances in genetics are playing an increasing role in our understanding of patterns of biodiversity and underlying evolutionary processes. It describes approaches that (a) document biodiversity, relationships and identity, (b) explore processes such as dispersal and population dynamics and (c) provide data on species, populations and individuals to inform management and conservation planning of Australian freshwater fishes. These approaches are readily translated into questions. When did these species diverge? Are these two forms members of the same species? Is this species diadromous? How many subpopulations, stocks or conservation units are present? Is there evidence of current or historical dispersal between these populations or regions? What larva is that, and what are these fish remains in the gut contents? Is there a different way to tag fish? Did the colourful or the subordinate male sire these offspring? Which broodstock should we spawn? Genetic data can help to answer questions like these.

Some understanding of genetics is becoming essential for ecology, conservation and natural resource management. The explosion of related publications over the last two decades, including 100+ journal papers on genetic research applied to Australian freshwater fishes in 1990–2011 and the emergence of journals such as *Molecular Ecology* and *Conservation Genetics*, is testament to the trend. However, ecologists without formal training in genetics may struggle to appreciate the insights that genetic data can provide and, equally, may not know the limitations of those data. Technical terms, methods, concepts and jargon all can complicate the learning process. We hope that this chapter will diminish some of these obstacles, explain the relevance of genetic data to the study of freshwater fishes in Australia and encourage broader collaborative approaches in freshwater ecology. With those goals in mind, we also highlight some issues for future research and management.

The chapter includes two approaches. First, we focus on basic ideas relevant to freshwater fishes (and many other organisms). Second, we review the background information needed to interpret information from the genetic literature. The chapter has three main sections: to consider the evolutionary processes which shape biodiversity in aquatic environments, different genetic approaches to investigations of biodiversity, and some practical examples of ways that genetic methods and ideas can assist researchers and managers. Readers need not aim to develop complete familiarity with the subject matter, as much as to build sufficient expertise to become a good collaborator – most ecologists will obtain their genetic data through collaboration with a geneticist. For those keen to explore further afield, we offer a supplementary guide to further reading.

EVOLUTIONARY PROCESSES

There are more than 15 000 species of freshwater fishes, an astonishing 25% of all vertebrates (Lévêque *et al.* 2008). They typically show high levels of within-species genetic diversity and much greater levels of genetic divergence between populations than marine fishes (Ward *et al.* 1994). In this section we consider three drivers of evolution and diversity in fresh water, namely speciation, the nature of habitats and the intrinsic characteristics of the fishes themselves.

Isolation and speciation

A basic scenario in evolution is that of a single ancestral population splitting into two or more daughter populations, each set on its own evolutionary trajectory. Over time, these groups diverge as they accumulate changes in their gene pools through mutation, genetic recombination, natural selection and adaptation and genetic drift. Levels of divergence can vary along a continuum from localised interbreeding populations or distinct subpopulations, to separate species, to deep ancestral relationships (Fig. 3.1).

A key question for biologists is, 'At what point in their evolution do different groups become distinct species?' For sexually reproducing organisms, including most freshwater fishes, it is when they become reproductively isolated – they no longer produce fully

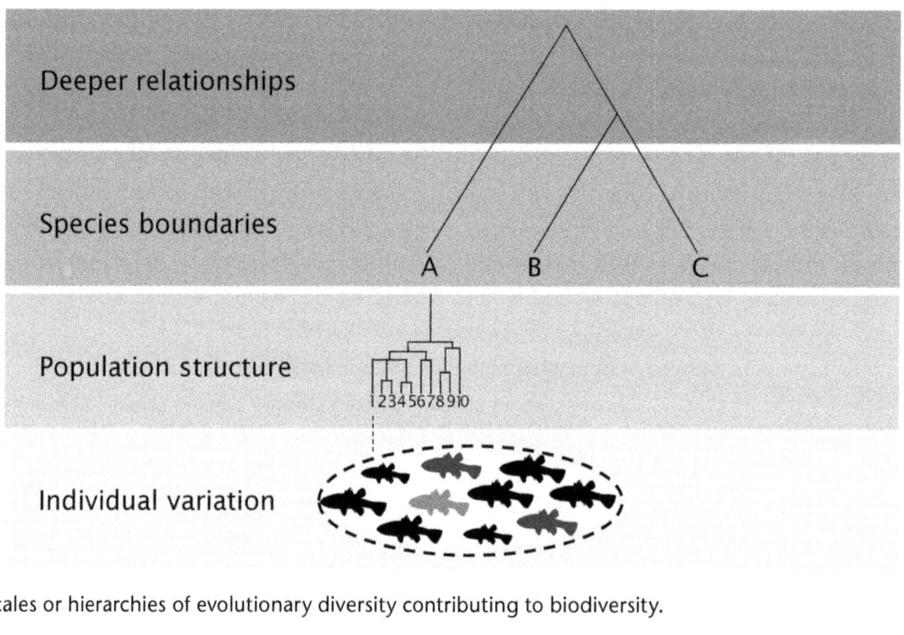

Figure 3.1: Scales or hierarchies of evolutionary diversity contributing to biodiversity.

fertile offspring as a consequence of some morphological, behavioural or genetic incompatibility (see 'Investigating biodiversity'). Reproductive isolating barriers may operate before mating occurs (pre-zygotic isolating barriers) or they may affect the fertility of offspring (post-zygotic isolating barriers). Thus, individuals in a population may change in shape, colour, fin pattern, age and breeding displays to a point where they are no longer recognised or preferred by the sister group (e.g. Pacific blue-eye, *Pseudomugil signifer*: Wong *et al.* 2004b) or where genetic mismatches prevent interbreeding (e.g. Murray cod, *Maccullochella peelii*; Clarence River cod, *M. ikei*: Rowland 1993). Reproductive isolating barriers can develop through geographic isolation, through ecological pressures and sexual selection or through hybridisation.

Geographic speciation

This form of speciation occurs when groups become isolated via fragmentation of their range or after dispersal. There are three modes, namely allopatric speciation (or vicariance speciation), where the groups have exclusive geographic ranges, parapatric speciation, where partial isolation restricts gene flow, and peripatric speciation, where one group is isolated at the periphery of another's range (Fig. 3.2).

Allopatry may be caused by geological barriers (e.g. catchment divides, waterfalls), physical or chemical barriers (e.g. seawater), biotic barriers (e.g. predators) or behavioural barriers (e.g. unsuitable habitats and specialisations for particular habitats). Obligate freshwater species are most vulnerable, as they may become locked in different areas, whereas separation may be less significant for species able to move between fresh water and the sea. Peripatric speciation may occur in diadromous species such as Tasmanian galaxiids that become landlocked by changes in geography or migratory behaviour (Ovenden *et al.* 1993; Waters and Wallis 2001).

Although permanent isolation of daughter populations will inevitably result in speciation, the time taken to achieve reproductive isolation can vary enormously. Typically, it takes thousands to millions of years (e.g. 1.5 million years for sister species of smelt (Retropinnidae) in the Lake Eyre and Murray-Darling basins: Hughes and Hillyer 2006; Hammer *et al.* 2007). The rate and degree of diversification are influenced by several processes. Fishes in contrasting environments are likely to experience greater selection pressures, especially in peripheral populations (e.g. Yarra pygmy perch, *Nannoperca obscura*: Hammer *et al.* 2010). Rapid divergence is likely to occur when only a few individuals are involved (the founder effect) or when there is a major population contraction or collapse that lasts for a few generations (a genetic bottleneck).

Ecological speciation

Gene pools can diverge even within an area of geographic overlap (sympatry). Opportunities arise when groups display specialisations for habitats, spawning grounds or food resources that polarise morphological, behavioural or reproductive characteristics within the population. In response, natural selection may rapidly change body shape or behaviour for traits such as feeding efficiency and predator avoidance, to a point where each ecotype no longer comes into reproductive contact with the other(s) or hybrids/intermediate types are less suited to either habitat, reinforcing their isolation (e.g. pelagic and demersal sticklebacks, *Gasterosteus* spp.: Rundle 2002). Likewise, a split may develop in the timing of migration or reproduction (e.g. spring and autumn spawning periods), driving the temporal isolation of two groups. Alternatively, strong sexual selection for mate traits such as colouration may promote sympatric speciation without any obvious relationship to habitat (e.g. Midas cichlid, *Amphilopus citrinellus*: Wilson *et al.* 2000).

While related species often co-occur, it is difficult to confirm that speciation has occurred in sympatry rather than in allopatry, followed by secondary contact, when the time-scale of these events may far outreach the observation spans of scientists! Some argue that most claimed instances of sympatric speciation are special cases of parapatric speciation, in that the two diverging genetic lineages occupy different microgeographic habitats. Freshwater fishes do provide examples of sympatric speciation, however, in restricted, isolated and geologically stable environments such as some lakes (Salzburger and Meyer 2004; Barluenga *et al.* 2006). Starting points for studies of sympatric speciation in Australian fishes could be rainbowfishes (Melanotaeniidae) and blue-eyes

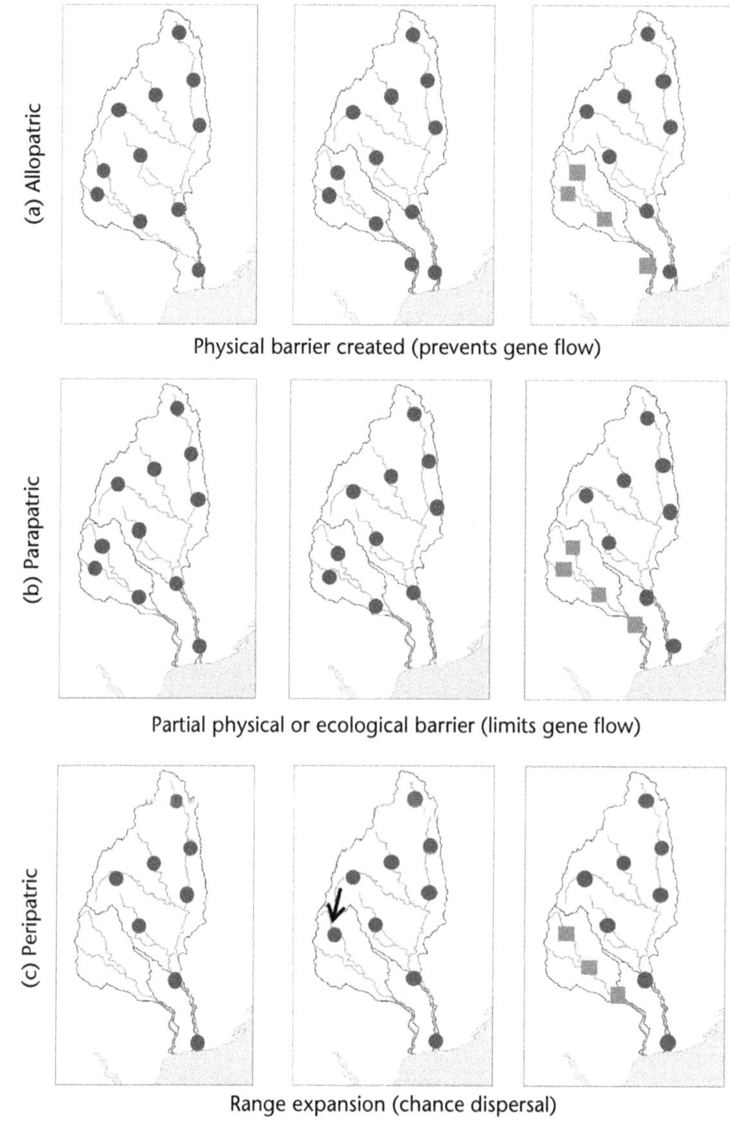

Figure 3.2: Hypothetical examples of ways that geographic speciation could occur in a stream habitat. (a) In allopatric speciation a population is fragmented by a complete barrier, promoting divergence. (b) Parapatric speciation involves a near-complete barrier to gene flow between groups. (c) Peripatric speciation involves colonisation of a new habitat and subsequent isolation.

(Pseudomugilidae) (Wong *et al.* 2004b), perhaps emulating work in Sulawesi on sailfin silversides (*Telmatherina* spp.: Coyne 2007; Herder *et al.* 2008).

Hybridisation and introgression

Strictly, hybridisation is when individuals of different species successfully produce offspring. A second, more liberal definition applies to any genetic cross between members of distinct populations. Where hybrids between species are not infertile, an opportunity exists for the transfer of alleles from one species into the gene pool of another (introgression). Hybridisation and introgression often are natural components of allopatric and sympatric speciation. In cases where two once-isolated groups come back into contact and barriers to reproduction are only partly established (i.e. the offspring are part-fertile), hybridisation can lead to a feedback loop known as reinforcement, driving further divergence towards complete speciation. In rare cases, highly successful hybrid forms may have adaptive advantages over their parental forms (thus, hybrid vigour) and may diverge as separate species (Herder *et al.* 2006).

Natural hybridisation is rare in most vertebrates but fishes are a notable exception, with a propensity to hybridise and introgress even across genera (e.g. goldfish, *Carassius auratus*; common carp, *Cyprinus carpio*: Hubbs 1955; Verspoor and Hammar 1991). This might be because the confined but variable nature of freshwater environments impels reproductive interactions and these may be further enhanced by external fertilisation in some groups. Hybridisation occurs in numerous Australian genera including *Craterocephalus* (Adams *et al.* 2011), *Galaxias* (Raadik 2011), *Maccullochella* (Wajon 1983; Douglas *et al.* 1995), *Macquaria* (Jerry *et al.* 1999; Shaddick *et al.* 2011b), *Nannoperca* (Hammer *et al.* 2010) and *Philypnodon* (Hammer 2008) and appears to be common in rainbowfishes and carp gudgeons (*Hypseleotris* spp.: Bertozzi *et al.* 2000; McGuigan *et al.* 2000; Unmack 2005; Schmidt *et al.* 2011a). Hybridisation and introgression are likely to be reflected in morphological and molecular characters and in behavioural traits, and may hamper identification of species and interpretations of ecological data.

Crosses between distinctive populations can have various outcomes, illustrated by human-mediated translocations. A simple consequence may be extinction of one form, as occurred in Queensland following translocations of freshwater shrimp (*Paratya australiensis*: Hughes *et al.* 2003). If the populations are reproductively compatible, there can be both deleterious consequences (outbreeding depression) and improved fitness (outbreeding enhancement). For example, many species of Salmonidae in the Northern Hemisphere are adapted to local catchments and crossing populations can lead to outbreeding depression, where the offspring are less able to survive in any catchment (e.g. Gilk *et al.* 2004). An infamous example of outbreeding enhancement concerns the invasive common carp in Australia. Where other strains of this alien species remained 'benign', the 'Boolarra' strain, a cross between separate stocks of carp, rapidly colonised the Murray-Darling Basin and other regions (Arthington 1991; Koehn 2004; Ch. 11).

Habitat complexity

Freshwater habitats are hotspots for biological diversification. By nature, they are spatially bounded, divided, heterogeneous and inherently variable over wide ranges of spatial and temporal scales (Ch. 4). These attributes provide many opportunities for reproductive isolation and eventual diversification of species.

Rivers, lakes and wetlands are divisive features of the landscape (Pringle 2001; Turner *et al.* 2001). Most fish cannot move between isolated habitats even if one is nearby, although there are exceptions such as eels (Anguillidae) and other species with climbing abilities (Ch. 5). Different systems often are completely independent, so that even a small topographic lift may separate catchments and send runoff in different directions. For instance, the headwaters of the upper Murray and Snowy rivers in south-eastern Australia abut but their mouths are separated by well over 1000 km, on opposite sides of the Great Dividing Range. The same is true of systems on the Atherton Tablelands, Cape York, the Kimberley and central Tasmania, and there are many more examples at smaller, local scales. Further segregation is imposed by physical barriers such as waterfalls, dry reaches, big rivers, wetlands and open water.

There are many facets to habitat diversity or heterogeneity in freshwater systems. First, drainage lines transcend clines in climate (e.g. temperature, rainfall, humidity) and geology (e.g. topographic relief, soil types), both of which influence aquatic habitats. Second, there are broad habitat types within drainage systems, including cool, fast-flowing headwaters and warm, slow-flowing lowland reaches, offstream wetlands and estuaries, and there are also localised habitats such as springs, sinkholes and perched swamps. A third facet of habitat diversity is linked to local geomorphic features and microhabitats (Ch. 4). These three layers offer a multitude of habitats for fishes.

Hydraulic connectivity within or between systems can be cyclical or stochastic, reflecting temporal variations in climate (i.e. seasonal flow regimes and periodic floods and droughts). Many aquatic systems in desert landscapes are the remnants of more extensive systems that are reconnected in wetter times (e.g. Johnson 2002). These remnant systems may be joined only periodically, like the rivers flowing to Lake Eyre in central Australia. These only occasionally reach the lake (once every three to five years) and rarely fill it (once every 50–100 years), and even more rarely do they join to allow movements of aquatic fauna from east (Cooper Creek) to west (Neales, Diamantina rivers). Thus, Lake Eyre golden perch (*Macquaria* sp.) populations in this region show an east–west genetic subdivision, implying low or episodic gene flow (Faulks *et al.* 2010b).

Long exposure to freshwater environments has promoted extensive radiations in fishes throughout the world. The diverse habitats of tropical Amazonia harbour the world's richest freshwater fish fauna, including 60 families and 3000+ known species and perhaps another 1000–1500 species yet to be described (Reis *et al.* 2003). There are multiple radiations among the Amazon catfishes (Siluriformes), tetras (Characidae) and dwarf cichlids (Cichlidae). The African Rift Valley lakes are famous for some 700 cichlid species derived from a few common ancestors, forming species flocks with incredible diversity in colour, shape, feeding methods and other specialisations (Johnson *et al.* 1996; Salzburger and Meyer 2004). The Wet Tropics region of Queensland has exceptional species richness by Australian standards (Ch. 2) and in Tasmania there are highland lakes with nine endemic species of galaxiids. Curiously, even deserts harbour unusual fishes, notably the pupfishes (*Cyprinodon* spp.) of North America and gobies (*Chlamydogobius* spp.) in central Australia (Larson 1995).

Intrinsic characteristics

The intrinsic characteristics of species provide a functional perspective for understanding diversification in freshwater fishes (Table 3.1). At a molecular level, genetic diversity and rates of evolution are seen to at least partly reflect differences between species – and populations within species – of characteristics including vagility, reproduction and behaviour (Nevo *et al.* 1984; Bromham 2009).

Vagility

The life histories of fishes broadly indicate their vagility (mobility or dispersal ability: see Chs 5, 7). Obligate freshwater fishes are more likely to have limited vagility (and more between-site genetic divergence) than diadromous or potamodromous species, as are philopatric species which have a limited home range or return to natal sites to breed. There may also be differences between the sexes. Among birds and mammals, dispersal tends to be mainly by females or males, respectively (Avise 2000; Petit and Excoffier 2009); it would not be surprising to discover that fishes also tend to gender-biased dispersal in some instances, but this awaits investigation.

Other factors affecting general mobility (Ch. 5) include the ability to overcome barriers. Fish with this ability include the aptly named climbing galaxias (*Galaxias brevipinnis*) with strong, wide pectoral fins and a flattened ventral surface, the lampreys (Geotriidae, Mordaciidae) with oral disks for attaching to surfaces, enabling them to climb over barriers, and the southern purple-spotted gudgeon (*Mogurnda adspersa*) with a remarkable ability to leap, as anyone who has kept them in aquaria will confirm (McDowall 1996a; Boxall *et al.* 2002).

Reproduction

The mode of reproduction influences how well eggs, free embryos and larvae (i.e. genes) are dispersed or mixed (Ch. 7). Clearly, the eggs of broadcast spawners are more likely to be genetically mixed than those of species with sticky eggs or parental care, or those that give birth to live young. A protracted spawning season or larval period increases chances for breeding to coincide with events favourable to dispersal, as would association with components of the flow regime (e.g. high flow and floods). Moreover, short-lived species appear to show more rapid rates of molecular evolution than those with longer generation times (Thomas *et al.* 2010).

Behaviour

Species-specific responses are also likely to depend on behavioural characters such as the habitat preferences and interactions with the environment (e.g. physical barriers). Habitat specificity provides an avenue for geographic or ecological isolation, whether at broader zones such as highlands, specific rare habitats such as groundwater springs or certain microhabitats such as

Table 3.1: Biological traits of freshwater fishes and matching hypothetical levels of genetic divergence

Intrinsic characteristics	Genetic divergence Strong ———	——→ Weak
Vagility		
Life history strategy	Obligate freshwater No spawning migration	Euryhaline/diadromous Potamodromous
Adult mobility	Small body size Benthic Weak swimming ability	Large body size Pelagic Strong swimming ability
Adaptations	Not able to climb Dispersal only in water	Morphological features for climbing Dispersal in water and over land
Reproduction		
Egg/free embryo/ larvae characteristics	Demersal/sedentary Eggs attached, adhesives	Pelagic/mobile Eggs dispersed
Strategy	Short spawning period Low–flow recruitment Short generation time (*r*–selected species) Small populations	Protracted spawning period Flow recruitment Long generation time (*K*–selected species) Large populations
Nesting behaviour	Brood care/ovivipary Sheltered position Specialist	No parental care Exposed position Generalist
Biological characteristics		
Habitat specificity	Narrow One habitat zone (e.g. upland) Particular microhabitat	Broad More than one habitat zone General habitat
Physical, chemical	Intolerant (e.g. salt, temperature, flow)	Wide tolerance
Metabolism	Fast metabolism	Slow metabolism
Behaviour	Natal homing/philopatry Innate response to avoid flood dislodgement	Opportunistic colonisation No flood-avoidance mechanisms
Biotic interactions	Prey Weak competitor	Predator Strong competitor

aquatic vegetation, riffles or vertical surfaces. Species that have a low tolerance for, or avoid, certain conditions such as high or low salinity, temperature and flow can also become restricted to certain areas (e.g. limits of distribution of tropical and temperate species).

Other behaviours that can help shape dispersal and genetic structure include innate responses to flooding, philopatry and interactions with other species. For example, the southern pygmy perch (*Nannoperca australis*) moves into riparian vegetation to avoid being dispersed by high flows, which may account for strong genetic structure in this species (Humphries 1995; Hammer 2001). Some diadromous salmonids in Europe and North America return to their natal sites to spawn, a phenomenon that promotes divergence by minimising interbreeding between catchments and stocks (e.g. Ferguson 2004). Strong genetic structure in the Cairns rainbowfish (*Cairnsichthys rhombosomoides*) may reflect the influence of predators that prevent dispersal between habitat patches (Thuesen *et al.* 2008).

Figure 3.3: 'Why did the fish cross the road?' 'Because it's in their genes!' Australia's most widespread freshwater fish, the spangled perch (*Leiopotherapon unicolor*), has a strong urge to disperse with the onset of flooding. Its dispersal ability is matched by relatively minor differences in genetic structure in populations across the continent (photo: R Pedler).

Ecological insights

Fish ecology is an ideal platform for testing hypotheses regarding patterns of gene flow within species or across landscapes (e.g. Tibbets and Dowling 1996; Neville *et al.* 2006; Hammer 2008; Hughes *et al.* 2009). For example, two distinct movement strategies have revealed different contemporary levels of gene flow. The spangled perch (*Leiopotherapon unicolor*), the most widespread of all Australian freshwater fishes, shows a strong urge to disperse with rising floods (Fig. 3.3). This enables colonisation of new habitats but requires generalist traits, including broad environmental tolerances (Merrick and Schmida 1984; Wager and Unmack 2000; Ch. 4). Pygmy perches (Percichthyidae), on the other hand, are sedentary and occupy much more specific habitats (Kuiter *et al.* 1996). The spangled perch displays little genetic variation between regions throughout Australia (Bostock *et al.* 2006) whereas pygmy perch populations show strong genetic structure at local scales, even down to individual streams (Hughes *et al.* 1999; Cook *et al.* 2007; Hammer *et al.* 2010). This association also works in reverse – genetic data can be used to infer the vagility of a species and to test the interplay between different traits.

INVESTIGATING BIODIVERSITY

Genetic data have four main ecological applications, namely (a) delineation of species boundaries, (b) documentation and analysis of genetic diversity within species, (c) reconstruction of evolutionary relationships and (d) comparative genetic profiling where unknown samples are compared genetically with reference material. The theme that unites these applications is systematics, and we begin this section by acknowledging systematics as the foundation for biodiversity research.

Systematics and genetic data

Systematics is a discipline concerned with hierarchical relationships among organisms (Fig. 3.1). The framework of relationships for each group of related organisms (species, taxa or operational taxonomic units) is a context for interpretation and comparison of all kinds of biological data. It follows that confusion over taxonomy can lead us to underestimate the true nature of fish assemblages, confound ecological data or act inappropriately in management or research, and could lead to losses of species (Bortolus 2008).

Among Australian freshwater fishes there are several examples of potential taxonomic traps. The presence of multiple species (unisexual lineages and/or hybrids) in carp gudgeons (*Hypseleotris* spp.) complicates research on these common, widespread fish (Bertozzi *et al.* 2000; Unmack 2000; Schmidt *et al.* 2011a). Previously unrecognised species in another widespread taxon, Australian smelt (*Retropinna semoni*), mean that comparisons of ecological data within and between regions could be misleading (Hammer *et al.* 2007). Research on the taxon golden perch (*Macquaria ambigua*), a popular angling species, has shown genetic divergences between three geographic populations (Murray-Darling Basin, Lake Eyre Basin, Fitzroy-Dawson river system in Queensland) and failure to recognise this in stocking programs could mean losses of genetic diversity or species (Musyl and Keenan 1992; Faulks *et al.* 2010b). The trout cod (*Maccullochella macquariensis*) was not formally recognised as a species distinct from Murray cod (*M. peelii*) until it was on the brink of extinction and it has required major investment for recovery (Trout Cod Recovery Team 2008).

Species may become extinct even before we know of their existence. This was true of an undescribed species in the taxon Macquarie perch (*Macquaria australasica*). Fish from the Shoalhaven river system in New South Wales show significant genetic and morphological divergences from other likely species in the Hawkesbury-Nepean river system and the Murray-Darling Basin. The Shoalhaven form declined and had disappeared in the wild by the late 1990s and the last known representative died in captivity in 2008 (Dufty 1986; Faulks *et al.* 2010a). This is the first documented extinction of an Australian freshwater fish, albeit of an undescribed species.

Taxonomic uncertainties exist in most groups of Australian freshwater fishes and a true measure of species richness could see a doubling of the number of currently recognised species. Thus, it is unwise to have complete faith in current species-level taxonomy (or the field guides they engender) and instead to think of it as work-in-progress. A valid systematic framework is a prerequisite in a research program and, in its absence, progress must be cautious.

Morphology and genetics

Traditionally, the data used in systematics have been limited to what is presumed to be heritable variation in morphological characters, supplemented occasionally by ecological, behavioural or physiological information. The emergence of molecular methods over the past 60 years has provided systematists with opportunities to include various kinds of genetic data in their investigations. Molecular and morphological appraisals are complementary, and each approach has strengths and weaknesses. There is now a broad consensus that (a) neither is innately more important than the other, (b) systematic studies incorporating both approaches are far superior to those that rely on morphology or molecules alone, (c) morphological datasets may need reappraisal where they appear to contradict molecular data, (d) existing systematic frameworks should not be overturned until a reappraisal has been made, (e) additional molecular data (perhaps a different kind of molecular marker) may be needed where there is an apparent conflict between preliminary molecular data and morphological data, and (f) the quality of each data type (e.g. how many genetically independent characters, their heritability) is far more important than whether the data are morphological or molecular. A major advantage of complementary approaches is that genetic evaluations can generate testable hypotheses for traditional methods based on morphology (Berendzen *et al.* 2009) (see 'Ecological insights').

Documenting species

Individually and collectively, species are the currency for comparative studies in ecology, conservation and management. They are indicators for environmental monitoring and reporting, the focus of threatened-species legislation and planning, monitors of site and ecosystem condition, the basis for environmental units in research and management (e.g. biogeography, bioregionalisation) and the building blocks in ecosystems.

The number of known freshwater fish species is growing rapidly, by about 200 species per year (Lundberg *et al.* 2000; Lévêque *et al.* 2008). Seldom-sampled regions, including parts of Africa, Asia and Central and South America, are important sources of new species. Recent discoveries in the Wet Tropics region of Queensland include Bloomfield River cod (*Macquaria wujalwujalensis*: Pusey and Kennard 2001) and colourful cling-gobies (*Stiphodon* spp.: Ebner *et al.* 2010). Even in well-sampled areas there may be new discoveries, such as the variegated pygmy perch (*Nannoperca variegata*) from Ewens Ponds (South Australia) and the Glenelg River (Victoria) in the mid 1980s (Kuiter and Allen 1986) and a new pygmy perch (*Nannoperca* sp.) from the Hay river system in south-western Western Australia (Morgan *et al.* 2011b). A second source of discoveries, one that challenges our views of fish faunas, is the new species revealed by taxonomic revisions. This can be from a simple lack of rigour in past studies or it may come from an enhanced ability, using genetic data, to detect cryptic (unrecognised) species.

Cryptic species may not be recognised by sight, as speciation can occur without changes in morphology. Molecular data expedite the search for cryptic species and, after rigorous study, many of these prove to be morphologically distinctive, albeit in sometimes subtle ways (Beheregaray and Caccone 2007). In New Zealand, for example, molecular systematics initiated a reinterpretation of local galaxiid taxonomy (Allibone *et al.* 1996), increasing the number of described species by 20–30% (some remain undescribed: Waters *et al.* 2010). Genetic data have complemented morphological reviews of Australian hardyheads (Crowley and Ivantsoff 1990), cod (Rowland 1993), purple-spotted gudgeons (Allen and Jenkins 1999) and galaxiids (Raadik 2011) and have revealed morphological characters for species and hybrid forms in the carp gudgeon complex (Bertozzi *et al.* 2000). Many other Australian species continue to be regarded as geographically widespread, despite morphological, ecological and genetic heterogeneity.

Species concepts

Species are arrayed on an evolutionary continuum, with older species having clearly diverged from related lineages and other, less-divergent species beginning the journey. Along the continuum, related groups become diagnosable species once they reach milestones that indicate differences in morphology, ecology, reproductive compatibility or genetic makeup. Among all living organisms (prokaryote versus eukaryote, plant versus animal, invertebrate versus vertebrate), among the variety of reproductive systems (clonal versus sexual, self-fertilising versus outcrossing, propensities for hybridisation) and other circumstances (e.g. sympatry versus allopatry), it is not surprising that there are many theoretical views of what it is that constitutes a species (e.g. 24 are listed by Mayden 1999). Some of these are attributable to the variety of ways that species are seen in different spheres of research (see Coyne and Orr 2004). The concepts can be assigned to one of five groups, namely typological concepts (morphological differentiation), evolutionary concepts (diagnosable lineages on separate evolutionary trajectories), biological concepts (reproductive isolation), phylogenetic concepts (diagnosed by shared, derived character states) and other concepts (based on assorted ecological or genotypic criteria).

Recent thinking is that traditional species concepts provide practical secondary evidence to identify species under a particular focus (de Queiroz 2007) but that the process of speciation warrants a unified theoretical concept, namely that species should be seen first as separately evolving metapopulation lineages. This view allows the criteria used to diagnose species to reflect a broader range of empirical data and focuses attention on whether the evidence presented in a particular study satisfies the operational criteria relevant to that group. Visual diagnosis has universal appeal and practical utility, of course, and most eukaryotic species are distinguished solely by morphological criteria. Yet evidence from genetics and ecology may help to overcome some of the inherent problems in reliance on morphological data. These include (a) the point that different species do not necessarily look different, (b) the criteria to distinguish species may be equivocal where characters are clinal, subjective or not consistent across all species/groups and (c) morphological divergence may not reflect divergence of lineages, owing to uncertainty over heritability, the number of contributing genes and the genetic independence of characters. The unified approach also counters the impression that 'good' species cannot hybridise – it takes account of the nature of the focal group (e.g. fishes show a propensity for natural hybridisation), its situation (e.g. it is difficult to assess reproductive isolation in allopatry, and breeding experiments decouple organisms from their natural biotic and abiotic environment), and the reality that incipient species may be at a point on the continuum where reproductive isolation is not yet complete.

This approach does not invalidate morphological taxonomy. Rather, it recognises that more lines of evidence now are available to review (a) new discoveries and information, and thereby more fully document divergences, (b) proposed species-level taxonomies as working hypotheses, (c) contentious or ambiguous situations (e.g. levels of divergence in allopatry), (d) assessments based on few characters (this is commonplace in fish taxonomy) and (e) identifications reliant on characters that, in hindsight, are not fully diagnostic.

Again, ecological researchers need to consider the soundness of the current systematic framework and allow for different interpretations that might apply under alternative taxonomic scenarios (i.e. a form of risk assessment). Ideally, researchers should test the validity of a putative 'species' by including in their program a molecular genetic component to search for cryptic species. Both responses are highly desirable in view of the continuing lack of taxonomic research on Australian freshwater fishes (Lundberg *et al.* 2000; Leis *et al.* 2007).

Operational criteria

The need for the research community to reach consensus on matters pertaining to the systematics of Australian freshwater fishes has never been more urgent, given predictions of imminent global extinctions. In the following we suggest an approach, including operational criteria and acceptable levels of rigour for species delineation, to guide this process.

Genetics offers a rapid, cost-effective way to scan groups and identify candidates for description. Levels of divergence beyond those normally encountered among conspecific individuals are used to flag populations that warrant taxonomic attention (e.g. overview study of Australian smelts, Retropinnidae: Hammer *et al.* 2007). Typically, this involves generating additional comparative molecular, morphological or ecological data for genetic markers, characters and traits not previously considered, plus a reappraisal of existing biological data, providing lines of evidence for distinct species (e.g. recent work on Australian freshwater cods, Percichthyidae: Nock *et al.* 2010). The more concordance there is among different types of data, the more rigorous the diagnosis, with obvious implications for the choice and quality of molecular data used to distinguish species (Table 3.2).

Where preliminary genetic data are available, they should be inspected for geographic patterns of heterogeneity unlike those expected for a single, substructured species. For example, a prominent genetic dichotomy between two adjacent groups of sites is often a good indicator of a species-level divide. Another indicator may be when a taxon is genetically homogeneous across most of its range but harbours a distinctive form in a region. Intensive sampling and *de novo* genetic analysis (using different markers) of the different forms are then clearly a priority. Sampling forms of unusual appearance or with other biological peculiarities, over the range of habitats and environmental conditions, increases the chances of detecting novel species. Sample sizes for genetic delineation need only be small, with a preference for low numbers of fish (three to five) across many sites, although bigger samples obviously provide more chance of detecting instances of multiple taxa occurring at a site. The protocol for genetic and morphological assessment should include (a) a working null hypothesis that two forms are the same species, to be rejected only where the forms are delineated by fixed differences in multiple characters (morphological and/or molecular characters), (b) a variety of taxonomic characters that, as best as can be determined, are genetically independent of all others, are biparentally inherited and display co-dominant character states, (c) no fewer than the absolute minimum numbers of characters (two in sympatry and three in allopatry) (again, more fixed differences mean a more robust diagnosis) and (d) a near-fixed taxonomic character that is effectively diagnostic of species even if the two forms exhibit a low frequency (e.g. $\leq 20\%$) of shared character states (this is standard practice for

Table 3.2: Attributes of genetic technologies and morphological data

Assessment	Morpho-logy	Karyo-types	Allo-zymes	AFLPs	mtDNA	nDNA (*introns*)	nDNA (*exons*)	Micro-satellites	SNPs
Technical issues									
Source of tissues/specimens (see Notes, below)	1	2	3	4	4	4	4	4	4
Technical expertise	L	M	H	M	M	M	M	M	M
Operating costs	L	[illegible]	M	M	M	H	H	M	H
Salary costs	M	M	L	H	M	H	H	H	H
General attributes									
Chars/genes screened: fully independent chars	30:? (< 30)	3:3	50:50	200:? (< 200)	2:1	5:5	5:5	20:20	200:200
Do characters display discrete states?	Y (some)	Y	Y	Y	Y	Y	Y	Y	Y
Are character states codominant?	Y (some)	Y	Y	N	N	Y	Y	Y	Y
Do characters have high heritability?	Y (some)	Y	Y (most)	Y	Y	Y	Y	Y	Y
Likely influence of selection on characters	H	L	M	L	L	L	M	L	L
Are data absolute (A) or relative (R)? (see Notes)	A,R	A	R	R	A	A	A	A	A
Are data comparable across different life stages?	N (rare)	Y	Y	Y	Y	Y	Y	Y	Y
Subjective decisions in data generation (+ often, – minor)?	+	–	+	–	–	–	–	–	–
Problems likely with character homology?	Y	N	N (rare)	N	Y/N	Y/N	Y/N	Y/N	N
Expected level of homoplasy in dataset	H	L	H	M	L	L	L	M	M
Utility									
Genetic identification	L	L	M	H	L	L	L	H	H
Extent of within-population variability	L	L	M	H	M	M	L	H	H
Extent of between-population divergence	L	L	M	H	H	H	L	H	H
Ability to diagnose similar species	M	L	H	H	M	H	L	M	L
Ability to detect hybrids, introgression	L	L	H	M	L	M	L	H	M
Phylogeography/landscape genetics	P	L	M	M	H	H	M	H	M
Phylogenetic affinities of congeneric species	L	M	M	L	M	H	M	L	L
Phylogenetic affinities of genera	M	M	L	P	M	M	H	P	P
Phylogenetic affinities of families and above	M	L	P	P	M	M	H	P	P

Attributes are shown for two classes of DNA (introns = more variable; exons = less variable) to highlight the potential of nuclear DNA (nDNA) sequencing. Comparisons are for the same numbers of individuals: Morphology = 30 meristic and/or measured characters; Karyotypes = gross, C-banded and G-banded; Allozymes = 50 loci; Amplified Fragmented Length Polymorphisms (AFLPs) = 200 loci; mtDNA sequencing = 2000 bp for two different genes; nDNA sequencing of introns = 2000 bp for 5 genes; nDNA sequencing of exons = 3000 bp for 5 genes; Microsatellites = 20 loci; Single Nucleotide Polymorphisms (SNPs) = 200 loci. It is assumed that 'informative' characters (chars) or genes have been chosen (those likely to distinguish individuals and/or taxonomic units).

Sources of tissue/specimens are coded 1 = museum voucher specimens, ideally formalin-fixed with matching genetic voucher; 2 = fresh or specially treated cells; 3 = fresh, or frozen while fresh; 4 = fresh, or frozen while fresh, or in ethanol, or dried.

L = low, M = medium, H = high; Y = yes, N = no; P = poor.

Data are absolute if character states are measured objectively; data are relative if character states are compared with each other.

traditional morphological characters and allows for the tendency of congeneric fish species to hybridise). Characters that do not fulfil criterion (b) above (e.g. mitochondrial DNA (mtDNA) genes, some morphological traits) nevertheless add rigour to a diagnosis where they provide concordant support for rejecting the null hypothesis. One important exception to guideline (c) is that genetic diagnosis of closely related species in sympatry can succeed in the absence of near-fixed differences provided that a sufficient number of partially informative genetic markers (the cumulative percentage of shared alleles at each marker is well above 10%) also indicate a genetic discontinuity.

Genetic diversity within species

Ecologists typically are interested in using genetic data to explore four broad topics: (a) the geographic distribution of variation, (b) assessments of dispersal ability, (c) times since populations split and (d) levels of within-population diversity. This section is concerned with genetic diversity within and between populations.

Some basic ideas

Genetic variation is ubiquitous in nature and the extent of variation itself varies substantially. Nevertheless, many genes, particularly those targeted in population-level studies, occur as multiple alleles (or haplotypes for DNA segments) in the gene pool of any given species. Different principles apply to different classes of eukaryotic genes, depending on whether they are single-copy nuclear genes, part of a gene family, sex-linked genes or mitochondrial genes. In this section, the focus is on single-copy nuclear genes, including the vast majority of genes and the most commonly used genetic markers.

All diploid organisms (i.e. most fishes) possess two non-identical copies of each autosomal chromosome and hence two copies of each autosomal locus. In species displaying well-differentiated sex chromosomes (e.g. mammals, birds, some fishes), most sex-linked loci occur as single copies in one sex and two copies in the other (the respective genders depend on the mechanism of sex determination). Diploid individuals therefore are either homozygous (two copies of the same allele) or heterozygous (copies of two different alleles) and occasionally hemizygous (one copy of a sex-linked allele) for every polymorphic locus in a population. Further, the number of potential genotypes at a polymorphic locus always exceeds the number of alleles, often considerably so (e.g. autosomal alleles *a*, *b*, *c* will produce genotypes *aa*, *ab*, *ac*, *bb*, *bc*, *cc*; 10 alleles = 55 genotypes; 40 alleles = 820 genotypes etc.).

The principle that underpins population genetic theory is the Hardy-Weinberg Equilibrium (HWE), which affirms that allele frequencies in a sexually reproducing population will always determine genotype frequencies and that both allele and genotype frequencies will remain unchanged over time, provided that six conditions are met. The conditions are that there is standard Mendelian inheritance, no mutation, no selection, random mating, a large population (thus, no genetic drift) and no migration. The HWE provides the null expectations for populations (i.e. genotype frequencies should match those expected for a given set of allele frequencies; allele frequencies should remain static across generations) and signposts for investigations where these expectations are not met (i.e. which one or more of the six assumptions does not apply; random mating and no migration are the conditions most likely to be violated).

A basic analytical procedure for all Mendelian loci is to make statistical comparisons of observed genotype frequencies with those predicted by the HWE for a given set of allele frequencies. Where there are statistically significant departures from HWE one of three patterns will be evident, each providing clues as to which assumption is violated. These patterns are that only a single locus departs from expectations, that multiple loci are involved but are otherwise discordant, and that multiple loci are involved and demonstrate a significant excess of certain genotypes, or heterozygotes or homozygotes. The last is of special significance as the signature of a Wahlund Effect, suggesting the presence of two or more subpopulations (or species), each with different allele frequencies at the loci deviating from the HWE. Assessments of linkage disequilibria among all supposedly independent loci are no less important but are less often applied. Statistical evidence of an association between genotypes at different loci in a single, wild population indicates either that the loci are tightly linked (and thereby not independent tests of population diversity), that there is non-allelic variation or that there is a more subtle Wahlund Effect.

Beyond these two basic analyses, there is a bewildering variety of measures available to assess genetic diversity. Discussion of these is beyond the scope of this chapter, but ecologists ideally should consider where each measure or analysis fits in the three key aims of population genetics, namely (a) to determine whether there is statistical evidence to reject a null hypothesis of panmixia (random mating across the entire population), (b) to quantify the diversity within populations and (c) to quantify the levels of divergence between populations. There are many computer programs available to analyse genotypic and sequence data; many are reviewed by Excoffier and Heckel (2006) and there is a list on the internet (<http://www.nslij-genetics.org/soft/group.html>).

Models of population genetic structure

Conceptual models help in understanding patterns of genetic structure among populations (Fig. 3.4). All require genetic data suitable for statistical examination, with consideration of issues such as setting levels of Type I and Type II errors (Richardson *et al.* 1986) and correcting for multiple tests of significance (Rice 1989). More than one model may be appropriate for one species, often at different geographic scales but occasionally at one scale, and different genetic datasets for one species may lead an analyst to refute or support alternative models, depending on the statistical analysis and the strengths and weaknesses of various molecular methods (Table 3.2).

The simplest model of population structure is panmixia (Richardson *et al.* 1986) (Fig. 3.4a). Although this is unlikely to represent any widespread, obligate freshwater species, it is the default model for assessments of population structure. Among Australian species, the spangled perch shows apparent panmixia across several major drainage divisions (Bostock *et al.* 2006). The potamodromous golden perch also varies little across its range (Faulks *et al.* 2010b). Panmixia is more common in diadromous or estuarine/marine species, provided that there are no physical barriers to migrations. For example, populations of common galaxias (*Galaxias maculatus*) in New Zealand are a single well-mixed stock and might be so in Australia (Waters *et al.* 2000a), Australian grayling (*Prototroctes maraena*) in Victoria (Schmidt *et al.* 2011b) and empire gudgeons (*Hypseleotris compressa*) and estuarine glassfish (*Ambassis marianus*) in eastern Australia all indicate high genetic connectivity (McGlashan and Hughes 2001; Mills *et al.* 2008).

For mobile species, but ones where dispersal is limited by distance and panmixia applies only within a single genetic neighbourhood, an isolation-by-distance model applies (Fig. 3.4b). Modest genetic divergence is possible between non-adjacent neighbourhoods, creating subtle clinal substructure (Wright 1969). Under this model there are no definable subpopulations but a continuum of overlapping neighbourhoods. This structure is seen in diadromous species where there is an association with habitat, or limited vagility, to prevent mixing across the range. This applies to Australian bass (*Percalates novemaculeata*) on the east coast (Jerry 1997) and barramundi (*Lates calcarifer*) in the north (Chenoweth *et al.* 1998). Less mobile, obligate freshwater species may be isolated also by distance within river systems, as for the western rainbowfish (*Melanotaenia australis*) in the Kimberley region, Western Australia (Phillips *et al.* 2009a).

A modification of the isolation-by-distance model, the stream-hierarchy model has been proposed to reflect two complexities related to the nature of freshwater systems (Meffe and Vrijenhoek 1988) (Fig. 3.4c). First, locations in near geographic proximity may not be closely connected by a stream network and it is necessary to consider specific stream distances. Second, the different levels of organisation (individual streams, groups of streams, subcatchments, whole systems) should also influence genetic structure in a corresponding hierarchical manner (Hughes *et al.* 2009). The Pacific blue-eye exhibits this pattern of genetic structure (McGlashan and Hughes 2002).

Where two or more genetically and geographically diagnosable subpopulations are present in a species, a bounded discrete-subpopulation model applies (Richardson *et al.* 1986) (Fig. 3.4d). Subpopulations are almost invariably separated by physical or biological barriers that limit gene flow, causing divergence. Diadromous species, for example, form subpopulations on regional (e.g. Tasmanian whitebait, *Lovettia sealii*: Pavuk 1994) or continental scales (e.g. common galaxias: Waters and Burridge 1999), unless they display natal homing. Most obligate freshwater fishes are likely

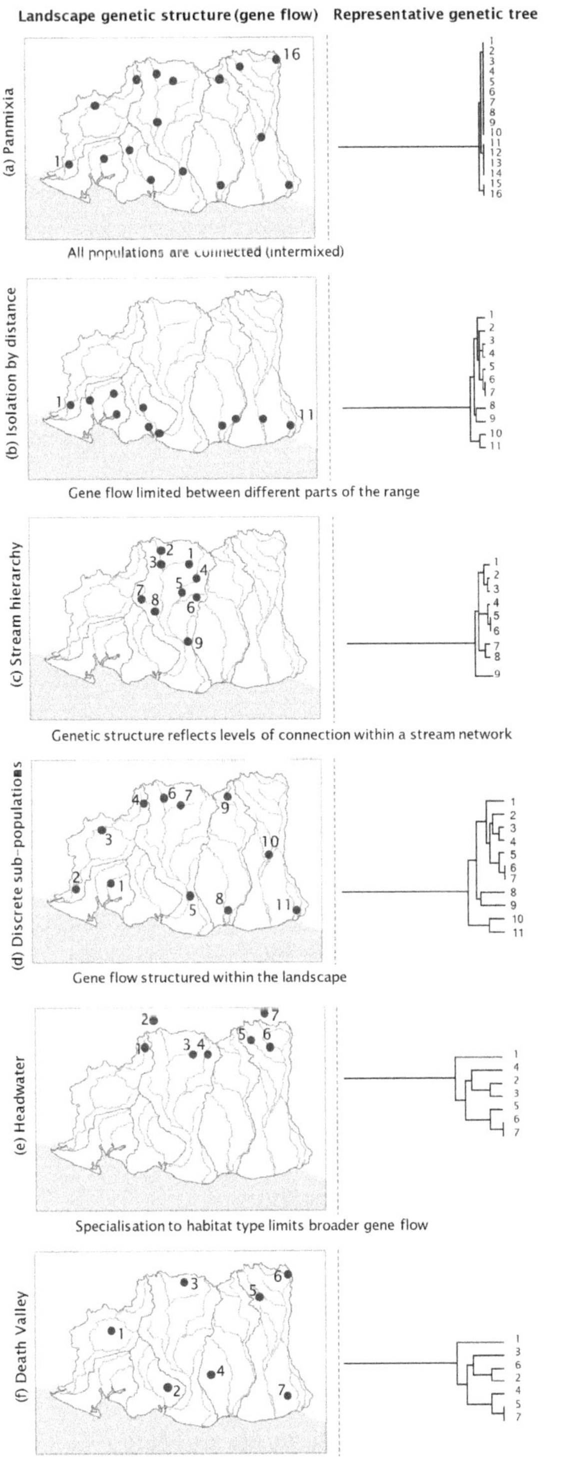

Figure 3.4: Hypothetical models of population genetic structure in a stream habitat. The maps show gene flow between populations in a physical landscape under different scenarios (see text) and the population genetic structure (trees) that could result.

to show some level of substructure, whether in systems separated by marine barriers (e.g. Yarra pygmy perch, *Nannoperca obscura*: Hammer *et al.* 2010), in specialised habitats (e.g. Oxleyan pygmy perch, *N. oxleyana*: Hughes *et al.* 1999; Knight *et al.* 2009), in subdivided subterranean habitats (e.g. cave gudgeon, *Milyeringa veritas*: Adams and Humphreys 1993) or in systems divided by waterfalls (e.g. fly-specked hardyhead, *Craterocephalus stercusmuscarum*: McGlashan and Hughes 2000) or predators (e.g. Cairns rainbowfish: Thuesen *et al.* 2008).

Headwater reaches are a common habitat type in stream systems, leading to a freshwater variant of the discrete-subpopulation model (Hughes *et al.* 2009) (Fig. 3.4e). Organisms whose biological requirements (e.g. fast flow, rocky habitat, cool well-oxygenated water) limit them to headwaters are likely to display little or no gene flow among subpopulations in different parts of a catchment, due to geographic fixation or intervening non-headwater habitat. Australian species with this type of genetic structure include distinct morphological/genetic forms of western galaxias in the Darling Range and Swan Coastal Plain, Western Australia (although this may be a species-level divergence: Watts *et al.* 1995) and divergent local and regional populations of the two-spine blackfish (*Gadopsis bispinosus*), an upland-stream specialist (Waters *et al.* 1994; Beitzel 2002).

Another variant of the discrete-subpopulation model is the Death Valley model, referring to localised remnants (Fig. 3.4f). This was conceived from studies of highly substructured desert fish, which persist as small isolated remnants from wetter historic times (Meffe and Vrijenhoek 1988). Rapid genetic drift, enhanced by the effects of isolation and small population size, causes strong divergence between subpopulations in which the pattern cannot be predicted from the spatial arrangement of sites (e.g. catchment boundaries or distance). This model remains to be tested on Australia's spring-dwelling desert fishes.

Just as population structure varies over time, it may vary spatially over the geographic range of a species. The pattern of genetic differentiation across categories of molecular markers can indicate the relative contributions and interplay of historical, recent and contemporary processes to the observed substructure (Hughes *et al.* 2009). The temporal scales can include evolutionary time (hence deep divides), more recent Earth history events, very recent anthropogenic changes (e.g. Clarence River cod: Nock *et al.* 2011) and current gene flow (e.g. Macquarie perch: Faulks 2010a). Often, more than one time scale can be discerned in patterns of gene flow (e.g. estuary perch, *Percalates colonorum*: Shaddick *et al.* 2011a).

Genetic insights into dispersal

Ecologists are likely to be interested in the actual or potential dispersal abilities of freshwater fishes. Genetic data provide independent information about dispersal in species for which ecological data are available (e.g. Australian smelt: Woods *et al.* 2010), and of course they are the only means to measure *in situ* reproductive exchanges (other techniques merely tell us where fish move). Genetic data confirm that limited gene flow may occur even among populations of highly vagile species (e.g. philopatric species, such as those salmonids which return to their natal stream). They also show that only a very small number of interbreeding migrants per generation is sufficient to prevent genetic differentiation.

There are numerous ways to assess gene flow (typically quantified as *Nm*, the product of the effective population number and rate of migration among populations, equal to the absolute number of genetic immigrants per generation). The most commonly used methods utilise either *F*-statistics (and related measures for multiple alleles), the frequencies of private alleles or gene genealogies (phylogenies for individual loci). All require data for multiple loci and all make explicit and implicit assumptions, some lacking biological realism. For this reason, estimates of gene flow are best viewed semi-qualitatively (thus, $Nm < 1$ = little or no gene flow; $2 < Nm < 10$ = low to moderate gene flow; $Nm > 10$ = moderate to high gene flow). Most phylogeographic and population-level genetic studies report these values (e.g. central Australian catfishes, Plotosidae: Huey *et al.* 2006), and studies in landscape genetics use these and other values to relate patterns of gene flow to anthropogenic changes, hydrology and other landscape features (e.g. Macquarie perch: Faulks *et al.* 2010a).

Having demonstrated an absence of contemporary gene flow between disjunct populations, ecologists may also wish to assess how long these populations have been separated. Two basic approaches are possible, namely the use of molecular clocks based on summary genetic distances for allozyme or DNA-based analyses (Hillis *et al.* 1996) or the dating of key nodes in individual gene trees based on DNA sequence data (Avise 2000). Both approaches are explored in the next section (see 'Evolutionary relationships').

Genetic diversity within a population

More insights into the ecology and biology of a population (and a species) can be obtained from comparative assessments of levels of genetic variation (e.g. diversity and frequency of loci). Large, stable populations (at historical and contemporary scales) are likely to maintain higher levels of within-population diversity than those that are smaller, restricted and prone to demographic fluctuations (e.g. Australian smelt versus pygmy perch: Hammer *et al.* 2007, 2010). Unusually high levels of genetic variability are a signature of localised introgressive hybridisation (e.g. Atherinidae: Adams *et al.* 2011) or, when accompanied by an excess of heterozygotes at multiple loci, of F_1 hybrids or unisexual lineages (e.g. carp gudgeons: Bertozzi *et al.* 2000). Trends in within-population variability inform conservation management because they may reveal the effects of fragmentation or reductions in population size (e.g. genetic bottlenecks: Frankham *et al.* 2010) and the potential for deleterious conditions such as inbreeding depression, reduced disease resistance and lowered ability to adapt to future changes. High-frequency private alleles may qualify for protection as a type of ecological asset, especially in stocking programs.

Evolutionary relationships

Another sphere of interest in systematics is the elucidation of evolutionary or phylogenetic relationships between taxa. As phylogenetic analyses are daunting, and constantly refined, we make no attempt to cover their terminology, methodology and controversies. Instead, we outline some basic principles and applications. Much of what follows also applies to assessments of relationships among populations within species, as the cognate disciplines of phylogeography and landscape genetics rely on phylogenetic analyses. A list of phylogenetic software is available on the internet

(<http://evolution.genetics.washington.edu/phylip/software.html>).

Basic principles

The evolutionary history of a group of organisms usually is shown as an evolutionary tree, obtained from phylogenetic analysis (overall genealogical relatedness) and/or cladistic analysis (relatedness based only on shared derived-character states). The raw data consist of either a matrix of character states or pairwise measures of genetic distance. Evolutionary trees include terminal nodes (which can be individuals, taxa or DNA sequences) and internal nodes (also known as hypothetical taxonomic units or coalescence points), all connected by branches whose relative lengths usually show the degree of relatedness between nodes. Once its topology has been determined, generally using computer software, a tree may thereafter be drawn as rooted (one internal node is designated as basal) or unrooted (a network). In either case, it is common practice to include one or more outgroup taxa to identify the basal node for the ingroup and to distinguish, if possible, ancestral (plesiomorphic) and derived (apomorphic) character states. Many trees also indicate the degree of statistical support (e.g. bootstrap percentages) contained in the raw data for each node.

As in other areas of systematics, the quality of data employed for phylogenetic analysis is as important as the choice of how to analyse the data. Ideally, the most suitable datasets for reconstructing phylogenies draw upon many genetically independent characters, each displaying high heritability, a low to moderate number of ordered character states per character (i.e. less than the saturation point), untainted by problems of strong selection, homoplasy (i.e. shared similarities through convergence, parallelism or secondary loss) and uncertain homology.

Three principles are relevant when assessing evolutionary trees. First, all tree-building methods have strengths, weaknesses, assumptions and limitations that influence the choice of a method and the confidence attached to the tree produced. Second, most methods have shortcomings that are biologically unrealistic, namely they generate a tree from any dataset, regardless of whether the raw data are 'tree-like', they assume that all evolutionary branching is dichotomous (they do not recognise polytomies) and they struggle to accommodate polymorphism. Third, while there is often concordance between the data for analysis and those that unite the taxa under study, there may be mismatches between gene-trees and species-trees (Waters *et al.* 2010).

Advantages of molecular data

Molecular data can provide insights into any revised evolutionary framework (e.g. galaxiids: Waters *et al.* 2000b). The best molecular datasets (i.e. unsaturated DNA sequences for multiple independent genes) offer two advantages over morphological datasets. First, they possess more of the features outlined (see 'Basic principles') as desirable for phylogenetic reconstruction. Second, they provide a more objective, readily quantified measure of genetic divergence across a standardised range of genes at all levels of the systematic hierarchy (until saturation is reached). It remains true, however, that phylogenetic studies that include both morphological and molecular characters (e.g. rainbowfishes: Sparks and Smith 2004) are likely to provide more powerful and robust descriptions of evolutionary relationships than those using only one data type. This integration is necessary, especially in systematics above the species level (Mooi and Gill 2010).

Estimating divergence times

The foregoing sections focused on reconstructing the sequence of cladogenic (branching order) events that led to the evolution of extant taxa. Another aim of phylogenetic analysis is to estimate the divergence times associated with clades thus identified. Rates of molecular evolution for a single class of molecular marker usually are stochastically constant within lineages (but often less so between lineages) and are far more predicable than rates of morphological evolution (Baverstock and Adams 1984). These observations have led to the concept of a molecular clock, where calculated levels of molecular divergence are translated into time elapsed since there was a common ancestor (e.g. pennyfish, *Denariusa australis*: Cook and Hughes 2010).

Molecular clocks are typically used to obtain three different types of dating for cladogenic events which are well-supported in a phylogeny, namely (a) relative ages (no explicit time scale), (b) absolute ages based on calibrating the divergence times for one or more key nodes, using fossil or other independent evidence and (c) absolute ages based on some nominated universal rate of molecular evolution for the organismal group or genes in use. Clearly, the number of underlying assumptions and the level of contention increase from (a) to (c). As rates of morphological evolution may vary enormously within and between lineages (Omland 1997), using comparative morphology to date cladogenic events is little better than guesswork in the absence of fossil data (i.e. for most Australian freshwater fishes). In these situations it is reasonable to use multi-locus molecular data to provide a first estimate of the age of the nominated lineages and to distinguish between competing biogeographic hypotheses where very different times of divergence are involved (e.g. relationships of mudfishes (*Neochanna* spp.) in Australia and New Zealand: Waters and McDowall 2005).

Limitations of single gene-trees

Gene-trees based on a single gene cannot be relied upon to evaluate phylogenies within or between species. Two failings are regularly observed. First, it is extremely common for single gene-trees to be discordant with the population- or species-tree for the individuals surveyed. For example, there is a 20–40% overall mismatch for mtDNA markers across a wide range of organisms (Funk and Omland 2003). Australian examples of species-pairs not diagnosable by mtDNA gene trees include Yarra and southern pygmy perch (*Nannoperca obscura* and *N. australis*, respectively: Unmack *et al.* 2011), the Finke and Lake Eyre hardyheads (*Craterocephalus* spp.: Unmack and Dowling 2010) and candidate species among the carp gudgeons (Schmidt *et al.* 2011a). Second, even where a single gene-tree is concordant with the underlying population- or species-tree, it does not follow that the timing of gene coalescence must coincide with the timing of the population split. In most instances, coalescence will pre-date isolation, often by a considerable margin (Knowles 2004); thus, the coalescence time is best regarded as an upper limit for speciation or population divergence.

Phylogenetics in practice

Phylogenetic analyses feature in some kinds of ecological research projects, particularly those with biogeographic associations. In addition, a researcher may wish to know whether particular attributes (e.g. aestivation, diadromy) evolved independently in congeneric species or whether species that share an attribute are all descended from a single ancestor with that attribute. These questions can help to identify the nearest living relatives, as a starting point for ecological research, but often stem simply from curiosity gained from working closely with a group. Local examples are studies showing that (a) the two diadromous members of the percichthyid genus *Percalates* (Australian bass; estuary perch), formerly regarded as species of *Macquaria*, are a separate lineage to the obligate freshwater species in that genus (golden perch, Macquarie perch), (b) sexually dimorphic galaxiids have a common ancestry (McDowall and Waters 2004) and (c) pygmy perches are indeed mini-percichthyids (Jerry *et al.* 2001). Often there are surprise findings – one example from deep-sea research concerns three fish of very different morphology, nominally assigned to three families, found to be different life stages and sexes of a single species (Johnson *et al.* 2009). Another pragmatic application of phylogenetic analysis is triage conservation assessments, making rapid identifications of species for priority protection, based on their evolutionary distinctiveness.

Comparative genetic profiling

The above-mentioned applications employ genetic markers to assess relatedness unconstrained by defined expectations and limits. Here, we touch upon applications in which unknown samples are compared with genetically defined known reference samples to obtain either/or genetic profiles to distinguish individuals, populations or species. These applications reflect the principle that most molecular genotypes are expressed in all life stages and tissues from an individual. Moreover, their utility does not rest upon a long list of theoretical and practical assumptions. Instead, they simply require that the necessary reference genotypes are known for a suitable suite of genetic markers. Examples of comparative profiling include (a) parentage testing of offspring (paternity or maternity assignment), where putative parents can be genotyped (e.g. Murray cod: Rourke *et al.* 2009), (b) pedigree analysis, gene mapping and kinship assessment in captive populations (e.g. rainbowfish: Evans *et al.* 2010), (c) assigning larval forms to their correct species (e.g. Murray-Darling fishes: Hardy *et al.* 2011), (d) identifying the species for unknown tissues such as fillets, scales or gut contents (e.g. lower Murray River food webs: Hardy *et al.* 2011) and (e) assigning individuals to their correct taxon in a complex of morphologically cryptic taxa (e.g. carp gudgeons: Schmidt *et al.* 2011a). From these kinds of outputs, specific projects can be developed, including projects in behavioural ecology (e.g. reproductive biology of Murray cod: Rourke *et al.* 2009) and issues of law-enforcement (e.g. confirming species identity).

Molecular genetic techniques

Molecular systematics obtains raw data from three broad kinds of technology (Hillis *et al.* 1996): methods for documenting karyotype variation, methods addressing genetically determined variation in proteins and methods which measure genetic variation in the nucleic acids (DNA, RNA). Key techniques are outlined below, and a comparison of their strengths, weaknesses and attributes is provided in Table 3.2.

Karyotype analysis

Karyotype analysis was used widely in years before the advent of DNA-based approaches. Observations on chromosome number, morphology and banding patterns were made for many species and groups of related species, providing insights into taxonomic ranks and acting as a stimulus for ideas about speciation. Although karyotype analysis is in decline it remains relevant as the simplest means of determining ploidy, as a useful method where there are narrow hybrid zones between apparently similar taxa and as a diagnostic character for cryptic species that are morphologically indistinguishable (e.g. North American madtom catfishes, *Noturus* spp.: Egge and Simons 2006).

Allozyme electrophoresis

This technique documents variation in more than 100 independent nuclear genes that encode soluble enzymes and non-enzymatic proteins in vertebrates. This is achieved by interpreting the relative mobilities and intensities of bands of enzyme activity displayed by tissue homogenates from different individuals following electrophoresis and histochemical staining (Richardson *et al.* 1986). Their strictly comparative nature, a requirement for fresh or frozen tissues, and dependence on correct interpretations of banding patterns have seen allozyme markers largely supplanted by DNA-based technologies. Given their speed of application, low cost, the number of genes examinable and the type of genetic data obtained (i.e. multiple, co-dominant, independent nuclear markers), allozyme data have continuing utility in molecular systematics, particularly as a method for delineating species and detecting hybridisation.

Polymerase chain reaction (PCR)

The following methods gather data directly from the primary sequences of bases in DNA (genes or gene portions) and all require a sufficient quantity and quality of DNA. The invention of the polymerase chain reaction (PCR) has greatly facilitated the use of DNA-based techniques, as PCR can amplify and purify thousands to millions of copies of a selected sequence from only small amounts of tissue. The required sequence is targeted using primers, short DNA fragments containing a sequence complementary to that in the target region.

Amplified fragment length polymorphisms (AFLPs)

This technique was introduced in the mid 1990s to compare the electrophoretic banding patterns of total genomic DNA from different individuals after digestion by two restriction enzymes, ligation with short sequences of synthetic DNA and selective amplification via PCR for randomly chosen pairs of DNA primers. The outcome is to indicate presence/absence at many hundreds or more anonymous AFLP loci. AFLP data can be generated rapidly and cheaply but the lack of co-dominant markers restricts their usefulness to certain kinds of molecular systematic investigations.

Mitochondrial DNA (mtDNA) sequencing

Unlike most other molecular technologies, DNA sequencing meets two ideals in molecular systematics as it indicates the actual raw genotypes at a locus and alleles/haplotypes that are ordered (i.e. their phylogenetic relationships can be readily assessed). Mitochondrial DNA sequences have been the most widely used molecular data since the early 1990s, encouraged by the availability of universal and taxon-specific primers, by decreases in complexity and cost, by the rise of phylogeography and by the enthusiasm that often accompanies rapid technological advances. Twenty years on, mtDNA sequence data are pre-eminent in many areas of molecular systematics (Beheregaray 2008), despite mounting evidence that the matrilineal gene trees they provide are subject to problems (e.g. nuclear paralogues, incomplete lineage sorting, selective sweeps, inability to diagnose hybrids), limiting their utility as a sole source of molecular information (Funk and Omland 2003). Sets of genetic markers that represent both mitochondrial and nuclear genomes are the best way to counter these gene-specific issues.

Nuclear DNA (nDNA) sequencing

Unlike mtDNA, the nuclear genome harbours thousands of potential genetic markers that occur as single copies and are independent, co-dominant and biparentally inherited. These two genomes also show substantial differences in mutation rates, the numbers of generations required for selectively neutral alleles/haplotypes to become fixed and the presence (in nDNA) versus absence (in mtDNA) of concerted evolution among near-gene families, partitioning of protein-encoding genes into introns and exons, and genetic recombination. These differences together render the two classes of DNA useful for differing applications in molecular systematics. In practice, nDNA sequence data are less utilised than might be expected; their uptake is hampered by a lack of suitable primers and the occasional need for molecular cloning to obtain separate sequences for the two alleles found in heterozygous individuals. Future developments in genomics (Avise 2010), such as next-generation sequencing (Glenn 2011), should overcome these difficulties. Many of the problems that confound use of mtDNA gene trees, however, also apply to single nDNA gene-trees (paralogues, lineage-sorting, selective sweeps), emphasising the need for concordance among multiple genetic markers.

Microsatellites

Microsatellites are repetitive regions in non-coding genomic DNA, consisting of variable numbers of a simple nucleotide motif (two to six base pairs in length), repeated in tandem. Their characterisation and use require the availability of PCR primers which target the conserved flanking regions adjacent to each microsatellite locus. Microsatellite loci have high heritability and show simple patterns of Mendelian inheritance, and display a number of advantages over other classes of co-dominant genetic markers (e.g. allozymes). In general, they occur in great numbers (typically, many thousands) in most organisms, are hypervariable (with up to about 50 alleles per locus), display high mutation rates, are more likely to be selectively neutral, undergo rapid evolution and harbour alleles that can sometimes be ordered phylogenetically according to simple mutational models (Oliveira *et al.* 2006). Together, these attributes make microsatellites ideal for all areas of molecular systematics at or below the level of species (e.g. assessing within-species variability, distinguishing individuals, populations and their offspring, assigning parentage, accessing gene flow). The ready availability of next-generation sequencing to identify hundreds of microsatellite loci, plus their associated primer pairs, assures continued widespread use in the future.

Single nucleotide polymorphisms (SNPs)

Once a considerable portion of a species' genome has been sequenced, it is possible to pinpoint genomic locations where different individuals display different genotypic combinations of the four bases (adenine, thymine, guanine, cytosine) that form part of the structure of DNA. When employed as a suite of genetic markers, unlinked SNP loci are useful in a variety of molecular systematic and population genetic applications, notably those focusing on within-population and within-species assessments (Brito and Edwards 2009). Technological advances in next-generation sequencing (e.g. Davey and Blaxter 2010) are likely to enhance their utility in well-studied organisms.

Obtaining genetic samples

A common concern among ecologists wishing to collaborate in a genetic study is how samples should be collected. There is a broad range of protocols for tissue collection, depending on taxonomic focus, ecological questions, methods of obtaining and analysing data, sampling factors (e.g. climate, transport and access to habitat, fish conservation status, abundance) and operational limitations (e.g. resources, funding). Thus, it is difficult to offer a detailed guide here; rather, discussions about these issues should form part of the collaboration. Some general considerations are the need to clearly define the questions to be answered, then to develop a sampling strategy with regard for the kind of samples required, geographic coverage, ideal sample sizes and appropriate field protocols. In the case of population studies, a pilot study to determine inherent variability, hence sample sizes required for statistical power, is more likely to succeed than a study based on *ad hoc* samples and guesswork (Richardson *et al.* 1986).

Continuously frozen tissue is the optimal method for preserving samples because, unlike most other methods, it preserves all macromolecules of interest (DNA, RNA, proteins) with high fidelity. Long-term storage is most effective in a cryogenic freezer at about -70°C; most museums and molecular research laboratories have these facilities. In practice, frozen tissue collection involves carrying a cylinder of liquid nitrogen, dry ice or, at a pinch, a portable car freezer into the field, or returning live fishes to the laboratory. Alcohol storage (which does not prevent long-term DNA degradation and limits projects to DNA-only methods) is best achieved using high-grade 90–100% ethanol, thinly sliced tissues (< 2 mm) and a generous liquid:tissue ratio (> 10:1), with samples stored in cold (to -20°C) or at least cool conditions, particularly in the first 24 hours after collection. A solution of saturated common salt is a temporary alternative where dangerous-goods transport is an issue. While providing no guarantee of DNA quality, other more readily available methods can be trialled when unexpected tissue-sampling opportunities arise, such as rubbing alcohol, isopropyl alcohol, high-strength spirits such as vodka, methylated spirits, preservation in washing powder and even sun-drying.

More sampled tissue rather than less, in the form of whole fishes or muscle fillets, allows for multiple projects and techniques. Some specific projects might target small, non-lethal samples of tissue (e.g. fin clips, scales or body mucus: Le Vin *et al.* 2011) or even historical fishery samples (e.g. Murray cod otoliths: Rourke *et al.* 2010). Experienced taxonomists maintain that fixation of voucher specimens in formalin is the preferred preservation method for morphological characters, as ethanol tends to render specimens brittle and misshapen. As formalin damages DNA, biodiversity assessments or taxonomic studies benefit from paired-tissue-plus-voucher specimens to enable matching of genetic and morphological data (sample tissue from right-hand side of fish).

CONTEMPORARY ISSUES

This section highlights some applications of genetics in the ecology of Australian freshwater fishes and suggests priorities for research and management.

Cryptic species

Cryptic species are a feature of faunal communities worldwide (Pfenninger and Schwenk 2007). They are prevalent in fishes but there is a high likelihood that they occur in other freshwater biota, particularly crustaceans (e.g. Chenoweth and Hughes 2003). The detection of cryptic species is important, for two reasons. First, similar-looking species may not necessarily have similar ecological characteristics, and there is potential for confounded data. Second, it is unlikely that management and conservation plans for a presumed single species will adequately protect true species diversity (Beheregaray and Caccone 2007).

These issues are reflected in the taxonomic history of four iconic species, the freshwater cods of the genus *Maccullochella*. Two species in particular, the sympatric Murray cod and trout cod, have had a long history of confusion. While the two are similar to an untrained eye, genetic data played a role in elevating trout cod to species status (Berra and Weatherley 1972) and physical and ecological differences between the two species now are well-documented (Lintermans 2007). Three allopatric species of cod have been identified using morphological, genetic and biological data, namely the true Murray cod in the Murray-Darling Basin, the Clarence River cod (*M. ikei*) from the Clarence River, New South Wales, and the Mary River cod (*M. mariensis*) from the Mary River, Queensland (Rowland 1993; Nock *et al.* 2010). The Richmond and Brisbane river basins in coastal eastern Australia formerly had resident cod, but these were lost before their specific status could be determined. All four extant species are nationally 'threatened' (Ch. 12).

The mountain galaxias (*Galaxias olidus*), carp gudgeons and smelts lack the iconic status of the cods but all are common in south-eastern Australia and provide striking examples of cryptic species. Australian smelts have been considered as two species, one on the mainland (*Retropinna semoni*) and another in Tasmania (*R. tasmanica*), but molecular data indicate at least five species, including allopatric species in distinct biogeographic provinces, multiple instances of sympatry along the east coast and a Tasmanian species more widespread than previously recognised (Hammer *et al.* 2007) (Fig. 3.5). These species now await morphological diagnosis. Mountain galaxias occur in upland habitats of the Great Dividing Range and a taxonomic review using a combined-lines-of-evidence approach has revealed 14 species, many with very restricted distributions (Raadik 2011). The complexity of the carp gudgeon genus *Hypseleotris* was revealed, in part, by a study on the lower Murray River in the early 1990s, but more research is needed to confirm the likely existence of unisexual forms and indications of cryptic taxa (i.e. allopatric populations of true western carp gudgeon (*H. klunzingeri*) and an informal taxon, 'Midgley's carp gudgeon': Bertozzi *et al.* 2000; Thacker and Unmack 2005; Schmidt *et al.* 2011a; Adams, unpubl.).

These few examples merely hint at future changes in the taxonomy of the obligate freshwater fish fauna of Australia. Molecular studies of species boundaries in freshwater catfish (*Tandanus tandanus*), eastern dwarf galaxias (*Galaxiella pusilla*), river blackfish (*Gadopsis marmoratus*), golden perch, southern pygmy perch and western pygmy perch (*Nannoperca vittata*), flathead gudgeon (*Philypnodon grandiceps*) and dwarf flathead gudgeon (*P. macrostomus*) all indicate cryptic species (Musyl and Keenan 1992, 1996; Hammer 2001; Miller *et al.* 2004; Coleman *et al.* 2010; Unmack *et al.* 2011). Deep divergences in phylogeographic studies of plotosids, galaxiids, atherinids, rainbowfishes, blue eyes, percichthyids and eleotrids also argue for cryptic taxa and a need for broad-scale, molecular systematic investigations (Watts *et al.* 1995; Hurwood and Hughes 1998; Page *et al.* 2004; Wong *et al.* 2004c; Thacker *et al.* 2007; Jerry 2008; Cook *et al.* 2010; Faulks *et al.* 2010b; Unmack and Dowling 2010; Unmack *et al.* 2011). There may be double the number of known freshwater fish species in Australia, including 'common' and 'threatened' species in all regions of the continent (Ch. 2). Some further examples of cryptic species are shown in Figure 3.6.

Systematic reviews that better represent the diversity of Australian freshwater fishes should promote conservation. This challenges researchers and managers to (a) encourage people to collect genetic material and voucher specimens, (b) allocate more funding for best-practice molecular systematics (i.e. multiple nuclear markers in combined lines of evidence) and (c) argue for more well-resourced taxonomists. If we do not promote biodiversity research and management, the pace of human-mediated change is likely to eliminate species even before they are documented.

Conservation units

A major component of biodiversity resides at systematic levels below that of species – subspecies, major lineages and subpopulations. To protect genetic diversity at these levels, a complementary approach is required to acknowledge the existence of conservation units aligned to major divergences within species (Ryder 1986; Vogler and Desalle 1994). This endeavour began with efforts to protect distinct stocks of salmon and trout in the USA. Patterns of genetic and ecological divergence were used to help identify discrete subpopulations for protection and management; these are termed evolutionarily significant units (ESUs) and are recognised, at least implicitly, in the US *Endangered Species Act* (Waples 1995). Defining criteria for identification of conservation units has proved controversial for the need to accommodate regional differences in management and legislation and differing opinions on appropriate biological and/or genetic criteria (Crandall *et al.* 2000; Wood and Gross 2008).

The most common approach has been to recognise deeper divides within species as ESUs and geographic subpopulations as management units (MUs: Moritz

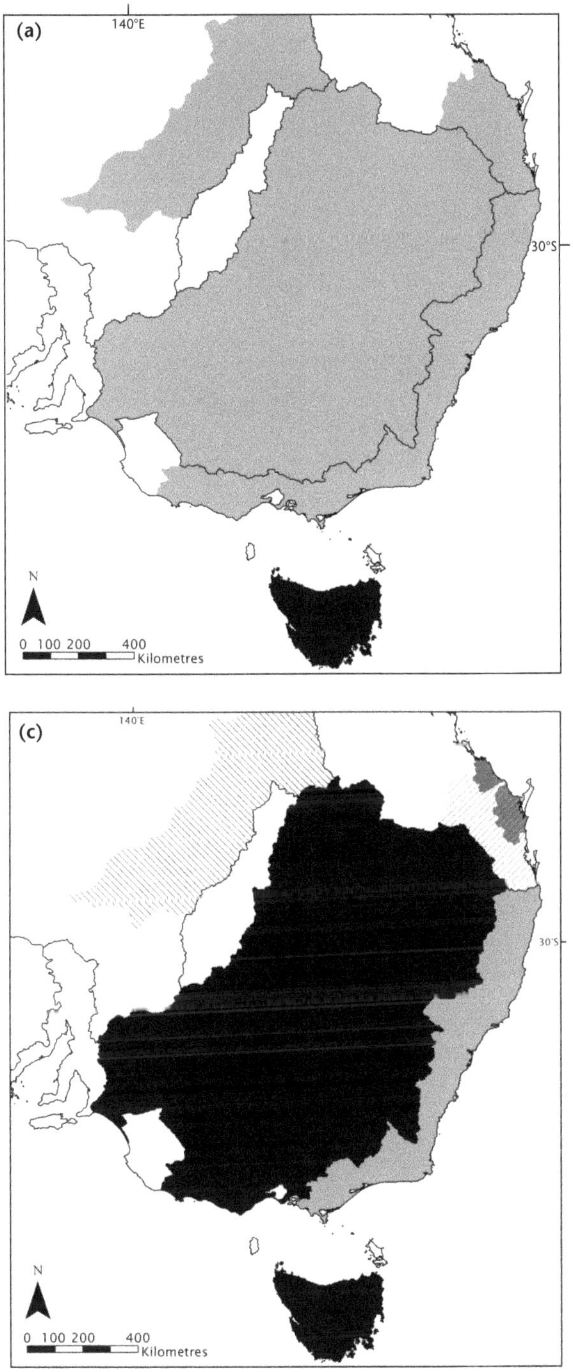

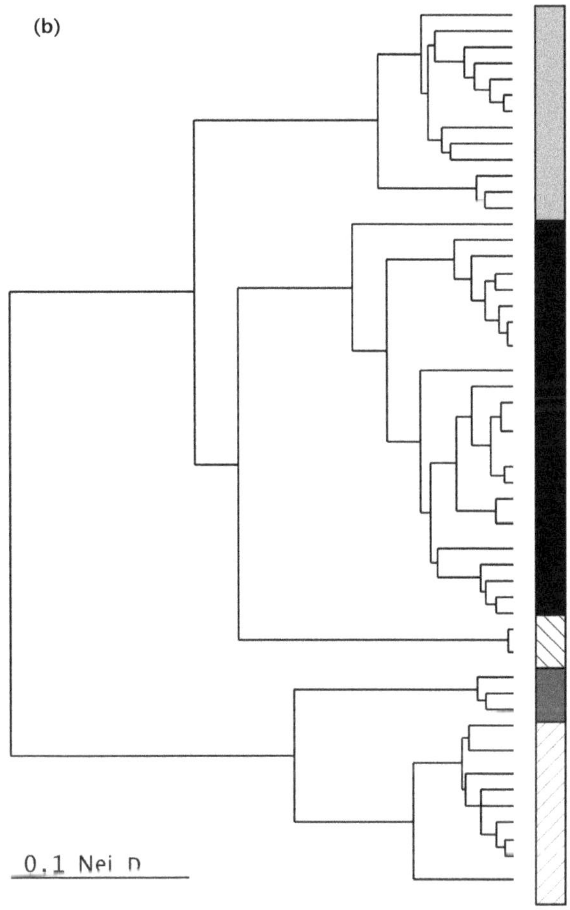

Figure 3.5: A study of species boundaries in Australian smelts (Retropinnidae), where molecular data have redefined the group (Hammer *et al.* 2007a). (a) Prior to the study, two species were recognised, one on the mainland and the other in Tasmania. (b) Species boundaries were investigated by screening for nuclear genetic markers (allozymes) at 62 sites across the range. (c) Five species were indicated, one of which occurs both on the mainland and in Tasmania. Nei's distance D reflects the difference in frequencies of one or more traits and assumes this has arisen by mutation and/or genetic drift.

1994, 2002). As with the combined-lines-of-evidence approach in systematics, comparative morphological, ecological and environmental assessments for different populations provide a more practical, holistic view of evolutionary potential (Moritz *et al.* 1995; Crandall *et al.* 2000; Palsbøll *et al.* 2007). Genetic assessments generally provide a first-stage overview of conservation issues and a basis to collate biological data for possible reinforcement or refinement. Attempts to define conservation units without reference to genetic

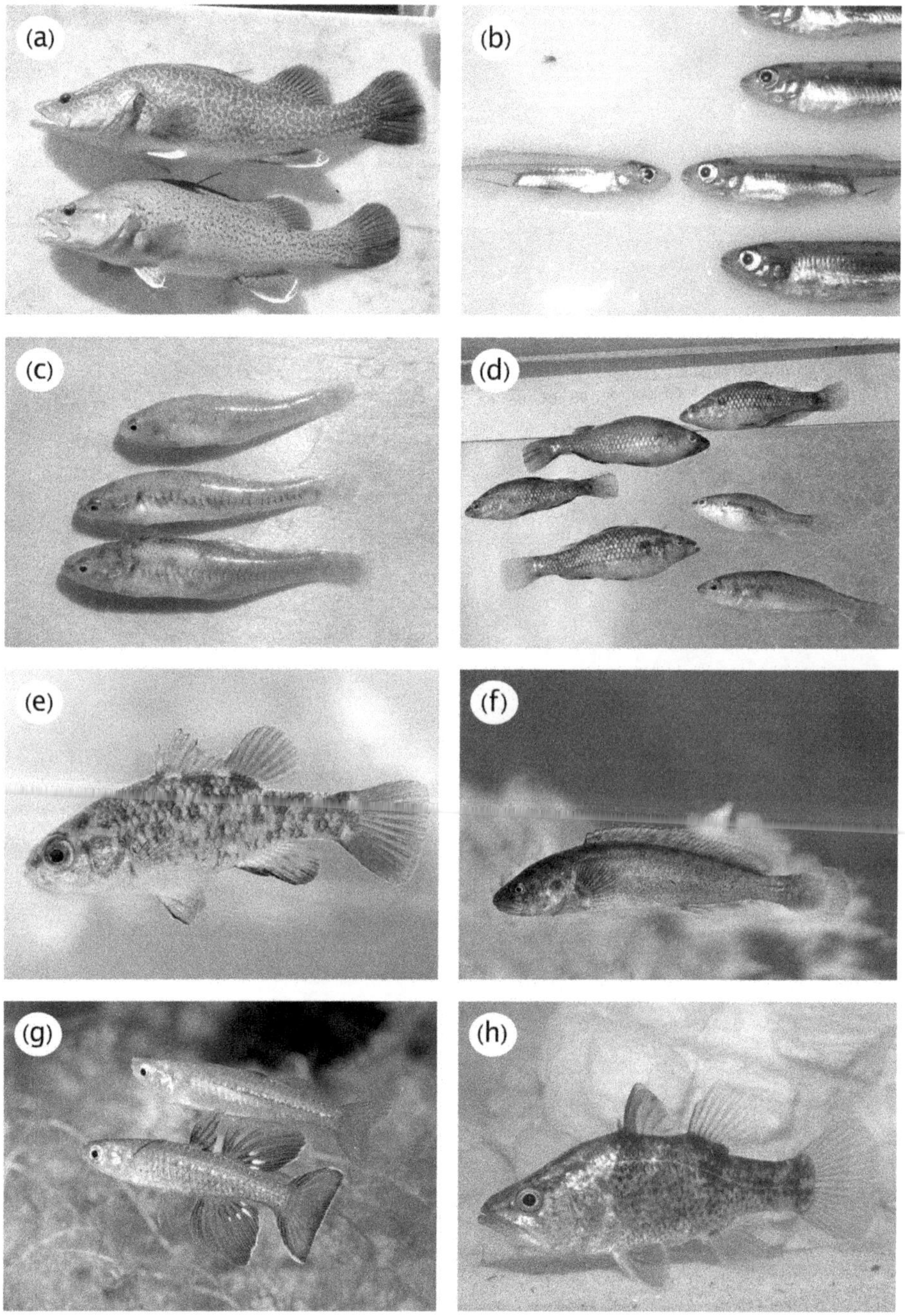

Figure 3.6: Examples of cryptic species among Australian freshwater fishes. In sympatry: (a) Murray cod (*Maccullochella peelii*, upper) and trout cod (*M. macquariensis*, lower) were once thought to be one species. (b) Australian smelt (*Retropinna* spp.) from the Mary River, Queensland (two species, facing different directions). (c) Carp gudgeons (*Hypseleotris* spp., all males) from the Murray River, South Australia – (top) Midgley's carp gudgeon, *H.* sp. 'Midgley'. (middle) True western carp gudgeon (*H. klunzingeri*). (lower) Murray-Darling carp gudgeon, *H.* sp. 'MD'. (d) Pygmy perches from south-west Western Australia – (bottom right) Balston's pygmy perch (*Nannoperca balstoni*). (middle right) Little pygmy perch (*N.* sp.). (rest) Western pygmy perch (*N. vittata*) (photo: M Allen). In allopatry: (e) Eastern species of southern pygmy perch (*Nannoperca australis*). (f) River blackfish (*Gadopsis marmoratus*) from the Murray-Darling Basin. (g) Delicate blue-eye (*Pseudomugil tenellus*), Daly River (photo: G Schmida). (h) Mouth almighty (*Glossamia aprion*), Carson River (photo: D Morgan).

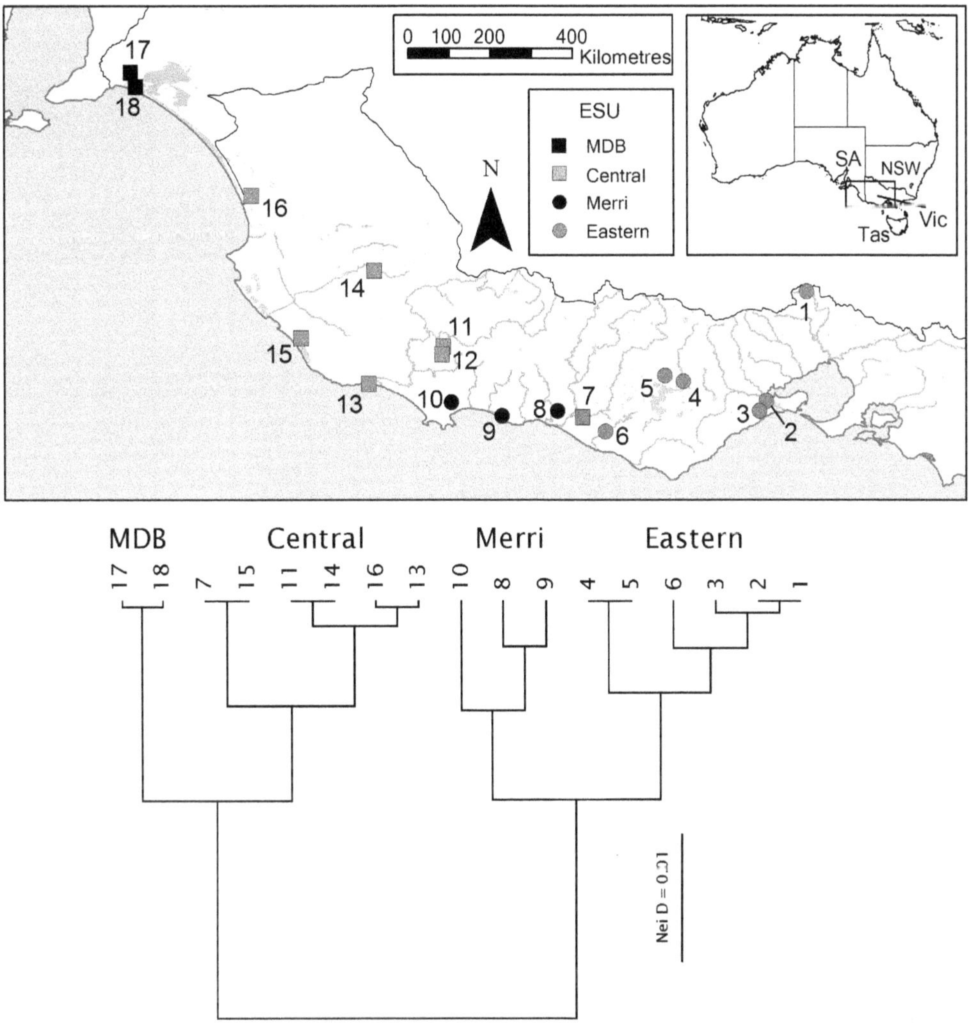

Figure 3.7: Conservation units in the Yarra pygmy perch (*Nannoperca obscura*). In this case, four deep divergences aligned with geographic boundaries provide a starting-point for protecting genetic diversity below the species level (modified from Hammer *et al.* 2010). Nei's distance D reflects the difference in frequencies of one or more traits and assumes this has arisen by mutation and/or genetic drift.

data (e.g. using arbitrary natural resource management boundaries) may misrepresent the evolutionary significance, hence conservation value, of regional populations (e.g. Firestone *et al.* 1999).

The spatial scale for conservation of diversity in obligate freshwater fishes is restricted, especially for small demersal species where conservation units may be confined to one or two river basins (Hammer *et al.* 2010). Conservation units have been defined explicitly for only a few Australian species (e.g. Yarra pygmy perch: Fig. 3.7) but they can be estimated from population genetic studies that identify major substructuring (e.g. for the Murray-Darling Basin: Moore *et al.* 2010). Genetic data were used to identify conservation units and action to protect threatened fish populations in Lake Alexandrina, near the mouth of the Murray River in South Australia, during the Millennium Drought of 1997–2010 (Hammer 2008; Hall *et al.* 2009a; Adams *et al.* 2011).

Should genetic subunits be mixed?

Having established that a threatened species includes one or more genetic subunits (whether ESUs or MUs), managers face the question of how best to act on this knowledge. Options include to maintain the *status quo*, protect and separately maintain each individual subunit, protect and separately maintain a few high-value subunits or deliberately mix different subunits. Aside from the obvious need to consider funding, managers ought to consider five questions: (a) how robust is the species-level taxonomy, (b) how genetically distinct are subunits, (c) how long have the subunits been genetically isolated, (d) how inbred are the individual subunits and (e) are there significant ecological differences between subunits? Hughes *et al.* (2009) and Frankham *et al.* (2010) provide guidance on how to address the first four questions, using genetic data.

The admixture of different ESUs should never be contemplated, as by definition these are major subdivisions and often equivalent to subspecies or phylogenetic species. Genetic rescue through admixture of subpopulations or MUs can be an effective strategy only where the subunits meet all the following criteria: they are demonstrably of the same species, they are genetically similar, they have not been historically isolated, they are themselves candidates for inbreeding depression and they do not display significant ecological differences (Frankham *et al.* 2010). A review of translocation as a tool for maintaining or improving genetic diversity and population security is provided by Weeks *et al.* (2011).

Identifying native and introduced populations

Whether a species is native, exotic or alien (Ch. 11) elicits different responses from legislature, management and research, and different responses from the community. The homogenisation of fish faunas through translocations of native species across natural barriers (Chs 11, 12) may cause uncertainty over whether or not a species is a local endemic. This is particularly so where there are few historical baseline data, as often applies for small or secretive species, or where there has been a long break in local records of a species. Doubts could be settled by genetic comparisons between possible source populations, ideally from across the species' range. Native populations can be identified quickly if they are distinctive (e.g. cryptic species, novel lineages, major differences in allele frequency, private alleles at multiple loci) or have relatively high levels of genetic diversity. In addition, rediscoveries could be matched to a historic reference population, where material is available and in good condition. Special protocols such as ancient DNA techniques can be used to obtain data from museum specimens or other tissues (de Bruyn *et al.* 2011). Fine-scale methods (e.g. microsatellites) may be needed when variability among and within populations is low.

Several Australian studies bear upon this issue. Investigations of climbing galaxias in the upper Murray River suggest that the population arrived via pipeline across the Great Dividing Range, rather than being a natural landlocked population or one derived from a diadromous migration (Waters *et al.* 2002). Genetic data have confirmed the introduction of populations of angling species such as freshwater catfish and Macquarie perch to some eastern coastal rivers and the presence of distinct native populations in some coastal systems (Jerry 2008; Faulks *et al.* 2010a, 2011). After a 30-year absence of records, and extensive searches, a single population of southern purple-spotted gudgeon was rediscovered in 2003 in a wetland on the lower Murray River, 2500 km from the nearest known extant population. Doubts were raised that this was a natural population and there was only minor interest from government until the wetland habitat began to dry. Genetic data showed the local population to be a genetically distinct MU (i.e. native), leading to plans to rescue the fish and develop a captive-breeding program (Fig. 3.8) in preparation for a recovery plan (Hammer 2008; Hall *et al.* 2009a; Carvalho *et al.* 2012).

DNA barcoding

DNA barcoding refers to applications which use DNA sequence data (for part of the mitochondrial cytochrome oxidase I gene) to assess the systematic affinities of individuals. It was proposed originally as an alternative, rapid method to distinguish species in situations where traditional keys are cumbersome or inapplicable (e.g. for larval forms). By inference, DNA barcodes that do not match those for known species are presumed to be undescribed species. The Barcode of Life project (Ratnasingham and Hebert 2007), including a program on fishes (FISH-BOL: Ward *et al.* 2009), plans to develop and maintain a global refer-

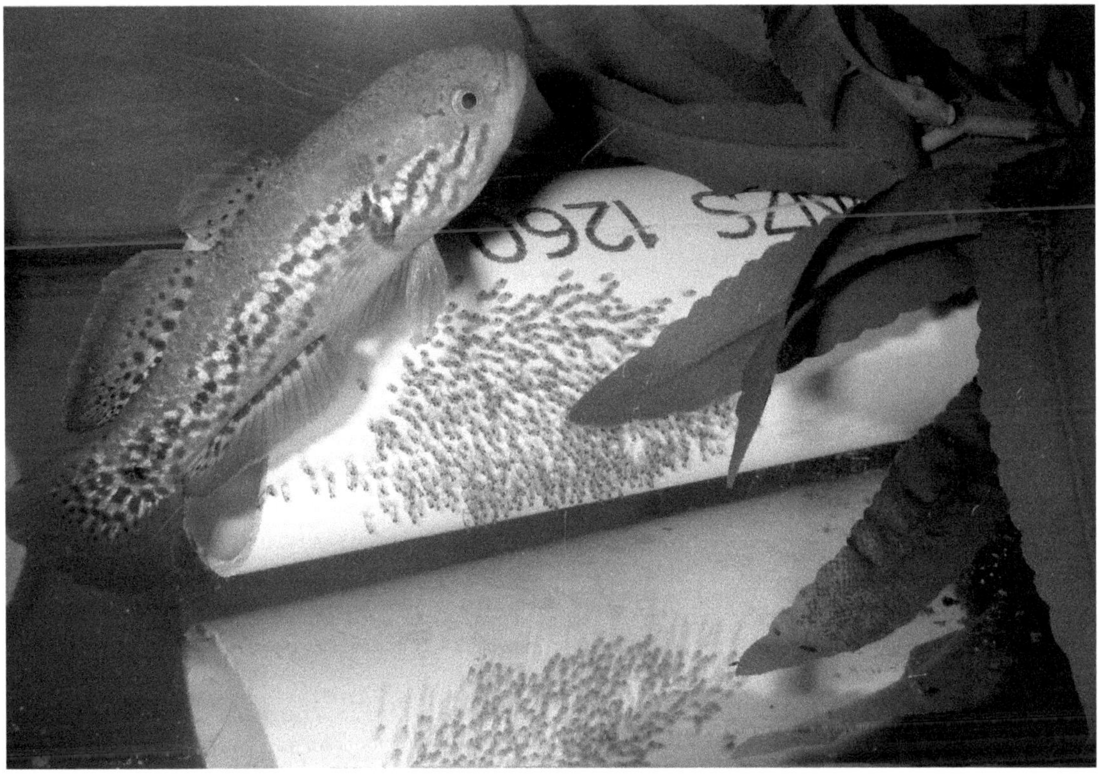

Figure 3.8: After many years when they were believed locally extinct, southern purple-spotted gudgeons (*Mogurnda adspersa*) were rediscovered in a wetland on the lower Murray River, South Australia (Hammer 2008). The wetland dried and fish were rescued into a captive-breeding program. The fish pictured is a male broodstock individual, guarding eggs. Genetic data helped to confirm that this was a native population and in urgent need of conservation action.

ence library of species to assist identification and promote taxonomy. It will include the freshwater fishes of Canada (Hubert *et al.* 2008) and later Australia.

This process remains relevant but it has been expanded to include any or all of the following: (a) using sequence data from other mitochondrial or nuclear genes, (b) examining genealogical concordance among small numbers of different genes, (c) using DNA sequence data to infer species boundaries in completely naïve groups and (d) defining species, genera or families in the absence of other data. Some of these pursuits are more laudable than others ((c) and (d) are particularly questionable) and DNA barcoding has become contentious. The strengths and limitations of this approach are considered in several reviews (e.g. Moritz and Cicero 2004; Will *et al.* 2005).

Locally, there are examples of both appropriate and inappropriate uses of DNA barcoding. In the former case, a barcode inventory has been produced for all native and alien fishes in the Murray-Darling Basin, for species identification and conservation purposes (Hardy *et al.* 2011), and the utility of different mtDNA genes for barcoding has been explored with regard for freshwater fishes in south-eastern Queensland (Page and Hughes 2010). In contrast, a revision of the rare cave gudgeon (Chakrabarty 2010) demonstrates the misuse of DNA barcoding. Three of four characters used to diagnose the southernmost populations as a new species were mtDNA genes (i.e. neither independent or biparentally inherited genetic markers: see 'Documenting species') and the fourth was a subtle morphological difference that applied only to some specimens.

A tag that lasts a lifetime (and beyond)

A genetic tag works like those used in traditional mark–recapture techniques except that it incorporates an ability to detect individuals from their DNA. Much the same technology is used forensically, to match suspects to evidence of crime. Genetic markers can distinguish individual fishes without issues of tag retention, for small and large individuals. For example, a small quantity of tissue (e.g. a fin clip) is returned to the laboratory, processed and the individual multi-locus genotype is fed into a database just as if it were a tag number. Where a match occurs it is used to assess growth rates of individuals, track distance moved, estimate recapture rates and population size and so on. More sophisticated analyses and larger samples can provide insights into issues such as gender, kinship analysis to determine whether parents, siblings or sired offspring have also been sampled, whether stocked individuals have reproduced with wild fishes, and how many individuals are successfully reproducing in a population (effective population size). They might eventually provide a non-lethal method for age determination. An innovative project to employ genetic tags is being developed in Australia for research into pelagic marine fishes. The GENETAG project has developed tagging-pole and baited-hook methods for procuring genetic material without the need to handle fishes, as a mark–recapture technique for use in management of the marine fishery for Spanish mackerel (*Scomberomorus commersoni*) (Buckworth *et al.* 2007).

Unisexuality in fishes

Hybridisation occasionally triggers events that lead to creation of so-called unisexual forms. These are rare among vertebrates but occur naturally in fishes (and amphibians and reptiles), and display some unusual modes of reproduction. Unisexual lineages reproduce without undergoing genetic recombination, and are so-named because almost all are females. Of three basic modes of reproduction among unisexual vertebrates, gynogenesis and hybridogenesis thus far are known only in fishes and amphibians, and apomictic parthenogenesis appears to be restricted to reptiles (certain sharks that reproduce biparentally may occasionally produce non-clonal parthenogenetic offspring). Parthenogenetic lineages can reproduce clonally and asexually, so that a single parthenogen is theoretically able to colonise new habitats. In contrast, unisexual fishes (and amphibians) can persist in nature only by co-occurring with (and effectively being sexual parasites of) their sexually reproducing (gonochoristic) relatives. This is because they require sperm either to stimulate clonal egg development (gynogenesis) or to provide a new haploid genome to wholly replace the paternal haploid genome from the previous generation (hybridogenesis). While all forms of unisexuality are novel, hybridogenesis is particularly bizarre, for two reasons. First, the maternal genome (which may be haploid or diploid) is transmitted from generation to generation without undergoing recombination (hemi-clonal inheritance). Second, while the sexual male becomes the genetic father of his hybridogenetic offspring, he is never their genetic grandfather! Among Australian freshwater fishes, unisexual lineages have been confirmed thus far only in carp gudgeons (Bertozzi *et al.* 2000; Schmidt *et al.* 2011a). As the molecular genetic signature of unisexuality is apparent only after applying a suite of nuclear genetic markers, its presence in other genera cannot be ruled out.

Behavioural ecology

Knowledge of parentage and kin relationships is a powerful tool for studying the reproductive biology and behaviour of fishes, especially considering that direct observations are often difficult. Genetic typing of offspring and putative parents can measure the contributions of different breeders and provide insight into sexual selection for certain traits (e.g. size, colour, aggression) and reproductive behaviour such as monogamy/polygamy, lekking, sneaker males, cuckolding, nest takeover, egg mimicry and sperm storage (Avise 2002; Ch. 7). This information is significant for evolutionary theory, for the ecology of species and for applications in management but there are few such data for Australian species. For example, female western rainbowfish prefer larger males, which are dominant in competition and sire the majority of offspring (Evans *et al.* 2010; Young *et al.* 2010). Investigations of Murray cod in aquaculture ponds have indicated polygamy and multiple spawning, two unexpected traits which have caused a rethink on broodstock management and hatchery protocols (Rourke *et al.* 2009).

Stocking and reintroduction

Releases of hatchery-bred fishes are made for various purposes beyond the creation of new fisheries; they may augment populations to offset decline, bolster low numbers of threatened species, re-establish previous distributions, provide additional population security or improve genetic diversity and adaptive potential. The process is widespread in Australia, especially with larger, recreationally important species, but also includes conservation-stocking or captive breeding of barred galaxias (*Galaxias fuscus*), several Tasmanian galaxiids, freshwater catfish, Macquarie perch, southern purple-spotted gudgeon, trout cod and Yarra pygmy perch (Dept of Primary Industries, Water and Energy Tasmania 2006; Hammer *et al.* 2009; Moore *et al.* 2010; Chs 11, 12). Without care, the positive intentions of such programs can be overshadowed by deleterious outcomes. Specific genetic consequences from captive breeding and translocation or release of hatchery-bred fish include (a) hybridisation with other local species (which may be morphologically cryptic), potentially leading to functional extinction of one or both species, wasted reproductive effort and additional competition or hybrid swarms, (b) swamping of local population diversity with loss of genetic diversity, adaptive potential and genetic heritage, especially with large stocking releases that include mainly brothers and sisters from a few broodstock, as has been common in Australia (e.g. over 5 million fish are released in the Murray-Darling Basin per year: Gillanders *et al.* 2006), (c) outbreeding depression and (d) failure to capture and maintain extant genetic variation in captive-breeding and release programs.

Locally, studies are beginning to investigate the genetic impacts of stocking; already swamping is documented by intensive stocking over a remnant population of Clarence River cod (Nock *et al.* 2011), intermixing of Murray cod (Rourke 2007) and the inadvertent release of hybrid offspring in stocking Australian bass (Shaddick *et al.* 2011b). Molecular techniques can help to minimise these threats to wild populations by guiding hatchery and release protocols with regard for the sources and numbers of broodstock (Rowland and Tully 2004), by designing stocking strategies to maintain and improve genetic diversity (Ingram *et al.* 2011), by monitoring hatchery production to avoid sibling mating (Carvalho *et al.* 2011) and by assessing the contributions of breeders to the numbers of juveniles produced (Rourke *et al.* 2009).

Threats and conservation

The global expansion of human populations and industry in the last two centuries has brought immense changes to the environment and to fauna and flora, on a scale like that of a mass extinction event but over a much shorter period of time. Indeed, the rate of extinctions has soared, especially among freshwater biota (Ricciardi and Rasmussen 1999), and may accelerate further as threats compound and climate change advances (Ch. 12). The threat of species extinction can be viewed on two levels, considering drivers (threatening processes) and genetic phenomena.

Broad-scale threats to aquatic ecosystems, such as altered hydrology, habitat degradation and alien fishes (Ch. 12), can drive the extinction of species, conservation units and unique genetic populations within short periods (e.g. habitat changes in the lower Murray River: Hammer 2008; Hammer *et al.* 2010; Adams *et al.* 2011). Artificial movements by various means (Lintermans 2004) can short-circuit and erode evolutionary processes, provide observable specific threats to genetic diversity (see 'Stocking and reintroduction') and potentially produce combinations of alien species or strains that are more tolerant or invasive (i.e. outbreeding enhancement, as for common carp). Anthropogenic changes can promote hybridisation through breakdown in behavioural or habitat separation. Eutrophication, for example, masks mate-recognition in swordtails (*Xiphophorus birchmanni*), leading to hybridisation between sympatric species (Fisher *et al.* 2006). Changed flow regimes may interrupt natural spawning cues and timing or alter habitats to force species interactions. For example, reduced river flows increase the chances of landlocking and reproductive interactions between yellowfin bream and black bream (*Acanthopagrus* spp.) in estuaries of eastern Australia (Rowland 1984).

The reduction and progressive elimination of genetic diversity may act insidiously to reduce both the ability to adapt to change and the resilience to cope with stochastic events, both natural and artificial. Losses of genetic diversity occur through local extinctions, fragmentation and geographic fixation, reduced habitat areas and diversity, greater interactions among species (e.g. increasing potential for competition and hybridisation), smaller and suppressed populations (e.g. inbreeding depression), losses of connectivity or gene flow and changed metapopulation dynamics (Frankham *et al.* 2010).

We see biodiversity within ecosystems as the products of evolution but view the overall process in only a split second of evolutionary time. How then should we try to manage species and habitats in face of current and future threats? Genetics provides tools to address some of these issues. Environmental degradation may divorce species and their genetic components from critical ecological and evolutionary processes which guide adaptation and evolution, and actions and measures will be needed from the top down to preserve and restore systems. Linking genetics to restoration and recovery planning may be as simple as cataloguing biodiversity to demonstrate the significance of a region, or as complex as maintaining a flow regime to maintain and stimulate ecological and evolutionary processes. In habitats changing more quickly under the press of human industry than species can adapt, actions will be needed from the bottom up to conserve gene flow and diversity within populations. These actions will impose a strong demand for knowledge of local environments and the genetic characteristics of species, populations and individuals. At an intermediate scale, between individuals and ecosystems, restoration of habitats and connectivity between habitats are proactive means to influence population processes and resilience (Cook *et al.* 2007).

Box 3.1: Further reading

Full bibliographic information appears in the References section at the end of this volume.

Avise (2002) A thought-provoking preface with case studies showing how genetic data have enhanced understanding of ecology and evolution.

Avise (2000, 2004) Reviews of species concepts, overviews of molecular markers and introductions to related theory.

Frankham *et al.* (2010) Issues in conservation of biodiversity, focusing on genetic processes, phenomena and threats.

Futuyama (2005) Overview of principles in evolutionary biology, with emphasis on systematics, genetics, biodiversity, biogeography and ecology.

Helfman *et al.* (2009) Starting point for readers interested in the taxonomy, evolution and ecology of fishes. Includes a chapter on fish genetics.

Hillis *et al.* (1996) Basic principles of molecular systematics and laboratory procedures for karyotype, allozyme and standard DNA analyses.

Holderegger and Wagner (2008) A review of the emerging field of landscape genetics.

Hubbs and Lagler (2004) A universal manual for fish taxonomists, with detailed methods for characterising physical variation.

Hughes *et al.* (2009) A review of gene flow and dispersal in stream habitats, with an Australian perspective.

Moore *et al.* (2010) A practical review of genetics for local resource management.

Page and Holmes (1988) An overview of principles and procedures in the use of molecular genetic data, with emphasis on phylogenetic reconstruction.

CONCLUSION

Evolution and adaptation underpin much of what we observe in ecology. In response to a diverse and changeable habitat template, freshwater biota display a huge variety of form and intrinsic characteristics. Genetics provide a powerful means of documenting this diversity and revealing patterns and processes. Various molecular techniques are available to test and critically evaluate hypotheses but there may be no one best choice – a recurrent message in this chapter is that multiple independent lines of evidence provide the most robust findings. Genetic data can support current practice, provide independent complementary analyses of ecological research and even challenge or enhance basic ideas and assumptions in ecology. This is especially relevant for how we view and recognise species, investigate dispersal, understand fish behaviour and test for patterns in the landscape. Genetic research has a formidable and exciting agenda that includes helping to review the species level taxonomy currently applied to many Australian freshwater fishes, assisting natural resource managers to appreciate and protect high levels of genetic heterogeneity in species and assemblages, continuing to explore and inform sometimes-desperate conservation scenarios (e.g. faunal rescues, translocations, *ex situ* conservation), refining theory (e.g. species concepts, mechanisms of speciation) and contributing to newer research areas (e.g. behavioural ecology of freshwater fishes). For readers wanting to explore these issues or to reinforce ideas visited earlier in this chapter, we conclude with some recommendations for reading (see Box 3.1).

4 Habitats

John D. Koehn and Mark J. Kennard

INTRODUCTION

There are no terrestrial fishes, of course. Some species have stretched the boundaries of an aquatic existence – for example, mudskippers (Gobiidae), salamanderfish (Lepidogalaxiidae) and lungfishes (e.g. Neoceratodontidae) – but none is entirely free from dependence on water as a medium. In their aquatic realm, fishes rely on water and all kinds of submerged structures to provide the conditions needed to live, grow and reproduce. To better understand them, we need to make a conceptual leap from life on land to life in the water.

Not all aquatic environments are similar. While some marine environments are comparatively stable, very different conditions prevail in rivers, lakes and other inland waters. These provide an array of dynamic habitats that present a great variety of physiological and ecological challenges. Even in their complex underwater world, freshwater fishes are exposed to the vagaries of a capricious climate (especially rainfall and runoff) and to the far-reaching consequences of human activities (e.g. dams, pollution, catchment clearing).

A habitat is more than a place where a species, assemblage or community is located, because fishes often occur in areas that could not sustain them for long periods. The core habitat provides the resources and conditions needed for them to complete their life cycles and to sustain viable populations. It includes abiotic (physical, chemical) and biotic (living) components. The size and shape of the water body, its hydraulics, the water chemistry and the nature of the substratum are all parts of the abiotic environment, many of them determined by the inherent properties of water (Gordon *et al.* 2004). Biotic components include living organisms linked by processes such as competition, predation and parasitism (Ch. 6). Some biotic and abiotic components, such as food or spawning sites, may be in short supply at different times or places and may thereby limit the value of the habitat and its capacity to support fishes. Although as ecologists we may try to isolate particular elements for study, like submerged wood, water velocity or chemistry, the true habitat – from a fish's viewpoint – is an inseparable combination of many such variables. Thus, the term 'habitat' refers generally to an area (or areas) within which the requirements of all life stages of a species are met. It may also refer to an area where recruitment occurs on a scale sufficient to sustain a population, and to the intersection of habitats for populations of other species (an assemblage: Ch. 10). The concept of habitat is therefore holistic, because it integrates many elements of the environment, and hierarchical, because it spans different levels of organisation.

In this chapter, we begin by outlining the diversity of Australian freshwater environments and the physical context for the habitats of fishes. We note some concepts relevant to fish habitats, consider the significance of spatial and temporal scale and describe the types of habitat needed by different species. We also appraise the state of knowledge of fish habitats, describe methods for study and identify opportunities for research. Readers will soon detect a bias toward work undertaken in south-eastern Australia. That is unavoidable, because it is a true reflection of the 'state of the art', and we trust that it will not disguise the need for work in other regions, particularly in tropical northern Australia.

FRESHWATER ENVIRONMENTS

Diversity

The sheer size of the Australian continent and its variety of geomorphic, climatic and hydrologic conditions (Ch. 10, Fig. 10.1) have engendered a diversity of freshwater habitats for fishes (e.g. De Deckker and Williams 1986; Boulton and Brock 1999). The most basic distinction – recognised by fishes as well as people – is between still and flowing water. Rivers and streams are *lotic* (running water) habitats; wetlands like billabongs, lakes, springs, swamps and artificial impoundments are *lentic* (standing water) habitats.

Lotic habitats

Rivers and streams flow downhill from their headwaters to the sea (or, in some cases, to a terminal lake or other wetland). The current is an all-pervasive force, compelling fishes to swim with or against it, or to seek shelter from it. At first sight, rivers and streams are longitudinal habitats, but they are also likely to have lateral connections to floodplain wetlands and anabranches, flood runners or other distributary channels.

Lotic habitats are governed by rainfall and runoff (hence flow), by altitude and gradient, geology and geomorphology (hence the substratum) and by submerged rocks, wood and plants. Their character changes along the channel, from upland streams (steep slope, fast and turbulent flow) through montane, lowland channels (gentle slope, slow flow) to estuaries and the sea. Where there is a floodplain, there are likely to be connections to many lentic habitats (see 'Hierarchies, patches and scale'). In both channel and floodplain habitats there are 'microhabitats' (small areas within sites) and 'mesohabitats' (e.g. pools or riffles) used by fishes and their young stages, including areas of open water, edges, backwaters, slack waters, benches, pools and riffles. Slack waters, for example, are a refuge from the current for young fishes and their invertebrate prey, and changed flows in these areas will affect their abundance (Humphries *et al.* 2006).

Many Australian rivers traverse arid or semi-arid areas and have highly variable flow regimes (Puckridge *et al.* 1998; Kennard *et al.* 2010), so the habitats available to fishes vary widely in space and time. They range from inundated channels and floodplains during floods, to isolated waterholes during drought (Balcombe *et al.* 2007; Figs 4.1, 4.2). In these dryland rivers, interactions between flow and habitat cover many spatial scales (e.g. floodplain, river, waterhole refuges and microhabitats). As the dry season progresses, bars, boulders, fallen wood and other structural elements are exposed as streams contract to waterholes that become vital refuges for fishes. The composition of assemblages is likely to change as drying proceeds. For example, there often are declines in common species such as northwest glassfish (*Ambassis* sp.), bony herring (*Nematalosa erebi*), silver catfish (*Porochilus argenteus*), Hyrtl's catfish (*Neosilurus hyrtlii*) and spangled perch (*Leiopotherapon unicolor*) (Arthington *et al.* 2005). In the Warrego River, a tributary of the Darling in north-western New South Wales, small shallow waterholes support more fishes and more species than do large deep waterholes, which are a refuge for large species such as golden perch (*Macquaria ambigua*) and freshwater catfish (*Tandanus tandanus*) (Balcombe *et al.* 2006). The persistence of waterhole refuges is governed mainly by channel morphology and evaporative losses: some disappear in the course of a season, others persist for two years or more without surface inflows (Bunn *et al.* 2006b) and some are boosted by subsurface flows (Boulton 1993).

Periodic flow pulses are needed to maintain habitats and sustain fish assemblages in these systems

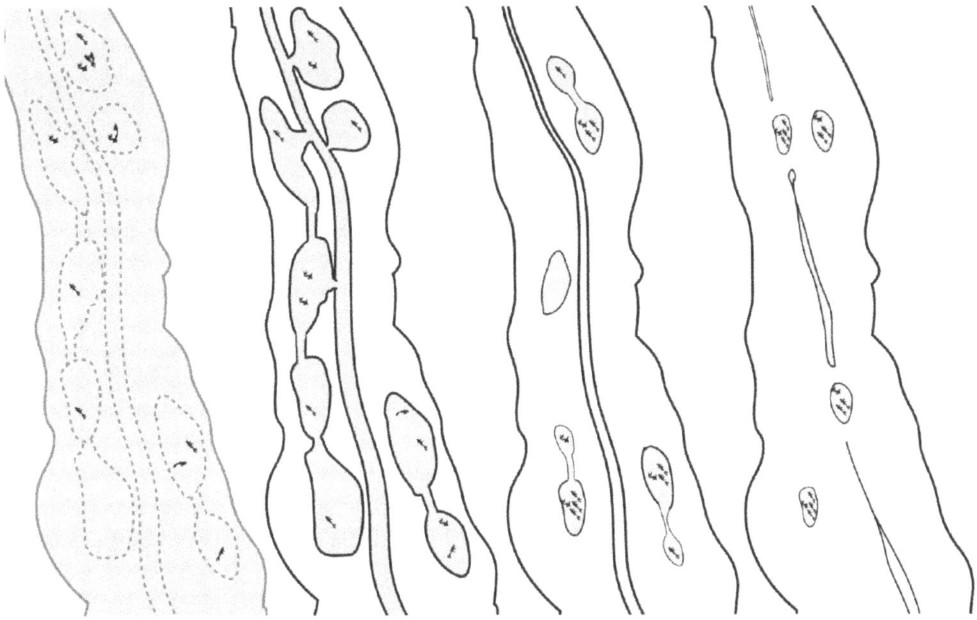

Figure 4.1: Dryland river changes to habitat area from (left to right) fully inundated floodplain to refuge pools. Shaded areas are water.

Figure 4.2: The Ovens and Murray rivers, south-eastern Australia, in low- and high-flow periods. (a) Ovens River in low flow at Peechelba Bridge. (b) Ovens River in flood around Peechelba Bridge. (c) Murray River in low-flow winter period near Cobrawonga. (d) Murray River in flood at Cobrawonga (photos: J McKenzie).

(Arthington *et al.* 2005; Bunn *et al.* 2006b). After flooding, fishes move from their refuges to exploit newly wetted habitats and increased productivity (Rayner *et al.* 2009a). The diets of bony herring, for example, change from filamentous algae and detritus during dry 'bust' periods to non-filamentous (benthic) algae during wet 'boom' times (Sternberg *et al.* 2008). Floods may cover vast expanses of floodplains, connecting lentic and lotic habitats (Fig. 4.3). These habitats are available only while water persists or while connected to the channel, so they undergo seasonal wetting and drying (Humphries *et al.* 1999). Many tropical Australian fishes, including the lake grunter (*Variichthys lacustris*), pennyfish (*Denariusa australis*) and delicate blue-eye (*Pseudomugil tenellus*), specialise in using these habitats (Pusey *et al.* 2011) (see 'Generalists and specialists'). These warm, shallow, food-rich habitats foster rapid growth and are used by most native fish species, either as juveniles and maturing fish or as spent adults (Balcombe *et al.* 2007). In the northern Murray-Darling Basin, floodplain habitats attract a variety of species including young-of-the-year golden perch and bony herring (Rolls and Wilson 2010), but they appear to be used less by fishes in the southern basin (Humphries *et al.* 1999).

Tropical floodplains, like those of the Amazon, Mekong and Zaïre rivers, have regular seasonal inundations that provide food, spawning sites and refuges for larval, juvenile and adult fishes (Goulding 1981a) that are a mainstay of fisheries in those regions (Welcomme 1985). The pulsing of flows between channel and floodplain drives the food web and its ecological role is encapsulated by the 'flood pulse concept' (Junk *et al.* 1989). Although this concept was conceived for rivers with seasonal flooding, flooding in many Australian rivers is much less regular. The emphasis on a regular pulse was because it was argued that species could not adapt to irregular events, but opportunistic traits, a capacity for rapid dispersal and broad tolerances all might be seen as evolutionary responses to unpredictability (Walker *et al.* 1995; Puckridge *et al.* 1998). Regular or not, the flood pulse appears to be no less significant for many Australian rivers than for rivers like the Amazon.

Fishes use many floodplain habitats, including anabranches, billabongs and flooded terrestrial areas

Figure 4.3: Aerial view of meanders in the lower Ovens River, Victoria, showing the expanse of inundated floodplain during flood (photo: J McKenzie).

(Ch. 5) but, as some habitats are temporary and others are permanent, they are used for different purposes and varying durations by different species or life stages (King 2004b). In temporary habitats, large-bodied species like Murray cod (*Maccullochella peelii*) may risk stranding if the water level falls rapidly. Low-flow periods occur in floodplain rivers throughout Australia and while the resident fishes have diverse life histories (Humphries *et al.* 1999) some (e.g. Murray cod) spawn only in the main channel and generally favour instream habitats (Koehn and Harrington 2005; Koehn 2009a).

Lentic habitats

Lentic habitats are characterised by a lack of flow, although there may be periodic flows, such as in billabongs connected to a river. Australia has few species specialised for these habitats (see 'Generalists and specialists'), compared to in other parts of the world. For example, Lake Victoria in Africa has about 300 fish species, many of which are lentic specialists (Bruton 1990). Natural lakes are not a significant habitat for fishes in most of Australia, and few of these were formed by glaciers, tectonic movements or volcanism except in Tasmania (Boulton and Brock 1999). There are coastal dune lakes, formed by aeolian (wind) and other processes (e.g. sand barriers) in regions like eastern Cape York Peninsula and on sand islands (e.g. Fraser, Moreton, Stradbroke islands) in south-eastern Queensland (Pusey *et al.* 2000c). These support diverse fish assemblages, including distinctive species. Billabongs (oxbow lakes) are another common lentic habitat, formed by migration of a meandering river channel. They may retain much of their riverine character or become like pools, depending on how often they are reconnected. Billabongs are important refuges for fishes and other aquatic organisms when floodplains are dry.

The most common lentic environments are those created by man-made storages. Impoundments, reservoirs and weir pools have greatly expanded the numbers of still and slow-flowing habitats throughout Australia, especially along rivers. Many are referred to as lakes, when they really are impounded rivers. They vary from small weir pools to large impoundments such as Lake Argyle, with a capacity (5800 GL) more than twice the mean annual flow of the Ord River in north-western Western Australia (Pusey and Kennard 2009).

Impoundments favour species typical of deep, still-water habitats (e.g. alien fishes such as tilapia, *Oreochromis mossambicus*, and common carp, *Cyprinus carpio*), or native species such as bony herring and blue catfish (*Neoarius graeffei*), and may cause a decline in local riverine species. In the Burnett River, south-eastern Queensland, 41% of river-channel habitats of the Australian lungfish (*Neoceratodus forsteri*) are affected by impoundments (Brooks and Kind 2002). A further 40% of spawning, larval, juvenile and adult lungfish habitats in the nearby Mary River would have been destroyed or fragmented by a proposed dam at Traveston Crossing (Balcombe and Arthington 2009), but construction was not allowed to proceed. On the Murray River floodplain at Chowilla, South Australia, there is concern about threats to anabranch habitats in an area where an environmental flow regulator is under construction (Mallen-Cooper *et al.* 2008).

Water levels in and downstream of impoundments can fluctuate on an hourly, daily, seasonal or annual basis, exposing marginal areas and plant beds that are habitat and potential nesting sites for freshwater catfish and other species. Dams and weirs are barriers to migratory fishes (Chs 5, 12), and poor water quality (e.g. cold, hypoxic water from stratified reservoirs) can cause fish kills. Barriers in estuaries, like those on the Fitzroy River, Queensland, and others near the Murray Mouth, South Australia, may limit the local tidal prism and reduce access by marine, estuarine and freshwater fishes to feeding and nursery habitats (Stuart and Mallen-Cooper 1999; Zampatti *et al.* 2010).

Small weirs, like those used to raise water levels to gravity-feed irrigation offtakes, can cause local problems, and greater problems where they occur sequentially along rivers. For example, weirs have transformed the Murray River in South Australia into a series of connected pools. The effects of the weirs, together with the barrages at the Murray Mouth, mean that no part of the 830 km Murray below the Darling confluence is unimpounded (Walker 2006). Deeper lakes and reservoirs provide a gradient of habitats from shallow littoral areas to deep-water habitats. While littoral zones provide submerged

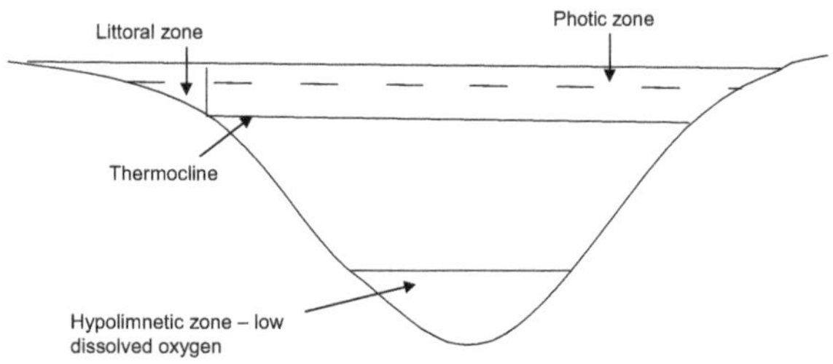

Figure 4.4: Habitat zones in lakes.

macrophytes as structural habitats, deep waters can stratify seasonally, forming hypolimnetic zones of cold hypoxic water (Fig. 4.4). Many impounded waters also retain old river-channel habitats (Fig. 4.5). The amount of structural habitat (usually wood) in reservoirs or weir pools depends on whether or not the site was cleared prior to filling.

Fishes in impoundments have received little scientific attention in Australia despite their importance for recreational fisheries. For example, Macquarie perch (*Macquaria australasica*) now rely largely on reservoir habitats (Cadwallader and Rogan 1977; Cadwallader and Douglas 1986; Ebner *et al.* 2007). Spawning areas for Macquarie perch are known in Lake Dartmouth on the Mitta River, Victoria (Tonkin *et al.* 2010), and artificial refuges for the species have been constructed in the Cotter Reservoir, Australian Capital Territory (Lintermans *et al.* 2010; Ebner *et al.* 2011). Smaller lentic species, including some galaxiids (Galaxiidae) and carp gudgeons (*Hypseleotris* spp.), have been investigated, particularly in relation to life histories and salinity tolerances (e.g. Chessman 1971; Pollard 1971b; Chessman and Williams 1974, 1975) and bioenergetics (Meredith 2005). The spawning and littoral-habitat associations of golden galaxias (*Galaxias auratus*) have been studied in relation to water-level variations in a Tasmanian lake managed for hydropower generation (Hardie *et al.* 2005, 2006, 2007).

Figure 4.5: Reservoirs on the Murray River. (a) Lake Hume full (photo: B Gawne). (b) Lake Hume cleared, partially empty (photo: B Gawne). (c) Lake Mulwala drained, showing old river channels (photo: J McKenzie). (d) Lake Mulwala drained, showing remaining timber (photo: J McKenzie).

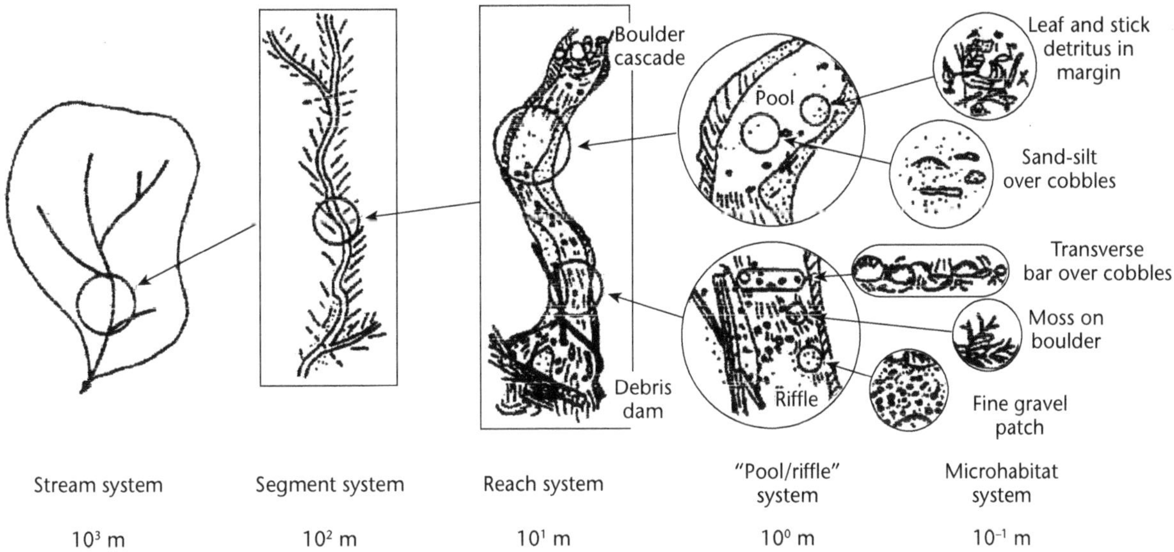

Figure 4.6: Spatially nested hierarchy of physical landscape units in a catchment (Frissell *et al.* 1986).

Fishes in lentic habitats may make daily movements up and down the water column. In billabongs, Australian smelt (*Retropinna semoni*) remain near the bottom during the day but rise through the water column at night, following the vertical migration of their zooplankton prey (Lieschke and Closs 1999) and avoiding predation by redfin perch (*Perca fluviatilis*) and other visual predators. Carp gudgeons also undertake vertical migrations, spending daytime near the surface and night time in deeper water (Meredith *et al.* 2003).

HABITAT FEATURES

Hierarchies, patches and scale

The structural arrangement of habitats, their connectedness and the scales at which these patterns and processes operate all influence the ways they are utilised by fishes. Freshwater habitats can be viewed as a hierarchy (Fig. 4.6), from regional features (e.g. climate, topography, geology, hydrology) down to the smallest micro-scales (Frissell *et al.* 1986). For example, a river reach (macrohabitat) may include riffles and pools. Within each riffle or pool (mesohabitat) there is variation in depth, flow velocity and substratum composition, and there may be woody debris, root masses or undercut banks (Fig. 4.7). These smaller elements (microhabitats) (Fig. 4.8) also contribute to physical diversity. In general, fishes are influenced by factors at each scale.

The habitat characteristics of rivers and streams are influenced by the landscapes through which they flow (Hynes 1975; Vannote *et al.* 1980); this is manifest in conceptual models describing the structure and function of natural river-floodplain ecosystems (Ch. 1). These models recognise in various ways the links between catchments, rivers and floodplains, the significance of spatial and temporal scales (Frissell *et al.* 1986; Amoros *et al.* 1987; Johnson *et al.* 1995a; Lowe *et al.* 2006), the dynamic nature of habitats for fishes and other biota (White and Pickett 1985; Pringle *et al.* 1988; Townsend 1989; Wu and Loucks 1995a; Thorp *et al.* 2006) and the role of disturbance (Resh *et al.* 1988). Together they describe how habitats are viewed, from the hierarchical perspective of landscape ecology (Wiens 2002b; Johnson and Host 2010; Winemiller *et al.* 2010). This approach is part of the emerging 'riverscape concept' (Fausch *et al.* 2002; Wiens 2002b).

Habitats may not be discrete areas. Rather, they are likely to occur as patches or mosaics that vary in size and quality, across the range of spatial and temporal scales. Patch boundaries, their positions in the environment and their connectivity with other patches are critical factors influencing their utility for fishes. This

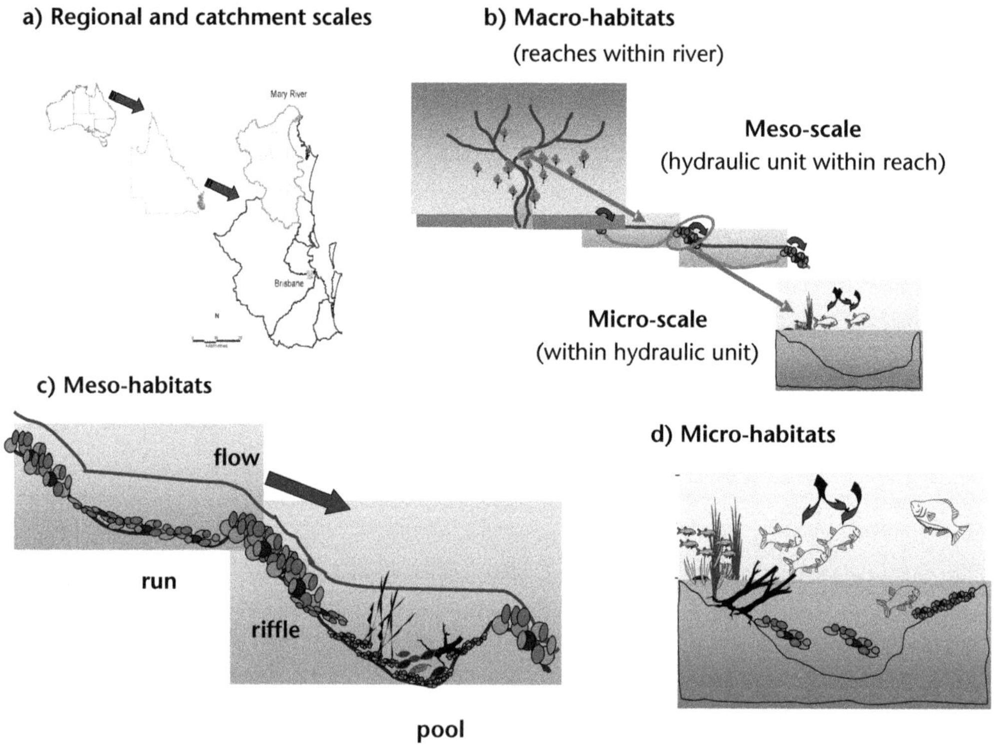

Figure 4.7: Habitat hierarchies for rivers and streams. (a) Regional and catchment scale. (b) Macrohabitat. (c) Mesohabitat. (d) Microhabitat.

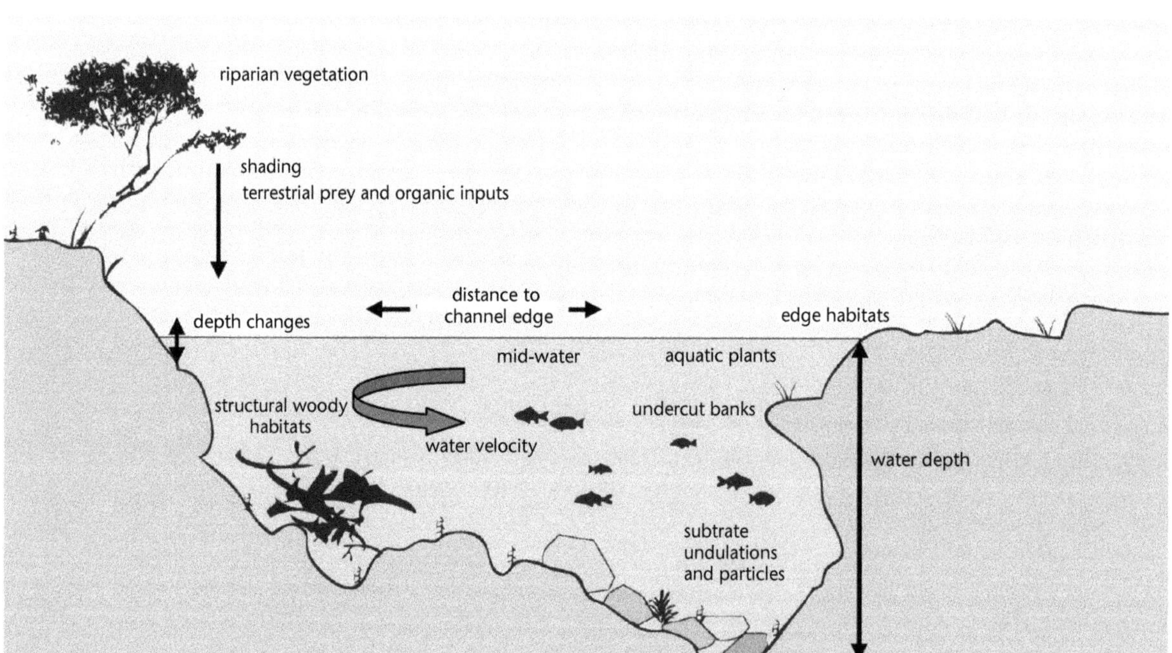

Figure 4.8: Microhabitat attributes in a pool.

is indicated by the ways that fishes move between habitat patches to access resources (Schlosser and Angermeier 1995; Ch. 5).

Hydrology

Patterns of flow at multiple scales govern river ecosystem structure and function. Long-term variations in the magnitude, timing, frequency and duration of flow events define the physical habitat template at the catchment scale, and short-term events influence habitat availability and connectivity at the reach scale. These hydrological factors – collectively, 'the flow regime' – affect refuges, food resources, opportunities for movement and migration and conditions for spawning and recruitment (Humphries *et al.* 1999; Kennard *et al.* 2007; Perry and Bond 2009). In essence, the flow regime determines the quantity and quality of the habitat. The different factors are significant for management of environmental flows and fish populations (King *et al.* 2010).

Australia's diverse climates and catchments provide a corresponding variety of river-flow regimes (Gordon *et al.* 2004; Kennard *et al.* 2010; Ch. 7, Fig. 7.10). Interannual flow variability is high, especially in inland regions (Walker *et al.* 1995; Puckridge *et al.* 1998; Kennard *et al.* 2010), so habitats can change dramatically in size and character from one year to another (Figs 4.1, 4.2). In northern Australia, summer wet-season flows are usually dominated by one or two large flood events associated with cyclones (Pusey *et al.* 2004; Warfe *et al.* 2011). In central Australia, rivers experience long periods with little or no flow, punctuated by occasional extreme floods. In south-eastern Australia, natural river flows peak in winter and spring but these regimes are modified by seasonal management for irrigation, so winter–spring flows are stored for release in summer and autumn. In ephemeral streams, with extended periods of zero flow, the availability of aquatic habitats depends on retention of surface water and maintenance of lateral, longitudinal and vertical connectivity (Bond *et al.* 2008).

Hydraulics

Hydraulics concerns the mechanics and movements of water, whereas hydrology considers the water cycle and regional patterns of runoff and flow. Thus, water-level changes, flow velocity and turbulence are in the realm of hydraulics. The hydraulic environment is critical for lotic fishes (Hynes 1970; Wiens 2002a), although there have been few Australian studies that have explicitly investigated eco-hydraulic relationships (e.g. Kilsby and Walker 2012). Local variations in flow (hydraulic diversity) affect water depth, instream cover and the nature of the sediments, and thereby provide habitats for different species and life stages.

Current in a stream channel depends on channel geometry (size, depth), gradient, sinuosity power (erosive capacity) and the substratum (e.g. rocks, boulders, wood). The velocity may be near zero at the stream bed and maximal just below the surface, but it also varies across the channel and is greatest at the thalweg, the deepest part of the channel and usually the line of fastest flow. For many fishes, the current determines the boundaries of the physical habitat. For example, currents limit the range of two-spine blackfish (*Gadopsis bispinosus*: Maddock 1999).

Flow near the stream bed forms complex microhabitats for many benthic and demersal species and their food organisms (e.g. algae, invertebrates). As flowing water meets a solid object, such as a rock, there is a stagnation point, where flow is arrested. Downstream, the flow is more nearly laminar, forming a smooth envelope over the surface of the object. Further downstream, the laminar sublayer becomes thin (depending on water velocity) and flow is turbulent. Shear stress (the force of water flow on the river bed) in the laminar sublayer is less than in the turbulent zone (Gordon *et al.* 2004), so the laminar region may be favourable for fishes.

The distribution of instream objects may cause small, low-velocity eddies and dead-water areas to form, providing refuges. Riffle-dwellers such as juvenile sooty grunter (*Hephaestus fuliginosus*) and demersal fishes like gobies (e.g. *Glossogobius* spp.) often rest in refuges in riffles and rapids in fast-flowing water (Pusey *et al.* 2004). In a stream bed with many obstructions, the roughness of the substratum is increased, the eddies interact and the turbulence intensifies (Gordon *et al.* 2004). Submerged wood also creates physical diversity, providing sheltered pockets, pools or scour holes that are favoured by many stream fishes (Wallace *et al.* 1995; Abbe and

Montgomery 1996; Crook and Robertson 1999). For example, the Oxleyan pygmy perch (*Nannoperca oxleyana*) favours habitats with abundant wood, plants or undercut banks (Pusey *et al.* 2004; Knight and Arthington 2008).

The vertical position of fishes in the water column is another expression of hydraulic preferences. For example, Murray cod favour demersal rather than mid-water or surface habitats (Koehn 2009b), while golden perch, a pelagic species, often lurk among submerged branches well above the bottom (Hughes *et al.* 2007). Species like congolli (*Pseudaphritis urvillii*) are also demersal, resting on the stream bed, while the ammocetes of lampreys (Geotriidae, Mordaciidae) are benthic, living in the sediments. The common galaxias (*Galaxias maculatus*) is a pelagic species, preferring open, slow-flowing channels, whereas the flathead gudgeon (*Philypnodon grandiceps*) is a demersal species that shelters among boulders. As flow increases, the common galaxias needs to change habitats, whereas the flathead gudgeon can remain in its preferred habitat (Kilsby and Walker 2012).

In streams with porous, coarse sand or gravel substrata, part of the flow enters the hyporheic zone (Boulton 1993), entering (downwelling areas) and leaving the substratum (upwelling areas). Downwelling areas usually occur at the downstream end of a pool, and upwelling areas occur downstream of riffles. Some fishes, notably Tasmanian mudfish (*Neochanna cleaveri*) and salamanderfish (*Lepidogalaxias salamandroides*), use the hyporheic zone as a refuge in dry periods and may aestivate in the moist sediments of dry pools. Some salmonids construct nests (spawning redds) over downwelling areas (Newbury and Gaboury 1993), where the flow of oxygenated water may promote egg survivorship. Freshwater catfishes construct large, saucer-shaped nests in coarse gravel, often along the thalweg at the downstream end of a run or pool. These nests may increase exchange between surface and subsurface water (Pusey *et al.* 2004).

Habitats obviously are of value to fishes only if they have access to them; movements are often blocked by artificial barriers or a lack of hydraulic connections due to road culverts, for example. Barriers are a threat to fishes, especially migratory species, and hydraulic connectivity is a factor of major importance in relation to movements undertaken by fishes (Ch. 5).

Physical structure

Habitat structure in lotic environments includes channel form, the nature of the substratum (e.g. rocks, gravel, sand, silt), instream wood and aquatic plants used by fishes for refuges and sites for feeding and spawning (see 'Habitat needs'). In lentic environments there is less diversity in substrata (e.g. sand, silt) and, while aquatic plants and submerged wood are significant in littoral areas, they are less so in deeper water.

Structural complexity greatly influences the capacity of aquatic environments to support fishes. It provides refuges for prey, limiting foraging by predatory fishes (Savino and Stein 1982; Johnson *et al.* 1988; Angermeier 1992; Everett and Ruiz 1993). It offers protection from sunlight and strong currents (Abbe and Montgomery 1996; Crook and Robertson 1999; Ch. 5), delineates territories (Fausch 1993; Crook and Robertson 1999), limits interactions between individuals (Dolloff and Warren 2003) and may provide spawning sites (Jackson 1978; Ch. 7). Boulders, rocks and cobbles provide habitats and spawning sites for species like two-spine blackfish (Koehn 1987; O'Connor and Zampatti 2006) and the Clarence River cod (*Maccullochella ikei*: Butler 2009b), although siltation – caused by catchment erosion, bushfires or bank slumping – may threaten the survival of eggs and adults (Doeg and Koehn 1994; Lyon and O'Connor 2008).

Submerged wood, or 'structural woody habitat' (SWH) (also known by the somewhat derogatory term 'snags' or 'large woody debris'), provides cover for fishes, roosts for birds, basking platforms for turtles and a surface for growth of biofilms. In streams, SWH creates variations in water depth and flow velocity (Richmond and Fausch 1995; Abbe and Montgomery 1996) and channel morphology (flow-mediated pools, scour holes: Wallace *et al.* 1995). Scour holes are particularly favoured by pool-dwelling species like Australian bass (*Percalates novemaculeata*: Borg *et al.* 2007).

Wood is delivered to the stream from the surrounding riparian zone (Koehn *et al.* 2004a) or swept from upstream, and is one of the most recognised and important structural habitat components, especially in lowland rivers where channels are more uniform and substratum particles are finer and provide less structure than in upland streams (Wallace and Benke 1984; Bilby and Ward 1991; Gordon *et al.* 2004). Instream wood also traps other organic matter (Bilby and Likens 1980) and provides a structurally complex, stable, hard substratum for biofilms, macroinvertebrates and other fauna (Benke *et al.* 1984; O'Connor 1991), all of which may be food for fishes.

Murray cod are consistently associated with wood more than any other habitat type (Boys and Thoms 2006; Koehn 2009a) and the same is true of two congeners, the Mary River cod (*Maccullochella mariensis*: Simpson and Mapleston 2002) and trout cod (*M. macquariensis*: Nicol *et al.* 2007). Habitats with wood may also be preferred by Murray cod, even in areas of low or no current (Koehn 2009a). This suggests that wood may offer advantages (e.g. territorial markers, feeding sites) other than shelter from the current (Crook and Robertson 1999).

The abundance and diversity of lotic fishes are strongly associated with SWH at micro- to macrohabitat scales, both in Australia (Koehn *et al.* 2004a; Boys and Thoms 2006; Nicol *et al.* 2007; Koehn 2009a) and in other parts of the world (Reeves *et al.* 1993; Inoue and Nakano 1998). One illustration of its significance is that a 34% increase in SWH in the upper Ovens River, Victoria, led to a six-fold increase in numbers of two-spine blackfish (Koehn 1987).

Aquatic plants are another form of structural habitat, used by many fishes for spawning and feeding and as refuges for all life stages. Along the Murray River in Victoria, and in lakes near the Murray Mouth in South Australia, the Murray hardyhead (*Craterocephalus fluviatilis*) is associated with submerged widgeon grass (*Ruppia*) or other plants, including milfoil (*Myriophyllum*) (Wedderburn *et al.* 2007). In south-eastern Queensland, Agassiz's glassfish (*Ambassis agassizii*), Marjorie's hardyhead (*Craterocephalus marjoriae*), fly-specked hardyhead (*C. stercusmuscarum*) and Oxleyan pygmy perch all favour beds of aquatic macrophytes (Pusey *et al.* 2004). Freshwater catfish (Clunie and Koehn 2001a), eastern dwarf galaxias (*Galaxiella pusilla*) and southern pygmy perch (*Nannoperca australis*) are other species often associated with aquatic vegetation (Humphries 1995). Australian lungfish spawn over plant beds in running water and use them as a nursery (Brooks and Kind 2002; Ch. 7).

Water quality

All species of fishes, and all life stages, have certain environmental tolerances, beyond which growth, reproduction and survival are compromised. Key variables include temperature, oxygen and salinity.

Water temperature affects metabolism, and spawning and larval development in many species are triggered when temperatures exceed a particular threshold (Pankhurst and King 2010). Depressed temperatures, as in low-level releases from stratified reservoirs, can make riverine habitats unsuitable for warm-water species (Sherman *et al.* 2007). Dissolved oxygen is inversely related to temperature (thus, warm water is likely to contain less oxygen than cold water). Oxygen levels may be depressed (hypoxia) or zero (anoxia) in low-level releases from stratified reservoirs (regardless of temperature), in still, warm, nutrient-rich water where there is abundant algal or plant growth, or in blackwater events (see below). Hypoxia may cause fish kills (Koehn 2005) and sublethal effects on growth and digestion (Wang *et al.* 2009), although some species are tolerant, including some in billabong assemblages (McNeil and Closs 2007).

Blackwater is dark brown or black water with high levels of dissolved organic carbon and low levels of dissolved oxygen. Blackwater events occur naturally in ephemeral streams, floodplain rivers and wetlands (Bishop 1980; Townsend and Edwards 2003; McMaster and Bond 2008; Watkins *et al.* 2010). They develop when water inundates dry channels or floodplain areas, leaching carbon from eucalypt leaves, bark and wood; the carbon is consumed by microbes, causing hypoxia and killing fish and other fauna (Townsend and Edwards 2003; McMaster and Bond 2008; Baldwin and Whitworth 2009).

Landscape-scale changes in water quality may reduce the ranges of some species. For example, rising salinities in south-western Western Australia, exacerbated by extensive land clearing, threaten the region's endemic freshwater fishes. While forested tributaries retain viable populations of western pygmy perch (*Nannoperca vittata*), nightfish (*Bostockia porosa*) and western dwarf galaxias (*Galaxiella munda*), elevated salinities in the upper Blackwood River system have caused declines, even local extinctions, of these species (Morgan *et al.* 2003). Salinisation also occurs along the Murray in south-eastern Australia, and salinities in riverine wetlands appear to be a factor in habitat separation of hardyheads (*Craterocephalus* spp.: Wedderburn *et al.* 2007).

HABITAT NEEDS

Quantity and quality

The numbers of fishes in an area depend on available habitat and free access (Hayes *et al.* 1996). When resources are in short supply the carrying capacity of the habitat is generally reduced, with effects on physiological condition, competition and recruitment. For example, pools with submerged structures have a finite capacity for fishes seeking refuge from predators (Johnson *et al.* 1988) and, if the numbers of prey fishes exceed the capacity of the refuges, their abundance may be limited by predation. Methods to assess habitats are established (see 'Habitat assessment') but habitat quality, hence carrying capacity, is more difficult to measure, partly because it varies with the innate characteristics of each species.

Generalists and specialists

The habitat requirements of a species are the abiotic features necessary for the persistence of individuals or populations (Rosenfeld 2003). They are subject to complex interactions with genetic traits, environmental tolerances, physiological needs, swimming capabilities, behaviour and other factors (Ch. 5). It is useful to distinguish 'habitat generalists' with flexible needs, able to exploit a wide range of environments, and 'habitat specialists' with particular needs that restrict the variety of suitable habitats. Generalists tend to predominate in Australia, broadly reflecting the nature of the environment. The fishes of dryland rivers, for example, tend to be generalists, with traits allowing them to persist in variable and often harsh conditions (Balcombe and Arthington 2009). More stable habitats foster trophic complexity (Menge and Sutherland 1976) and promote habitat specialisation (Pusey *et al.* 1995).

One habitat generalist, for example, is the spangled perch, found in desert springs and bores, floodplain wetlands and billabongs, lowland rivers, upland tributaries, temporary streams and artificial impoundments over most of Australia (Allen *et al.* 2002). Spangled perch are omnivorous and highly mobile, with broad tolerances and an opportunistic life-history strategy (Pusey *et al.* 2004; Olden and Kennard 2010). Other habitat generalists are southern shortfin eels (*Anguilla australis*), longfin eels (*A. reinhardtii*), bony herring, mountain galaxias (*Galaxias olidus*), barred grunter (*Amniataba percoides*), southern purple-spotted gudgeon (*Mogurnda adspersa*) and empire gudgeon (*Hypseleotris compressa*). An example of a habitat specialist is variegated pygmy perch (*Nannoperca variegata*) (Allen *et al.* 2002), found only in flowing, well-vegetated streams in a small region. Other examples are cited in Box 4.1.

Generalist traits are also associated with some invasive species, such as common carp (Koehn 2004). Common carp move freely through their habitat, sieving sediments to obtain food (Lammens and Hoogenboezem 1991). In contrast, golden perch are ambush predators that prefer deeper, flowing water and feed mainly on large, mobile prey such as shrimp and crayfish (Cadwallader and Backhouse 1983). Thus, common carp utilise a broader habitat than golden perch and are less strongly associated with particular needs (Crook 2004c). Feeding and nutrition, of course, are important determinants of habitat use (Ch. 6).

Habitat selection

Fishes often select particular habitat features over others, and are likely to cluster in patches that maximise individual fitness (Huey 1991; Krebs and Kacelnik 1991; Inoue and Nakano 1998). Riverine fishes, for example, may position themselves to maximise net energy gains (Fausch 1984; Grossman and Freeman 1987; David and Closs 2003).

Selection and settlement in a particular habitat may require some assessment of its suitability with regard for needs (spawning, food, refuge) and risks (predators, competitors, stranding). Habitat evaluation appears to occur in Panamanian catfishes (Loricariidae), for example, guided by memory and forays to assess alternative habitats (Power 1984). Losses in

Box 4.1: Habitat specialists

- The barred galaxias (*Galaxias fuscus*) is a small, threatened species confined to high-altitude (>400 m), low-order streams in the Goulburn River catchment, Victoria. These streams have waterfalls as barriers to predatory trout, and the barred galaxias thereby has some degree of protection. It prefers slow-flowing pools (<0.2 m s^{-1}) adjacent to fast-flowing reaches (0.55–2.01 m s^{-1}) (Shirley and Raadik 1997).
- The ornate rainbowfish (*Rhadinocentrus ornatus*) (Fig. 4.12a), honey blue-eye (*Pseudomugil mellis*) and Oxleyan pygmy perch (*Nannoperca oxleyana*) occur in the coastal lowland wallum (*Banksia* heathland; Fig. 4.12b) of south-eastern Queensland and north-eastern New South Wales. They inhabit lotic and lentic habitats, generally dystrophic, acidic, humic, nutrient-poor waters over siliceous sand with abundant plants (Pusey *et al.* 2004).
- Several species of freshwater fishes in northern Australia are floodplain specialists, including the pennyfish (*Denariusa australis*), delicate blue-eye (*Pseudomugil tenellus*) (Fig. 4.12c) and lake grunter (*Variichthys lacustris*). These are associated with densely vegetated billabongs and seasonally inundated wetlands (Pusey *et al.* 2011) (Fig. 4.12d).
- The threatened cave gudgeon (*Milyeringa veritas*) in Western Australia occupies an *anchialine* (groundwater estuary) habitat where salinities vary between fresh and sea water. Otolith microchemistry suggests that some cavefishes inhabit the sea or brackish waters; others remain in fresh water, and still others utilise these habitats opportunistically (Humphreys *et al.* 2006). Like the blind cave eel (*Ophisternon candidum*), this species lacks eyes or pigment and has well-developed sensory papillae (Humphreys 2001; Chakrabarty 2010). The patchy, discontinuous distributions of these species make them vulnerable to disturbance and potentially slow to recover.

dry-season waterholes and mass stranding on floodplains (Welcomme 1985; Lowe-McConnell 1987; Bishop and Forbes 1991) suggest that fishes may not always make effective assessments of long-term habitat suitability.

Interactions between individuals at intra- and interspecific levels affect the profitability of habitat patches (Natsumeda 1998) (see 'Species interactions'). As a consequence, individual patterns of utilisation differ among and between species (Lomnicki 1988; Nakano 1995b). This depends on the habitat needs, behaviour and mobility of species (Ch. 5). At another level, the preferences of lotic species may overlap (Baker and Ross 1981; Floyd *et al.* 1984; Koehn *et al.* 1994), perhaps as an adaptation to heterogeneous environments (Hill and Grossman 1993). Stable habitats allow more trophic complexity (Menge and Sutherland 1976) and more habitat specialisation (Pusey *et al.* 1995).

Spawning sites for some species can be highly specific (Ch. 7) and their availability and quality can have major effects on recruitment. For example, river blackfish (*Gadopsis marmoratus*) spawn in hollow logs, freshwater catfish lay their eggs in gravel nests and Australian lungfish lay their eggs among aquatic plants. Other species are less particular. Carp gudgeons, for example, spawn among rocks and cobbles, leaf litter, aquatic plants and root masses (Pusey *et al.* 2004).

Habitat preferences may vary over time. For example, Murray cod show stronger preferences in autumn, when they are at home sites, than in spring, when they migrate (Koehn 2009a). Golden perch demonstrate habitat shifts, where individuals move to new home sites (Crook 2004c; Ch. 5). Carp gudgeons become increasingly demersal as they grow, perhaps influenced by competition with eastern gambusia for space and food (Stoffels and Humphries 2003). There may also be diel changes in habitat selection. During the day, river blackfish favour slow-flowing, woody habitats or undercut banks, and at night they move into faster-flowing, open areas (Koehn *et al.* 1994; Khan *et al.* 2004; Koster and Crook 2008). Some individual golden perch also move between day and night habitats, whereas others do not (Crook *et al.* 2001).

Adult Murray cod (>450 mm total length, TL) and their young-of-the-year favour structural woody habitats. At a macrohabitat scale, the adults prefer channel habitats at high flows, and the floodplain is generally avoided (Koehn 2009a). Early life history also indicates that the species is a main-channel specialist (King

Figure 4.9: Habitats of Murray cod (*Maccullochella peelii*). (a) Lower reaches of the Ovens River, south-eastern Australia, exposed following lowering of Lake Mulwala downstream. (b) Ovens River, main channel.

2004b; Koehn and Harrington 2005, 2006). Similar microhabitats are selected regardless of hydraulic conditions, and choices are influenced by the presence of structural wood (Fig. 4.9) and other factors. Indeed, more than 90% of Murray cod may be associated with wood. The young-of-the-year select shallower habitats nearer the river bank than adult fish, but also require structural woody habitat (Koehn 2009a).

Nurseries

Many fishes need specialised nursery habitats to protect their early life stages. This may be related to competition, predator avoidance, trophic differences or phylogenetic factors reflected in physiological tolerances or morphological constraints (Ross 1986; Schlosser 1987b; Facey and Grossman 1991).

Bony herring and golden perch may use inundated floodplain habitats as nurseries (Pusey *et al.* 2004; Rolls and Wilson 2010). Many other species use still or slow-flowing littoral areas, backwaters and shallow embayments (Humphries *et al.* 1999); as the youngsters grow and develop their swimming ability they move into faster-flowing water (Pusey *et al.* 2002). These so-called 'slackwater' habitats are used by some species throughout their life cycle; others, such as Australian smelt (*Retropinna semoni*) and rainbow-

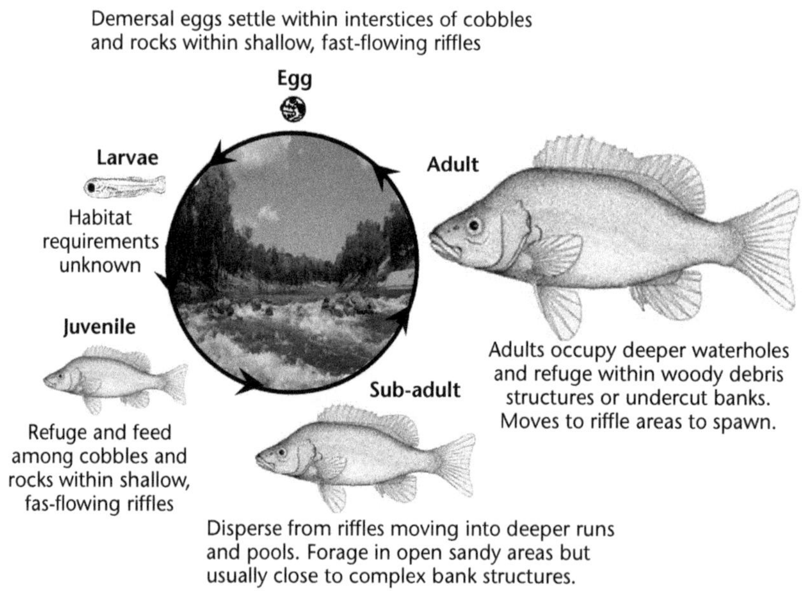

Figure 4.10: Ontogenetic changes in habitat requirements of sooty grunter (*Hephaestus fuliginosus*) in the perennial main channel of the Daly River, Northern Territory (B.J. Pusey and M.J. Kennard, unpubl.).

fishes (Melanotaeniidae) (King 2004b), use them only as nurseries. Still others, such as Murray cod, do not use them at all (King 2004b; Koehn 2009a).

Fishes tend to select different macro- and micro-habitats as their life cycle progresses (Sabo and Orth 1994) and striking habitat shifts may occur during ontogeny (Vilizzi and Walker 1999b; Pusey *et al.* 2002, 2004; King 2004b; Ch. 7). In rainforest streams in northern Queensland, larval fishes are restricted mainly to edge habitats with flows less than about 10 cm s^{-1}. Very young larvae remain close to cover and the stream bank in areas of zero flow whereas older larvae disperse into deeper areas with less cover and higher flows, further from the bank (Pusey *et al.* 2002). There is ontogenetic variation in the habitat requirements of sooty grunter in northern Australia (Pusey *et al.* 2004; Chan *et al.* 2012; Fig. 4.10). Shallow, fast-flowing, well-oxygenated riffles over coarse sediment are optimal conditions for the eggs, protecting them against large predators such as barramundi (*Lates calcarifer*) and providing aquatic insects as food for the juveniles. Adult sooty grunter occupy deeper, slow-flowing pools with complex structures and overhanging riparian vegetation that protect against avian predators and provide terrestrial fruits and insects as food (Pusey *et al.* 2010; Davis *et al.* 2011). Carp gudgeons and eastern gambusia are other species that show ontogenetic shifts in their use of food and other habitat resources (Stoffels and Humphries 2003).

There are size-dependent trade-offs between the risk of predation and opportunities for foraging (Werner and Gilliam 1984). Young-of-the-year Murray cod seek shallow habitats with woody debris, closer to the bank than adult fish (Koehn 2009a), offering more protection against predators such as golden perch, trout cod and redfin perch, and other Murray cod (Cadwallader and Backhouse 1983).

Riparian zones

Riparian areas are a transitional zone between channel, floodplain and terrestrial habitats, and mediate in the transfer of energy and materials between these areas. They may also buffer streams from catchment processes (e.g. agricultural chemicals, erosion).

The abundance of fishes is often associated with overhanging riparian plant cover (Koehn 2009a). Riparian vegetation is a source of plant material and terrestrial invertebrates as food for fishes and other biota, and it provides structural woody habitat in the form of logs, branches and leaves. Shading from sunlight and high temperatures is important for some fishes, and losses of riparian trees and shrubs through clearing and stock grazing are a common disturbance in streams. Shade protects the eggs of species such as the climbing galaxias (*Galaxias brevipinnis*), laid during high flows in upland streams and stranded out of the water until stimulated to hatch by another spate (O'Connor and Koehn 1998). Habitat alterations associated with riparian vegetation clearance may favour alien species such as eastern gambusia over native species such as rainbowfishes (King and Warburton 2007).

Refuges

Fish habitats can undergo substantial changes in quantity, quality and connectivity in time and space, often in response to variable flows (droughts and floods). Floods are 'pulse' disturbances; droughts are more protracted 'press' disturbances and the long-term impacts of climate change might be seen as a 'ramp' disturbance (Lake 2000). Habitats and habitat patches can vary from harsh and frequently disturbed (e.g. by drought or flood) to benign and comparatively stable, effectively meaning a gradient from abiotic to biotic controls (Schlosser 1987b; Townsend 1989). Thus, fish assemblages in unstable (unpredictable) environments are influenced by colonisation and extinctions, whereas those in stable (predictable) environments are influenced more by biotic interactions such as competition and predation.

One way for fishes to survive natural disturbances is by seeking refuges. Disconnection of riverine habitats during low flows, for example, can restrict movements and confine fishes to isolated pools for extended periods; these refuges are critical for maintenance of local fish populations (Bunn *et al.* 2006b; Beesley and Prince 2010). Refuge habitats are common on dryland-river floodplains (see 'Lotic habitats' above; Ch. 5) and in ephemeral channels (Bond *et al.* 2008). High summer temperatures and low oxygen levels can be stressful for trapped fishes (Lake 2003), and the volume of water (habitat area) in the system, seepage losses and the inherent 'carrying capacity' of a pool (the number of individuals it can support) are factors in survival (Perry and Bond 2009). The capacity of these pools to protect fishes from drought can be compromised by human interventions such as water extraction, erosion or sedimentation, or by allowing access by stock (Bond *et al.* 2008). In future, refuge pools will become even more vital in regions affected by climate change and lower rainfall (Morrongiello *et al.* 2011a).

Species interactions

Behavioural interactions influence habitat use, because many riverine fishes are territorial and defend their feeding and spawning areas (Grant 1997; Ch. 5). Dominance hierarchies can determine the spatial distributions of individuals in riverine fish communities. These are maintained by aggressive interactions; dominance is usually related to body size (Nakano 1995a). The dominant fishes tend to occupy the most energetically or reproductively profitable positions and subordinate fishes are obliged to occupy marginal habitats (Fausch 1984). Large river blackfish, for example, may gain priority access to preferred habitats through behavioural interactions (Koehn *et al.* 1994). In some riverine salmonids the population may be distributed as a loosely organised matrix of individual territories, although home ranges may overlap and spaces may be shared among individuals (Dill *et al.* 1981).

Predator-mediated impacts on habitat use are well-known (Schlosser 1987b). Behavioural responses to predators by potential prey fishes include aggregation (schools, shoals), movements to refuge habitats and reductions in foraging distances, feeding times and food intake. The consequences for prey may include intensified competition for resources among individuals, reduced fecundity and reductions in recruitment and survivorship (Mittelbach 1986).

In Australia, the distributions of small native fishes are often affected by the presence of alien trout (*Salmo trutta*, *Oncorhynchus mykiss*: Lintermans 2000). The presence of brown trout, for example, may be more important than habitat characteristics in determining the abundance in streams of spotted galaxias (*Galaxias truttaceus*: Ault and White 1994). In the absence of brown trout, climbing galaxias may inhabit slow-flowing pool habitats, but in their presence they are inclined to dwell in riffles (Koehn and O'Connor 1990b). Changes in the longitudinal distribution of mountain galaxias and brown trout in intermittent streams may also be related to their differential tolerances to habitat conditions in summer (Closs and Lake 1996). The upstream limits for brown trout may be set by low flows and high temperatures, whereas mountain galaxias can survive in small, scattered pools, especially in the absence of trout (Lintermans 2004). In other areas, small species such as galaxiids no longer co-occur with salmonids, owing to displacement or predation (McDowall 2006). In general, habitat use by small species is often consistent with the notion that large fishes prefer deeper water habitats to reduce predation risk. Small fishes typically are associated with shallow water, avoiding large predatory fishes, and with instream cover to protect them from avian predators (Koehn *et al.* 1994).

HABITAT ASSESSMENT

Identifying requirements

There is some ambiguity in the ways that terms like 'habitat availability', 'habitat use', 'habitat selection', 'habitat preference' and 'habitat requirements' are used. The *available* habitat is that component accessible to fishes; *usage* is the quantity of that component and the frequency of use over some period; *selection* is that component used disproportionately to its availability; *preference* reflects the likelihood of that component being chosen when offered on an equal basis with other components; and *avoidance* is the opposite of preference (Johnson 1980; Manly *et al.* 2002; Rosenfeld 2003). As mentioned earlier, habitat *requirements* are the abiotic features of the environment needed to ensure the persistence of individuals or populations. If changes to habitat features have little or no effect, they presumably are not critical (Rosenfeld 2003; Rosenfeld and Hatfield 2006; Fig. 4.12).

Our understanding of the habitat requirements of Australian species is patchy, being better for some species and some regions than for others. Another factor is that many different methods have been used to assess habitat requirements, and it is often difficult to compare data. These include direct observations (e.g. snorkelling, underwater video), sampling surveys (e.g. electrofishing) and more recently the use of telemetry (Ch. 5) and experimental approaches. Nevertheless, there are some useful compendiums containing information about habitats. They include national guides (Lake 1971, 1978; Merrick and Schmida 1984; Allen 1989; Allen *et al.* 2002), regional guides for north-eastern Australia (Herbert and Peeters 1995; Pusey *et al.* 2004), south-eastern Australia (Cadwallader and Backhouse 1983; Koehn and O'Connor 1990a; McDowall 1996b; Lintermans and Osborne 2002; Moffatt and Voller 2002; Lintermans 2007), inland Australia (Wager and Unmack 2000), Tasmania (Fulton 1990), Western Australia (Allen 1982) and northern Australia (Larson and Martin 1990; Bishop *et al.* 2001). There are also guides to particular families (e.g. rainbowfishes: Allen 1995).

Identifying habitat needs and understanding how they change in response to natural processes and human impacts are vital for effective management. Without that knowledge, managers must rely on guesswork, which often fails and can have economic and social consequences that are difficult to justify (Rosenfeld 2003). In the following, we outline some approaches to habitat assessment. We describe the kinds of data gathered by fish biologists and some of the methods employed, and we comment on progress thus far.

Methods of measurement

The most common analytical approaches to habitat requirements, aside from simple descriptions, include correlative habitat-association models (e.g. habitat-frequency curves or explanatory and predictive statistical models of species distribution or abundance), mechanistic models (e.g. bioenergetic models) and manipulative experiments. The strengths and weaknesses of each approach are described by Rosenfeld (2003), Lamouroux *et al.* (2010) and Lancaster and Downes (2010a, b).

Habitat use can be quantified by direct observations or indirect, capture-based methods at one or more spatial scales (macro-, meso-, microhabitat) and temporal scales (e.g. short-term behaviour, life span). Examples include visual observations by snorkelling or underwater video, standard sampling methods by electrofishing, hydroacoustics, sonar, mechanical tags or markers and telemetry (Table 4.1). The choice of method depends on the purpose of the study, the data required and logistical constraints (e.g. cost, expertise, occupational health and safety issues). Clear goals are needed if researchers and managers are to share and synthesise the data (Bain and Stevenson 1999).

Methods to assess habitat availability include qualitative surveys (visual assessments scored according to availability, quality etc.), quantitative surveys using standard stream-survey techniques (e.g. transect or random-point measurements of such variables as depth, velocity and channel dimensions) or a combination of the two (Table 4.1). Quantitative survey data can be used also to develop one- or two-dimensional hydrodynamic models that relate changes in habitat availability to variations in discharge. With these methods, habitat can be assessed at small scales (e.g. < 1 km) with variable accuracy, precision and cost. If broader scales are required, sonar, hydroacoustic and aerial or remote-sensing methods are more appropriate (Table 4.1).

Habitat frequency curves

Quantifying habitat requirements often involves relating the frequency of habitat use by fishes to the frequency of habitat components in the environment, and interpreting a mismatch as an indication of habitat selection. From this information, criteria can be devised to reflect the suitability of conditions for particular species. Frequency of habitat use and habitat selectivity may not truly indicate a species' preference for a particular habitat (or any other resource) and they should be interpreted cautiously (Pusey 1998). Only under experimental conditions, where all relevant states of a resource are equally accessible, can innate preferences be determined confidently (Baltz 1990). Even so, field observations of habitat use do provide useful insights, but data of this kind are lacking for many species.

There is a range of methods to compare habitat availability or use and to infer habitat selection or preference (Manly *et al.* 2002). One often used for stream fishes is the 'habitat suitability curve', a normalised frequency histogram relating utilisation or preferences to water velocity, depth, substratum composition or other habitat features (Johnson *et al.*

Table 4.1: Methods to assess habitat availability and habitat use by freshwater fishes, with notes on advantages and disadvantages

Method	Description	Advantages	Disadvantages	Sample references
Habitat availability				
Qualitative survey	Visual assessments, scored on availability, quality etc.	Rapid Low cost No special equipment Assess many variables Quantify at broad scales	Prone to bias Limited utility of data	Anderson (1993); Bain and Stevenson (1999); Kaufmann *et al.* (1999)
Quantitative survey	Stream surveys (transects, random points) to measure variables (depth, velocity) Data used in hydrodynamic models to predict changes with flow	Low-cost equipment Assess many variables Accurate, depending on observer, sampling intensity	Labour-intensive Prone to bias Requires expertise, software, equipment	Bain and Stevenson (1999); Kennard *et al.* (2006b); Grossman and Skyfield (2009); Chan *et al.* (2012)
Hydroacoustics	Use of sonar to detect depth and distribution of submerged structures (plants, substratum)	Quantify at broad scales Quantify rapid changes over long periods	Expensive equipment Requires expertise, software, equipment Limited features can be quantified	<http://www.biosonicsinc.com/>
Aerial photo, video, remote sensing (LIDAR, imagery)	Use airborne imagery to map distribution and characteristics	Quantify at broad scales	Coarse spatial resolution Limited features can be quantified	Koehn *et al.* (2004a); Marchamalo *et al.* (2007)
Habitat use				
Visual	Snorkel, bankside observation	Assess from fish perspective (snorkel) Rapid Non-destructive	Fright bias Conditions may limit efficacy (turbidity) May be unsafe for observer	Bishop *et al.* (1995); Grossman and Ratajczak (1998); Hardie *et al.* (2006)
Fish sampling	Electrofishing (boat, backpack) Netting (fyke, seine, gill) Trapping, poisoning, angling	Standard methods Equipment easily available Limited training required	Sampling bias Lacks microhabitat specificity (nets) Fright bias Conditions may limit efficacy (turbidity) May be destructive	Lockett (1988); Pusey (1998); Pusey *et al.* (1998b); Balcombe and Closs (2000); Arthington *et al.* (2005); Kennard *et al.* (2006b); Ebner *et al.* (2008b)

Method	Description	Advantages	Disadvantages	Sample references
Underwater video	Camera in waterproof housing Baited or non-baited systems	Continuous sampling Non-destructive	Conditions may limit efficacy (turbidity) Catch-selective (bait attracts different species) Large volumes of data Limited range	Ebner *et al.* 2(008a); Butler and Rowland (2009)
Hydroacoustics, Dual-frequency identification sonar (DIDSON)	Hydroacoustic systems provide data on abundance, distribution, size and behaviour Visual check on hydroacoustic detection in split-beam echograms	Continuous sampling Non-destructive	Expensive May be difficult to identify fishes Prone to interference (bubbles) Requires expertise, software, equipment Large volumes of data	Matveev (2007); Berghuis (2008)
Mechanical tags or markers	External or internal mechanical tags (dart and anchor tags, streamers, clips and discs) Marking (subcutaneous dye injection, branding, tattooing, spray-painting, fin-clipping, oxytetracycline, colour-coded plastic chips (microtaggants))	Low-cost equipment Monitor individual fishes	Time-consuming (mark–recapture) Limited number can be monitored May be destructive	Chapter 5
Sophisticated or electronic tags	Radio tags Acoustic tags Passive inductive transponder (PIT) tags	Continuous sampling Monitor individual fishes	Expensive Requires expertise, software, equipment Limited range (PIT) Limited number can be monitored May be destructive	Nicol *et al.* (2004); Butler *et al.* (2009); Ebner (2009); Koehn (2009a, b); Chapter 5

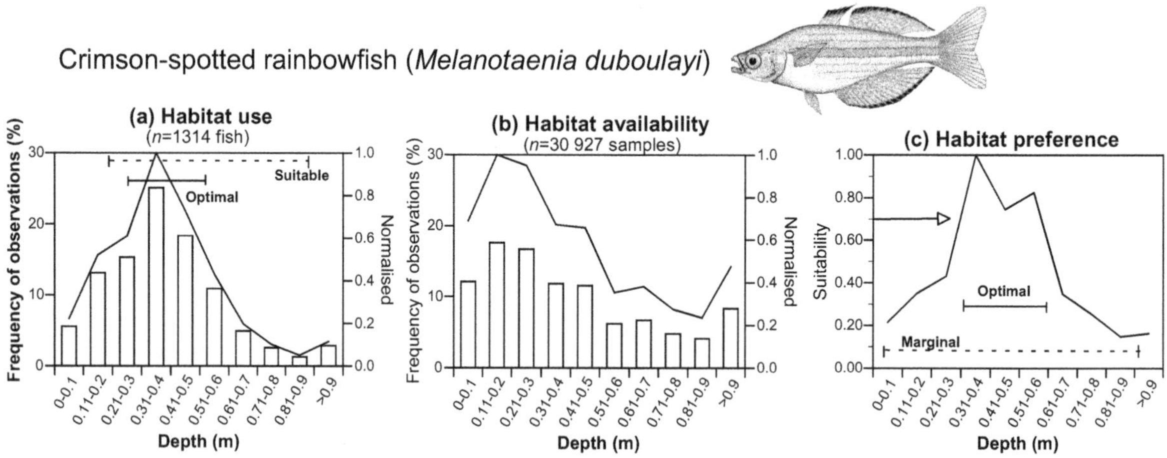

Figure 4.11: Habitat use and availability data for the crimson-spotted rainbowfish (*Melanotaenia duboulayi*). (a) Frequency distributions and normalised curves depicting habitat use (total water column depth) by crimson-spotted rainbowfish in the Mary River, south-eastern Queensland. (b) Frequency distributions and normalised curves of habitat availability (total water column depth) weighted by the relative abundance of crimson-spotted rainbowfish at each site and sampling occasion. Habitat availability and habitat use curves are used to derive the habitat suitability (preference) curves for this species. (c) Using habitat suitability criteria (Groshens and Orth 1994), depths are optimal for habitat suitability >0.7, marginal for >0 to <0.7 and unsuitable at 0.

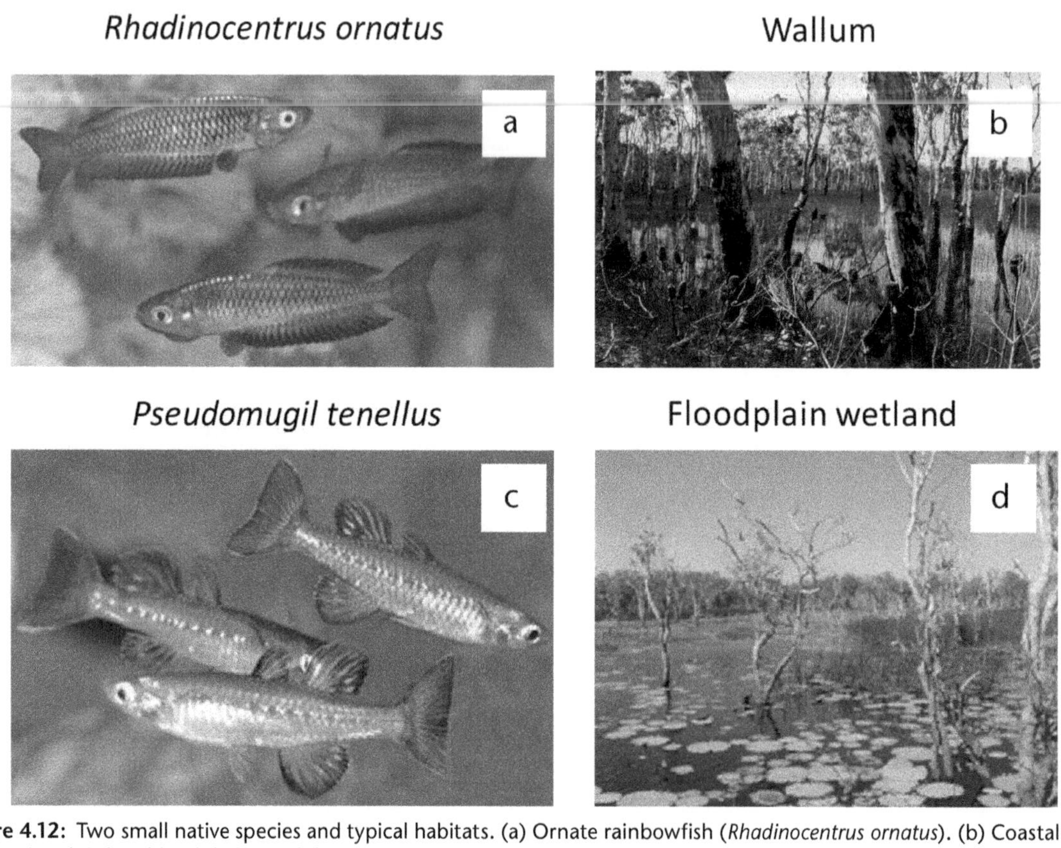

Figure 4.12: Two small native species and typical habitats. (a) Ornate rainbowfish (*Rhadinocentrus ornatus*). (b) Coastal wallum (*Banksia* heathland) habitat of the ornate rainbowfish. (c) Delicate blue-eye (*Pseudomugil tenellus*). (d) Floodplain wetland habitat of the delicate blue-eye.

1995b; Glozier *et al.* 1997). A simple way to characterise habitat suitability is to develop frequency distributions of habitat use for each particular feature (Category II curves of Thomas and Bovee 1993; Fig. 4.11a). Non-parametric tolerance limits are used to define habitat-suitability criteria and the optimal range for a variable is taken as the interval in the frequency distribution encompassing the central 50% of locations occupied by each species and/or size class. The suitable range for the variable is the interval containing the central 95% of occupied locations (Thomas and Bovee 1993; Fig. 4.11a).

Using Category II curves to depict habitat suitability is problematic because curves may not be transferable between sites, and they take no account of variations in habitat availability between sites. Further, suitability curves based on habitat use may be valid only if all fishes at any point in space and time are free to choose from the range of habitat conditions (Heggenes 1996). To offset these problems, it is necessary to standardise microhabitat use at each site with regard for habitat availability. Data from habitat surveys at each sampling location are pooled across sites, weighted by the relative abundances of species at each site, and the frequency values are normalised to the mode of the distribution (i.e. scored from 0–1). A new frequency distribution is then generated by dividing the frequency value for each interval of the normalised frequency distribution of habitat-use data (Fig. 4.11a) by the values for each corresponding interval of the distribution of available-habitat data (Fig. 4.11b). The new distribution is then normalised to the modal value and used to generate a habitat-suitability curve that reflects a species' preferences for instream habitat conditions (Fig. 4.11c). Habitat-suitability criteria (Groshens and Orth 1994) may then be used to define optimal and marginal habitat conditions.

Habitat association models

Another approach is to use correlative habitat association models (ecological niche models), usually at meso- to macrohabitat scales. These are explanatory or predictive statistical models based on co-variation between environmental gradients and the presence or absence of species, although abundance and biomass data can be modelled (Kennard *et al.* 2007). Several models have been developed for freshwater fishes in Australia (Kennard *et al.* 2006a, 2007; Stewart-Koster *et al.* 2007, 2011; Growns and West 2008; Bond *et al.* 2011).

Models of species presence/absence are commonly applied at broad spatial scales (e.g. distributions among river basins or within stream networks). They employ coarse-scale, readily measured predictor variables describing climate, catchment topography and flow regime, and indicate the *potential* distributions of species. Models of abundance or biomass are usually applied at smaller scales (e.g. within and among stream reaches), using landscape- and local-scale predictors (e.g. hydraulics, channel geometry, substratum composition, cover, water quality). These models elucidate the environmental factors that determine the carrying capacity of a habitat and they may have potential (depending on the predictor variables) to estimate total abundance in unsampled reaches or at larger scales (Rosenfeld 2003). Experiments are needed to distinguish causal and spurious relationships with environmental gradients and to test hypotheses about mechanisms and the trade-offs made by organisms confronted by interacting gradients (Lancaster and Downes 2010b).

Bioenergetic models

Bioenergetic models can yield insights into the causes that underpin correlations between environmental factors and the distribution or abundance of species (Rosenfeld 2003). They can predict individual habitat choices and potential growth rates based on the energetic costs and benefits of different habitats. Although these models are popular internationally (Rosenfeld 2003; Hartman and Kitchell 2008), there are few Australian examples (e.g. carp gudgeons: Meredith 2005; Murray cod: Whiterod 2010b). Limitations include the need for field and laboratory measurements to estimate parameters such as weight-dependent maximum consumption, respiration and activity, and to determine the thermal habitats occupied by fishes and how these vary with life stage (Hansen *et al.* 1993). Bioenergetic models also require accurate estimates of population sizes and survival rates (Perry and Bond 2009).

Manipulative experiments

Experiments can help to identify the critical habitat needs of species or life stages. They may involve manipulations of habitat features or the densities of predators and/or competitors in natural or artificial systems, and monitoring the abundance, habitat use or individual fitness of species (Rosenfeld 2003). These experiments typically show that preferences alone rarely control distribution or abundance because there are overriding factors that may be difficult to identify, control or manipulate (Lancaster and Downes 2010b). There are few Australian examples, and most have involved manipulations of submerged wood (Brooks *et al.* 2004; Nicol *et al.* 2004; Bond and Lake 2005; Howell *et al.* 2011), water velocity (Humphries *et al.* 2006) or streamflow (Tonkin *et al.* 2008a; King *et al.* 2009, 2010; Rayner *et al.* 2009a, 2011; Rolls and Wilson 2010).

MANAGEMENT

The rehabilitation of river habitats is receiving attention in Australia and elsewhere (Gore and Shields 1995; Murray-Darling Basin Commission 2004; Roni *et al.* 2005). Large investments are made to protect and restore freshwater habitats, yet management agencies often struggle to ensure that their expenditure is effective. Although there are clear, well-understood ecological guidelines for restoration (Bradshaw 1996; Lake *et al.* 2007), efforts are often hampered where there is too little knowledge or practical experience, inadequate monitoring or a lack of commitment to the principles of 'adaptive management'. These issues are addressed more fully in Chapter 12, but prior identification of habitat needs is critical, as access and availability are often cited as factors limiting fish populations in degraded ecosystems (Bond and Lake 2003b).

Area-based protection can be effective in isolated wetlands, mound springs and refuges (Lake *et al.* 2007) but it may not be feasible at larger scales in rivers or lakes, as they are vulnerable to events in catchments or watersheds remote from the site (Koehn 2003). In rivers, one option is to restore longitudinal and lateral connectivity to facilitate fish movements, including recolonisation from refuges (Ch. 5). This could involve preservation of refuges and removal of barriers that limit dispersal and recolonisation (Beesley and Prince 2010). There is also a need to protect floodplain and riparian habitats prone to damage from stock grazing, firewood collection and other impacts (Pusey and Arthington 2003; Koehn *et al.* 2004a).

There may be many potential degradation-recovery pathways, and recovered habitat can be quite different from the original habitat (Lake *et al.* 2007). The design of stream rehabilitation and enhancement works often borrows from civil engineering, fluvial geomorphology, open-channel hydraulics and other disciplines that complement ecology. These integrated disciplines provide a background for planning pools and riffles, channel realignments and fish passages, and require assessments of the present state of the stream, the flow regime and the appropriate channel geometry and slope (Newbury and Gaboury 1993).

Reinstatement of structural woody habitat (SWH) may be an effective, albeit costly, rehabilitation measure (Murray-Darling Basin Commission 2004) for native fish populations, and may help to offset the effects of the historical practice of de-snagging rivers (Treadwell *et al.* 1999). Recreating effective spatial arrangements of SWH is important (Koehn *et al.* 2004a) and very large snags should be included (Erskine and Webb 2003). Several case studies demonstrate how SWH can be used to good effect (Brooks *et al.* 2004; Nicol *et al.* 2004; Bond and Lake 2005; Howell *et al.* 2011).

Changed flow regimes have had major impacts on freshwater habitats, especially those associated with rivers in eastern and south-eastern Australia (Walker 2006), and these can override other issues. Too often, however, restoration of more nearly natural flow regimes is seen as a panacea rather than as merely part of the remedy. Restoration of flows without complementary restoration of habitats may do little to restore many native fish populations. This is not to deny the necessity for sympathetic flow management. On the contrary, we need to better understand and use environmental water allocations as a means to restore more natural flow patterns where possible. Recasting flow-restoration projects as experiments in restoration ecology is long overdue, as part of the adaptive management process (Arthington and Pusey 2003).

The future of Australia's freshwater fishes depends, in part, on the nature and extent of climate change, and our responses as resource managers. The impacts will vary among regions, species and habitats (Morrongiello *et al.* 2011a) and management will need to become more proactive than it has been in the past (Koehn *et al.* 2011). The combinations of direct and indirect effects will bring unexpected challenges requiring robust management and, above all, a sound knowledge of the science of ecology.

CONCLUSION

The rapid, widespread degradation of our freshwater habitats means that wise management will be needed to ensure the persistence of many populations and species of freshwater fishes. To manage their habitats effectively, we need to know more about:

1. the habitat requirements of each species and its life stages;
2. those habitat features that underpin spawning, recruitment, feeding, growth, maintenance and other processes that support populations and species;
3. the ways that climate change, flow alteration, water quality, alien species and other environmental impacts affect the extent and quality of available habitats;
4. the effects of patch size, quality, complexity and heterogeneity on the carrying capacities of habitats, and how these vary in space and time;
5. the restoration and maintenance of hydrological connectivity between habitats;
6. ways to restore habitats, learning and refining our approaches as we go;
7. the relationships between habitat features and riverscapes; and
8. how to protect key habitats and ecosystem-scale processes.

Rehabilitation is well-recognised as an option for management, and informed decisions in that regard could make substantial contributions to freshwater fish conservation. There is much that we do not understand, however, and there never was a greater need for research and management to combine, in the true spirit of adaptive management.

5 Movements and migration

John D. Koehn and David A. Crook

INTRODUCTION

Movement is integral to the ecology of fishes, enabling them to access resources such as food, shelter and potential mates and to avoid threats. Their movements range from lightning 'bursts', for ambushing prey or escaping predators, to sustained migrations that cover hundreds, even thousands, of kilometres up and down rivers. The patterns of movement reflect the diversity of behaviour, physiology, trophic roles and life history strategies.

The migrations of Australian freshwater fishes may be less conspicuous than the famous spawning runs of Northern Hemisphere salmonids, but they too are impressive and spectacular. For example, juvenile climbing galaxias (*Galaxias brevipinnis*) return from the sea to travel upstream, using their pectoral fins to climb waterfalls and other barriers and colonise inland habitats (McDowall 1988). Golden perch (*Macquaria ambigua*) move over distances up to 2300 km along inland rivers (Reynolds 1983). Southern shortfin and longfin eels (*Anguilla australis*, *A. reinhardtii*) spend five to 20 years in fresh water before metamorphosis and migration *en masse* more than 2000 km to deep-sea spawning grounds (Beumer 1996). These events are most obvious when there are aggregations of migrating fishes at natural or artificial instream barriers. They often are predictable, and render the fishes vulnerable to exploitation (Deap *et al.* 2003). In Australia, Aboriginal people over millennia developed an intimate knowledge of migrating fishes and gathered them as food (Humphries 2007). Stone fish-traps on the Barwon River near Brewarrina, New South Wales, were used for more than 40 000 years to harvest golden perch, Murray cod (*Maccullochella peelii*) and other fishes (Mathews 1903; Dargin 1976). At Lake Condah in south-western Victoria, Gunditjmara people used channels and traps to capture southern shortfin eels (Builth 2004; Box 5.1; Fig. 5.1). Migrations like these have also been exploited by commercial and recreational fishermen (McDowall 1988; Mallen-Cooper *et al.* 1996). For example, there was a commercial fishery based on spawning aggregations of Tasmanian whitebait (*Lovettia sealii*) from the 1930s to the 1970s (Fulton and Pavuk 1988). Commercial operations ceased in 1973 following population declines, but there is still a recreational fishery. Our understanding of the adaptive significance of these movements is poor, especially for smaller, non-commercial or recreational species (Koehn and O'Connor 1990b; Allen *et al.* 2002; Pusey *et al.* 2004). Most studies concern large-scale movements, but small-scale movements are also ecologically important.

Box 5.1: Eels at Lake Condah

Southern shortfin eels (*Anguilla australis*) were part of the diet of the Gunditijmara people of south-western Victoria, who developed one of the world's oldest aquaculture systems (Builth 2004). The eels are catadromous (Fig. 5.5), each life stage having a specific pattern of movement. The Gunditijmara constructed an intricate channel and trap network around the Mt Eccles (Budj Bim) lava beds and Lake Condah, so that in winter and spring glass eels and elvers moving upstream could be directed into penned-off swamps away from older, predatory eels. They also allowed mature, non-feeding silver eels, with high fat content needed for their long migration to oceanic breeding grounds, to make their way back to the river and the sea in autumn. Portable lattice barriers were used to control eel movements and some migrating silver eels were guided through rock raceways where they could be trapped in cone traps (Fig. 5.1) or caught in ponds. Eels trapped in these ponds could then be harvested as water levels fell.

In this chapter, we examine the movements made by different species, and methods for studying them. We also consider their evolutionary and ecological significance, and ways that this knowledge could be applied to conserve and manage Australian freshwater fishes. At the outset, we should acknowledge that most published studies refer to fishes in temperate regions, particularly the Murray-Darling Basin. This means there is some unavoidable bias in our choice of examples and, while that does a disservice to our tropical fishes, it also highlights the exciting discoveries that await researchers willing to brave the crocodiles and venture into the fresh waters of tropical Australia.

MODES OF MOVEMENT

Fishes move over a range of spatial and temporal scales (Table 5.1) and patterns of movement may be complex and difficult to predict (Simpson and Mapleston 2002). Movements may occur in longitudinal (upstream, downstream), lateral (offstream) and even vertical directions. Vertical movements involve fishes occupying different depths within the water

Table 5.1: Terms applied to different modes and scales of fish movements

Movement	Description	Functions
Micro-scale	Immediate, short bursts (< 15 s)	Predator avoidance/escape, prey capture
Meso-scale	Short-term, local, sustained	Home range, homing, exploration of habitats, diel, feeding/foraging excursions, avoidance of poor conditions, spawning
Macro-scale	Long-term, large-scale, prolonged	Migrations, exploration, homing, feeding, spawning, avoidance of poor conditions, home range shifts
Longitudinal	upstream, downstream	Spawning, exploration, migrations, access to resources, feeding, genetic mixing
Lateral	In channel, onto floodplain, in/out of lake littoral zones	Dispersal, exploration, between habitats patches, recolonisation
Vertical	Up/down water column	Feeding, surface breathing
Obligate	Necessary in life cycle	Spawning
Facultative	Not necessary in life cycle	Spawning
Active	Requiring energy expenditure	Homing, escape, prey capture, upstream movements
Passive	Not requiring energy expenditure	Egg, free embryo and larval drift, downstream movements
Seasonal	For part of the year	Spawning, feeding
Diel	Daily	Territoriality
Random	Non-directed	Dispersal (e.g. in some cases of drift)
Oriented	Directed (active)	
Non-oriented	Non-directed	Dispersal
Repeated	Undertaken more than once	Annual migrations, diel patterns, homing

Figure 5.1: (a) Richard Bamblett Jr demonstrating use of a traditional woven basket to direct eels to his spear. (b) Gunditjmara Tyson Lovett-Murray and Sean Bell at Lake Condah, south-western Victoria, demonstrating placement of a traditionally woven eel basket in a wooden eel trap. The baskets were constructed using reeds and placed in constructed channels used to direct migrating eels into the traps (photos: D Bell).

column, often over diel cycles (Jellyman and Tsukamoto 2002; Koehn 2009b; Ch. 4): these are little known and would be an interesting topic for research.

Upstream movements are undertaken to reach spawning grounds or to access new habitats (Lucas and Baras 2001). They have received most emphasis in the scientific and management literature, particularly with regard for fishway design, but downstream movements are no less important – indeed, fishes would all end up in the headwaters of streams if there were not compensatory downstream traffic! The up- and downstream movements of diadromous species are well-known (see 'Diadromy'), but mark–recapture tagging and telemetry show that many potamodromous fishes also move in both directions. Adult Murray cod, golden perch, spangled perch (*Leiopotherapon unicolor*) and common carp (*Cyprinus carpio*) all undertake bidirectional movements (Reynolds 1983; Pusey *et al.* 1994; McKinnon 1997; Koehn and Nicol 1998; Crook 2004c; Koehn *et al.* 2004b, 2008, 2009; O'Connor *et al.* 2005; Stuart and Jones 2006b; Sadlier *et al.* 2008; Koehn 2011).

Lateral movements are well-documented for fishes in tropical rivers (Goulding 1981b) but much less is known of those in Australian lowland rivers (King *et al.* 1999). In northern Australia, spangled perch, barramundi (*Lates calcarifer*), ariid catfish and various small-bodied species move onto inundated

Figure 5.2: Different conditions for studying movements on the floodplain of the Mitchell River, northern Australia. Fish capture trap in (a) the dry season and (b) the wet season (photo: B Pusey).

floodplains during the monsoonal wet season and return to river channels in the dry season (Fig. 5.2) (Bishop *et al.* 1995; Pusey *et al.* 2004; Jardine *et al.* 2011). In Cooper Creek, central Australia, most of the native fishes use inundated floodplains to gain access to new habitats and resources (Arthington *et al.* 2005). Their individual movements are not well-documented but they are revealed indirectly by studies of resource consumption (e.g. stable isotope analyses: Jardine *et al.* 2011) and gene flow within populations (Huey *et al.* 2008, 2011b; Hughes *et al.* 2009).

In lowland reaches of the Murray-Darling Basin, Murray cod and trout cod (*Maccullochella macquariensis*) may use floodplain channels, but their forays onto floodplains otherwise are limited (Koehn *et al.* 2008; Koehn 2009a). Common carp and golden perch, on the other hand, use both floodplain channels and the floodplain proper (Koehn and Nicol 1998; Stuart and Jones 2006a). Small-bodied species, including carp gudgeons (*Hypseleotris* spp.) and Australian smelt (*Retropinna semoni*), move in concert with changing water levels (Lyon *et al.* 2010). They move into offstream habitats as water levels rise, move about as levels peak and return to the channel as levels fall.

We should clarify some terms used to describe fish movements. The term 'movement' refers simply to any change in the location of a fish, whereas 'migration' and 'dispersal' are specific kinds of movement. Migration is movement that occurs with some periodicity, involves most of the population and generally is from one habitat or environment to another (Northcote 1978; Lucas and Baras 2001). 'Dispersal' describes movement from an area where fishes are concentrated to a dispersed (less-concentrated) state. The term 'obligate' indicates that a particular movement is necessary to complete the life cycle, and 'facultative' refers to movements that, while commonplace, are not vital.

ADAPTIVE SIGNIFICANCE

Movements are energetically expensive and incur certain risks, including greater vulnerability to predators (Switzer 1993, 1997; Alerstam *et al.* 2003) but these evidently are outweighed by the adaptive advantages. Here, we examine some of these trade-offs.

Resources at micro- and mesohabitat scales

Habitat quality in aquatic environments tends to be temporally variable and spatially patchy, and fishes must move between patches to optimise their use of resources. The 'ideal free distribution' concept (IFD: Fretwell and Lucas 1970) considers the adaptive significance of patch-scale resources (Milinski 1988; Tyler and Gilliam 1995). In its simplest form, the IFD concept describes a situation in which movement between habitats is unconstrained and individuals are free to colonise the habitat patch that brings most rewards. Not all individuals within a population can occupy one patch, because competition, resource depletion and other density-dependent mechanisms reduce patch profitability as the number of occupants increases. The IFD concept suggests that an individual should leave its habitat patch if profitability falls below that of other available patches – this is one explanation for small-scale movements by fishes (e.g. between microhabitats). At mesohabitat (within-reach) scales, however, movements are often restricted to discrete home ranges and there may be limited opportunities to visit other areas or respond to local changes in habitat quality (Armstrong and Herbert 1997). This limitation may be overcome if individuals make periodic excursions to assess the profitability of habitats outside their normal range, then either relocate or return to the home range (Fig. 5.3). The mechanisms influencing movement between habitat patches are complex (Tyler and Gilliam 1995; Ruxton *et al.* 1999), but it appears that a key function of small- to medium-scale movements is to enable fishes to optimise their use of the dynamic resources that characterise aquatic habitats.

Resources at macrohabitat scales

One explanation for the evolution of large-scale migrations in fishes is that they provide access to richer feeding areas than the natal habitat, and increase fitness and reproductive potential. Thus, the various forms of diadromy (see 'Diadromy') could have evolved in response to differences in marine and freshwater productivity that confer advantages to different life stages (Gross *et al.* 1988; McDowall 1988, 2007). Avoidance of harsh conditions in natal habitats (e.g. freezing in winter) could be a driver of diadromy in Northern Hemisphere fishes (Jonsson and Jonsson 1993). While freezing of streams is rarely an issue for Australian fishes, the differences between stream and ocean temperatures in winter could influence habitat profitability (McDowall 1988). In Australia, drying (rather than freezing) probably influences some potamodromous movements, where there is a need to avoid unfavourable conditions and colonise more profitable habitats.

In the Lake Eyre and northern Murray-Darling basins in central Australia, fishes move to exploit the boom–bust nature of arid environments (Puckridge *et al.* 1998; Balcombe and Arthington 2009). In dry periods, fish numbers decline and those individuals that remain are confined to disconnected waterholes (Arthington *et al.* 2005). In floods, the fishes move along the channel and onto the floodplain, taking advantage of the pulse of productivity that accompanies floods (Sternberg *et al.* 2008). Seasonal and inter-seasonal drying means that, for many species, a retreat to refuges and subsequent recolonisation is a basic movement pattern (Ch. 4, Fig. 4.1). These movements, with other dispersal abilities, contribute resilience in the face of drought and other disturbances (Crook *et al.* 2010a, b).

The extent of movements during floods in dryland rivers has been inferred mainly from population genetics (Huey *et al.* 2011b) and changes in assemblage structure in waterholes before and after floods (Balcombe and Arthington 2009). In the Lake Eyre and northern Murray-Darling basins, genetic structuring in subpopulations of catfishes (*Neosilurus hyrtlii*, *Porochilus argenteus*, *Tandanus tandanus*) (Huey *et al.* 2006, 2008, 2011b), golden perch (Huey *et al.* 2011b), bony herring (*Nematalosa erebi*) and Australian smelt is weak, indicating frequent, large-scale dispersal (Huey *et al.* 2006). Assemblage composition and the abundances of species in waterholes vary greatly after floods because many species move between refuges (Balcombe and Arthington 2009).

The Wet Tropics of northern Australia show a similar boom–bust dynamic, with predictable wet and dry seasons (Douglas *et al.* 2005). Again, many species have evolved movement behaviours to exploit this pattern. In the Mitchell River, northern Queensland, for example, many fishes move onto inundated floodplains during the wet season (e.g. barramundi; blue catfish, *Neoarius graeffei*; highfin catfish, *N. berneyi*; gizzard shad, *Nematalosa come*; spangled perch; glassfish, *Ambassis* sp.; common ponyfish, *Leiognathus equulus*; barred grunter, *Amniataba percoides*; toothless catfish, *Anodontiglanis dahli*; Hyrtl's catfish, *Neosilurus hyrtlii*; spotted scat, *Selenotoca multifasciatus*) (Jardine *et al.* 2011). Floodplain inundation persists for a few months each year, but the pulse of energy provided to fishes drives the food web and helps to explain the capacity of tropical rivers to support greater fish species diversity and biomass than temperate systems (Jardine *et al.* 2011).

Stress avoidance and recolonisation

As conditions change, fishes move to avoid acute or chronic stressors, and may later return. In some cases, avoidance of adverse conditions may not be possible and fishes die. For instance, bushfire is an acute stressor for fishes in many parts of Australia as sediment is flushed into rivers by following rains. The water may become de-oxygenated sludge and fishes that cannot escape face suffocation (Lyon and O'Connor 2008). Blackwater events are stressors (Ch. 4) that occur with flooding after extended dry periods, when the water becomes hypoxic due to decomposition of organic matter (Howitt *et al.* 2007). Alternatively, the drying of habitats due to seasonal or supra-seasonal drought may be a chronic stressor (Humphries and Baldwin 2003; Arthington *et al.* 2005). As habitats dry, water quality and food availability decline (Lake 2003). These events are usually transient, however, and the habitat is soon restored, opening up opportunities for fishes to recolonise. Carp gudgeons, for example, retreat from ephemeral creeks to permanent pools during dry periods, and disperse widely to recolonise vacant habitats once flows return (Perry and Bond 2009).

Metapopulation structure

So far, we have considered the adaptive significance of movements in terms of the fitness of individuals (i.e. classic Darwinian natural selection), but they are also part of the processes that maintain viability and persistence at a population level (Hughes *et al.* 2009). As the profitability of aquatic habitats is spatially and temporally patchy, many fish stocks have a 'source–sink' metapopulation structure, wherein losses of individuals from low-productivity patches (sinks) are offset by immigration of individuals from high-productivity patches (sources) (Woodford and McIntosh 2010; Huey *et al.* 2011b). Without movements between sources and sinks, species would eventually be lost from the sinks and the sources would become isolated. This is what happens when human activities (e.g. impoundments) fragment river systems (Hughes 2007; Huey *et al.* 2011b). Loss of connectivity between sources and sinks increases the vulnerability of populations to catastrophic disturbances and reduces their ability to respond to climatic variability (Fagan 2002; Schick and Lindley 2007; Huey *et al.* 2011b).

MECHANICS OF SWIMMING

Water is denser than air, providing buoyancy but also a more viscous environment in which organisms move. It constrains fish movements in ways unlike those experienced by terrestrial animals moving in air, and requires a different means of propulsion. Water may be still or in motion, with currents and turbulence that affect swimming in myriad ways. Active swimming movements have implications for energy expenditure, but in some situations passive drift (especially of eggs, free embryos and larvae) may be more energy-efficient.

Swimming ability

Fishes have diverse swimming abilities, depending on their developmental stage, size, morphology and physiology. In turn, the mobility and swimming ability of a species influence its ecology (e.g. mode of foraging, propensity to migrate: Videler 1993). Swimming is energetically expensive and fishes have evolved numerous ways to optimise their efforts in different surroundings (McGuigan *et al.* 2003). To understand the interactions between swimming ability and movements, it will help to touch briefly on some physiological matters.

The muscles of a fish must create enough thrust to overcome the opposing drag of the body mass and friction, and swimming ability is a physiological and morphological trade-off between these forces (Webb 1975; Beamish 1978). Trunk and caudal fin propulsion is the most efficient mechanical means of forward progress, enabling maximum acceleration, flexibility and sprint speeds. It involves undulation of the trunk and contraction of muscles on either side of the body to generate backward thrust and push the fish forward. Too much undulation can increase turbulence and drag but this can be offset by body scales, a streamlined body shape and reduced amplitude of body movement (Jobling 1995). Some species employ paired and unpaired fins, especially for fine-scale, directed movements. Fishes rarely rely on a single mode of swimming, but vary the mode in response to their needs. Remaining stationary and slow swimming have different physiological requirements, of course, compared to high-performance fast swimming. The relationship with oxygen consumption (or energy expenditure) varies with speed, and optimal swimming speeds generally are one to three body lengths per second (Jobling 1995).

Table 5.2: Comparison of the structure, physiological properties and functions of red and white muscle fibres in fishes

Red muscle	White muscle
Slow-contracting fibres	Fast-contracting fibres
Aerobic respiration	Anaerobic respiration
Abundant blood supply (hence colour)	Less blood supply
Small (100–200 μm) fibre diameter	Large (300–400 μm) fibre diameter
Store and use lipids and glycogen	Store and use mostly glycogen
Used for cruising speed	Used for acceleration and burst speed
Sustained speeds, long distances	High speeds, short distances
Typically higher in pelagic species	Typically higher in sit-and-wait predators

Source: After Jobling (1995).

Two types of muscle fibres influence swimming, namely *red* fibres ('slow' muscles that respire aerobically) and *white* fibres ('fast' muscles that respire anaerobically) (Table 5.2). The relative proportions and diameters of red and white muscle fibres help to determine swimming ability (Greek-Walker and Pull 1975), setting critical velocities and swimming speeds for the species. Even so, the proportions of these fibres (and other physiological features) can be plastic within species, depending on local conditions. For example, crimson-spotted rainbowfish (*Melanotaenia duboulayi*) in Queensland lakes have less caudal red muscle, lower swimming performance and different fin shapes and positions than their riverine counterparts, reflecting less need for swimming against strong currents (McGuigan *et al.* 2003). We therefore need to be cautious when generalising about the swimming capabilities of particular species.

Swimming is classified as *sustained* (more than 200 min), *prolonged* (15 s–200 min) or *burst* (< 15 s) (Blake 1983; Domenici and Blake 1997). Sustained swimming includes foraging, migration and schooling. Together with prolonged swimming (cruising, interspersed with bouts of more energetic swimming), this accounts for nearly all swimming time (Blake 1983). Prolonged swimming is influenced by body weight and temperature, and depends on physiological characteristics (Beamish 1978). A widely used measure of prolonged swimming ability is the 'critical swimming speed' (U_{crit}), determined by subjecting fishes to increasing current velocity until they become fatigued (Jobling 1995). These data can be useful to compare species or fishes of different sizes. For example, juvenile Murray cod are capable of moderately prolonged swimming, and have low-energy demands typical of ambush predators (Whiterod 2010b). Although Murray cod swim fastest at higher temperatures, their swimming capacity is maintained across a broad range of temperatures (16–25°C).

Burst swimming is an energetically expensive, anaerobic form of swimming that – as the name suggests – is invoked for short intervals to escape from predators, to capture prey or to negotiate strong currents (Webb 1984b; Ch. 6). Acceleration for burst swimming usually is achieved by large-amplitude tail movements (Jobling 1995). As in prolonged swimming, water temperature influences burst swimming performance because it affects water viscosity and muscle function (Mallen-Cooper 1992). Fast-start performance in juvenile golden perch includes a threshold, whereby performance is much reduced below 15°C and remains steady at 16–25°C (Lyon *et al.* 2008b). This is one of the less obvious ways that golden perch could be affected by cold-water releases from thermally stratified reservoirs.

Just as there are interspecific and ontogenetic changes in habitat use (Werner and Gilliam 1984; Ch. 4), there are developmental changes in the swimming capabilities of fishes (Kowarsky and Ross 1981; Stuart and Berghuis 2002). For example, juvenile barramundi and Australian bass (*Percalates novemaculeata*) have different abilities to negotiate high-velocity zones in fishways (Mallen-Cooper 1992). Ninety-five per cent of Australian bass can negotiate 1.02 m s^{-1} at 40 mm (total length, TL), 1.40 m s^{-1} at 64 mm TL and 1.84 m s^{-1} at 93 mm TL, whereas barramundi at 43 mm TL can negotiate only 0.66 m s^{-1}. Prolonged swimming speeds of 0.19–0.36 m s^{-1} and burst speeds of 0.47–1.35 m s^{-1} are reported for common galaxias (*Galaxias maculatus*) and apparently are little affected by turbulence (Nikora *et al.* 2003). Among larvae of Clarence River cod (*Maccullochella ikei*), burst swimming speeds increase daily until day 30, and stamina thereafter declines (Butler 2009a). This highlights an ontogenetic change in this species, coinciding with different phototactic responses and a switch from endogenous to exogenous feeding (Ch. 7).

The motion of a fish and its position in the water column are influenced by its buoyancy. Neutral buoyancy allows more energy to be available for forward motion. This is important for pelagic species because the thrust from swimming can also provide dynamic lift and assist mobility. Bony fishes generally have light skeletons and retain gas in a swim bladder that allows them to adjust their buoyancy, whereas demersal species often have small swim bladders and need to expend energy to rise into the water column (Jobling 1995).

Some fishes have a remarkable capacity for movement without swimming. For example, southern shortfin eels survive for some time out of water and glass eels and elvers are able to travel readily over wet surfaces and to climb wet surfaces (Beumer 1979a; Beumer and Harrington 1980b). The Australian lungfish (*Neoceratodus forsteri*) can slither down slopes and across wet grass (Illidge 1894, cited in Pusey *et al.* 1994). Pouch lampreys (*Geotria australis*), shorthead lampreys (*Mordacia mordax*) and juvenile climbing galaxias can climb wet vertical surfaces as they migrate upstream (Sloane 1984a, b, c; McDowall 1988). It is amazing to watch them climb the side of a bucket.

Drift

Drifting with the current is an efficient form of dispersal for the early life stages of many riverine fishes, including their eggs, free embryos, larvae and juveniles (Brown and Armstrong 1985; Robinson *et al.* 1998). Hatching from nest aggregations incurs risks from competition, cannibalism and agonistic encounters with siblings (Brown 1985; Henderson and Corps 1997), and drift may be an effective way to limit these and other density-dependent effects. While drift allows movement to more favourable rearing habitats it may expose larvae to other risks, such as predation and changed flow regimes in regulated rivers. Many (but not all) Australian fishes drift as eggs (e.g. Australian bass; barramundi; climbing galaxias; golden perch; grunters, *Scortum* spp.; silver perch, *Bidyanus bidyanus*) and free embryos or larvae (e.g. flathead gudgeons, *Philypnodon* spp.; Murray cod; trout cod) and the same is true of common carp (Humphries *et al.* 2002; Koehn 2004; Koehn and MacKenzie 2004; Koehn and McDowall 2004; Koehn *et al.* 2004b; Pusey *et al.* 2004; Humphries 2005; Koehn and Harrington 2005).

The longitudinal distributions of drifting eggs, free embryos and larvae are influenced by river flows during the spawning and hatching periods and drift distances may be affected by changes to the seasonality of flows in regulated rivers (Koehn *et al.* 2004b; Koehn 2011; Ch. 4). In summer, discharge under natural conditions is usually low or falling, limiting the distances that eggs, free embryos or larvae are carried by the current. High flows caused by releases for irrigation in spring, summer and autumn, however, can extend the drift distance before the young fishes settle. Adult fishes also can be displaced downstream by high-flow events. In the coastal rivers of Queensland, for example, freshwater catfish (*Tandanus tandanus*) and golden perch have been displaced by high flows over tidal barrages into estuarine habitats where they cannot survive (Stuart and Berghuis 2002).

Factors that stimulate movement

Movements are stimulated by physiological (e.g. sexual maturation, ontogenetic stage, body condition, hunger, metabolic balance, spawning condition), behavioural (e.g. homing), environmental (e.g. light, temperature, hydrology, weather, water quality, food availability, season, lunar phase) and biotic factors (e.g. species interactions, dominance, territoriality, predator avoidance), sometimes acting in concert (Lucas and Baras 2001). These factors are not well-understood, but seasonal flows and temperatures appear to be cues for migration by some potamodromous species (Mallen-Cooper 2000). Rises in flow have been suggested to initiate upstream (presumably pre-spawning) movements by several large Murray-Darling fish species (Cadwallader 1977; Llewellyn and MacDonald 1980; Reynolds 1983). The implicit hypothesis has been that such upstream movements counters, or is countered by, the downstream drift of free embryos following spawning (Lake 1967a; Humphries 2005).

The movements of freshwater sawfish (*Pristis microdon*) are influenced by high flows (Whitty *et al.* 2008; Phillips 2009b). Adult and subadult golden perch respond to changes in flow and temperature, including small (< 15 cm) rises in river level (Mallen-Cooper 2000) but cease their upstream movements when levels begin to fall. Upstream movements of silver perch are also stimulated by small rises in flow (Mallen-Cooper and Stuart 2003). Local movements of Clarence River cod are related to flow increases but apparently are not influenced by temperature (Butler 2009a).

Movement variability

There is wide variability in the movement patterns of individual fishes (Lucas and Baras 2001; Rodríguez 2002), posing a challenge for data analysis (White and Garrott 1990). Some variation may be attributable to differences between the sexes (Parkinson *et al.* 1999; Paragamian and Kruse 2001) and to differences in size (Auer 1996; Gaston and Blackburn 1996; Pyron 1999). The movements of Clarence River cod are predictable but also individualistic, and there may be no period when all individuals behave similarly (Butler 2009a). Mary River cod (*Maccullochella mariensis*) have relatively predictable, consistent patterns of behaviour within their individual home ranges but less predictable, highly individualistic long-distance movements (Simpson and Mapleston 2002). The movements of cyprinids have been described as 'an infinitely variable combination of active movements' (Linfield 1985).

Australian lungfish in free-flowing reaches of the Mary and Burnett rivers in south-eastern Queensland generally are sedentary; most individuals remain within one pool or adjacent pools and movements of more than 1 km are rare. In impounded reaches of the Burnett system, however, movements of up to 48 km are common, particularly during high flows in the spawning season (Brooks and Kind 2002).

The movements of Murray cod vary with body size and home location, and there may be big differences in the distances moved by fish resident in lakes and rivers (Koehn *et al.* 2009). The movements of Murray cod below and above 650 mm TL differ, perhaps due to immaturity (Jones and Stuart 2007). Fish from Lake Mulwala, on the mid Murray River near Yarrawonga, Victoria, travel greater distances than river fish. Some of the resident lake fish move only to the upper reaches of the lake (< 10 km), while others move well into the inflowing rivers (Koehn *et al.* 2009; Koehn 2011).

Variations can also occur in timing. In the middle reaches of the Murray River migrations of Murray cod begin asynchronously, usually from mid July until late October and sometimes through to December and January, and return movements are individualistic (Koehn *et al.* 2009). In the lower Murray, movements may occur several months earlier (Saddlier *et al.* 2008). Individual fishes may depart from the general movement pattern, some remaining at home sites and not migrating in some seasons, and others remaining for considerable times at presumed spawning sites and making unexpected stray movements (Koehn *et al.* 2009). Some Murray cod appear to switch between migratory and sedentary behaviours in different years. These kinds of variation are not unusual among fishes (Jackson 1980; Harcup *et al.* 1984; Gangl *et al.* 2000; Knights *et al.* 2002; Vokoun and Rabeni 2005) and may insure populations against changing environmental conditions.

LOCAL MOVEMENTS

An often-cited mark–recapture study by Reynolds (1983) in the lower Murray River reported that several tagged golden perch (11 fish, or 2% of all recaptures) were later recaptured more than 1000 km upstream (indeed, up to 2300 km) from their point of origin. This is justifiably used as evidence of migration in the life cycle of golden perch, but we should not overlook the fact that 86% of the golden perch in Reynolds' study were recaptured less than 60 km from the tagging site (cf. McKinnon 1997). In general, a proper understanding of the movements of freshwater fishes requires us to consider large- and small-scale movements among all individuals in the population and at all stages of ontogeny (McDowall 1988; Northcote 1992; Koehn and Nicol 1998; Lucas and Baras 2001). In this section, we consider the small-scale movements associated with day-to-day activities.

Restricted movement paradigm

Gerking (1953) suggested that stream fishes stay in the same area for most of their lives, and that pools or other preferred habitats are isolated units with their own distinct populations. This 'restricted movement paradigm' (RMP) has been influential in research and management, but it has been challenged (Gowan *et al.* 1994) by new knowledge of large-scale movements (Lucas and Baras 2001; Koehn *et al.* 2009). While Gerking's idea is no longer generally upheld, nonetheless for many species there is empirical evidence for restricted movements over extended periods (Rodríguez 2002; Crook 2004a; Koehn *et al.* 2008; Ebner *et al.* 2009a).

Home range concept

The area used by a fish (or other animal) in routine activities (feeding, mating, caring for young) is the 'home range' (Burt 1943). The term may be used loosely to describe the area through which an animal moves over the duration of a study, but it should be reserved for those that show 'site fidelity' (Spencer *et al.* 1990). Site fidelity is the tendency to remain within an area for an extended time, or to return to an area occupied previously (homing) (White and Garrott 1990).

Evidence of restricted movement and site fidelity by stream fishes dates from the early 20th century (Gerking 1950). The term 'home range' first appears to have been applied to fishes by Gerking (1953), in his mark–recapture studies of warm-water fishes in North America. He found evidence of restricted movement and a tendency for artificially displaced fishes to home to their site of capture. Since then, the existence of home ranges has been proposed for freshwater fishes throughout the world (Gerking 1959; Lewis and Flickinger 1967; Malinin 1969; Linfield 1985; Young 1999).

Mark–recapture studies generally show that most tagged stream fishes are recaptured at or near their original capture location. For example, Berra (1973) reported that 78% of tagged mountain galaxias (*Galaxias olidus*) moved less than 26 m in late winter and spring in a tributary of the Murrumbidgee River near Canberra, and concluded that the species' home range was 13–26 m. Studies of river blackfish (*Gadopsis marmoratus*: Koehn 1986), two-spine blackfish (*G. bispinosus*: Lintermans 1998), southern shortfin eels (Beumer 1979a) and alien brown trout (*Salmo trutta*: Jackson 1980) also indicate limited home ranges. Home ranges for radio-tagged Macquarie perch (*Macquaria australasica*) in Cotter Reservoir (Australian Capital Territory) are estimated as 0.2–18.4 ha, with mean lengths of 516 m (Ebner *et al.* 2010). Mark–recapture studies in other parts of the world suggest that populations include a behavioural dichotomy, with a dominant sedentary component and a smaller mobile component (Funk 1957; Hesthagen 1988; Northcote 1992; Smithson and Johnston 1999; Rodríguez 2002).

Homing

Return movements to home sites are typical of riverine fishes (Gerking 1953) and are associated with both small- and large-scale movements. For example, trout cod generally move within home ranges of less than 100 m and return home after moving further afield (Ebner and Thiem 2006; Koehn *et al.* 2008). Similar patterns occur in Murray cod (Koehn and Nicol 1998; Jones and Stuart 2007; Koehn *et al.* 2009), Mary River cod (Simpson and Mapleston 2002), Clarence River cod (Butler and Rowland 2009), golden perch (Crook 2004a; O'Connor *et al.* 2005), Australian lungfish (Brooks and Kind 2002), river blackfish (Khan *et al.* 2004; Koster and Crook 2008), southern shortfin eel (Jellyman and Sykes 2003) and common carp (Crook 2004a; Stuart and Jones 2006b). Homing may involve overshooting and backtracking before a fish resettles at home (Hasler and Scholz 1983; Koehn *et al.* 2009), owing perhaps to difficulties in relocating or re-establishing.

Mobility and home range shifts

Restricted movements and site fidelity do not necessarily imply that species with these attributes are sedentary over their entire life times. Even species that spend most of their lives in a sedentary state may migrate over large distances at times. Golden perch in the Murray River may remain in home ranges of less than 0.5 km but, with rising temperatures and river discharge in spring, many (15 of 19 fish) made rapid upstream or downstream movements of 16–290 km, possibly for spawning (O'Connor *et al.* 2005). Similarly, radio-tagged Mary River cod occupy restricted home ranges (70–820 m) for many months but occasionally undertake rapid, long-distance movements of up to 35 km (Simpson and Mapleston 2002). After three to nine months, most of these fish return to, or near to, their original locations.

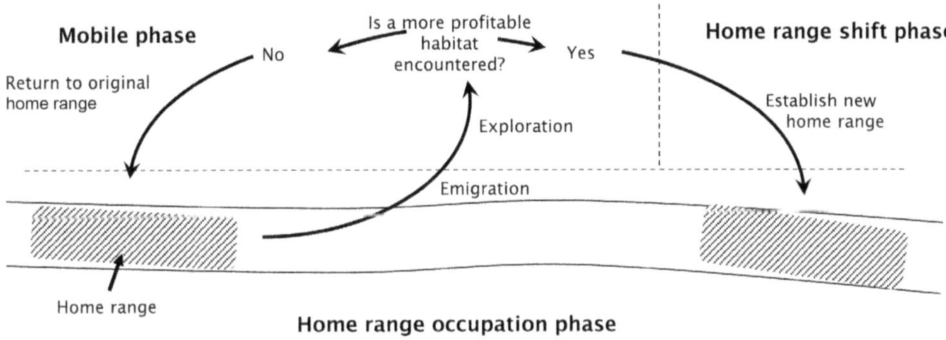

Figure 5.3: Home range shift model, showing fish movements to assess habitat profitability outside the normal home range. Shifts may also occur as a consequence of other movements, such as spawning migrations (modified from Crook 2004a).

Several theories have been proposed to resolve the apparent conflict in observations of strong site fidelity and high mobility in fish populations (Linfield 1985; Hesthagen 1988; Crook 2004c). The 'behavioural dichotomy' model (discussed above) suggests that populations are a mixture of sedentary and mobile individuals. Alternatively, the 'home range shift' model (Crook 2004a) proposes that individual fish spend most of their lives exhibiting sedentary behaviour but occasionally shift the locations of their home ranges. Periods of mobility include distinct stages of emigration, exploration and home range establishment (Fig. 5.3). The home range shift model differs from the behavioural dichotomy model in that individual fishes exhibit sedentary and mobile behaviours. It suggests that the mobility of a fish over its life time depends on the frequency and magnitude of home range shifts, as well as the size of its home range.

The energetic costs and risks associated with changing locations may be related to the level of site fidelity (Morris 1987; Morris and Brown 1992; Switzer 1993). Individuals that invest a lot of energy in the initial establishment of a home range might be expected to show strong fidelity to that site and to be less likely to change sites (Switzer 1993). Movements to new sites presumably would result in an overall energy gain, and the inclination to move should relate to the magnitude of the gain, habitat patch sizes and habitat stability.

Associated behaviours

Most studies of day-to-day movements by Australian fishes have overlooked the underlying patterns of behaviour. For instance, there has been little regard for intra- and interspecific interactions (e.g. territoriality, dominance hierarchies), despite research in other countries suggesting that these are drivers of demographic and distributional processes (Nakano 1995a; Armstrong *et al.* 1997; Armstrong and Herbert 1997; Gowan and Fausch 2002). This apparent oversight reflects difficulties in verifying activities such as spawning, which is often inferred from other observations such as the presence of larvae (Humphries 2005; Koehn and Harrington 2006; Koehn *et al.* 2009). One exception is the use of underwater video cameras to document the reproductive behaviour of Clarence River cod in the Mann and Nymboida rivers, New South Wales (Butler and Rowland 2009). Male Clarence River cod spent up to a week prior to spawning, cleaning and guarding nesting sites in slow-flowing pools. Females then entered the nesting sites to lay their eggs, and males remained for up to 24 days to guard and care for the eggs and larvae. Males did not feed or leave the nesting sites during the incubation period. Similar field-based observations have been made of nesting and spawning by freshwater catfish in southern Queensland (Merrick and Midgley 1981). These kinds of observations are a valuable complement to less direct methods such as telemetry and mark–recapture.

Where direct observation is not possible (e.g. in turbid water), indirect methods are used to infer the behaviour associated with movement. For example, reports of increased activity by fishes at night have been used to infer that the species concerned are nocturnal foragers that shelter during daylight hours. In northern Australia, eastern rainbowfish (*Melanotaenia splendida*), glassfishes (*Ambassis* spp.), spangled perch, barred grunter and sooty grunter (*Hephaestus fuliginosus*) are less active at night, possibly to avoid predators (Bishop *et al.* 1995). From variations in the strength of radio signals, tagged trout cod in the Murrumbidgee River apparently have crepuscular and nocturnal patterns of activity (Thiem *et al.* 2008). River blackfish in Armstrong Creek, Victoria, use various habitats at night, then return at dawn to the same undercut bank where they shelter through the day (Koster and Crook 2008). In contrast, Mary River cod may be much more active around dawn than during the day or night (Simpson and Mapleston 2002).

LARGE-SCALE MOVEMENTS

Potamodromy

'Potamodromy' refers to migrations that occur wholly within fresh water (Myers 1949). Although many Australian fishes are potamodromous, this pattern has received less attention than diadromy (Mallen-Cooper 2000). Large-scale migrations are often linked to reproduction (Wootton 1998) and many riverine and lacustrine species undertake pre-spawning upstream migrations in spring, followed by rapid, post-spawning, downstream returns (Scott and Crossman 1973; Silva and Davies 1986; Koed *et al.* 2000; Palstra *et al.* 2004; Koehn *et al.* 2009). The best-known potamodromous Australian species are golden perch (Llewellyn 1968; Reynolds 1983; Koehn and Nicol 1998; O'Connor *et al.* 2005) and Murray cod (Koehn and Nicol 1998; Koehn *et al.* 2004b, 2009; Koehn and Harrington 2006; Saddlier *et al.* 2008).

While there are early reports of long-range movements by Murray cod (1498 km in 488 days) and other reports of 48 km and 214 km (Llewellyn 1968; Anonymous 1971; Reynolds 1983), 69% of Murray cod tagged by Reynolds (1983) moved less than 10 km and he concluded that the species was sedentary, without a spawning migration. Radio-telemetry since has shown that although Murray cod move little for much of the year, they do undertake migrations (Koehn and Nicol 1998; Saddlier *et al.* 2008; Koehn *et al.* 2009). These occur prior to spawning and are followed by returns, presumably after spawning (Koehn 2006).

Australian lungfish are relatively sedentary, occupying linear ranges between 300 m and 47 km, and appear to have movement patterns like those of Murray cod. They are more active prior to spawning, and those in impoundments move further than those in rivers (Brooks and Kind 2002). Adult Macquarie perch migrate from lakes into inflowing rivers to spawn; spent fish then usually return to their lake location (Cadwallader 1976b; Cadwallader and Rogan 1977; Lintermans *et al.* 2010) and their pelagic larvae are swept downstream (Gooley and McDonald 1988; Tonkin *et al.* 2010). The movements of several other species (e.g. Australian smelt, carp gudgeons, bony herring) are being revealed by fishway monitoring (Barrett 2008), although the data have not yet shown the full extent or type of movements (Mallen-Cooper and Stuart 2003). Are these species also potamodromous?

Knowledge of migrations by adult Murray cod and the drift of free embryos and larvae has been incorporated into a simple model (Koehn and Nicol 1998). This raised the possibility of upstream colonisation by juvenile fish (less than three years), but later work suggested that movements by juveniles are limited (Mallen-Cooper 2000; Jones and Stuart 2007). There are also reports that some Murray cod move downstream prior to spawning (Saddlier *et al.* 2008). These untidy pieces of the movement 'jigsaw puzzle' emphasise the need to periodically review our models and incorporate new data.

Conceptual models synthesise our knowledge of potamodromous movements in ways that integrate the ecology of the species and point to strategies for managing populations, especially in a context of spatial and temporal scales (Koehn 2011). Models are needed to draw together evidence, often from disparate studies under dissimilar conditions. For example, we know that golden perch move in complex patterns that include upstream and downstream movements, home range shifts, lateral movements, egg and larval drift and upstream juvenile migration (Table 5.3; Fig. 5.4). Different perspectives of migrations by

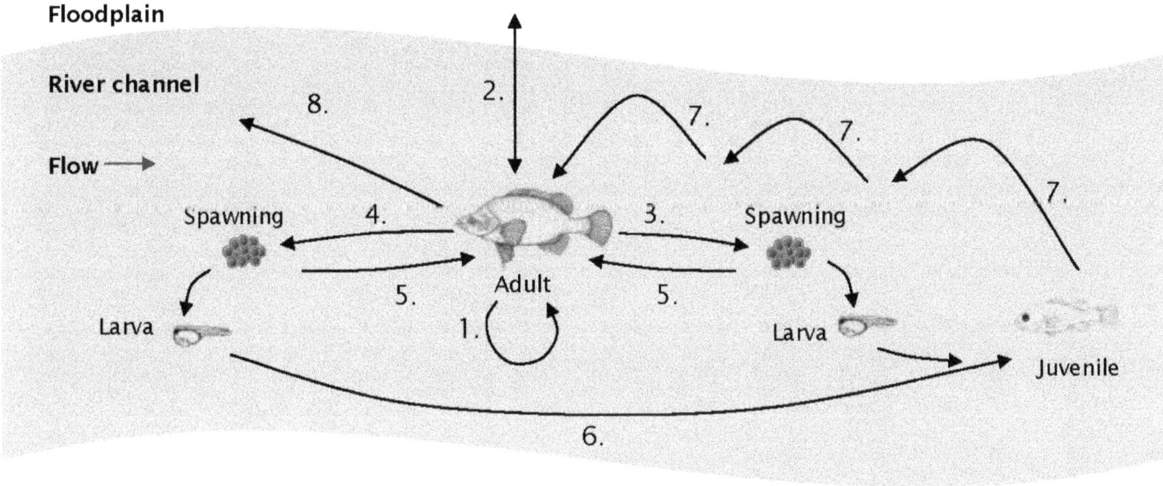

Figure 5.4: A model of movements by golden perch (*Macquaria ambigua*). (1) Local, home range movements. (2) Lateral movements to/from floodplain. (3) Downstream movements. (4) Upstream movements. (5) Return movements (homing). (6) Downstream drift. (7) Periodic upstream juvenile migrations in response to flow. (8) Large-scale movements and home range shifts. See also Table 5.3.

golden perch have come from studies in large rivers during floods (Reynolds 1983; O'Connor *et al.* 2005) and another over a short period in a small river (Crook 2004a, b, c). Despite extensive research over decades, the patterns for this widespread, common and popular species remain unclear. This shows how complex (and interesting!) the movements of potamodromous fishes can be.

Table 5.3: Components of a movement model for golden perch (*Macquaria ambigua*), with key references

	Component	Life stage	Method	References
1	Local, home range movements	Adults	Mark–recapture, telemetry	Llewellyn 1968; Reynolds 1983; McKinnon 1997; Crook 2004a, c; O'Connor *et al.* 2005
2	Lateral movements to/ from floodplain	Adults	Telemetry	Koehn and Nicol 1998; Rolls and Wilson 2010
3	Upstream movements	Adults	Mark–recapture	Llewellyn 1968; Reynolds 1983
4	Downstream movements	Adults	Mark–recapture, telemetry	McKinnon 1997; Koehn and Nicol 1998; O'Connor *et al.* 2005
5	Return movements (homing)			Koehn and Nicol 1998; Crook 2004c
6	Downstream drift	Eggs, free embryos, larvae	Driftnet capture	Koehn and Nicol 1998; Humphries *et al.* 2002; King 2004b
7	Periodic upstream juvenile migrations in response to flow	Juveniles	Fishway captures	Mallen-Cooper and Stuart 2003
8	Large-scale movements, home range shifts	Adults	Telemetry	Reynolds 1983; Crook 2004b, c; O'Connor *et al.* 2005

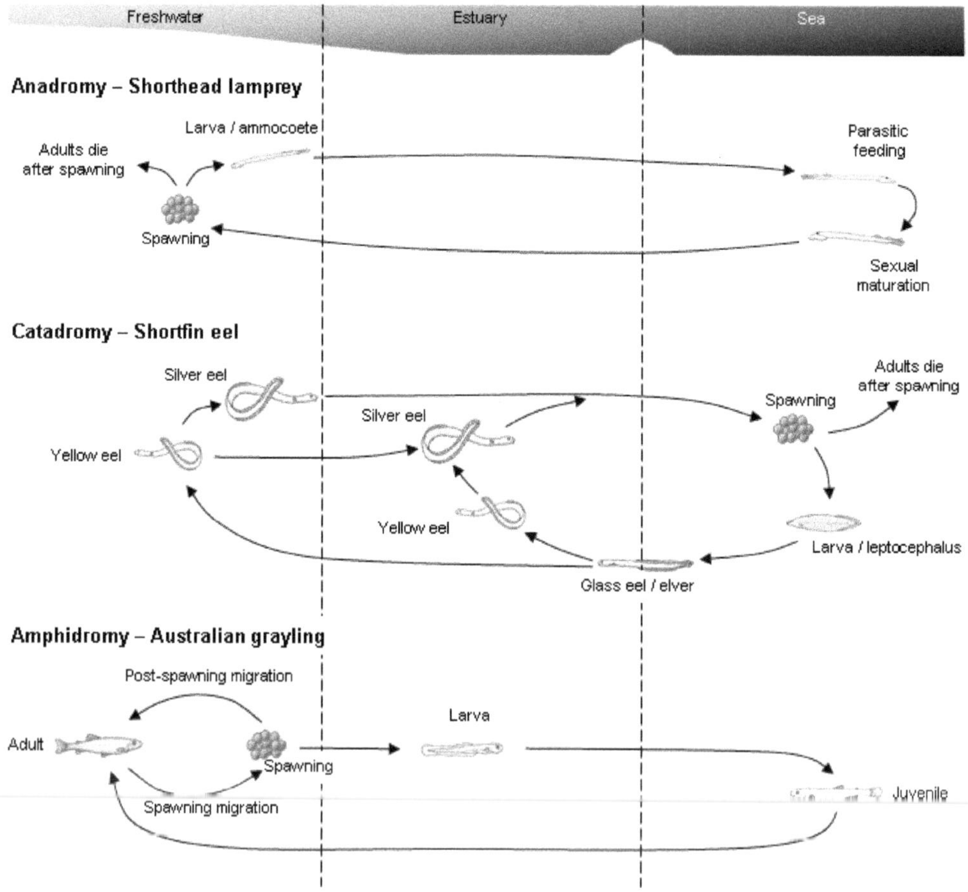

Figure 5.5: Examples of three modes of diadromy: shorthead lamprey (*Mordacia mordax*), southern shortfin eel (*Anguilla australis*) and Australian grayling (*Prototroctes maraena*).

Diadromy

Diadromy refers to migrations of fishes between freshwater and marine habitats. Many natural history documentaries have shown the upstream migrations of anadromous salmon as they leap and battle their way to spawning grounds, avoiding predation by bears, only to die an inglorious death after spending their energy in a frenzied spawning ritual. Diadromous migrations – albeit less spectacular – are also undertaken by many Australian fishes.

In Australia, many freshwater fish species are concentrated in coastal rivers, where they have evolved a variety of life history strategies to take advantage of access to estuaries and the sea. In the coastal rivers of south-eastern Australia, about 70% of fish species exhibit diadromy (Koehn and O'Connor 1990a, b, c). There are three modes, namely 'catadromy', 'anadromy' and 'amphidromy' (Myers 1949; Fig. 5.5). For more detailed information see McDowall (1988, 2007), particularly the seminal 1988 text *Diadromy in Fishes*.

Although diadromous fishes are grouped in categories, there often is flexibility in migratory behaviour within species and across locations. Where access to the sea is blocked, some diadromous species form landlocked populations, especially in lakes that provide suitable conditions for larvae and juveniles. Galaxiids are particularly flexible and commonly form landlocked populations (McDowall 1988). Landlocked common galaxias occur in natural coastal lakes (Pollard 1974; Chapman *et al.* 2006b), climbing galaxias and spotted galaxias (*G. truttaceus*) occur in Tasmanian highland lakes (Humphries 1989, 1990; Fulton 1990) and populations of climbing galaxias are established upstream of impoundments in Tasmania (McDowall 1996b).

Populations of some diadromous species include sympatric diadromous and non-diadromous stocks. This phenomenon is referred to as 'facultative diadromy' (McDowall 1988; Closs *et al.* 2003) or 'partial migration' (Jonsson and Jonsson 1993; Kerr *et al.* 2009). Diadromous and non-diadromous contingents are known in populations of barramundi (Pender and Griffin 1996), Australian smelt (Crook *et al.* 2008) and southern shortfin eels (Arai *et al.* 2004).

Catadromy

Catadromy is a form of diadromy in which mature adults migrate from fresh water to spawn in the sea (Fig 5.5). Catadromous migrations are made famous by anguillid eels, which undertake arduous journeys from the upper reaches of freshwater streams to spawn in the open ocean. Eels may migrate for 2000 km or more to reach their spawning grounds (Jellyman and Tsukamoto 2002) – a remarkable feat of endurance and navigation! Like the anadromous salmonids, the eels have a semelparous life history, dying once they have spawned (Ch. 7). Their *leptocephalus* larvae migrate from the spawning grounds with the assistance of ocean currents, back to rivers and streams, attracted by the scent of fresh water at the entrances to estuaries (Sola 1995; Jellyman and Bowen 2009). By the time the eels are ready to begin their upstream migration they have developed elongate, unpigmented bodies and are known as 'glass eels'. In estuaries, they develop brown pigmentation and are known as 'elvers'. Some elvers migrate upstream where they grow to become 'yellow eels'. Otolith chemistry has shown that many eels never actually enter fresh waters but remain in estuaries and/or coastal seawater throughout their lives (Arai *et al.* 2004). Thus, at least some anguillid eels exhibit facultative catadromy.

After settlement, yellow eels lead a sedentary existence for five to 20 years before metamorphosing into 'silver eels' (De Silva *et al.* 2002; Jellyman and Sykes 2003). They cease to feed and undergo physiological changes, including enlargement of the eyes and development of silver colouration on the flanks and dorsal surface. Silver eels return to the sea, where they congregate and spawn *en masse* in particular areas.

Four species of anguillid eels occur in Australian waters, including the southern shortfin eel, longfin eel, Pacific shortfin eel (*Anguilla obscura*) and the Indonesian shortfin eel (*A. bicolor*). All appear to be catadromous. The southern shortfin, longfin and Pacific shortfin eels probably spawn in the Coral Sea (south-western Pacific Ocean) and the Indonesian shortfin eel spawns in the northern Indian Ocean (Allen *et al.* 2002; Jellyman and Tsukamoto 2002).

Other Australian fishes have catadromous life histories. The common galaxias matures in fresh water but spawns in the upper tidal reaches of estuaries (Burnet 1965; Fig. 5.5). The eggs are laid during spring tides among fringing vegetation and the clutches develop above the water line once the tide recedes (Benzie 1968; Barbee and Swearer 2007). This species is often described as 'marginally catadromous' because spawning occurs at the freshwater–estuarine interface rather than at sea (McDowall 1988). Hatching occurs when the eggs are inundated by subsequent spring tides and the larvae spend four to six months in the sea before returning to fresh water as juveniles (McDowall 1994; Barbee and Swearer 2007). Juvenile common galaxias are apparently attracted upstream by the odour of conspecifics and, to a lesser extent, by the natural odour of stream water (Hale *et al.* 2009). The dispersal of larvae and juveniles during the marine phase facilitates gene flow, so that local populations tend to be genetically similar (Waters *et al.* 2000a).

Congolli (*Pseudaphritis urvillii*) exhibit an unusual form of catadromy in that the sexes are segregated for most of the life cycle, with males living in estuarine and marine habitats and females in the lower

Figure 5.6: Female congolli (*Pseudaphritis urvillii*, 200–350 mm), a catadromous species, from a downstream migration in winter. This photo was taken after a boat lock was opened to allow fish to enter the Murray River estuary at Goolwa, South Australia (photo: B Zampatti).

Figure 5.7: Photos of 114-day-old Australian bass (*Percalates novemaculeata*) held for the preceding 72 days in (a) sea water (38 g L^{-1}) and (b) fresh water (0.5 g L^{-1}), showing differences in pigmentation (Macdonald and Crook 2010). Juveniles develop in estuaries and are tolerant of salinity (photo: T Raadik).

freshwater reaches of rivers (Hortle 1979; Hortle and White 1980; Crook *et al.* 2010a, b). The females appear to live in fresh water for three to five years (B.P. Zampatti, South Australian Research and Development Institute, pers. comm.) before migrating rapidly downstream in winter, spawning presumably at sea (Crook *et al.* 2010a, b; Figs 5.6, 5.8).

The life history of Australian bass may be superficially like that of congolli, with females inhabiting the middle to upper freshwater reaches of coastal streams and males in the lower freshwater and upper estuarine reaches (Harris 1986). The females move downstream during high flows in late winter, spawning in estuaries. Unlike congolli, spawning and juvenile development occur in estuaries rather than at sea (Figs 5.7, 5.8). Thus, Australian bass should be considered marginally catadromous (McDowall 1988). There is considerable morphometric variation and genetic structuring in populations of Australian bass, suggesting that movement between river systems via the sea is limited (Jerry 1997; Jerry and Baverstock 1998; Jerry and Cairns 1998).

Barramundi show a unique version of catadromy (Fig. 5.8). The species is sequentially hermaphroditic; most individuals mature as males at around three years of age and develop into females after spawning (although some fishes, known as 'primary females', develop directly into females: Davis 1982). Spawning occurs in sea water and the larvae inhabit coastal swamps. The life history model for this species suggests that after about a year the juveniles migrate upstream to fresh water, where they spend two to three years before returning downstream to spawn, first as males and later as females (McDowall 1988; Allen *et al.* 2002). Recent research, however, has shown that – at least in some rivers – many juvenile barramundi remain in the sea, with no freshwater phase at all (Pender and Griffin 1996). Further, many of the barramundi that enter fresh water become sexually mature but fail to migrate downstream to participate in spawning (Milton and Chenery 2005). It appears that the behaviour of barramundi is a complex form of facultative catadromy. Clearly, the species' life history and migratory behaviours are highly adaptable (Robins *et al.* 2005).

Several other species in coastal Australian freshwaters are likely to be catadromous, but confirmation is needed. In eastern Australia, freshwater herring (*Potamalosa richmondia*) and bullrout (*Notesthes robusta*) may undertake spawning migrations from fresh water to estuaries or the sea (Pidgeon 1989; Briggs and McDowall 1996; Pollard and Parker 1996; Pusey *et al.* 2004), but the evidence is weak. Otolith chemistry has shown that both species move between waters of varying salinities at different life stages, but there is no clear evidence to distinguish true catadromous migrations and irregular, euryhaline movements (Miles *et al.* 2009). Thus, while the migrations of some catadromous species are reasonably well-known, research is lacking for many others.

Anadromy

'Anadromy' is a form of diadromy in which mature adults migrate upstream from the sea to spawn in

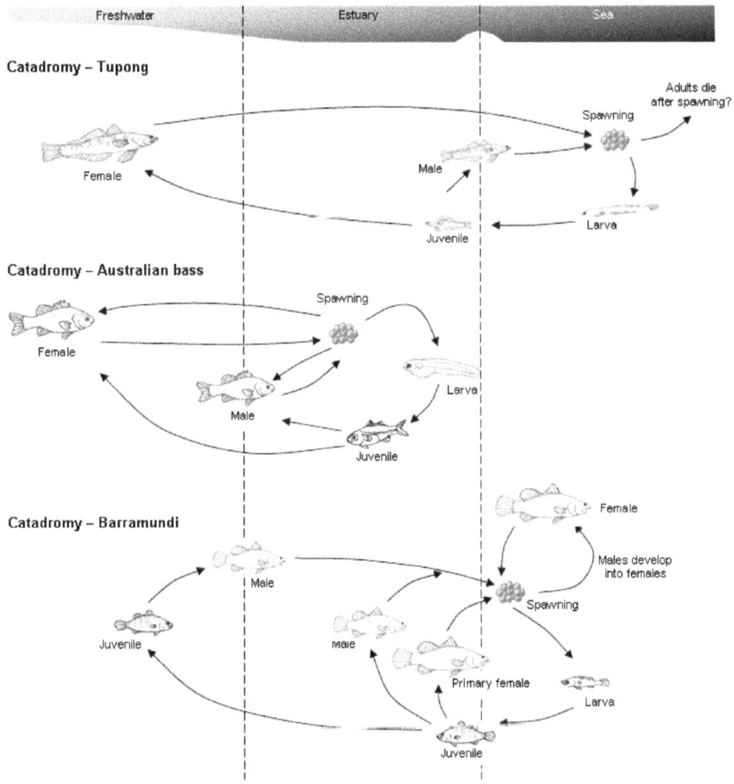

Figure 5.8: Variation in catadromous life histories: congolli (tupong) (*Pseudaphritis urvillii*), Australian bass (*Percalates novemaculeata*) and barramundi (*Lates calcarifer*).

fresh water. After hatching, the young develop for a few months to several years before moving downstream to the sea. The juveniles spend several years in the sea before sexual maturation, when they begin their return migration. Many anadromous fishes (particularly salmonids) exhibit natal homing, using olfactory cues to return to the location where they were spawned (Hasler and Scholz 1983). Many Northern Hemisphere species are anadromous, including salmonids (Salmonidae), sturgeons (Acipenseridae), lampreys (Petromyzontidae) and smelts (Osmeridae). In Australia, anadromy is comparatively rare – only the lampreys exhibit classical anadromous life histories.

Three species of lamprey are known from Australian fresh waters. The non-parasitic lamprey (*Mordacia praecox*) occurs in a few rivers in south-eastern Australia and completes its entire life cycle in fresh water (Hughes and Potter 1969; Potter 1996). The shorthead and pouch lampreys are anadromous. They spawn in fresh water; the juveniles (ammocoetes) remain there for about three years, as detritivores, before metamorphosing to adults (gaining eyes and teeth) and migrating to the sea, where they spend several years as parasites on other fishes before returning to fresh water to spawn (Potter 1996; Fig. 5.5).

The Tasmanian whitebait has also been tentatively classified as anadromous (McDowall 1988), with mature adults migrating from the sea to spawn in upper estuaries (Blackburn 1950; Fulton 1990). This species rarely penetrates far inland, however, and probably should be considered as marginally anadromous or as an estuary-dependent marine species.

Amphidromy

'Amphidromy' is a form of diadromy in which migrations between fresh water and the sea are not for breeding purposes but occur regularly during other phases of the life cycle (Myers 1949; McDowall 1988; Fig. 5.5). As spawning is in fresh water, amphidromy superficially resembles anadromy and the differences are often unrecognised (McDowall 2007). The salient distinction is that amphidromous fishes migrate as juveniles from the sea to fresh water, whereas anadromous fishes develop to maturity in the sea.

The geographic distribution of amphidromous fishes centres on oceanic islands of the tropics and subtropics, southward towards temperate regions in Australia and New Zealand (McDowall 2007). McDowall (1988) described more than 60 amphidromous species including galaxiids (Galaxiidae), sculpins (Cottidae), southern graylings (Retropinnidae), gudgeons and bullies (Eleotridae), gobies (Gobiidae) and ayu or sweetfish (Plecoglossidae). In contrast to catadromous and anadromous fishes, most amphidromous species are small-bodied, typically with sizes up to about 300 mm TL.

Amphidromy is common among the Galaxiidae. In Australia, there are three amphidromous galaxiids: the climbing galaxias, spotted galaxias and Tasmanian mudfish (*Neochanna cleaveri*). Spawning of climbing galaxias and spotted galaxias occurs in freshwater habitats (Humphries 1989; O'Connor and Koehn 1998) and the larvae are swept to sea soon after hatching. They develop into juveniles in the sea and after five to six months return to fresh water as unpigmented juveniles (whitebait) (McDowall 1996b). The life history of the Tasmanian mudfish is little known but is likely to be broadly similar (Fulton 1990).

Australian grayling (*Prototroctes maraena*) is an amphidromous species in coastal streams of south-eastern Australia (Bell *et al.* 1980; Berra 1982; Bacher and O'Brien 1989). Mature fish undertake rapid downstream migrations in autumn, laying their eggs in the lowermost freshwater reaches (Koster 2011; Fig. 5.5). The larvae grow into juveniles at sea, returning after four to six months in spring and early summer, often among galaxiid whitebait runs (Bishop and Bell 1978; Fulton and Pavuk 1988; Crook *et al.* 2006). Unlike many of the diadromous galaxiids, the Australian grayling appears to be strictly amphidromous, with all individuals spending their larval and early juvenile phases at sea. No landlocked populations are known (Bell *et al.* 1980; McDowall 1996b), and the apparent lack of genetic structure among populations indicates wide marine dispersal and genetic mixing.

While the migrations of Australian grayling and most amphidromous galaxiids are reasonably well-documented, other less-studied fishes may exhibit amphidromy. Otolith chemistry suggests that Australian smelt in coastal catchments exhibit facultative amphidromy (Crook *et al.* 2008). Freshwater mullet (*Trachystoma petardi*) may also be amphidromous, based on chemical analysis and reports of newly spent fishes in fresh water (Miles *et al.* 2009). Other amphidromous species could include Cox's gudgeon (*Gobiomorphus coxii*), striped gudgeon (*G. australis*), roman-nose goby (*Awaous acritosus*) and several species of *Glossogobius* (Pusey *et al.* 2004; Miles *et al.* 2009).

STUDYING MOVEMENTS

There has been a surge in research on the movements of Australian freshwater fishes in recent decades, with the emergence of new technologies for monitoring (Fig. 5.9). Prior to 1975 there were few studies; most information came from anecdotal and sometimes contradictory observations (Cadwallader 1977). Until recently there was little information even for widespread, familiar species such as Murray cod and golden perch. With new techniques available, there are exciting prospects for enterprising researchers.

Direct observation

At the most basic level, some information on small-scale fish movements can be gained from direct observations via underwater diving or cameras. In many Australian waterways direct observations are precluded by high turbidity but, where it is possible (Hardie *et al.* 2006), it can yield detailed information about small-scale movements and associated behaviour (Butler and Rowland 2009) and migrations past fixed points (Bishop *et al.* 1995). Potential disadvantages are that behaviour may be modified if fishes are disturbed, repeated observations of individuals may not be possible and observations can be made only over small areas and short periods (Dolloff *et al.* 1996).

Nets, traps and fishway monitoring

Nets and traps can provide information on the direction and timing of movements and migrations. Most such studies have set nets or traps in fishways or weirs to monitor the passage of diadromous (Beumer and Harrington 1980a; Sloane 1984a, b, c; Stuart and Berghuis 2002; Morgan and Beatty 2006a) and potamo-

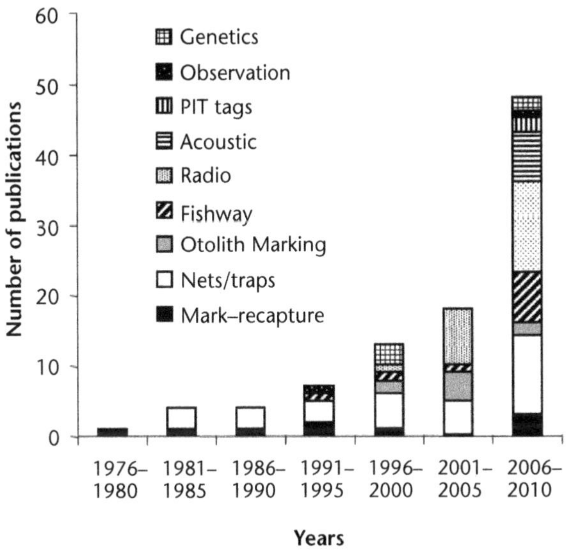

Figure 5.9: Numbers of 'movement' publications on Australian freshwater fishes 1976–2010, according to study method. Sources: Internet-based search engines (e.g. Scopus, Google Scholar) and references in published papers, book chapters and unpublished reports.

dromous fishes (Mallen-Cooper *et al.* 1996; Mallen-Cooper and Stuart 2003; Mallen-Cooper and Brand 2007). Indeed, monitoring at fishways, using PIT tag technology (see Box 5.2 and 'PIT' tag detection arrays'), has provided new ways to gather a wide array of data (Barrett 2008). Tag data indicate the effectiveness of the passage, diel movements, size variations and the scale and timing of seasonal migrations, including migrations by small fishes and alien species (Mallen-Cooper and Brand 2007; Baumgartner *et al.* 2008; Stuart *et al.* 2008a, b). Directional traps have been used to document movements between channel and floodplain habitats (Lyon *et al.* 2010). While nets and traps indicate the direction and timing of movement, they usually record only the presence of fishes at a single location, not the distance moved nor the proportion of the population that is moving.

Mark–recapture

'Mark–recapture' was a first choice in early studies of movements and is still widely employed (Fig. 5.9). As the name suggests, it involves collecting and marking or tagging fishes then setting them free. Information on movements (and other variables, such as growth) can be obtained if the fish is subsequently recaptured.

Box 5.2: Monitoring fishways

Monitoring is an important source of information about fishes passing through fishways. Although these data indicate movements at only one point in time, they may provide new information, especially for tropical coastal species (Pusey *et al.* 2004). With each new fishway there are questions such as 'How effective will it be in allowing fishes to pass the barrier?' 'Which species?' 'How many?' 'What size ranges?' 'What proportion of the populations?' 'Will it allow fishes safe passage over the range of flows, seasons and environmental conditions?'

There is a need for appropriate design, testing and refinement of fishway efficiency at each site. This is an issue especially for attraction flows to the fishway entrance – if fishes cannot find the entrance, they cannot pass through it.

Large numbers of fishes may be involved. An assessment of a new vertical-slot fishway on the Burnett River, Queensland, showed 52 000 fishes (34 species) passing, at up to 4500 individuals per day. This was far more than the 2000 fishes that daily passed a previous weir-and-pool type fishway at the same site (Kowarsky and Ross 1981; Stuart and Berghuis 2002). In the tropical Fitzroy River, Queensland, 17 species (13 potamodromous, two catadromous, two amphidromous) of fishes from 35 mm to 710 mm long passed over the weir (maximum of 3317 fishes per day), in a fish lock, during flows up to 80 000 ML d^{-1}. This was despite weir gates not being integrated with the entrance, and a narrow operational range unsuited to variable flows (Stuart *et al.* 2007). In a fishway on the Murray River, even when operating in less variable flows, up to 4415 fishes (13 species) passed per day (Stuart *et al.* 2008a). Comparisons of fishes trapped in the entrances and exits show that some species (or body sizes) are better at negotiating fishways than others, and various modifications have been made to resolve these problems (Baumgartner *et al.* 2008). One challenge is to determine the proportion of the population that needs to pass a barrier for long-term population benefits. PIT tags provide valuable data in this regard. For many diadromous species, it may be desirable for all the migrating population to pass and complete part of their life cycle. It may be less important for potamodromous species, or those where only some populations are migratory.

Mark–recapture studies often rely on recreational or commercial fishers to report recaptures, although targeted surveys may also be conducted.

There are various ways to mark or tag fishes (Hancock 1989; Guy *et al.* 1996; Neilson *et al.* 1999; Hancock *et al.* 2000). These fall into two main categories, namely *batch marks* and *individual tags*. Batch marks are quicker and less expensive to apply to large numbers of fishes. They include fin clips (Baumgartner 2006), dyes (van der Walt and Faragher 2003), chemical markers (Munro *et al.* 2008), thermally induced marks (Willet 1996), injectable elastomers (Gallagher and Hutchison 2004), tattoos (Lucas and Batley 1996), freeze-branding (Kumar and Chapman 1998) and some types of coded wire tag (Ingram 1993). Individual tags have a tag-specific identifier. They include external dart and T-bar types (Stuart and Jones 2006b), visual implant tags (Crook and White 1995), PIT tags (Barrett and Mallen-Cooper 2006) and individually coded wire tags (Crook and White 1995). Photographic identification of individual fishes using pigment patterns is also possible, especially for small numbers (Bansemer and Bennett 2008).

Knowledge of movements from mark–recapture methods typically relies on discontinuous observations of recaptured fishes over time, and detailed movements usually cannot be mapped (Priede 1980). Tag losses, low recapture rates and the potential effects of tags on fish behaviour are problems for data collection and interpretation. Further, many tag–recapture studies are biased against detecting large-scale movements because sampling intensity (hence the likelihood of encountering tagged fishes) decreases with distance from the original capture site (Gowan *et al.* 1994; Rodríguez 2002).

PIT tag detection arrays

'Passive integrated transponder' (PIT) tags are implanted and triggered by energised antennae that cause the tag to emit a coded radio signal. As PIT tags do not rely on an internal battery they have an indefinite life expectancy and, when combined with tag detection arrays, allow data to be collected automatically without the need to recapture tagged fish (Prentice *et al.* 1990). A system of automated PIT tag detectors and data loggers has been installed at fishways along the Murray River as part of the Murray-Darling Basin Authority's Sea to Lake Hume fishway program (Ch. 12; Barrett and Mallen-Cooper 2006). In 2008, more than 20 000 fishes were PIT tagged as part of that program (Barrett 2008). PIT tags have a limited detection range (< 500 mm), so are suitable in situations where fishes must pass in close proximity to detectors, as in fishways (Castro-Santos *et al.* 1996) and small streams (Lucas and Baras 2000).

Telemetry

The emergence of telemetry in research on Australian fishes in the early 1990s has provided an invaluable tool for use on many species (Koehn 1996). Telemetry allows individual fishes to be tracked in the wild, with high spatial and temporal resolution. Radio and acoustic transmitters may be surgically implanted or attached externally (Crook 2004a, b, c; O'Connor *et al.* 2009). They are battery-powered and emit coded signals that are detected by electronic receivers. In addition to signalling location, sensors can be installed to provide *in situ* data about behaviour and physiology (e.g. temperature, depth, heart rate, activity: Winter 1996; Eiler *et al.* 2000; Cooke *et al.* 2004; Nielson *et al.* 2009). Radio-tagged fishes can be tracked by foot, boat or aircraft, or recorded by automated logging stations. It is possible to determine the locations of radio-tagged fishes to within 5 m by triangulation using a directional antenna (Crook *et al.* 2001), or to within 1 m using aerial-reduction techniques (Koehn *et al.* 2012). The reception range of radio transmitters varies with environmental conditions, transmitter and receiver type, but ranges of 1 km by foot and boat and several kilometres by air are achievable (Koehn 2009a, b). The signal range decreases with increasing conductivity (salinity), particularly at conductivities above about 600 $\mu S\ cm^{-1}$ (Winter 1983; Koehn 2006).

Acoustic transmitters (pingers) emit low-frequency sound (pings) that are individually coded in patterns analogous to Morse code. Unlike radio transmitters, these acoustic signals travel effectively only through water and an underwater hydrophone or logging station is needed for detection. It is more difficult to accurately determine the fine-scale locations of fish using acoustic telemetry, although listening station arrays can be designed to provide an accurate three-dimensional location. Most acoustic telemetry studies use an array of fixed logging stations to continuously monitor movements between stations. Acoustic signals are unaffected by salinity (conductivity) and are especially useful for species that traverse freshwater, estuarine and marine boundaries. These arrays are also becoming popular in studies of wholly freshwater fishes because of their ability to monitor movements continuously at relatively low cost.

Otolith chemistry

In recent years, measurements of the chemical concentrations of trace elements in otoliths (ear bones) have become widely used for reconstructing the movements of fishes (Gillanders 2005; Ch. 8). Otoliths grow continuously throughout life and are not remetabolised once deposited (Campana 1999). As otolith material accretes, trace elements are incorporated, more or less in proportion to their concentrations in the environment (Macdonald and Crook 2010). Fine-scale measurements of trace elements can be used to reconstruct the ambient water chemistry experienced by a fish through its life. For example, variations in the ratios of strontium:calcium, barium:calcium and ^{87}strontium:^{86}strontium between the cores and edges of otoliths have been used to infer movement between marine and freshwater environments (McCulloch *et al.* 2005). Otolith chemistry has great potential but it is an indirect method, reliant on the existence of predictable relationships between otolith and water chemistry. Accurate interpretation of otolith chemistry data requires validation of these relationships, but in many studies, such information is lacking (Gillanders 2005).

Population genetics

Genetic data provide information about historical connectivity between geographically distinct populations of a species, enabling inferences about how fishes move across the landscape (Hughes 2007; Hughes *et al.* 2009). These inferences are based on the fact that isolated populations become genetically distinct over time due to mutation and genetic drift. Genetic differences can be used to estimate connectivity between populations (Slatkin 1985), and are often measured using analyses of allozymes, mitochondrial DNA and microsatellite DNA (Ch. 3).

Analysing movement data

Analyses of complex data often require non-standard statistical approaches. This is a rapidly developing field and is likely to benefit from parallel developments in terrestrial ecology and telemetry technology. Computer animations can visually illustrate the more complex movement patterns over time in relation to environmental conditions. The development of state-based movement models and their application to landscapes, linking movements to resource selection, will prove to be an important advance (Patterson *et al.* 2007). There is a need to maintain collaboration between data collectors, analysts and modellers, using a shared language to ensure appropriate analyses and robust interpretations.

The time frame over which analyses are undertaken is a critical part of the interpretation of results as the probabilities of detecting some patterns, and overall movement distances, increase over time (Gresswell and Hendricks 2007; Roberts and Angermeier 2007). Movement patterns should be examined over a full complement of seasons, using frequent sampling, to ensure that false interpretations are avoided. For example, the movements of a Murray cod over a 12 month period show travel in upstream (A to D) and downstream (D to E) directions (Fig. 5.10). A simple tag–recapture or short-term study could have provided different interpretations depending on when the fish was recaptured. For example, if the fish tagged in August at A was captured at B, C or D it would be presumed to have moved 35 km or 60 km, whereas a capture at E would have suggested that the fish did not move at all!

APPLYING KNOWLEDGE

Patterns and models

An understanding of the movements of stream fishes is needed to manage populations at an appropriate 'riverscape' scale (Fausch *et al.* 2002; Wiens 2002b).

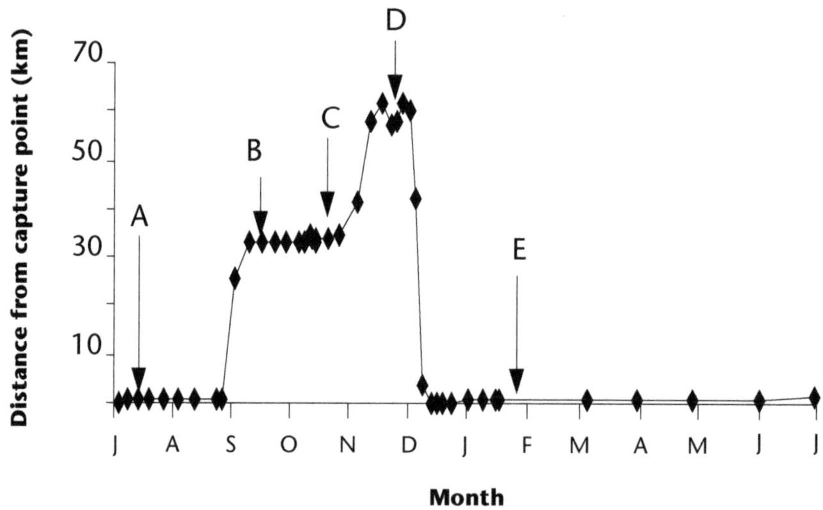

Figure 5.10: Pattern of movement locations over a 12-month period for a radio-tagged Murray cod (*Maccullochella peelii*) (modified after Koehn 2006).

This will provide scientific support for conservation actions such as provision of adequate fish passage (Mallen-Cooper 1999; Barrett and Mallen-Cooper 2006), and is a key element in river restoration (Palmer *et al.* 2005). Understanding movement patterns and rates is vital for pest management, to determine pathways for colonisation and invasion (Koehn *et al.* 2000; Koehn and MacKenzie 2004). Models are usually constrained by knowledge of species, and new findings have often overturned previous concepts (O'Connor *et al.* 2005; Koehn 2009a; Koehn *et al.* 2009). Models relating to movements are described earlier in this chapter (Figs 5.3–5.5), and these can also incorporate temporal components (movement calendars: Koehn and O'Connor 1990a) for each species.

While models are important for ecological understanding and guiding management (e.g. ensuring fish passage) they do have limitations (different techniques, times, sites, biases) and generalisations may obscure variations that may occur in the movements of individual fishes (White and Garrott 1990). In some circumstances it may be the minority 'non-conformist' fishes that are most important to the viability of a population (e.g. maintaining the links in metapopulation structure). The timing, spatial extent and variability of movements (e.g. between individuals, sexes, sizes, seasons, life stages) need to be incorporated into more complex models.

Barriers to movement

River regulation and associated infrastructure (dams, weirs, levees and other structures) impede fish movements and threaten species in Australia (Cadwallader 1978; Harris 1983; Koehn and O'Connor 1990a, c; Ch. 12) and elsewhere in the world (Cowx and Welcomme 1998). There are more than 4000 instream barriers in the Murray-Darling Basin (Murray-Darling Basin Commission 2004) and perhaps 10 000 structures when small regulators and road crossings are included. These barriers disrupt all forms of connectivity between populations, including longitudinal migrations and lateral connections between river and floodplain habitats.

Obstructed movements can lead to declines and localised extinctions of migratory species above and below barriers (Bishop and Bell 1978; Brumley 1987; Gehrke *et al.* 2002). For example, 10 diadromous species, including the threatened Australian grayling, are now extinct above Tallowa Dam on the Shoalhaven River, New South Wales (Gehrke *et al.* 2002). Barriers restrict access to large areas of habitat, especially in upland regions. More than half the areas of coastal stream habitats in south-eastern Australia are closed to migratory species (Harris 1983). The proposed (later abandoned) Traveston Crossing Dam on the Mary River, Queensland, was likely to fragment habitats of the threatened Australian lungfish; this was one of the factors that prevented dam construction (Brooks and Kind 2002; Arthington 2009).

The most common main-channel barriers to movements are low-level weirs, which have most impact during low-flow periods (because they are submerged by high flows). Floodplain regulating structures prevent fishes moving to and from floodplain wetlands, and many fishes can become trapped when floodplain waters recede (Jones and Stuart 2008). There is significant mortality of larvae passing over weirs (Baumgartner *et al.* 2006a). Weirs and other barriers also prevent recolonisation after catastrophic events such as drought, but ironically they may slow the colonisation of alien species such as common carp (Koehn *et al.* 2000). Artificial barriers have been used to protect threatened barred galaxias (*Galaxias fuscus*) and Swan galaxias (*G. fontanus*) from predatory brown trout (Crook and Sanger 1998; Lintermans 2004).

Barriers are likely to have intensified impacts when associated with other disturbances such as modified flows, predation on fish aggregations (including take by anglers), hydropower stations (Stuart *et al.* 2010) and poor water-quality conditions (Koehn 2005). Non-physical barriers are less noticeable, of course, than physical structures. Temperature affects the swimming abilities of fishes, and cold-water releases from dams may limit movements by some species. Similarly, culverts and pipes, lack of contiguous habitats and even channelised streams may restrict movements, producing water velocities or turbulence beyond the swimming capabilities of some species. Behavioural changes have been observed that restrict golden perch migrating downstream over low-level weirs (O'Connor *et al.* 2006), but little is known of other factors that may restrict movement behaviour such as changes in light (e.g. darkness in culverts).

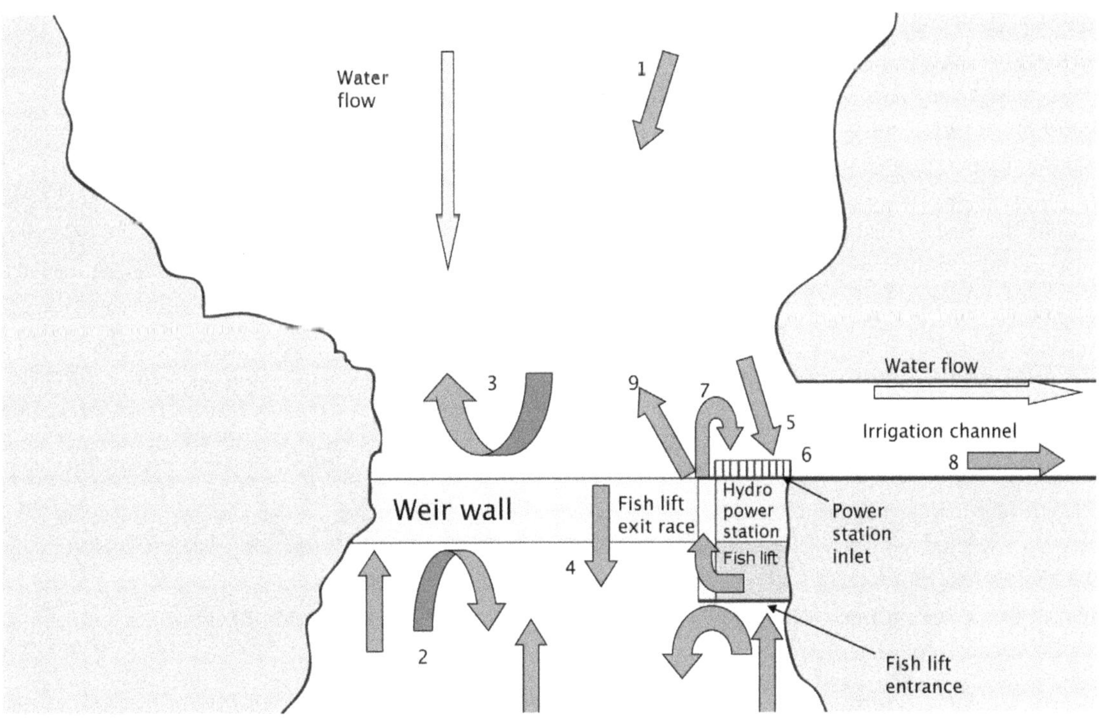

Figure 5.11: Impediments to fish movements at Yarrawonga Weir (Lake Mulwala, Murray River, south-eastern Australia). (1) Larvae may remain in the lake and not pass downstream. (2) The weir wall may stop adults moving upstream. (3) The weir wall may be a behavioural barrier to adults moving downstream. (4) Fishes (including larvae) may suffer mortality or physical damage if they pass through the undershot weir gates. (5) Fishes (including larvae) may suffer physical damage or mortality if they pass through the hydropower station turbines. (6) Fishes moving downstream may be caught on the trash racks of the hydropower station. (7) Fishes moving up in the fish lift may be swept downstream through the hydropower station or onto the trash racks. (8) Larvae (and some juveniles and adults) may enter irrigation channels and other suboptimal habitats. (9) Fishes move from a lotic to a lentic habitat.

Any barrier may have a multitude of impacts on fish populations. An example is Yarrawonga Weir on the Murray River (Fig. 5.11).

Passage

Providing passage past dams and other barriers is an essential step towards rehabilitation of native fish populations (Ch. 12; Murray-Darling Basin Commission 2004; Barrett 2008). Most concern has been directed at provision of upstream passage (Mallen-Cooper 1999; Stuart and Berghuis 2002) and modified designs may be needed to accommodate movements by large fishes such as Australian lungfish, freshwater sawfish and Murray cod. Downstream passage is also important, however, and solutions are potentially more difficult as fishes moving with the flow have less time to assess cues and bypass options (Williams *et al.* 2011). Safe passage is needed to prevent damage to adults and juveniles (O'Connor *et al.* 2006; Baumgartner *et al.* 2006b, 2009) and losses to irrigation channels (King and O'Connor 2007) and hydropower stations (Stuart *et al.* 2010). Downstream fish-passage technology (screens, bypasses, fish barriers) is well-advanced in Europe and the USA (Lucas and Baras 2001) but work in Australia is still preliminary (Blackley 2003). Automated fish-counting technology can also be useful in monitoring and evaluating fishway performance (Baumgartner *et al.* 2006a).

The Murray-Darling Basin Authority's 'Sea to Lake Hume' fish passage program is one of the largest programs to remediate fish passage anywhere in the world, restoring longitudinal connectivity to over 2000 km of the Murray River (Barrett and Mallen-Cooper 2006; Ch. 12). It will help to rehabilitate native fish populations and has been instrumental in increasing our knowledge of fish movements (Barrett 2008). A hundred or more other fishways have been constructed elsewhere in Australia (e.g. Burnett and Fitzroy rivers, Queensland; Nepean River system and the 'Fish Superhighways' program in the Macquarie River and Macquarie Marshes, New South Wales). Vertical-slot fishways are more effective than pool-and-weir designs (Stuart and Mallen-Cooper 1999) for large and small fishes (Morgan and Beatty 2006d), although options such as rock ramps and modified lock operations have also been trialled (Stuart *et al.* 2007, 2008a, b).

The remediation of fish passage in these ways requires clear goals, accounting for issues such as priority species, the range of sizes and swimming abilities to be accommodated and interbarrier distances to maximise benefits (Mallen-Cooper 2000; Baumgartner *et al.* 2010b). The concept of 'passage for the whole river fish community' is becoming popular (Stuart *et al.* 2008a, b) and the aim of providing passage throughout whole catchments is another laudable goal.

Knowledge gaps

The movements of many species – even common species – are little understood and there are many opportunities for research. In particular, we need to know more of small species and the fishes of tropical regions. Mere descriptions or quantification of movements are helpful but they may not elucidate the ecological purposes that underlie movements. Cues for movement are often obscure, especially flow cues for floodplain use and movements to and from the sea. Life cycles and associated movements need to be clarified. There is little quantification of swimming abilities, velocity tolerances and preferences, energy costs and swimming behaviours of different life stages, although these govern colonisation and dispersal. Larval behaviour, including vertical migrations and flow relationships, is little known. Knowledge of recolonisation rates (including from stocking) is important for management of threatened species. Understanding the movement capabilities and pathways of invasive fishes is another vital issue. Finally, there is scant information about the role of movements in ecological processes such as the contributions of migratory fishes to food webs and the mass transport of organic carbon (Jardine *et al.* 2011).

CONCLUSION

Movements are an integral part of the ecology of fishes, although patterns vary among species, individuals, life stages, sizes and sexes, and between seasons and locations. In some cases, the adaptive value of these movements is unknown save that they enable fishes to access and utilise resources needed for survival. We need to know more of the ecological and evolutionary roles of movements – their causes, mechanisms, complexities, behaviours and patterns – before our knowledge can be integrated as a framework to assist management and conservation (Nathan 2008). The burgeoning literature on the movements of Australian freshwater fishes is a rich source of the information needed for such a framework. We trust that this chapter has given a new appreciation of fish movements and, above all, shown the need for more knowledge to sustain wise management.

6 Trophic ecology

Rick J. Stoffels

INTRODUCTION

English mathematician and logician Augustus De Morgan (1806–71) had sequences in mind when he penned this verse, fashioned after one by Irish satirist Jonathan Swift (1667–1745):

> *Great fleas have little fleas upon their backs to bite 'em,*
> *And little fleas have lesser fleas, and so ad infinitum.*
> *And the great fleas themselves, in turn, have greater fleas to go on;*
> *While these again have greater still, and greater still, and so on.*

Freshwater fishes might appear to have little in common with fleas but, like us, they are part of a web of organisms that could not survive, let alone grow and reproduce, without consuming food from another part of the web. How much food do fishes need to survive? How do they find and capture food? What effects do fishes have on the food web? The aim of this chapter is to venture answers, or part-answers, to questions like these.

Trophic ecology is about food and feeding. It is a big topic, because freshwater fishes have diverse diets and modes of feeding, and it is integrative, drawing on multiple disciplines and levels of biological and ecological organisation. A comprehensive discussion could command its own book; the chapter has, of course, inclined to matters of most import for freshwater fishes in Australia. The aim is to outline a conceptual framework that is both an overview and a template for ideas and information that readers may wish to add. The chapter begins by considering how fishes use their senses to detect prey or other food items, then examines their modes of feeding and how these interact with bioenergetics and nutrition. Then comes a more strictly ecological perspective, considering trophic guilds and, finally, the effects that fishes have on food webs. Again, each topic is potentially a specialisation and some biologists spend their careers within those confines. Ecologists, above all, must keep in mind 'the big picture'.

SEEKING PREY

Aquatic vertebrates seek food in a world with very different physical properties from the one populated by terrestrial vertebrates. As water is denser than air, fishes literally can feel the medium in which they move. They may need to rely more on touch and taste than sight if the underwater light environment is poor, or more on touch and sight than taste if water quality is poor. The differences in and out of water are so profound that a biologist trying to understand the perceptions of fishes may do best to ignore the human sensory experiences and instead try to 'think like a fish'. It is a world where someone seated in their own room might 'feel' a friend walk past in the hallway, where they can 'taste' what friends are eating in the tearoom down the corridor. It is a world where a sense of sight may be a luxury.

The primary senses used by freshwater fishes to locate food are vision, mechanoreception and chemoreception. An appreciation of these senses helps us to understand how the trophic niches and foraging behaviour of fishes have evolved, within their particular phylogenetic and environmental constraints. Their behavioural decisions use information transmitted to the brain by the sensory systems. Thinking about this sensory information and how it is processed and applied leads to interesting questions and hypotheses about fishes in their environment. For example, will a significant increase in turbidity reduce the predatory efficiency of a fish? Do fishes that rely on vision for prey capture have different feeding modes from those that depend on a mechanosensory lateral line (see 'Mechanoreception')?

In the following section the focus is on predatory fishes – those that consume other living organisms. Of course, not all fishes are predators. Detritivores (see 'Trophic guilds'), for example, are not necessarily predators but they do have similar senses, such as chemoreception, that lead them to food. The sensory ecology of predatory fishes leads to some important insights.

Vision

Two processes govern the passage of light through water. Light photons are absorbed, so that energy is transferred to matter, and photons are scattered, deflected from their paths without loss of energy. Together, scattering and absorption control the quantity of light available for images and the quality of images that an eye can form (Lythgoe 1979). They determine three optical properties, namely intensity, colour and contrast, that determine the ability of fishes to visually detect their food.

Light intensity is affected by both absorption and scatter. Absorption is illustrated by the way that chlorophyll absorbs energy from certain wavelengths of light as part of the mechanism of plant photosynthesis. Scattering occurs as particles suspended in the water deflect photons through the water column as upwelling light.

The colour of light is related to its wavelength. Different wavelengths are affected differently by absorption and scattering, so that deeper into the water the light becomes increasingly monochromatic. Short-wavelength light (< 500 nm) is blue, intermediate wavelengths (500–600 nm) are green and long-wavelengths (> 600 nm) are reddish. The colour of water in a river or lake, on the other hand, has four components (Lythgoe 1979):

1. Rayleigh scattering (the differential scattering of short wavelengths by microscopic particles with diameters smaller than the wavelengths). This form of scattering deflects short-wavelength light, creating the blue colour of clear water (and the sky);
2. selective absorption of short- and long wavelengths by water molecules. Pure water is most transparent to wavelengths around 475 nm;
3. chlorophyll, which is most transparent to wavelengths of 560–750 nm (it preferentially absorbs long and short wavelengths, and so appears green);
4. 'yellow substance', essentially dissolved organic matter, which is transparent to long wavelengths and absorbs short wavelengths; thus it appears as yellow, brown and red.

Suspended organic and inorganic particles scatter light of all wavelengths, creating a veil of bright light between the image and the eye and reducing the contrast. This broad-spectrum scattering is measured as turbidity, in nephelometric turbidity units (NTU). Contrast is defined as $(T - B)/B$, where T is the spectral radiance of a target and B is the spectral radiance of the background. Contrast increases, therefore, as the target's radiance increases over that of the background and it decreases as the background's radiance increases over that of the target. The veiling radiance, V, caused by light scattered between the target and the eye, produces a reduced contrast of

$(T - B)/(B + V)$ (Lythgoe 1980). The loss of contrast caused by turbidity can significantly impair the prey-detection abilities of fishes (Utne-Palm 2002).

Contrast and colour are relevant optical properties for visual detection of prey. Fresh waters experience episodic, even sustained, levels of turbidity far beyond those in the marine environment. Further, high concentrations of organic matter mean that fresh water transmits long-wavelength light better than it does short and intermediate wavelengths. What features of a freshwater fish's eye improve image reception under these conditions?

Fishes generally have retinal cones that are most sensitive to the wavelengths of light that occur in their natural habitat (Lythgoe and Partridge 1989). The sensitivity of freshwater fishes to the long-wavelength reddish light in their habitat is enhanced by cones that mostly absorb red light (Lythgoe 1980; Bowmaker *et al.* 1994; Carleton *et al.* 2005). In contrast, many marine fishes are most sensitive to the short-wavelength blue light that is typical of their environment. Interestingly, freshwater fishes see well into the infrared (long wavelengths that humans cannot see), which penetrates fresh waters coloured by dissolved organic matter (Lythgoe and Partridge 1989). Not surprisingly, freshwater fishes have eyes well-adapted to detect prey in a reddened world.

The match between ambient lighting and the spectral sensitivity of the eye is less strong in freshwater fishes than it is in marine fishes. Figure 6.1 shows how well-aligned the absorption maxima of deep-sea fishes are to their ambient lighting; obviously, if light is limited, capturing sufficient light to form an image is paramount. Yet in turbid water, fishes in both coastal marine and freshwater environments have absorption spectra skewed towards wavelengths shorter than expected if maximal sensitivity was the only issue (Fig. 6.1). Their eyes must separate the noise (the veiling light produced by turbidity) from the signal (the light reflected by prey). This left-shift of absorption spectra away from background light improves the contrast of prey (or other objects) in turbid environments (Lythgoe and Partridge 1989).

Other features of freshwater fish eyes that enable them to improve the contrast of images in turbid environments include ultraviolet (UV) sensitivity and yellow filters in the cornea and/or lens. UV sensitivity may improve detection of prey such as zooplankton, which appear as silhouettes against a bright UV background (Reckel *et al.* 2002). Yellow filters help to separate scattered light from turbidity (Lythgoe 1980).

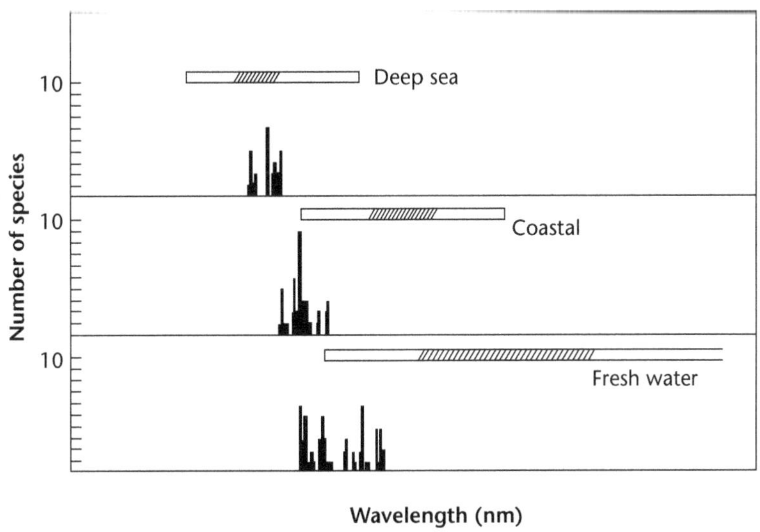

Figure 6.1: The relationship between spectral bandwidths available for vision in deep-sea, coastal marine and freshwater environments and the numbers of species with absorption maxima at certain wavelengths (vertical bars). The hatched areas of the horizontal bars represent the bandwidths that, if absorbed by visual pigments, would confer most sensitivity to ambient light (Lythgoe and Partridge 1989).

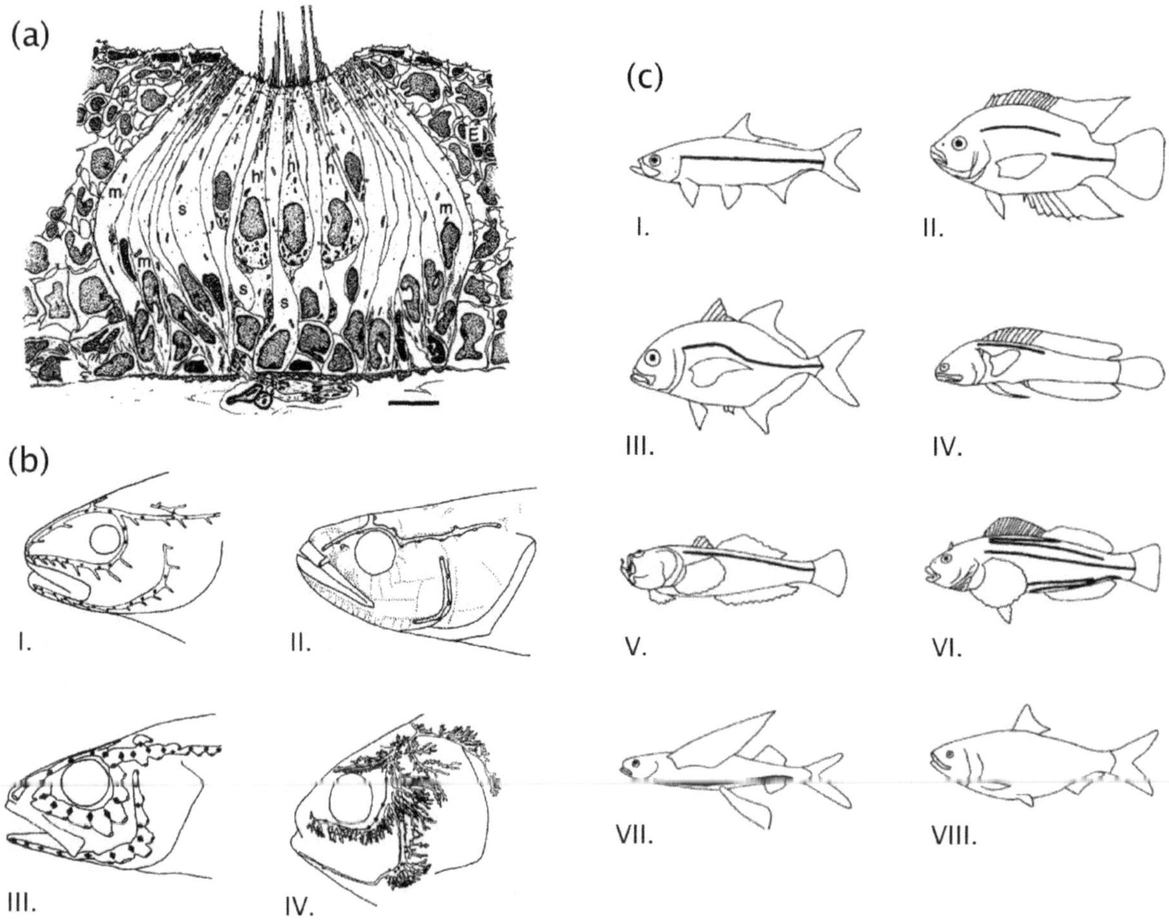

Figure 6.2: Gross morphology of the neuromast and canal lateral-line system in fishes. (a) Schematic longitudinal section of a superficial neuromast (h = hair cells; s = supporting cells; m = mantle cells). Scale bar = 10 µm. From Munz (1979). (b) Cranial lateral-line canal patterns. (I) Simple, narrow canals (Gadidae). (II) Reduced canals (Gobiidae). (III) Widened canals (Percidae). (IV) Branched canals (Clupeidae). From Webb (1989). (c) Trunk lateral-line canal patterns. (I) Complete (straight; oxeye herring). (II) Disjunct (Cichlidae). (III) Complete (arched: jack, Carangidae). (IV) Incomplete (jawfish, Opistognathidae). (V) Complete (dorsal displacement: stargazer, Uranoscopidae). (VI) Multiple (greenling, Hexagrammidae). (VII) Complete (ventral displacement: flying fish, Exocoetidae). (VIII) Absent (herring, Clupeidae). From Webb (1989).

Mechanoreception

All fishes have a lateral-line system of hair-cell receptors that enable them to detect motion of objects relative to the body. The lateral-line system enables a fish to feel its immediate environment, including movements by prey, predators, mates, currents and eddies, and to sense submerged logs, rocks or other inanimate objects which distort the currents around them (Montgomery *et al.* 1995; Coombs and Braun 2003). This sense is well-suited to animals that live in a dense medium like water, where disturbances to ambient currents are created by virtually anything that moves.

The structure that confers sensitivity to water movements is the neuromast, a group of sensory hair-cells surrounded by one or more non-sensory support cells (Fig. 6.2a). The hair-cells are covered by a gelatinous cupula, which transmits forces generated by fluid flow to the hair-cells, which relay the stimulus to the central nervous system. Superficial neuromasts occur in lines, as clusters or individually on the skin of the head or trunk; they generally are small (10–50 µm) and round or diamond-shaped. In contrast, canal neuromasts are larger (up to 400 µm) and of variable shape. Canal neuromasts are in either the cranial or trunk lateral-line canal systems, beneath the skin (Webb *et al.* 2008; Figs 6.2b, c). The differential morphology of these two kinds of neuromasts and the structures around them render the superficial neuromasts sensitive to low frequencies and canal neuromasts sensitive to high frequencies. Low-frequency signals typically are emitted by abiotic flows (current in streams) while high-frequency signals come from biotic flows (prey movement, wakes of swimming fishes etc.). In effect, superficial neuromasts are fluid-velocity detectors (velocimeters) and canal neuromasts are fluid-acceleration detectors (accelerometers) (Webb *et al.* 2008). This means that superficial neuromasts play a role in rheotaxis (orientation of the fish in the direction of flow), and canal neuromasts function in prey detection and the orientations adopted by predators prior to striking at prey (Montgomery *et al.* 1995; Coombs and Braun 2003).

The lateral-line system is important to nocturnal fishes and to fishes that feed in turbid water or other low-light habitats where vision is impaired. The morphology of the system varies among species, in ways determined more by habitat than phylogeny (Montgomery *et al.* 1995). For example, fishes with larger canal diameters and pore sizes can detect prey at greater distances (Janssen 1997; Janssen *et al.* 1999). Fishes that hunt prey in turbulent water may have few superficial neuromasts but a well-developed canal system, while those in still water or with a sedentary life style may have a greater density of superficial neuromasts on the skin and a less-developed canal system (Vischer 1990; Engelmann *et al.* 2002). Fishes hunting prey in turbulent streams may have few superficial neuromasts and a well-developed canal system that probably decreases their sensitivity to abiotic flows but increases their sensitivity to biotic flows. For fishes in still water, there may be enhanced sensitivity to slow flows and rheotactic behaviour to assist predators to intercept drifting prey. These species may be ultra-sensitive to fluid motion; while this has advantages, they may have trouble differentiating between patterns in water velocity and acceleration caused by their own swimming (essentially 'noise') and that of potential prey (Coombs and Braun 2003). Predatory fishes may eliminate this noise by adopting an ambush mode of feeding, where the predator is motionless before accelerating rapidly to capture its prey, or by saltatory behaviour, when the predator cruises with short pauses to feel for prey nearby (see 'Chemoreception'). Interestingly, the motionlessness of ambush predators would reduce the ability of prey fishes to detect the predator with their lateral lines. Staying still, then, is a form of lateral-line camouflage (Montgomery *et al.* 1995).

Chemoreception

Some fishes have very strong chemoreceptive abilities. For example, chemoreceptors on the barbels of certain North American catfishes (Ictaluridae) detect amino acids at concentrations as low as 10^{-11} moles per litre. This may be the lowest electrophysiological threshold reported for chemoreception in any vertebrate and is the equivalent of detecting less than one-hundredth of a teaspoon of alanine dissolved in an Olympic swimming pool (Schmidt-Nielsen 1997). Sensitivities of this magnitude are difficult to comprehend.

Fishes receive and process information about their surroundings using olfaction and gustation. In terrestrial vertebrates, olfaction is the sense of smell, enabling animals to detect airborne odorants, and gustation is the sense of taste, enabling animals to detect chemicals released by substances in solution (Moyes and Schulte 2008). This need not apply to all aquatic vertebrates, as the chemicals they need to detect are already in solution. Nevertheless, they do have distinct senses of olfaction and gustation. For fishes in particular, these senses are differentiated not by the medium that carries odorants to receptors but by the sensory receptors and organs that receive odorants and the associated neural pathways (Derby and Sorensen 2008). The neurological circuits associated with olfaction are separate and structurally different from those involved in gustation.

In olfaction, odorants are received by sensory cells in the nasal cavity and the sense is transmitted by the olfactory nerve (cranial nerve I). Fishes (unlike many terrestrial vertebrates) have an anterior and posterior nostril on either side of the ventral surface of their snout (Fig. 6.3a). Water flows into the anterior nostril, over the olfactory epithelium and out through the posterior nostril (Figs 6.3a, c). It is not yet clear how features of the nasal cavity affect the sensitivities of

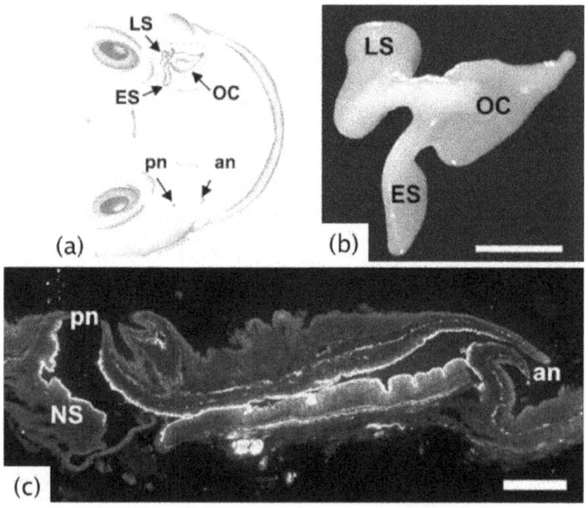

Figure 6.3: The nasal cavity of a Eurasian fish, the round goby (*Neogobius melanostomus*). (a) Location of nasal cavity and nostrils (an = anterior nostril; pn = posterior nostril). The area bounded by the dashed line shows three compartments of the nasal cavity (OC = olfactory chamber; and two accessory nasal sacs, arrowed). (b) A mould impression of the nasal cavity showing the olfactory chamber and two accessory nasal sacs. The ethmoidal sac (ES) is located medially and the lachrymal sac (LS) is positioned laterally. Scale bar = 1 mm. (c) Longitudinal section of the nasal cavity showing olfactory epithelium (bright layers) on the dorsal surface from the anterior (an) to posterior nostril (pn). The epithelium is prominent on the ventral surface and absent from the accessory nasal sac (NS). Scale bar = 500 μm. From Belanger *et al.* (2003).

fishes to odorants, or how sensitivity affects their predatory habits or capacity for prey-detection. Accessory nasal sacs influence water flow across the olfactory epithelium (Fig. 6.3b), enabling a fish to 'sniff' the water by rapidly opening and closing the operculum ('coughing'), changing the volume of the sacs and increasing flow across the epithelium (Nevitt 1991). The morphology of the sensory epithelium varies among taxa – it may be smooth, or folded into elaborate rosettes – but the degree of folding is not necessarily related to odorant sensitivity or mode of feeding (Zielinski and Hara 2007).

In gustation, odorants are received by sensory cells at various locations on the body and transmitted to the central nervous system via the facial (cranial nerve VII), glossopharyngeal (cranial nerve IX) and vagal nerves (cranial nerve X). Taste buds, the structures containing the sensory cells of gustation, are in five regions on the body (Hara 2007): oral, palatal and laryngeal, branchial (gills), cutaneous (epidermis) and barbels. The density and distribution of taste buds in these regions vary among species, depending on feeding mode and behaviour (Hara 2007). For example, in the East African cardinal fishes (Apogonidae), nocturnal species have higher densities of taste buds on the lips and in the mouth than species that are visual predators (Fishelson *et al.* 2004). Compared to fishes from clear waters, those in highly turbid waters have larger facial lobes (parts of the brain that receive and process gustatory stimuli: Huber and Rylander 1992). Thus, fishes in conditions where vision is limited are likely to have a well-developed sense of gustation.

The density of cutaneous taste buds on cyprinid fishes is related to their feeding habitat. Bottom-living suction feeders (see below) have more cutaneous taste buds than pelagic or littoral species (Gomahr *et al.* 1992), suggesting that low-light demersal environments should attract fishes with strong chemosensory abilities, because vision is limited. Like fishes in turbid water (Huber and Rylander 1992), the taste lobes in the brains of demersal cyprinids are larger than those of their pelagic relatives.

Multimodal integration in prey detection

Occasionally, a single sense is sufficient to detect a prey item. More often, input from several sensory modalities is integrated by the fish's central nervous system. For example, in European catfish (*Silurus glanis*), a nocturnal species, gustation may be essential for finding local concentrations of prey, but in complete darkness the fish use their lateral lines to track the wakes of swimming prey, and in pursuit and capture (Pohlmann *et al.* 2004). Two ambush predators, largemouth bass (*Micropterus salmoides*) and muskellunge (*Esox masquinongy*), both North American species, use vision in initial detection and orientation towards prey, then use their lateral lines to determine the optimal distance and angle for a strike (New 2002). Thus, the capture and ingestion of prey are an integration of multiple modalities, with sensory inputs varying through the process.

Sensory modalities of Australian species

Despite the integration of sensory modalities in prey detection and capture, there are strong associations between the sensory systems of freshwater fishes and their feeding modes or feeding habitats (Kotrschal and Palzenberger 1992; Huber *et al.* 1997). Little is known of Australian freshwater fishes in this regard, but there is no reason to suppose that there is not a similar match between senses and environment. For example, larval golden perch (*Macquaria ambigua*) and silver perch (*Bidyanus bidyanus*) are most responsive to long-wavelength light (Gehrke 1994). This would be expected if their retinal cones are most sensitive to the spectral properties of habitats in the Murray-Darling Basin, where water may be turbid and rich in dissolved organic matter, favouring transmission of long-wavelength light.

Many *Galaxias* species have an accessory lateral line of superficial neuromasts on the dorsal body surface (McDowall 1997). These species often are nocturnal, whereas those lacking accessory neuromasts tend to be visual, diurnal predators. The accessory lateral line improves a fish's ability to track nocturnal prey that is moving above the fish, in the overlying water or at the surface.

How might turbidity affect the feeding behaviour of juvenile Murray cod (*Maccullochella peelii*), a species typical of turbid lowland rivers in the Murray-Darling Basin? It appears that high turbidity (150 NTU) has no significant effect on the shape of the functional response of cod (the number of prey eaten as a function of prey density) and that the rate of consumption is little affected by darkness (Allen-Ankins *et al.* in press). As cod are able to locate inanimate food pellets on the sediment in complete darkness, they must have well-developed chemosensory abilities.

Robust, controlled experimental analyses of the sensory physiology of Australian species are scarce, and there are many opportunities for research. There are some salient questions:

1 As many Australian fresh waters are turbid (Kirk 1985), do the freshwater fishes tend to have enhanced mechanosensory and chemosensory systems? Murray cod, trout cod (*Maccullochella macquariensis*) and nightfish (*Bostockia porosa*) have large, regularly arranged slits over their heads that look as if they were cut by a scalpel. These open to the canal neuromast system and their large size may enable these species to detect prey at a distance, without vision (Janssen 1997; Janssen *et al.* 1999). The canal pores of the nightfish are very large, perhaps as an adaptation to nocturnal feeding on mobile prey such as shrimp or other fishes (in fact, the species name is from the Greek *poros*, 'bearing pores').
2 Australian rivers are often languid, meandering waterways unlike the boisterous, high-discharge rivers of New Zealand, for example. Do the fishes in our lowland rivers have high numbers of superficial neuromasts and an atrophied canal system? These features could be adaptive in a low-noise environment.
3 The barbels around the mouths of catfishes (Ariidae, Plotosidae) are gustatory organs, laden with taste receptors (Hara 2007). A freshwater catfish (*Tandanus tandanus*) will home in on a prey item that is buried, then use its snout to shovel the sediment aside and extract the prey. Common carp (*Cyprinus carpio*) and goldfish (*Carassius auratus*) feed in a similar manner and they too have extremely well-developed senses of gustation (Hara 2007). Is their sense of gustation stronger? If so, has this been a factor in their success as invasive species in Australia and, by association, a factor in the decline of the freshwater catfish?

As a further guide to research questions, Figure 6.4 is a graphical model showing ways that sensory modalities could be associated with habitat. The model is no more than a first draft, and like all models it is open to review and refinement. In its present form, one question is whether, in the turbid waters of the lower Murray-Darling Basin, demersal carnivores such as catfish rely more on chemoreception for prey detection than, say, carnivores that ambush mobile prey from cover, such as Murray cod.

FEEDING MODES

Freshwater fishes may be sorted into feeding modes based on mechanisms used to capture prey. Most attempts to do this have relied on either the functional consequences of predator morphology and

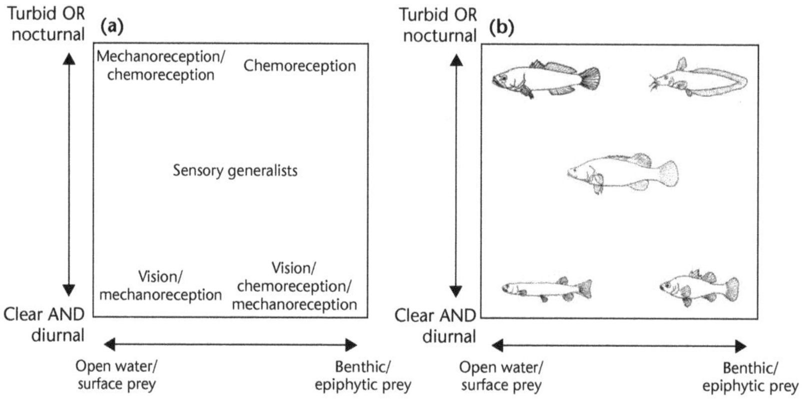

Figure 6.4: Dominant sensory modalities in relation to feeding habits. (a) A hypothetical classification. (b) Australian species that accord with the scheme in (a). Clockwise from top left: nightfish (*Bostockia porosa*), freshwater catfish (*Tandanus tandanus*), western pygmy perch (*Nannoperca vittata*), mountain galaxias (*Galaxias olidus*), Murray cod (*Maccullochella peelii*) (centre).

biomechanics (Liem 1980; Webb 1984a; Wainwright *et al.* 2004) or on predator behavioural ecology and 'optimal foraging' theory (Huey and Pianka 1981; Caraco and Gillespie 1986; Cooper 2005). Here, the former approach is adopted but behavioural patterns in foraging mode are not overlooked – some follow as a consequence of predator morphology.

Fishes have a wide repertoire of feeding morphologies, in a medium that is 900 times more dense than air and 80 times more viscous (Lauder 1985). It follows that the hydrodynamic properties of water have influenced the evolution of functional feeding morphologies. For example, the high density of water relative to air means that it is relatively incompressible. This could make it difficult for a fish to approach a prey item without affecting the position of that item, because the approach produces a bow wave that pushes the prey away (Lauder 1985). A solution to this problem is demonstrated by suction feeders.

Feeding modes are of three broad types: suction feeding, ram feeding and manipulation. Each has associated morphological and behavioural traits. These modes are discrete endpoints in a three-dimensional continuum; while some specialised species are close to those endpoints, most fishes are generalists to some degree and employ a combination of feeding modes under different circumstances.

Suction feeders

Suction feeding is the most common mode in fishes (Lauder 1985). Such fishes are able to rapidly increase the volume of the mouth, creating a sudden, strong pressure gradient that draws water and food into the mouth (buccal cavity). The pressure gradient may be gentle or strong, and the velocity of ingestion correspondingly is slow or rapid. Suction feeding is an adaptation to a high-density medium and is unique to aquatic vertebrates (Liem 1980; Lauder 1985).

As the prey moves and the fish does not (Norton and Brainerd 1993), a specialist suction feeder typically eats non-evasive, slow-moving or sedentary prey. There are two modes. Discriminate suction feeders target their prey items individually; their diet might include insects, snails or other invertebrates that cling to aquatic plants or woody debris. Indiscriminate suction feeders are less attracted to targets; they feed on benthic invertebrates or organic detritus, filtering items from mouthfuls of sediment.

What morphological features of fishes enhance suction-feeding ability? In the centrarchid fishes of North America, for example, there are four factors (Carroll *et al.* 2004; Higham *et al.* 2006; Fig. 6.5a): the surface area of the buccal cavity, cross-sectional area of the epaxial muscles (dorsal to the vertebrae), the distance between the centroid of the epaxial cross-section to the S-PT joint, L_{in}, and the distance from the S-PT joint to the centroid of the buccal area, L_{out}. Species with smaller buccal surface areas, longer L_{in}, shorter L_{out} and larger epaxial muscles produce the greatest pressures and therefore the best suction-feeding performance. Consequently, typical suction-feeding body shapes include smaller mouths and

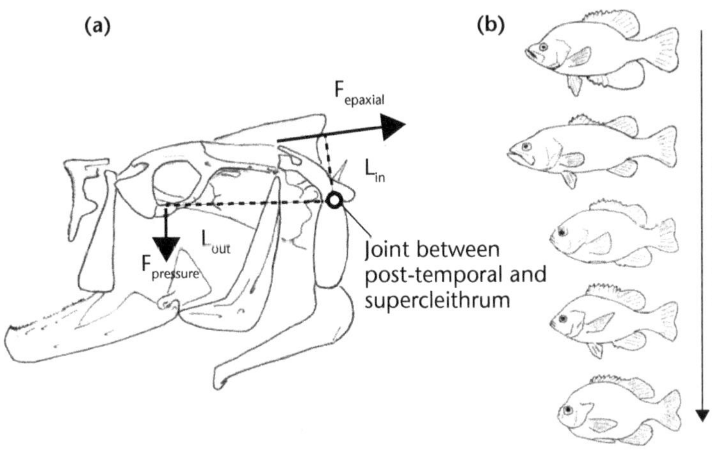

Figure 6.5: Suction feeding. (a) Model of a fish skull showing forces that affect suction-feeding performance. $F_{epaxial}$ is the force from epaxial muscles, behind the head, related to their cross-sectional area. $F_{pressure}$ is the force induced by the buccal cavity, related to its surface area. L_{in} is the distance from the centroid of the epaxial cross-section to the supercleithrum-post-temporalum (S-PT) joint. L_{out} is the distance from the S-PT joint to the centroid of the buccal area. (b) A morphological gradient in suction-feeding performance of five North American centrarchid fishes. From top: black crappie (*Pomoxis nigromaculatus*), largemouth bass (*Micropterus salmoides*), spotted sunfish (*Lepomis punctatus*), redear sunfish (*L. microlophus*) and bluegill sunfish (*L. macrochirus*). From Carroll *et al.* (2004).

deeper bodies (Fig. 6.5b). By extension, what are the locomotor features of a suction-feeding fish?

In a classic study linking morphology, locomotion and the foraging modes of fishes, Webb (1984a) suggested that discriminate suction feeders have adaptations for manoeuvrability, including propulsion by median/paired fins (MPF) rather than body/caudal fins (BCF) (Fig. 6.6). BCF swimmers move by undulations of the posterior part of the body and caudal fin, whereas fishes reliant on MPF locomotion use

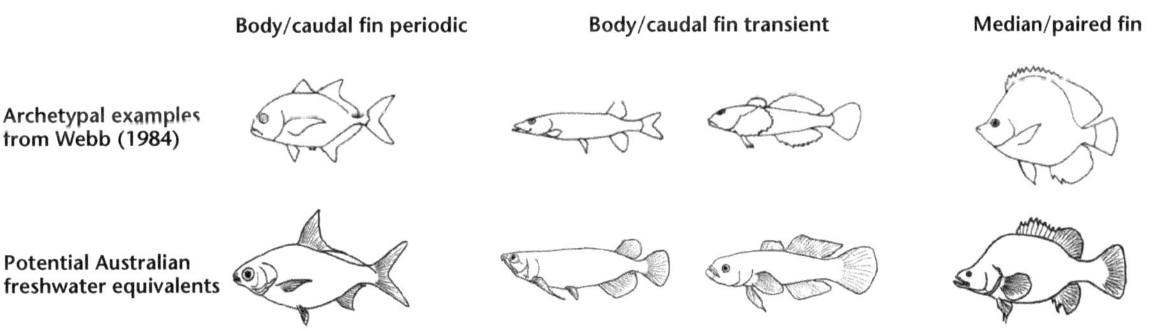

	Body/caudal fin periodic	Body/caudal fin transient	Median/paired fin
Design features	Tail of high aspect ratio (large span, fine chord); narrow caudal peduncle; anterior body depth; stiff body.	Posterior body depth; posterior dorso-ventral depth extended by enlarged soft-rayed dorsal and anal fins; flexible body; large muscle mass relative to whole body mass.	Soft-rayed, dorso-ventrally symmetrical dorsal and anal fins, extending caudally; deep, laterally compressed body.
Locomotor functions	Endurance swimming and sprints for sustained chases, patrolling, drift feeding, searching etc.	Fast-starts and powered turns for high acceleration prey capture.	Slow swimming and precise manoeuvres.

Figure 6.6: Archetypal forms of locomotor propulsion. Top row, from left: trevally (carangiform fishes), pike (esociform fishes), sculpin (cottiform fishes), butterfly fish (chaetodontiform fishes) and their morphological characteristics. Lower row: Australian fishes that may correspond to each form. From left: bony herring (*Nematalosa erebi*), northern saratoga (*Scleropages jardinii*), southern purple-spotted gudgeon (*Mogurnda adspersa*), golden perch (*Macquaria ambigua*). From Webb (1984b).

paired fins such as the pectoral fins and/or a combination of soft-rayed dorsal and anal fins. MPF swimmers also have a laterally compressed body (Fig. 6.6). These features facilitate accurate manoeuvres and braking, so that the mouth is positioned optimally for sucking up prey, such as insects and microcrustaceans, that cling to surfaces.

Not all suction feeders are MPF swimmers. Detritivorous fishes, and indiscriminate suction feeders in general, need to forage widely to meet their energy requirements owing to the low nutritional quality of their food. They require efficient means of locomotion over distances, and manoeuvrability, so that there may be compromises between BCF and MPF locomotion.

Ram feeders

Unlike suction feeders, ram-feeding fishes ingest their prey by accelerating towards it, essentially overtaking it. The predator moves faster than the prey (Norton and Brainerd 1993). Ram feeders pursue evasive, active prey such as shrimp and other fishes and they need to maintain strike accuracy while accelerating. They commonly have a large mouth, making it easier to catch prey while moving at speed (Higham *et al.* 2006).

The locomotor properties of ram-feeders, in contrast to suction feeders, include BCF propulsion to generate speed and acceleration. Manoeuvrability is less important than power, and they control their motion by movements of the body and a large caudal fin. There are two modes, each associated with a style of ram feeding (Webb 1984a).

Fishes using BCF periodic propulsion exhibit cyclical swimming kinematics, with regular, repeatable body movements. This mode is assisted by a tail with a high aspect ratio (large span but thin, often forked), a narrow caudal peduncle and a stiff, streamlined anterior body (Fig. 6.6). These fishes spend much of their lives swimming, covering long distances or perhaps swimming against a swift current, and are therefore likely to be cruising or drift-feeding predators.

Fish using BCF transient propulsion exhibit brief, non-cyclic kinematics, a large body depth and caudal area and a flexible body, enabling fast starts (Webb 1984a; Fig. 6.6). These typically are sit-and-wait ambush predators that rely on wood, rocks and other structural complexity to conceal their position; although they accelerate rapidly to capture prey, they are generally sedentary and have small home ranges.

Manipulation feeders

There is scant literature on the morphological and locomotor properties of manipulation-feeding fishes, perhaps because there is such a variety of adaptations that generalisations are elusive. Manipulation feeders either physically remove their prey from some kind of substrate prior to ingestion or take bites from larger prey without ingesting the whole. Examples include fishes that scrape algae from hard substrates (grazers), fishes that shoot jets of water to dislodge terrestrial prey from riparian substrates (e.g. archerfishes, Toxotidae: see below), fishes that scavenge on aquatic carrion (e.g. piranhas, Characidae: South America) and fishes that bite scales or nip fins of other fishes (e.g. some cichlids in Lake Malawi, Africa).

Manipulation feeders have some remarkable adaptations for specialised feeding. Grazing African cichlids, for example, have teeth adapted for scraping algae from hard surfaces (Fig. 6.7). They also have long intestines, like other grazing fishes, to enable them to extract nutrients from low-quality food. Another example is the archerfishes (*Toxotes* spp.) of northern Australia, with extraordinary anatomical and neurological adaptations that enable them to capture insects in flight above the water or clinging to overhanging branches. Archerfishes spit jets of water at prey, causing them to fall to the water surface. They must compensate for refraction at the air–water interface, making their target seem higher above the surface than it really is (Dill 1977; Fig. 6.8a). Within 100 ms of hitting the target, the predator initiates a rapid turn to align the body axis towards where the prey will land, and not to its position when it was hit (Rossel *et al.* 2002). The point of hitting the water surface is predicted from rapid (0.1 s) observation of the three-dimensional movement of the target rather than an extrapolation of the spatial trajectory of the falling prey. This high-speed neurological processing enables archerfishes to intercept their prey before other predators can reach it. Their capacity for rapid acceleration to the point of interception is enhanced by their morphology, suited for BCF transient propulsion and high velocities over short distances (Figs 6.6, 6.8b).

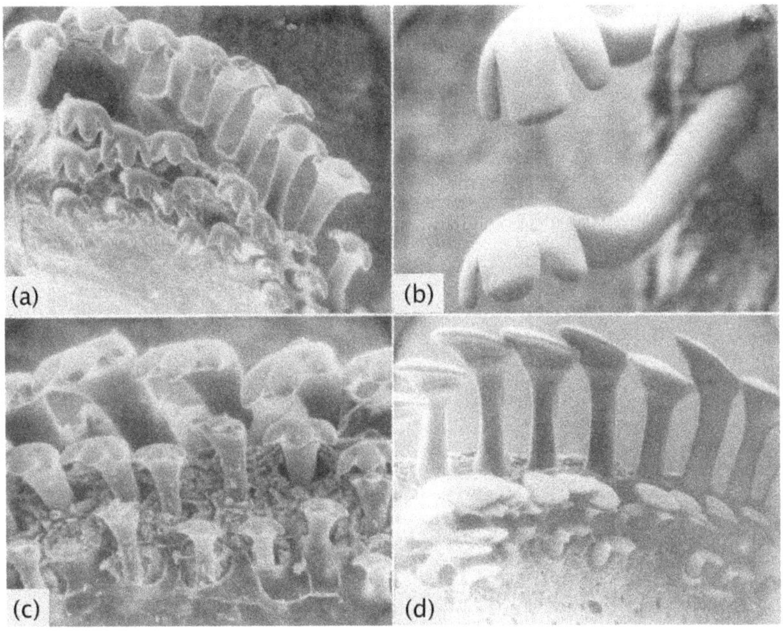

Figure 6.7: Scanning electron micrographs of dentition on the mandibles of grazing African cichlid fishes, showing how well they are adapted to scrape algae from hard substrates. (a) *Petrotilapia tridentiger*, Lake Malawi. (b) *Petrochromis polyodon*, Lake Tanganyika. (c) *Pseudotropheus zebra*, Lake Malawi. (d) *Hemitilapia oxyrhynchus*, Lake Malawi. From Liem (1980).

Modes of Australian species

If feeding modes are characterised in terms of the percentages of prey captures using suction, ram or manipulation methods, a group of species (e.g. size classes or populations) can be portrayed in a ternary plot (Fig. 6.9). The plot shows the hypothetical feeding modes of 12 Australian species, all near one of the three apexes. There are four species near each apex, as exemplars of each feeding mode, yet their behaviour is not wholly inflexible – species that employ one mode exclusively may be very rare.

Four candidates as ram feeders are barramundi (*Lates calcarifer*), mouth almighty (*Glossamia aprion*), southern saratoga (*Scleropages leichhardti*) and

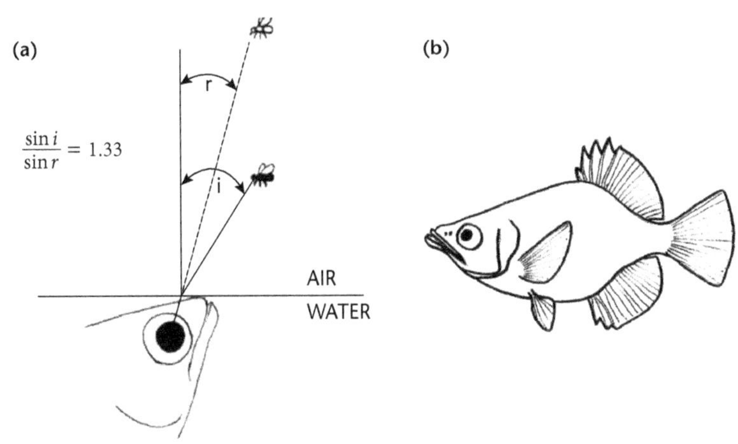

Figure 6.8: Archerfishes (Toxotidae) at the air–water interface. (a) Refraction. The incident ray is bent and the target, a fly, appears along the projection of the refracted ray (dashed line), above its true position. The angles of incidence (i) and of refraction (r) are related (sin i / sin r = 1.33), according to Snell's Law. From Dill (1977). (b) Body morphology, showing posterior body depth and fin surface area, adaptive for rapid acceleration.

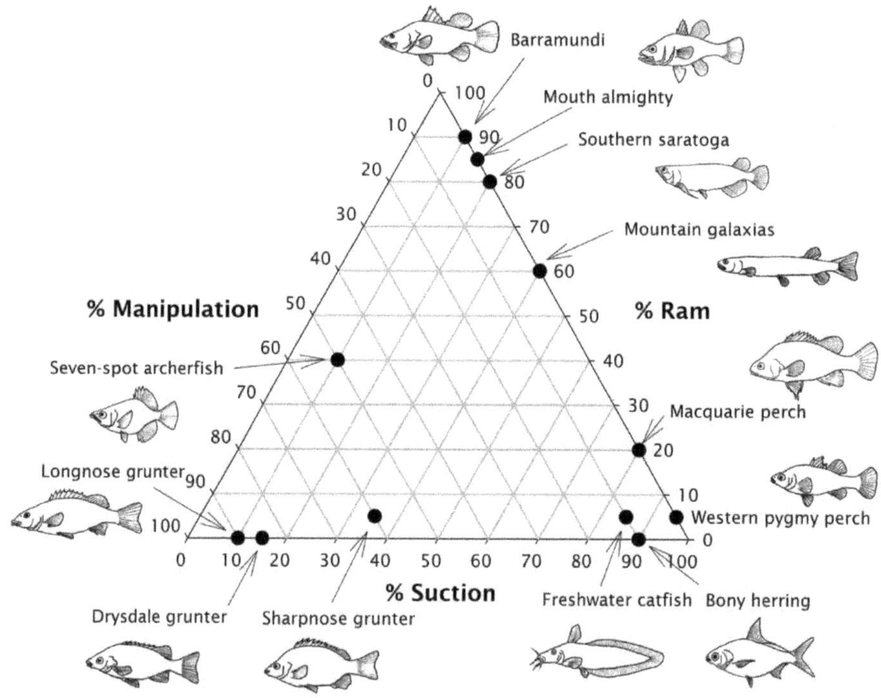

Figure 6.9: Ternary plot of the feeding modes of 12 Australian freshwater fish species (suction = right-sloping lines; ram = horizontal lines; manipulation = left-sloping lines). The sharpnose grunter (*Syncomistes butleri*), for example, might utilise suction feeding 35% of the time, ram feeding 5% of the time and manipulation 60% of the time.

mountain galaxias (*Galaxias olidus*). Each has a large gape relative to maximum body depth, typical of ram feeders (Higham *et al.* 2006). Barramundi and southern saratoga could be classic ambush predators, judging from their large gape and posterior body depth and from observations of their feeding behaviour. The mouth almighty is an ambush predator that lurks in dense cover (Allen *et al.* 2002; Pusey *et al.* 2004) and its morphology is suited to this mode of feeding. These three species have a capacity for rapid acceleration, and move by transient BCF propulsion. In contrast, the mountain galaxias has a low epaxial muscle mass and streamlined body, implying that it could use periodic BCF propulsion for ram feeding in streams, where it captures drifting aquatic and terrestrial insect prey.

Macquarie perch (*Macquaria australasica*), western pygmy perch (*Nannoperca vittata*), bony herring (*Nematalosa erebi*) and freshwater catfish have morphological attributes of suction feeders, including a small mouth relative to maximum body depth and a large ratio of epaxial muscle mass to length. The freshwater catfish could be an exception, as the epaxial muscle mass to length ratio probably is no higher than typical ram feeders. Yet catfish feed by sucking up soft sediment and filtering it for invertebrate prey, and could be regarded as an indiscriminate suction feeder. The same applies to the bony herring, with a body form adapted for endurance swimming (a tail with high aspect ratio and a slender caudal peduncle). As a detritivore (Allen *et al.* 2002; Pusey *et al.* 2004; Rayner *et al.* 2009b), the bony herring needs to range widely and process large quantities of detritus to meet its energy requirements. Detritivores are likely to use BCF periodic propulsion to cover distances efficiently (Webb 1984a).

The two remaining examples of suction feeders, Macquarie perch and western pygmy perch, have a morphology suited to discriminate suction feeding (small mouths, strong epaxial muscles) and to MPF swimming, giving them manoeuvrability to enable high-velocity suction feeding on prey that clings to surfaces.

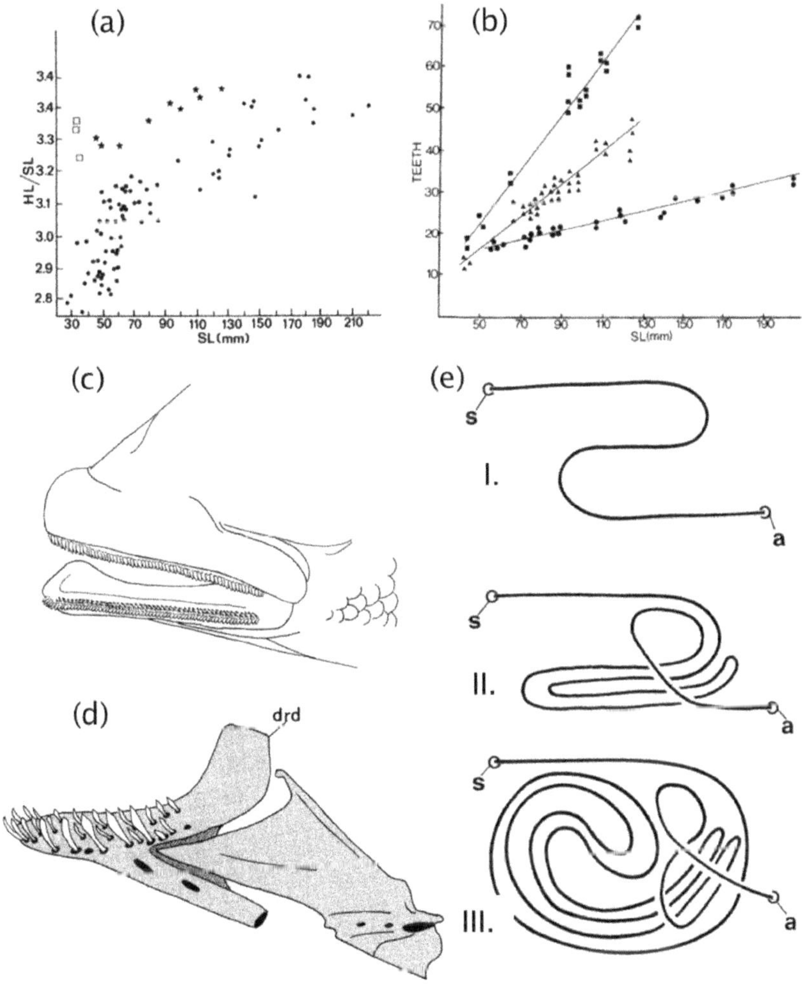

Figure 6.10: Grazing adaptations of grunters (Terapontidae) from northern Australia (from Vari 1978; Vari and Hutchins 1978). (a) Relationship between standard length (SL) and head length (HL) relative to standard length in sharpnose grunter (*Syncomistes butleri* – circles), longnose grunter (*S. trigonicus* – stars) and another grazing species, Kimberley grunter (*S. kimberleyensis* – squares). (b) Relationship between standard length and number of teeth on each side of the upper jaw of longnose grunter (squares), Drysdale grunter (*S. rastellus* – triangles) and sharpnose grunter (circles). (c) Lateral head view of longnose grunter, showing laterally directed lower jaw dentition. (d) Jaw form and teeth position in sharpnose grunter. (e) Intestinal convolution patterns in Terapontidae. (i): *Leiopotherapon, Amniataba, Hannia, Terapon, Lagusia, Pelates, Pelsartia, Rhyncopelates, Mesopristes.* (ii) *Hephaestus, Bidyanus, Scortum.* (iii) *Pingalla, Syncomistes.*

Manipulation may be the least common feeding mode among Australian freshwater fishes (Fig. 6.9). In addition to the archerfishes there is the longnose grunter (*Syncomistes trigonicus*), Drysdale grunter (*S. rastellus*) and sharpnose grunter (*S. butleri*), all from northern Australia. These are algal grazers. Diet and morphology both indicate that the longnose grunter is most specialised, followed by the Drysdale grunter and the sharpnose grunter (Vari 1978; Vari and Hutchins 1978; Bishop *et al.* 2001). Thus, the length of the head relative to body length is greatest in the longnose grunter (Fig. 6.10a), enabling access to algae in deeper crevices. The numbers of teeth on each side of the upper jaw are also greatest for the long-nose grunter, followed by the Drysdale grunter and sharp-nose grunter (Fig. 6.10b), and the shapes and positions of the teeth are suited for scraping algae from hard surfaces (Figs 6.10c, d). In general, the intestinal convolutions typical of this genus increase the length of the digestive tract and the surface area for nutrient absorption, facilitating digestion of algae (Fig. 6.10e).

ENERGETICS AND NUTRITION

We have considered the ways that fishes locate and capture their prey, and turn now to look at the physiological consequences for individual fishes. The focus is patterns of food ingestion and how they interact with the environment to affect the fitness of individuals, and the nutritional demands imposed by morphology and life style.

Bioenergetics

A fish is a ball of organised cells producing energy to maintain the fitness of the individual; indeed, energy must be expended to keep the fish alive. But what is 'energy'? The word is used freely in everyday conversation, but it is an abstract concept. It is the capacity to do work which, at a cellular level, involves maintenance of ionic concentrations in cells and molecular transport across membranes, repairing damaged proteins and DNA, fuelling electron-transport chains and so on. Collectively, these processes enable a fish to perform as an individual – to find food, build nests, conduct courtship, migrate and flee predators.

Much of the work involves the transformation of potential energy to kinetic energy and *vice versa*. Energy transformations are governed by the laws of thermodynamics, and the first law especially is key to understanding the bioenergetics of fishes. The first law states that 'energy can be neither created nor destroyed'. That is, it can be changed from one form to another, but the total energy remains the same. In biological systems, the efficiency with which energy is transformed from stored (potential) to usable (kinetic) forms is never greater than 30% and whenever this occurs some energy is lost as heat. A food web is like a network of energy transformations and the inevitable losses of energy as heat, in the course of transformations, determine its pyramidal shape. In other words, there is a negative relationship between biomass and height in the food chain (Begon *et al.* 1996).

The energy budgets of fishes are shaped by morphology, behaviour and the surrounding environment. A simple expression of mass balance is:

$$C = M + G + W \qquad (6.1)$$

where C is food consumed, M is metabolism due to maintenance, G is growth (or metabolism due to growth) and W is waste material excreted. In keeping with the first law of thermodynamics, there is a balance between what goes in, what stays in the body and what comes out.

Equation 6.1 can be expanded as:

$$C = M_s + M_a + M_{sda} + G_s + G_r + W_f + W_u \qquad (6.2)$$

where

M_s = standard metabolic rate for a resting fish that is not digesting food. This is the minimal energy needed by a fish of a given size to stay alive at a given temperature;

M_a = active metabolic rate, the metabolic rate due to mean activity per unit time;

M_{sda} = specific dynamic action, the metabolic rate required to digest a meal and synthesise its products (Secor 2009);

G_s = energy allocated to somatic growth;

G_r = energy allocated to reproductive organs;

W_f = energy lost as faeces;

W_u = energy lost as urine.

The units of energy are joules or calories, and terms in the equation are rates (e.g. cal h^{-1}). The cost of metabolism is estimated as the rate of oxygen consumption (respiration), converted to calories using the oxycalorific equivalent, 3.2–3.24 calories per milligram of oxygen (Adams and Breck 1990).

The goal is to understand how a fish's phenotype interacts with its environment to control the partitioning of energy, as in Equation 6.2. In fact, all terms in that equation are functions of body mass and temperature. First, standard metabolic rate scales as a power function of body mass:

$$M_s = M_{0,s} S^b \qquad (6.3)$$

where S is fish mass, $M_{0,s}$ is a normalisation constant and b is an exponent that, for teleost fishes, is between 0.4 and 1.29, with a mean of 0.79 (Clarke and Johnston 1999). Interestingly, this is near 0.75 (3/4), a theoretical value that may apply to all animals (West *et al.* 1999; West and Brown 2005) and is the basis for the so-called 'three-quarter power law'. Although this is highly controversial (Hochachka and Somero 2002), it is an elegant, mechanistic, unified explanation for

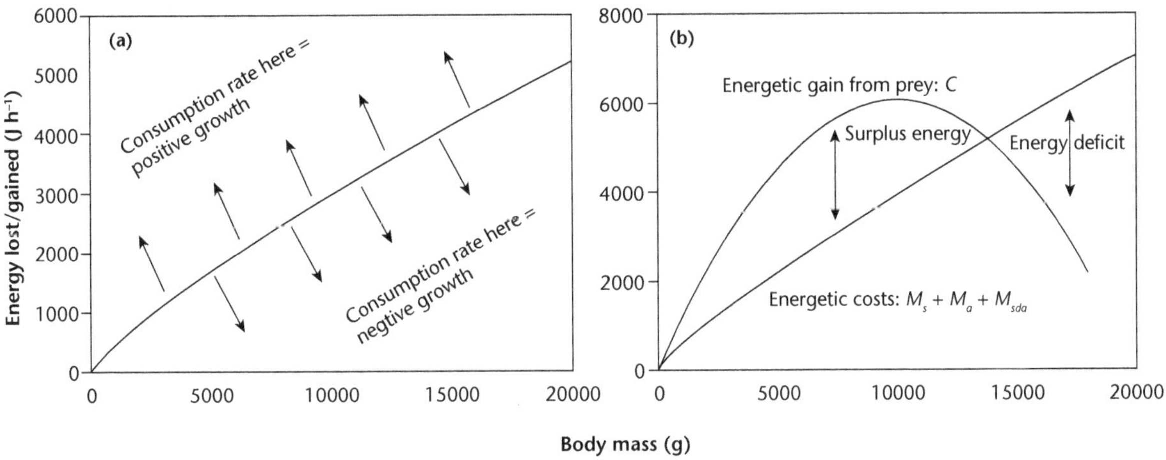

Figure 6.11: Effects of body mass on energy losses and gains. (a) Standard metabolic rate (joules per hour) as a function of body mass (g), showing a threshold rate of energy acquisition (consumption) that delineates regions of positive and negative growth. (b) Energetic costs of consumption (lower curve: the sum of standard and active metabolic rates and specific dynamic action) and energy acquired from consumption (upper curve) as functions of body mass. The difference between these rates is the surplus energy function.

inter- and intraspecific variation in metabolic rates in animals spanning many orders of magnitude in mass. If $b = 0.75$, the mass-specific metabolic rate of a fish is a negative function of body mass, suggesting that large fishes are more efficient and require less energy per unit mass than small fishes (West and Brown 2005). Thus:

$$M_s = M_{0,s}S^{3/4} \rightarrow M_sS^{-1} = M_{0,s}S^{-1/4} \quad (6.4)$$

The power-law dependence of metabolic rate on body mass (Equation 6.3) is a quantitative description of the effect of body size on the minimal energy requirement of a fish. Using parameter estimates from Clarke and Johnston (1999) and converting oxygen consumption rate to rate of energy consumption, we can plot the standard metabolic rate curve on a Cartesian plane. This delineates two regions of energy acquisition, one leading to positive growth and increased fitness and the other to negative growth and starvation (Fig. 6.11a).

The conditions for growth are described more explicitly by Equations 6.1 and 6.2:

$$G > 0 \leftrightarrow C > M + W \quad (6.5)$$

That is, growth is positive if (and only if) the rate of energy acquisition in consumption exceeds the rate of energy expended in maintenance metabolism and lost as waste. It is difficult to relate net energy gain to body mass because of uncertainty about how rates of energy consumption scale with size in ecosystems (i.e. under natural conditions). It is reasonable to assume power-law dependence between rates of metabolism and waste production and mass (Banavar *et al.* 1999; Gillooly *et al.* 2001; Savage *et al.* 2004; West and Brown 2005), but field data describing the scaling of consumption rates are lacking. We know that, under natural conditions, the relationship between consumption rate and body size in Arctic char (*Salvelinus alpinus*) is unimodal or peaked (Finstad *et al.* 2006). Figure 6.11b assumes such a relationship to illustrate how net energetic gain or surplus energy may vary with body size. Note that, given the assumption, surplus energy vanishes as body mass exceeds about 14 kg.

Most fishes are ectotherms – the temperature of their metabolism matches that of their environment. Biochemical reaction rates are temperature-dependent and it follows that the physiology of a fish, indeed all of its activities, is controlled by temperature. The metabolic rates of fishes increase exponentially with temperature, according to the Arrhenius Equation (Clarke and Johnston 1999; Gillooly *et al.* 2001; Hochachka and Somero 2002; Brown *et al.* 2004a):

$$M_s = Ae^{-E/kT} \quad (6.6)$$

where E is the activation energy of a reaction, k is Boltzmann's constant and T is the absolute temperature

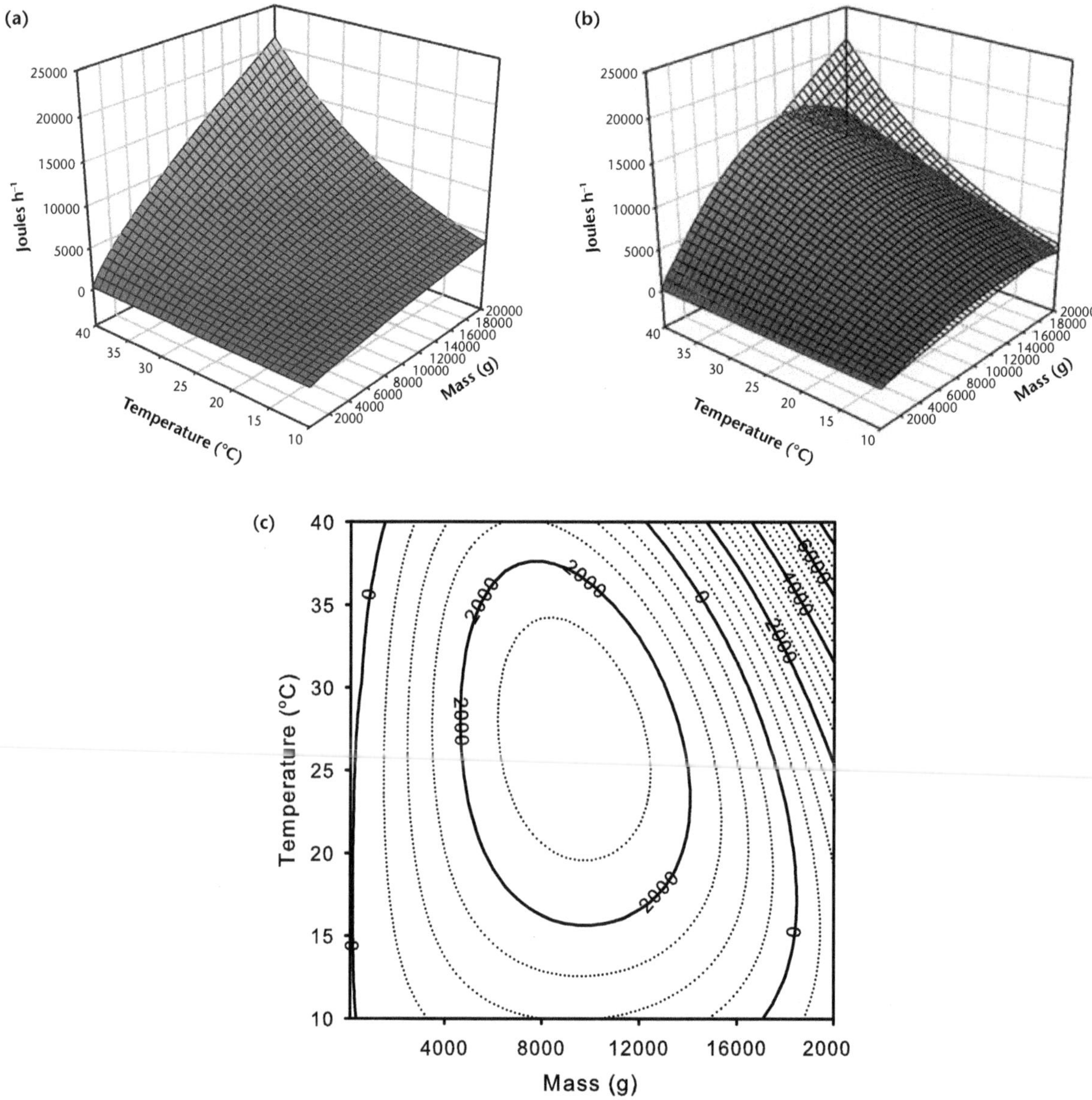

Figure 6.12: Effects of body mass and temperature on energy losses and gains. (a) Relationships between body mass, temperature and standard metabolic rate, generated using Equation 6.7 and parameter estimates from Downs *et al.* (2008). (b) A possible relationship between body mass, temperature and energy from consumption (solid surface). The difference between this surface and the standard metabolic rate surface (transparent) coarsely defines the surplus energy or net energetic gain function. (c) The surplus energy function from (b) as a contour plot (joules per hour).

(degrees Kelvin). A is a constant, representing the frequency of collision of particles (Chang 1994). This equation has real utility for freshwater fish ecologists as it is a close fit to empirical data for many species of teleosts (Clarke and Johnston 1999). For example, it suggests that, to sustain standard metabolism, a tropical fish at 30°C requires six times the energy input of a polar species at 0°C.

Equations 6.3 and 6.6 can be combined to create a general model showing how standard metabolic rates, hence the minimal energy requirements, of fishes scale with body mass and temperature:

$$M_s = aS^b e^{-E/kt} \tag{6.7}$$

where a is a normalisation constant. The scaling relationships between minimal energy requirements and temperature and body size are shown in Figure 6.12a (data from Gillooly *et al.* 2001; parameter estimates from Downs *et al.* 2008). This shows clearly that body size has a greater impact on energy demand in warm systems than in cool systems. Extending the above two-dimensional example, the plane surface in the graph (Fig. 6.12a) delineates two regions – above the plane, where consumption of energy leads to positive growth, and below the plane, where consumption results in negative growth. Ideally, we need to know, for a given species under natural conditions, how mean consumption rates vary as a function of temperature and body mass. That would allow the identification of a 'surplus energy' function to indicate thresholds for positive growth in freshwater systems.

Figure 6.12b shows how consumption might vary as a function of temperature and body mass, assuming a quadratic relationship between consumption and size (as above, following Finstad *et al.* 2006) and a linear relationship between temperature and consumption. The example is hypothetical but suggests that, as the system warms, the energy gain from consumption fails to keep up with metabolic costs, so that net energetic gain for a 12 kg fish declines above about 25°C (Fig. 6.12c). A similar result has been documented for terrestrial beetles (Vucic-Pestic *et al.* 2011) but there appear to be no examples from freshwater predator–prey systems. This is surprising, as freshwater food webs are dominated by ectotherms and are thereby strongly influenced by temperature.

Obviously, temperature and body mass are factors constraining the energy budgets of freshwater fishes, but what other factors are at play? We need to know more of the ways that morphology, feeding mode, phylogeny and the aquatic environment interact to shape energy requirements, how energy is partitioned into finding, catching and digesting food, and how energy is allocated to growth (surplus energy). Feeding mode, for example, will affect how much energy is allocated to locomotion (M_a). Ram feeders using BCF periodic propulsion will need to allocate more energy to M_a than will an ambush ram feeder, thus the latter may have more energy to allocate to digestion or other activities. Or perhaps ambush predators need less energy (less food) than ram feeders?

The energy allocated to digestion (M_{sda}) has been determined for four freshwater fishes (none native to Australia) with different foraging modes (Fu *et al.* 2009). These are an ambush predator, the southern catfish (*Silurus meridionalis*); an active grazer, the grass carp (*Ctenopharyngodon idella*); a sluggish omnivore, the dark-barbel catfish (*Pelteobagrus vachelli*); and an active omnivore, Crucian carp (*Carassius carassius*). The southern catfish has the lowest standard metabolic rate, allocates least energy to endurance swimming, eats large meals (as a percentage of body weight) and allocates most energy to digestion. The grass carp eats small meals, allocates most energy to endurance swimming and less to digestion. Eating small meals and allocating less energy to digestion seems a good strategy for a fish not protected by cover and needing to flee from predators. Thus, the energy budget is well-matched to the mode of feeding.

Bioenergetics of Australian species

The structures of the scaling equations above are universal, as the mechanisms underlying body mass and temperature effects apply to all fishes regardless of phylogeny or morphology (West *et al.* 1997, 1999; Banavar *et al.* 1999; Gillooly *et al.* 2001; West and Brown 2005). Using the equations and estimated parameters (Gillooly *et al.* 2001; Downs *et al.* 2008), we can estimate energy needs for Australian species. This is not to say that the scaling equations are complete descriptions of energy needs, or that the parameter values might not vary with morphology, phylogeny and environmental variables (e.g. oxygen tension in water). The result calculated from Equation 6.7 should be considered a minimal energy requirement, as part of the variance in energy needs across species is determined by morphology and other variables.

Studies of the bioenergetics of freshwater fishes in Australia are limited to research on spangled perch (*Leiopotherapon unicolor*: Gehrke 1988) and Murray cod (Clark *et al.* 2005) and two unpublished theses (Meredith 2005; Whiterod 2010a). Spangled perch is the most widespread of all Australian species but it does not occur south of about Condobolin (New South Wales) in the Murray-Darling Basin (Lintermans 2007), suggesting that its distribution may be temperature-limited. Although the species demonstrates a significant positive relationship between temperature and food consumption, the proportion of

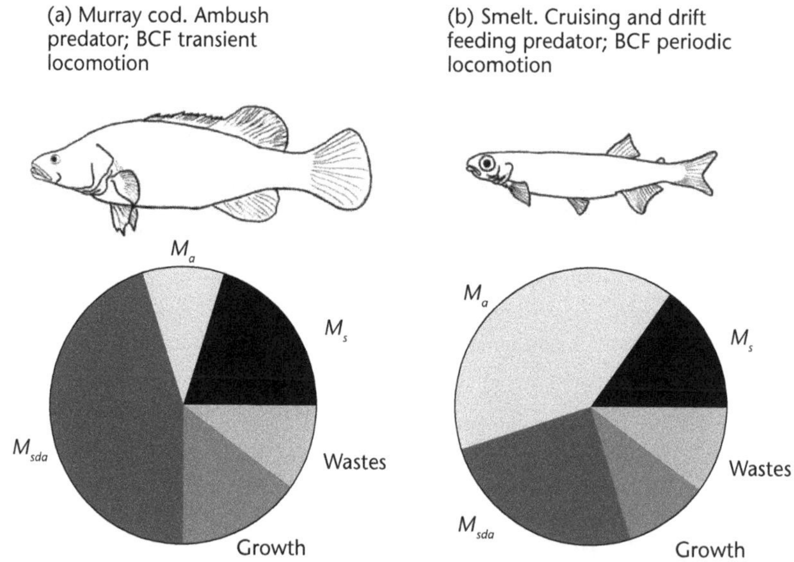

Figure 6.13: Hypothetical energy budgets for two species with different feeding modes. (a) Murray cod (*Maccullochella peelii*). (b) Australian smelt (*Retropinna semoni*). M_a = active metabolism; M_s = standard metabolism; M_{sda} = specific dynamic action.

energy allocated to growth does not vary between 16.8°C and 22.6°C (Gehrke 1988).

Meredith (2005) developed a bioenergetic model for carp gudgeons (*Hypseleotris* spp.), small fishes that are common throughout the Murray-Darling Basin. He showed that the normalisation constant and mass exponent for the standard metabolic rate are low ($a = 10^{-4}$; $b = 0.61$ at 30°C), suggesting that the minimal energy requirement in these species is low compared to other fishes ($b = 0.79$; Clarke and Johnston 1999). This could mean that carp gudgeons convert prey into biomass more efficiently than do other small-bodied species, but more comparative data are needed.

Whiterod (2010a) investigated the effects of temperature on the swimming energetics of Murray cod. He found that the swimming performance of Murray cod was like that of the northern pike (*Esox lucius*), a ram-feeding ambush predator from Europe and North America. Casual observations on the morphology of Murray cod, suggesting BCF transient propulsion, also fit the profile of an ambush predator.

Future research would benefit from comparative studies of the ways that differences in morphology and feeding mode interact with the environment to affect energetics. For example, we might compare Murray cod and Australian smelt (*Retropinna semoni*). The Murray cod has a large mouth and large posterior body depth, and its posterior dorso-ventral surface area is increased by enlarged dorsal and anal fins. These traits are suited to ambush piscivory – rapid acceleration, ram feeding (Webb 1984a; Higham *et al.* 2005, 2006). In contrast, the Australian smelt has a forked tail with higher aspect ratio and it uses BCF periodic propulsion in the drift or in the pelagic zone of lentic habitats, suggesting a cruising or drift ram feeder. We could hypothesise that the ambush predator consumes larger meals and allocates more of its energy to digestion (specific dynamic action) than does the cruising predator (Fig. 6.13; Fu *et al.* 2009). As the Australian smelt is an active species, using BCF periodic propulsion and feeding in habitats where it is exposed to predators, smaller meals allow it to maintain swimming performance during and after feeding and to allocate more energy to activity than the Murray cod.

Nutritional ecology

The aforementioned models are like nutrient budgets, or single-currency nutrient budgets, where the currency is energy (Raubenheimer and Simpson 1994). Describing the nutritional value of food in these terms alone is problematic, for two reasons (Clements and Raubenheimer 2006). One is that each species has

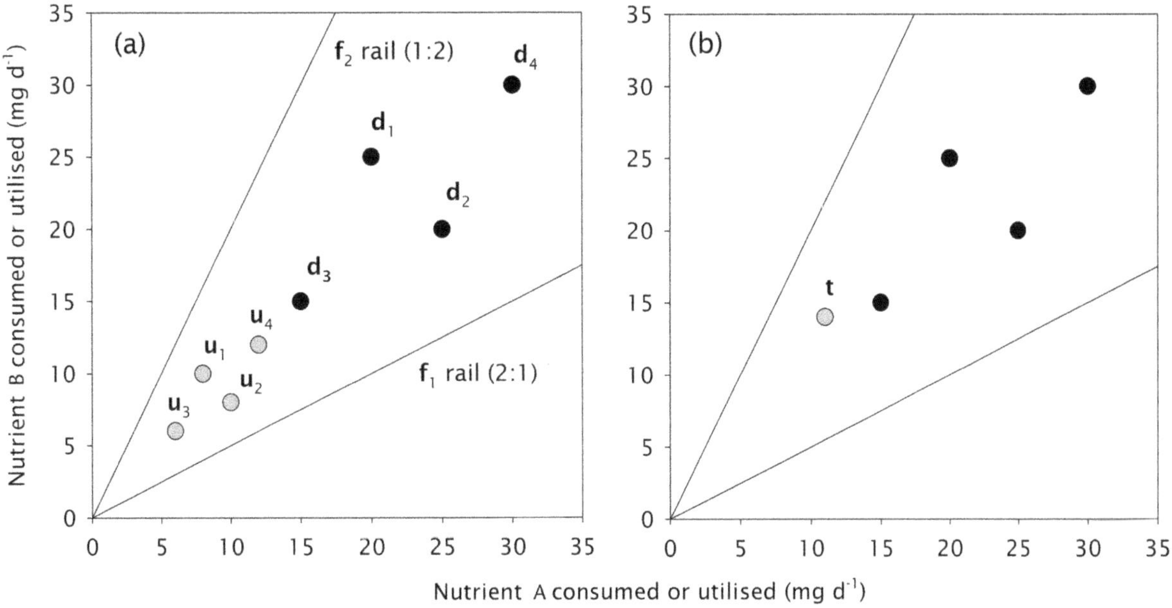

Figure 6.14: Two-dimensional nutrient spaces defining (a) the relationship between the nutrient utilisation of four groups of fishes (grey), each ingesting a unique diet (black), without homeostatic regulation of nutrient composition. Diets $\mathbf{d_1}$ to $\mathbf{d_4}$ are formed by taking linear combinations of the two foods, $\mathbf{f_1}$ and $\mathbf{f_2}$. The nutrients utilised when feeding on these four diets are defined by $\mathbf{u_1}$ to $\mathbf{u_4}$ respectively. (b) Nutrient utilisation of fishes fed different diets is similar when there is homeostatic regulation of nutrient composition. The nutrient composition of this homeostatic consumer is given by the target vector, **t**. Lines here are 'food rails' defining the proportions of nutrients A and B in each food type.

specific requirements for different nutrients, and aggregate measures of food quality, such as energy, could suggest that all species and life stages require the same thing. The second is that using energy as a measure of food quality implies that all the energy in a food item can be assimilated; often, that is not true.

Models in nutritional ecology differ from models in bioenergetics in that they are nutritionally explicit or multi-currency models. A nutritional ecologist is interested in how multiple nutrients flow between organism and environment. The goal is a quantitative understanding of how the spatial structure and temporal dynamics of the nutrient composition of the food (the nutritional environment) interact with the consumer's phenotype to shape its fitness. Nutritional ecology is therefore a core issue in fundamental and applied fish ecology.

There are two prominent theoretical frameworks for nutritional ecology. Ecological stoichiometry (ES: Elser *et al.* 2000; Sterner and Elser 2002; Anderson *et al.* 2004) concerns the flow of multiple elements (e.g. C, N, P) between consumer and the environment. The geometric framework (GF: Raubenheimer *et al.* 2009; Simpson *et al.* 2010) considers the flow of multiple macronutrients (proteins, carbohydrates, lipids) and micronutrients (amino acids, fatty acids) between consumer and environment. These frameworks have much in common (Anderson *et al.* 2004; Boersma and Elser 2006; Raubenheimer *et al.* 2009), including four principles that are tenets in nutritional ecology:

- prey species differ in nutrient composition and therefore their nutritional value to consumers;
- consumers have a characteristic stoichiometry, maintained by behavioural or physiological means;
- defence of characteristic stoichiometry may incur costs to fitness, hence the nutritional balance between consumer and resource affects the ecology of populations; and
- there is variation in the stoichiometry of consumers, hence the nutrients they require.

The first is self-explanatory, but the second and third principles are best understood using the nutrient space approach of ES and GF. Consider a fish feeding on prey that contains two nutrients, A and B.

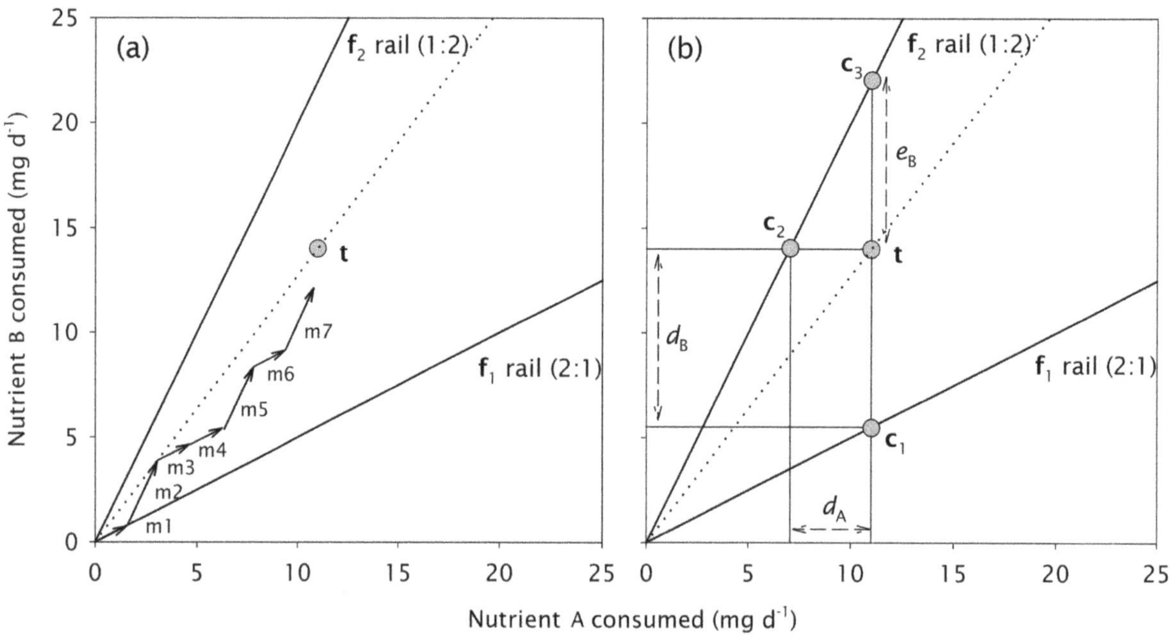

Figure 6.15: Two-dimensional nutrient spaces showing (a) how a fish may select different prey types to approximate its daily target nutrient intake, t, i.e. behavioural regulation of nutrient intake. Four meals (m) of prey type f_1 and three meals of prey type f_2 daily are sufficient to approximate the target. (b) Nutrient deficiencies (d_A, d_B) and an excess (e_B) resulting from three daily consumption rates (c_1, c_2, c_3). When consumption is at c_3, the fish may excrete an amount e_B to reach t, i.e. physiological regulation of nutrient intake. The target vector falls between the two food rails showing the proportions of nutrients A and B in two foods.

These might be individual elements (e.g. N, P), micronutrients (amino acids, fatty acids) or macronutrients (carbohydrates, proteins). We can now define the composition of a food type using a vector (n_A, n_B), where n_i is the quantity of nutrient i in milligrams.

Suppose we set up an experiment wherein we offer two foods, $\mathbf{f}_1 = (2, 1)$ and $\mathbf{f}_2 = (1, 2)$. That is, $\mathbf{f}_1$ has 2 mg of nutrient A for every 1 mg of nutrient B, and the reverse is true for $\mathbf{f}_2$. These two vectors now span a space containing all linear combinations of $\mathbf{f}_1$ and $\mathbf{f}_2$. The scalar multiples of these vectors are termed 'food rails' by Raubenheimer and Simpson (2004; Fig. 6.14), the two founders of nutritional ecology.

Now suppose that we assign fishes to four treatments, where those in each treatment are daily fed diets of different quantities of $\mathbf{f}_1$ and $\mathbf{f}_2$. Let diet 1, $\mathbf{d}_1$, be $5\mathbf{f}_1 + 10\mathbf{f}_2 = 5(2, 1) + 10(1, 2) = (20, 25)$, in units of milligrams per day. Three other diets are constructed, namely $\mathbf{d}_2 = (25, 20)$, $\mathbf{d}_3 = (15, 15)$ and $\mathbf{d}_4 = (30, 30)$. The positions of these diets in a two-dimensional nutrient space are shown in Figure 6.14.

Not all ingested nutrients are retained, as parts are egested or excreted. Let $\mathbf{u}_i$ be the vector describing nutrients actually utilised by a fish in growth and metabolism (fitness generation) when feeding on $\mathbf{d}_i$. If the fishes exposed to each diet do not regulate their stoichiometry or nutrient composition, through homeostasis, the **u** vectors will not all fall on the same spot in the nutrient space (Fig. 6.14a). If the fishes do regulate their nutrient composition then, no matter what food is available, the **u** vector will fall on the same point, known as the nutrient target, **t** (Fig. 6.14b; Raubenheimer and Simpson 2004). The nutrient target is the vector describing the nutrient composition of prey that should be ingested to maximise fitness of a particular consumer at a particular stage of development. It includes nutrient requirements for growth and other metabolic costs needed for fitness generation (e.g. swimming, active metabolism). The homeostatic regulation of a target nutrient utilisation vector is typical of fishes and other vertebrate consumers, and is a key principle in nutritional ecology.

To regulate their nutrient intake and achieve their target, fishes can be selective with the foods they ingest (behavioural mechanisms) or they can excrete or egest nutrients that are in surplus (physiological mechanisms). If a prey type has a food rail in the same direction as the consumer's nutrient target, the consumer needs to ingest sufficient food to reach its target (dotted line, Fig. 6.15). If several food rails lead to the target and the contributing prey types are sufficiently abundant, the fish can combine food types. Figure 6.15a shows a two-dimensional example, where a fish approximates its nutrient target within seven meals. Whenever x mg of each food type is consumed, the cumulative food intake moves x units along the food rail corresponding to the food eaten. It is not hard to understand why Raubenheimer and Simpson called this 'a geometric framework'!

If a fish is to regulate its nutrient intake by behavioural means, it must be able to discern the nutrient composition of foods (see 'Seeking prey') and to find and capture the food (see 'Feeding modes'). For example, the rainbow trout (*Oncorhynchus mykiss*) can select foods to meet specific amino acid requirements (Yamamoto *et al.* 2000). The cost of behavioural regulation is that fishes must spend energy searching for and/or discriminating between foods ($\mathbf{f}_1$ and $\mathbf{f}_2$ in Fig. 6.15). They could be exposed to predators during a search or, if the species involved is an ambush predator, they may need to spend time defending a territory.

Physiological mechanisms, and the problem posed by nutritionally homogeneous prey, are described in Figure 6.15b. Consider three vectors, $\mathbf{c}_1$, $\mathbf{c}_2$ and $\mathbf{c}_3$, describing the daily consumption of fishes feeding on only one of the two available foods, $\mathbf{f}_1$ and $\mathbf{f}_2$. Thus, the $\mathbf{c}_i$ vectors must fall on the food rails. If the fishes have access only to $\mathbf{f}_1$ and the rate of consumption falls on $\mathbf{c}_1$, then at this point sufficient nutrient A has been obtained (see the line joining $\mathbf{c}_1$ and $\mathbf{t}$) but there is a deficiency d_B of nutrient B. The analogous case, for fishes feeding only on $\mathbf{f}_2$, is shown by $\mathbf{c}_2$, where there is sufficient nutrient B but a deficiency d_A of nutrient A. Because d_A is less than d_B, in this example, it would be better to have access to $\mathbf{f}_2$ than to $\mathbf{f}_1$ if only the one prey type was available.

Suppose now that fishes have access only to $\mathbf{f}_2$ and that they consume $\mathbf{f}_2$ at a rate $\mathbf{c}_3$. At this point, the intake of nutrient A aligns with $\mathbf{t}$, but an excess e_B of nutrient B is consumed. The excess can be stored, excreted or egested (Rodehutscord *et al.* 2000; Vielma *et al.* 2002). Thus, physiological mechanisms of regulation, like behavioural mechanisms, incur a cost to the fitness of consumers (Boersma and Elser 2006; Raubenheimer *et al.* 2009), including freshwater fishes (Fu and Xie 2004).

The fourth principle of nutritional ecology is that consumer species differ in their nutritional requirements, implying that the nutritional value of a prey community (the nutritional landscape) varies among species of fishes. The nutrient requirements of fishes do vary among species (Bowen *et al.* 1995; Clements and Raubenheimer 2006; Hendrixson *et al.* 2007) and changes in prey communities could thereby lead to changes in fish community structure.

Thus, the GF is a very useful framework within which to consider the nutritional ecology of freshwater fishes. However, although we know that stoichiometric imbalances between fishes and their prey can reduce their fitness (Malzahn *et al.* 2007), the GF has not yet been applied to freshwater fishes. Raubenheimer *et al.* (2005) were first to apply the GF to a marine fish, and their lead appears not to have been followed.

Nutritional ecology of Australian species

Although there have been no studies of Australian freshwater fishes in terms of ES and the GF, one nutritional study concerns feeding and growth of larvae following yolk absorption. The larval stages of Murray cod and trout cod retain certain highly unsaturated fatty acids (HUFA) for an extended time after first feeding (Gunasekera *et al.* 1999). This appears to be a common trait in large, freshwater carnivores, such as percichthyids, and it is true also of juveniles of the northern pike, which may encounter a growth bottleneck if their microcrustacean prey are patchily distributed in space and time. The retention of HUFAs after first feeding may be an adaptation to an uncertain food supply (Desvilettes *et al.* 1997).

In aquaculture, the growth of silver perch (*Bidyanus bidyanus*) is maximal at 0.71% dietary phosphorus (Yang *et al.* 2006). This is consistent with a proposition (Boersma and Elser 2006) that too little or too much dietary phosphorus constrains growth, owing to the physiological costs of regulation (see 'Nutritional ecology'). Has eutrophication (nutrient enrichment) of rivers and lakes constrained the growth of freshwater fishes in Australia? Have the prey communities supporting native fishes changed, in terms of nutrition? We cannot answer questions like these without knowing more of the nutritional ecology of Australian species.

Many other questions await research. In keeping with the above discussion, one question could be: how common is behavioural regulation of nutrient intake in freshwater fishes, as opposed to physiological regulation? It is difficult to imagine ambush predators, reliant on rapid acceleration for ram feeding, being able to discriminate between prey of different nutrient composition. Perhaps the relative importance of behavioural and physiological regulation depends on the feeding mode, or membership of a trophic guild.

TROPHIC GUILDS

Liem's Paradox

It seems a straightforward matter to classify freshwater fishes into trophic guilds, or groups of species that exploit the same food resources (Begon *et al.* 1996). US ichthyologist Karel Liem spent many years pursuing this issue. He was particularly interested in African cichlids, a group of fishes with skulls modified for highly specialised modes of feeding including scraping algae from rocks, manipulating and crushing snails and removing the scales of other fishes. Liem (1980, 1990) found that, although skull morphology implied very narrow trophic niches, there was a lot of dietary overlap among the species. This became known as Liem's Paradox. Similar observations are reported for marine fishes (Wainwright *et al.* 2005; Bellwood *et al.* 2006).

A possible solution to Liem's Paradox rests on the premise that some resources are easy to obtain and do not require specialised adaptations, while others are intrinsically difficult to obtain and do require specialised adaptations (Robinson and Wilson 1998). The specialist phenotypes may have evolved so that resources could be partitioned at times when more easily obtained, preferred resources become scarce. That is, when food is hard to find, it pays to have a specialised morphology that enables exploitation of an alternative, less easily obtained food source. The morphological specialisation is more adaptive still if it does not restrict feeding on easily obtainable prey in times of plenty. Thus, resource partitioning among closely related fishes of different morphology may be apparent only when food is scarce. There is support

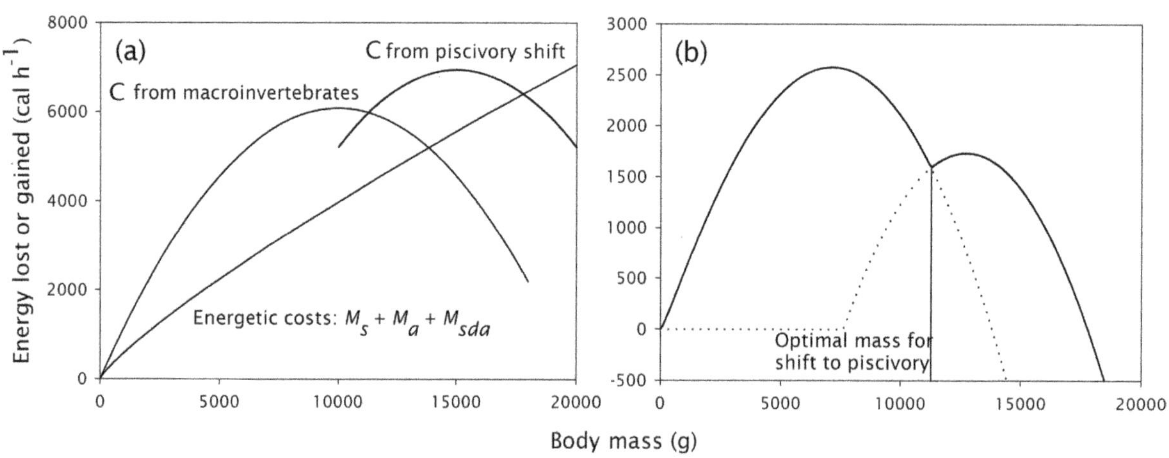

Figure 6.16: Hypothetical energy gains and losses associated with an ontogenetic dietary shift. (a) Energetic costs (M_s, = standard metabolism; M_a = active metabolism; M_{sda} = specific dynamic action) and energy gain from consumption (*C*) in a fish that makes a dietary shift from macroinvertebrate prey to piscivory. (b) The net energetic gain from the difference between costs of feeding and gains from feeding on different prey in (a). The dotted curve shows that net energetic gain from piscivory is suboptimal below a body mass of about 11 000 g and that piscivory is optimal above this mass. The reverse is true for aquatic insectivory.

for this hypothesis from investigations of an Arctic char community (Nakano *et al.* 1999).

Ontogenetic dietary shifts

It is axiomatic that in freshwater fishes a highly specialised diet is the exception rather than the rule – fishes are generalist feeders, most of the time. A second principle is that the diet of a fish is likely to change as it grows – the diet undergoes an ontogenetic shift (Ch. 7). This is not surprising, as the body mass and dimensions of fishes change by up to four orders of magnitude over the course of a life time and body size is a determinant of resource use in ecological communities (Wainwright 1996). An overview of this topic is provided by Werner and Gilliam (1984). Here, the discussion is limited to two consequences of ontogenetic dietary shifts, namely diminishing returns and species interactions.

As discussed above, the energy demand of a growing fish increases approximately as a three-quarter power of body mass (Equations 6.3, 6.4). It could be difficult for a fish to meet this demand by consuming the same prey throughout its life. If it did persist in feeding on small prey, the net energy return to the fish is likely to diminish as it grows, because the energy needed to hunt and consume small prey becomes greater than the energy returned to the fish. This pattern of diminishing returns determines the unimodal shape of a net energy gain function described by Finstad *et al.* (2006) (Fig. 6.16). By switching to larger prey, a fish avoids diminishing returns and maintains a net energy gain as it grows. Diminishing returns may drive an ontogenetic shift in carp gudgeons, from microcrustacivory (foraging on microcrustaceans) in open water to macroinvertebrates in benthic and littoral habitats (Meredith *et al.* 2003; Stoffels and Humphries 2003).

Ontogenetic dietary shifts can confound interspecific interactions. For example, Werner (1977) compared fish- and invertebrate-predators in a group of North American freshwater fishes. The juvenile piscivores (fish-eaters) inhabit the same littoral habitat as adult invertivores (invertebrate eaters) and compete for the same macroinvertebrate prey, but their morphology renders them less proficient at capturing and handling food and they are at a competitive disadvantage. Once they pass through this juvenile bottleneck they grow large and prey on the invertivorous fishes. Thus, dietary shifts can change the 'sign' of interspecific interactions.

Trophic guilds in Australia

Trophically, the most diverse family of Australian freshwater fishes, the Terapontidae or freshwater grunters, includes 21 species in seven trophic guilds (Davis *et al.* 2011). These are lepidophagous carnivores (scale-eaters), macrophagous carnivores (fishes, macrocrustaceans), meiophagous omnivores (aquatic invertebrates, plant material), macrophagous omnivores (larger fishes, macrocrustaceans, terrestrial plant material including fruits, flowers and seeds), herbivores (filamentous algae, macrophytes), detritivores/algivores and aquatic invertivores. Scale-eaters and terrestrial herbivory are unusual methods of nutrient acquisition among freshwater fishes.

How does the trophic guild structure in Australian rivers and lakes compare with that in freshwater fish communities elsewhere in the world? One way to explore this question is to make a meta-analysis of the diets of freshwater fishes from throughout Australia. Figure 6.17 shows the analysis for 92 species including only data reported as percentage composition by number; where there were multiple data for one species the mean was used, weighted by sample size. The analysis was restricted mainly to adults, as there is a lot of overlap in juvenile diets (Werner and Gilliam 1984). The percentage composition was recorded for detritus, algae, aquatic macrophytes, terrestrial plant matter, aquatic insects, microcrustaceans (Amphipoda, Cladocera, Copepoda, Isopoda, Ostracoda), macrocrustaceans (Decapoda), molluscs, fishes, fish scales and terrestrial animals.

The analysis implemented the SIMPROF routine (Clarke and Warwick 2001), using Bray-Curtis similarity to compare species. Six trophic guilds were identified, namely piscivores, algivores/detritivores, surface carnivores, microcrustacivores, aquatic insectivores and omnivores (Fig. 6.17). This aligns closely with guilds identified for freshwater fishes in Europe and North America (Matthews 1998). Putting aside variations in the names assigned by different investigators, six broad guilds are recognised: piscivores,

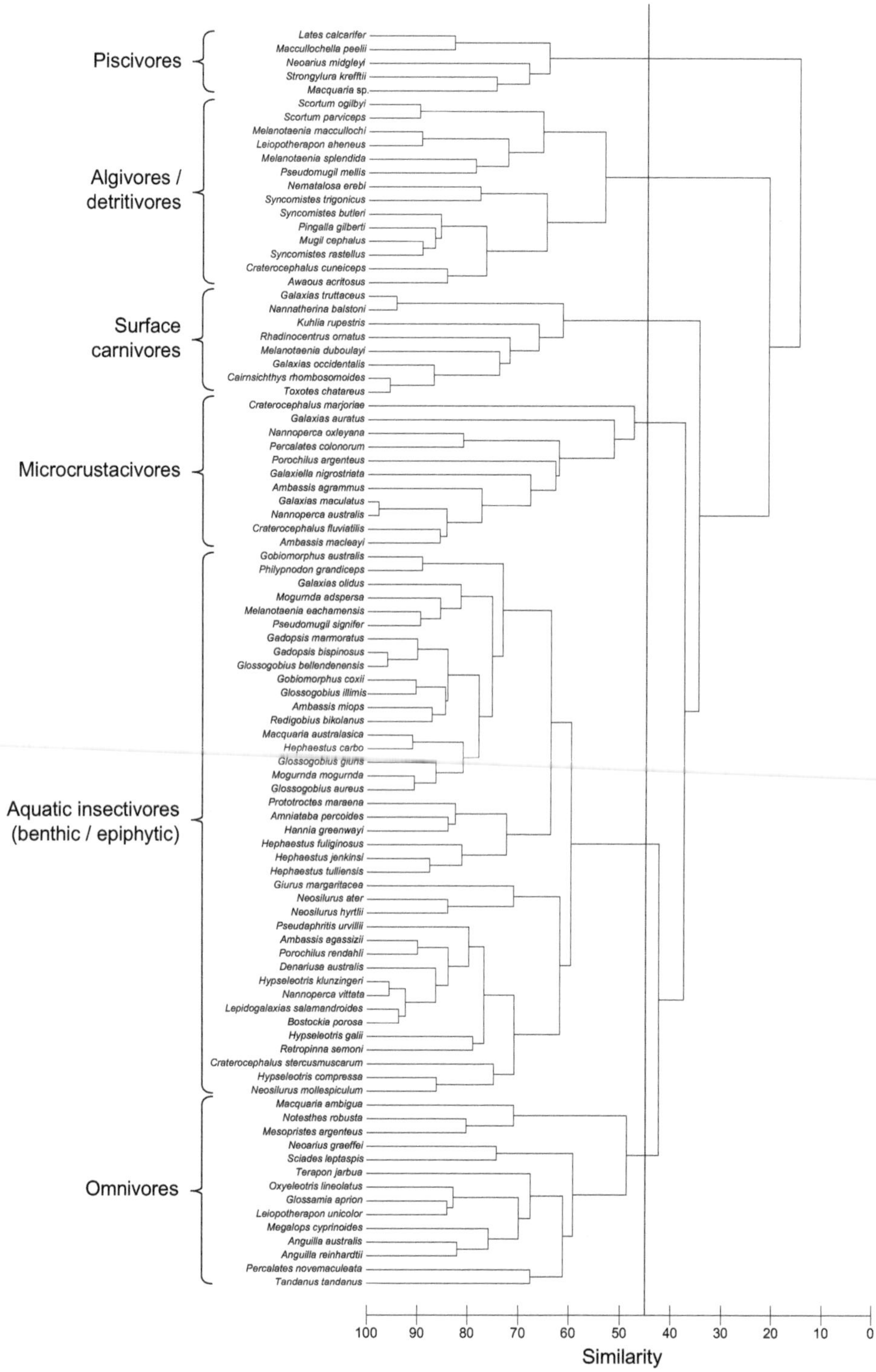

Figure 6.17: Gut-content similarities among 92 species of Australian freshwater fishes. The line at 45% similarity delineates major differences and suggests six trophic guilds (data from Pollard 1972; Jackson 1976; Cadwallader 1979; Cadwallader *et al.* 1980; Hortle and White 1980; Harris 1985; Cadwallader and Douglas 1986; Berra *et al.* 1987; Pen and Potter 1990, 1991a, b; Pen *et al.* 1993a, b; Humphries 1995; Morgan *et al.* 1995; Lintermans 1998; Gill and Morgan 2003; Morgan 2003; Howell *et al.* 2004; Pusey *et al.* 2004; Allen *et al.* 2005; Balcombe *et al.* 2005; Ebner 2006; Stuart-Smith *et al.* 2006; Gregory *et al.* 2007; Rayner *et al.* 2009b; Davis *et al.* 2011).

herbivores/detritivores, planktivores, benthic invertivores, midwater-surface feeders and omnivores.

One apparent difference, however, is that piscivory may be less common in Australian species. In North America, the ratio of piscivores to prey varies from 1:1 to 1:5 (Matthews 1998). In the lower Mulgrave River, north-eastern Queensland, the ratio of piscivores to potential prey is about 1:7.5 (Rayner *et al.* 2009b). In the Murray-Darling Basin fish assemblage, the meta-analysis suggests that there are about eight potential prey species per piscivore. These two estimates are outside the range reported for North American freshwater fishes (Matthews 1998) and they imply less diversity of piscivores in Australia. A meta-analysis is vulnerable to the biases associated with different methods, however, and the results are suggestive rather than conclusive. With that caveat, why might Australia have relatively few piscivores? That is a question for future research.

EFFECTS OF PREDATION ON PREY COMMUNITIES AND FOOD WEBS

Predation by freshwater fishes affects the structure and dynamics of prey communities. Four kinds of effects are illustrated.

Size structure and life history

In North America, populations of the amphipod *Hyalella* exposed to predation by centrarchid fishes have a smaller mean adult size, smaller size at maturity, smaller egg size and higher size-specific fecundity than *Hyalella* in habitats without fishes (Wellborn 1994). In effect, the fishes' preference for larger amphipods as prey encourages a shift in the size structure of the amphipod population. A similar shift has been demonstrated in long-term experiments (Reznick and Endler 1982; Reznick *et al.* 1990). Guppies (*Poecilia reticulata*) were subject to predation by millet (*Crenicichla alta*), a cichlid that preys on large guppies, and a killifish (*Rivulus hartii*) that preys on small sizes. The guppies exposed to *Crenicichla* predation matured at an earlier age, with a higher reproductive effort and more, smaller offspring per brood than populations subject to *Rivulus* predation. These selective pressures caused heritable shifts in the life history of the guppy population within 60 generations (11 years; Reznick *et al.* 1990).

Composition and diversity of prey

A classic study by Brooks and Dodson (1965) in the USA concerned the zooplankton in a lake before and after a natural invasion by a herring (Clupeidae), the alewife (*Alosa pseudoharengus*). Alewife are ram feeders with closely spaced gill rakers that selectively remove large zooplankters. Prior to the invasion, the zooplankton was dominated by large species but as these were removed by predators the community shifted to dominance by small species. Another example is an experiment in a Venezuelan river, where large piscivores were denied access to their prey and the abundance of the preferred prey (fishes 40–110 mm) in exclosures was significantly higher than in control treatments (Layman and Winemiller 2004).

Frequency-dependent predation (also known as 'switching behaviour,' wherein predation rates are related to the frequency of encounters with particular prey species) may promote prey biodiversity (Chesson 2000). As a species in a competitive prey community begins to dominate, its density is reduced by predation, nullifying its negative effect on others in the community. Although frequency-dependent predation has been demonstrated experimentally in fishes (Hughes and Croy 1993), its role in maintenance of prey biodiversity is unclear. For example, predation by redfin perch (*Perca fluviatilis*) may increase species evenness in the prey community, but not species richness (Diehl 1992). Another example showed that small insectivorous fishes increased the Shannon diversity of the entire guild of detritivores that they prey upon in the littoral zone of a volcanic lake in central Italy (Brooks and Dodson 1965; Mancinelli *et al.* 2002).

In Australia, selective predation by mountain galaxias on active, conspicuous invertebrate predators, dytiscid beetles, may release the beetles' prey from predation (Closs 1996). Similar direct and indirect effects of a carnivorous fish (*Hypseleotris*) on zooplankton have been demonstrated (Nielsen *et al.* 2000a, b; Ning *et al.* 2010).

Pelagic and benthic trophic cascades

A 'trophic cascade' occurs when a consumer has an indirect effect on resources that are more than one link below the consumer in the food web (Estes *et al.* 2011). These resources could be populations of animals or plants, or nutrients or detritus (Woodward *et al.* 2008).

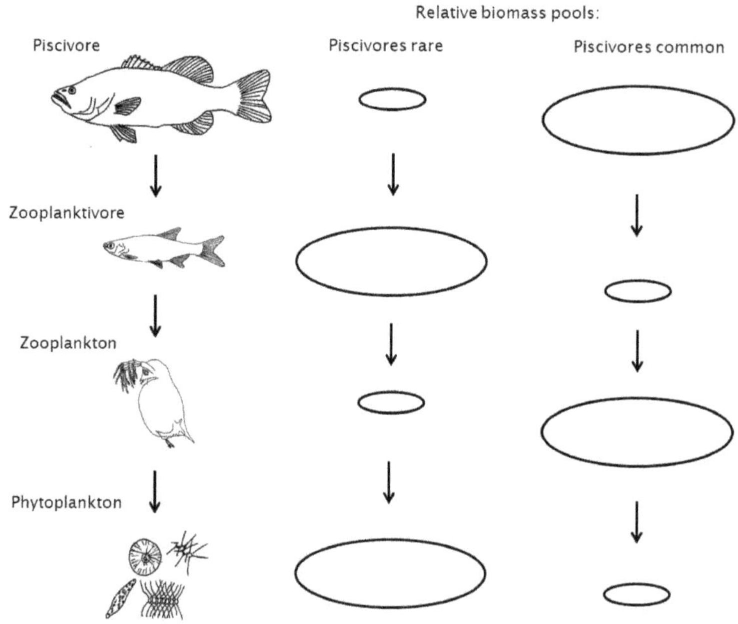

Figure 6.18: Basic model of the trophic cascade in a four-level pelagic food chain, showing contrasting distributions of biomass across trophic levels in lakes with high and low densities of top piscivores. When piscivores are rare (small oval), zooplanktivore biomass is high (large oval), resulting in low biomass of their prey (zooplankton), which in turn releases phytoplankton from predation pressure from zooplankton.

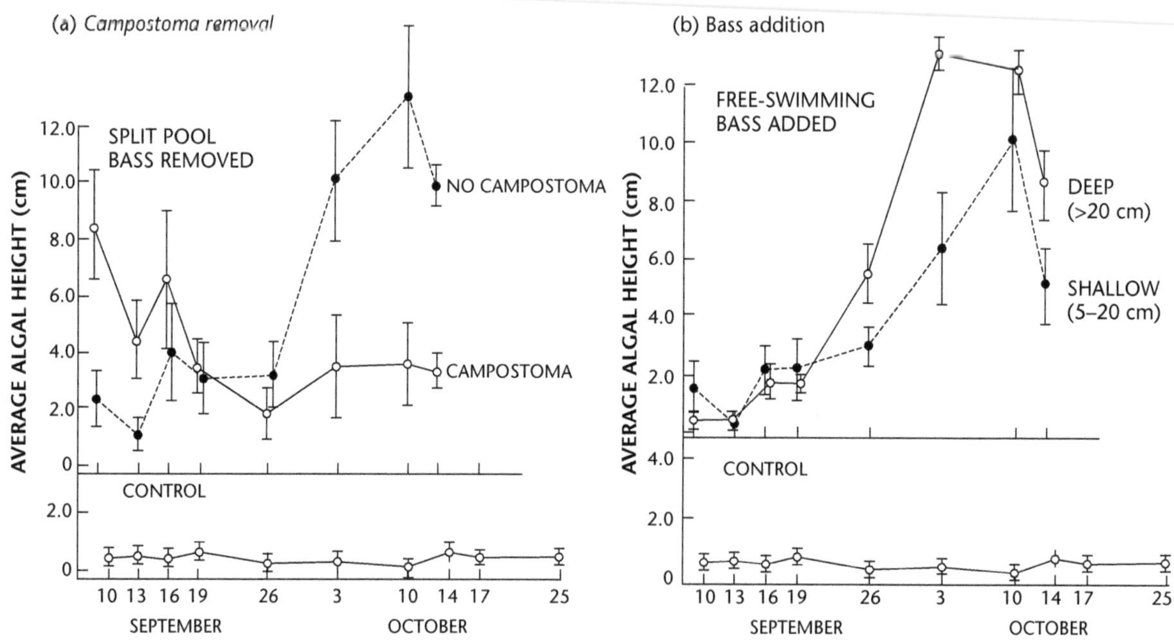

Figure 6.19: Relationships between algal standing crops and the presence of fish (modified from Power *et al.* 1985). (a) Trends in mean (±2 SE) algal standing crops in two sides of a pool (one with stoneroller minnows (*Campostoma anomalum*), the other without minnows) following manipulation of minnow abundance. The trend in mean algal standing crop in a control pool is also shown. (b) Trends in mean (±2 SE) algal biomass from shallow and deep sections of a pool to which piscivorous largemouth bass (*Micropterus salmoides*) were added. Again, the trends in a control pool are shown.

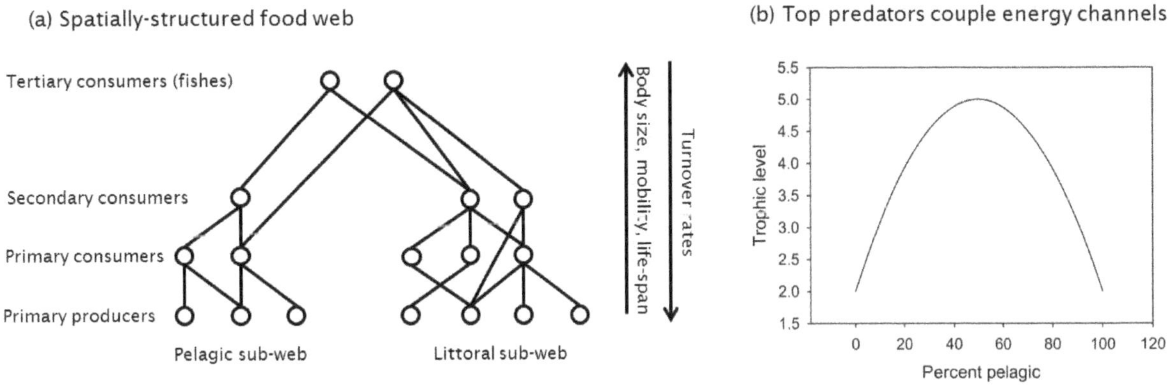

Figure 6.20: Spatially structured food webs and coupling by top (apex) predators. (a) A spatially structured limnetic food web with two distinct energy channels, one from the pelagic and the other from the littoral. Note how top predators couple the subwebs from distinct habitats. (b) The quadratic relationship between the proportion of energy derived from each of the energy channels in (a) and trophic level (see text).

Whole-lake experiments were performed by Stephen Carpenter and colleagues in Wisconsin (Carpenter and Kitchell 1993; Carpenter *et al.* 2001), manipulating the densities of a piscivore, largemouth bass, to determine the effects on the pelagic food chain. The food chain has four levels, with a top piscivore feeding on planktivorous fishes, which prey on zooplankton, which eat phytoplankton (Fig. 6.18). When bass are rare, planktivorous fishes thrive in the absence of predation. High densities of planktivorous fishes then depress zooplankton densities, freeing phytoplankton from grazing pressure. In contrast, when bass are abundant the densities of planktivorous fishes are reduced, increasing the abundance of zooplankton and thereby suppressing phytoplankton. Thus, the effects of predatory fishes cascade downward in the food chain, and the addition of a piscivore may lead to lower algal biomass. This is a 'pelagic cascade'.

A 'benthic cascade' is illustrated by a study of largemouth bass (piscivore), stoneroller minnows (*Campostoma anomalum*; algal grazer) and benthic algae in pools in an Oklahoma stream (Power *et al.* 1985). The abundances of largemouth bass and stonerollers were inversely related and there was a similar inverse relationship between stonerollers and algae, hence a positive relationship between bass and benthic algae. Pool experiments showed that, in the absence of bass, stonerollers depress the biomass of algae (Fig. 6.19a) but, when bass are added, stoneroller abundance declines and algal biomass increases (Fig. 6.19b). Thus, piscivorous fishes may have an indirect effect on benthic algae by directly reducing grazer abundance.

Spatially structured food webs

Food webs can be thought of as spatially structured channels for energy. That is, unique basal resources may be associated with discrete habitats, and the limited mobility of primary – and often secondary – consumers results in energy channels that are spatially decoupled (Fig. 6.20a) (Rooney *et al.* 2006). Organisms at lower trophic levels have higher turnover rates, hence populations of lower consumers may exhibit very rapid increases (booms) as basal energy flux increases and equally rapid reductions (busts) as those basal resources are depleted. These boom–bust cycles are not necessarily synchronised in space: while lower consumers in one habitat may be going through a boom, consumers in a neighbouring habitat may be in a bust. Theoretical and empirical research has shown that without predators to couple these spatially distinct energy channels, food web structure is unstable. That is, consumer densities, when perturbed, fluctuate strongly and return more slowly to equilibrium densities (Rooney *et al.* 2006).

However, large predators have two traits that may restore stability to spatially structured food webs: they are more mobile than lower consumers (McCann *et al.* 2005; McCann and Rooney 2009) and they often exhibit switching behaviour (frequency-dependent predation; Rooney *et al.* 2006). These traits enable them to couple subwebs (Fig. 6.20a) and respond to spatially asynchronous fluctuations in energy flux, dampening instability at large spatial scales. Indeed, meta-analyses of food webs show that predators with the highest trophic position obtain their energy from multiple habitats while lower-order consumers do not, resulting in a quadratic relationship between percentage of energy obtained from a habitat and trophic level (Fig. 6.20b).

A need for research

Australian researchers lag behind those in North America and Europe in understanding how freshwater fishes affect prey populations and food webs. How do Australian species compare with those documented elsewhere? It is widely accepted that trophic cascades occur in terrestrial and aquatic food webs (Estes *et al.* 2011), but we could expand trophic cascade theory through experimental manipulations of fish densities. Do cascades in Australia compare with those elsewhere, or does the variability and boom–bust nature of energy pulses in many inland rivers (Bunn *et al.* 2006b) result in food webs shaped more by external than internal factors? Further, if large mobile fishes are required to stabilise spatially structured food webs in riverine systems, how does foraging mode affect this process? For example, many large-bodied Australian fishes are sedentary (ambush predators with BCF transient locomotion) with a small home range (Crook *et al.* 2001). Does this foraging mode reduce their stabilising effect on freshwater food webs?

CONCLUSION

This chapter considers some elements of an integrative trophic ecology of Australian freshwater fishes. Some topics have been deliberately overlooked, including optimal diet/foraging theory (Macarthur and Pianka 1966), the consequences of nutrition for population dynamics (Persson and De Roos 2006) and the role played by phenotypes in nutrient cycles (Vanni 2002). These are homework for interested readers. Three points stand out. First, the trophic ecology of fishes is much more than the ecology of food webs – that is clear from the many references to cognate areas of science, including the above-mentioned omissions. Second, the simple questions posed in the Introduction (e.g. 'How much food does a fish need?') do not have simple answers – again, they draw upon skills from many related disciplines, especially physiology. Third, many exciting and testable ideas await the next generation of freshwater fish ecologists.

7 Reproduction and early life history

Alison J. King, Paul Humphries and Nicole G. McCasker

INTRODUCTION

The sheer diversity of reproductive and life history strategies in fishes is astonishing, reflecting the size, shape, behaviour and life styles of about 32 000 species that swim in the world's oceans, lakes and rivers. Some species breed within months of hatching and live for perhaps a year; others take 10 years or more to mature and live for more than a century. Some produce one or two young, cocooned in elaborate egg envelopes; others produce millions of microscopic eggs that appear as planktonic clouds in the water. The parents of some species spend months caring for their eggs and young, using sophisticated methods such as mouth brooding or even head brooding; others broadcast their eggs into the currents like seeds in the wind, and have no further role in their fate. Fishes are unique among vertebrates in producing hundreds, thousands and even millions of young, even in a single breeding event. In so large a brood the chances of survival for individuals are small, yet some clearly do survive. Their numbers determine the strength of their cohort and the number of recruits to the adult population. Understanding how different combinations of biological characteristics and environmental conditions affect survival and recruitment is the holy grail of fisheries research, and many scientists have dedicated their careers to that end.

Clearly, for a population to persist the fishes must produce viable young that grow to sexual maturity and reproduce themselves. Most fish begin life as a *zygote* (a fertilised egg); the eggs may be attached to surfaces, deposited in nests crafted by their parents or cast into the water to fend for themselves. Life cycles vary among species (Fig. 7.1) but most include an *embryo* (from zygote to first feeding, with all nutrition from the yolk sac), the *larva* (from first external feeding to juvenile metamorphosis), the *juvenile* (from when fins and other structures are fully differentiated to sexual maturity), the *adult* (the period dominated by reproduction) and *senescence* (the post-reproductive period, characterised by slow growth, reduced intensity and frequency of spawning and degeneration).

During the life cycle, the most profound changes are in the early, post-hatching stages of life and are reflected in morphology, physiology, behaviour and ecology. In this period, the weight of a growing individual can increase one million times! Despite the importance of this period for conservation and management, there is a great deal we do not know about the early lives of Australian freshwater fishes. How and when do they spawn? How many eggs do they produce? Do they spawn once only in a season and then die, or do they spawn multiple times? Which species care for their young by guarding them, or by head- or mouth-brooding? What constraints do the early stages encounter as they develop into adults? And how do flooding, food supply, habitat availability and other environment factors affect survival and recruitment?

This chapter explores the life history attributes, reproductive styles and spawning behaviours of Australian freshwater fishes, their early life stages and factors affecting mortality and recruitment. We borrow facts and ideas from research in other parts of the world, but use Australian examples where possible. Many of our examples come from south-eastern Australia, where most research has been done, but we have tried to minimise that bias. We conclude with an appraisal of progress so far, and some ideas that we believe are worthy of future study.

LIFE HISTORY

Models and theory

Life is an evolutionary game of chance, and fishes, like all organisms, 'gamble' on ways to ensure that their offspring have the best chance of survival to maturity. The variables include 'decisions' about age at first breeding, what time to breed, whether to breed once or more in a life time and how to package and disperse the progeny. Life history theory attempts to understand the factors responsible for the evolution of particular traits that maximise the potential for survival of offspring to reach reproductive age. Life history theory has been a popular topic for many years, but it was boosted 40-odd years ago by ideas about *r*-selection and *K*-selection (MacArthur and Wilson 1967; Pianka 1970). The term *r* is the intrinsic rate of population increase, and *K* refers to the carrying capacity of the local environment. Species that are *r*-selected are governed by density-independent factors (e.g. floods, droughts) and tend to mature early and live for short periods, to have high fecundity and to make a relatively small investment of energy per offspring. Species that are *K*-selected are affected by density-dependent factors (e.g. competition, predation, disease) and typically have delayed reproduction, long lives, low fecundity and a large investment of energy per offspring.

'Bet-hedging' was proposed later as an alternative idea relating life history traits to the survival of juveniles and adults (Stearns 1992). For instance, if juvenile mortality is greater than that of adults it is advantageous for the juvenile stage to be short and for individuals to reproduce early. Alternatively, if juvenile mortality is less than that of adults it may be better to remain a juvenile for longer, mature later and larger, then reproduce with those advantages. These 'strategies' may be adaptive if, for example, the effects of predation, food availability or competition are stronger at one life stage than another.

A triangular model of life history evolution in fishes (Winemiller and Rose 1992) was developed from these ideas and others concerning plants and insects (Grime 1977; Southwood 1977; Kawasaki 1980; Greenslade 1983) (Fig. 7.2a). The model attempts to explain the responses of organisms to a changeable environment and how, in fishes, this relates to the time taken to reach maturity. The *r*- and *K*-selection continuum is elaborated as follows:

- *opportunistic species* (*r*-selected species) – small, short-lived species that mature early and make a large reproductive effort, but with low fecundity per batch and a small investment of energy per offspring (e.g. carp gudgeons, *Hypseleotris* spp.; Australian smelt, *Retropinna semoni*; Agassiz's glassfish, *Ambassis agassizii*);
- *periodic species* – relatively large, long-lived species that mature late and make a moderate reproductive effort, with high fecundity per batch and a small energetic investment per offspring (e.g. bony herring, *Nematalosa erebi*; spangled perch, *Leiopotherapon unicolor*; lampreys: Geotriidae, Mordaciidae);
- *equilibrium species* (*K*-selected species) – species of variable size, maturing at medium to late stages, with low fecundity per batch and a large energetic investment per offspring (e.g. Murray cod, *Maccullochella peelii*; river blackfishes, *Gadopsis* spp.).

This model was developed for the Americas but applies broadly to marine fishes (King and McFarlane 2003) and to freshwater fishes in North America (Mims *et al.* 2010), Europe (Vila-Gispert and Moreno-Amich 2002; Eros 2005) and south-eastern Australia (Growns 2004). Analyses of the life history traits of Murray-Darling Basin species (Humphries *et al.* 1999) and another 53 Australian species (Fig. 7.2b) indicate general agreement with the model.

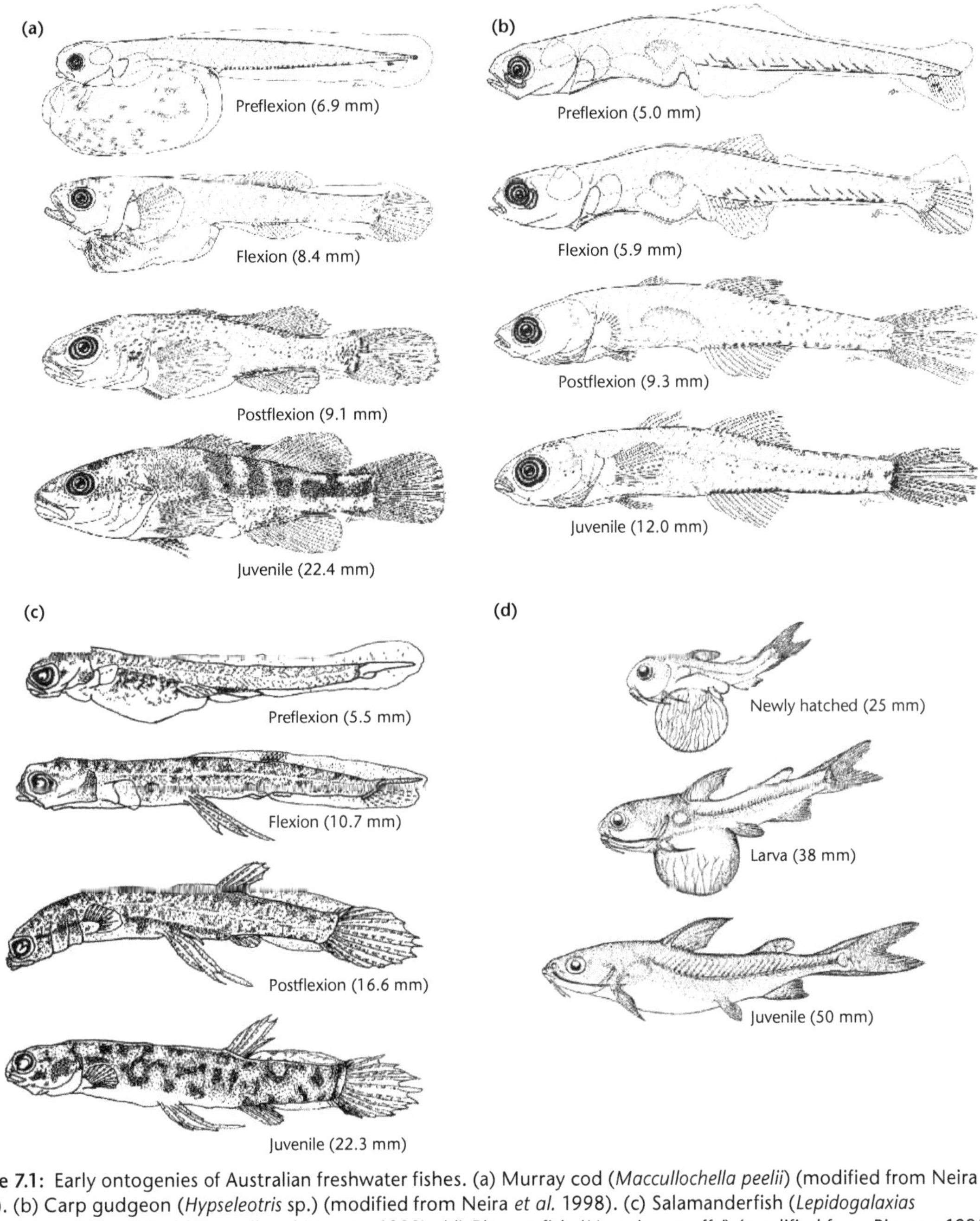

Figure 7.1: Early ontogenies of Australian freshwater fishes. (a) Murray cod (*Maccullochella peelii*) (modified from Neira *et al.* 1998). (b) Carp gudgeon (*Hypseleotris* sp.) (modified from Neira *et al.* 1998). (c) Salamanderfish (*Lepidogalaxias salamandroides*) (modified from Gill and Morgan 1999). (d) Blue catfish (*Neoarius graeffei*) (modified from Rimmer 1985a).

Species with different strategies may co-occur and can persist for long periods. For example, Murray cod, an equilibrium species, bony herring, a periodic species, and eastern gambusia (*Gambusia holbrooki*), an opportunistic alien species, all cohabit in the Murray-Darling Basin. Each is able to move between favourable and unfavourable habitats during different flow conditions (Schlosser 1991; Fausch *et al.* 2002; Lake *et al.* 2007). For fishes in highly variable, unpredictable inland rivers (Puckridge *et al.* 1998), mobility is a necessity.

An extension of this idea could suggest that recruitment in species from one strategy would be more favoured under particular climatic and flow conditions than other strategies (Fig. 7.3). For example, in drought,

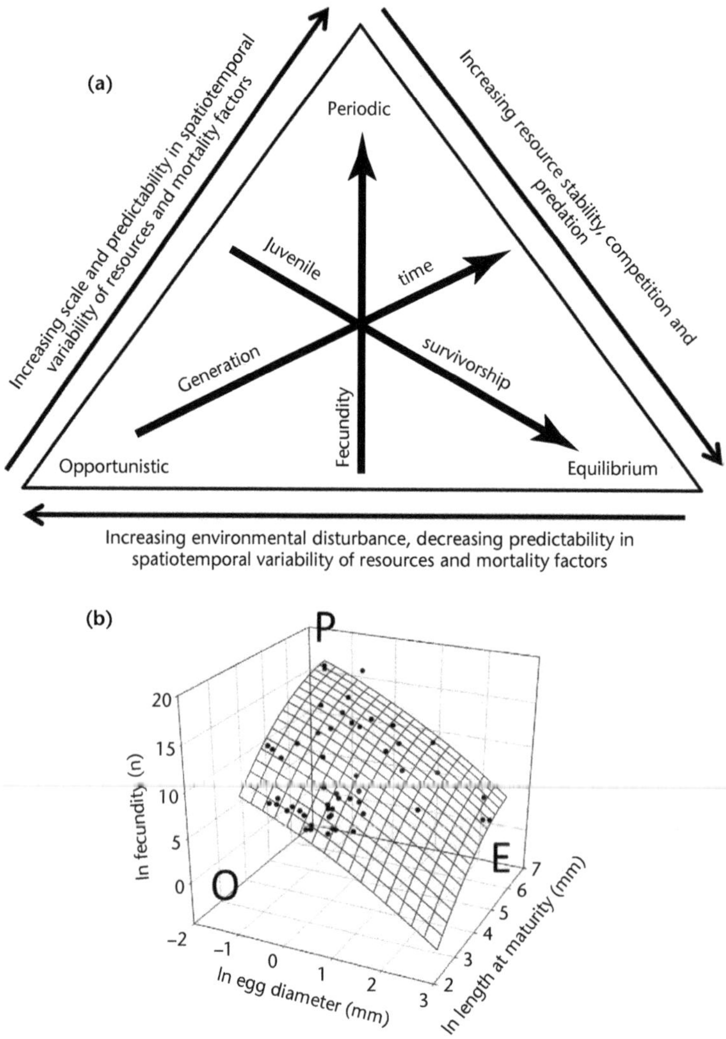

Figure 7.2: Triangular model of life history strategies of fishes. (a) General model of life history strategies (modified from Winemiller 1995). (b) Surface plot of length at maturity, intra-uterine egg diameter and fecundity and approximate location of equilibrium (E), opportunistic (O) and periodic (P) strategies for 53 species of Australian freshwater fishes. Non-linear 3D paraboloid analysis: $F = 50.28$, $R^2 = 0.79$, $P < 0.001$.

when rivers contract to pools and movements are extremely limited, equilibrium species with a big energetic investment per offspring but with low fecundity may recruit better than periodic species needing specific conditions (e.g. a flood) to cue reproduction. Opportunists may be compelled to breed, because of their short lives, and may have moderate recruitment success. At the other extreme, when rivers are in flood, large-scale movements are possible longitudinally and laterally and food and habitat resources are not limiting. This could favour periodic species, producing many young which disperse widely and potentially breed in response to rising flows. Thus, opportunists may breed and recruit moderately well under most conditions, whereas equilibrium and periodic species may assume dominance at different times. The comparative success of different life history strategies under various flow conditions would be an interesting focus for further research, requiring long-term studies conducted during contrasting flow conditions (but see Cadwallader 1976a; Balcombe *et al.* 2006; Balcombe and Arthington 2009; Kerezsy *et al.* 2011).

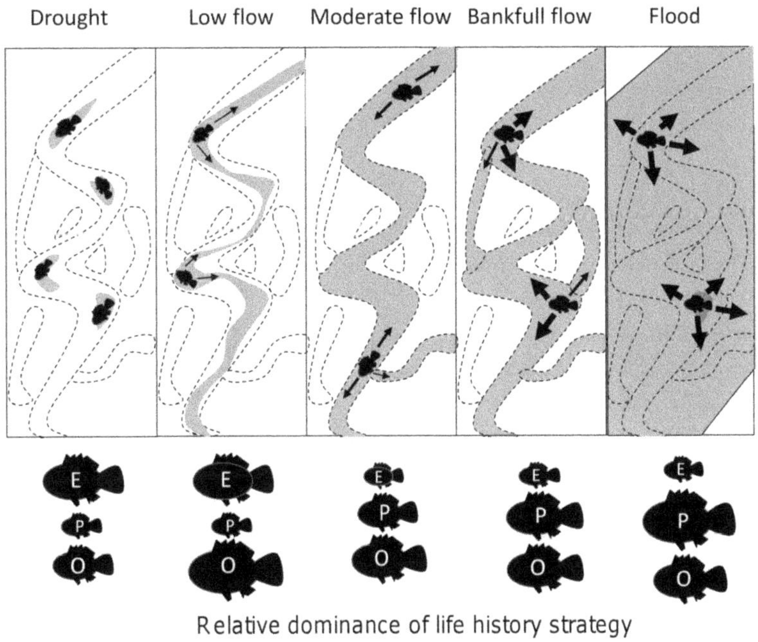

Figure 7.3: Speculative relative recruitment success of equilibrium (E), periodic (P) and opportunistic (O) species under different flow conditions (drought–flood). The size of the fishes represents the relative dominance of life history strategies under each flow condition. For example, equilibrium strategy species (e.g. Murray cod, *Maccullochella peelii*) should recruit well when flow conditions are relatively stable whereas periodic species (e.g. golden perch, *Macquaria ambigua*) should recruit well under periodic flooding.

Age and size at maturity

Fishes have limited energy to invest and a key issue in life history theory is to consider how that energy should be deployed – whether to favour maintenance and somatic growth, or reproduction (see Berg and Finstad 2008; Fig. 7.4). The reproductive investment includes not just gonad development but secondary sexual characters, spawning and post-spawning migrations and behaviours associated with spawning, such as nest building, mate attraction and guarding.

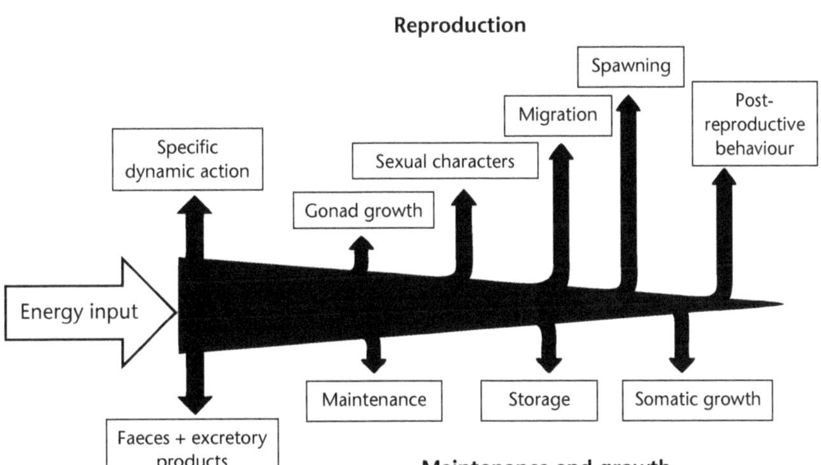

Figure 7.4: Conceptualisation of energy-use trade-offs for adult fishes. After loss of energy through specific dynamic action and waste excretion, energy ingested is allocated broadly to reproduction or maintenance and growth.

Fishes have *indeterminate* growth (Ch. 8). Thus, the bigger a female grows the more eggs she can carry, and the bigger a male fish the bigger his home range and the chance that he can defend a territory and attract a mate (Kramer and Chapman 1999). Being larger also brings a capacity to migrate longer distances, a significant advantage for diadromous species such as salmonids (Salmonidae), eels (Anguillidae) and lampreys (Geotriidae, Mordaciidae) (Ch. 5; Bird and Potter 1981; Fleming 1996). Even so, females generally mature at a larger size than males of the same species, and dimorphism is common (see 'Secondary sexual characters').

The longer that maturation in females is delayed, the larger they grow and the more eggs they produce (Reznick 1983). Then, the decision 'when to breed' depends on the probability of survival (Cole 1954; Stearns 1992). If there is strong larval and juvenile (non-reproductive) survival, selection should favour delayed breeding and a large body, but if there is weak larval and juvenile survival, selection should favour a small size and early maturation. For example, many hardyheads (Atherinidae) mature when small and young (Allen *et al.* 2005). On the other hand, males and females of the Australian lungfish (*Neoceratodus forsteri*) mature at 17 and 22 years, respectively (Pusey *et al.* 2004). Eels mature at similar ages (Allen *et al.* 2002).

Variations in size and age at maturity are often reported (Humphries 1989; Rowland 1998b; Hughes and Stewart 2006; Newman *et al.* 2008) but the cause is rarely identified. *Ultimate factors* ('evolutionary' factors) that influence age at maturity can be investigated by modelling and by interspecific or phylogenetic comparisons. Most Australian freshwater fishes mature by their first year; the percentage is slightly less for females than males (Fig. 7.5a). Size at maturity increases logarithmically with age and females tend to be slightly larger than males (Fig. 7.5b). *Proximate factors* are more easily studied. For example, intraspecific and even intrapopulation variations in age and size at maturity may be related to population density (Rodd *et al.* 1997), temperature (Morgan *et al.* 2002), food (Jonsson and Jonsson 2009), migration distance (Jonsson 2006), water-level changes (Chappaz *et al.* 1989), fishing pressure (Post *et al.* 2003) and lake acidification (Heibo and Vøllestad 2002). Thus, ultimate factors may determine whether there is scope for variation (plasticity) in body size and age at maturity in populations or species, but proximate factors determine the extent of variation. Size-selective predation, for example, is a proximate factor that could affect age at maturity in a population (Rodd *et al.* 1997).

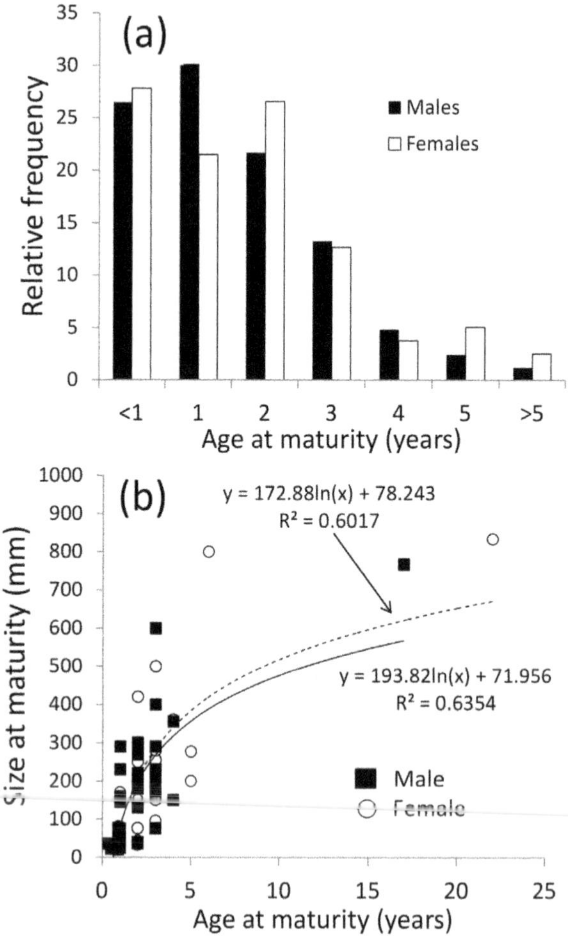

Figure 7.5: Age at maturity for Australian freshwater fishes. (a) Relative frequencies of age at maturity (males, 83 species; females, 70 species). (b) Relationship between age and size at maturity (males, 63 species, solid line; females, 64 species, dashed line).

Breed once or breed often?

Some fishes breed once in the course of their life cycle (semelparity) and others breed many times (iteroparity). Life history traits like these may be grouped as elements of strategies. Opportunists, for example, tend to be small and short-lived, to mature early and to be semelparous, whereas late maturers tend to be large, long-lived and usually iteroparous. There is a third, intermediate strategy that we will call 'quasisemelparity'. Most representatives are small species allied to the opportunists and their populations usually are dominated by 0+ individuals because, while individuals can live for more than one year, most do not. The contribution that these older, quasi-semelparous individuals make to their populations is not known. All three breeding types occur in Australia, but quasi-semelparous and iteroparous species predominate (Table 7.1). There are exceptions, of course, as nature often defies simple classifications. Lampreys, for example, are relatively large and long-lived but breed only once, and *Paragalaxias* species are small but live and breed over several years.

Breeding frequency and investment

Some fishes are 'single spawners' (total spawners) that spawn once per breeding season, and others are 'repeat spawners' that spawn on several occasions

Table 7.1: Breeding strategies and spawning patterns in Australian freshwater fishes (* alien species)

Strategy/pattern	Definition	Examples
Semelparous	Spawns once in life time	Australian eels (Anguillidae) (not confirmed) Some hardyheads (Atherinidae) Some *Galaxiella* spp. Tasmanian whitebait (*Lovettia sealii*) Lampreys (Geotriidae, Mordaciidae)
Quasi-semelparous	Small part of population lives for more than one year	Australian smelt (*Retropinna semoni*) Some carp gudgeons (Eleotridae) Some *Galaxias* spp. Salamanderfish (*Lepidogalaxias salamandroides*) Rainbowfishes (Melanotaeniidae) Some pygmy perches (Percichthyidae)
Iteroparous	Spawn more than once in life time	Ariidae Some *Galaxias* spp. *Paragalaxias* spp. Percichthyidae Plotosidae River blackfishes Common carp (*Cyprinus carpio*)*
Single – synchronous spawners	Spawn once per breeding season	Galaxiidae (many, incl. *Galaxias maculatus*, *Galaxiella pusilla*) Southern pygmy perch (*Nannoperca australis*) Nightfish (*Bostockia porosa*) Freshwater catfish (*Tandanus tandanus*) Kimberley archerfish (*Toxotes kimberleyensis*) Redfin perch (*Perca fluviatilis*)*
Single – asynchronous spawners	Individuals spawn at different times	Flathead gudgeon (*Philypnodon grandiceps*) Common carp (*Cyprinus carpio*)*
Repeat – serial spawners	Individuals spawn on several occasions, via incremental batches	Deep hardyhead (*Craterocephalus cuneiceps*) Jungle perch (*Kuhlia rupestris*) Barramundi (*Lates calcarifer*) Oxleyan pygmy perch (*Nannoperca oxleyana*) Murray cod (*Maccullochella peelii*) Australian smelt (*Retropinna semoni*)
Repeat – batch spawners	Individuals spawn on several occasions via several clutches	Agassiz's glassfish (*Ambassis agassizii*) Carp gudgeons (*Hypseleotris* spp.) Salamanderfish (*Lepidogalaxias salamandroides*) Cairns rainbowfish (*Cairnsichthys rhombosomoides*) Western pygmy perch (*Nannoperca vittata*) Eastern gambusia (*Gambusia holbrooki*)*

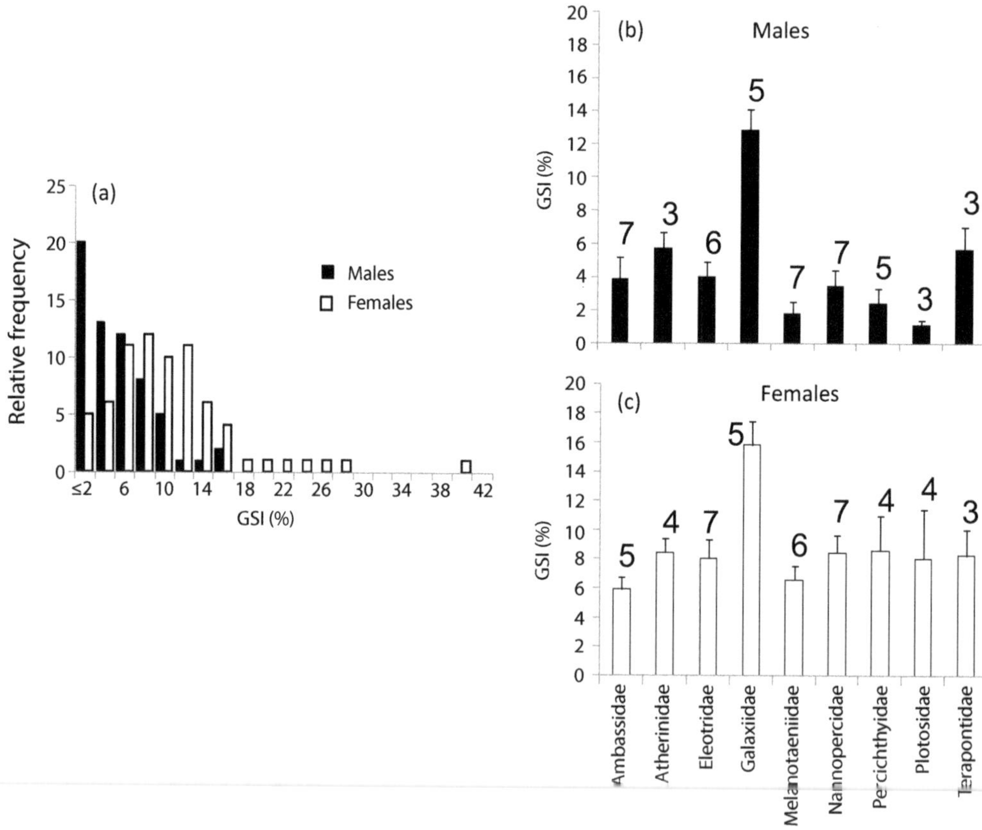

Figure 7.6: Records of Gonadosomatic Index (GSI) values in males and females of selected families of Australian freshwater fishes (numbers above columns show number of species). (a) Relative frequencies of GSI values (males, 62 species; females, 72 species). (b) Mean (+1 SE) GSI for males. (c) Mean (+1 SE) GSI for females.

during the season. Repeat spawners can be divided further as 'serial spawners', with incremental releases of eggs at different stages of development, and 'batch spawners', producing several batches or clutches of eggs. In some single spawners, some individuals may spawn at different times (asynchronously), so that the breeding season for the population as a whole is extended (Table 7.1).

Single spawning is advantageous when conditions are predictable and egg and larval mortality are low, or when eggs and/or larvae are abundant and readily dispersed. Thus, species such as Clarence River cod (*Maccullochella ikei*: Butler and Rowland 2009) spawn once per season and exhibit parental care and other sophisticated behaviour, while others, such as golden perch (*Macquaria ambigua*: Lake 1967c), broadcast many small eggs freely into the water. Repeat spawning or asynchronous single spawning may be advantageous when conditions are not predictable, because (excuse the pun) not all their eggs are 'put in the one basket' (Humphries *et al.* 2002). If the progeny from one batch die, the next batch may succeed.

It is interesting to consider the implications of the energetic investment in reproduction (mentioned earlier). The most common measure (albeit an index) is the weight of gonad as a percentage of total body weight, prior to spawning. This is the *gonadosomatic index* (GSI).

Females tend to have larger GSIs than males (Fig. 7.6). For example, blue catfish (*Neoarius graeffei*), whose females have ovaries 57 times the weight of the males' testes (GSI 16 v. 0.28%, respectively; Pusey *et al.* 2004); barramundi (*Lates calcarifer*), with female GSI eight times that of the males (GSI 40 v. 5%, respectively) and black catfish (*Neosilurus ater*), with female GSI six times that of the males (GSI 10.5 v. 1.7%, respectively; Orr and Milward 1984). Among Australian freshwater fishes, males typically have GSIs less than 2% (mode of 62 species: Fig. 7.6b) and females have GSIs of 6–8% (mode of 72 species: Fig. 7.6c), although values of 20–40% do occur (e.g. landlocked spotted galaxias, *Galaxias truttaceus*). Male GSIs tend to be uniform within and between taxonomic families, whereas female GSIs are more variable (Figs 7.6b, c). Members of the Galaxiidae have the largest GSIs, both for males and females.

The magnitude and timing of peak GSIs vary with local environmental conditions (Humphries 1989; Pusey *et al.* 2001). In Tasmania, landlocked populations of the spotted galaxias, normally an amphidromous species, occur in streams and lakes. The lotic females spawn in late autumn with GSIs peaking at 18%, but lentic females continue to develop until spring and spawn with GSIs of 25–30% (Humphries 1989). GSIs can vary with food supply – under experimental conditions, golden perch deprived of food and then fed may attain higher GSIs than others fed continuously to satiety (Collins and Anderson 1999).

Fecundity and egg size

Fecundity (the number of eggs produced per female per season) varies from a few hundred in river blackfishes and Swan River goby (*Pseudogobius olorum*) to hundreds of thousands in estuary perch (*Percalates colonorum*) and bony herring, to millions in sea mullet (*Mugil cephalus*), southern shortfin eel (*Anguilla australis*) and barramundi. One female barramundi is reported to have contained 46 million eggs (Davis 1984)!

Fecundity generally is related to the cube of length (Wootton 1998). Figure 7.7 illustrates this relationship for 16 Australian species. In the plot, the slopes and intercepts of the relationships vary considerably but the overall trend is similar, except for a group including Murray cod, freshwater catfish, blue catfish and river blackfish (*Gadopsis marmoratus*). These are large, relatively long-lived species that mature at 4+ years and have large eggs (3–12 mm), whereas most others mature earlier and have smaller eggs (typically < 3 mm).

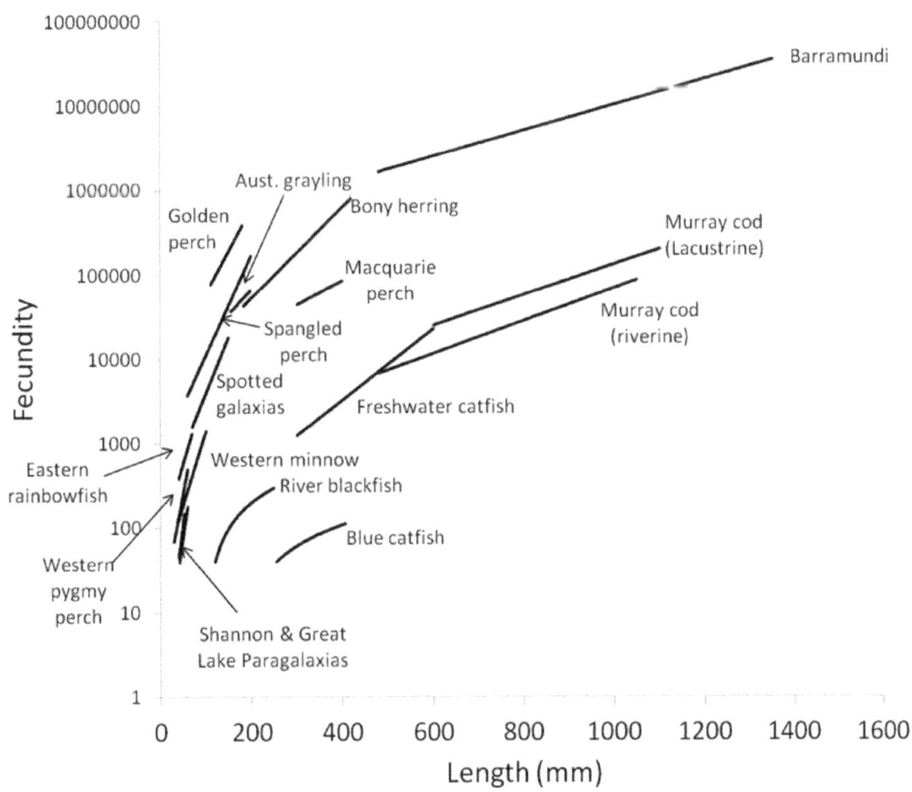

Figure 7.7: Length–fecundity relationships from published studies for 16 species of Australian freshwater fishes. Relationships may not span the full length range of the species.

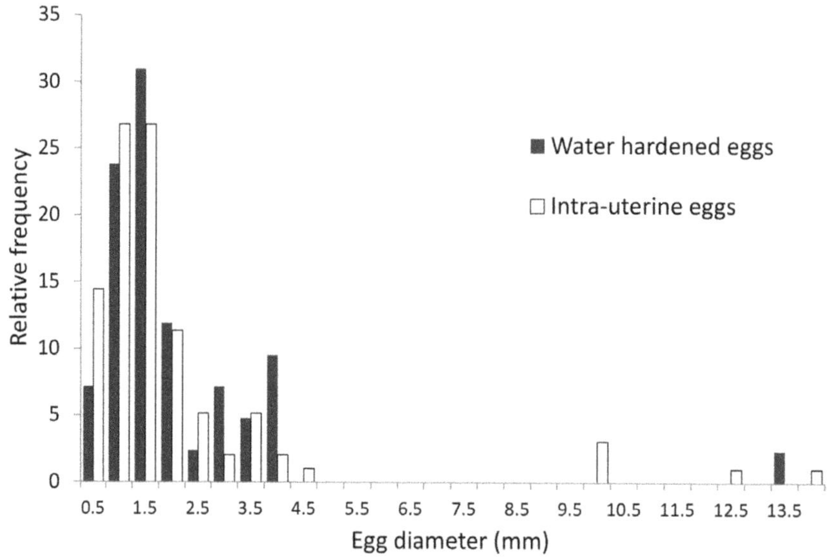

Figure 7.8: Frequency histogram of intra-uterine and water-hardened egg diameters for 45 species of Australian freshwater fishes.

Fecundity may vary between locations and habitats. For example, the fecundity of eastern rainbowfish (*Melanotaenia splendida*) varies among locations although the mean numbers of eggs per batch does not (Pusey *et al.* 2001). Another example is the difference in fecundity of lentic and lotic populations of spotted galaxias (Humphries 1989), mentioned above (see 'Breeding frequency and reproductive investment').

Egg size is usually measured as the diameter of eggs in the body cavity prior to the hardening and swelling associated with fertilisation and contact with water. In some species, egg diameters scarcely increase after fertilisation, but those of silver perch (*Bidyanus bidyanus*), Hyrtl's catfish (*Neosilurus hyrtlii*) and Macquarie perch (*Macquaria australasica*) can increase two- or three-fold (Appleford *et al.* 1998; Lintermans 2007). These changes may confound comparisons of intra-uterine data (Balon 1999) and although a more meaningful measure is to determine the density and volume of yolk, few researchers do that. Even so, there is information to be gleaned from coarse measurements of egg diameter. Most Australian freshwater fishes produce pre-fertilised eggs with diameters of 1.5 mm or less, a few have eggs of 1.5–5 mm and three species (blue catfish, boofhead catfish (*Sciades leptaspis*) and southern saratoga (*Scleropages leichhardti*)) have eggs exceeding 5 mm (Fig. 7.8).

Although species with high fecundity often have small eggs and species with low fecundity often have large eggs, there is not a statistically significant relationship between fecundity and egg diameter. This is not surprising, as fecundity is related to body size and egg diameter is not. In fact, the fecundity–egg size relationship is wedge-shaped, showing that larger species can produce larger eggs but do not necessarily do so. Few Australian studies have explored relationships between egg diameter and the length or weight of female fishes, although the topic of maternal effects on reproductive traits has received much attention overseas (Matthews 1998).

Parental care

Most species of fishes release their eggs or sperm into the water and do little more to care for the fertilised eggs or embryos, but 20% of all teleosts exhibit 'parental care', a form of behaviour that enhances the survival and development of the offspring (Sargent 1997). Of all species that show parental care, worldwide, about 50% are paternal care (male only), 18% are maternal care (female only) and 32% are biparental care (both parents) (Sargent 1997). The common mode is for one or both parents to guard the eggs from predators as they develop, and they may also fan the eggs to prevent siltation and maintain oxygen levels

(Sargent and Gross 1993). Other, more elaborate types include internal brooding, mouth-brooding and carrying the eggs externally (e.g. forehead brooding) or in a brood-pouch. Some species care for their developing young, including cichlids (Cichlidae), sunfishes (Centrarchidae), sticklebacks (Gasterosteidae) (Magnhagen 2008) and a number of Australian species (Tables 7.2, 7.3).

Parental care is well-studied by evolutionary biologists (see reviews by Sargent 1997; Magnhagen 2008). It is suggested to enhance the survival of eggs and larvae in situations where predation on adults is low (Wisenden 1999), and therefore is more likely to develop in larger species than in smaller species (Magnhagen 2008). In experiments with the biparental convict cichlid (*Archocentrus nigrofasciatus*), an alien species in Australia, removal of the male reduces the survival of young, although the effect is less at later stages of development (Keenleyside and Mackereth 1992). The costs of parental care to the parent can be significant. The consequences may be to prejudice survival of the adult, increase the time to next breeding and reduce fecundity and the numbers of future offspring (Sargent 1997). As carers generally do not feed, there is a trade-off, especially for females, between current and future reproduction. For males, there may be no opportunity to mate again during the current season. Little is known about the role of parental care in Australian freshwater fishes. There is scope for testing alternative models to evaluate the costs and benefits of parental care to the parents, their young and the population (Sargent and Gross 1993; Sargent 1997).

Of 92 Australian freshwater fish species for which we have data, about 35% show some form of parental care, 54 species show no parental care, 29 guard eggs only (mostly Eleotridae, Gobiidae), five tend eggs, free embryos and larvae in nest-like constructions (mostly Gadopsidae, Percichthyidae, Plotosidae), four brood eggs, free embryos and larvae orally (Ariidae,

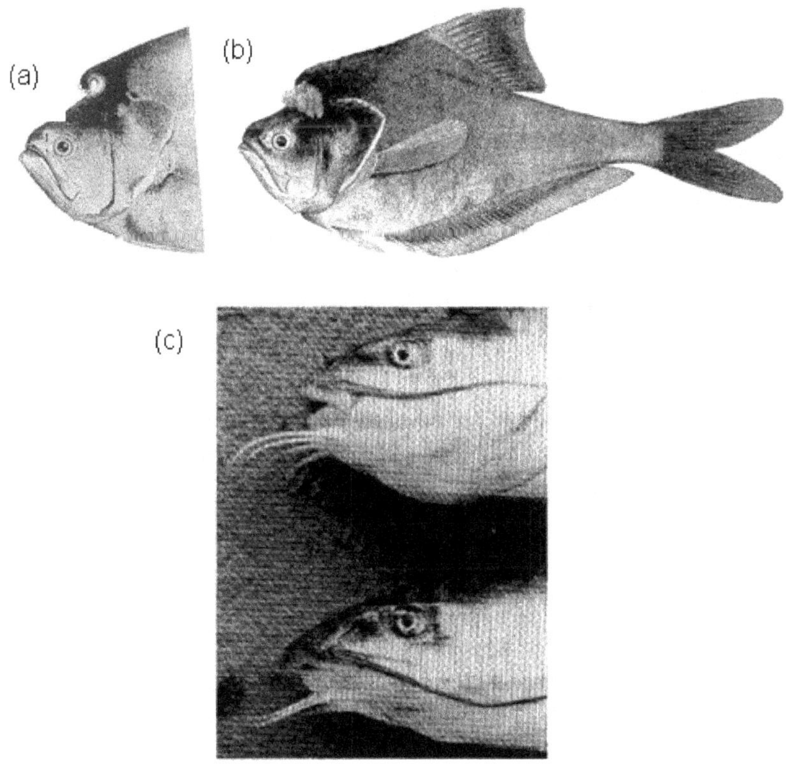

Figure 7.9: Male nurseryfish (*Kurtus gulliveri*). (a) Forehead hook. (b) With egg mass (from Berra and Humphrey 2002). (c) Brooding (top) and non-brooding (bottom) male blue catfish (*Neoarius graeffei*), showing distended branchial region (from Rimmer 1985a).

saratogas) and one, described below, broods its eggs on its head. Parental care in Australian freshwater fishes almost always is paternal, and there may be only one case (Great Lake galaxias, *Paragalaxias eleotroides*) where females guard the eggs (Fulton 1982). Three examples of paternal care are described below.

- The male nurseryfish (*Kurtus gulliveri*), from the lower reaches of rivers in northern Australia, develops a bulbous, scale-less, hook-like extension of the supra-occipital bone in the head, where it holds a cluster of eggs (Berra and Humphrey 2002) (Fig. 7.9a, b). Little is known of how the egg mass is fertilised or attached to the hook, but it is likely that the male uses his hook to move the fertilised egg mass from the female's vent and the eggs then adhere (Berra 2007). It is not known how long they remain attached, or whether the male performs other parental-care behaviour. This bizarre adaptation is thought to enhance egg survival in habitats that are prone to tidal disturbances and hypoxia.
- Fork-tailed catfishes (Ariidae) are oral brooders (Rimmer and Merrick 1983). Male blue catfish usually carry about 10 large (12–14 mm diameter) eggs but can carry as many as 123 eggs (Lake and Midgley 1970; Rimmer 1985a; Fig. 7.9c). The males do not feed during brooding, as the palatine teeth are covered by epithelial tissue. Eggs hatch after four to five weeks and the larvae are retained in the mouth at least until the yolk is absorbed (and may be retained up to 50–100 mm). The brooding period is usually six to nine weeks. Larvae begin feeding on plankton soon after hatching, while still in the mouth of the male parent.
- A revealing study of parental care involved underwater video-recording of spawning and post-spawning behaviour by Clarence River cod in rivers of north-eastern New South Wales (Butler and Rowland 2009). There is elaborate pre-spawning behaviour before egg deposition, but the female then leaves and does not return. The male remains to patrol and defend the nest, and ceases feeding. He guards the eggs and larvae for up to 24 days, initially protecting and cleaning the eggs then fanning them continuously after hatching. For five to six days after hatching the larvae remain near the substratum, then move about with their male parent still present. At 12–14 days after hatching the larvae leave the nest, and the male does likewise.

BREEDING SYSTEMS

For favourable genes to be passed to future generations, an individual must allocate time and resources to reproduction and produce offspring that, in turn, survive to reproduce. Fishes breed in a multitude of different ways; indeed, they display the greatest variety of reproductive styles of all vertebrates. Reproduction is a resource-intensive period, however, and patterns of investment vary with the breeding system employed by the respective species. In this section we consider breeding systems in fishes and highlight some Australian examples (Table 7.2).

Gender

In fishes, as in other vertebrates, gender is determined by primordial germ cells in the embryo. Sex determination is complex and subject to genetic, environmental and other factors that vary among species. Around 88% of the world's fishes are *gonochoristic* (Patzner 2008), meaning that gender is genetically determined and individuals possess only ovaries or testes. Gonochorists may develop either from an indeterminate gonad developing directly as ovary or testis (primary gonochorists) or from a gonad that initially is hermaphroditic (intersexual) and later becomes either ovary or testis (secondary gonochorists).

Individuals that possess both male and female sex organs at some stage of their life cycle, simultaneously or sequentially, are 'hermaphrodites'. Three types occur in fishes (Table 7.2), with a few examples in some Australian freshwater fishes. The barramundi is a *protandrous* hermaphrodite, commencing life as a male and becoming a female after spawning (Moore 1979; Davis 1982), although a few individuals may be synchronous hermaphrodites, containing functional male and female organs. In most barramundi, sex inversion (the transition from male to female) occurs after spawning in estuarine or marine environments, when the fish are about 800 mm in length (Moore 1979; Davis 1982). Protandrous hermaphroditism also occurs in some sparids (a mainly estuarine family), including western yellowfin bream (*Acanthopagrus latus*: Hesp *et al.* 2004) and tarwhine (*Rhabdosargus sarba*: Hesp and Potter 2003).

Unlike barramundi, most of the world's hermaphroditic fishes display *protogyny*, where an individual changes from female to male (Pavlov *et al.* 2009). Typically, this involves younger fish functioning as females until spawning, when testicular tissue develops and they become males. Despite its global prevalence, there are few examples in Australia. Swamp eels (*Ophisternon* spp.) potentially are protogynous but little is known of their reproductive biology (Pusey *et al.* 2004).

Modes of fertilisation

The majority of the world's fishes (94% of species: Patzner 2008) display *external* fertilisation, where sperm are mixed with eggs in the external environment (oviparity). Only 6% of fishes demonstrate *internal* fertilisation (Wootton 1998), where sperm are transferred to the female, the eggs are fertilised internally and develop in the ovarian follicles to hatch as free-swimming young (viviparity). This is true of salamanderfish (*Lepidogalaxias salamandroides*), where male and female remain coupled for up to 10 minutes (Pusey and Stewart 1989)! Some female viviparous fishes reach the pinnacle of parental care, catering for every need of their growing young, including supply of oxygen and disposal of wastes. The best-known examples are the Poeciliidae (all alien species in Australia), particularly the abundant and widespread eastern gambusia. Viviparity energetically is costly but it may confer a competitive advantage over native fishes when coupled with high reproductive output (McDowall 1996a).

In a few cases, eggs and sperm are transferred to the mouth of the male or female for fertilisation and rearing. An example is tilapia (*Oreochromis mossambicus*), an alien species, where the females take in eggs and sperm and the young hatch, in the mouth, after three to five days. Some native Australian species are mouth brooders (Table 7.2), but there is little information about how they fertilise their eggs.

Mating systems

The term 'mating system' refers to the number of mates an individual has per breeding attempt (Berglund 1997). 'Monogamy' applies when one male and one female mate only with each other, and 'polygamy' applies when individuals have several mates. Polygamy includes three patterns: *polygyny* (a single male mates with several females, but each female mates only with a single male), *polyandry* (a single female mates with several males, but each male mates only with a single female) and *polygynandry* (or promiscuity; both males and females mate with several partners).

Promiscuity may be an ancestral mating system in fishes, with external fertilisation of eggs involving many males and females simultaneously. It includes many mating scenarios, including passive broadcast-spawning of eggs and sperm with no active mate selection, and cases where an individual solicits several partners and those mate with several others. Mating systems may vary also depending on local conditions; for example, individuals may be polygamous when resources are abundant but tend to monogamy when resources are limited (Taylor and Knight 2008). The choice of mating systems also depends on phylogeny, sex-ratio imbalances and intersexual conflict, and access to resources (e.g. food, spawning and/or nesting sites).

The mating systems of Australian freshwater fish are little known. Some studies examine the breeding biology and reproductive behaviour of species but few have been able to determine their mating system. Following an example from above (see 'Parental care'), the female Clarence River cod selects a male at a prepared nesting site, spawns and leaves the nest while the male remains to guard the eggs, free embryos and larvae for up to 24 days (Butler and Rowland 2009). It is uncertain, however, whether the females mate with other males and thus are polyandrous (and polygamous) rather than monogamous. Captive Murray cod are polygamous (polygynous and polyandrous); there is multiple spawning by both sexes within a season and repeated mating between pairs across several seasons (thus monogamy: Rourke *et al.* 2009). Similarly, female empire gudgeons (*Hypseleotris compressa*) mate with more than one male then defend the spawning site (Auty 1978), but it is not known whether the male spawns with multiple partners over the breeding season. This species (and perhaps other eleotrids) appears to be polyandrous but could be promiscuous. The biology of Pacific blue-eye (*Pseudomugil signifer*) is

Table 7.2: Teleost breeding systems with Australian examples

Component	Explanation	Examples
1. Sex determination		
(a) Genetic	Gender determined by genetics	Most teleosts
(b) Environmental	Gender determined by environment	–
2. Gender system		
(a) Gonochoristic	Individuals with either ovaries or testes	Most teleosts
(b) Hermaphroditic	Individuals with ovaries and testes at different stages	
(i) Simultaneous	Ovaries and testes present at same time	–
(ii) Protandrous	Males change to females	Barramundi (*Lates calcarifer*), western yellowfin bream (*Acanthopagrus latus*),[1] tarwhine (*Rhabdosargus sarba*)[1]
(iii) Protogynous	Females change to males	Swamp eel (*Ophisternon gutturale*)[1]
3. Mode of fertilisation		
(a) External	Fertilisation in environment	Most teleosts
(b) Internal	Fertilisation in female	Eastern gambusia (*Gambusia holbrooki*),[2] salamanderfish (*Lepidogalaxias salamandroides*)
(c) Buccal	Fertilisation in mouth of male or female	Ariidae, tilapia (*Oreochromis mossambicus*)[2]
4. Mating systems		
(a) Polygamy	Individual has several mates	
(i) Polygyny	Males mate with several females, but females mate with one male	Murray cod (*Maccullochella peelii*)
(ii) Polyandry	Females mate with several males, but males mate with one female	Murray cod
(iii) Polygynandry (promiscuity)	Males and females mate with several partners	–
(b) Monogamy	Mates stay together	–
5. Secondary sexual characters		

Component	Explanation	Examples
(a) Monomorphism	Sexes indistinguishable even when mature	Most teleosts
(b) Sexual dimorphism		
(i) Permanent dimorphism	Sexes distinguishable after maturity	Empire gudgeon (*Hypseleotris compressa*), many Percichthyidae
(ii) Temporary dimorphism	Sexes distinguishable only in breeding season	Blue catfish (*Neoarius graeffei*), many Eleotridae (e.g. sleepy cod, *Oxyeleotris lineolata*; western carp gudgeon, *Hypseleotris klunzingeri*; firetail gudgeon, *H. galii*)
(c) Sexual polymorphism	One or both sexes as more than one distinguishable form	–
6. Spawning-site territory defence		
(a) Site not defended		Most teleosts
(b) Site defended		Many Eleotridae, freshwater catfish (*Tandanus tandanus*)
7. Parental care		
(a) No care (oviparity)	Eggs released by female, fertilised, then parents leave	Most teleosts
(b) Male parental care	Eggs and/or young cared for by male	Fork-tailed catfishes (*Neoarius graeffei*, *Sciades leptaspis*), many Eleotridae
(c) Female parental care	Eggs and/or young cared for by female	Great Lake galaxias (*Paragalaxias eleotroides*)
(i) Oviparity,with post-spawning care	Eggs released by female, but with parental care	Southern saratoga (*Scleropages leichhardti*), roman-nose goby (*Awaous acritosus*)
(ii) Ovoviviparity	Fertilisation internally, eggs retained until hatching or beyond, embryos reliant on yolk for nutrition	–
(iii) Viviparity	Fertilisation internally, eggs retained until hatching or beyond, embryos receive nutrient from female in addition to yolk	Eastern gambusia (*Gambusia holbrooki*)[2]
(d) Biparental care		–
(e) Juvenile helpers		–

1 = estuarine–freshwater species; 2 = alien species; – = no known Australian examples.
Classification modified after Wootton (1998, p. 151).

well-studied, particularly its elaborate mating and courtship behaviour (Wong *et al.* 2004a), but its mating system is unknown. Here is another potential research area, describing variations in the mating systems of related species, or individuals within species, across a spectrum of variable resource type and quality.

Secondary sexual characters

Secondary sexual characters enhance mating success. Charles Darwin (1871), for example, suggested that the brightly coloured, elaborate tail of the male peacock (*Pavo cristatus*) may have evolved as a consequence of female mate choice. Fishes also display these characters, and it is the males that are most likely to invest energy in secondary sexual characteristics and reproductive behaviour (Berg and Finstad 2008) whereas females invest more in gonad development. Secondary sexual characters can be used either for 'intrasexual selection', such as male–male competition, or mate selection ('intersexual selection'), usually by the female (Dugatkin and FitzGerald 1997). There may be associated body colour patterns (sexual dichromatism), body size differentiation, morphological differences or other ornamentation such as the hooked snout of male salmonids (this is what the generic name *Oncorhynchus* refers to).

Some Australian freshwater fishes display sexual dimorphism (Table 7.2). In some, such as the empire gudgeon (Auty 1978), the sexes are always distinguishable after maturity; in others, such as the western carp gudgeon (*Hypseleotris klunzingeri*), the dimorphism is noticeable only at the time of breeding. Many Australian percichthyids are dimorphic in that the bodies of mature males generally are smaller than those of females (Mallen-Cooper and Stuart 2003). In golden perch, size differentiation may or may not occur (Battaglene 1991; Anderson *et al.* 1992b; Mallen-Cooper and Stuart 2003). Eastern gambusia show conspicuous morphological differences, with males having a modified anal fin (gonopodium) that is used to transfer sperm in mating.

Sexual dichromatism (differentiation of sexes by colour patterns during the breeding season) is common in small-bodied Australian freshwater fishes such as galaxiids, gudgeons (Eleotridae), pygmy perches (*Nannoperca* spp.), blue-eyes (Pseudomugilidae) and rainbowfishes (*Melanotaenia* spp.). In these species, the males have fin and body colouration that develops during breeding, probably to attract females and perhaps threaten other competing males (Table 7.2). Many species are also capable of changing their body or fin shape during the breeding season only, such as honey blue-eye (*Pseudomugil mellis*), blue catfish, Agassiz's glassfish (*Ambassis agassizii*) and sleepy cod (*Oxyeleotris lineolata*). Female sleepy cod develop an enlarged triangular genital papilla only during the breeding season (Herbert and Graham 2004) and female blue catfish (*Neoarius graeffei*) develop long, round, thickened pelvic fins adapted for grasping (Rimmer 1985c). Two sympatric catfishes, black catfish and Hyrtl's catfish, normally look similar and spawn at the same time but develop different body and fin colouration that would guard against inappropriate mate choices at spawning (Orr and Milward 1984).

REPRODUCTIVE BEHAVIOUR

The reproductive behaviour of fishes may include movement to a spawning area, choice, preparation and defence of a site, courtship, mate selection and parental care. Investment in these behaviours may be energetically costly and can limit the time and energy available for gamete production (number and size), somatic growth and body maintenance (Wootton 1998). This is another significant trade-off for fishes.

Adult movements

Adults in spawning condition are likely to select the best environment that will maximise the survival of their young. Most Australian freshwater fishes appear to undergo localised movements only, to spawning sites within the same lake or river habitat (Ch. 5), but some make spawning migrations (regular, coordinated movements associated with breeding), sometimes over large distances. For example, Murray cod, a *potamodromous* species, may undertake pre-spawning migrations of up to hundreds of kilometres upstream to a spawning area, then spawn and return rapidly downstream, often directly to their home territory (Koehn and Harrington 2006). Long-distance potamodromous spawning migrations may also occur for golden perch and silver perch (Reynolds 1983; O'Connor *et al.*

2005) but the evidence for all these species is inconclusive and requires further investigation.

Diadromous fishes move between marine and freshwater environments for breeding. The transition involves great physiological and physical challenges, like the need to maintain water balance and other demands on general body condition, but these rigours clearly are outweighed by the adaptive value of bearing young in the best possible nursery environment. This may be why many diadromous fishes are semelparous (reproduce once then die).

Among diadromous fishes, *anadromous* species are rare in Australia. They include lampreys, which spawn in fresh water, undergo a larval metamorphosis and migrate to the sea as adults. Eels are the best known *catadromous* species; they spawn at sea, where the young develop before returning to fresh water. Many galaxiids and Australian grayling (*Prototroctes maraena*) are *amphidromous* (Ch. 5). The adults of these species move downstream to spawn in fresh water, sometimes within tidal limits, and the larvae spend time in estuaries or the ocean before returning to fresh water as juveniles or adults. Some diadromous populations that normally require access to the ocean to complete their life cycle are able to live and breed solely in freshwater habitats, having become landlocked (e.g. climbing galaxias *Galaxias brevipinnis*, common galaxias *G. maculatus*, spotted galaxias: Pollard 1971b; Humphries 1989; Chapman *et al.* 2006b).

Some Australian species form spawning aggregations (e.g. sooty grunter, *Hephaestus fuliginosus*: Pusey *et al.* 2004; silver perch; Macquarie perch). For example, there is a report of 50–70 silver perch moving rapidly in eddies, before spawning (NSW Fisheries 1915, cited in Clunie and Koehn 2001b). Macquarie perch also congregate in large groups in tributaries after spawning migrations from impoundments such as Lake Eildon and Lake Dartmouth in Victoria (Cadwallader and Rogan 1977; Tonkin *et al.* 2010). Aggregations of common carp (*Cyprinus carpio*) are a familiar sight in many rivers (Ch. 11).

Spawning site preparation

Pelagic (open water) spawners may select where they spawn but they do little more than cast their eggs into the water, leaving the young to fend for themselves. Other fishes, however, carefully select a spawning site and prepare it to receive their eggs. Males of some species may use the preparation of the spawning site to show their suitability as a potential mate to the opposite sex (e.g. Eleotridae). Preparations can be simple, like sweeping away debris (e.g. some gudgeons, Anderson *et al.* 1971; Auty 1978), or more sophisticated, like the nests constructed by freshwater catfish (*Tandanus tandanus*: Merrick and Midgley 1981).

Mate selection and courtship

Mate selection is behaviour that leads an individual to preferentially mate, ideally with the fittest partner. Female western rainbowfish (*Melanotaenia australis*), for example, deposit more eggs with large males than they do with smaller males (Evans *et al.* 2010). Mate choice is not exclusively female behaviour; male Pacific blue-eyes and desert gobies (*Chlamydogobius eremius*) expend more effort courting large females in preference to smaller females (Wong *et al.* 2004a; Wong and Svensson 2009).

Courtship fulfils several needs, including species recognition, mate choice, spawning site recognition and synchronisation, to ensure that both partners release gametes together (Wootton 1998). In fact, courtship can determine whether or not spawning occurs. Female Pacific blue-eyes apparently prefer males that engage in longer courtship (possibly inferring their paternal competence), leading to greater hatching success, and are not influenced by such traits as body size and fighting ability (Wong 2004). Courtship tends to be more elaborate – and more costly in terms of time, energy and risk of predation – in fishes that prepare their spawning sites and/or show parental care (Wootton 1998). The Australian lungfish has a complex courtship, including searching (pairs undertake circling movements and noisy breathing at the water surface) and follow-the-leader (as many as eight fish, males following and nudging the females). The paired fish then dive together, shedding eggs and milt over aquatic plants (Kemp 1984). Freshwater catfish also have an elaborate courtship (Merrick and Midgley 1981): the male and female approach each other, weave, nudge, arch/agitate, separate and inspect the nest (and probably spawn). This sequence may be repeated six times in 15 minutes, before the female swims away.

Territorial defence of spawning site

A spawning site is valuable property in the highly competitive task of rearing progeny. Not surprisingly, many of the species that exhibit parental care also defend their young against predation and disturbance. Male Clarence River cod vigorously defend their nest, eggs, free embryos and larvae from male competitors and potential predators (Butler and Rowland 2009). Aggressive 'bourgeois males' secure the premium nesting sites but there may also be 'sneaker fish' (males that move in to spawn with the female at the last minute) and 'parasitic males' (males that establish nest sites nearby, in the hope of winning an uncommitted female). Territorial defence of nesting sites may also occur in other freshwater cod (Rowland 1983; Pusey *et al.* 2004; Butler and Rowland 2009).

Small freshwater fishes may aggressively defend spawning sites. Male Oxleyan pygmy perch (*Nannoperca oxleyana*) defend their breeding territories against other males and court any female that swims near (Knight *et al.* 2007). Male firetail gudgeons (*Hypseleotris galii*) show similar behaviour, with darting movements, flaring fins and opercula (Anderson *et al.* 1971).

ENVIRONMENTAL REQUIREMENTS OF SPAWNING

Eggs, free embryos and larvae are vulnerable life stages and specific adaptations such as parental care, spawning site selection and timing of reproduction enhance their chances of survival. Their lack of mobility also makes them vulnerable to local environmental conditions.

Location

Freshwater fishes can choose spawning locations at two scales: the 'riverscape' (heterogeneous spatial features along a river) and the spawning site. Some species are able to move long distances to access floodplain wetlands or other areas likely to promote survival of the young (Bishop *et al.* 2001). The location of a spawning site should ensure that larvae and juveniles have easy access to nursery habitats and should limit the kind and intensity of hazards for eggs and young. Predation, disease, hypoxia, smothering by silt and poor water quality generally are significant hazards in early development (Wootton 1998). However, sites with abundant food and low predation risk, such as shallow, slackwater areas in rivers and inundated floodplains (King 2004a, b; Humphries *et al.* 2006), also tend to be more prone to poor water quality (Gehrke 1991).

Australian freshwater fishes represent a wide range of reproductive guilds, defined by Balon (1975), based on spawning site characteristics and types of parental care (Table 7.3). Some, like Australian grayling, golden perch and silver perch, are pelagic spawners that broadcast eggs and sperm into the water and show no parental care. These species are thought to select a breeding location, perhaps to maximise downstream dispersal of eggs and larvae (Humphries and King 2004; Tonkin *et al.* 2007), and the young develop as they drift downstream. Other species are demersal (or benthic) spawners over gravel or rock substrata (e.g. two-spine blackfish *Gadopsis bispinosus*: O'Connor and Zampatti 2006), on plants (e.g. Eleotridae) or at the water's edge. In the latter case, the common galaxias spawns in bankside vegetation submerged by a high tide; the eggs develop out of water then hatch on a subsequent tide (McDowall and Charteris 2006).

Timing of spawning

Gonad maturation is largely under endocrine control but spawning is driven mainly by external stimuli. For example, spawning in salamanderfish depends partly on lipid content at completion of aestivation (Pusey 1990), but the species also tends to spawn in late autumn and winter when temperature and day length are near their seasonal minima, enabling their young to grow before their pools dry (Morgan *et al.* 2000). For freshwater fishes in general, the timing of spawning must allow adults to spawn successfully and the young to survive, hence the timing of spawning is often linked to day length, temperature and food availability.

Annual cycles of photoperiod (the duration of daylight) and temperature are the most commonly linked cues for spawning, particularly in temperate regions. In these regions at least, the two are linked, as average water temperature generally increases as daylight hours are prolonged. However, few studies have attempted to tease these factors apart, as this requires long-term field data or laboratory experiments; most researchers are content to correlate environmental factors with some measure of spawning. Gonad maturation and onset of spawning may be linked to rising photoperiod and temperature but cessation of spawning is probably linked to critical temperature thresholds. Clearly, the timing of reproduction is influenced by more than one factor.

Differences in spring and summer temperatures across Australia lead to variations in the spawning times of widely distributed species. For example, the spawning time of bony herring varies geographically (Murray River, New South Wales, summer: 18–20°C; Burnett River, Queensland, spring–summer: 22°C; Alligator Rivers region, Northern Territory, year-round, peak in wet season: 24–28°C; Pusey *et al.* 2004). Variations in spawning time can occur at much smaller spatial scales; for example, the spawning

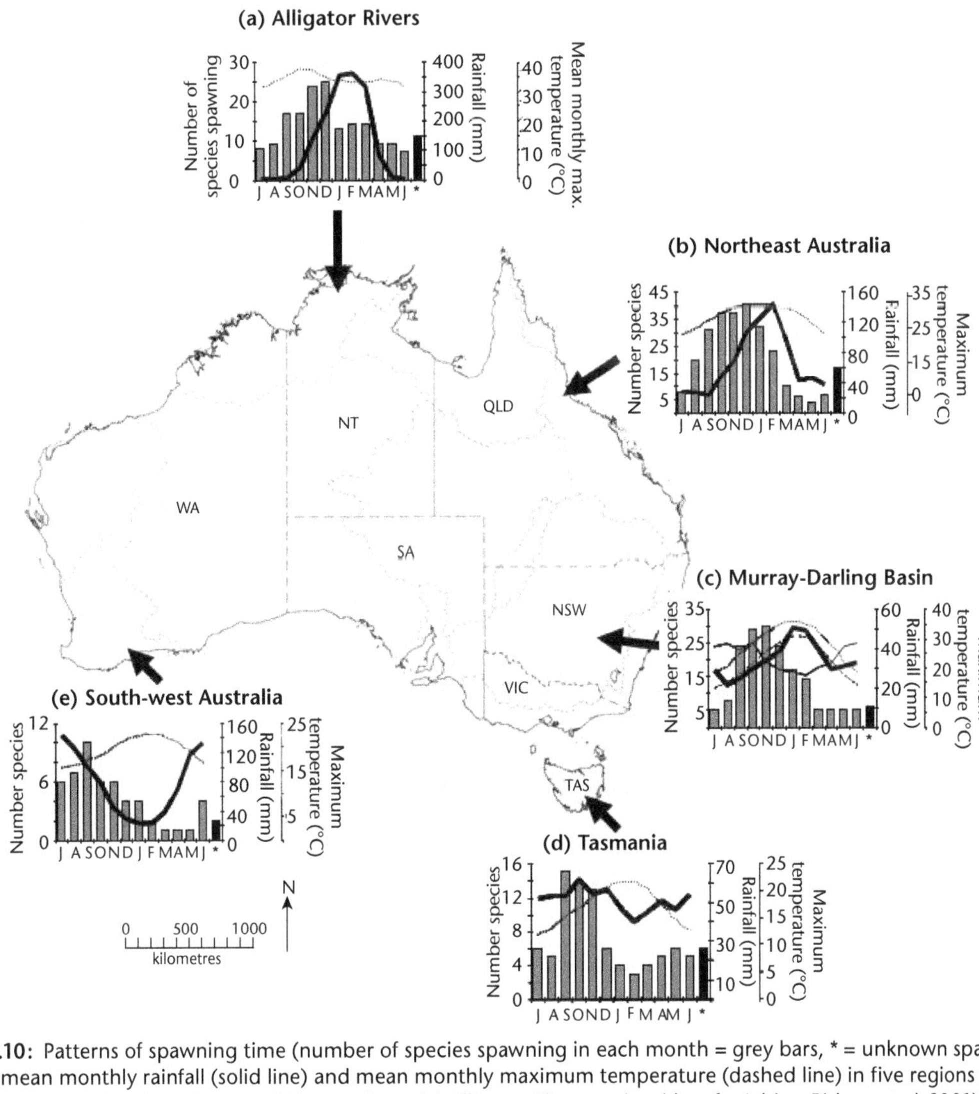

Figure 7.10: Patterns of spawning time (number of species spawning in each month = grey bars, * = unknown spawning period), mean monthly rainfall (solid line) and mean monthly maximum temperature (dashed line) in five regions of Australia. Climatic data from Bureau of Meteorology. (a) Alligator Rivers region (data for Jabiru, Bishop *et al.* 2001). (b) North-eastern Australia (data for Rockhampton, Pusey *et al.* 2004). (c) Murray-Darling Basin (data for Brewarrina, rainfall = heavy solid line, temperature = dotted line) and Echuca (rainfall = thin solid line, temperature = dashed line) (McDowall 1996b; Lintermans 2007; King *et al.* 2009). (d) Tasmania (data for Hobart, McDowall 1996b). (e) South-western Australia (data for Albany, Morgan *et al.* 1998).

Table 7.3: Reproductive guilds of teleost fishes, based on Balon's (1975, 1981) classification and potential Australian examples (*alien species)

Group	Guild	Examples
Non–guarders of eggs and young		
Open substrate spawners	Pelagic spawners	Australian grayling (*Prototroctes maraena*) Golden perch (*Macquaria ambigua*) Silver perch (*Bidyanus bidyanus*)
	Benthic spawners	
	Rock and gravel spawners with pelagic larvae	Golden galaxias (*Galaxias auratus*) Mountain galaxias (*G. olidus*)
	Rock and gravel spawners with benthic larvae	Brown trout (*Salmo trutta*)* Macquarie perch (*Macquaria australasica*)
	Non-obligatory plant spawners	Atherinidae Melanotaeniidae Agassiz's glassfish (*Ambassis agassizii*)
	Obligatory plant spawners	Australian lungfish (*Neoceratodus forsteri*) Macleay's glassfish (*Ambassis macleayi*)
	Sandy substrates	
	Terrestrial spawners on humid riparian fringe	Climbing galaxias (*Galaxias brevipinnis*) Common galaxias (*G. maculatus*)
Brood hiders	Benthic spawners	
	Cave spawners	Cave gudgeon (*Milyeringa veritas*) Blind cave eel (*Ophisternon candidum*)
	Annual spawners, eggs aestivate	
	Beach spawners, humid conditions	
	Spawners in live invertebrates	
Guarders		
Substrate choosers	Pelagic spawners, at surface in hypoxic waters	
	Terrestrial spawners	
	Plant spawners	Honey blue-eye (*Pseudomugil mellis*) Oxleyan pygmy perch (*Nannoperca oxleyana*) Many Eleotridae

Group	Guild	Examples
	Rock spawners	Sleepy cod (*Oxyeleotris lineolata*) Two-spine blackfish (*Gadopsis bispinosus*)
Nest spawners	Rock and gravel nesters	Freshwater catfish (*Tandanus tandanus*) Two-spine blackfish Some Eleotridae
	Sand nesters	
	Plant material nesters	
	Froth nesters	
	Hole nesters	Murray cod (*Maccullochella peelii*) River blackfish (*Gadopsis marmoratus*) Swamp eels (*Ophisternon* spp.)
	Miscellaneous material nesters	
	Anemone nesters	Not applicable
Bearers		
External bearers		
	Transfer brooders (eggs carried before deposition)	
	Auxiliary brooders (eggs carried on skin under fins etc.)	
	Forehead brooders	Nurseryfish (*Kurtus gulliveri*)
	Mouth brooders	Fork-tailed catfishes (*Neoarius graeffei, N. midgleyi, Sciades leptaspis*) Mouth almighty (*Glossamia aprion*) Southern saratoga (*Scleropages leichhardti*)
	Gill-chamber brooders	
	Pouch brooders	
Internal brooders	Ovi-ovoviviparous (retain internally fertilised eggs and release at early stage)	
	Ovoviviparous (one or few eggs develop at expense of others)	
	Viviparous (nutrition from female)	Eastern gambusia (*Gambusia holbrooki*)*

periods of carp gudgeon and Australian smelt differ substantially between the Campaspe and Broken rivers in north-eastern Victoria (Humphries *et al.* 2002).

While climatic conditions vary substantially across Australia, most freshwater fishes spawn in spring–summer, even in the tropics (Fig. 7.10). The warmer temperatures and longer photoperiod in spring–summer in temperate regions favour faster gonad, egg, embryo and larval development. But tropical regions show only minor seasonal changes in temperature and photoperiod throughout the year, hence other factors – like the onset of the wet season – are more influential. In the Alligator Rivers region of the Northern Territory, fishes such as sleepy cod, fly-specked hardyhead (*Craterocephalus stercusmuscarum*) and mouth almighty (*Glossamia aprion*) spawn year-round (aseasonal spawners), but most regional species breed at or during the onset of the wet season (November–December) (Bishop *et al.* 2001). Although very little is known of the early life stages of fishes in tropical regions of Australia, breeding at this time presumably enables young fishes to take advantage of the increased food supplies (e.g. plankton) and habitat (e.g. aquatic plants) that accompany extensive flooding, and promote rapid growth and dispersal (Bishop *et al.* 2001).

Fluctuations, usually rising rivers or water levels, induce spawning in many Australian freshwater fishes. Of 76 freshwater fishes in north-eastern Australia, flows may influence the timing of spawning for 10 species, although information for all species is poor (Pusey *et al.* 2004). One of the best-known and most controversial examples is golden perch. Lake (1967c) first reported that spawning of this species in artificial ponds is stimulated by water flooding dry ground at temperatures above 23°C, and Mackay (1973b) reported that the gonads were resorbed if those conditions did not occur. Golden perch thereafter were often described as 'flood recruitment specialists' (Berra and Cadwallader 1983; Harris and Gehrke 1994), but the evidence and its broad applicability have since been disputed. In fact, recent studies have shown that spawning and recruitment can occur across the flow spectrum (Mallen-Cooper and Stuart 2003; King *et al.* 2005; Balcombe *et al.* 2006; Roberts *et al.* 2008; King *et al.* 2009) and it appears that golden perch are highly flexible in their requirements. More extensive, long-term studies are needed to elucidate the factors that induce strongest spawning and recruitment (see 'Mortality and recruitment').

Rising rivers and the onset of the wet season stimulate spawning in barramundi, but the only participants initially are fish with easy access to their spawning environment in the sea. Later in the season, rainfall and flooding facilitate the movements of previously landlocked fish, triggering a second spawning peak. The timing, duration and intensity of the wet season determine the time of spawning and, in some cases, whether fish spawn at all (Bishop *et al.* 2001). Spawning of barramundi may also be linked to coincidence of high water temperatures and high tides associated with the full or new moon (Garrett *et al.* 1987).

EARLY LIFE HISTORY

Concepts and controversies

The early lives of fishes commence at activation (fertilisation of the egg) and finish when they become juveniles. This is a deceptively simple statement about a life stage that involves myriad morphological, physiological and behavioural changes; indeed, there are more changes in this period than at any other stage of the life cycle. The type, rate and evolutionary and ecological significance of these changes have fascinated many embryologists and ichthyologists. Their ideas are encapsulated in studies of *ontogenesis* (events during development of an individual) and *epigenesis* (the underlying mechanisms). This section focuses on concepts of ontogenesis.

There is a bewildering variety of terms and protocols used to describe phases in the lives of fishes. As ichthyologist Eugene Balon (1999, p. 18) said, 'biases have been caused by "fishery biologists" who, in the necessary attempt at cooperation, adjust their "terminology" to the jargon of their non-biologist counterparts in management, trade and politics. The result is a ridiculous life history sequence: egg–larva–fry–fingerling–subadult–spawner– (and ultimately) trophy, where "embryo", a term from the ivory-tower discipline of developmental biology has no place'.

The main point of contention is defining when the larval period begins. Larvae generally are characterised by elongated fin spines, stalked eyes, adhesive

Table 7.4: Stages in the early life history of fishes

Stage	Definition	Characteristics
Egg	Unfertilized oocyte	Egg envelope, perivitelline space, yolk and blastoderm before activation
Embryo	Commences with egg activation	Embryo enclosed within egg envelope; ontogenesis begins
Free embryo	Commences with hatching	Post-hatching; embryo is free to swim but feeds endogenously
Larva	Commences with ability to ingest and digest exogenous food	Exogenous feeding begins; temporary larval structures present
Juvenile	Commences when most/all embryonic and larval structures are lost; some adult structures present	Phenotype develops, completed at maturation

organs, pigmentation or other features that are lost on metamorphosis to a juvenile. Balon (1999) argued that in fishes, as in all vertebrates, larvae feed externally (*exogenous* feeding) whereas embryos draw nourishment from a yolk (*endogenous* feeding). Although hatching is a significant event, it does not necessarily occur at a fixed development stage, age or size, nor does the embryo change its source of nourishment. Therefore, according to Balon, this stage is not special enough and the embryo does not warrant the name 'larva'. Rather, it is often termed a 'free embryo' (Peňáz 2001). Others contend that 'hatching is a major turning point from ecological, physiological, and behavioural points of view … is easy to recognise, and is of importance in hatchery practice' (Kamler 2002, p. 19) and that Balon's protocol has not been widely used and is impractical (Fuiman and Werner 2002). While this may not justify rejection of Balon's arguments (Copp *et al.* 1999), it highlights a mismatch between theory and application. We lean to Balon's view and recognise the early life stages as egg, embryo, free embryo, larva and juvenile (Table 7.4).

Debate continues, however, because deciding when larvae become juveniles is problematic (Copp *et al.* 1999). A working definition (after Balon 1990; Peňáz 2001) is that 'the start of the juvenile period is when most or all of the embryonic and larval temporary structures have been lost and when some permanent adult structures are present'. There are variations on this theme, and a consensus might never be achieved.

Another debate is whether ontogenesis occurs gradually or in steps (saltation). Balon (1986) championed *saltatory ontogeny*, the idea that embryos, free embryos and larvae undergo rapid, stepwise changes interspersed with periods of little change. Others contend that ontogeny is continuous (Gorodilov 1996), that changes (albeit slower changes) do occur within each step and that the saltation model makes it difficult to quantify ontogenetic changes. Kamler (2002, p. 81) argued for a compromise, suggesting that 'ontogeny is a continuous process with temporary accelerations'.

Ontogenesis

Ontogeny begins with fertilisation of the egg, causing uptake of water (forming a perivitelline space), making it swell and 'harden', and closure of the *micropyle* (a pore that admits the sperm). The egg typically, but not always, becomes more buoyant as a result of fertilisation and reduced density. The eggs of golden perch, for example, are broadcast into the water column and float downstream; others, like those of many eleotrids and galaxiids, adhere to rocks, wood or plants. Still others are in nests, or in their mother until hatching. Thus, the environment in which an egg (hence the embryo) develops varies as widely as the diverse habitats of older stages of fishes.

Embryos hatch when they reach a particular size (and a limit imposed by the rate of oxygen diffusion) or in direct response to environmental hypoxia. Other factors affecting the timing of hatching can be temperature, pH and light. Free embryos emerge in any of a variety of states, from some that are tiny and blind with little capacity for movement (e.g. carp gudgeons: *Hypseleotris* spp.), to others that are relatively large with good visual acuity and swimming ability (e.g. Murray cod; Fig. 7.1a).

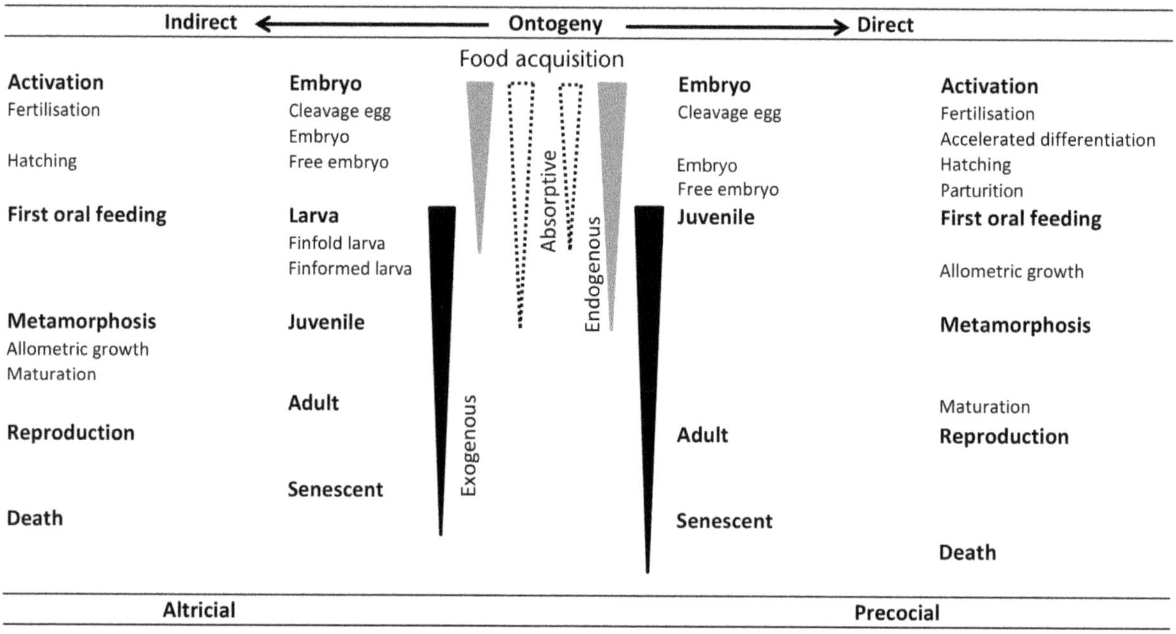

Figure 7.11: Balon's feeding and ontogeny model for fishes with indirect and direct development. Key events in bold. Black wedge = exogenous feeding; grey wedge = endogenous feeding; dotted wedge = absorptive nutrient uptake. Examples of indirect development include galaxiids, carp gudgeons (*Hypseleotris* spp.) and golden perch (*Macquaria ambigua*). An example of direct development includes freshwater catfish (*Tandanus tandanus*) (modified from Balon 1986).

There are two broad classes of ontogenesis (Fig. 7.11). *Indirect development* species, also known as 'altricial' species, feed endogenously (on yolk reserves) as embryos and free embryos, and the larvae commence exogenous feeding (with perhaps some overlap as yolk reserves are depleted), developing fins and organs as they metamorphose into juveniles. *Direct development* species, or 'precocial' species, bypass the larval stage and feed endogenously throughout the free embryo stage before switching to exogenous sources as juveniles. There is also an intermediate group, exemplified by Murray cod, in which there is a brief but distinct period of exogenous feeding by the larvae.

Indirectly developing species tend to produce many small eggs with little yolk and to be broadcast spawners with little or no parental care (*r*-selected species). Directly developing species tend to have fewer, larger eggs with more yolk and often are demersal spawners with some form of parental care (*K*-selected species). Thus, there are evolutionary links between reproductive modes, life history strategies and ontogeny.

Habitat and movement

The early life stages of fishes depend on nursery habitats where they can develop, grow and mature. For larvae to access these habitats, spawning must occur there or the eggs and/or larvae must move from elsewhere. The swimming abilities of free embryos and larvae vary greatly and typically improve with increased body size and ontogeny, but most can undertake large-scale movements only with the assistance of a current. In some species, the adults spawn upstream and rely on the flow of rivers to disperse the young. The free embryos and/or larvae of Murray cod, trout cod (*Maccullochella macquariensis*), golden perch and silver perch are thought to be obligate drifters (Humphries and King 2004). The eggs and free embryos of golden and silver perch drift (King *et al.* 2009), but only the free embryos and early larvae (at least seven days post-hatching) of Murray cod and trout cod drift (Humphries 2005; Koehn and Harrington 2006). In fact, Murray cod drift mainly at night (Humphries 2005) and, far from being passively entrained by the current with no control over their destination, they seem to be able to regulate when, where and for how long they drift. Presumably they must settle in suitable rearing habitats, but the nature and location of such habitats are not known.

The free embryos and larvae of many riverine fishes often utilise slackwaters or near-shore nursery habitats where there is less current, abundant prey, warm water and shelter from predators (Puckridge and Walker 1990; Humphries *et al.* 1999; King 2004b; Price 2007). Different species and life stages have their own microhabitat preferences in these slackwater patches (Price 2007), but it is unclear whether the spawning adults or larvae actively choose their nursery habitat or if the observed distribution of nursery habitats is simply a result of increased larval survival in favourable conditions (Werner 2002). This issue could be resolved by field investigations of spatial distribution of larvae through time, matched with experimental tests of habitat preferences, which has been investigated only in a limited manner for Australian freshwater fish. However, larvae and juveniles of common galaxias show apparent preferences for particular river systems, whereas those of climbing galaxias show no preference (Hale *et al.* 2008). Similarly, golden perch larvae actively avoid poor water quality conditions, such as hypoxia and high tannin loads, associated with blackwater events on inundated floodplains (Gehrke 1991; Ch. 4). Experimental studies of environmental preferences and tolerances are required for many Australian species.

Fishes of all life stages appear to select habitats based on their biological, physical and chemical requirements (Ch. 4). The rapid changes in the early life stages, however, oblige many species to make ontogenetic habitat shifts whereby young move between habitats as development progresses and their resource needs change. Diadromous species are an obvious example (Ch. 5) but habitat shifts occur also for potamodromous species, and often differ markedly among species. For example, four of five species in the Broken River, Victoria, employ different strategies (King 2004b; Fig. 7.12). It is clear that early life stages require very different habitats from adults, but surprisingly little is known of these habitats for the majority of Australian species.

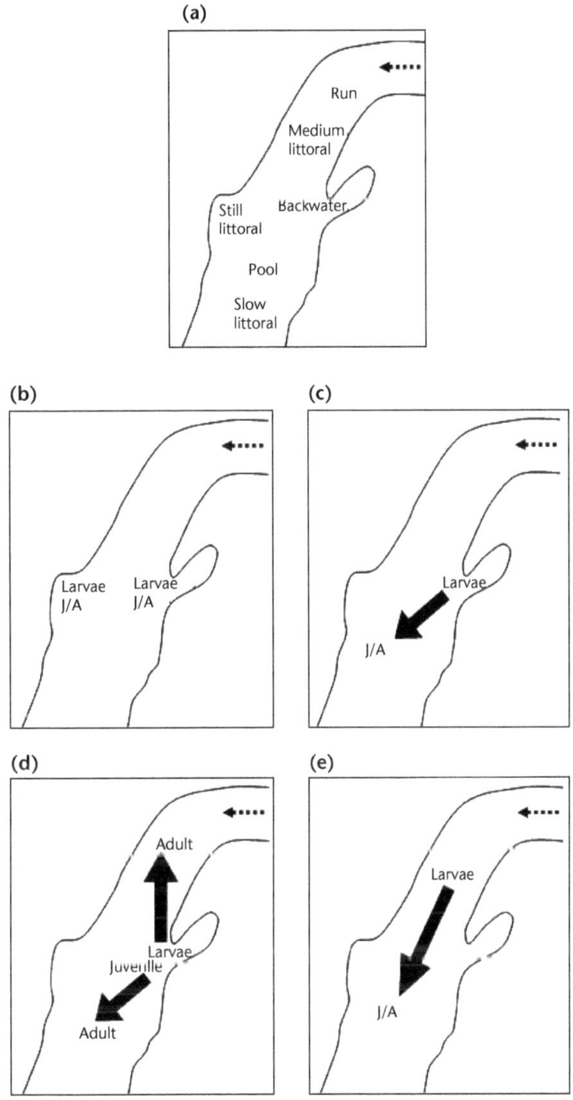

Figure 7.12: Four ontogenetic habitat strategies in fishes of the Broken River, Victoria. Solid arrows = direction of ontogenetic habitat change; dotted arrows = flow direction. From King (2004b). (a) Idealised location of sampled habitats. (b) Strategy 1 (e.g. eastern gambusia, *Gambusia holbrooki*; Murray River rainbowfish, *Melanotaenia fluviatilis*). (c) Strategy 2 (e.g. Australian smelt, *Retropinna semoni*). (d) Strategy 3 (e.g. common carp, *Cyprinus carpio*). (e) Strategy 4 (e.g. Murray cod, *Maccullochella peelii*). J/A = Juvenile/adult.

Larval feeding

First feeding and the risk of starvation

At hatching, fish rely on endogenous feeding; that is, they retain part of their yolk sac to sustain themselves. The transition from endogenous to exogenous feeding marks the critical first feeding period (Hjort 1914; May 1974), when larvae need abundant prey of the appropriate size to ensure their survival. The critical period is thought to be when most natural mortality occurs, and is one of the best predictors of year-class strength (see below). Golden perch in aquaculture ponds commence feeding five days after hatching; survival declines significantly if first feeding is delayed until day 8 and all larvae die if they fail to feed by day 10 (Rowland 1996). Other species, including Australian smelt, Murray cod and common carp, have larvae capable of feeding externally before the yolk sac is fully absorbed (King 2005) and thereby have some flexibility to develop their feeding skills and adapt to patchy prey distributions.

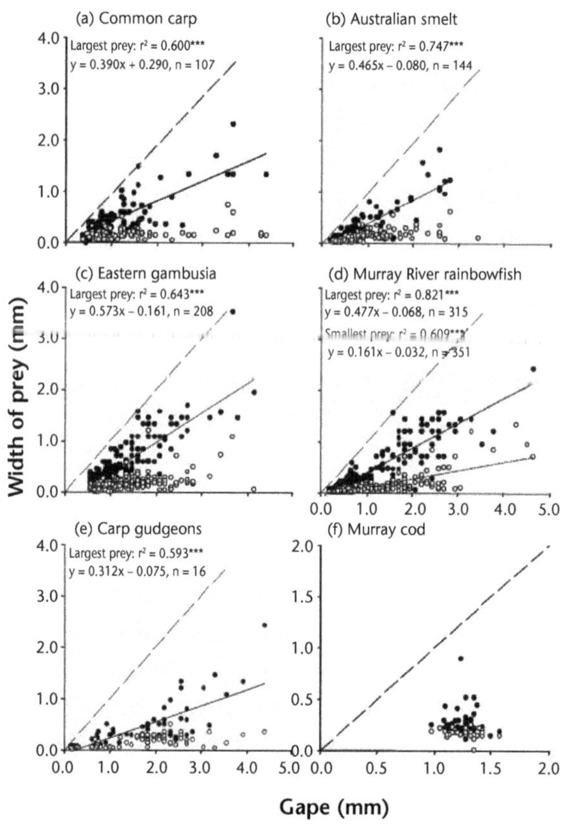

Figure 7.13: Linear regressions of gape size (mm) v. maximum width of largest (black circles, solid line) and smallest (white circles, dotted line) consumed prey for larval stages of (a) common carp (*Cyprinus carpio*), (b) Australian smelt (*Retropinna semoni*), (c) eastern gambusia (*Gambusia holbrooki*), (d) Murray River rainbowfish (*Melanotaenia fluviatilis*), (e) carp gudgeons (*Hypseleotris* sp.) and (f) Murray cod (*Maccullochella peelii*). Dashed line represents 1:1 ratio, where prey size = gape size. Note different *y*- and *x*-axis scales for Murray cod. Regression lines and equations shown only for significant results (overall linear regression models: *** $P < 0.001$; NS = not significant at $\alpha = 0.05$). From King (2005).

The 'critical period hypothesis' is an important concept concerning recruitment in fishes but supporting evidence is limited and often controversial, and it varies across species (Leggett and Deblois 1994). For example, recruitment strength in populations of carp gudgeons is probably determined during this critical period but this appears not to be true for fly-specked hardyhead (*Craterocephalus stercusmuscarum fulvus*) (McCasker 2009). Several hypotheses invoke the risk of starvation during early life, with the majority suggesting that cohort survival is enhanced when larvae encounter high densities of appropriately sized prey. There is little indication of the natural zooplankton densities required by the larvae of Australian fish, although densities of 20–1000 individuals L^{-1} have been proposed for marine species (Theilacker *et al.* 1996). Aquaculture facilities in Australia use a minimum of 500 individuals L^{-1} (King 2004a). Yet floodplain rivers contain abundant suitable prey for larvae (King 2004a; McCasker 2009; Cheshire 2010), with highest densities in epibenthic (demersal) environments. These Australian studies also support other studies, suggesting that food availability limits larval survival and growth (McCasker 2009). Given the lack of knowledge for predicting recruitment, we consider this another important area for research.

Diet, ontogeny and competition for food

Although the diets of adult freshwater fishes are reasonably well-known (Ch. 6), less is known of the natural diets of early life stages (but see Gehrke 1992; Gill and Morgan 1999, 2003; Vilizzi 1998b; King 2005; Kaminskas and Humphries 2009). Clearly, a fish can eat only what will fit into its mouth, thus a key factor limiting the diets of adults and larvae is gape size – mean prey size generally increases with gape size (and body length). Australian smelt, Murray River rainbowfish (*Melanotaenia fluviatilis*), common carp and eastern gambusia all consume prey as big as their respective gapes, but they also select smaller items (Fig. 7.13). Murray cod, and to some extent carp gudgeons, eat prey much smaller than their gape sizes. At first feeding, the larvae of Australian smelt and Murray River rainbowfish (mean gape 0.27 mm and 0.31 mm, respectively) eat rotifers, copepod nauplii and phytoplankton whereas older, larger larvae (mean gape 0.76 mm and 0.65 mm, respectively) prey on crustacean zooplankton, macroinvertebrates and terrestrial invertebrates (King 2005).

As larvae grow and develop, visual acuity and sensitivity, prey detection, prey capture ability and swimming performance improve. Thus, it is rare for a fish to specialise on one particular prey type throughout its life, giving rise to 'ontogenetic dietary shifts' (Werner and Gilliam 1984). Ontogenetic changes in the diets of Australian freshwater fishes from early life stages to adulthood have been described for only a few species (Gehrke 1992; Morgan *et al.* 1995; Gill and Morgan 1999, 2003; King 2005).

Different fish species and their ontogenetic stages often cohabit (Mark *et al.* 1987; King 2004b) and there is likely to be interspecific and intraspecific competition for food. Cohabiting species might thereby encounter a bottleneck that could influence recruitment strength (Werner and Gilliam 1984; Persson and Greenberg 1990), although there is little evidence that this occurs in Australian inland waters. In the Broken River, north-eastern Victoria, the larval diets of cohabiting Australian smelt and common carp in one case, and Murray River rainbowfish and eastern

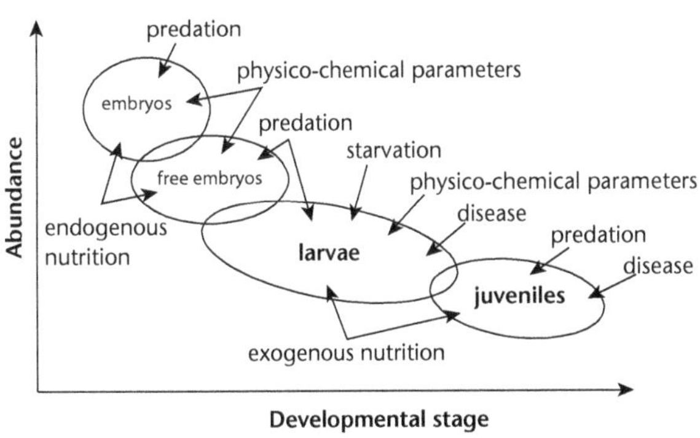

Figure 7.14: Conceptualisation of the recruitment process in fishes, including the sources of nutrition and probable extrinsic sources of mortality (modified from Houde 1987).

gambusia in another, did not significantly overlap (King 2005). In south-western Western Australia, ontogenetic dietary shifts by Balston's pygmy perch (*Nannatherina balstoni*) and dietary differences among other co-occurring species lessen the likelihood of competition for food (Morgan *et al.* 1995). Under laboratory conditions, however, there may be competition for small prey (< 500 µm) among larval golden perch, Murray cod and common carp, particularly in the first few days of exogenous feeding (Tonkin *et al.* 2006).

MORTALITY AND RECRUITMENT

Causes of mortality

Mortalities in the early lives of fishes can be caused by *extrinsic* or *intrinsic* factors (Fig. 7.14). Extrinsic factors include starvation, predation, disease or environmental stress, whereas intrinsic factors are morphological or physiological faults during larval development. Extrinsic factors are generally better understood than intrinsic factors. While few studies have shown unequivocally that one or other of the sources of mortality predominates in a particular situation, either starvation or predation is thought to drive the severity of mortality in young fish. Here, we outline these causes and some related concepts.

Starvation

Fish larvae are more vulnerable to starvation than juveniles and adults, owing to their smaller energy reserves, undeveloped hunting abilities and gape-limited foraging (Miller *et al.* 1988; Johnson and Dropkin 1995). Food limitation can ultimately cause death but it can also cause behavioural changes such as reduced swimming speed, feeding activity and stamina, or can affect sensory abilities, influencing the capacity to learn behaviours needed for social interactions, predator avoidance and food capture (Jonas and Wahl 1998). Starving larvae may be more vulnerable to predators (Cushing 1990).

Predation

The vulnerability of larvae to predation depends on size, development and growth rate (Shepherd and Cushing 1980; Bailey 1994), the relative size of predators (Pepin *et al.* 1992; Cowan *et al.* 1996) and the capabilities and number of predators and prey (Grimes and Isely 1996). The direct effects of predation are obvious, but the mere presence of predators may impede larval growth and development, limiting food intake and modifying behaviour. Turbidity, temperature, the availability of refugia and other environmental factors influence the chances of larvae encountering predators (Jacobsen and Berg 1998; Fiksen *et al.* 2002). There appear to be few quantitative estimates of mortality by predators in fresh water, but Houde (1989) ventured that the predation component of mortality in marine fishes could be up to 87% of total mortality.

Disease

Disease is a common agent of mortality in aquaculture (Katharios *et al.* 2008; Bigarre *et al.* 2009) but its influence on wild larvae is not clear. It may be a significant cause of mortality only when densities are high and infections or pathogens are readily passed between individuals (Wootton 1998).

Environmental causes

Recruitment in freshwater fishes can be affected dramatically by episodic physical or chemical conditions (Houde 1989; Fuiman and Werner 2002). Adverse water quality is a threat to free embryos and larvae, made vulnerable by their limited mobility and small size (large surface area to volume ratio). Changes in current velocity (Mion *et al.* 1998), temperature (Bartolino *et al.* 2008), oxygen (Fiksen *et al.* 2002), pH (Peterson *et al.* 1982), salinity (Jobling 1995), turbidity (Gadomski and Parsley 2005) and contaminants (Jobling 1995) may affect larval survival. However, the sublethal effects from these and other variables may be even more significant (McPeek and Peckarsky 1998). Interactions between abiotic and biotic factors also may intervene; for example, high turbidity can affect the feeding success of larvae and their exposure to predators (Fiksen *et al.* 2002).

Intrinsic mortality

Intrinsic factors in early mortality include physiological or morphological defects or constraints (Vladimirov 1975; Li and Mathias 1987). The vulnerability of larvae to these factors is influenced by a species' life history and by heritable and non-heritable parental traits, including condition and age at breeding (Cardinale and Arrhenius 2000). The larvae must develop adult structures for respiration, feeding, vision and locomotion before there is a critical need for them (Wieser 1995); defects in morphology of physiology, inherited or derived, may become lethal (Vladimirov 1975). Although intrinsic causes of mortality are important for larval condition and survival (Browman 1995; Burggren and Blank 2009), they have received little attention from ecologists.

Quantifying early mortality

Obtaining estimates of any of the vital rates (mortality, births, immigration and emigration) of animal populations is inherently challenging, and studying mortality during the early life history of fishes is especially difficult. Fish larvae are small, mobile, easily dispersed and patchily distributed, and a minute part of the biomass of organisms in their habitat (Houde 1989). The critical events that cause mortalities in larval populations occur over hours to days, making them hard to detect (Wootton 1998).

The rate of mortality (M) is the proportion of a population that dies in a given period. Conversely, the survival rate (S) is the proportion of initial numbers that remain after some mortality (thus, $S = 1 - M$). Estimates of mortality are made from age- or size-frequency distributions, known variously as catch curves (Ricker 1975), age-frequency curves (Beverton and Holt 1957) or abundance-at-age curves (Chambers and Leggett 1989). If the mortality rate *per capita* is constant, the plot of the logarithm of abundance-at-age (or size) data will yield a straight line, indicating that the risk of an individual dying in a given time interval does not change with age (or size). This involves two main assumptions, namely that mortality rates are constant over the time that individuals were collected, and that fishes are equally susceptible to mortality regardless of age, developmental stage or size. In fact, these assumptions are rarely met (McCasker 2009) and they may not be appropriate for early life stages, when mortality is both maximal and rapid (i.e. days). For example, small-bodied, opportunistic species like carp

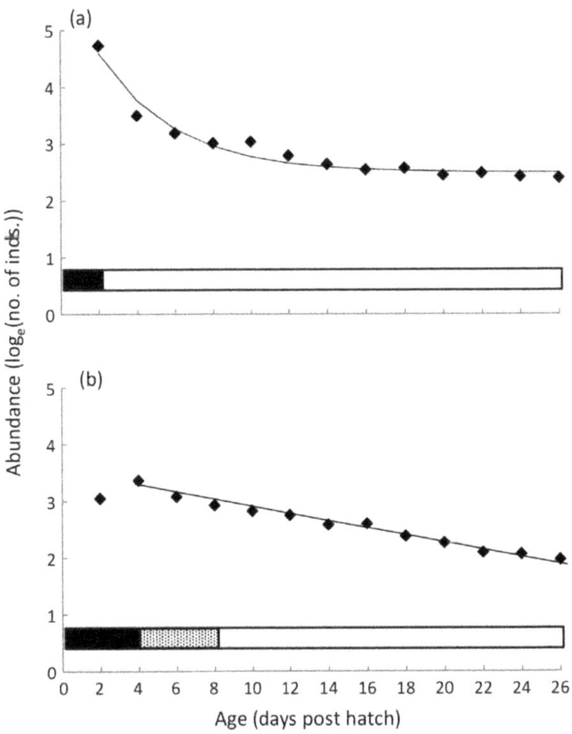

Figure 7.15: Survivorship curves during the free embryo and larval phase of two Australian species. (a) Carp gudgeon (*Hypseleotris* sp.). (b) Fly-specked hardyhead (*Craterocephalus stercusmuscarum*) (from McCasker 2009). The black solid line represents the mortality model of best fit, where mortality is the rate of decline between any two ages/stages; a straight line indicates that mortality is constant throughout development; a curved line indicates that mortality is greater at particular times during development. Horizontal bar: black = endogenous feeding; grey = transition between endogenous and exogenous feeding; white = exogenous feeding.

gudgeon and fly-specked hardyhead show clear differences in larval survivorship (Fig. 7.15).

Catch curves can provide a link between the mortality rates of free embryos and larvae and variations in recruitment (Miranda and Bettoli 2007). Their value is limited, however, because catch curves for fishes with protracted breeding seasons are likely to include multiple cohorts, so that age-abundance data are averaged across cohorts. These artificially smoothed survival curves may not be a true indication of mortality risks for individual larvae or individual cohorts (Fortier and Leggett 1985). The remedy is to follow individual cohorts through the larval period, to estimate mortality in the stages from hatching to juvenile metamorphosis and finally to link cohort age/stage mortality data to environmental conditions (Fortier and Leggett 1985, Dorsey *et al.* 1996; McCasker 2009).

Recruitment ecology

In fisheries science, *recruitment* is generally defined as the number of fishes in a year class as it enters the adult (or other segment) population (Heath and Richardson 1989). Recruits therefore are mature individuals, potentially able to reproduce and contribute to the population. In many species, the 'recruitment strength' of populations is established during the free embryo or larval periods (Pope and Willis 1998; Mukhina *et al.* 2003) and ecologically it may be more meaningful to focus on the status (age, size) of cohorts when the fishes have passed through the major survival and developmental bottlenecks. From that point, year-class strength is more or less fixed (van der Veer *et al.* 2000). Recruitment ecology has developed largely because of the importance of understanding the factors that explain or predict recruitment strength for managing commercially harvested marine fish stocks.

Recruitment strength is determined by the magnitude of spawning and by the survival and growth of early life stages. The first of these processes depends on the numbers of reproductive individuals (standing stock biomass) and appropriate environmental cues to trigger spawning and hatching. The second process is determined by individual fitness, habitat suitability and water quality, the quantity and quality of food and predation (Houde 1997). Most research concerns factors which affect spawning, hence reproductive success, but the factors causing variations in recruitment strength are less understood (King 1995; Cushing 1996). The indications are that variations in mortality during early life stages are a major contributor (Leggett and Deblois 1994; Pope and Willis 1998; Pine *et al.* 2000).

Several well-known hypotheses have been advanced to implicate food limitation (starvation) in the early lives of fishes in recruitment success (Table 7.5) but the evidence is inconclusive. In the late 1980s these ideas were challenged by claims that predation, rather than starvation, is the driver of early mortality.

Table 7.5: Definitions, proponents and comments on starvation-related recruitment hypotheses for fishes

Hypothesis	Definition	Proponents	Comment
Critical period	In the wild, mass mortality typically occurs towards the end of yolk-sac absorption, during the transition from endogenous to exogenous feeding	Domergue and Bietrix (1897); Hjort (1914)	Remains important, but evidence is inconclusive (Leggett and Deblois 1994)
Match/mismatch	If there is a match, when peaks of larval production and food production coincide, survival should be high and recruitment strength strong. If there is a mismatch and peaks do not coincide, survival should be low and recruitment strength poor	Cushing (1990)	Survival at high levels of feeding increases in the laboratory (Taylor and Freeburg 1984; Gotceitas *et al.* 1996), but evidence in nature is inconclusive (Leggett and Deblois 1994; Sirois and Dodson 2000)
Stable ocean	The maintenance of 'threshold for feeding' concentrations of food is associated with maximum chlorophyll layers in a stratified water column, formed during stable ocean conditions	Lasker (1978)	Inconclusive. In rivers, residence time influences the productivity of zooplankton and is similar to the 'inshore retention concept' (Schiemer *et al.* 2001) and the 'low-flow recruitment hypothesis' (Humphries *et al.* 1999)
Contact	Microturbulence increases encounters between larvae and prey, enabling larvae to survive at low prey densities	Rothschild and Osborn (1988)	Also proposed that turbulence slows the rate of starvation and mortality in a cohort because weaker individuals benefit from the effect of turbulence

The change of paradigm was based on arguments that starvation cannot explain the magnitude of mortality in the egg and free embryo periods, that there is little evidence of high rates of starvation in marine larval fishes and that larvae feeding under natural conditions, in the absence of predators, show high survivorship (Bailey and Houde 1989). Predation is now considered by many ecologists to be the main cause of early mortality (Houde 2002; Gisbert *et al.* 2004) but again the evidence is inconclusive.

This is a comparatively new field of research in Australia and it has focused mainly on the success and magnitude of reproduction (Humphries and Lake 2000). The 'flood recruitment model' (Harris and Gehrke 1994) proposed that flooding in rivers promotes larval survival and recruitment regardless of whether spawning was dependent on flooding. This built upon the 'flood pulse concept' (Junk *et al.* 1989) that recognised floods as driving variables in river ecosystems, and sparked interest in the recruitment ecology of Australian freshwater fishes. However, it emerged that many fishes spawn irrespective of flow conditions, including low-flow periods in summer, leading to the 'low-flow recruitment hypothesis' (Humphries *et al.* 1999). This recognised the unpredictable nature of floods in some regions of the Murray-Darling Basin and suggested that, in some species, spawning and recruitment are independent of floods. Indeed, it suggested that spawning and subsequent recruitment in small, short-lived, fecund species (opportunistic species) would be favoured under warm, low-flow conditions. This idea has gathered support (Humphries *et al.* 2002; King 2004b; Vilizzi *et al.* 2007), including a study that showed that flooding may have little effect on which species used floodplain habitats for spawning and recruitment on the Ovens River, Victoria (King *et al.* 2003a). In arid zone floodplains, however, the larvae and juveniles of many native fishes appear to recruit successfully in both flood years and non-flood years (Balcombe *et al.* 2006, 2007). While flooding clearly enhances spawning and recruitment of some native species (Roberts et al. 2008; King *et al.* 2009, 2010; Rolls and Wilson 2010), the mechanisms are still under investigation (Kaminskas and Humphries 2009).

Recently there has been a flurry of field-based studies investigating how environmental variables affect the recruitment ecology of Australian freshwater fishes, particularly in the Murray-Darling Basin (Douglas and Brown 2000; Humphries *et al.* 2002; Mallen-Cooper and Stuart 2003; Balcombe *et al.* 2006; Crook and Gillanders 2006), in south-eastern coastal rivers (Hale *et al.* 2008; Zampatti *et al.* 2010), arid regions (Balcombe *et al.* 2007) and tropical northern Australia (Staunton-Smith *et al.* 2004; Russell and McDougall 2005). The variability at all scales in the factors that influence mortality, such as climate and hydrology, river geomorphology, regional faunas, life history traits, prey availability and predators, challenges researchers to understand the mechanisms governing recruitment and, above all, to develop more ideas.

LARVAL ASSEMBLAGES

A 'larval fish assemblage' is a suite of species whose larvae occur at the same place and time (Miller 2002) but with no implied limit to spatial or temporal scales. For example, a larval assemblage could occur at a particular location and time (e.g. seven months in Wilson Inlet, south-western Australia: Gaughan and Potter 1997), or across a region over a longer time (e.g. seven years in the Campaspe and Broken rivers, Victoria: Humphries *et al.* 2002). The salient difference between a larval assemblage and general fish assemblage (Ch. 10) is that it is transient, being confined to one relatively short period in the life cycle. Importantly, this could include species that interact as part of a larval assemblage but may not interact as juveniles or adults. Understanding why larval assemblages occur together, the nature of their interactions and how these are related to recruitment is another of the challenges that confront recruitment ecologists (Humphries *et al.* 2002; King *et al.* 2003a; Vilizzi *et al.* 2007).

Both biotic and abiotic factors affect the formation, maintenance and disruption of larval assemblages (Miller 2002). Three factors in the formation of assemblages and their spatial and temporal patterns are the spawning time, location and behaviour of

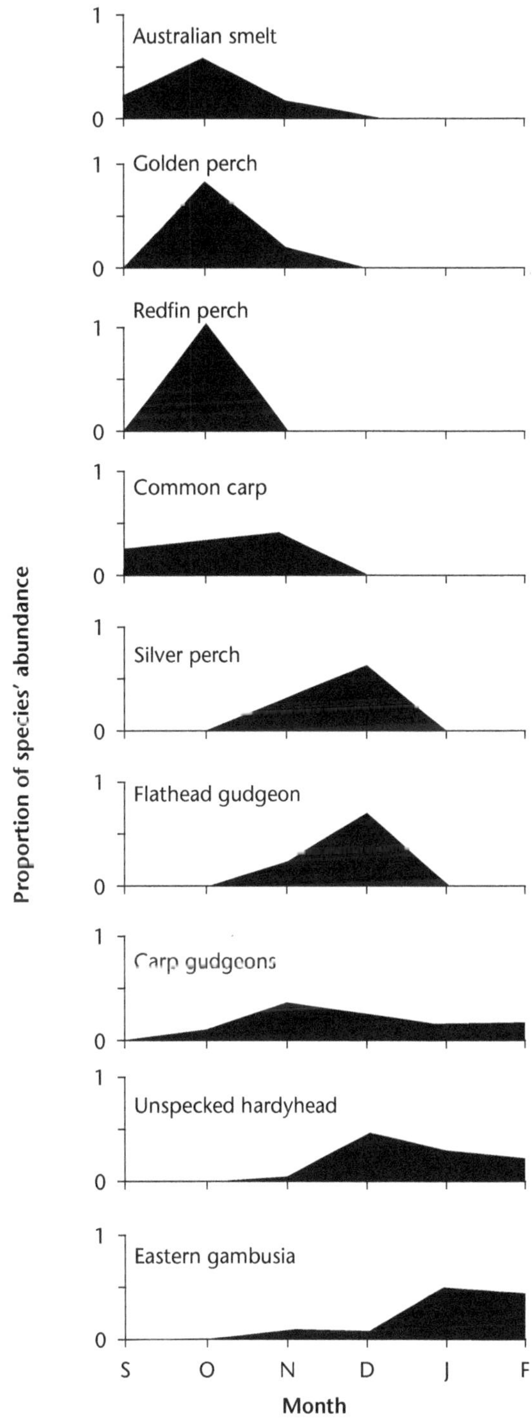

Figure 7.16: Temporal distribution (monthly across spawning season) of the nine most abundant species (embryos, free embryos and larvae) collected at Barmah-Millewa Forest, Murray River, Victoria, from 2003–08 (King, unpubl.). Sampling was by standardised drift net, sweep net, electrofishing and light-trap methods at 14 sites.

Table 7.6: Anthropogenic disturbances and associated factors, and their potential effects on reproduction and early life history

Disturbance	Associated factors	Potential effects on reproduction	Potential effects on early life history	Sample Australian references
Flow regulation: altered flow regime	Timing, magnitude, duration, rate of change of flow Flooding (minor, moderate, major) does not occur, or minor flooding occurs more often Greater incidence of moderate–high, stable, instream flows in summer/autumn	Spawning cues lost Lost access to spawning habitat and cues for movement	Altered timing, magnitude and type of food production Free embryos and larvae lose access to rearing habitat Eggs, free embryos and larvae entrained in current, swept downstream Increased mortality	Cadwallader (1976a); Humphries *et al.* (2002, 2006, 2008b) Humphries and Lake (2000); Humphries (2005) O'Connor and Mahoney (2004) Roberts *et al.* (2008); Zampatti *et al.* (2010)
Flow regulation: disconnection from floodplain	No flooding	Spawning and/or rearing habitat lost or degraded	Potential food resources from floodplain lost or diminished Rearing habitat lost Growth suppressed Increased mortality	King *et al.* (2003a); Vilizzi *et al.* (2007); Tonkin *et al.* (2008a); King *et al.* (2009, 2010)
Flow regulation: thermal pollution	Lower water temperature	Growth suppressed Spawning delayed or prevented	Growth suppressed Altered timing, magnitude, type of food production; increased mortality	Todd *et al.* (2005)
Flow regulation: impoundment	No flow, generally deep, stable water levels	Loss of spawning habitat	Food of lentic origin, increased with discharge from impoundment Drifting larvae encounter lentic habitat and sink Increased mortality	Humphries *et al.* (2002)
Flow regulation: barriers to movement	Dam wall Overshot and undershot weirs	Spawning movements prevented	Downstream drift prevented or damaged due to undershot weir	Baumgartner *et al.* (2006b)
Irrigation extraction	Water diverted from river to irrigation channel Pumping from irrigation channel to fields		Eggs, free embryos, larvae pumped from river into irrigation channels and fields Increased mortality	King and O'Connor (2007)
Habitat modification	Increased sedimentation Habitat removal	Sedimentation in spawning sites	Lost sites for egg deposition, larval rearing	Cadwallader and Backhouse (1983); O'Connor and Zampatti (2006)

Disturbance	Associated factors	Potential effects on reproduction	Potential effects on early life history	Sample Australian references
Pollution (e.g. endocrine disruptors, hydrocarbons)	Catastrophic or prolonged pollution events	Secondary sexual characters affected (e.g. reduced gonopodium in eastern gambusia) Altered sex ratios	Reduced survival, hatchability of eggs Larval deformities	Pollino and Holdway (2002); Pollino *et al.* (2007a) Batty and Lim (1999); Rawson *et al.* (2008)
Harvesting (commercial or recreational fishing)	Removal of mature individuals	Reduction in total spawning output		Walsh *et al.* (2010)
Climate change	Increased temperature, altered flow regime, more frequent hypoxia	Changes to seasonal timing of reproduction Interference with reproductive cues Reduced fertility and spawning	Reduced egg survivorship Changed growth, metabolism, development and stage duration Increased susceptibility to starvation	Pankhurst and Munday (2011)
Alien species	New predators New competitors		Increased predation, especially from early-spawning species Increased competition for food Increased mortality	Humphries *et al.* (2002); Tonkin *et al.* (2006)

adults. For example, in Barmah-Millewa Forest on the Murray River in Victoria, the composition of the larval assemblage differs between years owing to variations in the timing and duration of spawning by adult fishes (Fig. 7.16). Predation and competition also affect larval assemblage composition. Interspecific competition has been confirmed by research overseas (Miller 2002) but has received scant attention in Australia (King 2005; Tonkin *et al.* 2006). This could be important for conservation of native species, considering the dominance of alien species like common carp. Among abiotic factors, differences in the composition of assemblages have been reported between high- and low-flow years, between rivers and between hydraulic habitat patches (Humphries *et al.* 2002; King *et al.* 2003a; Vilizzi *et al.* 2007; Cheshire 2010).

ANTHROPOGENIC IMPACTS

Human disturbances have overwhelmingly been detrimental to native freshwater fishes (Olden 2006; Jelks *et al.* 2008; Ch. 12). Spawning adults and early life stages are particularly vulnerable to modifications to their environment because of their specific habitat and food needs and their comparatively narrow environmental tolerances. Across the spectrum of disturbances that could affect the early life stages of Australian freshwater fishes, the effects of flow regulation and diversions are the most pervasive and well-studied (Table 7.6).

Flow regulation is arguably the most severe and widespread disturbance to Australia's freshwater environments. In the Murray-Darling Basin alone there are more than 4000 dams and weirs and thousands of kilometres of floodplain levees, testimony to human industry and symbolic of profound changes to the ecological integrity of floodplain–river ecosystems. Flow regulation alters the frequency, magnitude, timing, duration and rate of change of flow; it promotes cold-water pollution through releases from deep, thermally stratified reservoirs; it increases the incidence of stable in-channel flows, particularly during summer and autumn; it replaces lotic reaches with lentic habitats; it alienates floodplain environments and introduces barriers to the passage of fishes and other biota. Changed flow regimes may remove cues for spawning of species such as golden perch and silver perch (Tonkin *et al.* 2008a; King *et al.* 2010), impede movements to and from spawning areas (Conallin *et al.* 2010) and limit the dispersal of young fishes (Table 7.6). For free embryos and larvae, rearing habitat may be lost through enhancement (e.g. higher than normal flows flushing Murray cod free embryos from the nest: Humphries 2005), reduction in flow (King *et al.* 2009, 2010) or drowning of favourable low-velocity habitats (Humphries *et al.* 2006). Flow regulation also tends to promote reproduction by alien species such as common carp and eastern gambusia (Gehrke and Harris 2001; Stuart and Jones 2006a; Bice and Zampatti 2011). Many of these changes may result in increased mortality of the young stages of native fishes and in poor recruitment. Although patterns are emerging, we are still a long way from understanding the mechanisms involved.

CONCLUSION

Knowledge of reproduction and early life history underpins the conservation and management of Australia's freshwater fishes; we have made substantial gains in knowledge but we still have much to learn. For example, our analysis of life history traits shows that even three basic elements of population biology – egg size, fecundity and length at maturity – are reasonably known for less than a fifth of all Australian freshwater fishes. The species for which we have most data are the large commercial or recreationally important species, while small-bodied species have received much less attention. In addition, there is a strong imbalance in favour of information from south-eastern Australia and too little information from the tropical north.

In this chapter we have highlighted several other issues that warrant further investigation and attention. It is ironic that knowledge of the processes governing early survival and mortality is so poor, as they control recruitment and, at least for larger species, the sustainability of commercial and recreational fisheries. Studies of spawning and recruitment in relation to biotic and abiotic factors especially are needed, to refine management tools such as the environmental flow allocations designed to maintain and enhance populations. These studies need have no geographic bounds – they are needed in all regions, in all climates and at all scales of time and space. Scale especially is an issue that threatens to confound many ecological investigations. For example, if temperatures and food resources fluctuate with daily flows, monthly sampling may be inadequate. We need to consider whether the survival of fish larvae also varies at this scale.

Finally, the reproductive behaviours and life history traits of our freshwater fishes are in many cases uniquely Australian. Our species have evolved under highly variable climatic conditions that greatly influence the diversity, extent and permanence of our aquatic environments, and their adaptations to these environments are surprising and instructive. Some species are distributed across vastly different climatic regions and have flexible life histories that allow them to reproduce under such variable conditions. Others have much more restricted distributions and live in some of the harshest conditions of any freshwater organism, yet still they survive, reproduce and carry out their life cycles successfully. Understanding how they have evolved to maintain viable populations under these circumstances, and how they are responding to challenges from human industry and alien species, may yield insights into ways that they might respond to the impacts of global climate change. We therefore have an opportunity to improve our understanding of the life histories of our fishes and to contribute to the understanding of freshwater fishes worldwide.

8 Age and growth

David A. Crook and Bronwyn M. Gillanders

INTRODUCTION

Every angler dreams of hauling in a really big fish, but how many would ponder the ramifications of taking, say, a 30 kg Murray cod from the river? How old is a fish that size, and how important might it be in maintaining the population? The answers to these apparently simple questions strike at the heart of this chapter.

Why do we need to understand the relationships between the age and growth of fishes? The answer is that our understanding of the ecology of fishes depends on this information, as do the conservation of fish populations and management of fisheries. Age and growth data underpin fisheries regulations used to control the numbers of fish that are taken and the methods used to harvest them. Unsustainable fishing in the absence of adequate regulations (and policing of regulations) has contributed to a global decline of fisheries in recent decades (Myers and Worm 2003; Walters and Martell 2004; Zeller and Pauly 2005). A famous case study is the fishery for a marine species, the orange roughy (*Hoplostethus atlanticus*). In the late 1970s, when exploitation of this species began in earnest, it was not known that it reaches maturity at about 30 years of age and may live for 150 years or more (Tracey and Horn 1999). Early management strategies assumed much faster growth and earlier maturation, and allowed big catches that resulted in the collapse of orange roughy fisheries (Clark 2001).

Australian freshwater fishes do not have the commercial value of many marine fisheries but information on age and growth is no less vital for their conservation and management. This information is only gradually becoming available. It was not until the 1980s that researchers were able to quantify the age and growth of Murray cod (*Maccullochella peelii*), and to discover that some cod live for 48 years or more (Anderson *et al.* 1992a). Only recently have we learned that the Australian lungfish (*Neoceratodus forsteri*) may live for more than 65 years (James *et al.* 2010). It is amazing to consider that some of the big fishes in our waterways were spawned in the days when Sir Robert Menzies was Prime Minister of Australia!

Although most scientific literature on age and growth in fishes concerns commercial marine species (Hilborn and Walters 1992), there are abundant data for freshwater fishes (reviews by Weatherley 1972; Summerfelt and Hall 1987; Weatherley and Gill 1987; Mommsen 1998). In this chapter, we outline some salient features of age and growth and their significance for the ecology of Australian freshwater fishes. We describe methods for ageing fishes and quantifying growth and consider the longevity of Australian species, before examining factors that characterise, regulate and influence growth. We conclude with some ideas for future directions in research.

VALUE OF AGE AND GROWTH DATA

Age and growth data, in combination with information on reproductive biology, are widely used as a basis for fisheries regulations. For example, size limits and closed seasons for Murray cod in New South Wales have been in force since the early 1990s (Rowland 1985, 1998a, b) and this appears partly responsible for the significant recovery of Murray cod stocks there over the last 15–20 years (Rowland 2005).

The value of age and growth data is exemplified by the use of age-structured statistical models to inform stock assessment and management (Haddon 2001; Walters and Martell 2004). These models divide populations into age classes (cohorts) to reveal demographic parameters such as growth rates, mortality and recruitment. They range from simple deterministic relationships (ones with fixed variables, e.g. von Bertalanffy growth curves: see 'Age–size relationships') to complex stochastic models that allow for variability in such factors as recruitment, natural and fishing mortality, environmental conditions, predation and density-dependent competition (Haddon 2001; Walters and Martell 2004). For example, the effects of minimum legal lengths and recreational fishing mortality on Murray cod have been explored using an age-structured model that refers to age-specific natural mortality, harvest and catch-and-release mortality, vulnerability to capture and fecundity (Allen *et al.* 2009). The model was used to evaluate regulation scenarios and to recommend sustainable levels of exploitation (Fig. 8.1a).

Aside from providing essential information for management, analyses of age and growth have contributed to specialised disciplines including ecology, evolutionary biology, conservation biology, comparative physiology and climate-change biology (Weatherley and Gill 1987; McDowall 1994; Perry and Bond 2009; Morrongiello *et al.* 2011c). Population viability analysis, for example, uses age-structured demographic models to assess the probability of extinction of species under different management scenarios (Possingham *et al.* 1993). The population dynamics of the threatened trout cod (*Maccullochella macquariensis*) have been modelled using this approach (Bearlin *et al.* 2002; Todd *et al.* 2004). These models included mortality rates, fecundity, intraspecific density dependence and habitat carrying capacity, and were used to estimate population sizes under different restocking scenarios (Todd *et al.* 2004; Fig. 8.1b). Similar models have been used to explore the effects of cold-water releases from reservoirs on the dynamics of Murray cod (Todd *et al.* 2005; Sherman *et al.* 2007) and brown trout (*Salmo trutta*: Brown 2004).

ESTIMATING AGE

We now consider methods to obtain age and growth data. One approach is to monitor fishes in captivity and record age, age versus length or weight and age at maturity. The growth of larvae, fingerlings and larger fish in hatcheries and aquaculture facilities has been described for some native fishes, including silver perch (*Bidyanus bidyanus*), golden perch (*Macquaria ambigua*), Murray cod, trout cod and Macquarie perch (*M. australasica*) (Rowland *et al.* 1994; Rowland 1996, 2004, 2005; Ingram 2009). Data from some life cycle stages, particularly free embryos, larvae and juveniles (Ch. 7) are often difficult or impossible to collect in the wild, whereas in aquaculture the absolute age to the day usually is known, the rearing conditions are known and controlled and the fish can be sampled with ease. However, because patterns of mortality and growth are highly sensitive to environmental conditions, data from aquaculture are not always readily transferable to wild populations (see 'Factors affecting growth'). For example, silver perch males and females reared in captivity on artificial feed mature sexually at two and three years of age, respectively (Rowland 2004); in the wild, the respective ages are three and five years (Mallen-Cooper and Stuart 2003).

Length-frequency analysis

A common technique is length-frequency analysis wherein peaks, or modes, in the frequencies of size classes are used to discriminate between yearly

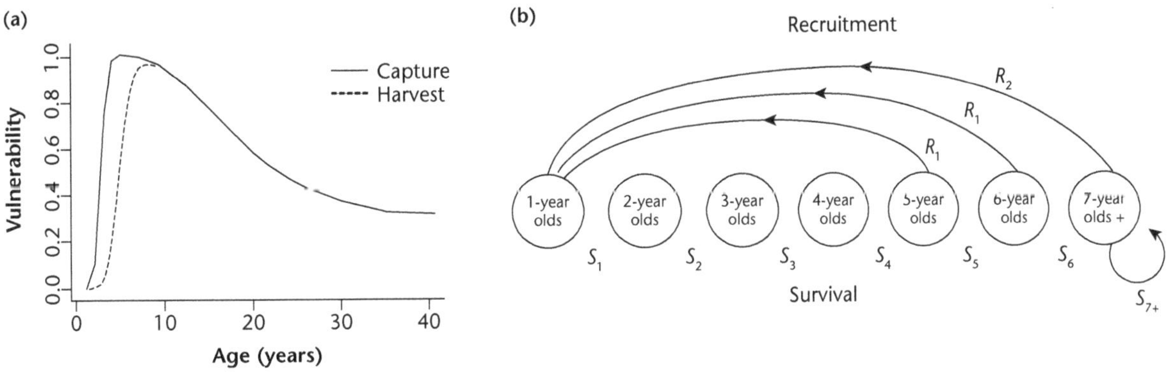

Figure 8.1: Use of age and growth data for fishery management and conservation modelling of Murray cod (*Maccullochella peelii*). (a) Modelled age-specific vulnerability of Murray cod to capture (solid line) and harvest (dashed line) under a 500 mm minimum length limit fisheries regulation scenario (from Allen *et al.* 2009). (b) Schematic diagram of age-based population model for Murray cod. The model splits populations into age classes with specific annual survival rates (S_1–S_7) and contributions to recruitment (R_1–R_2). Sexual maturity in the model is assumed to be reached at five years and fecundity increases with age (from Todd 2009).

cohorts (Fig. 8.2). This provides a snapshot of a population's size structure based on measurements of a sample. By sampling the population repeatedly over time, the progression of modes can be tracked as the fish in each cohort grow larger. Length-frequency analysis, usually in combination with other methods, has been applied to barramundi (*Lates calcarifer*: Dunstan 1959), spotted galaxias (*Galaxias truttaceus*:

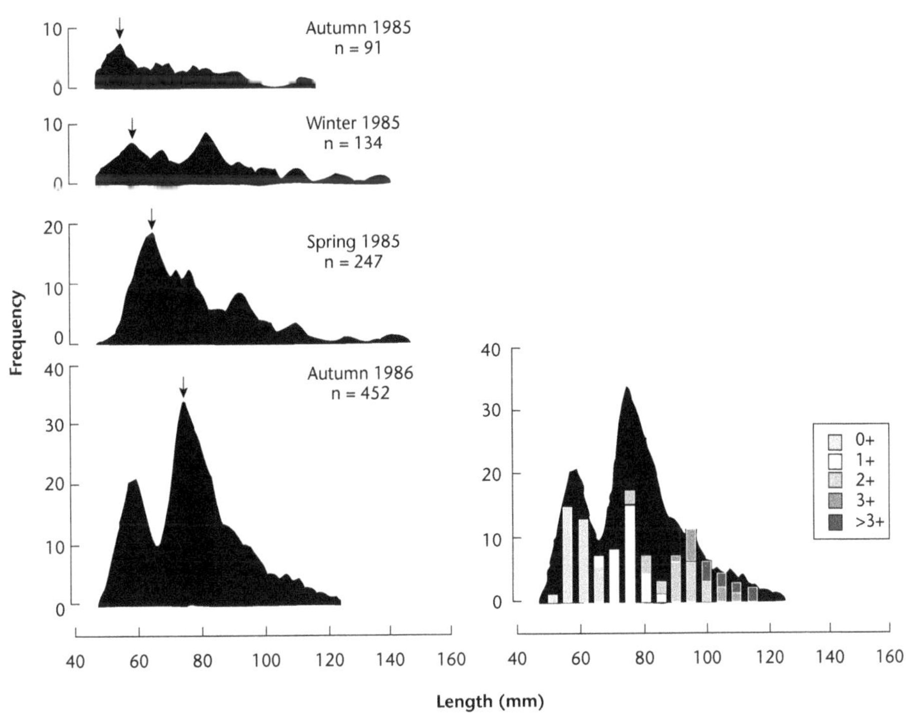

Figure 8.2: Length–frequency analysis of spotted galaxias (*Galaxias truttaceus*) from Tasmanian streams. The arrows show the average length of a single age-class over time. The graph at right shows the length–frequency plot for autumn 1986 with the ages of fish superimposed. Note the overlap of ages in the length–frequency modes. A three-point rolling average was used to smooth the data, and the fish were aged from annuli in the sagittal otoliths (modified from Humphries 1989).

Humphries 1989) (Fig. 8.2), golden perch (Anderson *et al.* 1992b; Balcombe *et al.* 2006; Ebner *et al.* 2009b), freshwater catfish (*Tandanus tandanus*), bony herring (*Nematalosa erebi*) and common carp (*Cyprinus carpio*) (Balcombe *et al.* 2006).

Although length-frequency analysis indicates the size structure of populations, its utility for estimating the ages of individuals is much reduced if there is wide variability in growth rates. This causes the modes to become indistinct – a phenomenon known as 'modal extinction' (Weatherley and Gill 1987). In most long-lived species, usually only the first one or two age classes are distinguishable in length-frequency analyses. Golden perch, for example, may live for more than 20 years (Stuart 2006); two-year-old fish from the Murrumbidgee River in 2005–09 were 120–405 mm total length (TL), while fish of 420–430 mm TL were between three and 18 years old (Crook and Gillanders, unpubl.). Such variability in individual growth rates means that length-frequency analysis is most useful for short-lived species or for identifying young-of-year cohorts in populations of long-lived fishes.

Calcified structures

Fortunately for ecologists, fishes have calcareous structures (in the case of rays and sharks, cartilaginous structures) with growth increments that are analogous to the growth rings in trees, allowing us to estimate the ages of individuals (Campana and Neilson 1985; Campana and Thorrold 2001). Increments are laid down in scales, vertebrae, cleithra (part of the pelvic girdle), fin spines, opercular bones and otoliths (earstones; Fig. 8.3).

The use of scales dates from the 19th century (Carlander 1987) and is still commonplace (Helle and Hoffman 1995; Friedland *et al.* 1996). The scales of most fishes have circuli, laid down in patterns determined by the growth characteristics of the fish. In many species, the circuli form an annual growth band or *annulus* (Wootton 1998; Fig. 8.3b) so that it is possible to estimate age by counting the annuli. These can be difficult to interpret, however, and they tend to under-represent the age of older fishes because material is resorbed during periods of slow growth. Further, false checks are often hard to distinguish from annuli, checks near the outer edges are close together and difficult to distinguish in older fish, and damaged scales are replaced with new scales that lack the older annuli (Linfield 1974; Carlander 1987). The main advantage of using scales is that they can be sampled without killing the fish. Scales are also used to back-calculate size-at-age relationships (see 'Age–size relationships').

Despite these issues, annuli on scales have been used to estimate the ages of barramundi (Davis and Kirkwood 1984), freshwater herring (*Potamalosa richmondia*: Pidgeon 1989) and, with some difficulty, golden perch (Llewellyn 1966). Annuli in skeletal and opercular bones have also been used (Llewellyn 1966; Rowland 1998a; Butler 2009a). For example, consistent annuli occur in sectioned dorsal spines in freshwater catfish (Davis 1977a) and the opercular bones of Murray cod (Rowland 1985, 1998a) but not in the spines or opercular bones of Clarence River cod (*Maccullochella ikei*: Butler 2009a). Even in species with consistent annuli, caution is needed when using dorsal spines for ageing since the inside of the spine may hollow out over time, resulting in loss of the early growth increments and potential underestimation of age (Davis 1977a).

For most species, otoliths are the most reliable structure for age estimation (Anderson *et al.* 1992b; Gooley 1992; Stuart and McKillup 2002; Butler 2009a). Otoliths are paired structures located within membranous sacs posterior to the brain, and collectively are part of the vestibular apparatus. They are analogous to the *otoconia* (earstones) of other vertebrates and they function in hearing, balance and acceleration (Popper and Lu 2000). In teleost fishes there are three types of otoliths: the *sagitta* (pl. sagittae), *lapillus* (pl. lapilli) and *asteriscus* (pl. asterisci) (Fig. 8.4).

Growth in otoliths is continuous throughout life and, because the otolith material is not remetabolised once it has accreted, otolith increments represent an uninterrupted record of the life of the fish (Campana 1999). They are generally more reliable than other calcified structures for annual ageing and may even reveal daily growth increments (Pannella 1971; Campana and Jones 1992), providing fine-scale information about early life history, including growth rates and back-calculated dates of hatching (Ch. 7; Campana and Jones 1992). Although whole otoliths have been used for annual ageing (Lake 1967a; Rowland

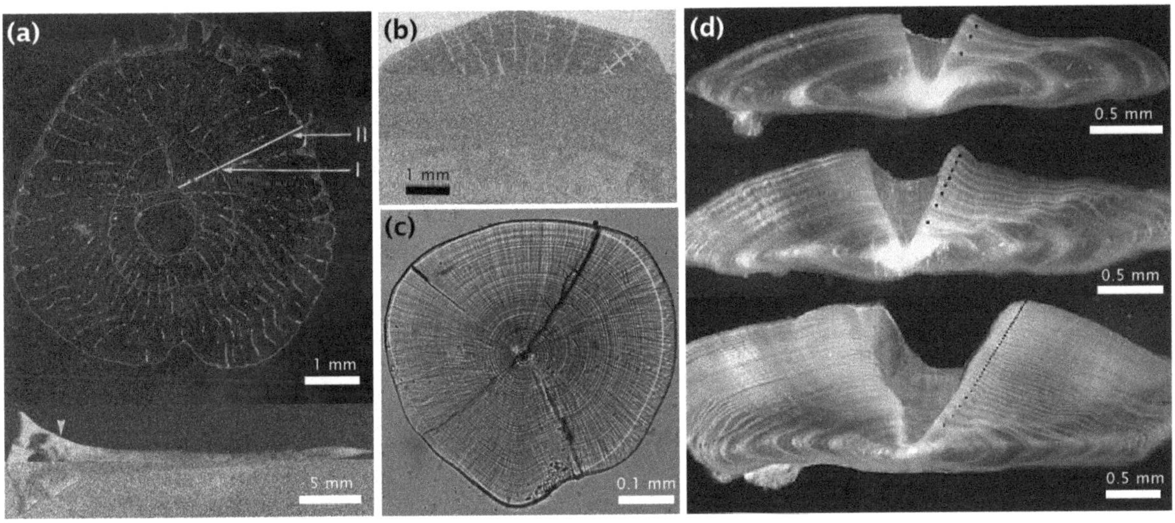

Figure 8.3: Examples of calcified structures used to age Australian freshwater fishes. (a) Transversely sectioned dorsal spine of a 2+-year-old freshwater catfish (*Tandanus tandanus*) showing the locations of the two annuli (top), and whole dorsal spine showing the location of the section (bottom) (modified from Davis 1977a). (b) Scale of a 6+-year-old bony herring (*Nematalosa erebi*) with locations of annuli marked (modified from Pidgeon 1989). (c) A 63-day-old Australian smelt (*Retropinna semoni*) sagittal otolith, showing daily increments (photo: Z Tonkin). (d) Transversely sectioned sagittal otoliths of estuary perch (*Percalates colonorum*), showing annuli (top: 3+ years, middle: 8+ years, bottom: 35 years) (photo: C Walsh).

1985, 1998a), they are usually sectioned and polished to reveal fine details (Secor *et al.* 1992; Fig. 8.3).

The otoliths of most fishes are composed of aragonite, a crystalline form of calcium carbonate that grows in association with a proteinaceous matrix. Other forms of calcium carbonate occur, namely vaterite (relatively common) and calcite (rare). The asterisci of common carp are composed of vaterite, whereas the lapillae and sagittae are aragonite (Li *et al.* 2009). Further, vateritic inclusions may occur as discrete regions within primarily aragonitic otoliths (Gauldie 1996; Brown and Severin 1999; Tzeng *et al.* 2007; Veinott *et al.* 2009). Vaterite is more translucent than aragonite (Tomás *et al.* 2004) and tends to lack incremental structure (Li *et al.* 2009), making vateritic otoliths less useful for age estimation (but see Vilizzi and Walker 1995; Brown *et al.* 2004b).

In most Australian freshwater fishes, the sagittae are most useful for annual ageing (Fig. 8.4) but there are exceptions among non-Australian taxa. In the Cyprinidae, for instance, the asterisci have been used for annual ageing of common carp (Vilizzi and

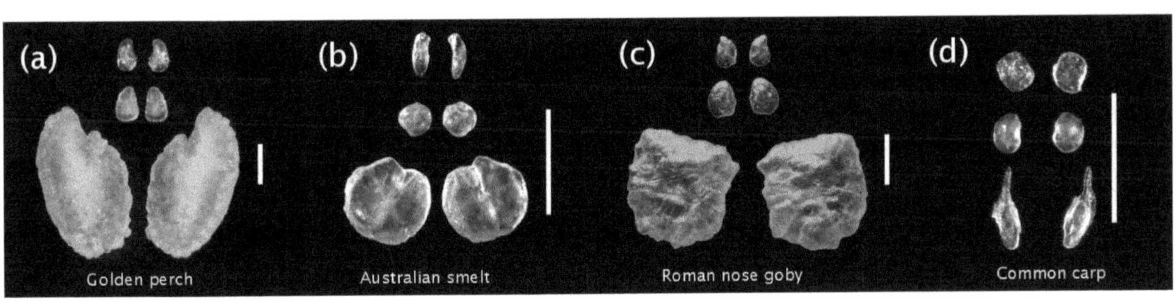

Figure 8.4: The three types of paired otoliths in teleosts: (top row) asterisci, (middle row) lapilli and (bottom row) sagittae. (a) Juvenile golden perch (*Macquaria ambigua* – 64 mm TL). (b) Adult Australian smelt (*Retropinna semoni* – 40 mm TL). (c) Adult roman-nose goby (*Awaous acritosus* – 100 mm TL). (d) Post-larval common carp (*Cyprinus carpio* – 18 mm TL). The asterisci in common carp become much larger relative to the lapilli and sagittae after the post-larval stage. Scale bars = 1 mm (photos: J Macdonald, D Crook).

Walker 1999a; Brown *et al.* 2004b, 2005; Crook and Gillanders 2006). The lapillae are often used to examine daily increments, particularly for species whose sagittae and asterisci are unsuitable due to the presence of secondary primordia, or because of unusual morphology (Campana and Jones 1992; Vilizzi 1998a; Smith and Walker 2003; Li *et al.* 2009).

Increment formation in otoliths

Daily increments in otoliths reflect diel variation in crystallisation processes caused by endogenous circadian rhythms, often reinforced by regular patterns in behaviour (e.g. feeding) and environmental conditions (e.g. light, temperature) (Campana and Neilson 1985; Morales-Nin 2000). Daily increments are laid down over 24 hours and appear as alternating translucent (light) and opaque (dark) bands when viewed under transmitted light (Figs 8.3c, 8.5a). An opaque zone will look dark under transmitted light, and light under reflected light. For consistency, it is best to use the terms 'opaque' and 'translucent' when discussing increment formation. The opaque zones represent discontinuities in otolith accretion and contain more organic material than the translucent zones; thus, they absorb more light when viewed with transmitted light (Morales-Nin 2000). Very fine-scale temporal variations in crystallisation patterns may also cause subdaily rings to appear within daily increments, but these usually are easy to distinguish from daily increments (Campana and Jones 1992). Daily increments are most easily observed during the first year or so of life and are used primarily to determine the ages of free embryos, larvae and early juveniles. Although daily increments may continue to form throughout life, they become less distinct as fishes (and their otoliths) grow (Campana and Jones 1992).

The processes that lead to annual increment formation in otoliths are less well-understood than those responsible for daily increments, although they appear to be linked to variations in somatic growth associated with endogenous factors (e.g. gonadal development) and environmental conditions (e.g. low temperatures) (Campana and Neilson 1985; Secor *et al.* 1992). Although annuli are much larger features than daily increments, they also consist of sequential opaque and translucent zones that vary in protein content (Fig. 8.3d). Because the processes responsible for annulus formation can be complex and variable among species, a good understanding of the timing and consistency of annulus formation is required to reliably estimate the ages of different species.

Validation of ageing methods

The usefulness of calcified structures for ageing fish depends on the regular and predictable formation of recognisable growth increments (Beamish and McFarlane 1983; Campana and Thorrold 2001). Several empirical approaches can be used to validate increment formation in otoliths. To validate daily increments, fishes usually are held in laboratory aquaria or (preferably) natural conditions, and either reared from hatching or marked using fluorescent dyes, such as oxytetracycline (OTC), and reared for several days or more. Double-marking methods (e.g. a week or so apart) may be useful where it is difficult to distinguish the edge-increments of the otoliths. Validation of daily increments is undertaken by comparing increment counts against the known post-marking or post-hatching age of the fish (Fig. 8.5a). These techniques have been used with the otoliths of Murray cod (Humphries 2005), golden perch (Brown and Wooden 2007), Australian smelt (*Retropinna semoni*: Tonkin *et al.* 2008c), eastern rainbowfish (*Melanotaenia splendida*: Humphrey *et al.* 2003) and common carp (Vilizzi 1998a).

The most common technique for determining the timing and consistency of annulus formation in otoliths and scales is 'marginal increment analysis', where the distance between the most recent annulus and the outer margin of the otolith or scale is measured (e.g. Butler 2009a) or classified qualitatively according to the positions of the opaque and translucent zones (Stuart and McKillup 2002). Samples of fish are collected regularly (usually monthly) to examine annulus-to-margin distances throughout an annual cycle. Following annulus formation, annulus-to-margin distances increase as new material is gradually deposited. Immediately after the deposition of a new annulus, the annulus-to-margin distance is much reduced. Evidence of valid annuli is inferred if rapid decreases occur once per year over a similar period for all individuals (Fig. 8.5b).

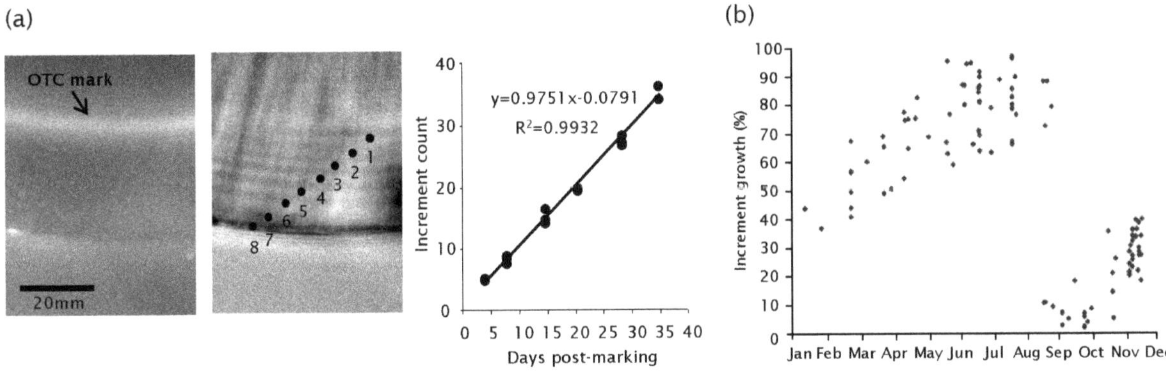

Figure 8.5: Examples of otolith increment validation techniques. (a) Use of chemical tagging to validate daily increments in sagittal otoliths of Australian smelt (*Retropinna semoni*) (modified from Tonkin *et al.* (2008b). The photographs show the oxytetracycline (OTC) mark under ultraviolet light and the locations of eight daily increments in the post-marking period. The graph shows the relationship between increment counts and the known ages of fish sampled 4–35 days post-marking. (b) Marginal increment analysis for validating annual increments in Clarence River cod (*Maccullochella ikei*) sagittal otoliths, showing annulus formation in August (from Butler and Rowland 2008).

Marginal increment analysis, often in combination with other techniques, has been used to validate annulus formation in nightfish (*Bostockia porosa*: Pen and Potter 1990), western pygmy perch (*Nannoperca vittata*: Pen and Potter 1991a), western galaxias (*Galaxias occidentalis*: Pen and Potter 1991a), golden perch (Anderson *et al.* 1992b; Mallen-Cooper and Stuart 2003; Roberts *et al.* 2008), Murray cod (Anderson *et al.* 1992a; Gooley 1992), Clarence River cod (Butler 2009a), barramundi (Davis and Kirkwood 1984; Stuart and McKillup 2002), salamanderfish (*Lepidogalaxias salamandroides*: Morgan *et al.* 2000), spotted galaxias (Morgan 2003), estuary perch (*Percalates colonorum*: Walsh *et al.* 2010), freshwater herring (Pidgeon 1989) and common carp (Vilizzi and Walker 1999a). It is important to minimise the subjectivity of marginal increment analysis by randomising samples, examining a minimum of two complete cycles, restricting the analysis to a few age groups at a time and using an objective approach to determine whether differences occur among seasonal groups (e.g. months; Campana and Thorrold 2001). Marginal increment analysis is often limited to a few early age classes, and other methods may be required to validate the ages of older fish.

The use of fishes of known age is another common approach to validating annual otolith increments, and is the most rigorous method (Campana and Thorrold 2001). In instances where the actual ages of fish (e.g. from a hatchery) and stocking history are known, and there is no chance of natural recruitment, information from the known-age fish can be used to validate annual increment formation (Davis 1977a; Anderson *et al.* 1992b; McDougall 2004; Stuart 2006). For example, Rowland (1985, 1998a) and Gooley (1992) used Murray cod from hatcheries, stocked into earthen ponds and farm dams, to validate annulus formation in otoliths over periods of three and four years, respectively. Stuart (2006) validated otolith annulus formation in golden perch stocked as fingerlings into a Victorian lake in 1982 and recaptured in 2002 and 2004.

Where the stocking history is unknown, mark–recapture techniques are often used. Fishes are marked with a fluorescent dye (e.g. OTC) and released back into the wild or held under semi-natural captive conditions for an extended period. The number and location of annuli laid down after marking are then used to validate the annual formation of increments. Variations of this approach have been used with the otoliths of barramundi (Stuart and McKillup 2002), longfin eels (*Anguilla reinhardtii*: Pease *et al.* 2003), estuary perch (Walsh *et al.* 2010) and common carp (Brown *et al.* 2004b).

Other age-validation methods (e.g. bomb radiocarbon dating, radiochemical dating) exist but they have not been widely applied to Australian freshwater fishes (Campana and Thorrold 2001). The bomb radiocarbon technique, in particular, has been used extensively to validate the ages of long-lived marine fishes (Kalish 1993; Campana and Thorrold 2001). It measures radiocarbon (C^{14}) concentrations in calcified structures and relates these to the pulse of radiocarbon that was released into the atmosphere from nuclear bomb tests in the 1950s and 1960s (Kalish 1993). Fishes must have been alive in 1958–65 to take advantage of the unique C^{14} values that occurred following the bomb tests (Campana 1997). Thus, the technique is useful only for validating the annuli of old fishes or archived samples. Radiocarbon concentrations in the scales of a 1.25 m long Australian lungfish have shown an age of 65–70 years (James *et al.* 2010). Other methods (e.g. elemental and isotopic cycles, length-frequency analysis) can be used to corroborate age interpretations but, strictly speaking, they are not equivalent to age validation (Campana and Thorrold 2001).

Otolith chemistry

Chemical analyses of otoliths are often integrated with age information to infer the environmental conditions experienced by fishes (Gillanders 2005, Fig. 8.6). As mentioned, otoliths grow continuously throughout life and the chemical structure is not remetabolised once it has been deposited. Dissolved trace elements in the surrounding water, such as strontium (Sr) and barium (Ba), move into the endolymph and become chemically bound within the calcium carbonate matrix as it accretes. Analysis of these elements can be used to reconstruct the ambient water chemistry at different stages of an individual fish's life.

Variations in otolith Sr, Ba and $^{87}Sr{:}^{86}Sr$ are often used to make inferences about diadromous migrations between freshwater and estuarine or marine environments, based on known relationships between salinity and trace-element concentrations or isotope ratios in otoliths (Gillanders 2005; Fig. 8.6). Otolith material formed while a fish is living in saline water typically has high Sr and low Ba concentrations relative to calcium (Ca) (i.e. high Sr:Ca, low Ba:Ca). The opposite typically is true for otolith growth in fresh water (low Sr:Ca, high Ba:Ca). $^{87}Sr{:}^{86}Sr$ ratios can be used to identify movements between saline and fresh water because seawater has a global $^{87}Sr{:}^{86}Sr$ value of 0.7091, whereas $^{87}Sr{:}^{86}Sr$ in fresh water reflects the underlying geology of the local catchment and therefore is highly variable (Kennedy *et al.* 2000; Gillanders 2005; McCulloch *et al.* 2005).

Otolith chemistry analyses have been used to examine diadromous migrations in barramundi (McCulloch *et al.* 2005; Milton and Chenery 2005), Australian smelt (Munro *et al.* 2008), Australian grayling (*Prototroctes maraena*: Crook *et al.* 2006), common galaxias (*Galaxias maculatus*: Hale and Swearer 2008; Hicks *et al.* 2010), congolli (*Pseudaphritis urvillii*: Cheshire 2005), Australian bass (*Percalates novemaculeata*: Macdonald and Crook 2010), southern shortfin eel (*Anguilla australis*: Arai *et al.* 1999, 2004), freshwater mullet (*Myxus capensis*), striped gudgeon (*Gobiomorphus australis*), Cox's gudgeon

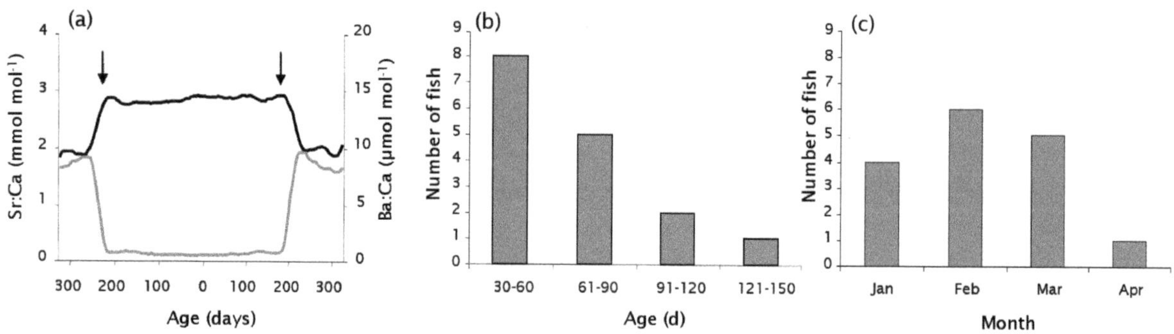

Figure 8.6: Example of the combined use of otolith chemistry analysis and daily ageing. (a) Transects of Sr:Ca (black line) and Ba:Ca (grey line) across sectioned otoliths were used to identify migrations from saline water into fresh water in diadromous populations of Australian smelt (*Retropinna semoni*) (see text). Arrows show the approximate timing of upstream migration. This information was combined with information from the otoliths to estimate (b) the age and (c) the timing of upstream migration (modified from Crook *et al.* 2008).

(*Gobiomorphus coxii*), freshwater herring and bullrout (*Notesthes robusta*) (Miles *et al.* 2009). Otolith chemistry analysis has also been used to discriminate between wild and hatchery golden perch (Munro *et al.* 2008, 2009; Woodcock *et al.* 2010) and Murray cod (Woodcock *et al.* 2011), to identify recruitment sources in common carp (Crook and Gillanders 2006) and to examine connectivity among coastal populations of common galaxias (Barbee and Swearer 2007) and inland populations of Australian smelt (Woods *et al.* 2010). Chemical analyses of other calcified structures (e.g. scales, fin spines) have been used for similar purposes as those mentioned above but are generally considered less precise and reliable than otolith chemistry analysis (Campana and Jones 1992). Nonetheless, freshwater and marine phases in barramundi have been successfully distinguished by analysing the chemistry of scales (Pender and Griffin 1996; Cappo *et al.* 2005).

LONGEVITY

Australian freshwater fishes have diverse physiological and life history characteristics, reflected in a range of growth characteristics and life expectancies. To understand why some species live longer than others, we need to examine life history strategies in terms of evolutionary adaptations and trade-offs (Ch. 7). Most small-bodied species (e.g. Atherinidae, Chandidae, Eleotridae, Galaxiidae, Melanotaeniidae) mature in their first year and live for one to five years (McDowall 1996b; Pusey *et al.* 2004). Large-bodied fishes mature later and live longer than smaller species. The Australian lungfish, for example, matures at 15–22 years of age, lives for 65 years or more and grows to more than 1.4 m and 25 kg (Brooks and Kind 2002; Arthington 2009; James *et al.* 2010). Other large species also have long maximum recorded life spans (e.g. barramundi, 32 years: Staunton-Smith *et al.* 2004; longfin eels, 52 years: Walsh *et al.* 2004; Murray cod, 48 years: Anderson *et al.* 1992a).

Not all species with long life spans grow to extreme sizes like Murray cod (purportedly to 113.5 kg: Rowland 1989). Australian bass and estuary perch, for example, are slow-growing, estuarine species that can attain 600 mm and 4 kg (Harris and Rowland 1996). Australian bass of about 400 mm from the Sydney Basin in New South Wales may reach 22 years (Harris 1987a) and estuary perch of 400 mm from the Bemm River, Victoria, can attain 41 years (Walsh *et al.* 2010; Fig. 8.3d). Several other species with only moderate body sizes are long-lived, including golden perch (oldest recorded individual 26 years, 533 mm, 1.8 kg) and silver perch (27 years, 440 mm, 1.2 kg) (Mallen-Cooper and Stuart 2003).

Fishes in other parts of the world have similar relationships between body size, age at maturity and longevity. For North American fishes, it has been proposed that species' life histories lie somewhere on a triangular continuum between three strategies: 'opportunistic' (small, rapidly maturing, short-lived), 'periodic' (larger, highly fecund, longer life spans) and 'equilibrium' (intermediate size, often with parental investment, fewer but larger offspring) (Winemiller and Rose 1992; Ch. 7). A similar, trait-based approach has been proposed to classify Australian freshwater fishes into five reproductive guilds (Growns 2004).

Aside from the characteristics of a species' life history, life expectancies can be strongly influenced by the ambient environment. For example, under conditions with high mortality rates (e.g. due to high predator densities) the probability of living to old age is much reduced (Stearns and Koella 1986). Fishes also exhibit (non-genetic) plasticity in their physiology and behaviour in response to environmental conditions (McDowall 1994). As a consequence, growth rates, energetic investment in reproduction and longevity within a species may vary considerably (Stearns and Koella 1986; McDowall 1994; Morrongiello *et al.* 2011b). For instance, Murray cod may grow faster in impoundments than in rivers (Rowland 1998a) and golden perch up to 27 kg have been recorded from impoundments (Harris and Rowland 1996), whereas fish weighing more than 5 kg are rare in natural riverine habitats.

CHARACTERISTICS OF GROWTH

Definition

For most practical purposes, growth is the change in length and/or weight of a fish over time. Growth in fishes sometimes is referred to, rather counter-intuitively, as positive (increase in size) or negative (decrease in size) (Busacker *et al.* 1990). While positive growth is part of the normal development of organisms, negative growth is a temporary condition during periods of food deprivation or physiological stress (Wikelski and Thom 2000). The lengths of juvenile salmonids, for example, may decrease by up to 10% during winter because of a lack of food and shelter in streams of northern Finland (Huusko *et al.* 2011).

From a metabolic perspective, growth is the change in the energetic (caloric) content of an organism over time (Mommsen 2001). It can be considered as the sum of somatic anabolism (G_s: production of body tissue) and gametic anabolism (G_g: production of reproductive tissue), and the energy balance of growth can be expressed as:

$$G_s + G_g = I - M - E \tag{8.1}$$

where I is ingested energy in the form of food, M is energy expended on metabolic processes (body maintenance, active metabolism, digestive metabolism) and E is energy excreted as waste (primarily as urea, ammonia and faeces). The faeces contain mucus and sloughed epidermal cells that are products of growth (Weatherley and Gill 1987), but for present purposes we can consider this as waste.

Indeterminate growth and growth plasticity

A feature of growth in many ectothermic vertebrates (fishes, reptiles, amphibians) is that they tend to continue to grow after sexual maturity (Sebens 1987; McDowall 1994). As a consequence, adult size is not fixed within a species. In contrast, birds and most mammals usually cease to grow once they reach maturity, resulting in a characteristic adult size for each species (Stamps 1993). The adaptive advantages of post-maturation growth for ectotherms appear to relate to increases in fecundity that a larger body allows – that is, larger fishes are able to carry more eggs and hence have greater reproductive output. In contrast, reproductive output in homeotherms is less related to maternal body size because litter or clutch size is limited by the need for parental care until weaning or fledging (Okuda *et al.* 1998).

Large, long-lived species such as Murray cod and barramundi demonstrate ways that fishes may continue to grow after sexual maturation. Sexual maturity in female Murray cod occurs at about four to seven years of age and 2–3 kg mass and in males at three to four years and 0.5–1 kg, yet this species may live for more than 45 years and attain more than 80 kg (Anderson *et al.* 1992a; Gooley 1992). Barramundi are protandrous hermaphrodites that mature as males at three to four years of age and 60–70 cm, and change into females after six to eight years of age and 80–100 cm (Davis 1982). The females may live to more than 30 years of age and may reach 1.5 m and more than 40 kg (Davis and Kirkwood 1984; Griffin 1987; Staunton-Smith *et al.* 2004).

Continued growth after maturation is often referred to as 'indeterminate' or 'continuous' growth (McDowall 1994). The use of these terms as interchangeable descriptors of growth in fishes is common in the literature, but requires some clarification. In this context, continuous growth refers to situations where growth occurs throughout life (McDowall 1994; Mommsen 2001). It is not strictly synonymous with indeterminate growth, which refers to species with no fixed adult size, but does not specify continuous growth throughout life. Post-maturation growth in fishes that display indeterminate growth typically is asymptotic in nature, with the change in length per time-step decreasing with age towards a finite maximum length (McDowall 1994). This type of growth is often numerically represented by plotting age against length for an individual fish or sample population and applying an asymptotic growth equation, such as the von Bertalanffy growth model (see 'Age–size relationships').

As mentioned earlier, the growth rates of fishes are highly flexible compared to other vertebrates, and are strongly influenced by environmental conditions and available resources (Weatherley and Gill 1987; McDowall 1994; Ali *et al.* 2003). The consequence of this phenotypic plasticity is that populations of the same species living under different conditions may exhibit disparate growth trajectories and different size–age relationships. Growth in spatially discrete stocks of golden perch demonstrates the potential for plasticity in asymptotic maximum sizes within a species (Mallen-Cooper and Stuart 2003). Length-at-age relationships in four stocks of golden perch from the Murray-Darling Basin are shown in Figure 8.7, and described using the von Bertalanffy growth model,

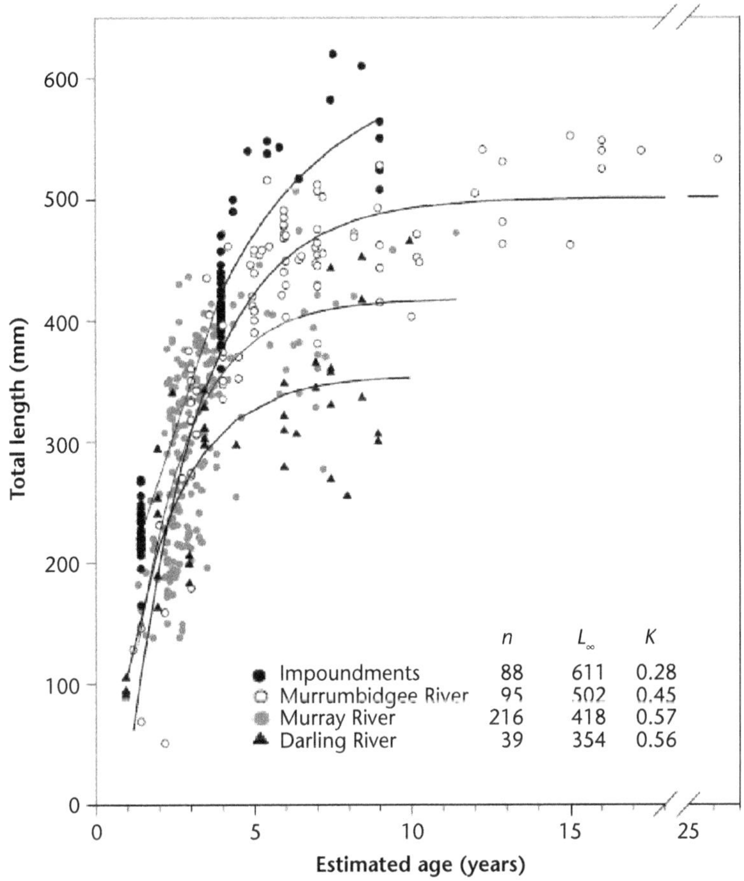

Figure 8.7: Growth-rate plasticity in golden perch (*Macquaria ambigua*) in the southern Murray-Darling Basin. Length-at-age relationships varied substantially in four sample populations, with faster growth in impoundments than rivers. The number of fish from each location (*n*) and estimates of L_∞ and *K* from fitted von Bertalanffy growth curves are shown. Annual increments in the sagittal otoliths were used to estimate age (modified from Mallen-Cooper and Stuart 2003).

with asymptotic average maximum sizes (L_∞) of 354–611 mm TL. Similar intraspecific variability occurs in other species (e.g. barramundi: Davis and Kirkwood 1984; Murray cod: Anderson *et al.* 1992a; Australian bass: Harris 1987a; Wilde and Sawynok 2005; longfin eels: Walsh *et al.* 2004).

Finally, it should be noted that some fishes exhibit determinate growth (Okuda *et al.* 1998; Magellan *et al.* 2005). Male cardinal fish (*Apogon doederleini*), for example, are an iteroparous, mouth-brooding species from coastal marine waters of northern Australia and other parts of the western Pacific. As in birds and mammals, post-maturation growth in males of this species ceases due to constraints imposed by a high level of parental care (Okuda *et al.* 1998). Several species of Australian freshwater fishes exhibit similar levels of paternal care of offspring (e.g. mouth-brooding ariid catfishes: Pusey *et al.* 2004; head-brooding nurseryfish, *Kurtus gulliveri*: Berra and Aday 2004; Ch. 7). In nurseryfish, males carry fertilised eggs using a hook located on their heads, and there is some evidence that following maturation they cease to grow, or at least grow more slowly than females (Berra and Aday 2004). Evidence for determinate growth in male salamanderfish is clearer, with fish maturing in the first or second year of life and continuing to breed up to four years of age with no further increase in size (Morgan *et al.* 2000). Female salamanderfish, on the other hand, exhibit indeterminate growth. The extent to which determinate growth occurs among Australian freshwater fishes, and its adaptive basis, are topics awaiting research.

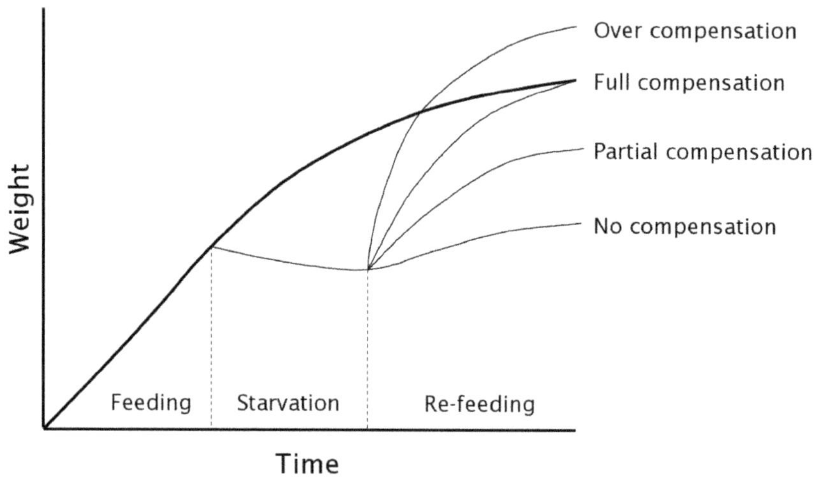

Figure 8.8: Schematic representation of compensatory growth following starvation. Note that full compensation restores the original growth trajectory (bold line), whereas fishes exhibiting partial compensation have a period of increased growth but never catch up to the growth trajectories of control fish. Fishes that exhibit no compensation resume a trajectory that is parallel with control fishes but at a lower weight-for-age. Over compensation has been demonstrated in some circumstances (modified from Ali *et al.* 2003).

Compensatory growth

An interesting consequence of the plasticity in growth characteristics is 'compensatory growth', a period of accelerated growth that occurs when favourable conditions are restored following a period of growth depression (Ali *et al.* 2003). This phenomenon occurs in a wide range of vertebrate and invertebrate taxa and is likely to be especially important for fishes, given their indeterminate, plastic, growth characteristics. In experimental situations, fishes that exhibit compensatory growth following growth depression grow significantly faster than control specimens under the same conditions. Thus, compensatory growth allows fishes to at least partially catch up to the trajectory they were on prior to the period of growth depression (Fig. 8.8). Compensatory growth is documented for Murray cod (Turchini *et al.* 2007), southern shortfin eel (Graynoth and Taylor 2000), barramundi (Tian and Qin 2003) and Australian smelt (Tonkin *et al.* 2008c).

Most studies of compensatory growth in fishes regard starvation as the growth-depressing mechanism, although other factors can be responsible (e.g. low oxygen, suboptimal temperatures). The mechanism responsible for compensatory growth following starvation is hyperphagia, a significantly higher rate of food consumption than in fishes fed *ad libitum* (free access to food; Ali *et al.* 2003). In other words, following growth depression, fishes are able to gorge themselves in order to achieve higher than normal growth rates. Increases in food conversion efficiency and changes to metabolic rate may also play roles in compensatory growth (Dobson and Holmes 1984; Wieser *et al.* 1992).

The obvious question that arises from studies of compensatory growth is that, if fishes are capable of increasing their growth rates following growth depression, why do they not grow at these higher rates all the time? Clearly, there must be fitness costs associated with higher than normal growth. These appear to be primarily associated with mortality risk (Priede 1985). For example, the increased foraging activity required to achieve hyperphagia may expose fish to higher rates of predation (Ali *et al.* 2003). While compensatory growth is generally treated as a distinct type of growth, periods of unusually high growth are likely to occur within the normal growth trajectories of fishes as individuals with highly flexible growth characteristics respond to environmental changes over time (e.g. gorging when prey items are abundant). Thus, growth trajectories in wild populations of fish reflect an optimal fitness strategy that balances the advantages of maximising growth against mortality risk and other factors that affect fitness.

Aestivation

Whereas compensatory growth represents higher than normal growth, some fishes have a capacity to reduce their rates of activity and metabolism and enter a state of dormancy known as aestivation, as habitats contract during seasonal dry periods or drought. Fishes stop feeding during aestivation and thus growth ceases. The salamanderfish is one of the best-studied examples. This small fish occurs in ephemeral streams in Western Australia, and avoids dehydration by burrowing into damp, sandy substrata, forming a thick layer of mucus around the body and reducing the rate of metabolism (metabolic depression) (Pusey 1986, 1989; Thompson and Withers 1999). While in this state of torpor, salamanderfish utilise reserves of lipid (Pusey 1990) and store urea (Pusey 1986). Respiration occurs through the skin in a manner similar to that of lungless salamanders (Berra 1997). When the rains finally come and re-wet the habitat, the salamanderfish emerge within minutes and resume feeding and growth (Berra and Allen 1989; Pusey 1990). Several other species of Australian freshwater fishes appear capable of aestivation, including the Australian lungfish (Guppy and Withers 1999), blackstriped dwarf galaxias (*Galaxiella nigrostriata*: Thompson and Withers 1999), eastern dwarf galaxias (*Galaxiella pusilla*: Koster 2003) and Tasmanian mudfish (*Neochanna cleaveri*: Koehn and Raadik 1991). Many Northern Hemisphere species also reduce feeding and activity during periods of extreme cold (overwintering: Metcalfe and Thorpe 1992) but this phenomenon is little studied among Australian freshwater fishes.

Allometry

'Allometry' refers to proportional changes in the parts of organisms due to growth. Body parts or bodily dimensions, whose sizes relative to the whole body change with growth, exhibit allometric growth, and those whose relative size does not change exhibit isometric growth. Analysis of relationships between length and weight is the most widely used method for exploring the allometry in fishes (but see Elliot 1976; Weatherley and Gill 1983).

As fishes increase in size, more growth goes towards increasing their bulk rather than length. Increased weight-to-length ratios will occur in any organism (or any object, for that matter) as it increases in size, provided that its relative dimensions (shape) and density remain the same (Huxley 1932; Weatherley and Gill 1987). The reason is that growth manifests as an increase in weight which, in turn, is the product of density and volume. Because objects grow in three dimensions (length, width, height) there is a cubic relationship between an object's volume and its linear dimensions. Therefore, assuming constant density, the weight (W) of an object is proportional to the cube of its length (L):

$$W \propto L^3 \qquad (8.2)$$

The relationship between weight and length in a sample of fish thus can be expressed as:

$$W = aL^b \qquad (8.3)$$

where a is the intercept and b is the slope, or 'allometric constant' (Weatherley and Gill 1987). The parameters a and b can be derived empirically by plotting length against weight and fitting a power curve to the data. If b is close to 3, the sample population closely conforms to the cubic relationship between weight and length, suggesting isometric growth (Isley and Grabowski 2007). Departures from $b = 3$ represent allometric growth – high values of b occur in fishes that become progressively fatter as they grow (positive allometry) and low values of b occur in fishes that become skinnier as they grow (negative allometry) (Zafar *et al.* 2003).

With the possible exception of live-bearing fishes (e.g. Poeciliidae) that give birth to miniature versions of the adult form, most fishes exhibit periods of allometric growth during their lives. The free embryos and larvae of most species have elongate bodies and gradually become bulkier as they develop into juveniles (Wootton 1998). Although changes to body dimensions tend to become much less obvious after larval metamorphosis, some fishes continue to grow allometrically throughout their lives (Wootton 1998). Isometric growth, or slight to strong positive allometry, occurs in juveniles and adults of most Australian freshwater fishes (Pidgeon 1989; Anderson *et al.* 1992a; Butler 2009b; Fig. 8.9).

Somatic versus gametic growth

In early ontogeny, energy for growth is expended almost entirely on somatic anabolism. As fishes

Figure 8.9: Photographs showing development of body shape in golden perch (*Macquaria ambigua*). Early larval stages are relatively elongate; the body becomes more massive relative to length during development. In many populations, this species continues to exhibit positive allometry as adults, becoming progressively more bulky (photos: Z Tonkin, W Koster, D Crook).

approach sexual maturity, investment in gametic anabolism increases rapidly and gonad growth is usually isometric relative to body size (Davis 1977c; Wootton 1998). As a consequence, larger individuals typically have higher fecundities, and often larger eggs, than smaller ones. From a life history perspective, fishes face a trade-off between maturing early and investing in gametic production at a small size and delaying maturation until a larger size and higher initial fecundity are attained (Roff 1983; Stearns and Koella 1986; Heino and Kaitala 1999; Ch. 7). For many species, the age and size at which maturity is reached are flexible and may be influenced by environmental factors (e.g. density-dependent competition) and the risk of mortality (Stearns 1984). In some species, very slow growth rates may result in populations of stunted individuals that mature at a much smaller than normal size. The stock of Murray cod in Lake Charlegrark, Victoria, is an example (Gooley 1992).

At the population level, the trade-off between investments in somatic and gametic growth strongly influences the way that the reproductive investment is apportioned between individuals. In long-lived species, continued isometric growth of the gonads results in very large fishes with a high reproductive output (increased fecundity, large egg size) compared to smaller fishes (Parker 1992; Birkeland and Dayton 2005). The fecundity of female Murray cod, for instance, may range from about 7000 eggs soon after maturation (2–3 kg) to more than 100 000 eggs in large (> 30 kg) fish (Koehn and O'Connor 1990a; Rowland 1998b; Stuart and Koehn 2007). In addition to their reproductive potential, older fish are the most experienced breeders in a population and tend to have more mating success than younger, less experienced fish (Birkeland and Dayton 2005). In male Clarence River cod, failure to secure spawning sites and attract females, and abandonment of nests prior to completion of hatching, may reflect the relative inexperience of younger individuals (Butler and Rowland 2009). There is, however, some evidence of reduced reproductive performance in older, large (> 40 kg) Murray cod (Rowland 2005). In contrast to long-lived fishes, the reproductive output of early maturing, short-lived species varies within a much narrower range. Australian smelt, for example, live for one to two years and fecundity is 100–1000 eggs per individual (Milton and Arthington 1985; Pusey *et al.* 2004).

Sexual dimorphism

Sexual dimorphism in body size occurs in many fishes and results from selective forces that influence fitness differentially between the sexes (Darwin 1871; Parker 1992). In most taxa that exhibit sexual dimorphism, females reach larger sizes than males. In fishes generally, it is rare for males to be twice as big as females, whereas females may be orders of magnitude larger than males (Parker 1992). Within the limits imposed by a species' life history strategy, larger size may be adaptive in females because fecundity increases with body size. On the other hand, male body size appears to involve trade-offs between several factors that may affect fitness, including the influence of size on the outcomes of male-to-male behavioural interactions, the degree of mortality associated with finding and attracting females and the amount of competition among male gametes to fertilise the available eggs (sperm competition) (Parker 1992).

There are numerous examples of sexual size dimorphism among Australian freshwater fishes. Female congolli may grow to more than 350 mm TL whereas the maximal size for males is 150 mm TL (Hortle 1979). Other examples of large female sizes relative to males include golden perch (Mallen-Cooper and Stuart 2003), longfin eel (Walsh *et al.* 2004), southern shortfin eel (Beumer 1979a), freshwater herring (Pidgeon 1989), estuary perch (Walsh *et al.* 2010) and Australian bass (Harris 1985). In contrast, the males of several species of rainbowfishes (Melanotaeniidae) and carp gudgeons (Eleotridae) grow larger than females (Pusey *et al.* 2001, 2004), although the differences are less than for most of the above species.

Hormonal regulation

Hormones are central to the physiological regulation of growth. Numerous hormones are involved, in a maze of pathways and feedback loops that is only partly understood (Mommsen and Moon 2001). Amid this complexity, growth hormone, insulin-like growth factor I (IGF-I) and insulin itself have pivotal roles (Duan 1997; Mommsen 1998; Fig. 8.10).

Growth hormone is produced in somatotrophic cells of the anterior pituitary gland and its actions are complex and multi-faceted (Mommsen and Moon 2001). Production is influenced by food intake, exercise, stress, ovulation, temperature, day length and salinity (Mommsen 1998). In terms of growth, probably the best-known effect is the stimulation of muscle growth via myogenesis and protein synthesis. This process is strongly mediated by IGF-I, but growth hormone appears also to act directly upon receptors in muscle and other tissues (Mommsen 1998; Fig. 8.10). Another critical action of growth hormone is to stimulate the breakdown of adipose (fatty) tissue, freeing amino acids (for growth) that otherwise would have been utilised for oxidative processes. Growth hormone may influence growth in fishes by increasing appetite or aggressive behaviour, which in turn results in higher food intake (Mommsen and Moon 2001).

IGF-I is produced in the liver and muscle tissue, mainly in response to growth hormone (Duan 1997).

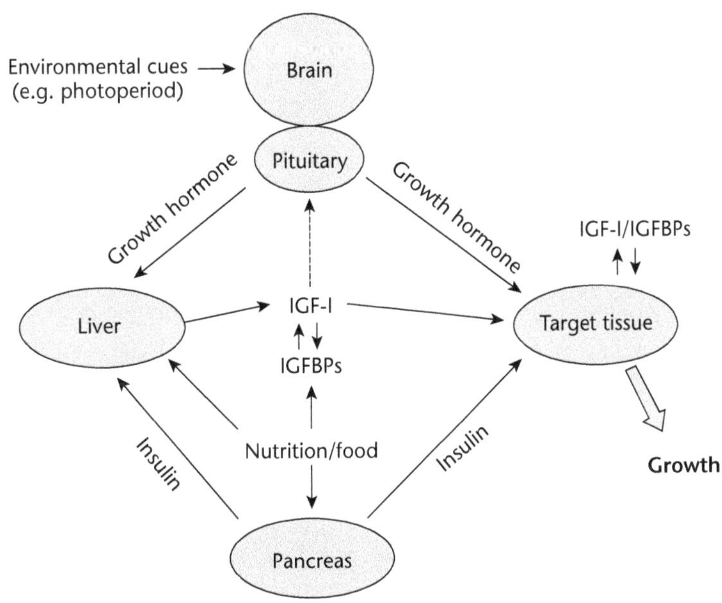

Figure 8.10: Schematic of interactions between growth hormone (GH), insulin-like growth factor-I (IGF-I), insulin-like growth factor binding proteins (IGFBPs) and insulin in regulation of fish growth. Solid lines represent stimulatory effects and the dashed line represents a negative feedback mechanism (modified from Duan 1997; Mommsen and Moon 2001).

Among its many functions, IGF-I stimulates the proliferation and differentiation of myoblasts (precursors of muscle cells), increases protein synthesis, suppresses protein degradation and accelerates glucose and amino acid uptake (Mommsen and Moon 2001). IGF-I tends to occur in association with IGF-binding proteins (IGFBPs) that regulate its functions (Duan 1997). IGFBPs originate from a number of sources, including the myoblasts, and appear to be influenced by the nutritional status of the fish (Mommsen and Moon 2001; Picha *et al.* 2008). Increases in IGF-I concentrations act as a negative feedback to growth hormone secretion in the pituitary gland (Fig. 8.10). In larval barramundi, IGF-I reactivity is tissue-specific and age-dependent, and IGF-I levels in the musculature and pectoral fins peak in newly hatched free embryos and decrease over time (Richardson *et al.* 1995). IGF-I increases in other tissues, including in the retinas, kidneys, gills and brain, as the larvae develop into juveniles.

As its name suggests, insulin is related to IGF-I and is produced by the β cells of the pancreas. In mammals, insulin mediates carbohydrate and lipid regulation, causing cells in the liver, muscle and adipose tissue to remove glucose from the blood and store it as glycogen, and to inhibit the use of lipids as an energy source. Insulin has similar actions in fishes but its role in regulating glucose is less prominent (Mommsen 1998). In terms of growth, its primary action is to accelerate protein synthesis and amino acid uptake in muscle tissue and thereby to stimulate growth in a manner similar to growth hormone and IGF-I. Insulin also has a lipogenic action, causing lipid to be deposited in liver, red muscle and subcutaneous and abdominal fat deposits (Mommsen 1998).

Hyperplasia and hypertrophy

A feature of growth that is unique to fishes is that 'hyperplasia' (increase in cell number) and 'hypertrophy' (increase in cell size) both contribute to growth after early development (Mommsen 2001). In most other vertebrates, including humans, hyperplasia ceases after early ontogeny, and increases in overall body size result from increases in the sizes of existing cells. The increase in muscle size of humans that occurs following weight training, for example, is due to an increase in the size of the cells rather than to an increase in cell numbers. In fishes that grow to large sizes, however, new muscle cells may continue to be created from undifferentiated myoblasts throughout life (Rowlerson *et al.* 1995; Mommsen 2001). There have been few studies of muscle growth in Australian freshwater fishes, but it is interesting that the Australian lungfish – the closest living relative of the tetrapods (four-legged animals) among fishes – has features intermediate between other fishes and amphibians (Kacperczyk and Daczewska 2008).

Energy storage

Another characteristic aspect of fish growth is the use of muscular protein as energy storage tissue. Fishes grow by increasing the mass of muscle tissue to a higher degree than most other vertebrates; they are able to do so because water provides buoyancy to support a much greater mass of tissue, relative to skeletal structure, than is possible for land-dwelling vertebrates. Fishes are also very efficient at excreting the nitrogenous waste products of protein turnover via the gills and skin. Thus, even though protein has a relatively low energy-to-mass ratio, fishes are able to use it effectively as an energy source for body movement (Mommsen 2001). Aside from protein, fishes store energy as lipid in the liver and red muscle and as subcutaneous and abdominal deposits (Sheridan 1988). Carbohydrate in the form of glycogen is stored in the brain, liver and muscle tissue as a short-term energy reserve (Mommsen 1998).

The ability to store large amounts of energy enables fishes to survive prolonged starvation, with different energy reserves used sequentially as the period of food deprivation continues. For example, sockeye salmon (*Oncorhynchus nerka*) that undertake non-feeding, upstream spawning migrations initially utilise lipid reserves then switch to protein metabolism as other energy sources are depleted (Mommsen *et al.* 1980). Starved golden perch in captivity mobilise lipid, protein and glycogen in the liver within 30 days after food deprivation, and mobilisation of abdominal lipid deposits and glycogen stored in the muscle occurs after 30–60 days (Collins and Anderson 1995). There may be no reduction in the protein content of the muscle of starved fish even after 210 days, suggesting that a longer period of starvation is required before golden perch rely on protein as the primary energy source (Collins and Anderson 1995).

FACTORS AFFECTING GROWTH

A variety of strongly interacting 'endogenous' and 'exogenous' factors determine rates of growth in fishes. Endogenous factors are those mediated by innate behavioural and physiological responses, and exogenous factors are imposed by the environment and interactions with other organisms (Wootton 1998). In this section, we outline the mechanisms by which these factors influence growth in fishes.

Sexual development and reproductive behaviour

As we have seen, growth in fishes after maturation occurs via somatic and gametic anabolism. Energy expended on gametic anabolism is lost to the individual as eggs or sperm during spawning, limiting the energy that can be devoted to somatic growth by sexually mature fishes (Hirshfield 1980; Day and Taylor 1997). The loss of energy to reproduction may be responsible for asymptotic growth. Thus, growth in many fishes is a two-stage process, including a linear pre-maturation growth phase and an asymptotic post-maturation growth phase (Lester *et al.* 2004; Fig. 8.11). Interestingly, when the environmental cues for spawning do not occur, some fishes are able to resorb their eggs and retain their energetic content. This process of 'ovarian involution' occurs in golden perch, Australian bass and Australian grayling when river flows are insufficient to initiate spawning (Mackay 1973b; Harris 1986; O'Connor and Mahoney 2004).

Energetic investment in reproduction is not restricted to gonad development. Many fishes, including Murray cod and Clarence River cod, have energetically expensive behaviours associated with reproduction. These include migration, elaborate courtship rituals and displays, aggressive interactions between competing males, guarding nests and breeding territories, nest-building and parental care, including fanning and aggressive protection of eggs, free embryos and larvae (Rowland 1983, 1998a; Butler and Rowland 2009; Chs 5, 7). Male southern purple-spotted gudgeon (*Mogurnda adspersa*) develop intense colouration and changes to the shape of the head during the spawning season and they cease feeding while they guard the fertilised eggs, free embryos and larvae (Tappin 1997). Similarly, male freshwater catfish expend energy constructing gravel nests and guarding eggs, free embryos and larvae for extended periods (Merrick and Midgley 1981). While such behaviours are energetically costly, they have fitness benefits that outweigh the limitations they impose upon somatic growth (Endler 1986).

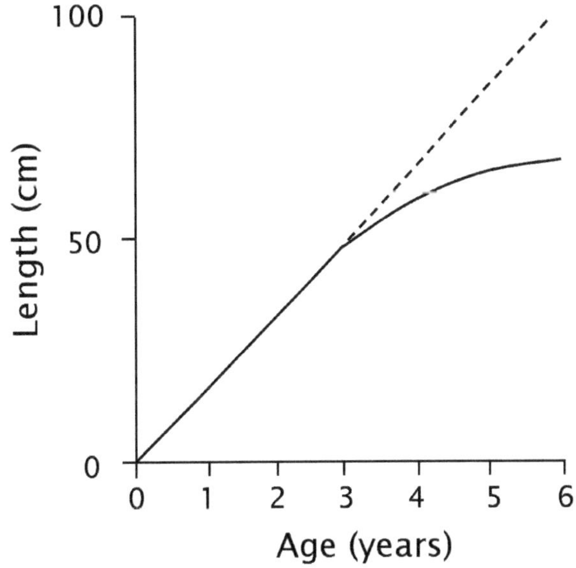

Figure 8.11: Schematic of the influence of reproductive investment on somatic growth for a species that matures at three years of age and 50 cm in length. The unbroken line shows the actual growth trajectory, with linear pre-maturation growth followed by asymptotic post-maturation growth. The broken line represents growth if all surplus energy had been allocated to somatic growth (modified from Lester *et al.* 2004).

Quantity and quality of food resources

Food is the source of energy for metabolism and the substrate for tissue synthesis, and a sufficient quantity and quality of food is needed to sustain growth (Condrey 1982; Bowen *et al.* 1995; Ch. 6). Fishes may use carbohydrate, lipids and protein as energy sources; protein in particular provides amino acids for synthesis of the proteins necessary for tissue growth and repair (Horn 1998). Essential amino acids cannot be synthesised by animals and must come from their dietary intake, and the same 10 amino acids are required for protein synthesis in fishes as in most other animals. The level of intake at which a fish neither gains nor loses weight is the 'maintenance ration'. Below this level, a fish will lose energy and exhibit negative growth. The food intake which maximises conversion of food into body tissue is the 'optimum ration'. Above this level, fishes feed with reduced conversion efficiency before reaching the maximum ration, then they cease to feed (Wootton 1998).

The nutritional quality of food is determined by its energy, protein and mineral content, and it varies between different food types (De Silva *et al.* 2002). For example, invertebrates generally have a much higher energy and protein content than algae and macrophytes which, in turn, contain more energy and protein than detritus (Bowen *et al.* 1995). Most fishes require diets with relatively high protein content for optimal growth: carnivorous and omnivorous fishes typically require dietary protein of 40–55% and 35–45% dry weight, respectively, compared with 15–25% in birds and mammals (Weatherley and Gill 1983; Horn 1998). The abilities of fishes to store protein as muscle tissue and to effectively excrete nitrogenous waste, and their ectothermy, account for their efficient utilisation of high-protein diets.

There have been many studies of the diets of Australian freshwater fishes (Davis 1977b; Cadwallader *et al.* 1980; Meredith *et al.* 2003; Balcombe *et al.* 2005; Balcombe and Humphries 2006; Ebner 2006; Baumgartner 2007; Sternberg *et al.* 2008; Davis *et al.* 2010a; Smith *et al.* 2011; Ch. 6), but little is known of the influence of food quantity and quality on growth rates in the wild. Most studies have concerned the search for effective and inexpensive artificial feeds and efficient feeding strategies in aquaculture (e.g. Murray cod: Gunasekera *et al.* 2000; Abery *et al.* 2002; De Silva *et al.* 2002; Francis *et al.* 2006; silver perch: Yang *et al.* 2006; barramundi: Tucker *et al.* 1988; Nankervis *et al.* 2000; Raso and Anderson 2003; Nankervis and Southgate 2006; freshwater catfish: Huynh 2010). Aquaculture studies are of limited use, however, for understanding the influence of diet on growth in the wild. Fishes in aquaculture are usually fed pelletised diets, and feeding can be manipulated to ensure that all individuals have access to sufficient feed. In the wild, competition for food can be intense and the profitability of particular prey items depends upon their nutritional value relative to the energy needed to detect, pursue, capture and digest them. Fishes in the wild have evolved divergent morphologies (e.g. increased intestinal length in herbivores) and behavioural strategies (e.g. predation versus grazing) to exploit different food resources (Bowen *et al.* 1995; Wootton 1998).

Biotic interactions

Interactions with other organisms may influence the growth of fishes via a number of mechanisms. An obvious example is predation – there is no more drastic effect on a fish's growth than being eaten! Interactions with predators also have more subtle effects. When faced with the threat of predation, fishes may be compelled to utilise habitats that offer suboptimal food resources but afford protection from predators, resulting in decreased foraging rates and slower growth (Werner *et al.* 1983). There is scant experimental evidence of the effects of predation on growth in Australian freshwater fishes, but observations of predator-avoidance behaviour suggest that predation is likely to influence growth in some species (Ault and White 1994; Herbert *et al.* 2003; Milano *et al.* 2010). Silver perch fingerlings may cease feeding when predatory cormorants (*Phalacrocorax* sp.) visit earthen ponds (Rowland *et al.* 1994), and sleepy cod (*Oxyeleotris lineolata*) reared in outdoor ponds increase in length more slowly in the presence of artificial shelter (plastic pipe) than fishes in ponds without shelter, presumably because the former spend less time feeding and remain in shelter to avoid predation (Herbert *et al.* 2003). Other agonistic behavioural interactions can affect growth. The alien eastern gambusia (*Gambusia holbrooki*), for example, negatively affects the growth and survival of some native species via interspecific aggression and fin-nipping (Howe *et al.* 1997; Morgan *et al.* 2004b; Keller and Brown 2008).

Competition for limited resources, particularly food and habitat, is another biotic factor that affects the growth of fishes (Werner and Hall 1977; Mittelbach 1988). Competition is a density-dependent mechanism whose effects become more intense as the number of individuals increases relative to the resources available. At high densities, growth rates may be depressed, resulting in populations of stunted individuals (Amundsen *et al.* 2007). Density-dependent stunting occurs in wild populations of the alien redfin perch (*Perca fluviatilis*) and tilapia (*Oreochromis mossambicus*) in Australia (Blühdorn and Arthington 1990; Morgan *et al.* 2002).

Inter- and intraspecific competition may also result in an unequal distribution of resources between individuals, with dominant fish gaining more access to resources and growing faster than subordinate ones (Wootton 1998). As the number of competitors increases relative to the available resources, the growth of the best competitors may be little affected, even though average growth rates decline across the population (Rubenstein 1981). This unequal distribution of resources is likely to account, at least in part, for variable growth rates in many populations. Evidence of the extent of density-dependent competition in Australian freshwater fishes is largely indirect and comes mainly from studies examining dietary and/or habitat overlap between species or ontogenetic stages (Pusey and Bradshaw 1996; Pusey *et al.* 2000a; Stoffels and Humphries 2003; King 2005; Tonkin *et al.* 2006). Density-dependent growth occurs in silver perch in aquaculture, with dominant individuals causing chronic stress and decreased feeding in subordinate fish (Rowland *et al.* 2006, 2009).

Fishes harbour parasites that may affect growth (Dove and Ernst 1998; Fletcher and Whittington 1998; Piasecki *et al.* 2004). Parasitic organisms utilise host-derived energy for their survival and growth and may infect vital tissues and organs including the gills, liver, gonads, muscle and intestinal tract (Langdon *et al.* 1985; Piasecki *et al.* 2004). As a consequence, fishes with high parasite loads often show reduced growth and poor body condition. For example, 50% of common galaxias in a lake in south-western Victoria are reported to have been infected with cestodes (tapeworm, *Ligula*) – the parasite load was up to half of the total body weight and the infected fish had reduced body condition (Pollard 1974). The ciliate protozoan *Chilodonella hexasticha* commonly infects large proportions of bony herring populations in winter, damaging the gill epithelia and eventually causing death (Langdon *et al.* 1985).

The cyclopoid copepod *Lernaea* (a crustacean, wrongly called 'anchor worm') is also common. This external parasite was introduced into Australia with common carp and it now infects a large number of native fish species (Piasecki *et al.* 2004; Ch. 11). *Lernaea* infestations cause inflammation around the site of attachment, and heavy loads cause loss of body condition and mortality (Rowland and Ingram 1991; Piasecki *et al.* 2004; Ebner 2006). Although the effects of parasite infection are usually negative, there are cases (none so far reported in Australia) where cestode infection enhances growth, causing the host to spend more time foraging and less time avoiding predators (Arnott *et al.* 2000; Loot *et al.* 2002). In addition to parasites, fishes are prone to fungal and bacterial diseases that may influence growth (Rowland and Ingram 1991).

Light and photoperiod

Light is essential for growth and development in most freshwater fishes: it allows them to use visual information to locate and capture food, it is essential for development of pigmentation and it influences the release of hormones that regulate growth (Boeuf and Le Bail 1999; Beckman 2010). Fishes detect light using photoreceptive cells in the retina of the eye and the pineal gland (a light-sensitive endocrine gland in the brain). Light information is transmitted to the brain via the hormone melatonin. Melatonin concentrations in the bloodstream increase at night and decrease during the day; this rhythmic variation acts as an internal clock, or *zeitgerber*, enabling fishes to use photoperiod information to perceive both time of day and season (Worrall *et al.* 2011). Melatonin affects growth primarily by acting upon receptors in the anterior pituitary gland that modulate the secretion of growth hormone and other associated hormones (Falcón *et al.* 2010). The daily and annual increments in otoliths appear to closely reflect the physiological responses of fishes to light/dark cycles and photoperiod (Mugiya 1987).

In aquaculture, manipulation of photoperiod has been investigated as a means to increase growth. Fishes typically grow faster when the photoperiod is artificially increased, because this encourages feeding activity and alters the secretion of melatonin and

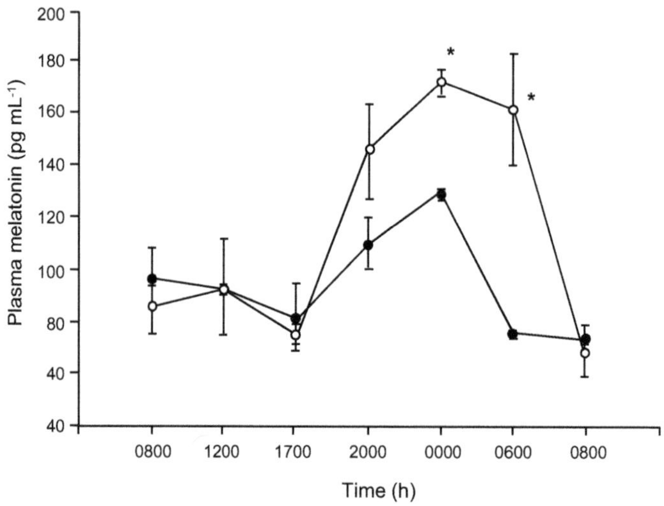

Figure 8.12: Diel periodicity in melatonin blood plasma concentrations (mean ± 1SE) in juvenile barramundi (*Lates calcarifer*) held for 20 days under 12 h light:12 h dark (open circles) and constant light (closed circles) following pre-trial acclimation to 12 h light:12 h dark. Note the increase in melatonin concentration during the pre-trial dark period (2000 h to 0600 h) and suppression of the diel melatonin cycle in fish held under constant light (* = significant difference between photoperiod treatments, $P < 0.05$). Modified from Worrall *et al.* (2011).

associated hormones (Boeuf and Le Bail 1999). For example, juvenile barramundi held for 20 days under a 24 h light regime grow significantly faster than fish on a 12 h light:12 h dark cycle, and the amplitude of melatonin cycling is suppressed under a 24 h light regime (Worrall *et al.* 2011; Fig. 8.12). In the wild, photoperiod oscillates as the seasons come and go, causing fluctuations in growth rates (Beckman 2010).

While most fishes require light for normal development, particularly during the larval period (Hart *et al.* 1996), many species are able to cope with very low light levels, as in turbid water. In fact, high-intensity light can be stressful, even lethal, to some fishes (Boeuf and Le Bail 1999). Some cave-dwelling species, such as the cave gudgeon (*Milyeringa veritas*) of Western Australia, live in the absence of light (Romero and Vanselow 2000).

Temperature

As ectotherms, the body temperatures of most freshwater fishes correspond closely to that of the ambient water, although they are able to partially regulate their temperature by moving between microhabitats (e.g. from warm, shallow margins to colder, deep water). The rates of chemical reactions associated with physiological processes are dependent on temperature and, as a consequence, the metabolic rate of a fish will increase as the ambient temperature rises (Katersky and Carter 2007b). At higher metabolic rates, fishes synthesise new tissue faster and forage more frequently and efficiently (Weatherley and Gill 1983; Wootton 1998). Provided that sufficient food is available to compensate for the energetic costs of increased metabolism, fishes grow faster with increasing temperature until an optimum level is reached (Beitinger and Fitzpatrick 1979; Katersky and Carter 2007b). Growth slows above the optimum temperature, as fishes are unable to deliver enough oxygen to respiring tissues to meet the increased demand. Eventually, they become heat-stressed and begin producing heat-shock proteins to protect vital organs (Iwama *et al.* 1999).

Different species have different temperature tolerances and optima according to their morphological and physiological adaptations (Beitinger and Fitzpatrick 1979). Spangled perch (*Leiopotherapon unicolor*), for example, occur in the hot, arid and semi-arid regions of Australia, and tolerate temperatures of 5–44°C (Gehrke and Fielder 1988). In contrast, some Antarctic marine fishes are unable to tolerate temperatures above 5°C (Macdonald *et al.* 1987). While many fishes are eurythermal (tolerating a wide range of

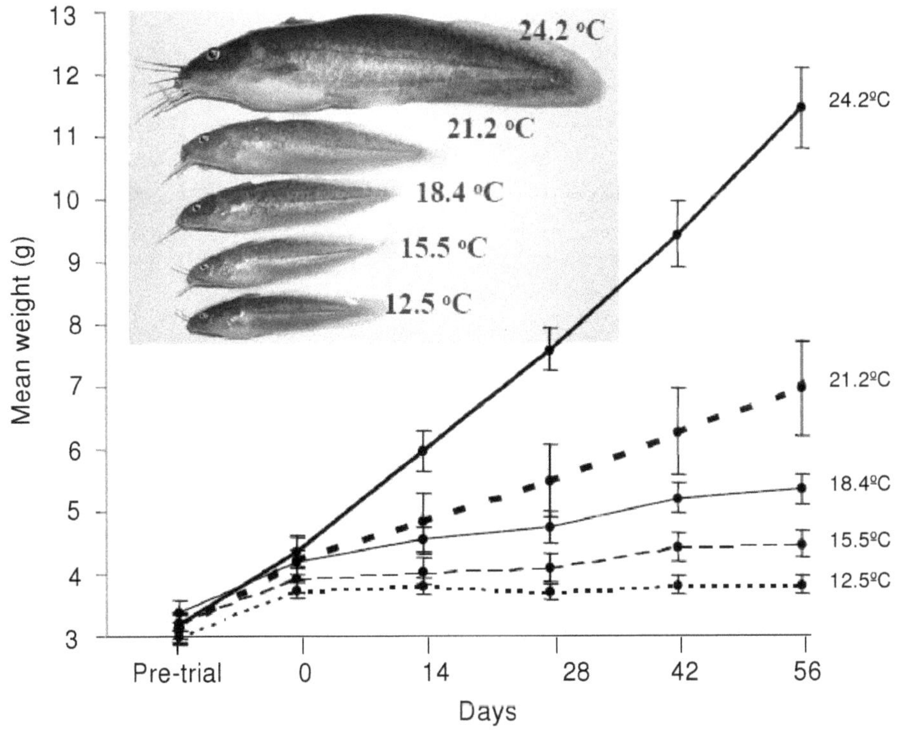

Figure 8.13: Effects of temperature on growth rate of freshwater catfish (*Tandanus tandanus*). Graph shows weight (mean ±1SE) of fish held at different temperatures on identical rations over 56 days. The inset shows examples of fish from each temperature treatment after 56 days (modified from Ryan *et al.* 2003).

temperatures), they usually exhibit distinct preferences within their tolerance range (Reynolds and Casterlin 1979). For example, golden perch tolerate temperatures of 4–37°C (Koehn and O'Connor 1990a) but show a strong preference for 24–31°C (Ryan *et al.* 2003). The tolerance range for silver perch is 2–38°C but the optimal range for growth in aquaculture is 23–28°C (Rowland 2009).

The effects of water temperature on growth rates have been studied under experimental or aquaculture conditions in Murray cod (Ryan *et al.* 2003; Vogel 2003), freshwater catfish (Ryan *et al.* 2003), barramundi (Katersky and Carter 2005, 2007a, b), Australian smelt (Tonkin *et al.* 2008c) and silver perch (Rowland 2009). Juvenile freshwater catfish in aquaria with an average starting weight of about 3 g reached an average 14 g after 90 days at 24.2°C, whereas the average weight of fish reared at 12.5°C was only 4 g (Ryan *et al.* 2003; Fig. 8.13). Similarly, juvenile barramundi held in aquaria for 20 days at four temperatures (27–39°C) increased their feed intake and grew faster with temperatures up to 36°C. The optimal temperature was exceeded at 39°C, indicated by reductions in feed intake, growth and food conversion efficiency (Katersky and Carter 2005).

Water chemistry

Dissolved oxygen is a key determinant of growth in fishes, due to its central role in tissue respiration and synthesis (Brett and Blackburn 1981). Under conditions of very low oxygen, growth slows and fishes often exhibit stress-related behaviours, including increased gill-ventilation rates and erratic movements (Brett and Blackburn 1981; McNeil and Closs 2007). Australian freshwater species have evolved a range of behavioural and physiological mechanisms to cope with hypoxia. For example, Australian smelt and flathead galaxias (*Galaxias rostratus*) ventilate their gills within the thin oxygenated layer near the water surface (McNeil and Closs 2007), and northern saratoga (*Scleropages jardinii*), oxeye herring (*Megalops cyprinoides*) and Australian lungfish gulp air and transport oxygen into the bloodstream using modified organs comparable to lungs (Grigg 1965; Wells *et al.* 1997).

In euryhaline and diadromous fishes (Ch. 5), rapid changes in salinity can affect growth because water and ions moving across cellular membranes must be regulated to maintain the ionic composition of body tissues (Langdon 1987). Osmotic and ionic regulation are energetically expensive and limit the energy that can be allocated to growth (Wootton 1998). Wholly freshwater species are exposed to relatively stable salinities, although there may be variation due to groundwater seepage and other factors; this may impose osmotic stresses which limit growth and development (Langdon 1987; Ingram *et al.* 2002). Fishes are also sensitive to variations in pH, alkalinity, hardness and turbidity (Rowland and Tully 2004). For example, very acidic (low pH) water disrupts the structure of gill membranes and oxygen cannot be effectively transported into the bloodstream (Fromm 1980). High turbidity reduces the ability of fishes to locate and capture prey (Sweka and Hartman 2003) and, in extreme cases, may cause thickening of the gill epithelium and reduced respiratory function (Henley *et al.* 2000).

As fishes consume high-protein diets, they excrete nitrogenous wastes including ammonia, nitrite and urea (Walsh 1998). Ammonia and nitrite can accumulate at high fish densities and negatively affect growth. For example, prolonged exposure of silver perch to un-ionised ammonia concentrations above 0.1 mg L^{-1} reduce growth and production in earthen ponds (Rowland 2009), and concentrations up to 0.36 mg L^{-1} of ammonia and above 1.43 mg L^{-1} of nitrite in aquaria cause discolouration and destruction of the gill epithelia, reduced growth and eventual mortality (Frances *et al.* 1998, 2000). Although most information on the effects of nitrogenous wastes comes from aquaculture research, high concentrations are likely to develop in the wild, particularly during drought, as habitats contract and fish densities increase (Crook *et al.* 2010b; Morrongiello *et al.* 2011a, c).

Pollutants from human activities may interfere with biochemical pathways that influence development and growth of fishes. Endosulfan, for instance, is an agricultural insecticide with highly toxic residues (Sunderam *et al.* 1992; Holdway *et al.* 2008) and endocrine disruptors from sewage effluent may affect sexual development (Batty and Lim 1999; Pollino *et al.* 2009). Short-term exposure to endosulfan, for example, reduces fertility and hatch rates in Murray River rainbowfish (*Melanotaenia fluviatilis*: Holdway *et al.* 2008) and oestrogenic compounds in human birth pills reduce the gonopodia (modified anal fins used to transfer sperm) of male eastern gambusia (Batty and Lim 1999).

Habitat structure

In inland waters, habitat structure may include structural woody habitat (snags, large woody debris), aquatic plants, geomorphic features (e.g. pools, riffles), boulders, cobbles and other benthic substrata (Ch. 4). In streams it can influence growth directly by creating areas of low velocity, reducing the energetic demands of swimming (Crook and Robertson 1999). Woody debris and plants act indirectly as a food resource for fishes, providing a hard substrate for biofilms and habitat for invertebrates (Scholz and Boon 1993; Lester and Boulton 2008). Habitat structural complexity also influences growth indirectly by modifying predator–prey and competitive interactions (Crowder and Cooper 1982; Schlosser 1987b; Crook and Robertson 1999; Quist and Guy 2001).

Climate

Although the factors that directly affect growth operate within the immediate vicinity of a fish, large-scale climatic variations also have effects. Correlations between the Southern Oscillation Index (SOI) and growth (measured using otolith increment widths; see 'Sclerochronology') are reported for golden perch in the Murray-Darling Basin, with faster growth during periods of positive SOI values (Pritchard 2004). Increased temperatures and longer growing seasons associated with human-induced climate change are likely to increase growth of golden perch in impoundments of the southern Murray-Darling Basin (Morrongiello *et al.* 2011c). Indeed, global climate change has pervasive implications for growth in freshwater fishes throughout the world.

QUANTIFYING GROWTH

Methods for measuring and modelling the growth of fishes range from simple length and weight measurements to examination of the internal structure of otoliths. Proxies of growth, such as otolith growth increments and measures of body condition, also indicate past growth and the potential for future growth. This section outlines some approaches used to quantify growth.

Growth from length–weight data

Growth in fishes often is measured in terms of length or weight, with reference to time incorporated to allow for calculation of a growth rate. Growth can be expressed as 'absolute growth' (length or weight at time 2 minus length or weight at time 1), 'relative growth' (difference in length or weight between two times divided by length or weight at time 1) and 'instantaneous growth' (difference in the natural logarithm of length or weight between two times). Instantaneous growth also is commonly referred to as 'specific', 'intrinsic', 'exponential' or 'logarithmic' growth.

Growth estimates from length and/or weight data can also be derived from mark–recapture techniques, through scientific or cooperative angler tagging programs (Wilde and Sawynok 2005; Walsh *et al.* 2006). Thus, the difference in size at tagging and recapture, divided by the time elapsed, indicates the growth rate of a recaptured fish. Graphical methods traditionally have been used to convert individual growth-increment data from tagging studies into descriptors of mean growth across a population (e.g. Walford plots: Walford 1946), and these can then be used to parameterise growth models like the von Bertalanffy growth equation (see 'Age–size relationships'; Francis 1988b; King 1995).

New statistical approaches provide more robust estimates of growth parameters from mark–recapture information (Haddon 2001). The computer program GROTAG, for example, uses a maximum-likelihood approach to estimate mean annual growth rates, growth variability and measurement error from mark–recapture data (Francis 1988b; Wilde and Sawynok 2005; Walsh *et al.* 2006). This can provide better descriptors of growth than estimates generated using von Bertalanffy growth curves (Francis 1988b, 1995).

Age–size relationships

As mentioned, age information from calcified structures can be plotted against information on the size of fish to estimate growth rates. One of the most-often used models for representing these relationships is the von Bertalanffy growth equation (von Bertalanffy 1938). An Austrian, Ludwig von Bertalanffy (1901–72) is regarded as the father of General Systems Theory but he is also famous for the model he developed originally for human growth, since widely applied to fishes.

There are several alternative parameterisations, but the model is functionally the same. The most widely used version is:

$$L_t = L_\infty(1 - e^{-K[t - t_0]}) \tag{8.4}$$

where L_t is the length at age t, L_∞ (*L-infinity*) is the average size at maximum age, K is a growth rate coefficient that describes how quickly the asymptotic size is approached and t_0 is the hypothetical age when the average size is zero (t_0 is a modelling artefact rather than a realistic biological parameter) (Beverton 1954; Beverton and Holt 1957). An illustration of the use of the von Bertalanffy equation is in Figure 8.7.

Although the von Bertalanffy model is widely used in fisheries management and research, it has important limitations (Knight 1968; Roff 1980; Sainsbury 1980; Haddon 2001) and we strongly advise against uncritical applications. Note that L_∞ is not the maximum size achievable in the modelled population; rather, it represents the average size of the oldest age classes. Thus, fish larger than L_∞ will often occur in a population; indeed, if no fish larger than L_∞ are present we probably should question the suitability of the data for estimating L_∞. Further, as L_∞ and K are often calculated by extrapolation from the available data, they are prone to yield unrealistic values when those data do not fully represent the population size structure. For example, unrealistically high estimates of L_∞ may arise when large size classes are under-represented (Knight 1968). Alternative parameterisations of the von Bertalanffy model overcome some of these problems (e.g. the Schnute parameterisation: Schnute 1981), and there are alternative statistical techniques (Gamito 1998; Haddon 2001). The Gompertz model, in particular, is commonly used to model growth in the early life stages of fishes (Tonkin *et al.* 2008c).

Aside from age–size at capture relationships, it is possible to obtain growth information from calcified structures by estimating the size of a fish at each age increment via 'back-calculation' (Quinn and Deriso 1999). Back-calculation uses measurements taken at one time to infer the size of a fish at an earlier time (Francis 1990), and relies on a predictable relationship between body length and the radius of the otolith, scale or other hard part as the fish grows. The earliest method for back-calculation is the Dahl-Lea method, using a proportional relationship between growth of the calcified structure and body length. Using this approach, the body length of a fish at the time of increment formation (L_i) is:

$$L_i = (S_i/S_c)\,L_c \qquad (8.5)$$

where S_i is the radius of the ith increment, S_c is the overall radius of the calcified structure and L_c is the length of the fish at the time of capture (Francis 1990). Various other mathematical approaches are used to back-calculate lengths, and their suitability depends on characteristics of the data and the hypotheses being tested (Francis 1990).

Validation is desirable in studies that estimate back-calculated lengths and is best conducted by making repeated observations of individual fish over time. This is difficult in practice, because it requires the recapture of individuals (Francis 1990). Group comparisons (e.g. between cohorts) are possible but it is difficult to know whether the comparisons are valid; for example, fish from older cohorts may not have the same first-year growth characteristics as one-year old fish collected at the same time, due to length-dependent mortality or temporal variation in growth rates (Francis 1990). Although back-calculated length-at-age data are often incorporated into von Bertalanffy models (Milton and Arthington 1985; Butler 2009a), caution is needed as estimates of size-at-age within fishes are serially correlated and not independent (Jones 2000). While the apparent simplicity of back-calculation is appealing, applications can be complex and require an appreciation of the strengths and weaknesses of the available approaches.

When growth increment information is not available, age–size relationships can be estimated using length-frequency data (see 'Length-frequency analysis'). The first methods were simple graphs of length-frequency data collected at a single time and they assumed that each mode represented an equal time interval. Parameters of the von Bertalanffy growth curve (K and L_∞) were estimated from Ford-Walford plots (a plot of length at age $t + 1$ versus length at age t) with t_0 set to zero (King 1995). Length-frequency samples at multiple times can be used to estimate growth. Depending on the frequency of sampling, several graphical methods are possible (Ford-Walford plot, von Bertalanffy plot, Gulland-Holt plot: King 1995). Increased computing power has made likelihood-based procedures possible. These methods simultaneously analyse the different samples of length-frequency distributions, yielding estimates of the number of cohorts in the population, growth parameters, the age of the first cohort and proportions at age (Fournier *et al.* 1990). The simplest model assumes that mean lengths-at-age lie on a von Bertalanffy growth curve and that the standard deviations of length-at-age are identical for all cohorts. More complex models allow for seasonally oscillating growth (e.g. fast growth in summer, slow growth in winter), sampling bias for the first cohort and age-dependent standard deviations in length-at-age. Although many studies of Australian freshwater fishes have plotted length-frequency distributions (Walsh *et al.* 2010), quantitative estimates of growth parameters generally are not attempted using length-frequency data.

There are important caveats on the use of age relationships to estimate growth in fishes. In particular, age–size estimates of growth derived using different approaches may not be comparable. For example, growth curves estimated from back-calculated lengths can be considered as true growth rates because they consist of multiple length–age estimates for individual fish (Campana 1990). In contrast, population growth rates estimated from the ages and lengths of fish at the time of capture represent a single age and length value for each individual; these are combined to estimate average growth for the sample population. Length-frequency estimates also produce averaged population growth rates. Population-averaged estimates should be treated as general descriptors of growth; they do not capture temporal variation in the growth rates of individuals. This may be important ecologically – it is highly unlikely that individual fishes grow in the manner described by the modelled growth curve for the population (Sainsbury 1980). A further caveat is that estimates of growth from age-based (e.g. calcified structures) and size-based (e.g. mark–recapture) growth models are not directly comparable (Francis 1988a, 1995). If comparisons are made, they should be in terms of growth rates rather than the parameters derived from fitted growth curves (Francis 1995).

Morphological condition indices

As fishes grow in both mass and length, indices that incorporate the relationship between body mass and body length provide an indicator of past growth as well as the potential for future growth. Fishes generally are considered to be in better condition (i.e. growing faster) if they have a higher weight relative to their length. Several morphological condition indices exist (Stevenson and Woods 2006; Peig and Green 2010) and there is no consensus over the best option (Peig and Green 2010).

Conventional methods that calculate ratios between mass and length include Fulton's index K, Quetelet's index and the body mass index (Peig and Green 2010). Fulton's Condition Factor (K) is widely used, and takes the form:

$$K = \frac{W}{L^3} \tag{8.6}$$

where W is weight/mass of fish (usually g × 10^5) and L is length (usually Total Length, mm). High values of K indicate better body condition. As length is raised to the third power, K assumes isometric growth (Suthers 1998). In many cases, however, K is correlated with length due to allometric growth. To account for this, relative condition (K_n) is calculated as:

$$K_n = \frac{W_i}{W_i^*} \tag{8.7}$$

where W_i is the observed weight of a fish and W_i^* is the weight predicted by the weight–length relationship for the sample population ($W_i^* = aL_i^b$; see 'Allometry').

Fulton's Condition Factor has been calculated for a number of Australian freshwater fishes (Milton and Arthington 1984; Harris 1985). Relative condition may vary between sexes or through time (Davis 1977c; Rimmer 1985b; Butler 2009b). For example, the relative condition of Clarence River cod may be less in summer (0.999) than in winter (1.026), following spawning (Butler 2009a).

Biochemical indices

Biochemical and physiological indices have also been developed (Bolger and Connolly 1989; Ferron and Leggett 1994; Stevenson and Woods 2006). These indicate the state of energy reserves and may be used as proxies of growth rate. Chemical assays of lipid and protein levels (proximate analyses) or tissue combustion (bomb calorimetry) traditionally are used to determine the energy contents of fishes.

Biochemical indices either quantify chemical constituents or measure physiological indicators. RNA/DNA ratios are widely used to indicate nutritional condition and, to a lesser extent, assess instantaneous growth in larval and juvenile fishes (Ferron and Leggett 1994; Bergeron 1997; Buckley *et al.* 2008). The premise for this approach is that while the amount of DNA per cell is relatively constant, the amount of RNA varies with physiological status and protein synthesis, reflecting recent growth and nutrition (Buckley *et al.* 1999). Thus, RNA/DNA appears to be a sensitive measure of short-term (days to weeks) growth rate, although few calibration experiments exist to confirm this (see Buckley *et al.* 2008 for examples). Many different analytical methods have been used, making comparisons between studies and species difficult (but see Caldarone *et al.* 2006). Most research has focused on bulk RNA analysis but specific nucleic acid-based biomarkers (e.g. heat-shock protein mRNA, metallothionein mRNA) may have applications in fish-growth studies (Buckley *et al.* 1999). Other potential biochemical measures include protein content, enzyme activity and lipid content (Suthers 1998; Peig and Green 2009). Endocrine biomarkers, such as IGF-I (see 'Hormonal regulation'), have also been used as measures of instantaneous growth (Picha *et al.* 2008).

Rapid-assessment methods for determining energy content also exist. Microwave transmission is used frequently in aquaculture to assess the fat content of fishes destined for markets, and it has been used on wild fishes (Crossin and Hinch 2005). The water content of fishes can be measured using a handheld electrical conductivity meter and, with knowledge of the relationship between water and lipid content, water-content values can be used to estimate lipid content. This technique requires calibration for individual species and has been applied to Murray cod (Whiterod 2010b).

Sclerochronology

Measurements of otolith-increment widths (sclerochronology) can be used in a manner similar to back-calculation to estimate somatic growth rates. Several studies have shown that somatic growth and otolith growth are correlated (Pritchard 2004; Tonkin *et al.* 2008c; Morrongiello *et al.* 2011c), although there is a time lag of days to weeks (Suthers 1998). Wide growth increments generally indicate periods of high growth rates and narrow increments indicate slow growth. In many sclerochronological analyses, the growth chronologies of multiple individuals in a population are combined to produce population growth chronologies (Black 2009). Analysis of correlations between environmental and/or biological variables and growth rates (i.e. increment widths) can be used to make retrospective inferences about the effects of environmental conditions on growth (Pritchard 2004; Morrongiello *et al.* 2011c).

CONCLUSION

Age and growth data have played, and continue to play, a central role in the management and conservation of Australia's freshwater fishes. The ecological understanding provided by analysis of these data can inform us about the rates at which we can sustainably harvest fishes and how we should regulate and manage fisheries, about strategies for rehabilitating threatened species, the consequences of anthropogenic change and numerous other issues.

Much of the research in this field to date concerns the validation of ageing methods and size-at-age relationships. This information is important, as there are no data for many species, but there is a need also for research into the ecological implications of age and growth at the levels of individuals and populations. For example, the effects on growth of density-dependent intraspecific competition are little known, although this underpins management activities such as the stocking of hatchery-reared fishes and development of regulations (e.g. bag limits).

A recurrent theme in this chapter is the plasticity of growth rates in fishes. Again, this has implications for the ecology, conservation and management of fishes. For example, maximum sizes and ages at maturation may vary widely across a species' geographic range, especially as many species reside in environments with highly variable hydrology and productivity (Puckridge *et al.* 1998; Bunn *et al.* 2006b; Balcombe and Arthington 2009). Given the extent to which humans have changed aquatic environments, it is essential that we increase our understanding of the drivers and ecological consequences of growth plasticity.

Integration of physiological and ecological data is another area that could provide insights into the growth of fishes. Given the boom–bust nature of many Australian freshwater habitats, we need to understand better how fishes cope with the bad times and take advantage of the good times. While there has been research on nutrition and growth for a few species (e.g. Murray cod, silver perch, barramundi), most relates to aquaculture rather than ecology. A notable exception is research on golden perch by Collins and Anderson (1995, 1997, 1999), where physiological data were used to develop hypotheses on how fishes cope with boom–bust environments.

New methods provide a glimpse of what the future may hold for age and growth research. For example, the detailed information available from otoliths has only recently become apparent, first with the use of growth increments to age fish then, more recently, with the advent of otolith chemical analysis to retrace environmental histories and validate the ages of long-lived species. Analysis of otolith-increment widths to hind-cast individual growth (sclerochronology) is a burgeoning area of research that, among its many potential applications, can inform us about the likely future impacts of climate change (Morrongiello *et al.* 2011c). There are also emerging techniques for analysing age and growth data that could help avoid some of the pitfalls associated with deterministic age/growth models.

We are now in a position to venture an answer to the question posed at the beginning of this chapter, about the age and importance of a 30 kg Murray cod. Thanks to careful validation and analysis of annual otolith increments in Murray cod (Anderson *et al.* 1992a; Rowland 1998a), we are able to say confidently that this fish would be more than 20 years old (and possibly 48+ years). We also know that older fish are the most experienced, most highly fecund breeders in a population. Although we would need to consider the abundance of other big cod in the population to assess the consequences of removing one fish, the evidence suggests that big fish play a vital role in contributing to future generations of their species. Armed with this knowledge, would you take that big cod home for dinner, or return it to the river?

9 Dynamics of populations

John H. Harris, Nick R. Bond, Gerry P. Closs, Peter C. Gehrke, Simon J. Nicol and Qifeng Ye

INTRODUCTION

Why are fishes more abundant in some areas than others? Why do abundances vary over time? What factors limit the sizes and productivity of populations? How do natural populations respond to harvesting, and where do we draw the line between sustainable and unsustainable fisheries? These are some of the questions that motivate researchers, and they underpin important issues in conservation and management. The answers might easily command a book but our focus, of course, is freshwater fishes and fisheries in Australia.

The size of a population is determined by interplay between the environment and four processes, namely recruitment, mortality, immigration and emigration. To understand how these interact, we need information about movements (Ch. 5), trophic dynamics (Ch. 6), life histories and recruitment (Ch. 7), age and growth (Ch. 8) and the influence of invasive species (Ch. 11). Here, we draw on these topics to illustrate the dynamic changes that occur in patterns of abundance, recruitment, mortality, migration, growth and production. We introduce some concepts, then consider population models and the impacts of fishing.

BASIC CONCEPTS

While the biology of some Australian freshwater fishes is reasonably well-known (e.g. Australian bass, *Percalates novemaculeata*; barramundi, *Lates calcarifer*; common galaxias, *Galaxias maculatus*; freshwater catfish, *Tandanus tandanus*; golden perch, *Macquaria ambigua*; Murray cod, *Maccullochella peelii*; river blackfish, *Gadopsis marmoratus*; silver perch, *Bidyanus bidyanus*), population dynamics have rarely been investigated other than in the context of applications in fish passage, aquaculture and fisheries. The term 'dynamics' here refers to the processes that affect the size of populations over the course of time, especially recruitment, growth, mortality, migration and their interactions.

Populations, stocks and metapopulations

The distributions of fishes within biogeographic boundaries are limited by their environmental tolerances and their abilities to compete with one another. Individual fishes rarely mate randomly with other individuals within the population's range; there is a tendency to form discrete stocks, or subpopulations, that to some degree are genetically isolated in space, time or both. As a consequence, subpopulations often show morphological and other differences. For example, there are genetically distinct stocks of Australian bass (Chenoweth and Hughes 1997), barramundi (Chenoweth *et al.* 1998), freshwater catfish, golden perch, silver perch and other species (Keenan *et al.* 1996) (Chs 2,3).

A metapopulation is a group of spatially separated populations of the same species (in effect, a population of populations). Each member population cycles independently of the others, and the chances of individuals mating with others from different populations are low, even over long periods. Although individual populations have finite life spans, a metapopulation as a whole is often stable for long periods of time because immigrants from one population are likely to recolonise habitats left open by the demise of another population. This highlights the importance of connectivity between apparently isolated populations, even at long intervals. No single population is likely to secure the long-term survival of a species, but many populations might do so in combination. There are many examples of species that occur as metapopulations (Woods *et al.* 2010), including barramundi (Pusey *et al.* 2004) and several Murray-Darling species (e.g. freshwater catfish, golden perch, silver perch: Keenan *et al.* 1996; Ch. 3).

Abundance

Three approaches are used to estimate the number of fish in a population: ratio methods, catch-effort methods and direct enumeration (Bagenal 1978; Gerking 1978; Southwood 2000). Ratio methods include mark–recapture and change-in-ratio methods. In a Petersen mark–recapture population estimate, for example, a sample of fish is marked using numbered or visible tags (plastic anchor or dart tags, coded-wire tags, PIT tags, visible-implant polymer tags) or chemical marks, and released back into the population. One or more random samples are then taken and the numbers of marked and unmarked fish are recorded. The proportion of marked fish in the recaptured sample is assumed to be the same as that in the entire population, so that a simple calculation indicates the size of the population. Various refinements are possible, but critical requirements generally are that there should be no recruitment or immigration, no change in mortality or behaviour from marking, random mixing of marked with unmarked fishes and no loss (or a known loss) of marks, that samples should be taken at random across the range of the population, and there may need to be allowance for size-selectivity of the sampling gear (Bagenal 1978). A modified Petersen method applied to Australian bass in a closed lagoon indicated an average of one fish per 5 m length of bank (Harris 1988).

The assumption in catch-effort methods is that, for a closed (contained) population, the catch per unit effort (CPUE, where effort is indicated by electrofisher 'on' time, gill-net hours, trap 'soak time', seine hauls etc.) is proportional to the size of the population. Thus, CPUE should decline as a series of samples is taken. In the commonly used Leslie method, added assumptions are that the probability of capture is equal for all fish and proportional to the sampling effort (Bagenal 1978). Other catch-effort methods have been applied, especially using electrofishing gear (Davies *et al.* 1988, 2010; Brown *et al.* 2003; Driver *et al.* 2005a). These methods are also used to monitor the relative population sizes of angling species using angler-recorded data, with or without the use of tags or other marks (Faragher and Gordon 1992; van der Walt *et al.* 2005a). A further variation, using repeated population-depletion sampling of enclosed segments of populations and recording the declining catches, was applied in the New South Wales Rivers Survey (Harris and Gehrke 1997); in that case, population estimates were obtained by applying removal, jackknife and mark–recapture estimators.

The constraints on interpretation of survey data and catch-effort results are sometimes difficult to convey. For example, when standardised electrofishing surveys sampling on 80 site–season combinations (20 sites, four seasonal samples over two years) in the Murray River catchment failed to detect freshwater catfish or Murray cod (Harris and Gehrke 1997), there was criticism from local anglers who were still able to catch these fishes by angling. They had failed to understand why sites were chosen at random (rather than merely selecting known fishing spots), so that results could be compared and statistically analysed over whole populations. The chosen sites obviously missed areas containing the remnants of once widespread, abundant populations. The two species were frequently sampled in other rivers during the study.

Direct counting is seldom possible in turbid Australian rivers and lakes, especially for cryptic species, but it can be efficient in shallow, clear systems in the tropics (Bishop *et al.* 2001) and in upland streams (Closs and Lake 1996; Harris *et al.* 2006). It is widely applied to migrating segments of populations, as when trout (Salmonidae) are caught in traps set to sample spawning runs (Tilzey 1977b; Davies and Sloane 1987). Fishway channels are often used to trap and count migrating fishes (Barrett and Mallen-Cooper 2006; Baumgartner *et al.* 2010b). Another variation involves sonar devices such as the sophisticated DIDSON (dual-frequency identification sonar) and split-beam hydroacoustics equipment (Berghuis 2006; Matveev 2007), enabling provisional identification, counts and approximate size classification for free-swimming fishes even in turbid water (Ch. 5). This technology offers exciting new research opportunities.

Recruitment

The number of new fish entering a population, or some segment of the population, in a given time is termed 'recruitment'. For the population as a whole, the term generally applies to new members entering via reproduction and immigration in a given year (Bagenal 1978; Walters and Martell 2004; Ch. 7). Recruitment may also be defined relative to some particular population segment, such as juveniles, the spawning stock (as in Ch. 7) or a fishable size-class. Assessments of recruitment are included in the Murray-Darling Basin Sustainable Rivers Audit (Davies *et al.* 2010) as an indicator of the resilience of fish populations (their capacity to remain viable in face of disturbances).

The process of recruitment is characterised by great variations among species and by order-of-magnitude fluctuations in annual rates within life stages and populations. There is a need for quantitative studies and meta-analysis of stage-specific mortality (Myers 1995), especially during the egg and larval stages, to better understand the sensitivities of species and to refine environmental management. These data are needed especially for better approaches to species conservation in regard to water resource management and hydrological conditions (Phillips 2001; Sherman *et al.* 2007; Ch. 7) and to the roles of invading species (Ch. 11). For instance, environmental water allocations to sustain recruitment have come under intense scrutiny as competition for water grows (Graham and Harris 2005; Murray-Darling Basin Commission 2005; King *et al.* 2009, 2010). Some financial imperatives have disappeared, because few Australian species are now commercially harvested in fresh water, but the conservation and ecosystem maintenance requirements remain as real as ever.

Growth and production

Populations grow through immigration and increases in biomass (increases in the numbers and/or sizes of individuals) (Ch. 8), offset by emigration and mortality. Concepts of population growth obviously are critical for fishery management, and fisheries operate by harvesting a portion of growth, the surplus production (Weatherley 1972; Gerking 1978; Walters and Martell 2004). Ecosystem production begins with plant photosynthesis (primary production) and continues through a number of trophic levels, each producing stocks of organisms that typically increase in individual size as they decrease in total biomass. Secondary production is derived, directly or indirectly, from consumption of plants, and includes fish production or the increase in fish biomass over a year (or other interval) in an area, including reproductive output and the biomass of fish that die (Bagenal 1978). The term 'productivity' here refers to the general extent of production.

There are two widely used methods of estimating production. In a landmark study of brown trout (*Salmo trutta*) production in the Horokiwi Stream, New Zealand, K. Radway Allen (1951) obtained a graphical estimate by plotting composite year classes in the population against their mean biomasses and integrating the resulting Allen curve, as it became known. W.E. (Bill) Ricker (1975) later developed a numerical method to determine the aggregated product of successive growth rates and biomasses in cohorts within a population.

Knowledge of production can provide insights into the roles of fishes in an ecosystem, especially when combined with data on other demographic parameters and diet. Table 9.1 includes examples of the

Table 9.1: Indicative demographic data for some representative Australian freshwater fishes. For many species there is wide variation in records of these values

Species	Longevity[1] (y)	Maximum size[2] (mm)	Age at maturity[3]	Mature size (mm)[4]	Fecundity (000s)[5]	Mortality[6]	P/B ratio[7]
Australian bass	22	600	2–4(m), 5–6(f)	18(m), 280(f)	50–150	0.28	0.19
Australian grayling	3	330	2		25–80		
Australian lungfish	65–70	1500	15–22				
Australian smelt	3	100	1	40	0.1–1		
Barramundi	>30	1800	3(m), 6–8(f)	300(m), 600(f)	5000–20 000		
Blue catfish		600		250–350	40–120		
Bony herring	5	480	1	183	40–400		2
Climbing galaxias		278	2		23		
Common galaxias	2	190	1	10	0.175–4		
Congolli	5	360	<4	100(m), 150(f)	400		
Cox's gudgeon		190					
Crimson-spotted rainbowfish		110			0.1–0.15		
Estuary perch	41	750	2(m), 3(f)	220(m)–280(f)	180–540		
Flathead gudgeon		115					
Fly-specked hardyhead		108			0.45–2.2		
Freshwater catfish	8	900	3–5		2–26		
Freshwater herring	11	320					
Freshwater mullet	14	810	4	300	1000–3000		
Golden perch	26	760	2–3(m), 4(f)	190(m), 400(f)	>500		0.73
Jungle perch	15	300(m), 450(f)	4–6(m), 2–5(f)	70(m), 210(f)	500/kg	0.3	
Longfin eel*		650(m), 1650(f)	>30	558(m), 931(f)			
Macquarie perch**		460	2(m)–3(f)	210(m)–300(f)	30–100		0.53
Mountain galaxias		135			0.4		
Murray cod	48	1800	3–4(m), 4–7(f)		20–200		
Southern purple-spotted gudgeon		120			0.1–1.3		

Species	Longevity[1] (y)	Maximum size[2] (mm)	Age at maturity[3]	Mature size (mm)[4]	Fecundity (000s)[5]	Mortality[6]	P/B ratio[7]
River blackfish		350–750	3–4	104	0.04–2.5		
Sea mullet		760	3	320	4000		
Silver perch	27	610	2(m)–3(f)	250(m)–290(f)	300–500		
Snub-nosed garfish		850		215			
Sooty grunter		450					
Southern pygmy perch	5	85	1	30(m)–33(f)	0.5–4		
Southern saratoga		1000	4		0.075–0.173		
Southern shortfin eel*		500(m), 1100(f)	6–12(m), 10–30(f)		500–10 000		
Spangled perch	3	250	1	58(m), 78(f)	24–113.2		1.16
Tasmanian whitebait		77	1		0.3		
Trout cod	25	850	3–5	350(m)–430(f)	1.2–11		
Two-spine blackfish	6	325	1	116(f)	0.15–0.35		
Western carp gudgeon	3	60	1	29(m), 27(f)	0.33–5.4		2.54
Western pygmy perch	5	80	1	43			

1. Longevity = maximum recorded life span (years).
2. Maximum recorded size (mm) may refer to total length, fork length or standard length.
3. Age at first sexual maturity for males (m) and females (f).
4. Smallest recorded size at sexual maturity for males (m) and females (f).
5. Total fecundity (thousands).
6. Instantaneous rate of natural mortality for adults.
7. Production/biomass ratio.

* Eels remained in Lake Eucumbene, New South Wales, in 2011, despite complete blockage of entry since 1954.

** Murray-Darling Basin species of Macquarie perch; eastern species much smaller.

Source: Collated from Allen *et al.* (2002); Froese and Paul (2010); Fulton (1990); Gehrke *et al.* (2011); Harris (1988); Hoyle and Jellyman (2002); Koehn and O'Connor (1990); Lake (1967, 1971); McDowall (1996); Pusey *et al.* (2004); Todd *et al.* (2004); Ch. 8.

production:biomass ratio (P/B) for a number of Australian species. Thus, the western carp gudgeon (*Hypseleotris klunzingeri*) is an abundant, small-bodied, short-lived species that feeds at low trophic levels, reflected in a high P/B (2.54). In contrast, the Australian bass is a larger, long-lived, high-order predator; the tabled P/B value (0.19) is for adult fishes in an upstream lagoon remote from estuarine habitats where younger, faster-growing individuals occur, and the estimate is extremely low by world standards (Harris 1988). Bony herring (*Nematalosa erebi*) have a moderate P/B (2.0). They are fast-growing, early maturing and relatively fecund, implying high productivity, and they are herbivores, feeding on attached algae and biofilms. Bony herring feature in the diets of predators (barramundi; freshwater longtom, *Strongylura krefftii*; golden perch; Murray cod; sooty grunter, *Hephaestus fuliginosus*), and are a cornerstone species in many lowland rivers and lakes in warmer areas of Australia.

Survival and mortality

How long do fishes live? How many members of a cohort will be alive in following years? Knowledge of mortality and its complement, survival, is fundamental to understanding the dynamics of populations. Survivorship curves typically show one of three patterns (Fig. 9.1), showing maximal mortality at older ages (Type I), a constant, proportional rate of mortality (Type II) or maximal mortality at young ages (Type III). Most fish populations have Type III curves, but many have a Type II curve as adults. In this case annual mortality is high among larvae and juveniles (Wootton 1998; Ch. 7), declines to become more-or-less stable in adulthood but is prone to peak again during spawning and migration. The number of adults in a cohort declines annually, in proportion to the number alive at the time (Wootton 1998).

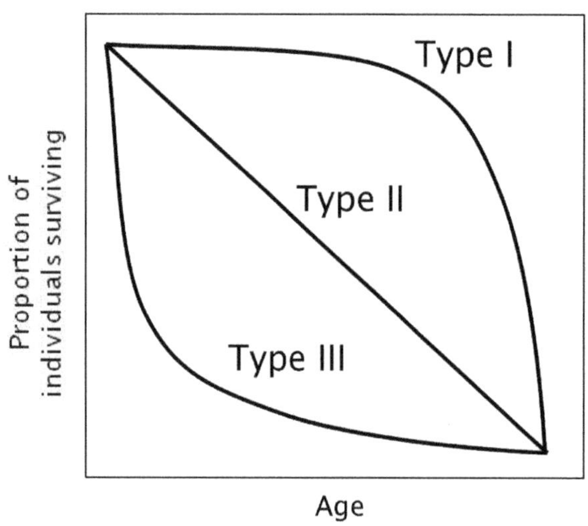

Figure 9.1: Theoretical survivorship curves for fish populations characterised by three broad types of life history strategies (see text).

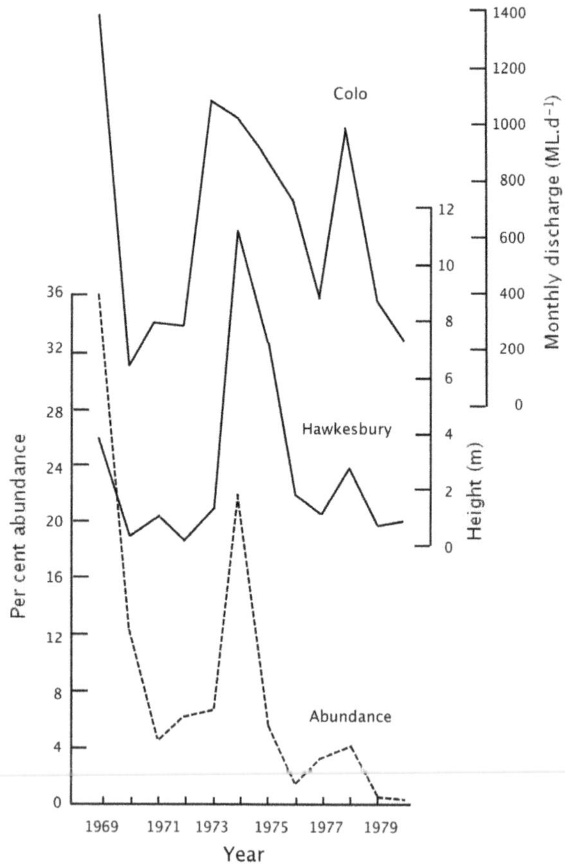

Figure 9.2: The relationship between recruitment (initial year-class abundance) of Australian bass (*Percalates novemaculeata*) during the July–August spawning season, shown by peak river height in the Hawkesbury River and daily discharges in the Colo River, New South Wales, averaged over those two months (Harris 1988).

Early mortality is usually associated with intensive predation, scarce prey (e.g. zooplankton or other invertebrates), disease or habitat changes (Ch. 7). Blackwater events, for example, can affect survival at all stages (Gehrke 1991; Dept of Industry and Development, New South Wales 2011b; Murray-Darling Basin Authority 2011). In the boom–bust flow regimes of many Australian rivers, mortality and survival may vary with erratic changes in the size and quality of habitats (Young 2001; Balcombe *et al.* 2007). In 1975, when Lake Eyre began to dry and become more saline after major flooding and high productivity, about 40 million bony herring and hardyhead (*Craterocephalus* spp.) were stranded on the shore (Ruello 1976). On the other hand, flooding may encourage high rates of survival and productivity of young and adults in many riverine fishes, including Australian bass (Harris 1988; Reid *et al.* 1997; Fig. 9.2).

Age data are usually a prerequisite for studies of mortality, survival, recruitment variation, growth and production, and the technology for age determination in fishes has a rich supporting literature (Bagenal 1978; Gerking 1978; Campana 2001). While an approximate idea of the age of young fishes can be gained from size measurements, more exact determinations require analyses of growth-related marks in hard body parts including scales, vertebrae, spines, fin rays and otoliths (Ch. 8).

With the capacity to age fishes comes an ability to determine total mortality, although length-based approaches also can be used, especially for young individuals and fast-growing species. A catch-curve estimate (Everhart *et al.* 1975; Fig. 9.3) is often used to assess the mortality of that part of the population that is large enough to be caught reliably by the particular fishing gear. This method plots the natural logarithm of the frequency of each age group in the catch against age, and the slope of the line indicates the total mortality in those age groups (Ch. 7).

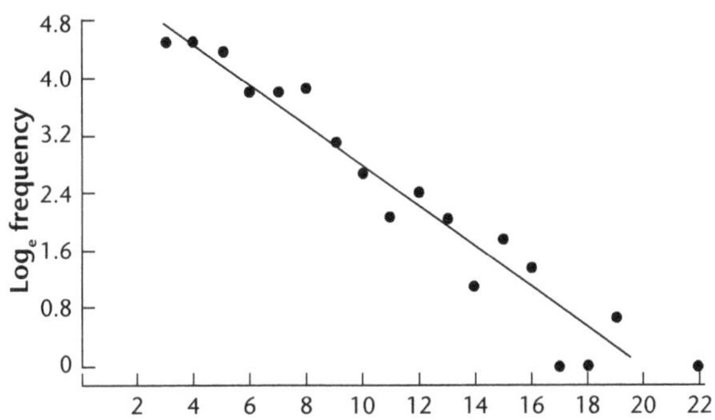

Figure 9.3: A catch-curve estimate of the mortality of Australian bass (*Percalates novemaculeata*) older than 2 years and large enough to be caught by the fishing gear. The slope of the line indicates the rate of total mortality (Harris 1988).

Harvesting adds to underlying natural mortality, so that total mortality is the sum of natural mortality plus fishing mortality. Fishing mortality is focused on particular sizes (hence age groups), of course. Fisheries regulations often involve limiting the numbers that may be caught and setting minimum and (occasionally) maximum sizes that may be captured (see 'Impacts of fishing'). In most harvested species, minimum legal length criteria are set to ensure that fishes reach maturity and have an opportunity to reproduce at least once before capture. For some species (Australian bass, barramundi, Murray cod) a maximum legal size is set, ensuring that the larger, generally faster-growing and most-fecund individuals are conserved. For Australian bass, barramundi and golden perch, where females are substantially larger than males, maximum size limits may be used to avoid distortion of population gender ratios through excessive captures of females. If there is strong selection for large sizes, the effective rate of mortality could be doubled.

Competition and predation

Competition and predation drive the dynamics of freshwater fish populations (Weatherley 1972; Wootton 1998; Walters and Martell 2004). Predation is a key component of total mortality and can be demonstrated through conventional dietary analyses and factored into models. Competition is more complex, however, and may be difficult to represent in equations. There are two forms.

- *Exploitation competition* is commonly associated with food, and occurs when individuals (regardless of species) compete for a resource that is in limited supply. It need not involve direct interactions between individuals, but nevertheless may inhibit the survival and growth of populations (Wootton 1998).
- *Interference competition* occurs when individuals aggressively defend a resource, denying access to other individuals. This may be associated with territorial behaviour over habitat patches, and typically involves direct interactions between individuals.

Debate over the role of competition has a long history in ecology (Krebs 2009). We could assume that populations are at equilibrium and that disturbances to predator or prey populations, perhaps through fishing pressure or stocking, will cause a shift to a new equilibrium. But assumptions of equilibrium are problematic, particularly in the variable conditions of Australian inland waters (Puckridge *et al.* 1998). Fish populations in these environments often undergo large fluctuations, owing to erratic changes in habitat availability and prey abundance (Balcombe *et al.* 2007), and the notion of an equilibrium is difficult to apply except perhaps at large scales of space and time or in a metapopulation context.

An illustration of competition for food resources is provided by the fish community in Lake Kununurra, a reservoir on the Ord River, Western Australia (Morgan and Gill 2004). The lake is impounded by a diversion dam, 55 km downstream of the much larger Ord River Dam. The diversion dam impedes fish passage and has caused the disappearance locally of barramundi and other migratory fishes. In response to proposals to stock barramundi in the lake, to create a recreational fishery, Morgan *et al.* (2004c) wanted to determine whether predation by barramundi on smaller fishes might increase competition among other piscivores in the lake. Most of 13 species examined had distinctive diets, different from that of barramundi, suggesting that competition would be low even when food resources were limited. The lake environment was relatively stable with abundant food for resident fishes, but competition was deemed possible if the barramundi were stocked at densities that could reduce the abundance of small fishes that are the prey of other piscivores such as freshwater longtom.

In contrast, dryland rivers such as the Moonie River in the northern Murray-Darling Basin typically experience boom–bust cycles (Puckridge *et al.* 1998; Arthington and Balcombe 2011). During flow events, when waterholes are connected, productivity booms and species such as bony herring and golden perch consume large quantities of non-filamentous benthic algae and crustaceans, respectively (Sternberg *et al.* 2008). During bust periods, when the river contracts to waterholes in which food and habitat availability are depleted, these species switch prey to filamentous algae and terrestrial insects, respectively. Switching thereby reduces competition when preferred prey are in short supply. The ability of many species to switch prey in this way suggests that interspecific competition will occur less often than might be predicted from assumptions of an environmental equilibrium (Walters and Martell 2004).

An example of interspecific competition is provided by an investigation of common carp (*Cyprinus carpio*) in paired lagoons in the northern Murray-Darling Basin (Gehrke *et al.* 2011). The biomasses of common carp in the lagoons were reduced experimentally by 33% and 43%, and subsequently the abundance of fishes competing with the common carp for zooplankton as food increased by 240% to 1130% in the case of bony herring, and by over 1600% for western carp gudgeons. The removal of common carp reduced the competition for zooplankton, so that the native fishes gained access to an increased supply of large zooplankters (*Boeckella*, *Calamoecia*, *Daphnia*).

The combination of predation and competition for food and habitat resources is exemplified by 'foraging arena' theory (Walters and Martell 2004). According to this theory, fishes spend most of their time avoiding being eaten by predators, perhaps by hiding among structural habitat. As the need for food increases, they must risk leaving shelter to find prey, and risk being eaten themselves (Ebner *et al.* 2011). The mere presence of predators, even in low numbers, can change the foraging behaviour of prey species (Ch. 11). Strategies for reducing predation risk and increasing feeding efficiency are being used increasingly in non-equilibrial ecosystem models designed to predict the responses of fish populations to disturbances such as fishing, stocking, invasive species and environmental changes such as improved catchment management or water resource development.

Dispersal, migration and population sources and sinks

Australian freshwater fishes rarely, if ever, pass through their larval, nursery, growth and mortality phases at one location. All species migrate between habitats but over very different spatial scales, from tens of metres to thousands of kilometres (Harris 2001; Ch. 5). Dispersal and migration are population movements distinguished by differences in predictability and directedness. Dispersal refers to non-directed movements away from an area, often in response to population density or declining resources. Migration is a more directed, predictable process that refers to 'movements resulting in an alternation between two or more separate habitats occurring with regular periodicity and involving a large fraction of the population' (Northcote 1978, p. 326). Young freshwater catfish, Murray cod and river blackfish leaving their nest sites demonstrate small-scale dispersal with little apparent direction, apart from downstream drift in currents, whereas the oceanic dispersals of larval and juvenile eels (*Anguilla* spp.) extend over enormous distances, assisted by the East Australian Current. The highly directed, predictable movement of larval and juvenile Australian bass, riding upstream gravity-circulation currents near the bed of estuaries (J.H. Harris, unpubl.), also exemplifies a migration.

For some species it is possible to identify zones that are population sources or sinks, where there is net recruitment or net mortality, respectively. The predominantly catadromous and amphidromous fishes of eastern Australian coastal rivers (McDowall 1996b; Allen *et al.* 2002; Ch. 10) have sources in oceanic or estuarine zones, by definition. Some potamodromous species recruit most strongly in lowland zones and travel upstream to grow in habitats where mortality of adults provides a population sink. This pattern may hold for smelt (*Retropinna* spp.), Cox's gudgeon (*Gobiomorphus coxii*) and empire gudgeon (*Hypseleotris compressa*). In other species (at least, those which are sufficiently well-known) there may be no clear pattern, perhaps because they occur in isolated wetland habitats (e.g. flathead galaxias, *Galaxias rostratus*; carp gudgeons). An ontogenetic record of the habitats occupied by fishes may be recorded in their otoliths and might be used to elucidate patterns of movement (Ch. 8).

An emerging hypothesis relates to large-scale movements in inland rivers. Research in the Murray-Darling Basin (Mallen-Cooper 1996; Barrett and Mallen-Cooper 2006; Baumgartner *et al.* 2010b) indicates size gradients in some potamodromous fishes, notably bony herring, golden perch and silver perch, at a river-valley scale. Samples of these species from the Murray River between the South Australian border and Lake Hume, at sites separated by hundreds of kilometres, indicate progressively greater body sizes. A hypothesis to explain this is that the young fishes move long distances upstream, growing larger as they go. Further, the numbers of fishes recorded moving downstream appear to be substantially lower than those moving upstream, although the data should be viewed cautiously because different sampling methods apply for upstream and downstream migrants.

These observations pose a tantalising opportunity for research to review the evidence for large-scale source-and-sink models of population dynamics in different systems. As above, these models suggest that some species are sourced from downstream recruitment zones and that the young fishes then migrate to grow and mature in upstream habitats (sinks), where their numbers decline through mortality. They assume that sufficient adult fishes remain in the recruitment zone and that they are capable of generating sufficient new recruits annually, as upstream recruitment zones are seldom significant among these particular species (Lake 1971; McDowall 1996b; Ch. 7). This pattern appears to be true of common carp as well as some native fishes (Crook and Gillanders 2006), and there is scope to consider wider applications.

The dependence of some freshwater fishes on migration pathways, and their sensitivity to loss of connectivity between habitats, are key vulnerabilities. Local extinctions above large dams (e.g. Australian bass; barramundi; bony herring; eels (Anguillidae); freshwater longtom; golden perch; jungle perch, *Kuhlia rupestris*; silver perch) underscore these dependencies and have encouraged developments in fish-passage restoration (Hancock *et al.* 2000; Keller and Peterken 2001; Barrett and Mallen-Cooper 2006). Increasingly, it becomes clear that changes in a particular habitat may produce population responses in quite another, distant habitat.

Influence of population density on demographics

How do population parameters change in response to changes in the space available to individuals? The mechanisms that regulate populations may be density-dependent or density-independent. Density-dependent mechanisms arise when there is competition for limiting resources between the individuals in a given habitat area, or when multiple species depend on the same resources. If the availability of food, spawning areas or other resources becomes limiting for a particular population density, then emigration and/or mortality will increase and/or growth and recruitment rates will decrease until a balance is restored. As population density increases there may be a corresponding increase in predation (Gerking 1978; Wootton 1998; Ch. 6). Conversely, decreased population density may lead to reduced predation, either through prey switching or less-successful foraging by predators. Further, predator density may modify prey behaviour. Predation pressure may cause prey to modify their habitat use and interfere with their foraging strategy (Pitcher *et al.* 1986; Walters and Martell 2004), so that productivity is reduced. Cannibalism may also be a factor, within or between cohorts. Barramundi, like many other piscivorous fishes, prey on younger, conspecific cohorts (Pusey *et al.* 2004), regulating total population size and modifying year-to-year variations in abundance. Within-cohort cannibalism is a form of density-dependent regulation, with larger fish preying on smaller ones, producing a faster-growing cohort of large individuals.

Fish diseases also may have density-dependent effects. In hatcheries, the frequency of encounters with infectious pathogens such as anchor worm (*Lernaea* spp.), gill parasites (e.g. *Chilodonella* spp.) or the agent of goldfish ulcer disease (*Aeromonas salmonicida*) affects the incidence of infection in the population (Humphrey and Langdon 1986; Rowland and Tully 2004) and clearly is a function of the density of susceptible fishes.

Density dependence may operate at low population densities via the Allee effect (Allee 1978), whereby low numbers of individuals reduce the likelihood of encountering a mate. These effects are difficult to demonstrate, but could limit population growth in newly colonised habitats and could accelerate population declines if densities fall to very low levels. Allee effects are relevant in conserving populations of rare species, but there appear to be no published Australian studies in this area.

Density-independent mechanisms are usually catastrophic, broad-scale events, such as severe drought, that reduce the carrying capacity of habitats (Ruello 1976; Balcombe *et al.* 2011a). Deteriorating water quality on floodplains through acid runoff and aluminium toxicity in the first rains of the wet season in the Northern Territory (Bishop 1980), and blackwater events (Murray-Darling Basin Authority 2011), are other causes of mass mortalities unrelated to population density. Thus, density-dependent and density-independent factors interact at various scales to regulate fish populations, and the dynamic nature of freshwater environments and the variety of trophic and other predator–prey interactions mean that these populations would rarely be in equilibrium (Wootton 1998).

MODELLING DYNAMICS

Fisheries studies were forerunners in quantitative ecology, leading to sophisticated methods for monitoring the status of exploited populations and modelling the potential effects of management interventions (Gulland 1977). For more than a century, driven by global commerce, methods have been developed, tested and refined (Beverton and Holt 1957; Nikolskii 1969), producing a store of knowledge on the dynamics of harvested species (Hilborn and Walters 1992; Walters and Martell 2004; Krebs 2009). While there are many international studies (Weatherley 1972; Bagenal 1978; Gerking 1978), Australian research is scarce (Harris 1988; Hancock 1994), partly because we have few significant commercial freshwater fisheries (Kailola *et al.* 1993). Nevertheless, elegant models are being developed with regard to conservation (Todd *et al.* 2005; Box 9.1), pest management (Brown and Walker 2004) and other issues (Gehrke *et al.* 2011). There is great potential for modelling to support management of large-scale drivers of recruitment such as flows in regulated rivers (Harris and Gehrke 1994; Gehrke and Harris 2001; Humphries *et al.* 2002), or to relate boom–bust populations to the highly variable flow regimes of inland rivers (Balcombe *et al.* 2007).

The conceptual basis for most population modelling is a simple equation in which the future population size (N_{t+1}) is expressed as a function of the current local population size (N_t), the numbers of births (B) and deaths (D) in the population and the numbers of individuals immigrating to (I) or emigrating from (E) the population:

$$N_{t+1} = N_t + B - D + I - E \qquad (9.1)$$

Box 9.1: Cold-water pollution and Murray cod

While cold-water releases maintain cold-adapted species such as trout, common carp and redfin perch downstream of reservoirs, they threaten the viability of native warm-water fishes. Todd *et al.* (2005) and Sherman *et al.* (2007) combined empirical data on survivorship at different temperatures in a stochastic population model to explore the possible mitigation options for the population viability of Murray cod and trout cod in the Mitta Mitta and Murray rivers.

The model included 10 age classes, with survivorship in the first age class incorporating spawning, egg, larval and early juvenile survival (less than one year old). Survival rates of eggs and larvae were determined from laboratory studies, and age-specific survival rates were based on age–length and length–frequency field data. Reproduction commenced at five years of age and increased until the 10th age class, and the model included stochastic variation in fecundity and survival. Only females were modelled (this is a common approach in populations with balanced sex ratios). Water temperature was modelled as a function of discharge in the Mitta Mitta River and using a hydrodynamic model for the Murray; in both cases, temperature scenarios based on historic, current and potential future management regimes were examined. For each scenario, stochastic population trajectories (n =10 000 replicate model runs of 25 years pre- and 23 years post-thermal pollution) were compared using interval quasi-extinction curves, which summarise the probability of the population size falling below a given threshold at any time during the model simulation. Scenarios also considered uncertainty in the timing of spawning by examining the effects on early and mid-season spawners. Results

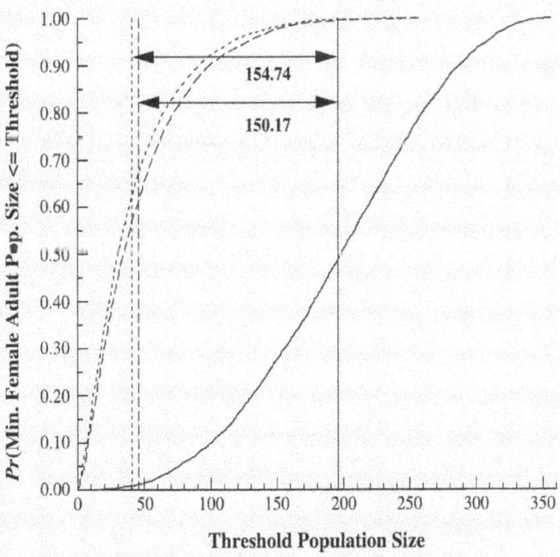

Box Figure 9.1: Effects of cold-water pollution on Murray cod, shown in quasi-extinction curves (female population). The solid line represents the NI (No Impact) model and the dashed lines represent realised thermal effects based on historical reservoir-release patterns and two different spawning schedules (SS): early (long dashed lines; SS1) and late (short dashed lines; SS2). The vertical lines indicate the average minimum population size for each model. The added risk attributable to impacts on post-spawning survival through the altered thermal regime is the difference between the vertical lines (from Todd *et al.* 2005).

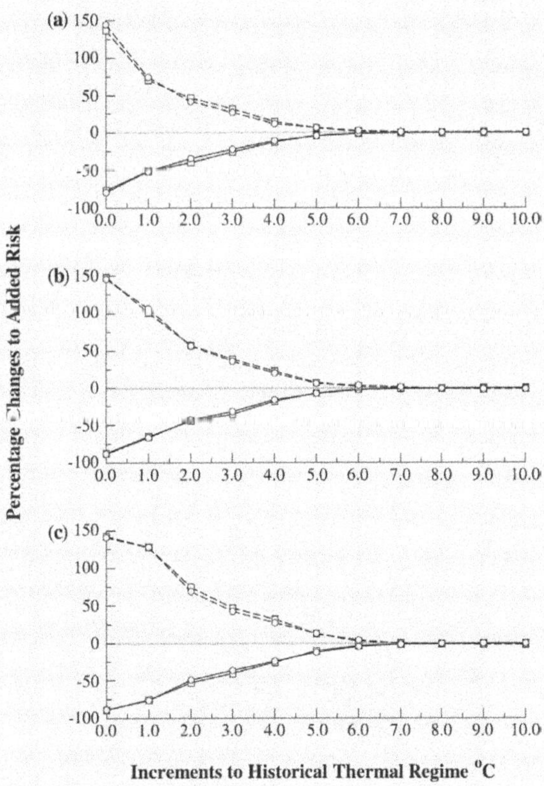

Box Figure 9.2: Effects of increasing the temperature of water releases. The solid lines indicate the percentage change in average minimum population size, and the dashed lines indicate the percentage change in the coefficient of variation (CV). Symbols represent results from models based on the two different spawning schedules (circles: SS1; squares: SS2). The critical period post-spawning period in each case was (a) n = 16, (b) n = 28, and (c) n = 40 days (from Todd *et al.* 2005).

from the Mitta Mitta showed a clear increase in the quasi-extinction risk as evidenced by the increased likelihood of populations falling to low levels (Box Fig. 9.1). Mitigation options considered the reduction in added risk associated with incremental increases in the temperature of water releases (Box Fig. 9.2) and allowed exploration of the potential outcomes of different approaches to mitigate cold-water pollution. Both studies highlighted the capacity to include uncertainty in population models via distributed parameter estimates as well as explicitly modelling scenarios based on plausible parameter values (in this case, timing of spawning and length of critical period).

This is often referred to as the BIDE (births, immigration, deaths, emigration) model. Other representations of population growth, such as the logistic growth model (Bagenal 1978; Hilborn and Walters 1992), effectively encompass the BIDE model parameters in a single population growth parameter (r), but are conceptually similar. The BIDE model is often applied to closed populations (e.g. in lakes, contained between barriers along rivers), or it may be used as a simple starting point, where immigration and emigration are ignored. For spatially complex metapopulations these two parameters, describing movements of individuals between local populations, may be very important.

Despite these simple concepts, the complex dynamics of wild populations reflect the fact that the parameters in the BIDE model are time-variant as a result of stochastic environmental and demographic changes and are often difficult to estimate because of sampling error. Nonetheless, the simplicity of the basic equation (Equation 9.1), and the fundamental processes it describes, should not be forgotten.

Because survival rates and reproductive output vary with age and/or developmental stage, most models further describe the abundance of individuals in the population in a structured form. They may be based on age (numbers of individuals in each age class) or stage (numbers of individuals present as juveniles or adults, or in different size cohorts), and are referred to as age- and stage-structured models, respectively. These can be shown pictorially in a life cycle diagram (Fig. 9.4) acknowledging that survival and reproductive output can each vary in structured population models. The decision about which model is best depends on the nature of the life history, the available demographic data and the questions being explored (Caswell 2000; Akçakaya 2005). Stage- or size-structured models are particularly suited to populations in which age is a poor predictor of demographic rates, or to examining the impacts of minimum-size limits of caught fish on population viability and growth.

Having established a representation of the structure of a population, it is necessary to estimate the

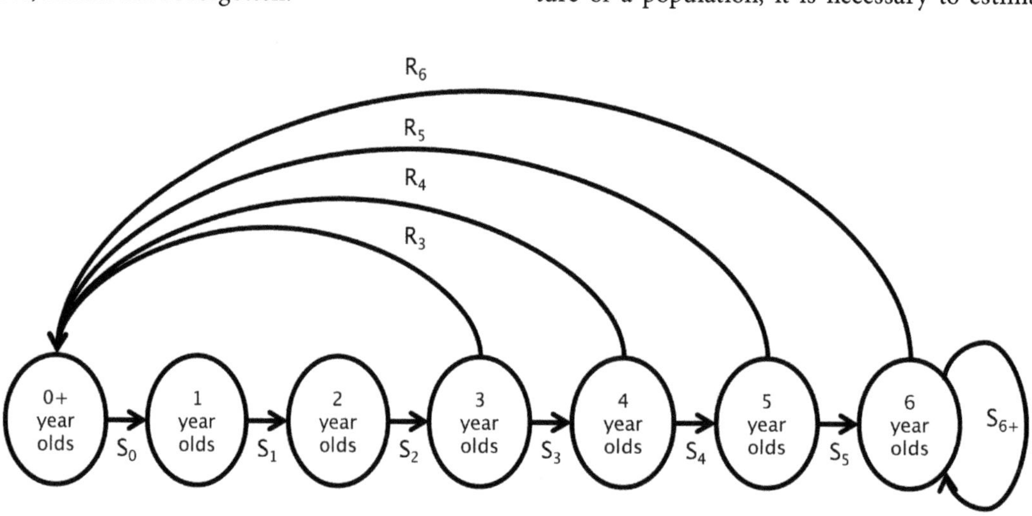

Figure 9.4: Example of an age-structured population life cycle diagram indicating the different survival and reproductive parameters ($S_{1...n}$ and $R_{3...n}$) required as input to a demographic model.

parameters in Equation 9.1. Although the adage 'garbage in, garbage out' does apply, imperfect knowledge is not a reason to forego modelling. Much can be learned by applying a best estimate of each of the key parameters together in a 'sensitivity analysis' where the values of parameters are varied systematically to see the effects (Burgman and Possingham 2000). Representing parameter estimates as distributions (e.g. normally distributed, with mean and variance) allows the uncertainty and stochasticity in parameter values to be monitored, and models are often run as repeated simulations to generate a distribution of possible population trajectories.

CPUE data alone are not sufficient to fully parameterise a population model, as carrying capacity, density-dependent effects and the relative abundances of different size classes may vary. In rare cases, estimates of population size come from a complete census of the population, or they may be derived from mark–recapture and capture–removal experiments. These methods are described by Southwood (2000) and supported by free software such as MARK (White and Burnham 1999). The same approaches can be extended to estimations of movement by adopting individual- or population-specific marks or tags (Hanski *et al.* 2000; Hanski and Gaggiotti 2004).

Estimates of survival and reproductive output come from life tables, which summarise the numbers of individuals surviving to a given age or stage (Hunt *et al.* 2011b). The most basic form is a static life table, which captures the abundance of each age- or stage class at a given time and assumes a stable age- and size-structure over time. A cohort life table is constructed by tracking the fate of cohorts over a number of years, and often includes data from age-growth analyses to estimate growth and survival (Ch. 8). Given sufficiently long time series, these approaches allow superior estimates of survival and fecundity and can be used to relate temporal changes in these parameters to environmental changes (Beissinger 1995; Van Winkle *et al.* 2011). For fishes where recruitment is sensitive to flow variability, it may be possible to relate year-class strength to annual streamflow (Reid *et al.* 1997), if mortality can be assumed constant from one year to the next (Jager *et al.* 1997; Sakaris and Irwin 2010).

The last critical parameter in population models considers the effects of density dependence on demographic processes and their parameters. Density-dependent processes reflect the internal feedbacks arising from depletion of resources such as food and space, and the increased risks of disease as populations grow. The mechanisms at play in any particular population will depend on the life history, physiology and behavioural characteristics of the species, and on the nature of the limiting resource (Quinn and Deriso 1999; Akçakaya 2005).

Having captured the structure of a population, most models incorporate a mix of deterministic processes and one or more of three main sources of chance variation, namely demographic stochasticity, environmental stochasticity and spatial variability (Akçakaya *et al.* 1997). Once the various parameters have been estimated, by implementing simulations in which demographic parameters or environmental conditions are systematically varied, the viability of populations may be assessed. So-called 'population viability analysis' is a powerful tool for projecting and evaluating the potential consequences of management actions, drought or other environmental perturbations (Burgman and Possingham 2000), and comparative studies among species with different life histories can be revealing even where parameter estimates have large uncertainties (van Winkle *et al.* 2011).

Individual-based models (IBMs) that track the fate of individuals through survival, reproduction and movement have been developed for fishes (DeAngelis *et al.* 1997, 2005; Perry and Bond 2009). These offer ways to explore the influence of fine-scale interactions and behavioural decisions but are more demanding computationally than population-level models (Scheffer *et al.* 1995) and often involve many more parameters.

The use of population models in fisheries management is varied but it may include assessment of different stocking regimes, the influence of alternative size or catch limits and other factors (Lorenzen 2012). Few examples exist from Australia, perhaps owing to the challenge of producing reliable BIDE parameter estimates in highly variable environments, undermining confidence in estimates and limiting model-based management for large stocking programs. Nonetheless, the BIDE modelling framework illustrates the basic drivers of population variability and there is considerable potential to expand on existing modelling studies – both in exploring natural population variability and alternative management scenarios – by ensuring that the needed data are collected as part of monitoring programs.

IMPACTS OF FISHING

Fisheries

Fishes and other freshwater biota sustained indigenous Australians for millennia before Europeans arrived and continued to support the early growth and development of the colonies. The cumulative impacts of exploitation and other pressures in the last two centuries are estimated to have reduced freshwater fishes to about 10% of their pre-European numbers (Murray-Darling Basin Commission 2004). In this section, we consider how fish populations respond to fishing and habitat changes.

How populations respond to fishing

At the simplest level, exploitation involves harvesting a proportion of a fish population. In many cases there may be unintended captures (by-catch) of other species. In recreational fisheries, even the increasingly popular, conservation-oriented practice of catch-and-release fishing has some unintended adverse effects and may cause delayed mortality through injuries and stress (Ch. 12). Thus, a fishery adds a population-mortality component to the natural mortality. The scope and implications of this added mortality have attracted vigorous scientific attention, and computer-based assessments and models have provided powerful tools for estimating population parameters, quantifying fishing mortality and exploring ways that management might be improved.

Removing fishes (usually the larger, older individuals) from a population produces some predictable results but the responses often are complex, causing shifts in various parameters, and each case is likely to differ from others. A generalised scenario of fishing impacts might include density-dependent changes to growth and production, recruitment, size and age distributions, inter- and intraspecific competition and predator–prey relationships. A fished population may attempt to restore lost biomass by increased growth, reproduction and/or survival whereas an unfished (minimally exploited) population approaching peak biomass will react by increasing mortality and/or decreasing growth and reproduction (Everhart *et al.* 1975; Ricker 1975).

Empirical studies of exploited fishes have been proving grounds for various kinds of dynamic models, such as surplus-yield and yield-per-recruit models, that quantify responses to harvesting by a variety of fishes and fisheries (Hilborn and Walters 1992; Haddon 2001; Walters and Martell 2004). These models have rarely been applied in Australian freshwater fisheries, as these generally have declined and were relatively small and of little economic significance. Most of the few scientific papers describing Australian freshwater fisheries are dated before 1990, and most refer to fishes in hatcheries or impoundments rather than natural riverine populations.

One response to removal of adult fish is for younger individuals to become relatively abundant. Younger fish are at a steeply rising stage of their growth curve and their rate of growth in a particular period will be greater than that of older animals, which may not grow at all. More younger, faster-growing fish implies that fishery production should increase in response to appropriate management. Managers exploit this characteristic and harvesting plans commonly identify maximum yields at around one-third to one-half of the initial, unfished virgin biomass (Everhart *et al.* 1975; Gulland 1977; Walters and Martell 2004).

Negative effects from fishing may appear as changes in patterns of growth or recruitment. Growth overfishing results from removal of larger fishes through size-selective harvest and/or fishing gear. Australian bass in the Hawkesbury River, for example, are heavily exploited through capture of larger fish, so that the population is dominated by smaller, young individuals (Harris 1988). Similarly, the recreational fishery for Murray cod in Lake Mulwala (Murray River) has cultivated a population in which the great majority of individuals are below the minimum legal length (J.H. Harris, unpubl.). Recruitment overfishing, on the other hand, occurs when the stock of spawners is so reduced that too few eggs are produced to provide for compensatory recruitment. Stock-recruitment curves

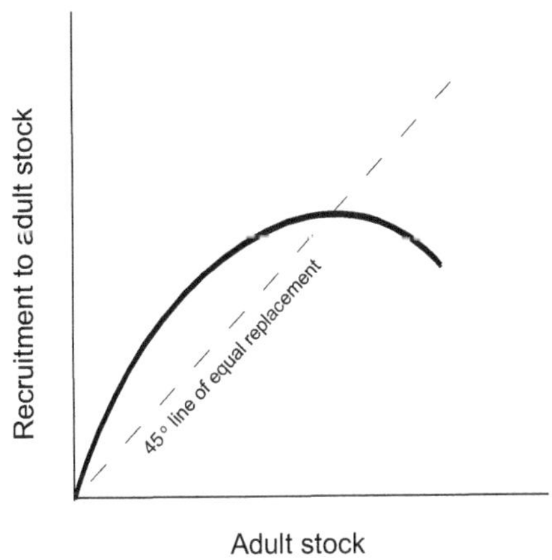

Figure 9.5: A conceptual stock-recruitment curve for a fish population experiencing density dependence that is driven by predation, cannibalism or resource limitation, and showing the relationship between the abundance of recruits and the size of the adult spawning stock.

(Gulland 1977; Fig. 9.5) typically are unimodal, with peak production of recruits at some point well below the population's maximum possible abundance. At lower abundances there are insufficient spawners and at high abundances density-dependence reduces total recruitment. An example of recruitment overfishing is excessive harvesting of mature eels (Hoyle and Jellyman 2002; Prosek 2010). Recruitment overfishing is a particular risk in sexually dimorphic species such as Australian bass, barramundi and golden perch, as the harvest favours larger individuals that, in these species, are females.

Exploited Australian freshwater fishes

Australia's total fishery production is small by world standards, ranked 55th among nations in 1989 (Kailola *et al.* 1993), and freshwater production contributes a very small proportion of the total. Commercial fisheries statistics are routinely compiled nationally (Bureau of Agricultural Resource Economics and Sciences 2010) and by the state and territory jurisdictions.

Of 92 exploited finfish species in fresh waters only five native (barramundi; golden perch; longfin eel, *Anguilla reinhardtii*; southern shortfin eel, *A. australis*) and two alien species (redfin perch, *Perca fluviatilis*; common carp) support commercial fisheries (Kailola *et al.* 1993). In aquaculture, several native species and three salmonids (Atlantic salmon, *Salmo salar*; brown trout; rainbow trout, *Oncorhynchus mykiss*) are produced. Species that were previously exploited commercially include Australian bass, estuary perch (*Percalates colonorum*), freshwater catfish, golden perch, Murray cod, Tasmanian whitebait (*Lovettia sealii* and associated galaxiids), pouch lamprey (*Geotria australis*) and shorthead lamprey (*Mordacia mordax*). This summary does not account for the substantial (unquantified) production that is generated in freshwater habitats by migratory and other species but is harvested in estuarine or coastal fisheries. Examples include barramundi, black bream (*Acanthopagrus butcheri*), eels (*Anguilla* spp.), mangrove jack (*Lutjanus argentimaculatus*), sea mullet (*Mugil cephalus*) and yellow-eye mullet (*Aldrichetta forsteri*). Freshwater fisheries production data ideally should embrace these species, and many more species that are targeted by indigenous and recreational fisheries (e.g. Australian bass; estuary perch; fork-tailed catfishes (Ariidae); freshwater catfish; freshwater cobbler, *Tandanus bostocki*; jungle perch; Macquarie perch, *Macquaria australasica*; mullets (Mugilidae); saratogas, *Scleropages* spp.; sooty grunter and other northern and inland Terapontidae). Fisheries development has had many impacts on the dynamics of these species, including growth overfishing, recruitment overfishing, species invasions and the displacement of native stocks by alien species or hatchery fishes, but these impacts are both compounded and confounded by habitat changes.

Australia has a poor record in early management of freshwater fisheries. In the Murray-Darling Basin, intensive exploitation by commercial and recreational fishing by the late 1800s had depleted the enormous stocks of Murray cod and other favoured species (Rowland 1989; Mackay and Eastburn 1990), leading to Royal Commissions of Enquiry in 1880 and the early 1900s. For example, in 1835 29 064 kg of fishes were sent to market in Melbourne from two mid-Murray ports (Koondrook, Moama) but in 1896 only 4273 kg were taken from those areas (Mackay and Eastburn 1990). The drastic reductions prompted New South Wales Commissioners of Fisheries to declare the whole of the Murray in that state closed to netting for a period. A commercial fishery on the Murray in South Australia was well established by

1850 (Ye and Hammer 2009). A reach fishery began there in 1923 (Poole 1984; Rohan 1989), taking Murray cod, freshwater catfish, golden perch and silver perch (Reynolds 1976). The catches varied greatly, peaking after floods, but there were strong downward trends (Fig. 9.6). The same pattern applied in New South Wales (Reid *et al.* 1997; Fig. 9.7). Commercial fishing for most native fishes in South Australia ceased in 2003, when the reach fishery was restructured as a fishery mainly for common carp, although bony herring and yabbies (crayfish, *Cherax destructor*) are also harvested.

Early fishway records provided clear evidence of severe declines in native fish populations. Data from five-year observations over a 50-year period (Mallen-Cooper and Brand 2007; Griffin 1993) show that the abundances of three native species passing through Euston Weir on the Murray River had declined by 95–100%. Four prized species, Murray cod, Clarence River cod (*Maccullochella ikei*), Mary River cod (*M. mariensis*) and trout cod (*M. macquariensis*) proved to be extremely vulnerable to overfishing (Cadwallader and Backhouse 1983; Rowland 1989; Simpson and Jackson 1996). Their large size, eating quality, market acceptability, slow growth and high vulnerability to fishing methods all combined to create precipitous declines in catches (Cadwallader 1978; Wager and Jackson 1993).

In a broad-scale study of fish as an indicator of river health, the New South Wales Rivers Survey (Harris and Gehrke 1997) reported that fish at 80 sites in coastal and inland rivers were in severe decline. Twenty-nine percent of the state's known freshwater fish species were not detected, and the status of many once-abundant species argued for their inclusion in conservation listings. Up to 25% of fish showed evidence of parasites, diseases or abnormalities and alien species dominated at a number of sites, being 57.5% of total fish abundance in Murray River sites. Common carp dominated over large areas; a population-depletion experiment in the Bogan River, New South Wales, indicated an average of one common carp for each square metre of water surface area. The Murray-Darling Basin Sustainable Rivers Audit (Davies *et al.* 2010) instituted comprehensive river health assessments using fish communities as one of several indicators. Other recent surveys in the Murray-Darling Basin

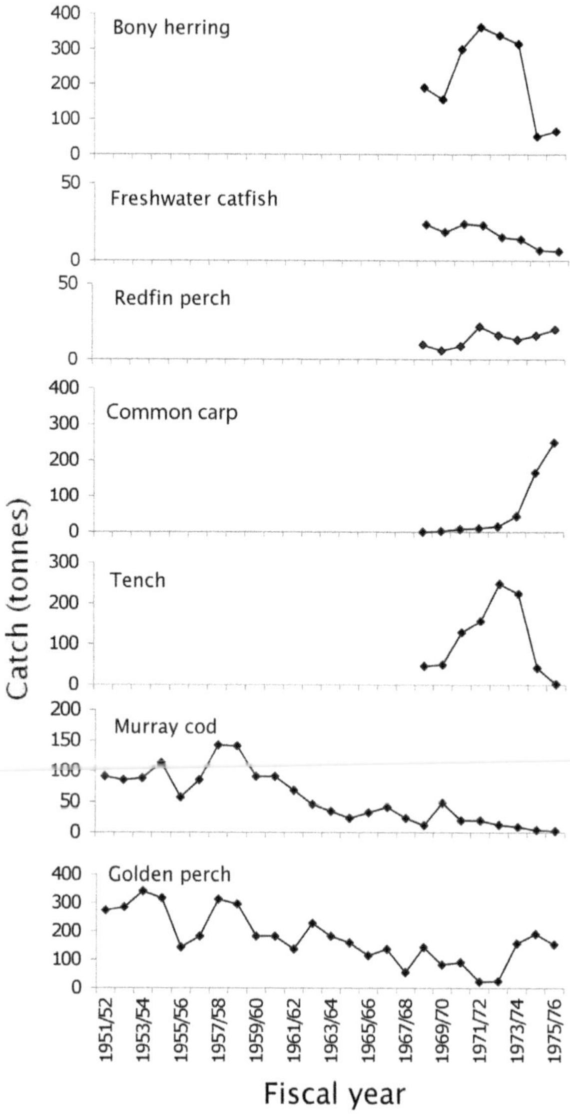

Figure 9.6: Annual commercial catches of key freshwater species from the South Australian Murray River and Lower Lakes 1951/52–1975/76 (Reynolds 1976). Modified from Ye and Hammer (2009).

emphasise downward trends in abundance and diversity. In the lower Murray-Darling rivers and the Murrumbidgee River, Gilligan (2005a, b) reported that many species were absent or in very low numbers.

These declines, most pronounced in the Murray-Darling Basin, are attributable to numerous influences although overfishing and habitat changes have been key contributors (Breckwoldt *et al.* 2004; Murray-Darling Basin Commission 2004; Ye and Hammer 2009). The roles of other influences are evident in

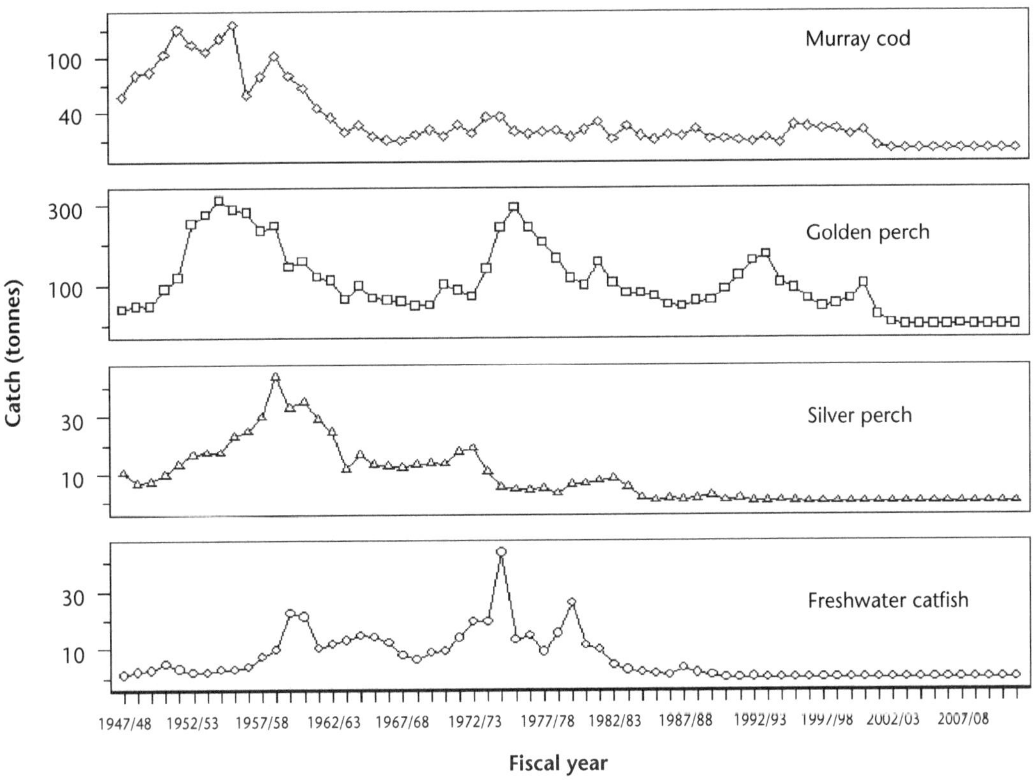

Figure 9.7: Declines in commercial fisheries catches of four species in New South Wales, 1947–2006 (modified from Reid *et al.* 1997).

similar declines apparent for unexploited species, especially smaller ones. Box 9.2 cites some examples.

In the late 20th century, most commercial freshwater fisheries were closed in the face of concern over the status of stocks. The concern was driven by research and catch data (Reid *et al.* 1997; Murray-Darling Basin Commission 2004) and by political pressure from recreational fishing groups demanding more of the resource, even though the commercial industry had long been constrained. Reid *et al.* (1997) reported that only 5% of the total linear extent of rivers in New South Wales was open to commercial fishing for native finfish, and there were other controls over gear and catch composition. Currently, the only commercial fisheries for native species are for bony herring and golden perch in the Murray River, South Australia, and for eels in several states. Illicit commercial harvesting has long been a substantial but unquantified problem, and probably continues. Common carp and redfin perch (and yabbies) continue to be commercially harvested in south-eastern mainland states.

Little is known of population impacts from recreational fishing, but they are likely to be substantial and should be considered as part of sustainable management (Kearney 1995; Reid *et al.* 1997).

Recreational fisheries

Fishing provided protein and leisure activity for early European settlers and recreational fisheries continue to flourish with increased human populations, more leisure time, economic independence, greater mobility on land and water and sophisticated fishing gear. As Kearney (1995) emphasised, catch data from recreational fisheries have almost always been inadequate, in contrast to commercial fisheries data, and despite the high employment (80 000–100 000 jobs) associated with the national recreational fishing industry. Until recently, the few available data were from a creel survey in Lake Keepit, Namoi River, New South Wales (Battaglene 1985), from South Australia (Phillipson *et al.* 1986), from roadside and creel surveys of the barramundi fishery in the Northern Terri-

> **Box 9.2: Declines of Murray-Darling Basin fishes**
>
> These are excerpts from the scientific literature (modified from Murray-Darling Basin Commission 2005).
>
> - In an oral history of fishes in the Gwydir River, people reported a massive drop in freshwater catfish numbers during the previous 50 years, in all but the most western regions. The decline in silver perch was more variable, but in some places equally severe.
> - There was a 95% fall in silver perch using the Euston fishway from the 1940s to 1990s.
> - There were only 92 silver perch caught from a total of 48 907 fishes in the Barmah Forest between 1990 and 1994.
> - Only nine silver perch were caught from a total of 8018 native fish in a survey of the Murray-Darling system in New South Wales over two years.
> - Two silver perch were caught from samples of 1556 native fish at Goondiwindi on the McIntyre River.
> - Seven silver perch were caught from a sample of 9178 fish above and below Bourke Weir on the Darling River.
> - No silver perch were caught from 2819 fish above and below the main weir at Menindee.
> - The New South Wales Rivers Survey, in which 50 438 fish were recorded at 80 sites across New South Wales, concluded that the condition of rivers in the Murray River system gave 'cause for particular alarm'.
> - Trout cod were once relatively common throughout much of the Murray-Darling Basin. There are now only two breeding populations in the wild.
> - Non-native fish species dominated the fish community in a survey at Barmah Forest, comprising 94% of the catch. Of these, common carp were the most abundant and made up 86% of the catch.
> - Four species are now absent from South Australia, namely trout cod, Macquarie perch, flathead galaxias and Agassiz's glassfish. Agassiz's glassfish was common historically; the others were rare.
> - Several species are listed as 'threatened' under the Environment Protection and Biodiversity Conservation Act 1999. For a current list, see http://www.environment.gov.au/biodiversity/threatened

tory (Griffin 1993) and from mail surveys of Tasmanian inland anglers (Davies 1995). Kearney (1995) attempted to estimate the national recreational fisheries catch and, after acknowledging assumptions and data deficits, concluded that it was of a similar order to the total commercial catch of the targeted recreational species.

A national, randomised telephone survey of recreational fishers (Henry and Lyle 2003) provided statistics on fish catch, effort, species composition and profiles of anglers and their expenditure. The survey indicated that 19.5% of people over five years of age fished annually, spending 20.6 million fisher-days (about 20% in fresh water) and harvesting 2.1 million common carp and 1.3 million redfin perch, plus trout and native fishes. Estimated total expenditures were $1.8 billion. Total recreational catches in the Murray River in South Australia in the 12 months to April 2001 were estimated at 2233 Murray cod, 211 299 golden perch and 421 790 common carp (equivalent to 23 t, 222 t and 239 t, respectively). Ten percent of Murray cod and 62% of golden perch reportedly were released after capture. In 2007–08, during drought in that region, the recreational fishing effort eased and catches of Murray cod and golden perch fell sharply.

Although data are scant, these levels of effort must have impacts, especially on less-abundant fishes, including growth overfishing and recruitment overfishing. Illicit and quasi-commercial fishing practices are another significant, but unquantified, impact (Reid *et al.* 1997). Sustainable management would involve intensive monitoring of recreational and other catches, allowing for temporal shifts in catches and population structure and the spatial and temporal distributions of fishing effort. Many tools exist to support monitoring and management of recreational fisheries and stocks (Cowx 1994; Pollock *et al.* 1994).

Indigenous fisheries

Freshwater systems are of major significance to indigenous people throughout Australia (Humphries 2007). They have strong cultural and spiritual values associated with fishes (Rowe *et al.* 2008), and fishing is strongly connected to traditional responsibilities for land management and kinship, maintaining social networks. In Henry and Lyle's (2003) national survey, participation in fishing among indigenous people was extremely high (91.7% of the indigenous population), mostly in northern Australia, with an estimated 420 000 indigenous fisher-days of effort nationally in the year of the survey.

Fish stocking

Stocking programs support recreational fishing or conservation (Cowx 1994; Phillips 2003). The aims include rehabilitation of threatened native species, rehabilitating fisheries or fish populations affected by bushfire, drought or flood and establishing or enhancing fisheries in waterways where target species are not recruiting well. Enhancing populations entails value judgements, a need for biological knowledge and a balance between competing interests. Stocking for conservation presents additional difficulties, including developing new propagation techniques, ensuring appropriate genetic quality and meeting the ecological requirements for success in the wild. Long-term commitments and resources are required.

Propagation methods for common carp, redfin perch and salmonids were introduced to Australia in the colonial era (Ch. 11), and hatchery production for native species has expanded greatly since the 1960s. Many millions of young fishes are liberated annually for recreational fishing in reservoirs and streams in Queensland, New South Wales, Victoria and Tasmania, guided by state policies and strategies (Phillips 2003). The main native species produced are Australian bass, barramundi, golden perch, Murray cod and silver perch, with smaller numbers of species such as southern saratoga and sooty grunter. In stocked fisheries, management is able to manipulate the species, sizes and numbers of fish released, together with regulations to control effort and take (e.g. seasonal and area closures, bag and size limits).

Aboriginal Australians manipulated freshwater habitats in a form of aquaculture based on eels (Ch. 5), and young eels are still captured during their upstream migrations and stocked in enclosed waters to support fisheries in Victoria and Tasmania (Sloane 1984c; Hall *et al.* 1990). Enhancements of existing populations usually succeed only in receiving waters that are below their carrying capacity. Occasionally, large hatchery fish may be released to create short-term, 'put-and-take' trout fisheries, which make fewer demands on habitat resources. Otherwise, adding extra fish to populations in habitats that are at or near carrying capacity is at least wasteful, and may damage existing stocks through competition, predation and displacement (Cowx 1994; Phillips 2003; Hutchison *et al.* 2006).

Multiple factors influence the success of fish stocking, some driven by conditions and events in the hatchery, others responding to factors in the release habitat. Key issues are summarised in Table 9.2. Many past stocking activities, particularly in trout fisheries, were founded more on customary practice and

Table 9.2: Factors in hatcheries and release habitats that influence the survival of stocked fishes

	Hatchery	Release habitat
Fish experience	Foraging for live prey, presence of predators, group dynamics and learning, use of structural cover, conditions in grow-out facilities, fish-training procedures	Hatchery experience and conditioning, abundance of prey and predators, group dynamics and learning, habitat quality, fitness for swimming
Genetics	Genetic adaptation of broodstock to hatchery conditions, inbreeding or outbreeding depression, group behaviour, growth rate	Hatchery adaptation, inbreeding or outbreeding depression, group behaviour, growth rate and other genetically determined fitness attributes
Size at release	Grow-out period, risk of accidental mortality, behavioural adaptation to hatchery conditions, cost per fish	Foraging ability, predator avoidance, swimming ability, capacity for fasting, density dependence, fishery exploitation
Body condition	Health, fat stores, immune status, density dependence	Health, fat stores, capacity for fasting, density dependence
Management procedures	Nutrition, design and operation of fry ponds and raceways, training for foraging and predator avoidance, vaccination, health certification	Transport, liberation technique, temporary artificial cover for release, release timing

supposition than on understanding of population dynamics. Anglers' expectations and notions that 'more and bigger fish are better' commonly drove stocking decisions and objective evaluations were rare until the 1990s. A key factor was observations of intensive predation on newly released hatchery fish by resident species, especially redfin perch. This led to an apparently reasonable – but flawed – notion that bigger fish, which swim faster, are more likely to survive. The flaws were that bigger fish would also be more adapted to predator-free hatchery conditions, with routine deliveries of artificial food, and that policies favouring stocking with larger fish mean order-of-magnitude reductions in the numbers of hatchery fish available, because of continual losses during the grow-out stages. In truth, bigger fishes means far fewer fishes.

Hatchery stocking has often been viewed as a panacea for fishery issues but stocking has often failed, especially in attempts to enhance existing populations (Cowx 1994; Lintermans 2007; Phillips 2003). A study of a Tasmanian brown trout population was repeated after a 30-year period when stocking was stopped (Davies *et al.* 1988). The abundance and sizes of fish increased by more than 50% in the absence of stocking, and angler catches increased significantly. In this case, abundance was driven by hydrological variation rather than by stocking. There has been similar research in streams, lakes and impoundments (Faragher and Gordon 1992; Faragher *et al.* 2007). Insights into the benefits of stocking native fishes, especially in impoundments, come from other studies (Phillips 2003; Hutchison *et al.* 2006).

Golden perch have been stocked in impoundments for several decades to enhance existing wild populations, but the effects were not clear until recently. Hunt *et al.* (2011a) introduced chemically marked juvenile golden perch to three lakes where natural recruitment was low. There were overall population increases of 47–90%, showing that the stocked fish had increased the abundance in each lake rather than replaced wild fish. On the other hand, the contributions of stocked fish declined with increasing connectedness to the Loddon River (Victoria), a probable source of recruitment of wild fish. Stocking may augment fish numbers when natural recruitment is low, but probably is wasted in populations that are recruiting naturally. Another stocking experiment (J.H. Harris, unpubl.) assessed the survival of larval and juvenile golden perch and silver perch over 12 months, following release to eight small reservoirs. The stocked larvae did not survive, but survival of juveniles was similar for both species, from 3–23% in reservoirs where there was predation by abundant redfin perch and 22–100% in low-predation habitats. This suggests that there may be benefits in stocking for fisheries enhancement even in habitats with strong predator populations.

Hatchery propagation and stocking are also practised for species conservation, including Clarence River cod (Rowland 1985), Mary River cod (Simpson and Jackson 1996), trout cod (Douglas *et al.* 1994) and Macquarie perch (Lintermans 2006). These programs face many challenges in re-establishing viable populations, possibly through Allee effects. Long lag-times are likely before results can be detected, especially for long-lived species. After several decades of trout cod stocking, for example, only a few signs of natural recruitment have been recorded (Lintermans 2007; Ch. 12). One factor may be the tendency for conditioning to occur among young fish reared under hatchery conditions; this militates against their survival in the wild (Table 9.2).

Sustainability issues affect fish stocking programs, and need to be managed (Ch. 12). Potential issues include effects on threatened species, other native species and communities, genetic swamping, spread of parasites and disease, loss of genetic diversity (hence less viability), inappropriate introductions and release of contaminating biota. Health-certification procedures underpin hatchery production standards and address many of the potential issues, and there is a need for continual upgrading of hatchery quality-management safeguards to meet national and international standards (Rowland and Tully 2004). Hatchery standards need to ensure optimal genetic quality of stocked fish. Routine genetic screening is required to eliminate hybrid parents and hatchery procedures must prevent repeat-breeding of propagated strains, inadequate mixing of progeny and other genetic threats. For example, research has identified genetic problems with Australian bass stocked in the Snowy River system (Schwartz and Beheregaray 2008), where almost half of sampled fish, some from hatchery broodstock and stocking programs, were hybrids of estuary perch and Australian bass. Hybridisation threatens the viability of parent populations, emphasising the need for genetic screening and broodstock management.

Hatchery propagation and stocking therefore can have mixed effects. Stocking may help to rehabilitate biodiversity by re-establishing threatened species and it generates economic and social benefits through support for recreational fisheries. This applies especially in impoundments, where many favoured angling species do not recruit well. There may also be negative effects from poorly managed stocking, with homogenisation of fish communities and displacement of wild native species and stocks through competition and predation by propagated fish. Notwithstanding the potential benefits of stocking, wild self-sustaining stocks have major advantages in terms of biodiversity value, angling quality and population viability. Wild fishes are genetically diverse and better able to respond to changing conditions, and they provide better angling. Wild stocks should always take precedence.

Sustainability

How can we ensure that populations targeted by fisheries will be managed sustainably? Legislative provisions are described in Ch. 12, and Australia's National Strategy for Ecologically Sustainable Development (ESD Steering Committee 1992) defines ecologically sustainable development as 'using, conserving and enhancing the community's resources so that ecological processes, on which life depends, are maintained, and the total quality of life, now and in the future, can be increased'. Many previously exploited species are now protected to redress population declines or threatened status (e.g. Australian lungfish, *Neoceratodus forsteri*; Clarence River cod, freshwater catfish, Mary River cod, river blackfish, silver perch, Tasmanian whitebait, trout cod and species of *Ambassis*, *Nannoperca* and *Mogurnda*). Other restrictions to improve the sustainability of recreational fishing include seasonal and area closures and minimum/maximum size and bag limits (Ch. 12). These controls are intended to reduce overfishing, to protect populations during vulnerable periods (especially spawning) and to avoid excessive mortality among large/female fish.

Thus, fisheries management legislation in the states and Commonwealth gives priority to conserving fishes and protecting their environment, and to ensuring that fishing and aquaculture are ecologically sustainable (Kearney *et al.* 2012; Ch. 12). An ongoing challenge is to ensure that there are adequate resources to service these ideals, given the declines that have occurred and the need for urgent and effective action.

Population effects of habitat change

Changes in freshwater habitats (catchment changes, sedimentation, altered flows, riparian modification, impoundment, contamination, salinisation, thermal pollution) are extensively documented (Young 2001; Breckwoldt *et al.* 2004; Murray-Darling Basin Commission 2005; Dept of Sustainability, Environment, Water, Population and Communities 2011a), and are discussed in other chapters. Only recently have the complex global interactions between fishes and forestry been recognised. These are summarised as 'multifaceted, dynamic processes involving most inland surface waters (streams, rivers, marshes, lakes, reservoirs, estuaries), forests, subsurface waters, geology and soils, climate and its changes, and the biotic components of the relevant ecosystems' (Northcote and Hartman 2004, p. 3). The accompanying text, including a review of Australian conditions, emphasises the profound consequences of forestry practices, as with agricultural land use, for freshwater habitats and fish populations. Among these multiple impacts, water resource development may have the most immediate and direct effects on the dynamics of fish populations.

Water resource development

Water resource development for agriculture, mining or urban use inevitably alters the flow regimes of rivers. The flow regime includes patterns of magnitude, frequency, duration, timing and rate of change of flood events, and virtually drives the ecology of freshwater systems. Spatial and temporal variations in the natural flow regime underpin the integrity of freshwater communities (Poff *et al.* 1997; Bunn and Arthington 2002; Poff and Zimmerman 2010) and the spawning, recruitment, growth, migration and mortality of fishes often can be related to one or more flow variables. Regulated regimes generally seek to remove interannual variability and perhaps ensure a minimum flow, causing a phenotype-environmental mismatch (Marshall *et al.* 2010) for indigenous species. Although patterns of discharge in Australian rivers are highly variable, flow regimes can be distinguished by key hydrological variables (Puckridge *et al.* 1998; Kennard *et al.* 2010) and are associated with the structures of fish communities (Ch. 10).

Several aspects of flow regulation cause population declines in native fishes (Poff *et al.* 1997; Bunn and Arthington 2002). Dams and weirs obstruct migrations while altering physical and chemical conditions (Todd *et al.* 2005; Walker 2006). Weir pools retard the downstream-drifting dispersal of larvae, and larval mortalities occur at weirs with undershot gates (Baumgartner *et al.* 2006b). Thermally stratified reservoirs cause cold-water pollution over many hundreds of kilometres of rivers (Phillips 2001; Sherman *et al.* 2007), lowering temperatures below the spawning thresholds and preferences of fish (Phillips 2001; Astles *et al.* 2003; Todd *et al.* 2005). Some impoundments may provide habitat for the threatened Macquarie perch (Ebner *et al.* 2011) and for the few other species (freshwater catfish, Murray cod) that can recruit in reservoirs. Water resource developments suppress flow variability (Poff *et al.* 1997; Balcombe *et al.* 2011b), absorbing the spring–early summer flood peaks that boost population recruitment in inland south-eastern Australia (Graham and Harris 2005; Humphries *et al.* 2002; King *et al.* 2009; Tonkin *et al.* 2008a), in summer in tropical rivers (Pusey *et al.* 2004) and in winter–spring in south-eastern coastal rivers (Harris 1988).

CONCLUSION

Knowledge of the dynamic processes in fish populations underpins biodiversity management, recreational use and harvesting for food. Few basic parameters are involved (recruitment, growth and production, mortality and migration), but their interactions and responses to human interventions are complex and varied. Traditional approaches to freshwater fisheries in Australia have changed in the aftermath of excessive pressure on favoured species, collapsed stocks, fishery closures and a new emphasis on conserving biodiversity. Most commercial harvesting of freshwater finfish is now for alien species only. Among recreational anglers there has been strong growth of a conservation ethic. Many anglers have come to value the benefits of their interaction with nature as a key part of the angling experience.

Several themes recur in this chapter. Many native fish populations in southern Australia have declined drastically, reflecting the deteriorating health of rivers. Changed river flow regimes and patterns of land use continue to threaten freshwater fishes in many regions. Recreational fishing is now the dominant form of exploitation, rather than commercial fishing, but there are deficiencies in monitoring, management and research. Applications of validated, predictive models of fish population dynamics would illuminate the long-running debate over the effects of water resource development on river flow regimes (Murray-Darling Basin Commission 2005), establishing a way to explain and predict the biological consequences of changed regimes.

Population ecology offers a bright array of opportunities and tools for scientists, managers and students interested in freshwater fishes. As the focus of research funding has shifted from declining commercial fisheries towards conservation, recreational fisheries and environmental management, there have been exciting advances in sampling, marking and imaging gear, computing hardware and software, hatchery facilities and modelling and laboratory techniques. The scientific literature in Australia, and the world generally, is growing stronger in all these areas. In addition, there are benefits in interdisciplinary studies alongside hydrologists, geographers, engineers, limnologists, chemists, water managers and other key professionals. Interactions with these associated disciplines bring stimulation, broader perspectives and enhanced productivity.

10 Assemblages

Angela H. Arthington, Mark J. Kennard, Bradley J. Pusey and Stephen R. Balcombe

INTRODUCTION

What controls 'who is where' in freshwater fish communities? Fishers often claim to know, and probably do most of the time (or often enough for fishing to be fun). Fish ecologists are also keen to know which species tend to occur together in wetlands, lakes and rivers, and why some groups, or assemblages, occur in some places and not others. In this chapter, we review the physical, chemical and biological factors that govern the structure of assemblages (the member species, their relative abundances, ecological attributes and interrelationships). We consider how and why they vary in space and time, the threats they face and their future prospects.

In Australia, as in other countries, freshwater fishes are affected by human activities in catchments, by impoundments, water diversions and altered flow regimes, by pollution, overfishing, alien species and other habitat changes (Boulton and Brock 1999; Arthington and Pusey 2003; Lintermans 2004). As the global climate becomes warmer and drier, we can expect changes to affect many, perhaps most, fish assemblages (Bond *et al.* 2008; Murphy and Timbal 2008). Ecologists and resource managers will be asked to develop practical solutions to problems of diminishing water resources, changes in the distribution of species and local extinctions. To underwrite those solutions we want to understand how environmental factors and processes affect fishes and the ecosystems that support them (Dudgeon *et al.* 2006; Palmer *et al.* 2008).

How do abiotic and biotic processes influence fish assemblages? Do their effects vary in space and time and among ecosystems, and if so how might we apply this knowledge? These questions have occupied many ecologists (Gido and Jackson 2010), and we now seek answers from the literature on fish assemblages in Australian inland waters. For ease of reference, Figure 10.1 shows the lakes, rivers and major drainage basins mentioned in this chapter.

WHAT IS AN ASSEMBLAGE?

The term 'assemblage' is used widely in fish ecology, but a rigorous definition is elusive. We accept the view of Matthews (1998), who regarded an assemblage as a group of species that occurs together in a single locality, where individuals have a reasonable opportunity for daily contact with each other. Matthews (1998) avoided the term 'community' because it may wrongly imply predictable interactions among species. Further, community is often used broadly to refer to all

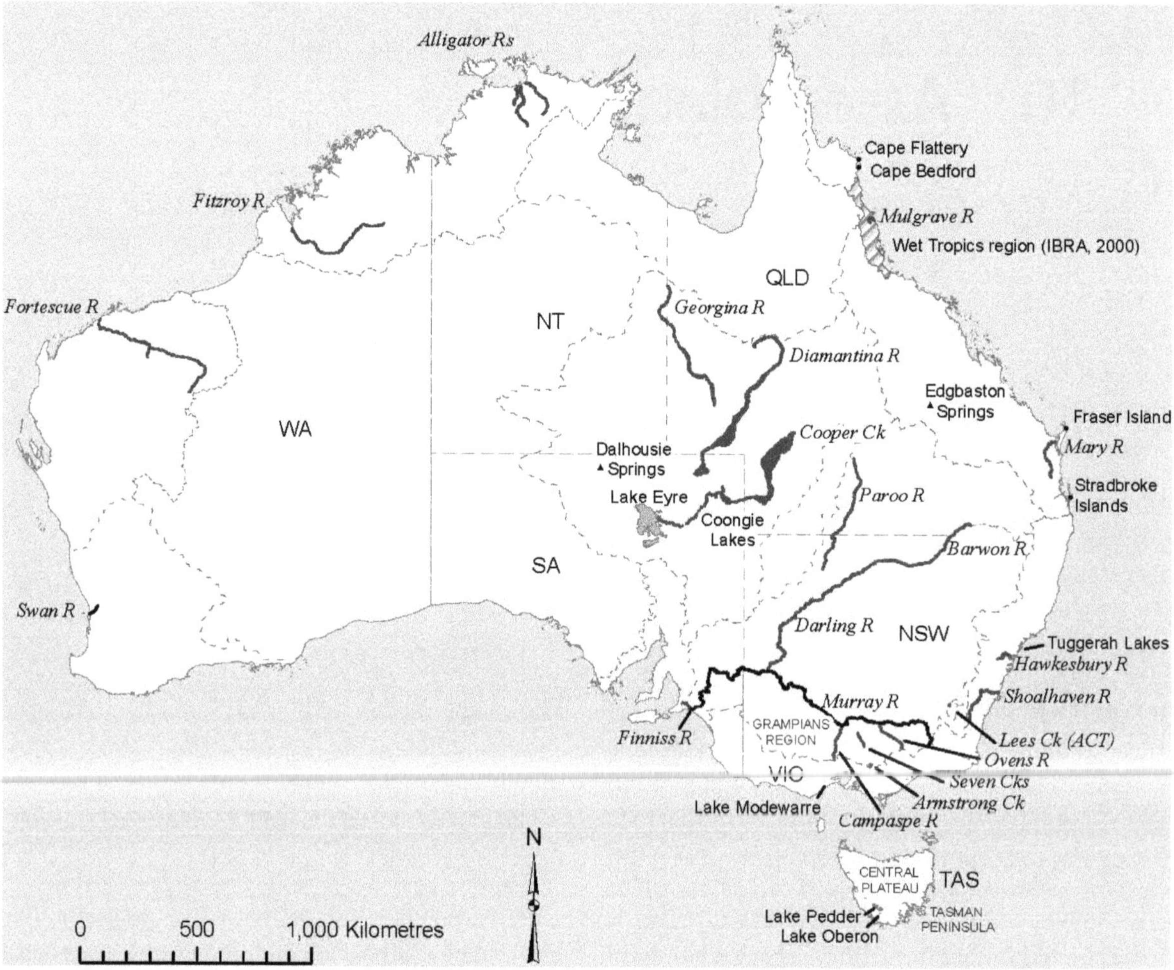

Figure 10.1: Australia, showing state and territory boundaries and some of the rivers and lakes mentioned in the text.

species in all taxa (fish, birds, plants etc.) in a particular place (thus, all living components of an ecosystem, including humans). Indeed, the concept of community can be extended further to a metacommunity, or a number of communities at different sites, linked by dispersal of multiple, potentially interacting species (Leibold *et al.* 2004). We consider this extended concept later (see 'Metacommunities').

The numbers of species regarded as part of an assemblage can depend on the study design, sampling gear and methods, the length of stream or shoreline, the types of habitat and the season of sampling. To make valid comparisons we need to be conscious of these effects and to standardise the sampling effort. It is common practice to employ a suite of methods in a replicable manner, so that entire assemblages (fishes of all body sizes and behaviours) are sampled within a defined location and period of time. The result is termed catch per unit effort (CPUE). Once standardised for a type of habitat (e.g. a riffle, pool or lake) and a suite of methods, CPUE data provide reliable, comparable estimates of species richness, taxonomic composition and relative abundance or biomass.

Some of the methods used in sampling fish assemblages include backpack and boat electrofishing, nets (gill, fyke, seine, sweep, trawl, drift and dip nets), fish traps (baited and light traps) and line fishing. Underwater observations by snorkelling, underwater photography, echo-sounding and fish tracking devices also contribute. In addition, rotenone (a naturally occurring, toxic ketone) may be used for quantitative sampling in structurally complex habitats such as among aquatic plants (Lockett 1988). Some comparisons of fishing methods can be found in Pusey *et al.* (1998b), Kennard *et al.* (2006b) and *Standard Methods for Sampling North American Freshwater Fishes* (Bonar *et al.* 2009).

ASSEMBLAGES

Species pools and environmental filters

Regardless of differences between sampling methods, the number of species in an assemblage at a particular location is a function of the number of species in the regional species pool and the effects of past and present processes that selectively remove species from the pool (Smith and Powell 1971; Jackson and Harvey 1989; Tonn 1990). Seen this way, abiotic conditions and biotic processes are like a series of filters through which species in the regional pool must pass to have a chance of persisting at a locality (Poff 1997). Species can pass through each filter only if they possess certain traits, such as tolerances to high temperatures or strong currents, and if their other habitat and reproductive needs are met (Poff and Allan 1995; Poff *et al.* 2006). Environmental filters operate over both historical and contemporary time frames and their effects vary in different regions (Unmack 2001b; Ch. 2). In the Wet Tropics of northern Australia, for example, riverine fish assemblages reflect a mix of historical influences such as the barrier effect of the Great Dividing Range, past climate change, drainage realignments and allopatric speciation in some taxa (e.g. freshwater catfishes, *Tandanus* spp.: Jerry 2008). Contemporary influences could include dispersal of fish from the Fly River (Papua New Guinea) at times of high discharge, and the transport of marine-dispersing larvae on the East Australian Current (Pusey *et al.* 2008). In other parts of Australia, glaciation and volcanism have shaped drainage basins and thereby influenced fish assemblages.

Lacustrine assemblages

The filter model originally applied to lacustrine (lake-dwelling) assemblages selected from a continental fauna by regional filters (Fig. 10.2). For example, regional lacustrine faunas may be shaped by climate, dispersal barriers, geomorphology, the size of lakes, water sources, chemical and thermal regimes, optical characteristics and other habitat features and biotic processes (Eadie and Keast 1984; Tonn 1990; Olden *et al.* 2001).

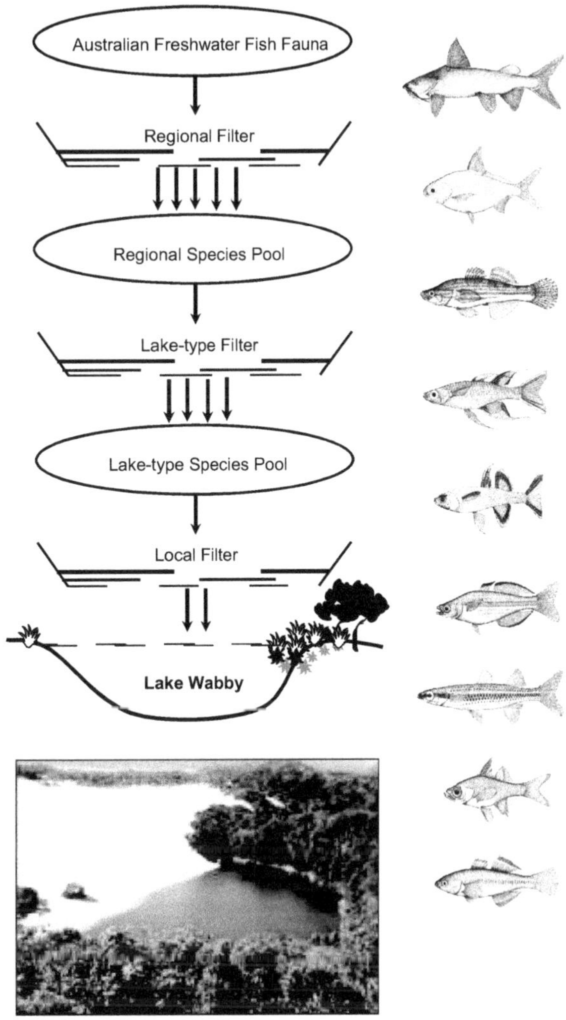

Figure 10.2: Conceptual model (from Tonn 1990) of environmental filters that influence fish assemblages in lakes, showing representatives of the assemblage in Lake Wabby, Fraser Island, Queensland. Species are (from top to bottom): blue catfish (*Neoarius graeffei*), bony herring (*Nematalosa erebi*), striped gudgeon (*Gobiomorphus australis*), Pacific blue-eye (*Pseudomugil signifer*), honey blue-eye (*Pseudomugil mellis*), crimson-spotted rainbowfish (*Melanotaenia duboulayi*), fly-specked hardyhead (*Craterocephalus stercusmuscarum*), Agassiz's glassfish (*Ambassis agassizii*), firetail gudgeon (*Hypseleotris galii*) (fish drawings by B.J. Pusey).

Australia's aridity precludes the large permanent lakes and extensive lakelands of other continents (e.g. English Lake District, North American Great Lakes), except for the glacial lakes, tarns and cirques of Tasmania and the coastal dune lakes of eastern Australia. As a result, Australia has few native fishes found only in lakes (Ch. 4), although when the lakes of a region are grouped as physiographic types they usually prove to have relatively discrete fish assemblages. The paucity of lacustrine species has changed, however, with the many artificial lakes created by dam and weir construction and with introductions of alien species adapted for lacustrine environments (Ch. 11). For example, extinction threatens one mainly lacustrine fish, the Pedder galaxias (*Galaxias pedderensis*), endemic to the former Lake Pedder and nearby waters in south-western Tasmania, where it cohabited with the swamp galaxias (*G. parvus*). In 1973–74 the region was flooded to create a new Huon-Serpentine impoundment, part of the Middle-Gordon hydro-electric power scheme. In the mid 1980s very few of either species could be found, and the Pedder galaxias was last recorded in its natural habitat in 1996 (Dept of Sustainability, Environment, Water, Population and Communities 2011b). Translocations of Pedder galaxias to nearby Lake Oberon and to Strathgordon Dam met with mixed success. Today, the Pedder galaxias is apparently extinct in what remains of its natural habitat, and alien brown trout (*Salmo trutta*) and a translocated native species, the climbing galaxias (*G. brevipinnis*), flourish in its place (McDowall 2006). Some ecologists believe that the endemic fauna of Lake Pedder might be restored if these introduced fishes, and the recently arrived alien redfin perch (*Perca fluviatilis*), could be suppressed (Lake 2001).

Another claim for extinction has proved wrong. The Lake Eacham rainbowfish (*Melanotaenia eachamensis*), once thought to be confined to Lake Eacham (Atherton Tableland, north Queensland), disappeared from the lake and was claimed to be extinct (Barlow *et al.* 1987) before it was rediscovered in nearby streams (Pusey *et al.* 1997).

Dune lakes in places such as Cape Bedford, Cape Flattery and Fraser Island, Queensland (Fig. 10.1) are perched or watertable window lakes, distinguished by their elevation relative to the regional watertable and a host of associated characteristics including size and depth, water chemistry, colour and transparency (Timms 1986; Pusey *et al.* 2000c). Some of these lakes lack native fish (e.g. Brown Lake, North Stradbroke Island) whereas others (e.g. Lake Wabby, Fraser Island) support 13 or more native species (Bensink and Burton 1975; Arthington *et al.* 1986; Fig. 10.2). Dune-lake species are adapted to low pH (< 7) and very fresh water (low ionic content), and their distributions vary with lake type and geographic location. Three such species in south-eastern Queensland and northern New South Wales have very restricted distributions, namely the honey blue-eye (*Pseudomugil mellis*), the Oxleyan pygmy perch (*Nannoperca oxleyana*) and the ornate rainbowfish (*Rhadinocentrus ornatus*: Pusey *et al.* 2004). All are threatened by wetland drainage, coastal development, pollution and alien species, especially eastern gambusia (*Gambusia holbrooki*: Arthington and Marshall 1999; Knight and Arthington 2008). Lacustrine fish assemblages generally are threatened by changes in the hydrology of inflowing rivers and by groundwater dynamics, habitat disturbance, alien species and climate change (Kingsford 2000; Pusey *et al.* 2004).

Australia has vast tracts of arid and semi-arid country where erratic low rainfall (< 500 mm annually) and occasional spates create rich mosaics of temporary lakes, wetlands, billabongs (oxbow lakes) and anastomosing stream channels. Some arid-zone wetlands are saline because they accumulate salt from rainfall, catchment soils and groundwater. One example is Lake Eyre, at the terminus of the Lake Eyre Basin in central Australia. This is Australia's biggest lake (9500 km^2) and the 18th largest in the world (Krieg 1989). Most of the time it is a dry saltpan, but it fills occasionally with floodwaters from Cooper Creek and the Diamantina and Georgina rivers (Fig. 10.1). Floods stimulate a massive boom of productivity that sustains huge populations of fish, drawing flocks of piscivorous birds from other regions (Kingsford *et al.* 1999). The lake fish fauna includes bony herring (*Nematalosa erebi*), Lake Eyre hardyhead (*Craterocephalus eyresii*), Australian smelt (*Retropinna semoni*), golden perch (*Macquaria ambigua*) and spangled perch (*Leiopotherapon unicolor*) (Glover 1982), eastern rainbowfish (*Melanotaenia splendida*), a glassfish (*Ambassis* sp.), tandans (Plotosidae) and grunters (Terapontidae) (Wager and Unmack 2000). When Lake Eyre dried in 1975, 40 million fishes were stranded on the shore, in lines reflecting their tolerances to increasing salinity as the water receded (Ruello 1976). The Coongie Lakes on lower Cooper Creek (Fig. 10.1) persist for seven years or so after flooding and host their own resident fauna comprising 10 native species including the dominant bony herring and Australian smelt, and two alien species (eastern gambusia; goldfish, *Carassius auratus*: Puckridge *et al.* 2000). During floods, lakes on the Paroo River floodplain in the north-western Murray-Darling Basin have been likened to a lakeland, although few are deeper than about 2 m and they persist for only a few months after filling (Timms 1999). Their fish fauna is more diverse than that of regulated rivers and floodplains in other parts of the Murray-Darling river system (Gehrke *et al.* 1995).

Mound spring assemblages

Mound springs are an extreme habitat for desert fishes. They occur particularly along the south-western margins of the Great Artesian Basin, where tectonic action and fractures allow artesian water rich in calcium and bicarbonate to escape to the surface (Habermehl 1982). Precipitation of carbonates builds up a characteristic mound and the water seeps into shallow marshes. The largest complex of active mound springs, Dalhousie Springs in South Australia, supports four endemic fish species (Wager and Unmack 2000), namely the Dalhousie catfish (*Neosilurus gloveri*), Dalhousie hardyhead (*Craterocephalus dalhousiensis*), Dalhousie purple-spotted gudgeon (*Mogurnda thermophila*) and Dalhousie goby (*Chlamydogobius gloveri*) as well as Glover's hardyhead (*Craterocephalus gloveri*) and spangled perch. Springs in the Lake Eyre supergroup support desert goby (*Chlamydogobius eremius*), Lake Eyre hardyhead, spangled perch and the alien eastern gambusia. Edgbaston Springs in the Springsure supergroup are mostly very small and shallow, without mounds, yet they support three endemic fish: redfin blue-eye (*Scaturiginichthys vermeilipinnis*), Edgbaston goby (*Chlamydogobius squamigenus*) and an undescribed hardyhead (*Craterocephalus* sp.) (Wager and Unmack 2000; Dept of Environment and Resource Management, Queensland 2011).

Mound spring fish assemblages are under threat. Since European settlement, thousands of artesian bores have reduced aquifer pressures and depleted surface-upwelling zones. Threats include the alien eastern gambusia, alien plants (e.g. pasture grasses, bamboo, date palms, rubber vine), disturbances from cattle, feral pigs and camels, excavations intended to enhance flows and improve stock access, impoundments and tourist visitations (e.g. trampling, inappropriate bathing). The ecological community dependent on discharge from the Great Artesian Basin is listed as 'endangered' under the Commonwealth *Environment Protection and Biodiversity Conservation Act 1999* and an associated recovery plan has prescribed actions to restore and maintain the springs and their fauna and flora (Fensham *et al.* 2010).

Coastal lake assemblages

Irregularities along coastlines favour the formation of sand bars and coastal lakes. Intermittently closed and open lakes and lagoons (ICOLLS), like those along the coasts of New South Wales and south-western Western Australia, have characteristic assemblages depending on their history of isolation, morphology and salinity regime. ICOLLS usually harbour fewer fish species than open estuaries but they can be much more productive for fishes that tolerate variable physical and chemical conditions (Pollard 1994). ICOLLS that occasionally open to the sea often support populations of commercial and recreational species of marine and estuarine origin, such as yellowfin bream (*Acanthopagrus australis*), dusky flathead (*Platycephalus fuscus*), sand whiting (*Sillago ciliata*) and sea mullet (*Mugil cephalus*), and some coastal lakes are nursery habitats for the young of these species (Hadwen *et al.* 2007). Threats to coastal lakes are intensifying with urban development and tourism, eutrophication from sewage (Hadwen and Arthington 2007) and alien species. Many are affected by artificial opening of lake entrances to allow flushing and prevent flooding of infrastructure, disrupting the movements of freshwater and estuarine species (Pollard 1994; Griffiths 1999).

Riverine assemblages

In rivers, environmental filters can be visualised as a spatially nested hierarchy of physical landscape units (Frissell *et al.* 1986) in which each level is nested within the level above, with levels from catchment (river basin) to valley, reach, channel unit (e.g. riffle, run, pool) and microhabitat scales (Fig. 4.6). The hierarchy means that processes within a riffle or pool are likely to be influenced by processes at the reach, valley and catchment scales. A corollary is that habitat restoration at a reach scale may not succeed if stressors impacting at larger scales are not addressed (Fausch *et al.* 2002; Howell *et al.* 2012). Spatial variations interact with temporal ones as rivers, wetlands and lakes may vary with changes in runoff or groundwater inflows, temperature and other seasonal factors. Collectively, these variations can be likened to the opening and closing of pores in environmental filters that allow species to pass and persist in local assemblages (Poff 1997; Wiens 2002a, b).

Many ecologists have sought to understand the spatial and temporal patterns of riverine fish assemblages and to interpret them in terms of underlying mechanistic processes. Correlations are often reported between environmental factors and assemblage structure (e.g. species richness, dominance patterns, traits) but they do not necessarily indicate causal mechanisms. Causality can be confirmed only by experimentation; this is difficult in many field situations and near-impossible at large spatial scales. Nevertheless, the associations occur repeatedly in freshwater environments between variables, and species responses can be used, through the veil of species traits, to infer biological mechanisms that affect fish within particular habitats (Townsend and Hildrew 1994; Wiens 2002b; Hoeinghaus *et al.* 2007).

Fish assemblages in coastal Australian rivers typically increase in species richness with distance downstream. For example, surveys in the Alligator Rivers region of the Northern Territory (Fig. 10.1) have recorded distinctive assemblages (up to seven species) in dry-season pools in upper escarpment habitats and richer assemblages (up to 21 species) in lowland creeks, main channels and billabongs (Bishop *et al.* 2001). In rivers of the Wet Tropics in Queensland, there may be two to three native species in upland reaches and tributaries and 20 or more species in lowland reaches (Hortle and Pearson 1990; Pusey and Kennard 1996). Streams in south-eastern Queensland show a similar pattern (Kennard *et al.* 2006a, 2007). In the Shoalhaven River system, New South Wales, upland tributaries have fewest species and lowland streams have most species, albeit with some anomalies (Bishop and Bell 1978). In Victoria, in the Seven Creeks system and streams in the Grampians, species richness generally increases with stream size (Cadwallader 1979; Jackson and Davies 1983). More species are recorded from the lower than the upper Barwon River in the Otway Ranges (Hall and Tunbridge 1988). In a surveyed stream on the Tasman Peninsula, Tasmania, the numbers of species increased downstream (Lake and Fulton 1981). In South Australia, surveys along the Murray River have shown variable numbers of small species in river edge, stream, billabong and backwater habitats, with 12 species in the lower reaches of the Finniss River, tributary to Lake Alexandrina, and fewer in the upper reaches (Lloyd and Walker 1986). The assemblages (23 freshwater species) of the Fitzroy River in the Kimberley region, Western Australia, are most speciose in lower reaches, moderately so in middle reaches and least so in the upper reaches (Morgan *et al.* 2004a).

The apparent downstream increase in species richness may arise partly because many of the so-called freshwater species in coastal rivers spend part of their lives in estuarine or marine environments, and because some marine and estuarine species enter the lower reaches and so contribute to freshwater assemblages. The flexible responses to salinity, and vagrant or diadromous (migratory) behaviour, make the distinction between estuarine and freshwater fishes somewhat equivocal (Chs 2, 5). Although 256 fish species inhabit Australian fresh waters for part or all of their life history, the lower reaches of coastal rivers attract many marine and estuarine fishes of ecological and commercial importance (Morgan *et al.* 2004a; Pusey *et al.* 2004) and their incursions into fresh waters foster a variety of biotic interactions. This point is discussed later (see 'Biotic factors').

The idea of a river channel as an environmental continuum with corresponding patterns in community composition is captured in the well-known 'river continuum concept' (Vannote *et al.* 1980). Fish received scant attention in the original formulation of the concept, but Décamps and Naiman (1989) proposed an overlying continuum related to fish assemblages. The distributions of fishes, mapped onto this continuum, link their responses to variations in energy sources and food web structure across a hierarchically arranged landscape (Romanuk *et al.* 2006). These linkages are further emphasised in the 'riverine ecosystem synthesis' (Thorp *et al.* 2008), which views a river system as a series of 'functional process zones' or hydrogeomorphic patches with distinct energy sources and food web structures. Thus, there is clear altitudinal zonation of fishes in rivers of New South Wales (Gehrke and Harris 2000), and variations in assemblages along the Barwon-Darling River (Murray-Darling Basin) correspond to functional process zones (Boys and Thoms 2006).

ABIOTIC FACTORS

Flow and habitat interactions

How do the myriad environmental factors interact across the spatial domains of a river system, and how much of the variation in assemblage structure is correlated with them? This question has been addressed in many Australian streams and rivers, as the studies cited above indicate. Another example is work by Stewart-Koster *et al.* (2007), who investigated fish assemblages in the Mary River, south-eastern Queensland, using canonical correspondence analysis – a multivariate technique that partitions patterns of ecological variability according to their main environmental correlates. Fish assemblages and habitat were sampled seasonally over four years at 17 relatively undisturbed river sites. The analysis associated 65% of spatial and temporal variation in assemblage structure with hydrologic variables (frequency, timing and duration of high and low flows and flow variability), landscape/habitat variables (distance to river mouth, channel width, depth, substrate characteristics, presence of large woody debris and litter and riparian cover) and interactions among flow and habitat variables.

Close scrutiny of direct and indirect interactions among flow variables, habitat attributes and fish traits is necessary to appreciate the importance of these assemblage–environment relationships (Bunn and Arthington 2002; Stewart-Koster *et al.* 2007). For example, low flows in rivers such as the Mary can reduce the system to isolated pools where fishes may be driven to local extinction or where their opportunities to disperse to more favourable habitats may be compromised by long periods of hydrologic disconnection (Pusey *et al.* 1993; Kennard *et al.* 2007; Stewart-Koster *et al.* 2007). Daily fluctuations in low flows over short periods (e.g. the season or year before sampling) can influence survival or inhibit recruitment by interfering with spawning and larval survival. In the Mary River, several small species (e.g. crimson-spotted rainbowfish, *Melanotaenia duboulayi*; Agassiz's glassfish, *Ambassis agassizii*) have a low-flow recruitment strategy (Milton and Arthington 1984; cf. Humphries *et al.* 1999) that involves spawning during steady low flows in spring, before the onset of flashy wet-season flows and floods. Too much variation in daily flows can strand fish egg masses or expose the river edge and slackwater habitats that provide refuges from strong currents, hence energetic advantages (Humphries *et al.* 2006). On the other hand, high flows can scour structural habitat elements such as aquatic vegetation, woody debris and soft substrata that provide food, cover and protection for many species (Pusey *et al.* 1993, 1995).

The effects of flow variability depend on the position of a reach within the catchment (Kennard *et al.* 2007; Stewart-Koster *et al.* 2007; Grossman *et al.* 2010). Fishes in headwater streams may be exposed to flashy flows and higher velocities, whereas those in lowland reaches experience more stable conditions (Horwitz 1978; Schlosser 1987a; Poff and Allan 1995). These hydrological disturbances have implications for biotic interactions, discussed below.

The flow of water along a river and through its diverse habitats is a driving factor in ecological processes and the activities of many species (Junk *et al.* 1989; Walker *et al.* 1995; Poff *et al.* 1997). Flow shapes the channel and drives sediment transport, it determines the type, extent and accessibility of habitat for riverine animals and plants, feeds energy into food webs through transport of carbon and is a major influence on animal and plant behaviour, life history strategies, recruitment and productivity. Fishes are particularly susceptible to flow patterns altered by regulation. The typical effect is to change the composition of assemblages, with intolerant species declining or becoming locally extinct and opportunistic native and alien species assuming dominance (Fausch *et al.* 2001; Lamouroux and Cattaneo 2006). Fish assemblages in regulated rivers of the North Coast, South Coast, Murray and Darling drainages have lost native species and gained alien species (Gehrke *et al.* 1995; Gehrke and Harris 2001). Over a seven-year study in the regulated Campaspe River, Victoria, six alien species comprised 64% of the total biomass of fish and only three of 10 native species were considered self-sustaining (Humphries *et al.* 2008a). Barriers to movements often exacerbate the effects of flow alteration, preventing movement into and out of regulated reaches and inhibiting recolonisation (Lucas and Baras 2001; Nilsson *et al.* 2005). When a deep-release dam develops thermal stratification (deep, cold water overlain by warmer surface water), the water released downstream is likely to be colder than natural and may discourage breeding of some fishes (Sherman *et al.* 2007; Olden and Naiman 2010).

The interplay of flow regime and habitat structure is recognised in the 'natural flow paradigm' and related ideas (Poff *et al.* 1997; Bunn and Arthington 2002; Naiman *et al.* 2008), and underpins most widely used methods for estimating the flow volumes required to maintain fish habitat in regulated streams (Tharme 2003). The concept of environmental flows provides opportunities to direct water into the right habitats at the right time of year to sustain fish reproduction, dispersal, assemblages and ecosystem processes (Arthington *et al.* 2010a and references therein). Increasingly, flow releases from dams are being treated as experiments to test concepts about the ecological benefits of flow releases as spawning triggers and for floodplain inundation (Poff *et al.* 2003; King *et al.* 2010). Flow releases can also be manipulated to reinstate variability in regulated rivers (Reich *et al.* 2010).

Riparian influences

Riverine fishes respond to the nature of the catchment, especially the corridor of riparian vegetation (Naiman and Décamps 1997; Allan 2004). The floristic composition, structure and condition of riparian vegetation influence fish assemblages by regulating shade and temperature, limiting instream primary productivity, consolidating the banks and providing instream habitat structure (aquatic plants, undercut banks, root masses, woody debris and leaf litter) and delivering organic material (Pusey and Arthington 2003; Davis *et al.* 2010a). Figure 10.3 shows a model of these processes and their effects on fishes.

The roles of flow regime, riparian condition and water quality may be clarified by so-called natural experiments. Sites are selected to compare reference (unimpacted, control) sites with treated (impacted) sites in terms of one or more factors of interest (Stoddard *et al.* 2006). The idea is to isolate the factors of interest, as in laboratory experiments. One such experiment in the Hawkesbury-Nepean River, New South Wales, showed changes in fish species richness and abundance in lowland reaches where overhanging riparian vegetation had been removed by sand and gravel mining, and the banks had slumped (Growns *et al.* 1998). Although some sites were affected by nutrients from treated sewage and urban runoff, and one species (sea mullet) responded to enrichment, other changes in the fish assemblages could be attributed to the loss of habitat complexity.

In the Wet Tropics of northern Queensland, clear streams from steep escarpments enter lowland agricultural areas where sugarcane and banana plantations and associated activities disturb the landscape and riparian corridors. In these degraded lowland areas, native species richness and abundance decline and alien species (e.g. platy, *Xiphophorus maculatus*; guppy, *Poecilia reticulata*; black mangrove cichlid, *Tilapia mariae*) become more prevalent (Pusey *et al.* 2008). The displacement of native species by alien species is a common pattern in streams subject to agricultural and urban impacts and it may be exacerbated by changed flow regimes below dams (Pusey *et al.* 1993; Gehrke *et al.* 1995; Kennard *et al.* 2005).

Inland riverine assemblages

Rivers in the arid and semi-arid areas of Australia differ from those in coastal regions in the extreme variability of their hydrologic regimes (Puckridge *et al.* 1998). For example, the flow regime of Cooper Creek in the Lake Eyre Basin includes long periods of zero flow, punctuated by seasonal in-channel flows and occasional large floods. Between floods, the system contracts to a mosaic of waterholes that may be disconnected for months and years (Hamilton *et al.*

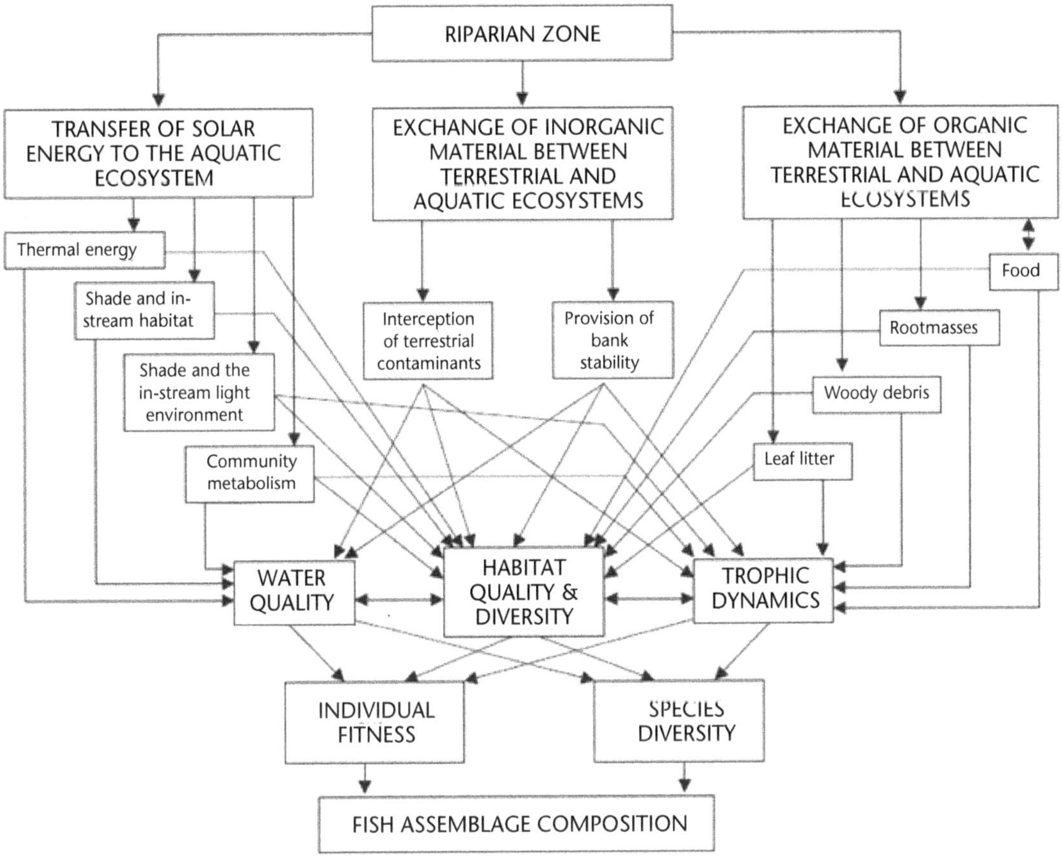

Figure 10.3: A conceptual model showing the influences of riparian processes on freshwater fish assemblages in streams (from Pusey and Arthington 2003).

2005). Species richness per water hole (up to 11 species) depends on another spatially nested habitat hierarchy – one that represents the shape and depth of waterholes, habitat diversity and the extent and type of floodplain (Arthington *et al.* 2005, 2010b). Between floods, the waterholes dry causing huge mortalities (e.g. 93% loss of individual fish in 15 isolated waterholes over one dry season in Cooper Creek: Arthington *et al.* 2005).

In hydrologically variable systems, high frequencies of disturbance favour tolerant species and habitat and trophic generalists (Poff and Allan 1995; Pusey *et al.* 2000b). Physiological tolerance, especially for dissolved oxygen, structures the fish assemblages of floodplain billabongs in the Ovens River, Victoria (McNeil and Closs 2007) and in pools in the episodic Fortescue River, north-western Australia, where only the most tolerant species (e.g. spangled perch; western rainbowfish, *Melanotaenia australis*) survive in pools (Beesley and Prince 2010).

Arid-zone fish assemblages respond strongly to floods. In Cooper Creek, for example, the fishes (except perhaps the endemic Cooper Creek catfish, *Neosiluroides cooperensis*) leave their dry-season refuges to live on inundated floodplains (Balcombe *et al.* 2007). Strong responses are consistent with the 'flood pulse concept', developed originally for tropical floodplain rivers (Junk *et al.* 1989). In tropical rivers, the annual flood pulse and associated productivity on floodplains underpin massive booms of fish biomass, in some of the world's most productive freshwater fisheries (Welcomme *et al.* 2006). In Cooper Creek, even though flooding is irregular, the opportunistic life history strategies of most fish species provide massive recruitment and booms of fish production, and the larger the area inundated, the greater the response

(Puckridge *et al.* 1998; Arthington *et al.* 2005). When the water recedes the fishes return to their waterholes, becoming food for the water birds (cormorants, darters, herons, pelicans and spoonbills) that track floods across the inland (Kingsford *et al.* 1999).

BIOTIC FACTORS

In his treatise on freshwater fish ecology, Matthews (1998) concluded that fish assemblages show evidence of both biotic and abiotic regulation, and that neither provides a complete explanation for the success of a fish assemblage in a habitat. Yet ecologists have emphasised the roles of predation, competition or both in the regulation of riverine fish assemblages and there is a continuing debate in the literature over which is more important in different circumstances (Jackson *et al.* 2001a; Grossman and Sabo 2010).

The original 'environmental filter' model (Smith and Powell 1971) has been modified to emphasise biotic interactions, especially predation (Jackson and Harvey 1989; Tonn 1990), and species traits have also been incorporated (Poff 1997). These concepts and lines of argument suggest that both historical (e.g. past competition and predation) and contemporary biotic processes contribute to the structure of freshwater fish assemblages (Oberdorff *et al.* 1997; Jackson *et al.* 2001a; Pusey *et al.* 2008). On the other hand, the riverine ecosystem synthesis suggests that stochastic factors are more important overall and throughout the riverine ecosystem (Thorp *et al.* 2008) because the type, diversity and distribution of habitat patches within functional process zones act to minimise the influences of competition and predation.

Predation

Predation may involve direct and indirect mechanisms. Size-selective predation on small individuals and species, for example, directly affect assemblage structure (Woodward and Warren 2007). Indirect top-down effects (where a species high in the food chain exerts control over species at lower levels) have been demonstrated experimentally in boreal temperate lakes (Carpenter and Kitchell 1993) and rivers (Power *et al.* 1988). These effects become apparent particularly when alien piscivores are introduced to streams and lakes. A dramatic example is the extinction of 200 species of endemic cichlid fishes in Lake Victoria, East Africa, following the deliberate introduction of a voracious piscivore, the Nile perch (*Lates niloticus*) (Goudswaard *et al.* 2008). The fate of Lake Victoria's cichlids became newsworthy at a time when one fish biologist was advocating the introduction of Nile perch to northern Australia for sport fishing; fortunately, the proposal lapsed (Barlow and Lisle 1987).

Another example of predator impact is the near-disappearance of some galaxiid fishes from streams in Australia and other southern landmasses after the introduction of alien salmonids (Lintermans 2000; McDowall 2006). Brown trout and rainbow trout (*Oncorhynchus mykiss*) were introduced to Australia more than a century ago and are responsible for declines of mountain galaxias (*Galaxias olidus*), river blackfish (*Gadopsis marmoratus*) and other native species in south-eastern Australia (Ch. 11). Although much of the evidence for direct impacts on native fishes is inferred, there are records of the demise of mountain galaxias after invasion by rainbow trout of a small stream in New South Wales (Tilzey 1976) and a recovery of mountain galaxias in Lees Creek, Australian Capital Territory, after eradication of rainbow trout (Lintermans 2000).

The introduction of redfin perch to Tasmania in 1862, and later to mainland Australia, also warrants mention. In Tasmania, the southern pygmy perch (*Nannoperca australis*) may be threatened by this species (Humphries 1995) and redfin perch has been implicated also in the decline of native fishes in Western Australia (Hutchison 1991; Morgan *et al.* 2002). The aggressive alien eastern gambusia, now ubiquitous in southern Australia, has also had major effects. In Western Australia, its impact is most evident in the shallow coastal lake habitats of the Swan Coastal Plain; coexistence with native species is possible during winter in streams with elevated discharge, where native species are more able to tolerate strong currents and breed successfully (Pusey *et al.* 1989). Common carp (*Cyprinus carpio*) is implicated in major declines in fish species richness in most of the Murray-Darling Basin (Schiller *et al.* 1997; Davies *et al.* 2010), especially in reaches with altered flow regimes (Gehrke *et al.* 1995, 1999). The effects of alien species are described further in Chapter 11.

Tropical food webs include relatively few large consumers that control energy flow (Douglas *et al.* 2005; Winemiller 2007). Piscivory, but not necessarily obligate piscivory, is typical of many northern Australian freshwater fishes (Ch. 6). For example, about a quarter of all fish species in a survey of three diverse northern rivers are reported to have diets in which fish contribute more than 10% (Pusey *et al.* 2004; Bishop *et al.* 2001). Predatory species appear to have most effect during base-flow conditions in the dry season, when predators and prey are concentrated by falling water levels and resources are in short supply. Similar patterns occur in floodplain rivers elsewhere (Welcomme *et al.* 2006).

What happens when predators are removed from aquatic ecosystems? 'Fishing down' the food web (Pauly *et al.* 1998) has become a global phenomenon in the oceans and in many river systems. It refers to the removal of large fishes at the top of the food web, until only small individuals and species remain. The ecological implications include effects on carbon and nutrient cycling, alterations to food web structure, losses of biodiversity and ultimately the endangerment of big fish species (Taylor *et al.* 2006; Winemiller 2007). In the southern Murray-Darling Basin the abundance of Murray cod (*Maccullochella peelii*), a large predator, has declined drastically since European settlement (Humphries and Winemiller 2009) and the species is listed as 'vulnerable' under the *Environment Protection and Biodiversity Conservation Act 1999* (Lintermans 2007). Other large native species similarly are threatened (Ch. 12).

Competition

The literature on competition and its effects is vast and often controversial. Matthews (1998) listed several lines of evidence to indicate competition among fishes. Many studies of resource partitioning are claimed as evidence of competition for shared resources as habitat, food or both (Ross 1986; Grossman *et al.* 1998). Yet when species forage for the same resources in the same habitat they are not necessarily competing, unless the resources are in short supply, and even then their numbers may be below the habitat's carrying capacity and not sufficient to cause competition (Matthews 1998). Many of the studies that invoke competition fail to record information about resource availability and other environmental conditions (Matthews 1998; Arthington and Marshall 1999). Although studies often invoke competition, predation can also influence patterns of habitat selection, resource use and diet composition in fishes (Balcombe and Closs 2004).

Harsh–benign habitat model

Streams and rivers are characterised by longitudinal and lateral habitat heterogeneity and by patterns of natural disturbance, particularly those related to stream-flow variability in space and time (Poff *et al.* 1997; Grossman and Sabo 2010). Fish assemblages in harsh, unpredictable environments (e.g. upland reaches) are viewed as products of colonisation and extinctions, whereas those in more benign, predictable environments (e.g. lowland reaches) are regulated by biotic interactions (Zalewski and Naiman 1985; Schlosser and Angermeier 1995). Under benign conditions, populations can grow until resources become limiting, and survival therefore is related to the ability to obtain resources (e.g. to compete for food) and/or avoid becoming food for predators (Beesley and Prince 2010). This is referred to as the 'harsh–benign habitat model'.

Biotic interactions are likely to be more important in downstream reaches, not least because greater species richness and higher abundances mean more interactions. There are often longitudinal progressions in fish trophic guilds, from generalised invertebrate-eaters in upstream areas to omnivores, detritivores, herbivores and piscivores in downstream areas (Romanuk *et al.* 2006; McGarvey and Hughes 2008). Similar trends are reported for Australian rivers (Gehrke and Harris 2000; Rayner *et al.* 2008, 2009b; Pusey *et al.* 2010; Ch. 6). In general, there is a relationship between water depth and fish body length (the bigger fish – deeper habitat pattern: Schlosser 1987a), so that more large fish species tend to live downstream than upstream. The river blackfish exhibits such a relationship in Armstrong Creek, Victoria, as its mode of habitat use changes with growth (Koehn *et al.* 1994). Fish living in structurally complex microhabitats lessen the risk of predation (Crowder and Cooper 1982). In billabongs of the Murray

River, small carp gudgeons (*Hypseleotris* spp.) and other species forage at night and take cover in littoral sedges during the day, avoiding predators such as redfin perch (Balcombe and Closs 2004).

The spatial scale of investigations influences our perceptions of biotic and abiotic factors as drivers of fish assemblage structure (Jackson *et al.* 2001a). From that perspective, small-scale studies typically indicate a greater role for competition whereas large-scale studies emphasise abiotic controls. This seems to be a consequence of the fact that experimental research on biotic interactions typically is small-scale (e.g. behavioural observations, fish tracking, removal plots, exclusion cages, introduction of predators). This is the scale at which individual fish interact and the scale that humans can readily observe or measure fish interactions (Wiley *et al.* 1997; Durance *et al.* 2006). Large-scale studies seldom incorporate experimental measurements of biotic processes (cf. Kennard 1995), but much can be learned from empirical studies in rivers with distinctive patterns of hydrologic variability and habitat disturbance.

The dynamics of fish assemblages in lowland reaches of the Mulgrave River, a small coastal river in the Wet Tropics, Queensland, provide a useful spatial–temporal model of abiotic–biotic processes (Fig. 10.4). In the dry season, the lowermost reach of the Mulgrave River is a single 'functional process zone' of repeated habitat patches with similar abiotic characteristics. Overlaying these patches are longitudinal gradients of habitat change, such as increasing water depth and conductivity (salinity), reflecting tidal penetration under low-flow conditions (Rayner *et al.* 2008). The interaction of gradients and habitat patches produces different conditions for individual species, hence spatial variations in assemblage structure, with eastern rainbowfish, freshwater catfish (*Tandanus tandanus*) and sooty grunter (*Hephaestus fuliginosus*) more common at upstream sites and barramundi (*Lates calcarifer*), sailfin glassfish (*Ambassis agrammus*), speckled goby (*Redigobius bikolanus*) and mangrove jack (*Lutjanus argentimaculatus*) dominating numerical abundance and biomass at downstream sites. These patterns correspond to a combination of the 'hierarchical patch dynamics' model (Townsend 1989; Wu and Loucks 1995b) and the river continuum concept (Vannote

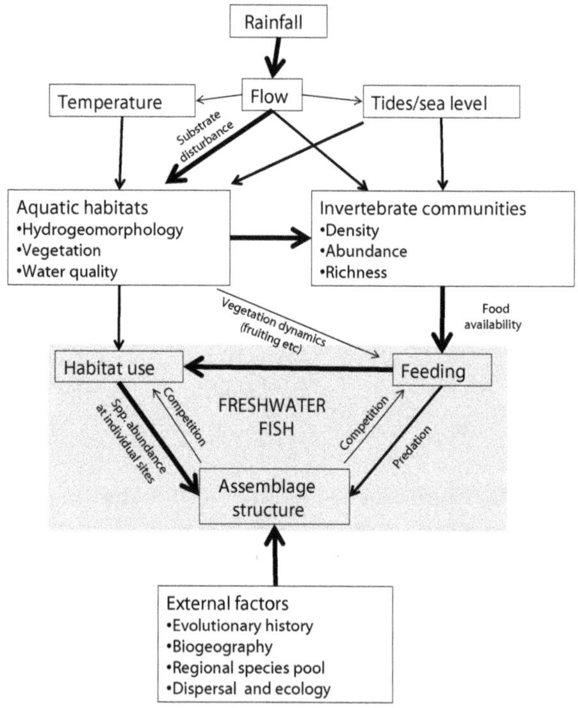

Figure 10.4: A conceptual model of influences on freshwater fish assemblages in the lower Mulgrave River, Queensland. Large arrows indicate key processes (from Rayner *et al.* 2009b).

et al. 1980). During the wet season in the Wet Tropics, pulses of fast-moving, turbulent water erode substrates and river banks and scour aquatic vegetation so that the habitat patches of the dry season are reduced to relatively deep water over areas of bare sand, lacking cover for fishes (Rayner *et al.* 2008). After the flood peak, habitat structure and fish assemblages in the main channel of the Mulgrave River enter a transition period, when the diversity of habitat patches increases with the growth of vegetation patches on mid-channel bars and there are other changes associated with falling water levels. At this time, fish quickly occupy the new habitat patches and biotic interactions, especially predation, regulate their recolonisation of sites and tributaries (Rayner *et al.* 2008). This modified 'flood pulse' model envisages a shift from stochastic processes during floods to deterministic, biotic interactions during dry periods. Predation (piscivory) is most significant during the wet–dry season transition and competition is less significant because in this assemblage prey selection is constrained by body and gut morphology, mouth size relative to body size and the ability to assimilate different food types. These constraints appear to reduce competition for food (Pusey *et al.* 2000b; Rayner *et al.* 2009b).

The idea that fish assemblages in harsh environments are governed by abiotic factors rather than biotic interactions has support from studies of arid-zone rivers. In Cooper Creek, as water levels decline in dry periods, there are predictable losses of fishes in response to water depth, temperature and habitat loss (Arthington *et al.* 2010b; Sheldon *et al.* 2010). Assemblages in individual waterholes are often reduced to half the original number of species, and to 10% by the end of the dry season (Arthington *et al.* 2005). In other rivers biotic processes, especially predation, contribute most when water is receding from the floodplain. This may be true of fish in Cooper Creek, where most piscivory occurs when fish are on the floodplain rather than in waterholes (Balcombe *et al.* 2005). As floods ebb, fish may be most affected by resource competition when the benthic algal band (the bathtub ring) that sustains waterhole food webs (Bunn *et al.* 2006a) is not fully developed, as fish numbers decrease in waterholes with low primary productivity (Fellows *et al.* 2009).

Disease agents and parasites may infect fishes confined to drying water bodies and rivers affected by drought or other forms of stress. In the Coongie Lakes region of Cooper Creek, several species (bony herring, eastern rainbowfish, golden perch and the alien goldfish) are regularly affected by dermatitis caused by the fungus *Achlya* (Puckridge and Drewien 1988). The Barcoo grunter (*Scortum barcoo*) is also susceptible to fungal infections under crowded conditions in drying waterholes (Balcombe and Arthington 2009). In the lower Murray River, winter stress and spring spawning stresses make bony herring prone to bacterial and fungal infections (e.g. *Aeromonas*, *Saprolegnia*) that cause skin lesions, scale loss, bleeding and muscle degeneration (Puckridge *et al.* 1989).

METACOMMUNITIES

Fish assemblages vary in richness and composition among coastal and inland lakes and mound springs, and along longitudinal and lateral floodplain river gradients. They also vary in time, in response to discharge and/or groundwater variations, chemical conditions and seasonal factors like temperature. Local habitat patches may be isolated or connected, depending on flow conditions. Hydrological connectivity provides opportunities for fishes to move from one patch to another in search of better habitats, food, shelter and breeding opportunities (Ch. 5). Dispersal ability underlies these movement patterns. Some species are highly vagile, such as the spangled perch, which takes virtually any opportunity to move, even along shallow wheel ruts, and is the most widespread of all native Australian species (Pusey *et al.* 2004). These movements may bring breeding opportunities and other rewards, or they may end in disaster if the new patch dries completely, as do many associated with dryland rivers. As species come and go in space and time, the membership of local assemblages changes such that the full suite of species persists as a composite entity only across space and time. These composites are termed 'metacommunities'.

Metacommunities are spatially distributed networks of local assemblages in which dispersal between assemblages and interactions within assemblages affect the persistence and turnover of species (Leibold *et al.* 2004), hence assemblage structure in space and time. In a metacommunity, local species richness and assemblage structure represent a balance between colonisation (from regional source habitats) and extinctions, depending on local habitat factors and competition, predation and disease. Four kinds of conceptual models describe metacommunity processes (Leibold *et al.* 2004). *Patch dynamics models* describe species composition in multiple patches, such as terrestrial islands (or 'islands' of water in a 'sea' of land), and emphasise trade-offs between colonisation and competition. *Species-sorting models* concern the responses of individual species to environmental heterogeneity, in that certain local conditions favour some species and not others, enhancing species coexistence within assemblages. *Source–sink* or *mass effects models* present a framework in which dispersal alters the species composition in heterogeneous assemblages. Finally, the *neutral model* regards species as essentially equivalent in their competitive and dispersal abilities, so that local and regional composition and abundance are determined mainly by stochastic, demographic processes and limits to dispersal.

The biota of rivers that become isolated in pools during dry periods are said to function as metacommunities, with the composition of assemblages being a function of upstream, downstream and lateral dispersal (Larned *et al.* 2010). The metacommunity concept also applies in dryland rivers with extensive floodplains and many temporarily isolated habitat patches (waterholes) that join during occasional large floods (Sheldon *et al.* 2010). The choice of model is significant. For example, if mass dispersal is the most influential process, we must be aware of human activities that constrain dispersal (e.g. weirs, dams, levees). If habitat heterogeneity is important, we must manage processes that alter the diversity of habitat patches. If competitive interactions and/or predation are significant, introduction of an invasive alien piscivore may be disastrous. Clearly, the relative strengths of different metacommunity processes and their effects on spatially fragmented assemblages will influence the ways that managers think about, and implement, river restoration (Jansson *et al.* 2007; Pires *et al.* 2010). The idea that elements of all four metacommunity models could apply to assemblages in various combinations at different times, depending on flow variability (Falke and Fausch 2010), offers a lot of scope for new research.

CONCLUSION

Fish assemblages in Australian fresh waters are not haphazard associations of species occurring in one place at one time. On the contrary, they are often highly structured by historic events, abiotic and biotic factors and ecological processes. Assemblages in coastal and inland rivers are structured in ways that are consistent with several popular concepts and models including the river continuum concept, the flood pulse concept and the natural flow paradigm. Although biotic processes have received less attention than abiotic variables, there appears to be more piscivory (and possibly competition) with increasing distance downstream in rivers and in drying waterbodies (streams, floodplain waterholes, lakes) where habitat and food resources decrease over time. Stochastic, abiotic factors tend to dominate during wet periods and on inundated floodplains. Ultimately, we may need to refine our models of metacommunity structure and adopt a more experimental (rather than observational) approach to determine the relative roles of abiotic and biotic processes, especially in the fragmented habitat mosaics typical of Australia's dryland rivers, lakes and mound springs.

Knowing that much spatial and temporal variation in assemblage structure is correlated with abiotic factors, is this sufficient to inform community/ecosystem restoration and the conservation of iconic and endangered species? We suggest that ecological studies of fish assemblages have much to offer when managers ask how a river's flow regime, aquatic habitat or riparian corridor should be restored to recover and maintain native biodiversity and ecosystem resilience. We know enough of abiotic processes to predict the likely outcomes of physical restoration efforts in many rivers, but biotic factors may ultimately determine the success of these projects, especially where native species have declined and alien piscivores and competitors are established (Ch. 11). To succeed, the restoration of ecosystem function must account for critical biotic components, some of which may have been lost and require reintroduction, if possible, whereas alien species may have to be eliminated.

With their multiple roles in assemblages, communities and ecosystems, fishes may be among the first, and the last, members to benefit from catchment management, environmental flows and habitat restoration. A better appreciation of the abiotic and biotic processes affecting fish assemblages would undoubtedly promote the protection and restoration of freshwater ecosystems and inform climate-change responses. We need to learn more about what controls 'who is where' in freshwater fish assemblages to manage our impacts and ensure that none of our native fishes becomes extinct as a result of human activities.

11 Fishes from elsewhere

John H. Harris

INTRODUCTION

If European colonists could return today, they would discover that Australian rivers now harbour many familiar fishes to supplement their rations. Rather than antipodean natives, they would find common carp (*Cyprinus carpio*), brown trout (*Salmo trutta*), rainbow trout (*Oncorhynchus mykiss*), goldfish (*Carassius auratus*) and redfin perch (*Perca fluviatilis*). They might question, now as then, the relative values of native fishes and those introduced from the Old World. Were these wise imports? Were some misguided choices, and what were the consequences of their introduction? Australian rivers have been transformed in the past two centuries, and the alien fishes are both agents and symptoms of change. The title for this chapter could apply to native species introduced outside their natural ranges or to species from other parts of the world. It could also refer to different genetic stocks of a native species introduced to another part of a drainage system, beyond a waterfall or other barrier.

In 200 years, technologies, trade and transportation have breached barriers that once isolated the freshwater biotas of catchments and continents. The study of biological invasions has become a theme in ecology, responding to trends towards globalisation and biological homogenisation of floras and faunas (Vermeij 1996; Strayer *et al.* 2006; Davis 2009). Extinctions and declining genetic diversity are associated threats (Moyle and Light 1996; Rahel 2002). Among the least-controlled and least-reversible of these changes are introductions of non-indigenous species (Strayer 2010).

This chapter considers the nature of freshwater fishes brought to Australia from elsewhere, how they have become established in the wild, and the effects they have had on native fishes and riverine ecosystems. It also considers the nature and effects of invasive species, exemplified by alien and translocated fishes, the pest status of several alien fishes and the positive and negative values attributed to some of these species.

A GLOBAL ISSUE IN AUSTRALIA

Global biodiversity is in decline (Butchardt 2010) and alien fishes are agents of change because they reduce native fish populations, degrade habitats, compromise gene pools and introduce diseases and parasites (Corfield *et al.* 2008). In North America, there have been 314 transplants of native fishes and 116 established alien species (Strayer 2010). In Europe, there have been 58 transplants and 95 alien species (Butchardt

2010). Alien fishes also dominate aquaculture and fishery enhancements (Davis 2009; Gozlan *et al.* 2010). For example, the production of African tilapias (Cichlidae) is seven times higher in Asia than in Africa; in Chile, non-indigenous salmon species (Salmonidae) provide 30% of world salmon production.

In Australia, alien fishes similarly threaten native biodiversity (Harris and Silveira 1999; Georges and Cottingham 2002; Olden *et al.* 2008). Of 73 feral vertebrate pests (Bomford and Glover 2004), 43 are freshwater fish and 34 continue to spread (Koehn and McDowall 2004; Lintermans 2004). The International Union for Conservation of Nature has nominated five of these species (common carp; tilapia, *Oreochromis mossambicus*; eastern gambusia, *Gambusia holbrooki*; rainbow trout; brown trout) among 100 of the world's most invasive species (Lowe *et al.* 2000). Another 30 small aquarium species occur in the wild, and many have potential to become serious pests (Corfield *et al.* 2008; Ch. 12). Virtually all regions of Australia now support fishes that originated elsewhere.

No part of Australia demonstrates the extent of alien fish invasions better than the Murray-Darling Basin, where about one-quarter of the fish fauna is non-indigenous species (Lintermans 2007). Fish communities are a foundation for river health assessments in the Murray-Darling Basin Authority's Sustainable Rivers Audit (Davies *et al.* 2010). In the first audit (2004–07), sampling yielded 38 fish species, 10 of which were alien (43% of total numbers; 68% of total biomass). Common carp, eastern gambusia and goldfish were present throughout the basin, and common carp was overwhelmingly dominant (87% of alien fish biomass; 58% of total fish biomass). In a survey of New South Wales rivers, Harris and Gehrke (1997) concluded that native fishes were in severe decline as a result of habitat degradation through sedimentation, river regulation, cold-water pollution, migration barriers and the effects of alien species (Gilligan 2005; Gilligan and Rayner 2007). The Murray-Darling Basin Native Fish Strategy (Murray-Darling Basin Commission 2004) includes controlling alien fishes among six 'driving' actions.

Assessments of the impacts of alien fishes in Australia need to be based on their ecology, their potential for spread and likely interactions with native biota and habitats. Even native fishes introduced outside their natural ranges may have effects like those of alien species (Phillips 2003). Further, the threat of more introductions today is by no means diminished (Ch. 12). Predicting the impacts from alien species and translocated native species is a major challenge for science and management. In the absence of a strong conceptual foundation for identification and management of potentially harmful species and their impacts, the need to implement the precautionary principle is paramount (Leprieur *et al.* 2009). Risk assessment procedures have a key role to play (Bomford and Glover 2004; Corfield *et al.* 2008; Keller *et al.* 2009).

A CHRONICLE OF INTRODUCTIONS

Non-indigenous fishes may originate on other continents, in other countries and in river basins in other parts of Australia. Although these origins have different implications for ecology and management, there is no consensus over categories and terms (Davis 2009; Gozlan *et al.* 2010). In this chapter, non-indigenous species and genetic stocks (a line having specific genetic traits) are classified as *alien*, *exotic* or *translocated*. These are defined in Table 11.1.

Figure 11.1 shows a short history of introductions. They began in the mid 1800s; there was little change in numbers of species in the first half of the 20th century, then a steep increase from 1960 (Lintermans

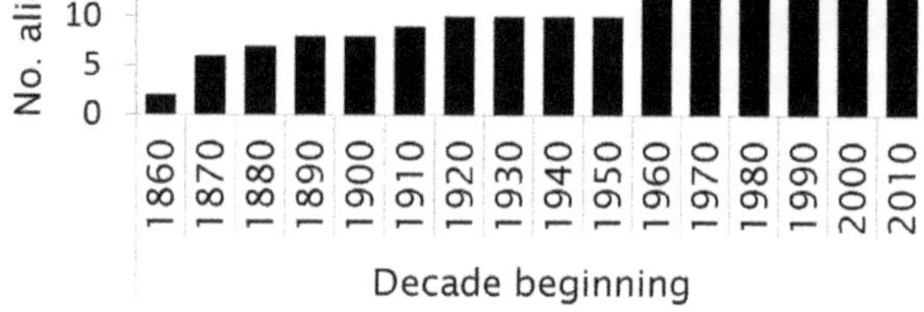

Figure 11.1: Numbers of alien fish species in the wild in Australian fresh waters at the beginning of each decade from 1860 (modified from Lintermans 2004).

Table 11.1: Classes (species and genetic stocks) of non-indigenous fishes in Australia

	Definition	Relevance	Examples
Exotic	Species that do not exist in the wild	Species held in captivity, potential alien species	Hundreds of ornamental species held in aquaria; many more potential imports through aquaculture, goods transport etc.
Alien	Species imported and now established	Previously exotic species now established in the wild. Implicated in biotic interactions, habitat changes	Currently 43 species (e.g. cichlids, cyprinids, poeciliids, salmonids)
Translocated	Native species introduced into areas outside their natural ranges	Potential for displacement of species and other biotic interactions and loss of genetic diversity through interbreeding. Impacts like those of alien species	Propagated native angling species, introduced forage species, species distributed in interbasin water transfers, hatchery contamination, relocations for conservation

2004; Ayres and Clunie 2010a). Three salmonids, four cyprinids and a percid were established before 1900; eastern gambusia was introduced before (and again during) World War II and numbers remained static for another 40 years. In two centuries, exotic fishes have established alien populations mainly in southern Australia but, from the late 20th century, the balance moved northward, with a surge in arrivals of tropical and subtropical aquarium species (Ch. 12).

ALIEN SPECIES

Most attention devoted to non-indigenous fishes has concerned alien species (Table 11.2). The exact number is uncertain as survey data are scarce and unevenly spread, some species may have died out and, in any case, numbers continue to change. Of 43 species listed in the table, there are 18 cichlids, eight cyprinids, six poeciliids and five salmonids, and the remaining families have only one or two representatives. There are no naturally occurring Australian members of these families, other than in the Gobiidae. All have crossed Wallace's Line, a biogeographic boundary in the Indonesian Archipelago that separates the Australian biota from that of the Old World (Van Oosterzee 1997).

Nearly 50 years ago, Alan Weatherley and John Lake (Weatherley and Lake 1967) observed wryly that 'amazingly little attention has been paid to the general biology of the unique fish fauna of (Australia's) rivers, estuaries, lakes and ponds. From the first, anglers and acclimatisationists were primarily concerned to establish here numerous fish species already familiar to them in the Northern Hemisphere'. Even in the late 1970s, one state fisheries department head was striving to introduce more salmonids, arguing that Australian species had limited value for recreational fisheries.

Salmonids

The first introductions of salmonids (and other alien species) were driven by zealous acclimatisation societies (Brinsley 2010), some of which persist today, albeit under government regulation. The colonists wanted to import familiar animals and plants from Europe and to enhance food production and recreational opportunities. With strong public support in Tasmania and Victoria, and a Salmon Commission, James Youl and others struggled to introduce salmonids. From 1852, ice and fresh water were used to cool and aerate fertilised eggs of brown trout (see Box 11.1) and Atlantic salmon (*Salmo salar*) during the three-month journey by sailing ship from England. After many failures, Youl applied new research from America (Halverson 2010), delaying ova development by packing them in layers of ice and moss (Roughley 1951). He succeeded in liberating trout and salmon near Hobart in 1864, and was knighted for his achievement.

Nearly 500 000 eggs of Atlantic salmon were imported in these Tasmanian endeavours and 3000 larvae were released (Weatherley and Lake 1967) but the species has not become established in the wild, possibly owing to insufficient propagule pressure (see 'Establishment') or unfavourable maritime conditions.

Box 11.1: Profile of brown trout, *Salmo trutta*

Alias: None.

Origin: Europe.

Introduced: Tasmania, 1864.

Distribution: Coastal south-eastern Australia, Murray-Darling Basin, Tasmania, south-western Australia. Regions: NEN, SEN, NTAS, STAS, SWWA, SWV, SEV, MDB, SAG (Fig. 2.2).

Potential range: Occupied in Australia, may decline with climate change.

Movement: Some Tasmanian fish are anadromous.

Size: <900 mm, <14 kg.

Diet: Predator on insects, crustaceans, fishes.

Habitat: Cool water (< 23–25°C). Favours higher altitude streams (e.g. > 600 m) on mainland, but down to sea level in Tasmania.

Reproduction: Spawns over gravel in flowing streams. Natural spawning augmented by stocking.

Impacts: Implicated in decline of some galaxiids through predation and competition. Habitat and/or dietary niche overlaps with upland native fishes. May cause trophic cascades (see 'Food webs').

Value: Premier angling and aquaculture species, popular table fish. Economically important.

Control: None.

Select references: Roughley (1951); Weatherley and Lake (1967); Scott and Crossman (1973); Tilzey (1976, 1977a); Jackson and Williams (1980); Crowl *et al.* (1992); Cadwallader (1996); Townsend (1996); Dept of Primary Industries, New South Wales (2006); McDowall (2006); Brinsley (2010).

From McDowall (1996b).

Nevertheless, some hatchery-bred salmon do escape from farms in Tasmania and the species could yet become established. In 1963, Atlantic salmon were reared in New South Wales from Canadian eggs and now supply regional aquaculture stocks (Brinsley 2010).

Eggs of rainbow trout were first imported to Australia from California through New Zealand in 1894 (Roughley 1951; Brinsley 2010). The species continues to support aquaculture and fisheries in Australia and around the world, wherever there are supplies of cool, high-quality water (Halverson 2010). Industry and government hatcheries routinely provide fish for the table, and young fish to stock upland streams and impoundments in cooler parts of southern Australia.

Common carp

Three strains of common carp are established in Australia (Shearer and Mulley 1978; Davis *et al.* 1999; Haynes *et al.* 2009). The Prospect strain was the first alien fish species imported to Australia, established near Sydney in 1850–60. It was followed by the Yanco strain of domesticated koi carp introduced to the Murrumbidgee Irrigation Area in 1876 (McDowall 1996b). Soon afterward, common carp and tench (*Tinca tinca*) were imported to Tasmania (Roughley 1951) and acclimatisation societies distributed them in Victoria. Ornamental koi carp continue to invade many new areas (Davis *et al.* 1999; Haynes *et al.* 2009). In 1961, an aquaculturist in south-eastern Victoria illegally imported the Boolarra strain of common carp from Europe; these escaped into the Murray-Darling Basin, with disastrous consequences despite attempts at eradication (see Box 11.2).

Redfin perch

Redfin perch was the second species introduced, with 11 fish liberated in Tasmanian streams in 1862 (Roughley 1951). This species tolerates temperatures up to 31°C (Weatherley and Lake 1967), enabling ready transport by sailing ship from England (Roughley 1951). Soon afterward, acclimatisation societies released redfin perch into waters on the mainland (see Box 11.3).

Eastern gambusia

Eastern gambusia was introduced to Australia as an aquarium species and later ostensibly to control

Box 11.2: Profile of common carp, *Cyprinus carpio*

Alias: Carp, European carp, mirror carp, koi carp.

Origin: Asia.

Introduced: Ornamental strains introduced to New South Wales in 1850s, 1878. Spread rapidly after Boolarra strain introduced in 1961, plus ornamental koi strain.

Distribution: Widespread in southern Australia. Successful, vagile invader. Prefers lentic habitats, especially well-vegetated, fresh–brackish sites. Regions: MDB, NEN, SEQ, SEN, SEV, SWV, NTAS, STAS, SAG, TORR, SWWA (Ch. 2, Fig. 2.2).

Potential range: Not yet occupied.

Movement: Potamodromous spring spawning movement to vegetated wetlands.

Size: <800 mm, <10 kg.

Diet: Omnivore/detritivore, benthivorous, plastic diet.

Habitat: Broad tolerances including poor water quality.

Reproduction: Highly fecund; spawns in rivers and floodplain wetlands.

Impacts: Affects water quality and vegetation; promotes algal blooms (see 'Food webs'); hybridises with goldfish.

Value: Declared noxious species. Major threat to biodiversity. Commercial fishery in SEV. Ornamental koi valued by aquarists.

Control: Some progress towards integrated control and recombinant genetic techniques for eradication. Various traps and exclusion devices.

Select references: Roughley (1951); Shearer and Mulley (1978); Hume *et al.* (1983); Fletcher *et al.* (1985); Roberts *et al.* (1995); Roberts and Tilzey (1996); Harris and Gehrke (1997); King *et al.* (1997); Roberts and Ebner (1997); Robertson *et al.* (1997); Davis *et al.* (1999); Koehn *et al.* (2000); Khan *et al.* (2003); King *et al.* (2003a); Brown and Walker (2004); Koehn (2004); Brown *et al.* (2005); Driver *et al.* (2005a, b); Smith (2005); Ansell and Jackson (2007); Gilligan and Rayner (2007); Stuart-Smith *et al.* (2008); Thresher (2008); Bax and Thresher (2009); Haynes *et al.* (2009); Ingram (2009); Jones and Stuart (2009); Froese and Pauly (2010); Gehrke *et al.* (2010).

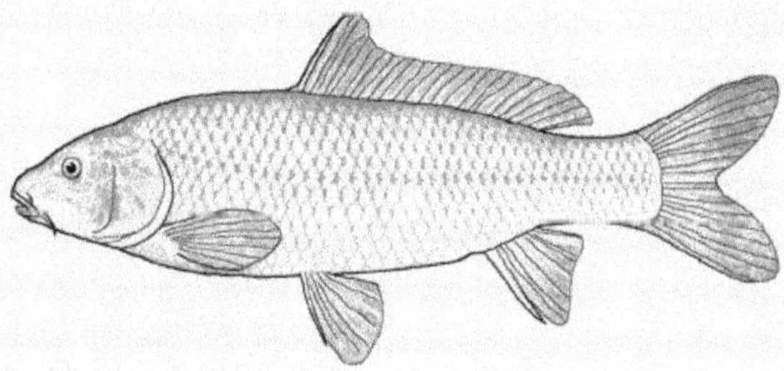

From McDowall (1996b).

malarial mosquitoes, although malaria is a problem only occasionally in some tropical northern areas. The name 'mosquitofish' has facilitated its dispersal and it is now ubiquitous in many regions, with the unfortunate implication that it is a specialised mosquito-predator, when it is no more effective than many native species (see Box 11.4).

Tilapia

The name 'tilapia' applies loosely to many members of the cichlid subfamily Tilapiinae but in Australia the species of most concern is *Oreochromis mossambicus*, which exists as invading populations in Queensland and Western Australia. The species is implicated in major declines in native fish populations and it continues to spread, potentially to include the Murray-Darling Basin where it could rival common carp as a dominant species (see Box 11.5).

Aquarium trade

After the 1960s, a booming aquarium industry, fuelled by air transport and international trade, brought a surge in exotic and alien fishes (McKay 1977; Arthington and McKenzie 1997; Lintermans 2004; Fig. 11.1). The species included poeciliids, cichlids and a cobitid, the oriental weatherloach (*Misgurnus anguillicaudatus*) (Table 11.2). At least 1181 exotic freshwater species have been imported for the aquarium trade (McNee 2002); 33 of these species have escaped into waterways and now dominate all alien species, being 77% of the total (Arthington and McKenzie 1997; Ansell and Jackson 2007). Fishes that have escaped

Table 11.2: Alien freshwater fish species in Australia, showing distribution and introduction vectors

Family	Common name	Species	Origin	Distribution[1]	Vectors
Cichlidae	Black mangrove cichlid	*Tilapia mariae*	W Africa	SEV, NEQ, SEN	Aquarium
	Blue acara	*Aequidens pulcher*		SEV, SEQ	Aquarium
	Convict cichlid	*Archocentrus nigrofasciatus*	Cen America	SEV, NEQ	Aquarium
	Firemouth	*Thorichthyes meeki*	Cen America	NEQ	Aquarium
	Green severum	*Heros severus*		NEQ	Aquarium
	Green terror	*Aequidens rivulatus*		NEQ	Aquarium
	Hybrid cichlid	*Labeotropheus* × *Pseudotropheus*?	?	SEV	Aquarium
	Jack Dempsey cichlid	*Cichlasoma octofasciatum*	Cen America	NEN, Vic, Qld	Aquarium
	Jewel cichlid	*Hemichromis bimaculatus*		ARNH, Qld	Aquarium
	Midas cichlid	*Amphilophus citrinellus*		NEQ	Aquarium
	Oscar	*Astronotus ocellatus*		NEQ	Aquarium
	Pearl cichlid	*Geophagus braziliensis*	S America	SWWA, NEN, FITZ	Aquarium
	Red devil	*Amphilophus labiatus*		NEQ, SEQ, SEV	Aquarium
	Redbelly tilapia	*Tilapia zillii*	Africa, Eurasia	PILB	Aquarium
	Redhead cichlid	*Cichlasoma synspilum*		Qld	Aquarium
	Three-spot cichlid	*Cichlasoma trimaculatum*		SEQ	Aquarium
	Tilapia	*Oreochromis mossambicus*	E Africa	SEV, SEQ, NEQ, BURD, FITZ, PILB	Aquarium, stocking
	Victoria Burton's Haplochromis	*Haplochromis burtoni*		SEQ, NEQ	Aquarium
Cobitidae	Oriental weatherloach	*Misgurnus anguillicaudatus*	Asia	MDB, SEN, SEV, SEQ, SAG, NEN, SEQ	Aquarium, bait-bucket
Cyprinidae	Common carp	*Cyprinus carpio*	Cen Asia	MDB, NEN, SEQ, SEN, SEV, SWV, NTAS, STAS, SAG, TORR, SWWA	Aquarium, stocking, aquaculture, bait-bucket
	Crucian carp	*Carassius carassius*	Europe	MDB	Aquarium
	Goldfish	*Carassius auratus*	Asia	MDB, NEN, SEQ, SEN, SEV, SWV, NTAS, STAS, SAG, TORR, SWWA, LEB	Aquarium, bait-bucket
	Roach	*Rutilus rutilus*	Europe	MDB, SWV, NEN	Stocking
	Rosy barb	*Puntius conchonius*	India	SEQ	Aquarium

Family	Common name	Species	Origin	Distribution[1]	Vectors
	Sumatra barb	*Puntius tetrazona*	Asia	Qld	Aquarium
	Tench	*Tinca tinca*	Europe	MDB, NTAS, STAS, SEV, SAG	Aquarium
	Whitecloud mountain minnow	*Tanichthys albonubes*	China	NEN, SEQ	Aquarium
Cyprinodontidae	American flagfish	*Jordanella floridae*	N America	Qld	Aquarium
Gobiidae	Yellowfin goby	*Acanthogobius flavimanus*	Asia	SEN, NEN, SWV	Ballast
	Streaked goby	*Acentrogobius pflaumii*		SEN, SWV	Ballast
Osphronemidae	Threespot gourami	*Trichogaster trichopterus*	Asia	BURD	Aquarium
Percidae	Redfin perch	*Perca fluviatilis*	Eurasia	MDB, TAS, SWWA, SEN, SWV, SEV, SAG	Stocking, bait-bucket
Poeciliidae	Eastern gambusia	*Gambusia holbrooki*	Cen, S America	NEQ, BURD, FITZ, SEQ, SECYP, SGC, ARNH, WKIM, LEB, BULL, PILB, WP, SWWA, NEN, SEN, MDB, SEV, SWV, SAG, TORR, NTAS	Pest control, ornamental
	Guppy	*Poecilia reticulata*	S America	NEN, SEQ, FITZ, BURD, NEQ, SECYP, ARNH, PILB	Ornamental
	One-spot livebearer	*Phalloceros caudimaculatus*	S America	SWWA, SEN, LEB	Ornamental
	Platy	*Xiphophorus maculatus*	Cen America	SEQ, NEQ, FITZ, BURD, ARNH	Ornamental
	Sailfin molly	*Poecilia latipinna*	Cen America	SEQ, NEQ, ARNH	Ornamental
	Green swordtail	*Xiphophorus hellerii*	Cen America	NEN, SEQ, NEQ, FITZ, SECYP, ARNH, PILB, SWWA, LEB	Ornamental
Salmonidae	Atlantic salmon	*Salmo salar*	Europe, N America	SEN	Stocking
	Brook trout	*Salvelinus fontinalis*	N America	SEN, TAS	Stocking
	Brown trout	*Salmo trutta*	Europe	NEN, SEN, TAS, SWWA, SWV, SEV, MDB, SAG	Stocking
	Chinook salmon	*Oncorhynchus tshawytscha*	N America, NE Asia	SWV	Stocking
	Rainbow trout	*Oncorhynchus mykiss*	N America, Siberia	NEN, SEN, TAS, SWWA, SWV, SEV, MDB, SAG	Stocking

1 See drainage region codes in Figure 2.2.

Sources: Allen *et al.* (2002); DPI (2001); Lintermans (2004); Corfield *et al.* (2008); Rowe *et al.* (2008); Ayres and Clunie (2010a); Froese and Pauly (2010); Industry and Development, New South Wales (2010a).

Box 11.3: Profile of redfin perch, *Perca fluviatilis*

Alias: English perch, Eurasian perch, redfin.

Origin: Eurasia.

Introduced: Tasmania from England, 1862.

Distribution: Widespread in cooler areas of southern Australia. Northern distribution limited to < 31°C. Prefers lentic waters, especially with abundant vegetation. Regions: MDB, TAS, SWWA, SEN, SWV, SEV, SAG (Ch. 2, Fig. 2.2).

Potential range: May not have reached southern limit, may contract southward with climate change; range expansion by deliberate introductions, hatchery contamination and bait-bucket transfers.

Movement: Not known to move large distances.

Size: < 400–450 mm, < 1–2 kg.

Diet: Predator on fishes, crustaceans, molluscs and insects.

Habitat: Tolerates poor water quality.

Reproduction: High fecundity, capacity to spawn when small can lead to large populations of stunted individuals.

Impacts: Threat to aquatic biodiversity; vector of epizootic haemopoietic necrosis virus (pathogenic to some native fishes).

Value: Declared noxious in some states. Popular recreational species and table fish.

Control: None attempted.

Select references: Roughley (1951); Weatherley *et al.* (1967); Baxter *et al.* (1985); Langdon (1989, 1990); Morgan *et al.* (2002); Ludgate and Closs (2003); Simon and Townsend (2003); Wilson (2005); Ansell and Jackson (2007); Corfield *et al.* (2008); Wilson *et al.* (2008).

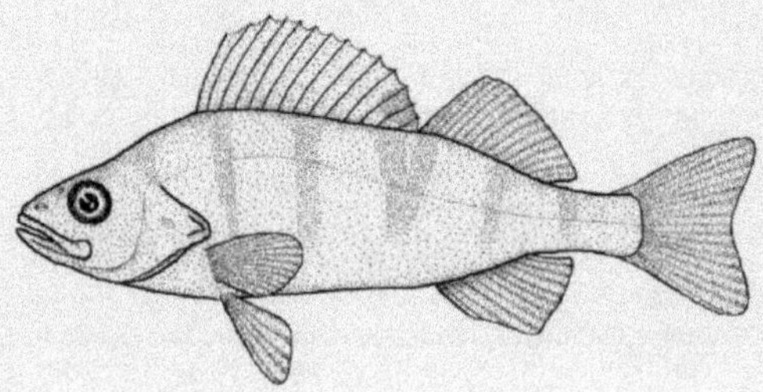

From McDowall (1996b).

from ornamental ponds to populate watercourses include goldfish, guppies (*Poecilia reticulata*), oriental weatherloach and koi carp. The Australian government now uses risk assessment modelling to monitor importations of alien species that have potential to become invasive pests (Bomford and Glover 2004; Ansell and Jackson 2007), but this does not affect the 1000+ species already present in aquaria.

Regulatory inadequacies among jurisdictions (the states and territories) have allowed many incoming species to bypass border controls (McNee 2002) and virtually any exotic fish species sought by aquarists has become available through one avenue or another. Of 1181 exotic species recorded in the last 40 years, only 481 (41%) were listed as permitted imports. In 2002, around 40% of 22 million pest fishes in Australia were imported rather than locally bred, and hundreds of species are smuggled in to avoid restrictions and quarantine (McNee 2002). Invasions continue, despite improved import regulations.

Ballast-water introductions

Australia is fortunate to lack major freshwater shipping ports. In the North American Great Lakes and many large European rivers, ballast-water discharges from ships have carried many alien organisms between continents (Lodge 1993; Moyle and Light 1996). Catastrophic invasions of North American waterways by zebra mussels and quagga mussels (*Dreissena* spp.) are compelling examples (Davis 2009). Although ballast water has introduced coastal invaders to Australia, only two euryhaline gobies have penetrated fresh water (Corfield *et al.* 2008).

TRANSLOCATED SPECIES

Translocated species are native fishes (or genetic stocks) introduced to areas outside their natural ranges. The initial introductions come from a variety of human activities and (rarely) natural processes (Table 11.3). Examples include Macquarie perch

Box 11.4: Profile of eastern gambusia, *Gambusia holbrooki*

Alias: Gambusia, mosquitofish, plague minnow, topminnow.

Origin: Central America: rivers flowing to the Gulf of Mexico.

Introduced: In 1920s for aquarium industry, later widely spread for mosquito control.

Distribution: Throughout most of Australia. Prefers warm, shallow areas in lentic fresh or brackish waters. Regions: NEQ, BURD, FITZ, SEQ, NEN, SEN, MDB, SEV, SWV, SAG, TORR, SWWA, TAS, NT (Ch. 2, Fig. 2.2).

Potential range: Not fully occupied; dispersed through introductions for mosquito control and bait-bucket transfers.

Movement: Limited movement.

Size: < 35–60 mm.

Diet: Predator on fish eggs and larvae, small invertebrates.

Habitat: Eurythermal (survives under ice, tolerates up to 44°C). Tolerates low water quality.

Reproduction: Sexually dimorphic, viviparous.

Value: Of little value in controlling mosquitoes; detrimental impacts on native mosquito predators, including small fishes.

Impacts: Declared noxious species. Threat to biodiversity through predation and competition, even against large fishes. Habitat and/or dietary niche overlaps with native fishes. Implicated in declines of native fishes and other biota.

Control: No effective control; genetic methods under investigation.

Select references: Lloyd (1990); Arthington (1991); Pen and Potter (1991b); Ivantsoff and Aarn (1999); Wilson (2005); Macdonald and Tonkin (2008); Pyke (2008); Rowe *et al.* (2008); Reynolds (2009).

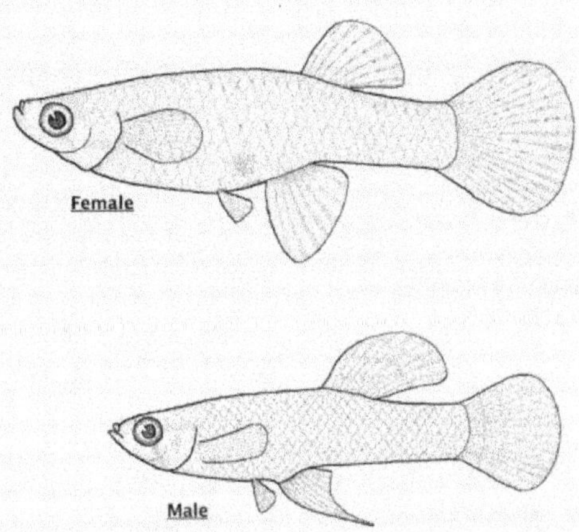

From McDowall (1996b).

(*Macquaria australasica*) in Victoria (Cadwallader 1981) and near Canberra (Lintermans 2007), *Maccullochella* species in Victoria (Cadwallader and Gooley 1984), climbing galaxias (*Galaxias brevipinnis*) in the upper Murray River and Murray cod (*Maccullochella peelii*) in the Gwydir River headwaters, New South Wales (Davies *et al.* 2010). Genetically isolated stocks, including golden perch (*Macquaria ambigua*), silver perch (*Bidyanus bidyanus*) and freshwater catfish (*Tandanus tandanus*: Keenan *et al.* 1996), Australian bass (*Percalates novemaculeata*: Chenoweth and Hughes 1997) and barramundi (*Lates calcarifer*: Chenoweth *et al.* 1998), highlight the risks of translocations. Many stocking programs have proceeded with little knowledge of the genetic character of the stocked fish or their relationship to wild populations (Rowland and Tully 2004). Rourke *et al.* (2010) provided an encouraging report that genetic diversity and the effective population size of Murray cod may be unchanged after prolonged stocking with local broodstock.

The attraction and robustness of native eels (Anguillidae), percichthyids and plotosids, and the proximity of different drainages, have facilitated relocations, often without documentation or monitoring. Most translocated native species are angling targets, forage species or bait. Murray cod, trout cod (*Maccullochella macquariensis*), golden perch, Macquarie perch, Australian bass and freshwater catfish have been released into new drainages in Victoria, Queensland and New South Wales (Berra and Cadwallader 1983; Cadwallader and Gooley 1984; McDowall

Box 11.5: Profile of tilapia, *Oreochromis mossambicus*

Alias: Mozambique tilapia, Mozambique mouth-brooder.

Origin: East Africa.

Introduced: In 1977 via aquarium trade, spread through illegal introductions.

Distribution: Patchy, near northern population centres in Queensland and Western Australia. Regions: SEV, SEQ, NEQ, BURD, PILB (Ch. 2, Fig. 2.2).

Potential range: Not fully occupied, may extend further through climate change; threatens Murray-Darling Basin, Gulf drainage and warmer coastal rivers.

Movement: Amphidromous, also through deliberate introduction and escape from spilling dams.

Size: < 350 mm, < 1.1 kg.

Diet: Omnivore, adults herbivorous, piscivorous and/or detritivorous, phenotypic and dietary plasticity.

Habitat: Euryhaline, in fresh–brackish water; tolerates low water quality; temperature tolerance 8–42°C. Prefers vegetated, warm, slow-flowing streams, lakes and impoundments.

Reproduction: Fecund mouth-brooder; hybridises with other cichlids.

Value: Popular table fish, aquaculture species that provides protein and food security in developing countries.

Impacts: Declared noxious species. Major threat to biodiversity through habitat modification, predation and competition for space and food.

Control: Public education program, eradication in ornamental ponds, barrier netting on high-risk irrigation channels.

Select references: Pollard (1990); Arthington (1991); Canonico *et al.* (2005); Ansell and Jackson (2007); Doupé and Burrows (2008); Doupé *et al.* (2009); Doupé and Knott (2010).

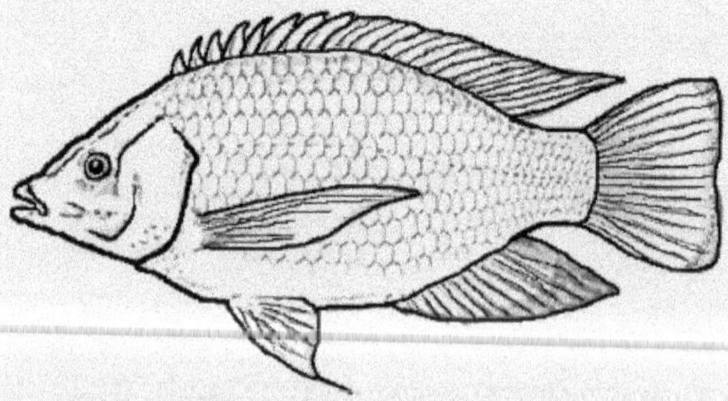

Drawing by JH Harris.

Table 11.3: Sources of Australian freshwater fish translocations, and human activities and natural processes potentially leading to redistribution and spread

Source	Activities and processes
Recreational fisheries	Fish stock introductions, maintenance or enhancements
Species conservation	Population rehabilitation using relocated or hatchery-bred fishes
Illicit stocking	Anglers making covert releases to waterways
Aquarium releases	Deliberate release of aquarium species
Mosquito control	Agencies distributing eastern gambusia (formerly 'mosquitofish')
Bait-bucket releases	Anglers releasing unused bait-fish
Bait-breeding releases	Anglers and/or entrepreneurs using waterways to breed bait-fish
Hatchery contamination	Unintended species in hatchery ponds and shipments
Fish farm spills	Flooding, pond overflows, inadequate screening
Farm dam spills	Flooding, overflows, lack of screening in stocked farm dams
Interbasin transfers	Unscreened hydroelectric and water-supply schemes
Ballast-water releases	Ship ballast taken aboard in remote regions
Inland marine aquaculture	Poor screening of marine species in inland saline aquifers
Rains of fishes	Natural uplift and redistribution through thunderstorm activity
Floodwater transport	Redistribution between catchments in flood waters

Source: Modified after Phillips (2003).

1996b). Murray cod are especially valued and there have been attempts to establish them outside their native range, especially in Western Australia (Dept of Fisheries, Western Australia 2003).

Unintended species have sometimes contaminated stocking shipments. Hatcheries have been under limited regulation until recent years, and many have drawn water directly from watercourses without effective filtration. The fertilised eggs, free embryos and larvae of local fish species readily become established in hatchery ponds and are then difficult to remove from stocking shipments. At least four species (gudgeons, Eleotridae; barred grunter, *Amniataba percoides*; redfin perch; common carp) have been spread by this route (Rowland 2001).

Conservation, water transfers and bait buckets

Three groups of fishes have been translocated. The first are listed threatened species relocated – usually as small numbers of wild-caught fish (Phillips 2003) – for protection or conservation (e.g. Australian grayling, *Prototroctes maraena*; Australian lungfish, *Neoceratodus forsteri*; southern purple-spotted gudgeon, *Mogurnda adspersa*; Tasmanian galaxiids). For example, a few Pedder galaxias (*Galaxias pedderensis*) were translocated to preserve the species following its decline in Lake Pedder after hydropower development (Dept of Primary Industries, Water and Energy, Tasmania 2006). Another example is the translocation of Macquarie perch to re-establish a population in the Queanbeyan River, New South Wales (Lintermans 2006). In addition, there are hatchery-based conservation programs for trout cod, Mary River cod (*Maccullochella mariensis*) and Clarence River cod (*M. ikei*) (Douglas *et al.* 1994; Rourke *et al.* 2009).

The second group of translocations of fishes (and other aquatic biota) is associated with interbasin water transfers. For example, the climbing galaxias occurs naturally in lowland tracts of the Murray River but water transfers from the Snowy Mountains Hydroelectric Scheme have introduced it to the Murray headwaters (Georges and Cottingham 2002). In Victoria, transfers from the Glenelg River to the Wimmera River were probably responsible for introducing common galaxias (*Galaxias maculatus*), alongside golden perch stocked for angling (Davies *et al.* 2008).

The third group includes small native and alien fish used as live bait, with escapees from so-called bait-bucket introductions. The scale of this process is undocumented, but anecdotal reports and unexplained occurrences of goldfish, oriental weatherloach, eastern gambusia, redfin perch and common carp suggest that it has been widespread (Keane and Neira 2004).

Stocking

Fish aquaculture has been practised in China since 2100 BC but artificial propagation of fishes began in France in 1850, when Joseph Remy first stripped eggs and milt from brown trout and hatched and reared the progeny (Halverson 2010). His techniques were adopted swiftly in the USA, with brook trout (*Salvelinus fontinalis*) propagated in 1853 and rainbow trout in 1872 (Halverson 2010).

Global fish stock enhancements yield about 2 million tonnes annually, mostly in freshwater fisheries, and account for 20% of captures (Lorenzen *et al.* 2000). Stocking is a favoured method for rehabilitating freshwater fisheries (Cowx 1994) but it is often applied uncritically (Welcomme and Bartley 1998). Despite long-standing arguments over the effectiveness of stocking programs, managers and politicians tend to pin hopes for the future of fisheries on stocking (White *et al.* 1995) rather than more challenging interventions such as habitat rehabilitation or fishing restrictions.

The term 'fish stocking' has quasi-agricultural connotations, reflecting both practice and perception. The practice is to breed many fish, liberate them in a waterway and later exploit the new-grown fishery resource. It began early in Australian colonial history, encouraged by the zeal of acclimatisation societies for Northern Hemisphere fishes. Later, river-dwelling native species not disposed to breed in reservoirs (Schiller and Harris 2001; King *et al.* 2003a; Ch. 7) often failed to colonise these artificial habitats. As most impoundments are near regional population centres, there has been a growing demand for recreational fisheries. Anglers seized on new propagation technologies, private and government hatcheries have proliferated and there have been routine mass releases of trout (in upland areas) and native species (in lowland areas). The acclimatisation societies still exist (Brinsley 2010), promoting anglers' interests, but state agencies carry most of the burden in stocking public waters. Private landowners also stock fish.

Warm-water fisheries in southern Australia are now reliant on golden perch, silver perch, Australian bass and Murray cod; those in northern Australia rely on barramundi, sooty grunter (*Hephaestus fuliginosus*) and saratoga (*Scleropages leichhardti*). Hatcheries in the eastern states routinely produce these native species and Clarence River cod, Mary River cod and trout cod. For others (e.g. eels, freshwater catfish), large-scale propagation has not succeeded because of the lower fecundity and/or specialised reproductive needs of those species.

Stocking is on a massive scale. From 1995–96 to 2000–01, nearly 37 million young native fishes of nine species were stocked in New South Wales, Queensland, Victoria and the Northern Territory (Simpson *et al.* 2002). In 2009, the New South Wales government released 5 million native fishes and trout. In Victoria, government stocking averaged 1.3 million native fishes and trout in the three years to 2009. In the Wet Tropics, Queensland, 2 million fishes comprising 36 native species were stocked in 1986–2004 (Burrows 2004). In the Murray-Darling Basin, more than 60 million native fishes were stocked over 30 years (Crook *et al.* 2009). The vast majority of these were to support recreational fishing.

Anglers tend to regard stocking as a panacea for fishery problems (Phillips 2003; Halverson 2010). Managers are also likely to favour stocking as an easy option, although hatchery-reared fish are prone to low fitness, behavioural problems and poor viability through a variety of genetic and developmental issues as well as predator naïvety (Phillips 2003; Rowland 2004). Stocking may obscure related environmental problems such as invasions, habitat degradation, biodiversity impacts, disease and overfishing (Ayres and Clunie 2010a). The real costs and benefits of stocking need careful analysis, but research on the long-term consequences is scant (Davies *et al.* 1988; Hutchison *et al.* 2006; Halverson 2010). Nevertheless, there is growing awareness that, in the long term, there is more benefit in managing wild stocks than in dependence on hatchery releases (Phillips 2003; Halverson 2010; Chs 9, 12).

Changed distributions

Native fishes in 28 families and 76 species have been translocated (Ayres and Clunie 2010a), mainly in eastern and south-eastern regions in the 1970s, 1980s and 1990s, although many transfers were not recorded. The natural distributions of species (and genotypes) have been shuffled and boundaries have become blurred (Burrows 2004; Closs *et al.* 2004; Olden *et al.* 2008). For native species, most translocations have been in Queensland (41 species), followed by Victoria (18), South Australia (14), New South Wales (11), Tasmania and the Australian Capital Territory (each nine), Western Australia (three) and the Northern Territory (one). For alien species invasions, Queensland (29 species) and Victoria (19) again lead, followed by New South Wales (18), Western Australia (12), Tasmania and the Northern Territory (each eight) and South Australia (seven). These data are shown in Table 11.2.

Ecologists consider that alien invasions are often facilitated by ecological disturbance (Harris and Silveira 1999; Kennard *et al.* 2005; Davies *et al.* 2010), although cause–effect relationships are usually not obvious because there are many confounding influences (see 'Invasion ecology'). Many kinds of disturbance are associated with human settlement and populous areas therefore are likely habitats for non-indigenous fishes (Arthington and McKenzie 1997; Olden *et al.* 2008).

As mentioned, alien fishes became established in southern Australia earlier than in the tropics and subtropics. Aquarium fishes are mainly warm-water species (McNee 2002; Corfield *et al.* 2008) and alien aquarium species have accompanied the spread of human populations towards the north. The distributions of 30 aquarium fishes mapped by Corfield *et al.* (2008) are clustered around population centres. At least 16 species are established near Townsville, 10 near Brisbane, five near Cairns, four near Darwin, four near Sydney and three near Canberra. The spread of aquarium fishes to remote areas has been limited (Ch. 2). Twelve of 14 drainages apparently free of non-indigenous fishes are in the Kimberley and Northern Provinces (Fig. 2.2). In the south, the remote, arid Bulloo-Bancannia Basin and Western Plateau appear not to have not been invaded.

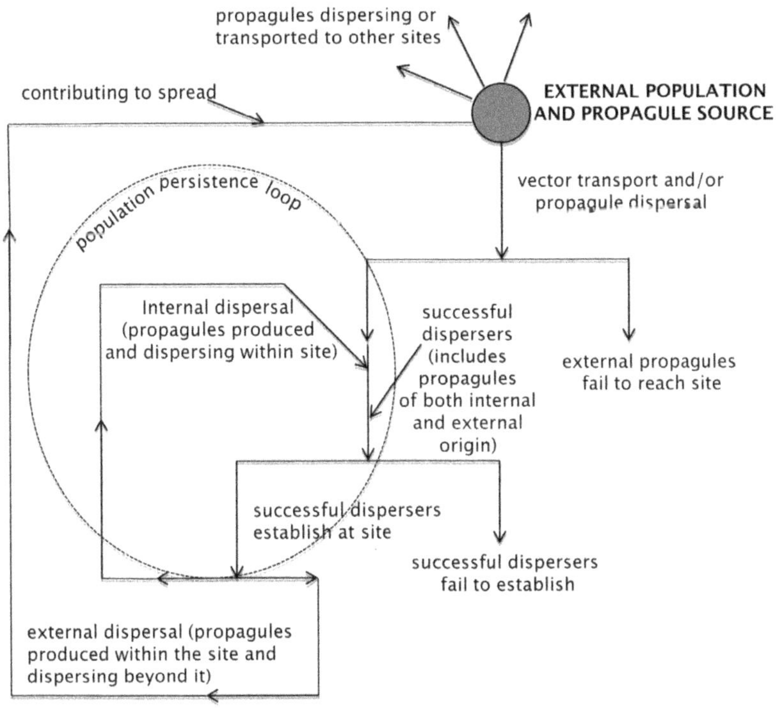

Figure 11.2: The sequence and iterations of dispersal or vector transport from a source population, and establishment of propagules giving rise to the establishment, persistence and spread of new populations (from Davis 2009).

INVASION ECOLOGY

Invasive fishes face many challenges in a new environment. Predation, nutrition, reproduction, dispersal and environmental tolerances all present challenges that take precedence at different points in the rolling, repetitive sequence of events that unfolds as a new species imposes itself on a community. The staged events in invasion include introduction, establishment, naturalisation, dispersal, population distribution and invasive spread, all part of an ongoing series of cyclical iterations. Mark Davis (2009) provided a thoughtful review in his book *Invasion Biology*. Two critical stages in invasion are establishment and dispersal. Both operate at the individual level, but ultimately lead to the persistence and spread of populations and metapopulations (Fig. 11.2).

Establishment and dispersal

For a population to become established, sufficient individuals must survive to reproduce within parameters of mortality and numerical strength. Mortality is a complex function of time, existing enemies (predators, parasites, diseases), food availability, water quality and sheltering cover and other habitat variables (Weatherley 1972; Gulland 1977; Wootton 1998) Abiotic factors are critical (Moyle and Light 1996; Bomford and Glover 2004) because if these suit the invading species, regardless of other biota, establishment is likely. Numerical strength is expressed as 'propagule pressure', the product of the number of individuals at each introduction and the number of events (Rahel 2002; Davis 2009; Strayer 2010). Invasions are more likely to succeed after multiple introductions of many individuals, leading to cumulatively increasing propagule pressure (Fig. 11.3).

About 50% of reported fish introductions result in establishment of new populations and the probability of success increases with releases of large numbers at different times and places (Bomford and Glover 2004). Worldwide, species with more than 10 introductions are likely to establish at least one population, especially when the climate in the new area is like that of

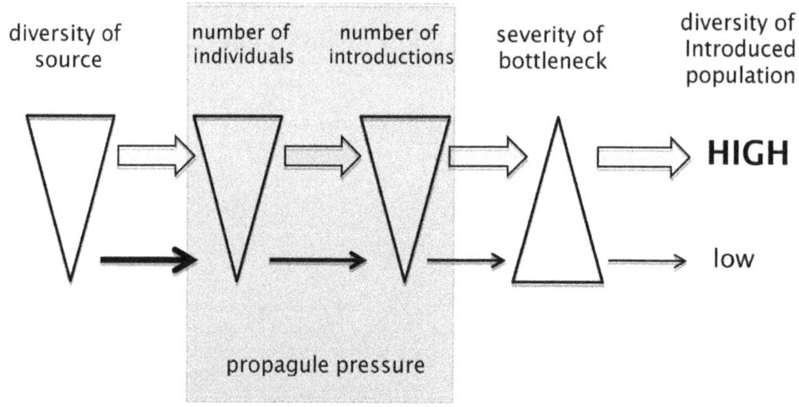

Figure 11.3: Factors in the genetic diversity of invading populations. Triangles indicate the magnitude of each factor, from high (base) to low (apex). Arrow thicknesses represent the proportions of original (source) genetic diversity preserved at each step. The grey box highlights factors generating propagule pressure (see text). Genetically diverse invaders are likely to be from high-diversity sources and associated with high propagule pressure from numerous individuals, or multiple introductions, or both. Invaders can avoid population bottlenecks if the founder population is large or if there is rapid population expansion following introduction (modified from Roman and Darling 2007).

the source area. In Australia, examples include rapid invasions of trout, introduced on multiple occasions to cool-water habitats. In contrast, the establishment of populations of threatened species such as trout cod can be problematic even after many large-scale releases (Douglas *et al.* 1994; Lintermans 2007).

Following establishment, individuals in the new population must disperse to consolidate the invasion and may spread to form a metapopulation (see 'Metapopulations'). This requires spawning and recruitment, generating new propagule pressure and reiterations of establishment and dispersal (Fig. 11.2). The limits for dispersal will be influenced by structural features (catchment boundaries, waterfalls, dams) and other physical and chemical barriers (hydraulics, temperature, salinity, oxygen).

Dynamic models are needed for predictions and analysis of establishment and dispersal. Different traits (longevity, fecundity, growth rate, tolerances, migratory behaviour) become important at different stages, and even over five to 10 years invasive populations can increase by orders of magnitude (Keller *et al.* 2009). It is likely that lessons from epidemiology can be applied to invasive species and there are refined statistical methods for preparing risk assessments. A species' capacity to be integrated into a resident community is enhanced in an environment lacking its natural enemies (diseases, parasites, competitors, predators). The success of introduced trout bred from transported eggs attests to the strength of this 'enemy release' phenomenon.

Metapopulations

Ultimately, the success of an invader hinges on evolutionary processes (Vermeij 1996; Mooney and Cleland 2001; Davis 2009). Its capacity to adapt to new conditions (e.g. climate change, interactions with other species) depends on its innate behavioural and physiological flexibility, its genetic diversity, its propensity to move and its ability to form a metapopulation (i.e. to maintain interactions with populations of the same species in different areas).

Common carp, for example, show exceptional plasticity, invading diverse communities, habitats and geography throughout the world. In Australia, the Prospect and Yanco strains established small populations but failed to spread (Shearer and Mulley 1978; Davis *et al.* 1999). But when Boolarra strain carp and later koi imports escaped into the wild their populations boomed, leading to rapid invasion (Hume *et al.* 1983; Roberts and Ebner 1997; Haynes *et al.* 2009). Presumably, the two early strains lacked genetic variability, phenotypic plasticity and a propensity to move large distances, but those attributes were very evident in the later introductions. Redfin perch also show phenotypic plasticity, with early maturation producing super-abundant populations of small individuals where cannibalism is common, with larger individuals eating smaller individuals (Morgan *et al.* 2002). On the other hand, global warming is likely to cause contractions in the geographic and altitudinal ranges of cool-water fishes such as redfin perch, trout and tench, and these may have limited adaptive capacity as a consequence of genetic bottlenecks caused by introductions of small numbers of fish on few occasions (Weatherley and Lake 1967).

An interesting aside to the invasion of common carp is that top-level native predators (e.g. cormorants, pelicans and herons; Australian bass, golden perch and Murray cod) may change their foraging behaviour to prey selectively on this now super-abundant fish. This factor, together with drought, may be implicated in an apparent widespread decline in numbers of common carp in 2000–10. Another behavioural adaptation that may affect the success of invaders is social learning by the invaded species (Duffy *et al.* 2009), where individuals learn by observing others. Other factors could include declining resource availability and pathogens introduced by invaders. The interplay of these factors is reflected in the cycles of irruption and decline demonstrated by many invaders. Rapid establishment and dispersal produce abundant populations that, over time, regress to lower abundances and become integrated into local communities (Strayer *et al.* 2006; Davis 2009).

Environmental invasibility

The invasibility of an environment, or its vulnerability to establishment of new species (or genotypes), is often invoked as a key to the success of an invader (Arthington and McKenzie 1997; Phillips 2003; Davis 2009). It interacts with the attributes of a species and its propagule pressure and varies over time, subject to environmental factors including droughts and floods, river regulation, catchment land-use and existing aquatic communities. Some striking examples of increased invasibility occur where reservoirs discharge cold, hypolimnetic water to rivers, fostering cold-tolerant trout, redfin perch or common carp above native warm-water species (Roberts and Tilzey 1996; Phillips 2001; Todd *et al.* 2005). Invasible environments also occur near urban centres, where degraded habitats and native fishes are vulnerable to introductions of aquarium fishes (Arthington and McKenzie 1997; Rowe *et al.* 2008).

Blackwater events in the Murray River are further examples of changes in invasibility. Flooding after protracted dry periods may wash organic detritus and leachate into stream channels, de-oxygenating the water for hundreds of kilometres downstream (Industry and Investment, New South Wales 2011; Murray-Darling Basin Authority 2011). Common carp tolerate poor water quality and may recruit in enormous numbers in these situations, where most native fishes do not survive (McNeil and Closs 2007).

In rivers of the Lake Eyre Basin in arid central Australia, the abundances of alien species decline in big floods when native species may recruit strongly (Costelloe *et al.* 2010). These rivers may have comparatively low invasibility, conferred by variable hydrological regimes and native fishes with life cycles attuned to boom–bust dynamics. In general, habitats prone to wide variations (e.g. flow, thermal regime, water quality) may be less prone to invasion because the resident native species are better able to cope with the extremes.

Processes affecting invasibility may also drive regional biodiversity (Davis 2009). Environments prone to disturbances at intermediate, rather than high or low, frequencies tend to have the highest species diversity because dominant species periodically decrease, allowing the persistence of rarer species (the intermediate disturbance hypothesis: Connell 1978). Natural events such as bushfires, storms, floods and droughts, and human impacts such as soil erosion, pollution, instream barriers, changed water regimes and alien species add to the range and severity of these disturbances (Connell 1978; Reice *et al.* 1990). Presumably, disturbed environments may be colonised by either indigenous or non-indigenous species, if they are present and suitably adapted.

Invasibility, biodiversity and river health

There are parallels in concepts of invasibility, biological diversity and river health in that all respond to ecological degradation and are emergent properties of systems. Nativeness (the relative contribution of alien fish species to riverine fish assemblages) has been used as an index of river health in the Murray-Darling Basin (Davies *et al.* 2010). The index of biotic integrity (Karr 1981; Harris and Silveira 1999) employs similar data to assess the condition of freshwater ecosystems.

Does good river health promise low invasibility? Certainly, the opposite can be true. Poor health (e.g. blackwater events, urban streams) may favour invasions by some species (e.g. common carp, eastern gambusia, guppies). The presumption, from traditional niche theory, that invasibility declines as species richness increases (because there are fewer available niches) is often cited but has little evidence in support (Bomford and Glover 2004; Davis 2009). Others suggest that the resistance of native fauna is not important in mediating fish invasions (Moyle and Light 1996; Kennard *et al.* 2005).

What makes a successful invader?

Narrow tolerances (e.g. temperature, salinity, oxygen, turbidity, water velocity) account for most of the many species introductions that fail to establish populations (Kolar and Lodge 2001). Further, physiological and other attributes may intervene at different stages in an invasion (introduction, establishment, dispersal, integration: Marchetti *et al.* 2004). Successful establishment is favoured if a species has broad physiological tolerances, cares for its young and has a long life span. Successful dispersal is favoured if a species has a long life span, is close to its native habitat and is a trophic generalist. High abundances of invaders tend to be associated with small body size, broad tolerances and proximity to their native habitat. The number of propagules introduced, and prior success of a species as an invader, also suggest success in establishment.

The physical and chemical nature of water, and climate, limit the success of non-indigenous fishes in Australia. Only eurythermal species such as goldfish, common carp and eastern gambusia cover broad latitudinal ranges and these same species have a capacity to prosper even in waters where oxygen falls to levels not tolerated by most fish (McNeil and Closs 2007). Others, such as tilapia, guppies and brown trout, tolerate saline water, unlike most freshwater species. Tolerances are summarised by McDowall (1996b), Allen *et al.* (2002) and Froese and Pauly (2010).

Current speeds also limit non-indigenous fishes. While lowland and slopes-region rivers are generally slow-flowing, fish with the modest swimming abilities of guppies, eastern gambusia or redfin perch are precluded from swift-flowing streams which they might otherwise penetrate. Even common carp, which are powerful swimmers, have limited success in headwater streams – their spawning hotspots are apparently in lowland reaches (Smith and Walker 2004; Gilligan and Rayner 2007) or impoundments, and often are separated by barriers. Nevertheless, many invasive fishes do have broad tolerances of current, temperature, salinity, oxygen and other variables, faciltating their success in diverse habitats.

Many alien species evolved in environments more predictable than most Australian freshwater systems and, as a consequence, their spawning tends to be seasonally limited (Ch. 7). Common carp is an exception, being able to delay spawning until flooded wetlands become available. This ability, and its longevity, has contributed to the success of its invasion (King *et al.* 2003a; Koehn 2004; Smith 2005; Gilligan and Rayner 2007). Eastern gambusia has a life cycle unlike any Australian species (maturing early, producing multiple batches of young and having internal fertilisation, giving birth to live young), making it perhaps a super-invader (Pyke 2008).

Successful invaders often are species that can use food resources that are underexploited by native species (Marchetti *et al.* 2004). Abundant aquatic plants, especially attached algae, probably have contributed to the success of tilapia, common carp and goldfish. Relatively few Australian species are herbivorous; those few include sea mullet (*Mugil cephalus*), freshwater mullet (*Trachystoma petardi*) and bony herring (*Nematalosa erebi*). Plants do contribute to the diets of other species, such as silver perch.

IMPACTS OF NON-INDIGENOUS FISHES

Environmental, economic and social impacts follow upon fish introductions (Davis 2009; Keller *et al.* 2009; Strayer 2010). An extreme example followed the introduction of the predatory Nile perch (*Lates niloticus*) to Lake Victoria, Africa, in the 1950s. Within a few decades more than 200 endemic fishes became extinct (Kolar and Lodge 2001; Global Invasive Species Database 2004) and there were disastrous impacts on lakeside human communities. These events derailed attempts to introduce Nile perch to Australia, to offset a decline in barramundi (Williams 1982).

Movements of species, facilitated by commerce and transport, have increased local species richness in many parts of the world even though most transported species fail to establish and few of those become pests (Mooney and Cleland 2001). Where invaders do succeed, they often integrate into existing assemblages without extirpations (Moyle and Light 1996). New assemblages come to resemble co-evolved communities. Where extirpations occur, they result mainly from piscivore introductions, perhaps because the prey are not adapted to the new predator's behavioural traits. This may explain the impact of trout on galaxiids, which previously experienced less predation pressure from other fishes (McDowall 2006).

Impacts of invasive species are mediated through both biotic and abiotic interactions (Table 11.4) but

Table 11.4: Recognised and predicted mechanisms of change through non-indigenous fish invasions in Australia

Mechanism	Impact	Alien examples	Translocated examples
Piscivorous (macrophagic predation)	Strong ecological effects at apex of food webs, replacing high-level consumers. Predation mortality, species displacement, reduced genetic diversity. Trophic cascades possible	Redfin perch, salmonids, oscar	Barramundi, *Maccullochella* spp., golden perch, northern saratoga, Australian bass
Mesophagic predation	Often important effects at centre of food webs – smaller fishes and macroinvertebrates. Predation mortality, species displacement and/or reduced genetic diversity. Trophic cascades possible	Eastern gambusia, redfin perch, tench, salmonids, cichlids esp. tilapia	Golden perch, silver perch, Australian bass, barred grunter, climbing galaxias, Macquarie perch, mouth almighty, sooty grunter
Microphagic predation	Affects bottom of food webs – small macroinvertebrates, zooplankton. Predation mortality, species displacement, reduced genetic diversity. Trophic cascades and algal blooms possible	Most cichlids, poeciliids, oriental weatherloach, small cyprinids	Young of many native species
Omnivory and herbivory	Transform organic resources into fish biomass and excreted wastes. Water-quality impacts. Loss of aquatic plants. May induce algal blooms	Goldfish, common carp, crucian carp, roach, cichlids esp. tilapia	Australian lungfish
Resource competition	Limited by food, shelter, space, spawning sites	Potentially all species	Potentially all species
Interference competition	Abundant or aggressive individuals and biomass interfere with native species' behaviour. Probable disruption of native species' nests	Common carp, eastern gambusia, cichlids esp. tilapia, salmonids, redfin perch, other alien species	Abundantly stocked angling species
Vectors for parasites and diseases	Introduce disease organisms and parasites to native fishes often lacking in experience and resistance. Increased rates of transmission	Goldfish, common carp, eastern gambusia, other ornamental fishes, redfin perch, roach, salmonids, other alien species	Australian bass, Murray cod, silver perch, barramundi, other native species
Genetic effects	Hybridisation, displacement of discrete genetic stocks, displacement of wild stocks by hatchery fish, evolutionary bottlenecks, possible increased fitness of alien hybrids	Cyprinids, cichlids	Australian bass, Murray cod, silver perch, golden perch, barramundi
Habitat disruption	Loss of aquatic plants, turbidity, eutrophication, algal blooms	Common carp, goldfish, tilapia	?
Evolutionary effects	High selection pressures, evolving adaptations of invading species and invaded communities	All species	All species

many reports are anecdotal or based on mere associations, due to the difficulty of establishing cause–effect relationships and through confounding the effects of introductions by habitat degradation (Bomford and Glover 2004; Rowe *et al.* 2008). The trend towards homogenisation through non-indigenous introductions has been responsible for losses of unique character in native communities (Ch. 10). Introductions have changed biogeographic patterns and increased faunal similarity among drainage systems, particularly those experiencing most human-induced change (Olden *et al.* 2008).

Bomford and Glover (2004) proposed a method for ranking the risks associated with introduced fishes, using Fishbase (Froese and Pauly 2010). Fishbase is a comprehensive resource but it reflects the limitations and inadequacies of available data. For example, common carp are described as 'never known to dominate the environment', a comment that – in Australia – could not be further from the truth!

Widespread stocking of both alien and native species can cause native-species displacement, loss of genetic stocks and introductions of parasites and diseases (Table 11.4; Clunie *et al.* 2002b; Phillips 2003; Gillanders *et al.* 2006). These threats have prompted agencies in Australia to develop guidelines for hatchery and stocking operations (Ansell and Jackson 2007; Ch. 12). Rigorous evidence for the impacts of alien fishes is often in short supply and pest potential is difficult to assess (Bomford and Glover 2004; Corfield *et al.* 2008). The data used for risk assessments are often anecdotal, speculative or scant, and invasiveness is not necessarily a useful indication of the potential impacts of fishes. For example, some aquarium species (goldfish, oriental weatherloach, guppies) are widespread without significant impacts having been detected. Because evidence of actual impacts may be slow to appear, assessments of risk potential should be based on prior knowledge of traits related to a species' likely impacts on native biota in addition to its invasive potential. The contrasting histories in Australia of two alien cyprinids, the destructive common carp and comparatively benign goldfish, show how different the impacts may be for superficially similar species. Further, the impacts of alien species are often synergistic, augmenting those of other stressors. In the words of Corfield *et al.* (2008, p. 56), 'exotic species might be a primary cause for decline, a contributing factor for a species already in trouble, the final nail in the coffin or merely a bouquet at the funeral'.

Predation and competition

Invading fishes may be at any trophic level, from herbivore to detritivore to top predator (Table 11.4), but most important alien invaders, aside from common carp, have been carnivores. The same is true of translocated native species. Regardless, the invaders play diverse roles in ecological modifications. Predator–prey interactions in particular are species- and habitat-specific and likely to vary with ontogenetic stage (Stoffels and Humphries 2003). For example, removal of redfin perch from ponds in New Zealand caused marked changes in the abundance of their prey (common bully, *Gobiomorphus cotidianus*: Ludgate and Closs 2003) whereas removal of the species from streams in South Australia had no effect on numbers of flathead gudgeon (*Philypnodon grandiceps*), although it did change their microhabitat use (Wilson *et al.* 2008).

Brown trout and rainbow trout introductions around the world generally have been detrimental to native fishes, leading to declines and even extinctions. Species displacement and fragmentation have also occurred in Australasia (Tilzey 1976; Jackson and Williams 1980; Crowl *et al.* 1992). Trout overlap the ranges of relatively few native fishes, notably river blackfishes (Gadopsidae), eels, a number of Tasmanian galaxiids, Australian grayling, trout cod and Macquarie perch. These species cohabit with trout in many areas (Harris and Gehrke 1997; Davies *et al.* 2008) although some, such as mountain galaxias (*Galaxias olidus*), require macrophytes, broad shallows or other structural cover to maintain populations in face of predation. In streams lacking cover, galaxiids become restricted to the upper reaches less accessible to trout, and their populations become fragmented (Lintermans 2000; Green 2008).

Trout predation undoubtedly occurs on other small fishes, where their ranges overlap. One region that needs close monitoring is the eastern Murray-Darling Basin tributaries near the lower altitudinal limit of trout (about 600 m), where there are endangered trout cod and Macquarie perch. In Lake Pedder,

Tasmania, following hydropower development, the combined effects of predation by brown trout, habitat degradation and competition and predation by translocated climbing galaxias led to a precipitous decline of Pedder galaxias and a declaration of the species being extinct in the wild (Dept of Primary Industries, Water and Energy, Tasmania 2006; Ch. 12). In general, there is no doubt that trout have had negative effects on some native fish populations (Cadwallader 1996; Erskine and Harris 2004) but they are only one of multiple stressors, which particularly include those related to land-use changes and habitat degradation. The interactions of trout, native fishes and environmental factors are dynamic – dry periods and warm temperatures displace the trout, allowing local recovery of native populations (Closs and Lake 1996). A challenge for research is to distinguish between the impacts of habitat change and those of trout and other alien species, and to evaluate interactions between the two (Ormerod *et al.* 2010).

Salmonids are by no means the only predatory alien species to warrant concern. Redfin perch is a declared noxious species in New South Wales and an 'exotic' species in South Australia (not to be transported or returned to the water) but is valued as an angling species in Victoria (Rowe *et al.* 2008). Eastern gambusia are implicated in declines in the range and abundance of many small and larger native fishes through fin-nipping, competition and predation on eggs and young (Lloyd 1990; Arthington 1991; Corfield *et al.* 2008; Macdonald and Tonkin 2008). Some of the other 32 escapees from the aquarium industry (Table 11.2) may be having comparable impacts, but information is lacking. Tilapia are aggressive and successful competitors for space, spawning areas and food, and have a wide diet that may include other fish (Ansell and Jackson 2007; Doupé *et al.* 2009; Doupé and Knott 2010; Froese and Pauly 2010). They presently occur in Queensland and Western Australia and there is a risk that they will spread into the Murray-Darling Basin and Gulf rivers. Their impacts on biodiversity elsewhere in the world (Canonico *et al.* 2005) earn them a high rank among threatening invasive species, despite their value as a source of protein in some Asian and Pacific Island cultures (Ansell and Jackson 2007; Doupé and Burrows 2008).

Food webs

Ecological communities are complex webs that link plants, herbivores, carnivores, parasites, detritivores and decomposers. Changes to the structure or dynamics of food webs through introduction of predators such as redfin perch and trout, or omnivores such as common carp, can permeate all levels of ecological organisation (individual, population, community, ecosystem) and alter nutrient and energy fluxes (Simon and Townsend 2003). The effects are amplified as a trophic cascade (Townsend 1996; Polis *et al.* 2000), where a change in some element of the food web has indirect consequences for organisms at other trophic levels (Table 11.4; Ch. 6). For example, removal of a top predator that preys on a zooplanktivore could increase the abundance of zooplankton and thereby decrease the abundance of phytoplankton. Equally, the introduction of a top predator might lead to an increase in benthic macroinvertebrates, mediated through increased predation on bottom-feeding fishes. A top-down cascade refers to the effects of removal of a high-order predator or consumer, whereas a bottom-up cascade might indicate removal of phytoplankton or a fish grazing on phytoplankton. Trophic cascades often indicate the presence of a keystone species, one whose removal has a disproportionate effect on its community (Mills *et al.* 1993).

Fish may trigger trophic cascades that culminate in nuisance blooms of Cyanobacteria (blue-green algae) by preying on zooplankton, or they may promote blooms through effects on nutrients and light (Gehrke and Harris 1994). Common carp contribute to these bottom-up cascades as juveniles preying on grazing zooplankton and by adults excreting nutrients, resuspending sediments and causing damage to aquatic plants (Gehrke and Harris 1994; King *et al.* 1997). There have been attempts to manipulate predators to control trophic cascades and reduce blooms (Gehrke *et al.* 2010), and their potential is demonstrated by enclosure–exclosure experiments (Khan *et al.* 2003; Schmitz *et al.* 2004).

Predator-avoidance behaviour, where prey employ adaptive foraging strategies to reduce the risk of predation, can reduce the impacts of herbivores, driving a trophic cascade leading to over-abundant plant growth (Schmitz *et al.* 2004). Thus, the presence of brown trout and galaxiids in a stream may cause mayfly nymphs to seek refuge, reducing their grazing impact on attached algae (McIntosh and Townsend 1996; Townsend 1996; Polis *et al.* 2000). Strategies such as this may reduce losses to predation but sublethal impacts may still occur; for example, brown trout may cause golden galaxias (*Galaxias auratus*) to avoid open-water habitat, reducing scope for feeding (Stuart-Smith *et al.* 2006).

Parasites and pathogens

Through geological isolation and the demands of a variable climate, Australia's native fish fauna lacks representatives of many of the world's freshwater fish families, and the same is true of many associated parasites and diseases. The native fishes are therefore vulnerable to introduced pathogens because they lack specific resistance (Lymbery *et al.* 2010). The 8–10 million ornamental fishes imported to Australia each year pose a particular hazard as vehicles for pathogens (McNee 2002; Ch. 12). In fact there have been many disease incursions, leading to establishment of viral, bacterial, fungal, protozoan and metazoan pathogens from ornamental fishes (Whittington *et al.* 2010) Past quarantine protection has been inadequate, although practices have improved in recent years. Disease incursions will continue, however, and exotic pathogens will continue to become established unless procedures, monitoring, research and policing are dramatically intensified.

The Asian fish tapeworm *Bothriocephalus acheilognathi* is a pathogenic, non-host-specific parasite that is new to Australia. It is usually associated with common carp and eastern gambusia but has crossed to at least one native fish (Dove *et al.* 1997) and threatens others (Henderson 2009). Four new monogenean tapeworms have been introduced with alien ornamental fishes (Dove and Ernst 1998). The parasitic copepod (anchor worm, *Lernaea cyprinacea*) is well-established, and apparently was introduced with common carp or redfin perch (Lintermans 2007). At least two new pathogenic parasites have been introduced to Western Australia by alien fishes (Lymbery *et al.* 2010). Epizootic haematopoietic necrosis virus is a virulent iridovirus, probably endemic to Australia, which causes severe mortalities in redfin perch and may also be carried by rainbow trout (Whittington *et al.* 2010). It is an indiscriminate pathogen and at least 10 native species (Australian bass, golden perch, Macquarie perch, mountain galaxias, Murray cod, silver perch) and other alien species have proved susceptible to infection under experimental conditions (Lintermans 2007). Aquarium gouramis (Belontiidae) carry Gourami iridovirus that, in 2003, caused 90% losses in farmed Murray cod (Whittington and Chong 2007). Goldfish ulcer disease, a bacterial infection, has caused outbreaks among goldfish, roach (*Rutilus rutilus*) and silver perch, and threatened the Atlantic salmon aquaculture industry (Humphrey and Ashburner 1993). These are a few examples among many (Whittington and Chong 2007).

Nodaviruses (single-stranded RNA viruses) have caused mortalities in freshwater and marine fish hatcheries. Nodavirus has infected Australian bass in hatcheries and threatened wild fish, leading to curtailment of stocking plans and compulsory screening of hatchery bass (Frances *et al.* 2005). Viral encephalopathy and retinopathy (a neuropathological condition) is virulent in barramundi and at least seven other freshwater fishes are susceptible, naturally or experimentally (Dept of Agriculture, Fisheries and Forestry 2007). These outbreaks emphasise the hazards of distributing mass-propagated fish and the need for intensive control of global trading, hatchery operations and stocking procedures (Gozlan *et al.* 2010; Whittington *et al.* 2010).

Engineering the environment

Some organisms are called 'ecosystem engineers' for their propensity to create or modify habitats and change the structure of communities, discouraging some species and favouring others (Jones *et al.* 1994; Crooks 2002). The term applies to humans, of course, and to many alien species. Common carp, in particular, consume benthic detritus and organisms, and their feeding habits change turbidity and nutrient levels, hence phytoplankton and zooplankton assemblages (Koehn *et al.* 2000) and fish communities (Gehrke *et al.* 2011). In lentic environments they may cause a shift from clear-water conditions, where macrophytes dominate, to turbid waters where phytoplankton dominate (Gehrke and Harris 1994; Matsuzaki *et al.* 2009; Weber and Brown 2009).

In Australia, the environments engineered by common carp may be beyond the tolerances of native fishes. The effects are especially dramatic in the Murray-Darling Basin where, after nearly 50 years' residency, common carp has come to represent nearly 60% of total fish biomass (Davies *et al.* 2008). In the lower Murray River, South Australia, common carp are likely to have been a major factor in the decline of many species of aquatic snails (Sheldon and Walker 1997). For these reasons, a primary concern in management of offstream wetlands is to exclude common carp (Hillyard 2011). A related problem is to avoid environmental flow allocations that benefit common carp above other species (Conallin *et al.* 2012).

The term 'ecosystem engineer' has no bounds but other alien fish species that could qualify, in some degree, are redfin perch, eastern gambusia, tilapia and trout. Each has a story to tell; their cumulative impacts also tell a story.

Long-term impacts

Breached barriers can have long-term biogeographic, ecological and evolutionary consequences for invaders and the invaded (Vermeij 1996; Strayer *et al.* 2006). Invasive species may alter the evolutionary trajectories of native species through competitive exclusion, niche displacement, hybridisation, introgression, predation and even extinction. Invaders themselves evolve in response to their interactions with native species and in response to new environments created by human interventions (Mooney and Cleland 2001; Davis 2009). Introduced and native fishes may hybridise. While low-level introgression occurs naturally and promotes genetic variation in closely related species, it becomes a concern for conservation when increased by human disturbances (e.g. species introductions, habitat modification, over-exploitation: Schwartz and Beheregaray 2008). The fitness of native species may be reduced by hybridisation, especially when stocked hatchery fishes swamp remnant native populations (Mooney and Cleland 2001; Phillips 2003), and population declines may erode genetic isolating mechanisms (Rowe *et al.* 2008).

Among native Australian freshwater fishes, where there are few family-level taxonomic overlaps (Table 11.1) between alien, exotic and native species, hybridisation is less of a risk than between translocated and indigenous native species. Hybridisation does occur, however, among alien species (e.g. common carp and goldfish; tilapiine cichlids, including *Oreochromis mossambicus*: McDowall 1996b) and may enhance their adaptation to local environments. Increased genetic variation and hybrid vigour probably facilitated the explosive spread of common carp in the Murray-Darling Basin in the 1970s, following introductions of the Boolarra and koi strains alongside the stable Prospect and Yanco strains (Davis *et al.* 1999; Haynes *et al.* 2009).

In translocated hatchery-bred fishes, there may be insidious threats from reduced genetic diversity, loss of co-evolved adaptations to specific habitats and swamping of indigenous populations through competitive displacement, hybridisation and introgression (Phillips 2003; Rowland 2004). Stocking incurs risks through interbreeding of hatchery progeny with conspecifics and hybridisation with congenerics, potentially affecting the genetic integrity and fitness of native species in degraded habitats or remnant populations. For example, parental lineages of Murray cod and trout cod were lost following stocking in Cataract Dam (Nepean River system, New South Wales) outside their native ranges, and the stocked population now consists only of fertile hybrids (Harris and Dixon 1988). Hybrids of Australian bass and estuary perch (*Percalates novemaculeata* and *P. colonorum*, respectively) occur in eastern Victoria in an area where their ranges overlap, and the genotypes of many individuals indicate hybridisation and back-crossing (Schwartz and Beheregaray 2008). Australian bass have been stocked repeatedly in the region and it is possible that the parental populations could be swamped through outbreeding depression or replacement by hybrids from hatchery broodfish and stocking programs.

Economic and social effects

Invasive species are part of international trade that may have irreversible side effects. In the USA, numbers of alien species have grown exponentially over 150 years in association with the importation of goods (Keller *et al.* 2009); in 2001 they caused estimated economic losses of $US137 billion (Kolar and Lodge 2001). With globalisation, non-indigenous species are now among the leading global threats to biodiversity (Strayer *et al.* 2006). In Australia, the Invasive Animals Cooperative Research Centre (<http://www.invasiveanimals.com>) has estimated that invasive vertebrates in Australasia cost at least $720 million annually. These costs should not, however, be assessed by dollars alone; there are other positive and negative values, and assessments should be informed by financial, scientific, social and ethical considerations (Koehn 2004; Keller *et al.* 2009).

There are compensations, in that the same species that damage ecosystems may bring social and economic benefits (Global Invasive Species Database 2004; Rowe *et al.* 2008; Gozlan *et al.* 2010). The National Recreational and Indigenous Fishing Survey (Henry and Lyle 2003) reported that, in one year, recreational or indigenous fishers among Australians over five years of age (19.5% of the population) spent about 5 million fishing-days, or 20% of total days fished, in fresh water. The harvested totals were 2.1 million common carp, 1.3 million redfin perch and other species, with associated expenditure of more than $1.8 billion (Chs 9, 12). Non-indigenous fishes may provide additional benefits through aquaculture and food security (Cadwallader 1996; Rowe *et al.* 2008; Halverson 2010). Some 17% of global finfish production is from non-indigenous species (Gozlan *et al.* 2010).

In Australia, the benefits from deliberate fish introductions include the recreational, social and economic values of trout fisheries, and fisheries for hatchery-bred native species in rivers and impoundments (Simpson *et al.* 2002; Ansell and Jackson 2007). Aquaculture of silver perch and barramundi is also economically significant (Australian Bureau of Agricultural and Resource Economics and Bureau of Rural Sciences 2010). Social impact assessments and cost–benefit analyses for six alien species (eastern gambusia; redfin perch; roach; tench; streaked goby, *Acentrogobius pflaumii*; yellowfin goby, *Acanthogobius flavimanus*) are provided by Rowe *et al.* (2008). Cost–benefit assessments generally would be improved by better estimates of the values of ecosystem services.

The social impacts of non-indigenous fish introductions are not well understood. Records of oral history from residents along the Lachlan River, New South Wales, refer to the disappearance of aquatic plants, loss of water clarity and invasion of common carp (Roberts and Sainty 1996). Pervasive environmental degradation – and presumably the presence of alien species – has negative effects on emotional well-being in rural communities (Rowe *et al.* 2008). Horwitz *et al.* (2001) commented that 'biodiversity, and its endemic features, contribute to a person's attachment to a particular place and become part of a person's identity. Loss, destruction or change in a location has the potential to affect an individual's psychological well-being and challenge a community's identity and image of itself.' Aboriginal communities suffer when their connections with native fishes, which are a vital part of traditional life styles and social interactions, are lost to environmental degradation and alien species (Rowe *et al.* 2008).

PEST CONTROL AND ERADICATION

State and Commonwealth governments have gazetted controls (Ch. 12) to manage pest species, and native fish translocations are better controlled now than in the past. The Invasive Animals Cooperative Research Centre is developing strategies and methods for integrated pest management of common carp and other pest fishes, including development of genetic technologies, biocides, pheromone-assisted trapping and environmental manipulations. The 'daughterless carp' program is a radical, still-unproven concept using recombinant DNA methods directed at common carp and eastern gambusia (Thresher 2007, 2008; Bax and Thresher 2009). It aims to manipulate the genotypes to produce heritable constructs that bias gender ratios towards male progeny without reducing fitness, thereby driving population fertility downwards. This program offers some hope as the first long-term, sustainable tool for control and possible eradication of pest animals, including fishes. A host-specific pathogen, the koi iridovirus, is also being investigated as a possible control agent for common carp.

Other, more direct tools are emerging. Williams' carp separation cages are being installed in fishways (Stuart and Jones 2002, 2006a; Stuart 2008) to exploit the propensity of common carp to leap over barriers, separating them from native fish which rarely leap, with a few exceptions (e.g. mullets, Mugilidae; spangled perch, *Leiopotherapon unicolor*). The separated common carp are then harvested.

Common carp spawn in floodplain wetlands (Smith 2005; Gilligan and Rayner 2007; Conallin *et al.* 2012) and flow manipulations, exclusion screens and innovative trapping techniques enable them to be targeted at those sites (Stuart and Jones 2006a; Hillyard *et al.* 2010). In Lake Crescent and Lake Sorell, Tasmania, common carp eradication is being attempted using containment, intensive multi-gear fishing, sterile Judas common carp and barrier-netting. This has succeeded in Crescent and there has been a major reduction in the population in Sorell (<http://www.ifs.tas.gov.au/ifs>). More information about control of alien species is given in Chapter 12.

VALUES OF ALIEN FISHES

Fishing is one of few activities in a modern, sedentary, post-agricultural society where people are free to express their hunter-gatherer nature. Recreational fishing is highly valued in Australia, with benefits across the spectrum from native to alien species. It enhances people's enjoyment of natural environments, contributes to personal health and well-being and is an outlet for emotional stresses (Horwitz *et al.* 2001; Henry and Lyle 2003; Rowe *et al.* 2008). Indeed, many scientists and managers are freshwater anglers. Fishing is a fundamental part of the traditional culture and modern-day economics of indigenous communities.

Yet people's attitudes to alien fishes range widely. There is general disdain for common carp, burgeoning enthusiasm for aquarium fishes from hobbyists, zeal for salmonids (especially trout) from anglers and aquaculturists, ambivalence about redfin perch, scant knowledge of eastern gambusia and tilapia and general ignorance of oriental weatherloach, gobies, platys and many more species. Issues related to native fish translocations have little currency in the general community, and arguments about the merits of wild stocks and hatchery fishes rarely enter the public arena. Each of these issues motivates polarised community groups and there is neither consensus (except for common carp) nor broad understanding. Nor is there consensus over what constitutes a pest fish, aside from common carp. This perception depends on the backgrounds and affiliations of individuals, it varies widely between regions and there is probably a divide between the attitudes of urban and rural Australians. As a consequence, relatively few alien species are gazetted as pests (see Ch. 12).

There is an entrenched divide between community attitudes to trout (and salmonids in general) and other alien species. Trout provide economic and social benefits and most, but by no means all, Australians would consider them desirable (Karolak 2006). Fishing may foster an affinity with nature that often seems denied to modern city-dwellers. In an ideal world, wild native fish stocks might meet these needs but the realities of a transformed environment and well-established trout fisheries suggest that this is unrealistic. Nevertheless, the divide reveals a double standard in public attitudes to native fishes and other matters of natural heritage.

CONCLUSION

Change is a constant in evolution, and there were invasions and extinctions long before humans intervened. Why be concerned, then, about homogenisation of the global fauna and flora? One response is that the changes now underway are happening faster than ever before. Another is that they are unfolding within a global context of profoundly altered natural systems whose resilience and resistance to change are severely diminished. A sad aspect of homogenisation, as Closs *et al.* (2004) noted poignantly, is that 'with each new introduction, somewhere different becomes just a little more like everywhere else'.

The impacts of fish introductions in Australia range from apparently benign species such as goldfish to habitat destruction by common carp, to competitive pressures from eastern gambusia and tilapia, to disruptions caused by predators such as redfin perch and trout, to the little-understood effects and continuing threats of guppies, oriental weatherloach and a host of other ornamental fishes. There are also the genetic and evolutionary implications of translocated native species, with growing concern over the low fitness and viability of hatchery fish compared to wild stock. The philosophical divides in perceptions of alien and native fishes are often based on value judgements rather than on objective data. Some favour a nativism paradigm, wherein native species are inherently more desirable than non-indigenous species, and 'non-indigenous' is synonymous with 'pest'. Others recognise the social and economic benefits of some non-indigenous species, particularly trout and most translocated native species. Is it realistic to aim for control or eradication of alien pest fishes? Is it realistic to protect all native fishes? Or should we simply accept that homogenised fish communities will persist, manage them as best we can and pragmatically appreciate their values? What are the respective returns on investment for habitat rehabilitation rather than species management? And, more generally, where do we draw a line between our regard for natural heritage and our aspirations for the quality of human lives?

Non-indigenous fishes may appear to be here to stay, although we should not underestimate the potential for modern technology to discover practical means for control or even eradication of common carp, eastern gambusia and tilapia. For the present, we must look for ways to limit and ameliorate their impacts, to avoid further introductions and to protect and promote native fishes. Perhaps the clearest lessons concern the divide between salmonids and other alien species, the recognition that – as a group – alien fishes have harmed native biodiversity and ecosystem services and that the issues are not merely scientific ones. Values held by the community at large, and the inevitable compromises that lead us to pluralist policies on issues such as this, argue for the pragmatic view that the role of science is to inform people about the implications of policy decisions made for reasons other than science alone.

12 Conservation and management

Mark Lintermans

INTRODUCTION

Globally, freshwater fishes and their habitats are imperilled (Malmqvist and Rundle 2002; Dudgeon *et al.* 2006; Jelks *et al.* 2008). The main threats are habitat loss or modification, altered flow regimes, changed water quality, barriers to fish passage, alien species, translocation and stocking and overexploitation. Freshwater fishes are about 40% of all fish species but their habitats – in rivers, lakes and wetlands – are only a tiny part (0.01%) of the world's water. Australia in particular has a comparatively small freshwater fish fauna of 256 recognised species, of which 74% are endemic (Ch. 2).

For many years, recreational fisheries, especially for alien salmonids, were the main concern in Australia. From the 1980s, reviews (Cadwallader 1978; Michaelis 1985; Harris 1987b) drew attention to threatened native species, regionally and nationally, and helped shift the focus to conservation, including the monitoring, maintenance and restoration of species, communities and ecosystems. This will seem curious to readers outside Australia, but we have few commercial freshwater fisheries and their yield is dwarfed by that from marine fisheries (Kailola *et al.* 1993). This reflects the nature of our freshwater fauna and the fact that most Australians live near the sea. Indeed, the pressures on native fishes are concentrated in the southern part of the continent, where most people live. Fish populations in the north generally are in better condition.

Australia's freshwater fishes are at a crossroads. Intense competition for water, proliferations of alien species and continuing habitat losses mean that, without concerted action, extinctions will occur within the next 30 years. What should we conserve, and how should we go about it? This chapter explores these issues, including discussions of threatened species, the nature of threats to fishes and responses by management. Although our context is Australian, the book by Helfman (2007) is a good introduction for those interested in a global perspective.

CONSERVATION

Legislation

The legislative basis for freshwater conservation in Australia is a recent development. Awareness of the decline of biodiversity increased in the 1970s and 1980s, with the Victorian *Flora and Fauna Guarantee Act 1988* in the vanguard. National approaches followed, with the *Endangered Species Conservation Act 1992* (ESC Act) supplanted by the still-current

Environment Protection and Biodiversity Conservation Act 1999 (EPBC Act). The EPBC Act is designed to protect and manage matters of national environmental significance, including nationally and internationally significant fauna, flora and ecological communities. The Act provides for listing these entities as threatened (vulnerable, endangered, critically endangered) and provides conservation advice and recovery plans, recognition of key threatening processes and, where appropriate, threat abatement plans. The provisions are explained on the website of the Department of Sustainability, Environment, Water, Populations and Communities (<http://www.environment.gov.au>). The EPBC Act was recently reviewed (Hawke 2009) and procedural matters may change, but the basic framework is likely to remain intact.

Some provisions of the EPBC Act apply only to the small areas of land and water that are under Commonwealth control; the states and territories (the jurisdictions) have their own legislation. Although the various governments are working to align procedures and criteria, inconsistencies are likely to remain (Hawke 2009). This chapter mainly concerns national legislation, with incidental references to jurisdictional matters.

Threatened species and ecological communities

The first official list of threatened fauna in Australia, compiled in 1980, included three freshwater fishes, namely trout cod (*Maccullochella macquariensis*), Macquarie perch (*Macquaria australasica*) and Australian grayling (*Prototroctes maraena*) (Burbidge and Jenkins 1984). In 1985, a national review listed 14 species of concern (Michaelis 1985), and in 1987 a workshop identified 13 species as endangered, vulnerable or potentially threatened (Harris 1987b). By 2011, 49 species had been listed as extinct in the wild, critically endangered, endangered or vulnerable. Of these, 36 are listed under the EPBC Act, 13 are on a list maintained since 1985 by the Australian Society for Fish Biology (ASFB) and an additional 25 species are listed under state or territory legislation (Table 12.1).

State and federal lists can have different emphases (Possingham *et al.* 2002), but most agencies have adopted or approximated the higher-threat categories of the IUCN, the International Union for the Conservation of Nature (Table 12.2). Only these higher categories are considered here.

Of 74 fish taxa listed as threatened under state or national legislation (including three significant populations: Dept of Industry and Development, New South Wales 2011a), the Galaxiidae has most threatened taxa (19 of 23 described taxa). Other conspicuous families are the Percichthyidae (15 of 26 taxa), Atherinidae and Eleotridae (six taxa each) and Gobiidae (five taxa) (Table 12.3). Many of these taxa occur in south-eastern Australia, where the pressures of human populations and agricultural development are greatest. Indeed, 46–55% of species in the Australian Capital Territory, South Australia, Tasmania and Victoria are listed as threatened (Fig. 12.1).

Listing under the EPBC Act does not prevent recreational fishing for a species in the states or territories. Threatened species not protected from 'take' (capture and retention) include Macquarie perch in Victoria, Mary River cod (*Maccullochella mariensis*)

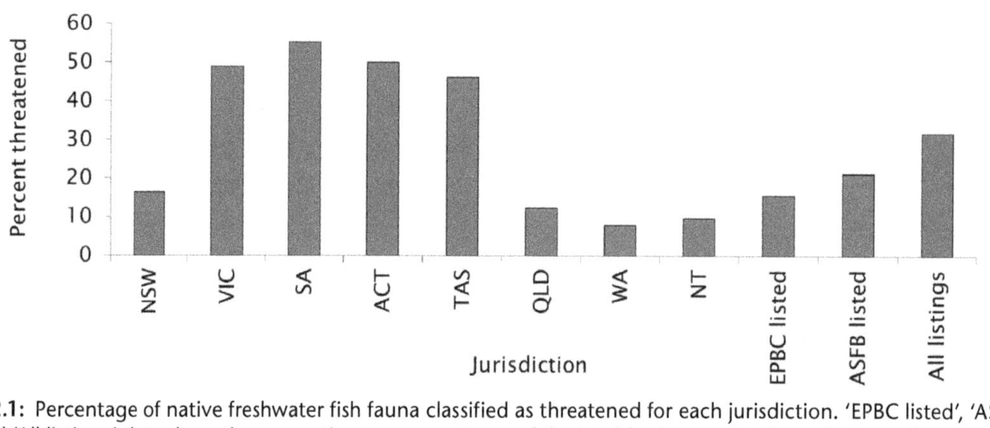

Figure 12.1: Percentage of native freshwater fish fauna classified as threatened for each jurisdiction. 'EPBC listed', 'ASFB listed' and 'All listings' data show the proportions as percentages of the total freshwater fish fauna in Australia.

Box 12.1: Conservation management of *Maccullochella*

All four species of the freshwater cod genus *Maccullochella* are listed as endangered or vulnerable under the EPBC Act. Two of three endangered species (trout cod, Clarence River cod) are totally protected and angler harvest is prohibited, but limited harvest is allowed for the third species (Mary River cod) in stocked waters (Lintermans *et al.* 2005). The fourth species, Murray cod, is vulnerable but widely distributed and an iconic target for anglers. Some 78% of captured Murray cod are released (Henry and Lyle 2003). The continued availability to anglers of Mary River and Murray cod has been a catalyst for engagement with stakeholders. In contrast, trout cod, which has been unavailable for harvest since the 1980s, has a reduced community profile that has contributed to difficulties in communication with anglers. For example, long-term restocking programs have increased local abundances of trout cod but have also raised difficulties in persuading some anglers that the prohibition on harvest should remain until natural recruitment is well-established. Some controlled recreational harvest of trout cod may help to boost community appreciation of the species, and a proposal to this end has been suggested (Native Fish Australia 2006; Trout Cod Recovery Team 2008).

Management arrangements for individual species often vary between jurisdictions. Where a threatened species is distributed across state boundaries, harmonised recreational and conservation management is desirable but regulations need not be identical (e.g. Murray cod, trout cod). Identification of well-defined and discrete spatial management units would promote local stewardship of threatened species.

From top: Murray cod, trout cod, Clarence River cod, Mary River cod (photos: G Schmida, E Beaton, J Mathews, Qld DEEDI respectively).

at some locations in Queensland and Murray cod (*M. peelii*) nationally (see Box 12.1). Other listed species are protected from take but not necessarily from capture (they must be returned alive to the water). The effects of capture and release are little known (Bartholomew and Bohnsack 2005; Cooke and Suski 2005), although many anglers believe that released fish are not harmed (Arlinghaus *et al.* 2007). A study of silver perch (*Bidyanus bidyanus*) showed that most shallow-hooked or lure-caught fish survive, but that more than 70% of deep-hooked fish die, regardless of the type of hook (Van der Walt *et al.* 2005b).

Fish species may be included in national- or state-listed ecological communities (Table 12.4). For example, native fish (Edgbaston goby, *Chlamydogobius squamigenus*; Elizabeth Springs goby, *C. micropterus*; redfin blue-eye, *Scaturiginichthys vermeilipinnis*) are part of an EPBC-listed endangered ecological community associated with mound springs in the Great Artesian Basin. There are no freshwater fish

Table 12.1: Freshwater fish species listed under the EPBC Act and the ASFB list, and the status of recovery plans for each species (0 = none, 1 = current, 2 = in preparation)

Species	Common name	EPBC Act	ASFB	Recovery plan status
Bidyanus bidyanus	Silver perch	–	Vu	0
Cairnsichthys rhombosomoides	Cairns rainbowfish	–	Vu	0
Chlamydogobius japalpa	Finke goby	–	Vu	0
Chlamydogobius gloveri	Dalhousie goby	–	Vu	0*
Chlamydogobius micropterus	Elizabeth Springs goby	En	C En	0*
Chlamydogobius squamigenus	Edgbaston goby	Vu	C En	0*
Craterocephalus amniculus	Darling hardyhead	–	Vu	0
Craterocephalus dalhousiensis	Dalhousie hardyhead	–	Vu	0*
Craterocephalus gloveri	Glover's hardyhead	–	Vu	0
Craterocephalus fluviatilis	Murray hardyhead	Vu	C En	1
Galaxias auratus	Golden galaxias	En	En	1
Galaxias fontanus	Swan galaxias	En	En	1
Galaxias fuscus	Barred galaxias	En	C En	1
Galaxias johnstoni	Clarence galaxias	En	En	1
Galaxias parvus	Swamp galaxias	Vu	En	1
Galaxias pedderensis	Pedder galaxias	Ex Wild	Ex Wild	1
Galaxias rostratus	Flathead galaxias	–	Vu	0
Galaxias tanycephalus	Saddled galaxias	Vu	Vu	1
Galaxias truttaceus hesperius	Spotted galaxias	C En	C En	1
Galaxiella pusilla	Eastern dwarf galaxias	Vu	Vu	1
Glyphis glyphis	Speartooth shark	C En	C En	0
Glyphis garricki	Northern river shark	En	En	0
Guyu wujalwujalensis	Bloomfield River cod	–	Vu	0
Himantura chaophraya	Freshwater whipray	–	Vu	0
Maccullochella ikei	Clarence River cod	En	En	1
Maccullochella macquariensis	Trout cod	En	C En	1
Maccullochella mariensis	Mary River cod	En	C En	1
Maccullochella peelii	Murray cod	Vu	Vu	2
Macquaria australasica	Macquarie perch	En	En	2
Melanotaenia eachamensis	Lake Eacham rainbowfish	En	En	0
Milyeringa veritas	Cave gudgeon	Vu	Vu	0
Mogurnda clivicola	Flinders Ranges purple-spotted gudgeon	Vu	Vu	0
Mordacia praecox	Non-parasitic lamprey	–	Vu	0
Nannatherina balstoni	Balston's pygmy perch	Vu	Vu	0
Nannoperca obscura	Yarra pygmy perch	Vu	Vu	1

Species	Common name	EPBC Act	ASFB	Recovery plan status
Nannoperca oxleyana	Oxleyan pygmy perch	En	En	1
Nannoperca variegata	Variegated pygmy perch	Vu	Vu	1
Nannoperca sp.	Little pygmy perch	–	CE	0
Neoceratodus forsteri	Australian lungfish	Vu	Vu	2
Neosilurus gloveri	Dalhousie catfish	–	Vu	0
Ophisternon candidum	Blind cave eel	Vu	Vu	0
Paragalaxias dissimilis	Shannon galaxias	Vu	En	1
Paragalaxias eleotroides	Great Lake galaxias	Vu	En	1
Paragalaxias mesotes	Arthurs galaxias	En	En	1
Pristis microdon	Freshwater sawfish	Vu	C En	2
Prototroctes maraena	Australian grayling	Vu	Vu	1
Pseudomugil mellis	Honey blue-eye	Vu	En	0
Scaturiginichthys vermeilipinnis	Redfin blue-eye	En	C En	2*
Stiphodon semoni	Neon goby	C En	C En	0

Ex = extinct; Ex Wild = extinct in the wild; C En = critically endangered; En = endangered; Vu = vulnerable; R = rare; T = threatened.
*: See Fensham *et al.* (2010).

Table 12.2: Categories and authorities for listing threatened freshwater fishes in national, state and territory jurisdictions. Lesser categories (e.g. near threatened) are not included

Jurisdiction	Authority	Extinct	Extinct in the Wild	Critically Endangered	Endangered	Vulnerable	Rare	Other
Commonwealth	*Environment Protection and Biodiversity Conservation Act 1999*	✓	✓	✓	✓	✓		
National#	Australian Society for Fish Biology	✓	✓	✓	✓	✓		
Queensland	*Nature Conservation Act 1992*		✓		✓	✓		
Victoria	*Flora and Fauna Guarantee Act 1989*							✓A, C
New South Wales	*Fisheries Management Act 1994*	✓		✓C	✓B, C	✓C		
Australian Capital Territory	*Nature Conservation Act 1980*	✓			✓	✓		
South Australia*	*National Parks and Wildlife Act 1972*				✓	✓	✓	
Western Australia	*Wildlife Conservation Act 1950*	✓						✓D
Northern Territory	*Territory Parks and Wildlife Conservation Act 2000*	✓	✓	✓	✓	✓		
Tasmania	*Threatened Species Protection Act 1995*	✓			✓	✓	✓	

A = listed as threatened (with non-legislative advisory lists to categorise species); B = includes populations; C = includes ecological communities; D = top category is 'rare or likely to become extinct'; # = non-statutory listing; * = no species currently listed, but the State Action Plan adopts IUCN categories (Hammer *et al.* 2009).

Table 12.3: Familial composition of 74 threatened freshwater fish taxa (species, subspecies and significant populations) at state or national level (Ex, Ex Wild, C En, En, Vu)

Family	Total Australian freshwater taxa	Threatened taxa
Ambassidae	8	1
Atherinidae	16	6
Bovichtidae	1	1
Carcharhinidae	3	2
Clupeidae	4	1
Dasyatidae	1	1
Eleotridae	31	6
Galaxiidae	23[a]	19
Geotriidae	1	1
Gobiidae	17[b]	5
Melanotaeniidae	21	3
Mordaciidae	2	2
Neoceratodontidae	1	1
Percichthyidae	26[a,c]	15
Plotosidae	20	2
Pristidae	1	1
Retropinnidae	7	1
Pseudomugilidae	6	2
Synbranchidae	2	1
Terapontidae	25	3

a: Taxonomy under review (T. Raadik, M. Hammer, pers. comms).
b: Likely to rise with new species of cling goby identified in northern Australia (B. Ebner, pers. comm.).
c: Includes former Gadopsidae, Nannopercidae.

communities listed nationally, although some (e.g. the Murray River below the Murray-Darling junction) have been proposed.

Extinctions

No Australian freshwater fish species is known to have become extinct since 1788, although this claim has been challenged by research using molecular methods (Hammer *et al.* 2007; Faulks *et al.* 2010a; Nock *et al.* 2011; Chs 2, 3). Even so, claims of extinction should be made with caution. Reports of the extinction in the wild of the Lake Eacham rainbowfish (*Melanotaenia eachamensis*) were proven wrong when the species was rediscovered in local streams (Pusey *et al.* 1997; Zhu *et al.* 1998).

Many so-called regional extinctions are better seen as range contractions, although they do represent significant losses of genetic diversity (Ch. 3) and possibly cryptic taxa (e.g. Macquarie perch in the Kangaroo River, New South Wales: Faulks *et al.* 2010a; Ch. 3). Examples include Agassiz's glassfish (*Ambassis agassizii*), southern purple-spotted gudgeon (*Mogurnda adspersa*) and freshwater herring (*Potamalosa richmondia*) in Victoria (Dept of Sustainability and Environment, Victoria 2007) and flathead galaxias (*Galaxias rostratus*), Macquarie perch and trout cod in South Australia (Dept of Sustainability and Environment, Victoria 2007; Hammer *et al.* 2009). For these species, recolonisation of former ranges seems unlikely

Box 12.2: Translocation of the Pedder galaxias

The Pedder galaxias (*Galaxias pedderensis*) is the only Australian freshwater fish listed as extinct in the wild, and its survival is a result of sustained conservation effort. It is a Tasmanian endemic, and naturally occurred only in Lake Pedder and nearby streams (Dept of Primary Industries, Water and Energy, Tasmania 2006). Following flooding of the region for hydropower generation in the early 1970s, the population appeared to have survived, but it declined dramatically after 1980 (Lake 1998). It is likely that interactions with alien brown trout and the native climbing galaxias (*Galaxias brevipinnis*) were responsible. Surveys in 1988–91 found few Pedder galaxias in streams flowing to the new Lake Pedder. Between 1991–97, 34 individuals were introduced to nearby Lake Oberon, and monitoring in 1997 showed evidence of successful recruitment (Crook and Sanger 1997). The species now has been introduced to two other localities.

From Fulton (1990).

Table 12.4: Threatened ecological communities and threatening processes listed in New South Wales and Victoria, and recovery plans or threat abatement plans (if any)

Community/process	Category (year)	RP/TAP (year)
New South Wales		
The aquatic ecological community in the natural drainage system of the lower Murray River catchment	Endangered EC (2001)	
The aquatic ecological community in the natural drainage system of the lowland catchment of the Darling River	Endangered EC (2003)	
The aquatic ecological community in the natural drainage system of the lowland catchment of the Lachlan River	Endangered EC (2005)	
Hook-and-line fishing in areas important for the survival of threatened fish species	KTP (2004)	
Removal of large woody debris from rivers and streams	KTP (2001)	2007
Introduction of fishes to fresh waters within a river catchment outside their natural range	KTP (2001)	
Degradation of native riparian vegetation along watercourses	KTP (2001)	
Installation and operation of instream structures and other mechanisms that alter natural flow regimes of rivers and streams	KTP (2002)	
Human-caused climate change	KTP (2010)	
Victoria		
Lowland riverine fish community of the southern Murray-Darling Basin	Threatened Community (2000)	
Alteration to the natural flow regimes of rivers and streams	PTP (1992)	2003
Alteration to the natural temperature regimes of rivers and streams	PTP (1992)	2003
Degradation of native riparian vegetation along rivers and streams	PTP (1996)	2003
Input of toxic substances into rivers and streams	PTP (1995)	
Prevention of passage of aquatic biota as a result of the presence of instream structures	PTP (1995)	2003
Introduction of live fishes into waters outside their natural range within a river catchment after 1770	PTP (1992)	2003
Removal of woody debris from streams	PTP (1991)	2003
Increase in sediment input into rivers and streams due to human activities	PTP (1991)	2003

RP = recovery plan; TAP = threat abatement plan.
EC = ecological community; KTP = key threatening process; PTP = potentially threatening process.

without human intervention. In Tasmania, the Pedder galaxias (*Galaxias pedderensis*) is extinct in the wild (Table 12.1), but persists as two translocated wild populations outside its natural range (Dept of Primary Industries, Water and Energy, Tasmania 2006; Box 12.2). Other species or populations may be in 'extinction debt', meaning that they have survived environmental change thus far but are unlikely to do so in the long term (Kuussaari *et al.* 2009). Better understanding is needed of the magnitude, probability and time scales of extinction debt in freshwater fishes to assist conservation programs (Olden *et al.* 2010).

Recovery plans and other provisions

A recovery plan is intended to counter threats, initiate monitoring or research, inform the public and generally improve the status of listed species or ecological communities (Taylor *et al.* 2005; Kerkvliet and Langpap 2007). The EPBC Act originally required such a plan for each listed entity, but in 2006 this became discretionary. If required, a plan must be in place within three years of listing, and reviewed within five years. Of course, mere preparation of a plan is not enough; it must be implemented (Lundquist *et al.* 2002), often by agencies with scant resources. Implementation is at state, regional or local levels and is not a federal responsibility other than for listed species or ecological communities on Commonwealth land.

An alternative is to promote regional and ecological community-based approaches (Bryant and Harris 1996) through multi-species recovery plans (Dept of Environment, Climate Change and Water, New South Wales 2010). Under the US *Endangered Species Act 1988*, species listed in this way may receive less attention than those with dedicated plans, and are more likely to decline (Lundquist *et al.* 2002; Taylor *et al.* 2005; Schwartz 2008). Ideally, there should be a mix of single- and multi-species approaches.

Ideally, a recovery plan should lead to downlisting or delisting of species, but this has not occurred for any freshwater fish in Australia, suggesting that threatened species could require conservation management indefinitely (Doremus and Pagel 2001; Scott *et al.* 2005). The primary goal is to prevent extinction, however, and without listing and recovery actions the Pedder galaxias and barred galaxias (*Galaxias fuscus*) arguably would be extinct, and the Mary River cod would be near extinction.

Few key threatening processes have been declared under the EPBC Act (or its predecessor: Woinarski and Fisher 1999) and threat abatement plans are discretionary. Neither of these provisions has realised its potential in regard to conservation of freshwater fishes, and the problem is only partly offset by state and territory legislation (Table 12.4).

Management of threatened fishes

Most EPBC-listed species are protected from take and have some provision for habitat protection or rehabilitation. Some large species (e.g. *Maccullochella* spp., Macquarie perch) have hatchery breeding and release programs, as their breeding biology is inferred from knowledge of congeners which sustain stocked recreational fisheries. These programs are a means to prevent extinctions and to establish new, self-sustaining populations within the historic distributions of the respective species. Most involve stocking with fingerlings, as this minimises the time (and cost) needed for holding and maintenance. Some programs have reported wild offspring from stocked fish (Douglas and Brown 2000; Lyon et al. 2008a), but claims of success are premature until there are several new generations in the restocked population.

In Australia, restocking programs rely on taking wild broodfish and releasing their captive-bred progeny back into the wild. These programs are an easier option than efforts to restore habitat, eradicate alien species or counter other large-scale threats (Philippart 1995). They provide life history data to support conservation efforts, but cannot wholly replace them (Rakes *et al.* 1999). Indeed, stocking can reduce genetic diversity in wild populations if it is based on few broodstock (Nock *et al.* 2011). Hatchery-based programs can fail for lack of understanding the causes of decline, vague or unrealistic objectives, lack of evaluation, failure to incorporate management tools and lack of cost–benefit analysis (Molony *et al.* 2003). In addition, the survival of hatchery-reared adult fish can be affected by behavioural deficits such as predator naïvety (Ebner *et al.* 2009a).

Reintroduction programs have commenced for some small fish species (e.g. southern purple-spotted gudgeon; southern pygmy perch, *Nannoperca australis*; Murray hardyhead, *Craterocephalus fluviatilis*: Australian Society for Fish Biology 2009; Hammer *et al.* 2009). The biology of some small species may be inferred from knowledge of related species in the aquarium trade, and others have been investigated specifically for the purposes of stock recovery (Ellis and Pyke 2010). This work has been spurred, in part, by prolonged drought in the southern Murray-Darling Basin in 1997–2010.

'Environmental flows' (environmental water allocations) are employed to compensate *in situ* for water shortages caused by drought or diversions for agriculture and industry. They are generally intended to preserve small or discrete habitats, offset poor water quality (e.g. blackwater events: see 'Management responses: water quality') or promote spawning. For example, environmental flows have been deployed to protect Murray hardyhead (Pritchard *et al.* 2009; Dept of Sustainability and Environment, Victoria 2010;

Ellis and Pyke 2010) and river blackfish (*Gadopsis marmoratus*: Hammer *et al.* 2009), and to promote spawning of golden galaxias (*Galaxias auratus*: Dept of Primary Industries and Water, Tasmania 2008).

Where environmental flows are not practical, fishes may be collected from the wild and maintained *ex situ* (rescues). Subjects have included include barred galaxias, golden perch (*Macquaria ambigua*), Macquarie perch, Murray cod, river blackfish, silver perch, southern pygmy perch and Yarra pygmy perch (*Nannoperca obscura*) (Australian Society for Fish Biology 2009; Hammer *et al.* 2009; Pritchard *et al.* 2009).

Challenges

From the viewpoint of freshwater fish conservation, the EPBC Act has shortcomings (Hawke 2009). For example:

- The Act requires nominations as a starting point for threatened species listings and thereby is a passive rather than strategic approach. There has been no review of the national status of freshwater fishes since the action plan of Wager and Jackson (1993), and a revision is needed urgently.
- The Act does not cover actions initiated prior to 1999, and long-standing, still-current threats (e.g. alien species, river regulation, vegetation clearance) are not addressed.
- The time to respond to issues is a limitation in that nomination, assessment and listing processes, preparation of recovery plans and engagement with communities and agencies may take years. The status of species may change on much shorter time scales (Pritchard *et al.* 2009). Responses are slowing under mounting pressures but the process needs to accelerate, with less emphasis on process and more on outcomes.
- Listing species without recovery plans, or without resources to prosecute plans, is unlikely to succeed (Kerkvliet and Langpap 2007). Delisting or downlisting species are desirable outcomes but they alone do not indicate the success of plans (Doremus and Pagel 2001); indeed, no EPBC-listed species has been permanently delisted as a result of conservation actions. Australia might emulate a requirement of the US *Threatened Species Act 1988*, namely that biennial measures of population trends are needed to indicate whether or not a species is recovering (Scott *et al.* 2005; Taylor *et al.* 2005).
- The Act provides for declarations of 'critical habitat' (habitat essential for conservation of a species) but this is not a mandatory part of the listing process. The only such declaration for a freshwater fish in Australia is for Oxleyan pygmy perch (*Nannoperca oxleyana*) (Dept of Industry and Development, New South Wales 2010b). This is significant because under the US *Endangered Species Protection Act 1988* the status of species with identified critical habitat is more likely to improve than for other listed species (Taylor *et al.* 2005).

Other challenges relate to strategic issues and learning from experience. There is vigorous debate about the best strategies to deploy scarce resources (Bottrill *et al.* 2008; Briggs 2009; Jachowski and Kesler 2009), given that resources for conservation are invariably inadequate to meet all needs (Schwartz 2008; Joseph *et al.* 2009). Funding for research and management is influenced by the level of endangerment of the target species (Possingham *et al.* 2002), but the degree of conflict that a species engenders, its perceived charisma and political factors are also important (Schwartz 2008). One issue, for example, is to determine whether we should invest heavily in protection of critically endangered species at the expense of vulnerable species. Another is that past experience is often not incorporated into proactive management, in the spirit of adaptive management. Environmental extremes (flood, drought, fire) often trigger novel crisis management approaches (e.g. refugia mapping, drought contingency plans, rescues, translocations), but it appears that lessons are soon forgotten.

ISSUES FOR MANAGEMENT

The principal threats to Australia's freshwater fishes (Allen *et al.* 2002; Lintermans 2007) are like those in other parts of the world (Malmqvist and Rundle 2002; Jelks *et al.* 2008) and broadly are of seven kinds: habitat modification, altered flow regimes, water quality, barriers, alien species, translocation and stocking, and overexploitation. The threats vary regionally and are related to patterns of human settlement and resource use. Each is considered below.

Habitat modification

The National Land and Water Resources Audit (2002) reported that more than 85% of assessed river reaches in Australia were significantly modified. Riparian zones and stream channels may be eroded or silted; they may be cleared of native vegetation and invaded by blackberries, mimosa, willows and other alien species; channels may be desnagged or dredged to improve water delivery; and wetlands may be drained, disconnected or flooded by dams, weirs, levees and other structures (Robertson and Rowling 2000; Pusey and Arthington 2003).

Siltation reduces instream habitat diversity by filling holes and smothering spawning sites, reducing light penetration and the diversity and abundance of invertebrates (Doeg and Koehn 1994; Davis and Finlayson 2000; Bartley and Rutherfurd 2005). Channelising and desnagging reduce hydraulic diversity (Hortle and Lake 1982), remove shelter from predators and strong currents and destroy breeding sites (Crook and Robertson 1999; Erskine and Webb 2003; Lester and Boulton 2008). The isolation of rivers and floodplains has reduced connectivity between nursery, feeding and spawning habitats (Kingsford 2000). Dams and weirs have altered sediment and flow regimes and may release cold, hypoxic water, degrading the habitats of native fishes. These effects may persist for hundreds of kilometres downstream (Preece and Jones 2002; Preece 2004).

Altered flow regimes

The natural flow patterns of streams, including flow magnitude, frequency, duration, timing, variability and rate of change, are critical for maintenance of biodiversity and ecosystem integrity, and changed regimes are a major threat to lotic species (Poff *et al.* 1997; Richter *et al.* 1997; Naiman *et al.* 2008). Patterns of flow in Australian lowland rivers, particularly inland rivers, are highly variable (Puckridge *et al.* 1998) and most rivers are now regulated by dams, weirs and other structures, changing flow regimes downstream (Kingsford 2000; Arthington and Pusey 2003). Reduced flooding has isolated floodplains, limiting dispersal opportunities and access to spawning and larval rearing sites (Jones and Stuart 2008; Tonkin *et al.* 2008a), and altered flow regimes have impacted on breeding and migration cues for native fishes (Gehrke *et al.* 1995; Tonkin *et al.* 2008a). Increased or more stable baseflow in regulated rivers may remove significant natural low-flow events, affecting the availability and quality of larval rearing habitats and influence recruitment (Humphries and Lake 2000; McMahon and Finlayson 2003). Altered flow regimes favour generalist alien species with broad tolerances and a capacity for rapid growth and reproduction (Stuart and Jones 2006a; Pyke 2008; Costelloe *et al.* 2010). Familiar examples are common carp (*Cyprinus carpio*) and eastern gambusia (*Gambusia holbrooki*).

There may be other, more subtle impacts on native fishes. Irrigation offtakes divert fish from rivers to unfavourable habitats, such as irrigation channels and flooded paddocks, where there may be no return. Pumps cause injury and mortality (King and O'Connor 2007; Baumgartner *et al.* 2009), and each year perhaps a million Murray cod larvae are extracted in irrigation water diverted from the Murray River (Gilligan and Schiller 2003).

Water quality

Broad-scale land clearing raises groundwater tables and may cause salinisation of land and water (Allison *et al.* 1990; Halse *et al.* 2003; Nielsen *et al.* 2003). Salinity is a major water quality issue in many river basins (National Land and Water Resources Audit 2001), and reduces fish habitat (Hart *et al.* 1990, 1991; Robson and Mitchell 2010). Its impacts are especially severe in streams and wetlands dependent on groundwater inflows, as in south-western Western Australia (Beatty *et al.* 2010). Adult fishes may tolerate a comparatively wide range of salinities but eggs, larvae and juveniles generally are more susceptible (Clunie *et al.* 2002a; Ye *et al.* 2010).

Many industrial and agricultural chemicals affect fishes (Jones and Reynolds 1997; Hose *et al.* 2003). For example, endocrine disruptors, including oestrogenic chemicals in discharges from sewage treatment plants and industries, are cause for concern (Milla *et al.* 2011). Endocrine disruption has been demonstrated for eastern gambusia in Australia (Batty and Lim 1999; Norris and Burgin 2010) but little is known of the effects on native fishes.

Elevated nutrient concentrations (eutrophication) from agricultural and urban runoff and sewage effluent can promote excessive plant growth and algal blooms, such as Cyanobacteria, with implications for water quality and aquatic biota, and for stock and human use (Boulton and Brock 1999). Turbid water, low oxygen levels and algal toxins can have severe effects on fishes (Engström-Öst *et al.* 2006).

Acidification of freshwater and estuarine habitats through oxidation of sulfidic soils is a potential problem in coastal and inland waters (Sammut *et al.* 1995; McCarthy *et al.* 2006). It is also associated with mining operations that mobilise zinc, cadmium, lead and copper and other heavy metals. Parts of the Molonglo River in New South Wales and the South Esk River in Tasmania have, or have had, impoverished fish and invertebrate communities as a result of contamination by acid mine drainage (Weatherley *et al.* 1967; Tyler and Buckney 1973; Norris *et al.* 1982).

Most dams have low-level offtakes, and deep water released downstream in summer and autumn is often cool and hypoxic. Cold-water pollution can change river temperatures for hundreds of kilometres downstream of large dams, delaying insect emergence and reducing fish growth, reproduction and swimming speeds (Astles *et al.* 2003; Ryan *et al.* 2003; Rutherford *et al.* 2009). Low oxygen levels (< 5 mg L^{-1}) are stressful and have caused numerous fish kills (Koehn 2005). Cold-water discharge favours alien cold-water species, particularly salmonids (Ryan *et al.* 2003; Brown 2004), and in the case of Dartmouth Dam may have been responsible for the disappearance of Murray cod and Macquarie perch from the Mitta Mitta River, Victoria (Koehn *et al.* 1997; Todd *et al.* 2005).

Barriers

Fishes move laterally (offstream) and longitudinally (instream) to access feeding and spawning areas, to avoid adverse conditions and to disperse and colonise new habitats (Ch. 5). Barriers to movements, including dams, weirs, levees and causeways, are a global concern. There are more than 45 000 big dams (> 15 m high) worldwide (Nilsson *et al.* 2005) and their effects on fishes can be compensated only by fishways or other engineered infrastructure. Even small structures, such as weirs, culverts, wetland regulators and causeways, can exclude fishes from large areas of potential habitat during low flows, although the smallest of these are drowned out during high flows.

In the Murray-Darling Basin there are more than 4000 regulating structures (Lintermans 2007) and hundreds of kilometres of floodplain levees. Some native fish, notably golden perch, can move thousands of kilometres in several months (Reynolds 1983), although this may not be typical of the species (Crook 2004a; O'Connor *et al.* 2006). Other species may move hundreds of kilometres in the absence of barriers (Kerezsy 2010a; Ch. 5).

Barriers affect downstream as well as upstream movements. Fishes are often reluctant to approach barriers when moving downstream (Lintermans and Phillips 2004; O'Connor *et al.* 2006) and passage over weirs is a cause of significant mortality. 'Undershot weirs' (where water is released at the bottom) may cause more injury than 'overshot weirs' (water overflows the weir). Mortalities of more than 90% of the larvae of golden perch and silver perch, and more than 50% of Murray cod larvae, are reported from undershot weirs. Mortalities of small species can also be severe (Baumgartner *et al.* 2006b, 2010a).

Alien species

Alien fishes threaten native species and ecological communities, nationally and internationally (Ch. 11). Australia has established populations of at least 37 alien species and the number continues to grow. The acclimatisation phase of sport fish introductions (e.g. salmonids; tench, *Tinca tinca*; redfin perch, *Perca fluviatilis*) has passed (Clements 1988) and the aquarium trade is now the main source of new introductions. Of the 37 alien species, 25 are believed to have arrived via the aquarium trade (Lintermans 2004; Corfield *et al.* 2008). Indeed, 24 of the last 27 species to become established originated from the ornamental trade (Fig. 12.2), whereas only one ornamental species (the goldfish, *Carassius auratus*) was established prior to about 1960. Another six ornamental species (blue tilapia, *Oreochromis aureus*; firemouth cichlid, *Thorichthys meeki*; green severum, *Heros severus*; redhead cichlid, *Vieja*

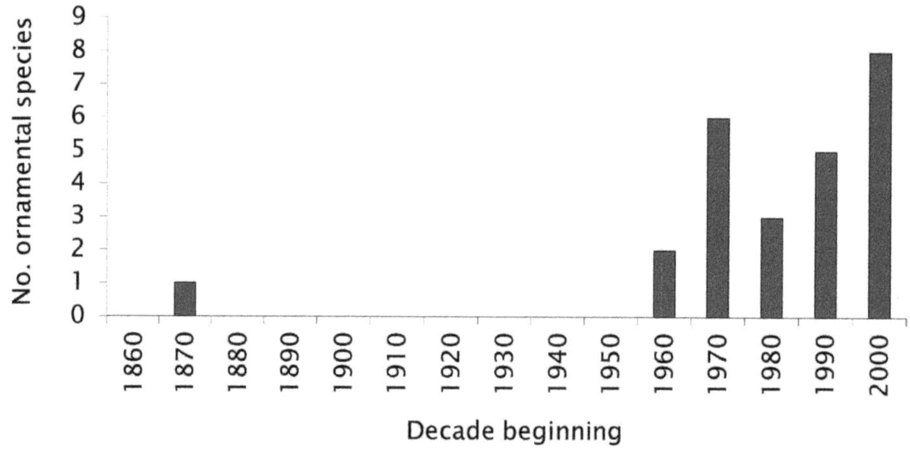

Figure 12.2: Number of ornamental species and the decade in which they became established (or were first recorded as established) in Australian fresh waters.

synspila; green terror, *Aequidens rivulatus*; Sumatra barb, *Puntius tetrazona*) are recorded or established, but their true status is unclear (Arthington *et al.* 1999; Kailola 2000; Corfield *et al.* 2008).

Some 1500 retailers (aquarium and pet shops) trade in ornamental fishes in Australia (Lintermans 2004; Dept of Agriculture, Fisheries and Forestry 2007); 12–14% of Australians keep ornamental fishes (Patrick 1998) and more than 16 million were imported in 2008–09. Smuggling of ornamental fishes is equivalent to perhaps 5–10% of legal imports (Australian Quarantine and Inspection Service 1999) or 740 000–1.48 million fish per year, based on average imports in 2006–10. Subtropical and tropical regions of Australia have borne the brunt of the 'ornamental invasion' (Webb 2007), but temperate regions are not immune and several species there are expanding their ranges (e.g. green swordtail, *Xiphophorus hellerii*; one-spot livebearer, *Phalloceros caudimaculatus*; oriental weatherloach, *Misgurnus anguillicaudatus*).

Alien species affect native fishes through predation, aggression, competition for resources, habitat changes, the spread of disease and parasites and, to a minor extent, hybridisation (Ch. 11). Predatory and aggressive interactions have affected small species such as galaxiids (Lintermans 2000; Morgan *et al.* 2002; McDowall 2006), eleotrids, melanotaeniids, pseudomugilids and small percichthyids (Barlow *et al.* 1987; Aarn and Ivantsoff 2001; Rowe *et al.* 2008). Habitat changes are attributed to common carp and tilapia (*Oreochromis mossambicus*) through feeding (Koehn 2004), nest-building (Maddern *et al.* 2007), destruction of plants (Roberts *et al.* 1995) and increased turbidity (King *et al.* 1997). Competition has been demonstrated with salmonids, eastern gambusia and tilapia (Howe *et al.* 1997; Lintermans 2006; Maddern *et al.* 2007; Stuart-Smith *et al.* 2008). Diseases, such as epizootic haematopoietic necrosis virus, a megalocytivirus and parasites such as *Lernaea cyprinacea*, *Ligula intestinalis* and *Bothriocephalus acheilognathi* are spread by alien fishes (Whittington *et al.* 1996; Chapman *et al.* 2006a; Go *et al.* 2006; Marina *et al.* 2008). Mortality of Murray cod in a fish farm in Victoria in 2003 was caused by a megalocytivirus imported with dwarf gourami (*Trichogaster lalius*), an ornamental species from south-east Asia (Go *et al.* 2006; Go and Whittington 2006).

Alien fishes in the USA cost about $US5.4 billion annually, excluding control costs (Pimentel *et al.* 2005). Common carp cause $16 million damage annually in Australia (McLeod 2004), and eastern gambusia and tilapia probably incur similar costs. However, salmonids and other alien species support recreational fisheries, and balancing environmental damage against social values can be problematic (Jackson *et al.* 2004). In south-eastern Australia, 4–5 million salmonids are stocked annually (Ch. 11).

Translocation and stocking

The global fish fauna is being homogenised by introductions of alien species and translocations of native species (Rahel 2007; Marr *et al.* 2010). Translocations can be accidental or deliberate (International Union for the Conservation of Nature 1987) and may involve interbasin water transfers, bait-bucket introductions, recreational stocking, escape from aquaculture and conservation activities (Arthington and McKenzie 1997; Lintermans 2004; Gozlan *et al.* 2010; Ch. 11).

In the Murray-Darling Basin, 3–4 million native fishes are stocked in public waters each year (Murray-Darling Basin Commission 2004); this is a significant increase from the late 1970s, when less than 0.5 million were stocked annually. Stocking sustains recreational fisheries, enables populations to persist in habitats (e.g. artificial lakes) that would not naturally sustain them, and augments natural populations (Molony *et al.* 2003). So-called 'forage fishes', including bony herring (*Nematalosa erebi*) and snub-nosed garfish (*Arrhamphus sclerolepis*), are stocked to sustain impoundment-based fisheries for Murray cod and barramundi (*Lates calcarifer*) (Moore 2007).

The impacts of stocking and translocation include hybridisation and other genetic issues with native species (Gozlan *et al.* 2010). For example, translocated trout cod and Murray cod have hybridised in Cataract Dam, New South Wales (Harris and Dixon 1988). Genetic swamping by hatchery-bred stock may reduce genetic diversity by mixing distinct stocks, underrepresenting true diversity, inbreeding, outbreeding and changing selection pressure (Ford 2002). Reduced genetic diversity is documented for hatchery broodstock of some species (Bearlin and Tikel 2003; Farrington *et al.* 2004; Rourke *et al.* 2009) and some Australian species show eroded genetic diversity from stocking (Nock *et al.* 2011). Several stocked species have 'genetic structure', indicating 'evolutionarily significant units' that warrant recognition in conservation (Keenan *et al.* 1996; Rourke 2007; Faulks *et al.* 2010b; Ch. 3).

Overexploitation

Overexploitation for commercial, recreational and artisanal use has resulted in the decline of freshwater fishes and fisheries throughout the world (Cooke and Cowx 2004; Lewin *et al.* 2006; Humphries and Winemiller 2009). In south-eastern Australia there are no longer significant commercial fisheries for native freshwater species, although small fisheries do exist for common carp (Koehn *et al.* 2000).

Overexploitation, for commercial and recreational purposes, has contributed to the decline of Murray cod (Rowland 1989, 2005; Fig. 12.3), Clarence River cod (*Maccullochella ikei*: Lintermans *et al.* 2005), trout cod (Berra 1974) and Macquarie perch (Cadwallader and Rogan 1977). In addition, fisheries may affect non-target species. For example, commercial barramundi gill-net fisheries in the lowland rivers of northern Australia have impacts on threatened species such as speartooth sharks (*Glyphis* spp.), bull shark (*Carcharhinus leucas*) and freshwater sawfish

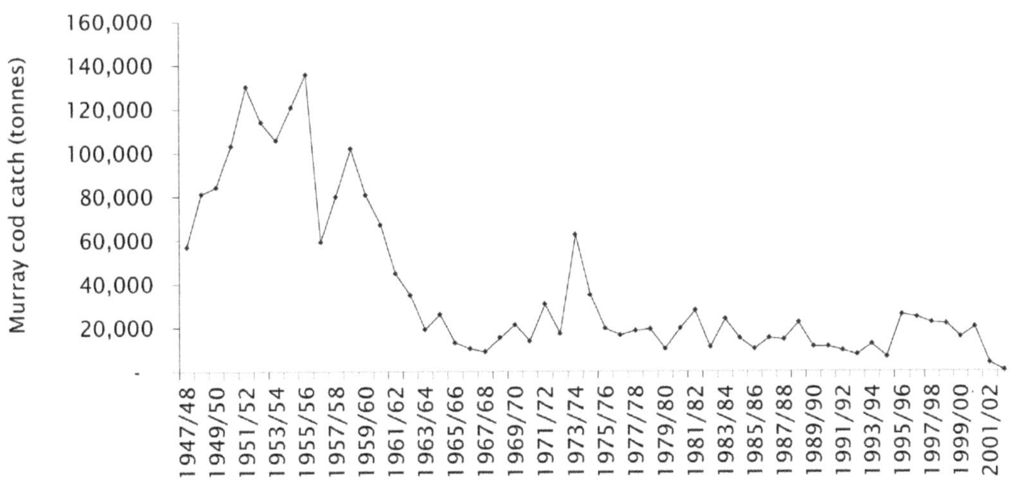

Figure 12.3: Annual commercial catches of Murray cod (*Maccullochella peelii*) in New South Wales, 1947–2001 (NSW Department of Industry and Investment/HCatch and ComCatch extracts 21-01-2011).

(*Pristis microdon*), which use fresh waters as a nursery (Peverell 2005; Field *et al.* 2008).

Overharvesting can change the size structure of populations, so that larger individuals become rare. The removal of large predatory species can affect food webs because top-down regulation of prey species is disrupted (Humphries and Winemiller 2009). Surveys in the Murray River have reported few Murray cod above the minimum size limit (Nicol *et al.* 2005; Allen *et al.* 2009) and similar effects are seen in populations of Australian bass (*Percalates novemaculeata*) in coastal New South Wales (Hall *et al.* 2009b). Clearly, if individuals are removed before they reproduce the population is likely to decline (Ch. 9). Overexploitation is also an issue where harvesting from the wild for aquarium display purposes affects small species (Ebner and Thuesen 2011).

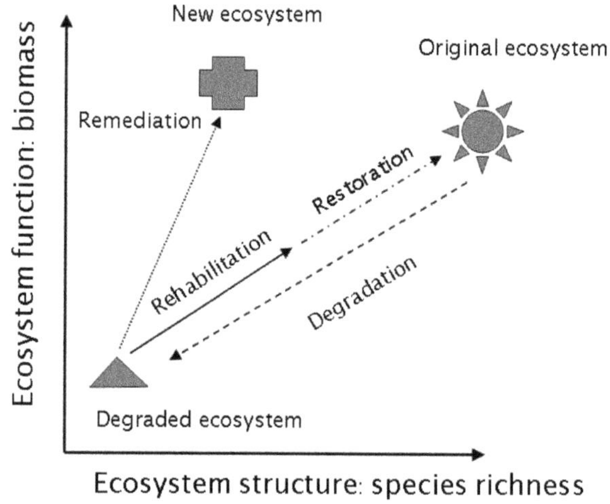

Figure 12.4: Stream management options (redrawn from Rutherfurd *et al.* 2000).

MANAGEMENT RESPONSES

Habitat modification

More than 10 000 rehabilitation projects were completed in Australia in 1995–2007, with annual expenditures of $100 million (Price *et al.* 2009). Rehabilitation works include riparian planting, bank stabilisation and provision of instream structural habitat such as woody debris (Brooks and Lake 2007). Another popular activity is release of water for environmental purposes (see 'Altered flow regimes'). 'Restoration' is often used loosely as a generic term, as it is here, but it can mean 'the return of an ecosystem to a close approximation of its condition prior to disturbance' (Brookes 1995; Higgs 1997; Rutherfurd *et al.* 2000). 'Rehabilitation' aims for a 'return to pre-disturbance for a limited number of attributes' (Brookes 1995) and 'remediation' improves the ecological condition of the waterway, not necessarily to resemble its original state (Rutherfurd *et al.* 2000; Fig. 12.4). 'Enhancement' is improvement in one or more attributes (Brookes 1995).

The attributes of a successful rehabilitation project are well-documented (Bond and Lake 2003a; Bernhardt *et al.* 2005; Palmer *et al.* 2005) and establishing clear goals is paramount (Lake 2005; Ryder *et al.* 2008). In one survey of river rehabilitation projects in the USA, only 46% had explicit goals and many were non-quantitative and difficult to measure in the field (e.g. 'establish a natural channel': Bernhardt *et al.* 2007). Most projects are poorly documented (Palmer *et al.* 2007) and failures are rarely reported (Smokorowski *et al.* 1998; Fryirs *et al.* 2007). A review of 2247 projects in Victoria showed that only 14% included provisions for monitoring (Brooks and Lake 2007). Resources are often not available for monitoring, 'before' data are often missing and control or reference sites are lacking, so that few conclusions can be drawn (Rutherfurd *et al.* 2004; Lake 2005). Fish do not always respond as expected. For example, Smokorowski *et al.* (1998) reported that while 27% of projects reported an increase in fish abundance or biomass, they generally could not distinguish an aggregation of fish from an increase in production. In such cases there is little scope for learning from the intervention, and adaptive management is precluded (see 'Monitoring and adaptive management'). There is a range of potential responses to the original stressor (Fig. 12.5), and a range of recovery trajectories. Hysteresis occurs where the pathway or thresholds for recovery are different from those for degradation, and the time frame for recovery is rarely short and often related directly to the scale of the intervention (Lake 2005; Fig. 12.5c). Bond and Lake (2003a) listed five potential reasons why anticipated fish responses to habitat rehabilitation may not occur: barriers to colonisation, temporal shifts in habitat use, introduced species, long-term and large-scale processes, and inappropriate scales of restoration. Many rehabilita-

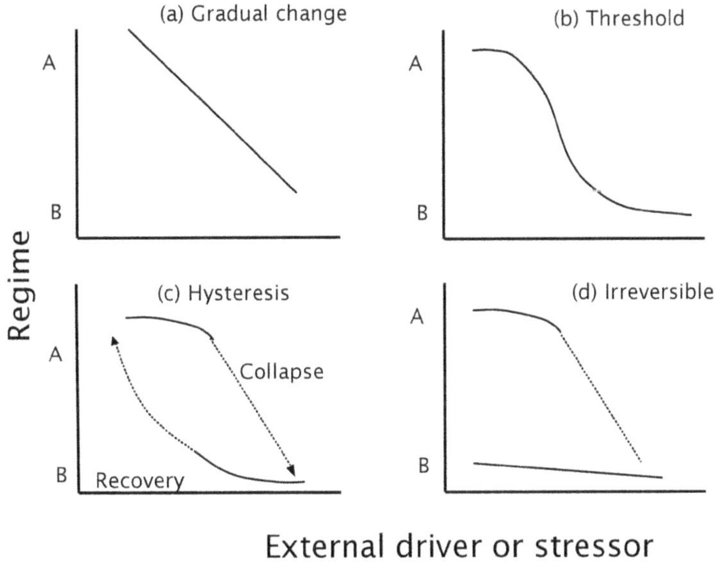

Figure 12.5: Conceptual models of system responses to stressors (redrawn from Davis *et al.* 2010b). (a) Gradual ecological change model where there is a linear relationship between two regimes (persistent ecosystem states A, B) and an external environmental driver or stressor. A gradual change in the stressor results in a gradual transition from regime A to regime B, which is reversible by reducing the stressor. (b) Threshold model with a nonlinear response to change in the external stressor. An abrupt change, or regime shift, occurs from A to B over a small change in the external stressor; this is reversible if the stressor is reduced below the threshold value. (c) Hysteresis model with a nonlinear response in which the threshold of change differs between the pathway of collapse from A to B and the pathway of recovery from B to A. (d) Irreversible change from A to B.

tion projects are based on the 'field of dreams' hypothesis ('If you build it, they will come': Palmer *et al.* 1997), which presumes that restoration of physical components is sufficient to initiate a return to the desired community or ecosystem. This is arguably one of the 'myths' of restoration ecology (Hilderbrand *et al.* 2005), although it does facilitate thinking about responses to interventions in complex systems.

Habitat management interventions

Habitat rehabilitation involving livestock exclusion, replanting and creation of riparian buffer zones is a common management intervention (Rutherfurd *et al.* 2000), but is often carried out on small scales with little support from monitoring data. Exclusion of stock can significantly increase the abundance of eucalypts and groundcover (Robertson and Rowling 2000). Guidelines are available for the amount of riparian zone rehabilitation required to ameliorate elevated water temperatures (Davies *et al.* 2004). In lowland streams, methods to reintroduce structural woody habitat are well-advanced (Nicol *et al.* 2001, 2002; Erskine and Webb 2003), but evaluations of outcomes are not (Lester and Boulton 2008). Increases in habitat heterogeneity are reported (Bond and Lake 2005; Lester and Wright 2009) and native fish diversity and/or abundance often increase (Koehn 1987; Brooks *et al.* 2004), but the responses are highly variable (Nicol *et al.* 2004; Bond and Lake 2005; Hindell 2007). A resnagging project on the Murray River between Lake Hume and Lake Mulwala has installed more than 4450 large snags in 194 km of river channel; while the snags have proved attractive to native fishes, it is too early to confirm whether populations have increased (J. Lyon, Dept. of Sustainability and Environment, Victoria, pers. comm.). It is often not clear whether additions of structural woody habitat increase fish productivity or merely cause more fishes to aggregate (Brickhill *et al.* 2005). The response of alien fishes to such additions is another neglected issue (Nicol *et al.* 2004).

'Sand slugs' (a pulse of mobile sand in streams) have reduced geomorphic complexity and habitat heterogeneity in many streams, widening channels and eliminating pool habitats (Davis and Finlayson 2000; Rutherfurd *et al.* 2000; Bartley and Rutherfurd 2005).

Interception or extraction measures can take years to have an effect, depending on the quantities of sediment. Time to recovery without management intervention is measured in decades to centuries, as a result of the large quantities of sediment in the channel. Efforts have been made to accelerate the process of recovery (Bond and Lake 2005; Lintermans 2005), but local efforts are overwhelmed unless large-scale processes are addressed (Lake 2005; Howson *et al.* 2009). Rock groynes or flow deflectors are widely used to increase scour and provide pool habitat in reaches affected by sand slugs (Roni *et al.* 2005), but their benefits are localised and they are expensive if applied at larger than site or reach scales (Lintermans 2005; Robson and Mitchell 2010). Structural woody habitat also promotes scouring in affected reaches but the benefits may be erased by siltation during high-flow events (Borg *et al.* 2007). The source and delivery of sediment to streams have been well-documented in south-eastern Australia, with forest roads and hill-slope, gully and channel erosion identified as major contributors (Davies and Nelson 1994; Croke *et al.* 1999; Prosser *et al.* 2001; Hairsine *et al.* 2002; Olley and Wasson 2003; Croke and Hairsine 2006; Lane *et al.* 2006). Improved models for amelioration of sediment delivery, the value of buffer strips and forestry codes of practice have all resulted in decreased sediment delivery to streams.

Many rehabilitation projects tend to focus on a single stressor (e.g. alien species, resnagging, erosion control, riparian rehabilitation), although it is well-understood that multiple stressors are usually involved (Davis *et al.* 2010b; Robson and Mitchell 2010). To emphasise the potential virtues of multiple actions, the 'demonstration reach' concept is being applied as part of the Native Fish Strategy for the Murray-Darling Basin (Murray-Darling Basin Commission 2004). This is a comprehensive program that addresses multiple threats in five states and territories over a decadal time scale (Koehn and Lintermans 2012). Demonstration reaches generally are 20–100 km long and subject to a variety of threats and ongoing management, with support from nearby communities (Barrett and Ansell 2005). Examples are established in all Murray-Darling Basin jurisdictions, and interventions variously include instream and riparian habitat rehabilitation, provision of environmental flows, amelioration of cold-water pollution, provision of fishways and alien fish control. Actions are tailored to each demonstration reach and monitoring is mandatory (Boys *et al.* 2008).

Rehabilitation of lentic habitats (other than wetlands) has received much less attention, although these may be critical habitats for threatened fishes (Dept of Primary Industries, Water and Energy, Tasmania 2006; Hammer *et al.* 2009; Lintermans 2012). In other countries, enhancements of lake and reservoir habitats do command attention, especially to promote recreational fishing (Bolding *et al.* 2004). As Australia has many artificial impoundments and river restoration necessarily is a slow process, we may need to look to the potential of reservoirs to help meet future conservation goals.

Altered flow regimes

In the 1990s, Australian governments agreed to pursue a water reform agenda, leading to 12 *National Principles for the Provision of Water for Ecosystems* (ARMCANZ and ANZECC 1996). As a result, governments now have legislative, policy and management mechanisms for provision of environmental flows (Brisbane Declaration 2007). The actual delivery of environmental flows has been slower than desirable, particularly in the major rivers, but this is a problem worldwide (Le Quesne *et al.* 2010). However, progress has been made and the environment now has a well-established 'water right'. More than 200 'environmental flow methodologies' (EFMs) have been described, grouped broadly as hydrological rules, hydraulic rating methods, habitat simulation methods and holistic approaches (Arthington and Pusey 2003; Tharme 2003).

Hydrological rules (rule of thumb) EFMs rely on generalisations that are of dubious value in rivers with high inter-annual flow variability (Arthington *et al.* 2006). They usually involve analysis of historical flow records and construction of a flow duration curve, with flow percentiles as targets. Although these rules are commonly used, they may result in long periods of constant flow, with negative consequences (Poff *et al.* 1997; Lytle and Poff 2004; Poff *et al.* 2010).

Hydraulic rating EFMs require analyses of channel cross-sections (often in riffles) at a range of discharges to measure hydraulic variables, such as maximum depth or wetted perimeter, used as surrogates for the habitats needed by fishes and other target biota. These approaches were once widely used in North America and less so in Australia; they have been superseded by habitat simulations (Gippel and Stewardson 1998).

Habitat simulation EFMs combine habitat suitability data for target species and habitat availability data from multiple transects under a range of discharges. The most widely used is the 'instream flow incremental methodology' (IFIM: Bovee 1982), in combination with the 'physical habitat simulation' (PHABSIM: Tharme 2003). Other methods include the 'river hydraulics and habitat simulation' (RHYHABSIM) from New Zealand (Jowett and Richardson 1995).

Holistic approaches are useful where there are many interacting elements, and are usually applied to river systems rather than reaches or sites (Arthington and Pusey 2003; King *et al.* 2003b; Tharme 2003). They consider all elements of the hydrograph (Fig. 12.6), along with geomorphic, hydraulic, biotic and ecological data (e.g. connectivity, nutrient processing, disturbance) and sometimes social issues. Holistic EFMs, including expert or scientific panel approaches, have been widely applied in Australia and South Africa and often incorporate quantitative flow ecology models (Tharme 2003). One such approach, the 'benchmarking' EFM, is used routinely in Queensland (Arthington and Pusey 2003). Other

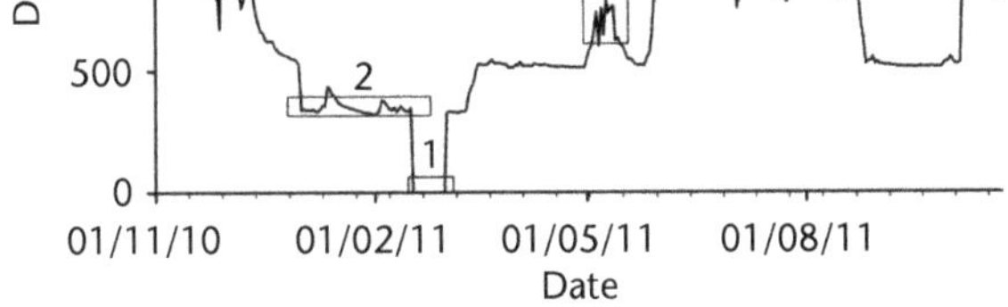

Figure 12.6: Typical daily hydrograph of a temperate river showing different flow components (1 = cease to flow; 2 = low flow; 3 = freshes; 4 = base winter flows; 5 = bankfull; 6 = overbank).

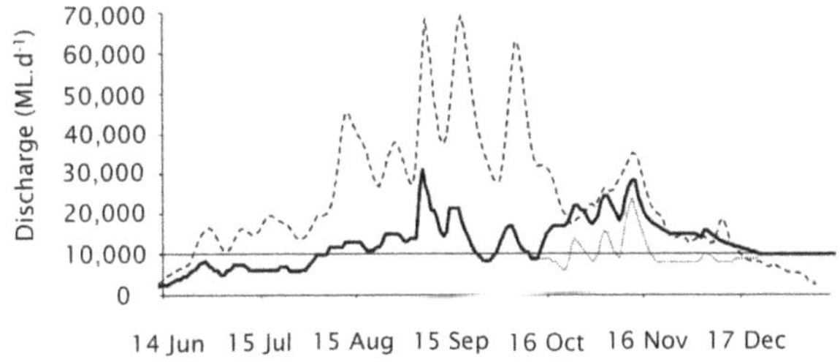

Figure 12.7: Actual mean daily discharge (bold solid line), simulated mean daily discharge without environmental water allocation (EWA, dotted line), simulated natural flows (dashed line) during the 2005 Barmah-Millewa EWA (see text). Horizontal solid line represents the discharge at which floodplain inundation commences. Flows are measured or simulated downstream of Yarrawonga, Victoria, on the Murray River upstream of Barmah-Millewa Forest. Simulated natural flows refer to modelled flows, based on tributary inputs and no upstream regulation (redrawn from King *et al.* 2009).

regional, generic approaches based on rivers in hydrological classes are exemplified by the 'ecological limits of hydrologic alteration model' (ELOHA: Poff *et al.* 2010). While the majority of environmental flow releases focus on returning water to regulated rivers, the impacts of elevated baseflows warrant more attention (McMahon and Finlayson 2003).

There are numerous reports detailing how environmental flow allocations or regimes have been determined, but few case studies of environmental flow outcomes for Australian fishes (Saddlier and Doeg 1997; Lind *et al.* 2007). One is that of King *et al.* (2009, 2010), relating to a release of 513 GL to augment a natural flood in spring 2005 in Barmah Forest on the Murray River, Victoria (Fig. 12.7). This promoted spawning of golden perch and silver perch, and although Murray cod and trout cod showed no evidence of enhanced spawning there was increased recruitment of young-of-year fishes following the event. The same flood enhanced the dispersal and recruitment of southern pygmy perch (Tonkin *et al.* 2008a) and its benefits extended hundreds of kilometres downstream, where there was enhanced spawning of golden perch and silver perch (Bucater *et al.* 2009). However, there were also benefits for the alien common carp and oriental weatherloach (King *et al.* 2007).

Environmental flows may be implemented for alien species. For example, baseflows from Parangana

Dam, Tasmania, are maintained to benefit populations of brown trout (*Salmo trutta*) (Dept of Primary Industries, Parks, Water and Environment, Tasmania 2009). Monitoring over 10 years showed that numbers of adult trout increased and stabilised about two years after the commencement of environmental releases. The abundance of young-of-year fish followed a similar pattern, but with a three-year lag. The increased abundance was attributed to increases in available wetted habitat and food resources (macroinvertebrates). There was also an increase in the local abundance of congolli (*Pseudaphritis urvillii*), possibly related to increased upstream migration. In Australia, monitoring of fish outcomes from the provision of environmental flows is still in its infancy, with few published studies. Improved monitoring is needed if true adaptive management is to be implemented (see 'Monitoring and adaptive management').

Water quality

Notwithstanding legislation to limit pollution of inland waters, enforcement is lacking and penalties often are paltry (Sinclair 2005). The development of National Water Quality Guidelines has brought more consistency across jurisdictions (Australian and New Zealand Environment and Conservation Council 2000), but the links between criteria and measurable outcomes are weak (Likens *et al.* 2009). Again, the need for better monitoring is indicated.

Salinisation is a priority in management of agricultural lands and domestic water supplies (Murray-Darling Basin Ministerial Council 2001), but concern at the impacts of salinity on biodiversity and aquatic ecosystems is usually secondary (Clunie *et al.* 2002a). A range of salinity mitigation activities is employed, including engineering (draining and pumping of saline water, capping of artesian bores, use of evaporation basins) and revegetation (planting of deep-rooted perennial species to lower watertables) (Bari and Schofield 1992; Murray-Darling Basin Ministerial Council 2001).

Many small-scale threats to water quality are often outside legislative control and responses are reliant on coordination between land, water and fish management agencies. For example, cessation of flows during drought can result in fish being restricted to shrinking habitats where local events such as dieback of aquatic plants can cause oxygen depletion and death. In this case, water, fishes and fish kills are all managed by different agencies (Koehn 2005)

Blackwater events occur when eucalypt leaves and other organic material decay in wetlands, consuming oxygen and rendering the water black with tannins, polyphenols and other chemicals. Low oxygen levels cause stress to fishes and other animals. Blackwater events in the Murray-Darling Basin have caused the deaths of several species, including Murray cod (Pritchard *et al.* 2009), and they undoubtedly also have sublethal effects (Morrongiello *et al.* 2011b). These events can sometimes be ameliorated if environmental flows are provided to dilute and reoxygenate the black water.

Cold-water pollution impacts on rivers are beginning to be addressed, but most large reservoirs still affect downstream thermal regimes. Modelling has elucidated the potential benefits of mitigation for fishes (Brown 2004; Todd *et al.* 2005; Sherman *et al.* 2007) and some local options have been costed (Sherman 2000; Boys *et al.* 2009; Raine *et al.* 2009). The costs of retro-fitting multi-level offtakes or other devices are sometimes prohibitive, but even low-cost options have received little attention. In 2004, however, the New South Wales government initiated a 25-year strategy for cold-water pollution mitigation (Raine *et al.* 2009). A surface offtake has been constructed at Tallowa Dam on the Shoalhaven River, where destratification using compressed air increased the outlet temperature (21 m depth) in January from 12.6°C to 21.9°C. Surface offtakes have also been retro-fitted at Tantangara Dam on the Murrumbidgee River and Jindabyne Dam on the Snowy River, and options have been explored at Burrendong (Macquarie River) and Keepit dams on the Namoi River (Raine *et al.* 2009).

Barriers

The need to ensure fish passage in rivers was recognised more than a century ago, and the challenge since has been to facilitate movements over dams, weirs and other structures without compromising their operations. The solution generally has been to construct artificial fishways; the first in Australia was constructed in 1913 (Mallen-Cooper and Harris 1990).

Transit through fishways involves attraction, passage and post-passage (Odeh 1999). Some early designs lacked 'attraction flows', thus fish might not detect the fishway entrance (Harris 1984), but they are now commonplace. Little is known of post-passage effects (for individual or sequential fishways) and they warrant more research. The issues in providing passage, however, are better understood.

For many years fishways were based on designs from Europe and North America, intended for migratory salmonids, and 91 were constructed in Australia until as recently as the 1980s (Mallen-Cooper and Brand 2007). These designs had steep slopes (1:5 to 1:9) with 300 mm steps and internal velocities up to 2.4 m s^{-1}. We know now that these specifications are not suitable for native fishes, and later designs typically have lesser slopes of 1:18–1:30 and smaller steps of 50–165 mm, depending on the target species (Thorncraft and Harris 2000). Some designs do retain steep slopes (e.g. Denil fishways, 1:12) but they also have more resting pools. These are intended to encourage fish to remain within the fishway, particularly those which (like bony herring) tend to retreat if they cannot complete passage in daylight hours (Mallen-Cooper 1999). Their value is not proven (White *et al.* 2011) and requires further study.

Research on fish movements has predictably concentrated on commercial and recreational species (Reynolds 1983; Mallen-Cooper 1992, 1994) but the needs of small species and juveniles, and the importance of lateral as well as longitudinal movements, are recognised (Barrett and Mallen-Cooper 2006; Tonkin *et al.* 2008a; Ch. 5). Modern designs are intended to facilitate the passage of multiple species and life stages (Roscoe and Hinch 2010).

Most fishways are associated with small structures like barrages and weirs (Harris 1984; Stuart and Mallen-Cooper 1999; Stuart and Berghuis 2002). The commonest are 'rock ramp' designs, suited to barriers up to 1.5 m high, and 'vertical slot' designs on barriers of 1–6 m (Harris *et al.* 1998; Thorncraft and Harris 2000; Figs 12.8, 12.9). Other designs include 'pool-and-weir', 'submerged orifice', 'Denil', 'bypass', 'eel fishways', 'fish locks', 'fish lifts' and 'trap-and-transport' (catch-and-carry) arrangements.

Fishways on higher structures (> 15 m) are becoming more common. A 43 m high fishway was constructed in 2005 at Paradise Dam on the Burnett River, Queensland, including a fish lift for upstream migration and, significantly, a fish lock for downstream migration. Another 37 m fishway was constructed in 2009 at Tallowa Dam on the Shoalhaven River, New South Wales, also with a fish lift for upstream migration. In Queensland, fishways are planned for Wyaralong Dam (34 m, with fish lift) on the Logan River and Hinze Dam (108.5 m) on the Nerang River. The Hinze fishway will be a trap-and-transport type, suited for upstream passage where water levels fluctuate, because the hopper can be lowered to the level of the receiving water. A number of smaller fish locks (< 15 m) have been installed on structures in north-eastern Australia (Gehrke *et al.* 2002; Stuart *et al.* 2007).

Provision of fish passage on existing barriers is now an accepted part of waterway management and in some regions there have been systematic efforts to identify and remediate barriers (Dept of Primary Industries, New South Wales 2006, Macdonald and Davies 2007; Dept of Industry and Development, New South Wales 2009). The importance of investigating (rather than assuming) the effectiveness of fishways has been recognised and is now regarded as an essential component of mitigating barriers to fish passage. Most remedial work has occurred in southern and eastern regions, where most of the barriers occur, but there are some installations in northern and western Australia (Stuart and Berghuis 2002; Morgan and Beatty 2006d; Beatty *et al.* 2007). In Victoria, nearly 7000 km of streams were reopened to fish passage in 2002–09 (Dept of Sustainability and Environment, Victoria 2010).

Few operational weirs are removed, because there often are social or economic implications, but many others are redundant and could be removed. Since inception of a weir-removal program in New South Wales, 14 structures have been removed (e.g. Muswellbrook Weir, Hunter River; Norco Weir, Richmond River: C. Copeland, Dept of Industry and Development, New South Wales, pers. comm. 2010).

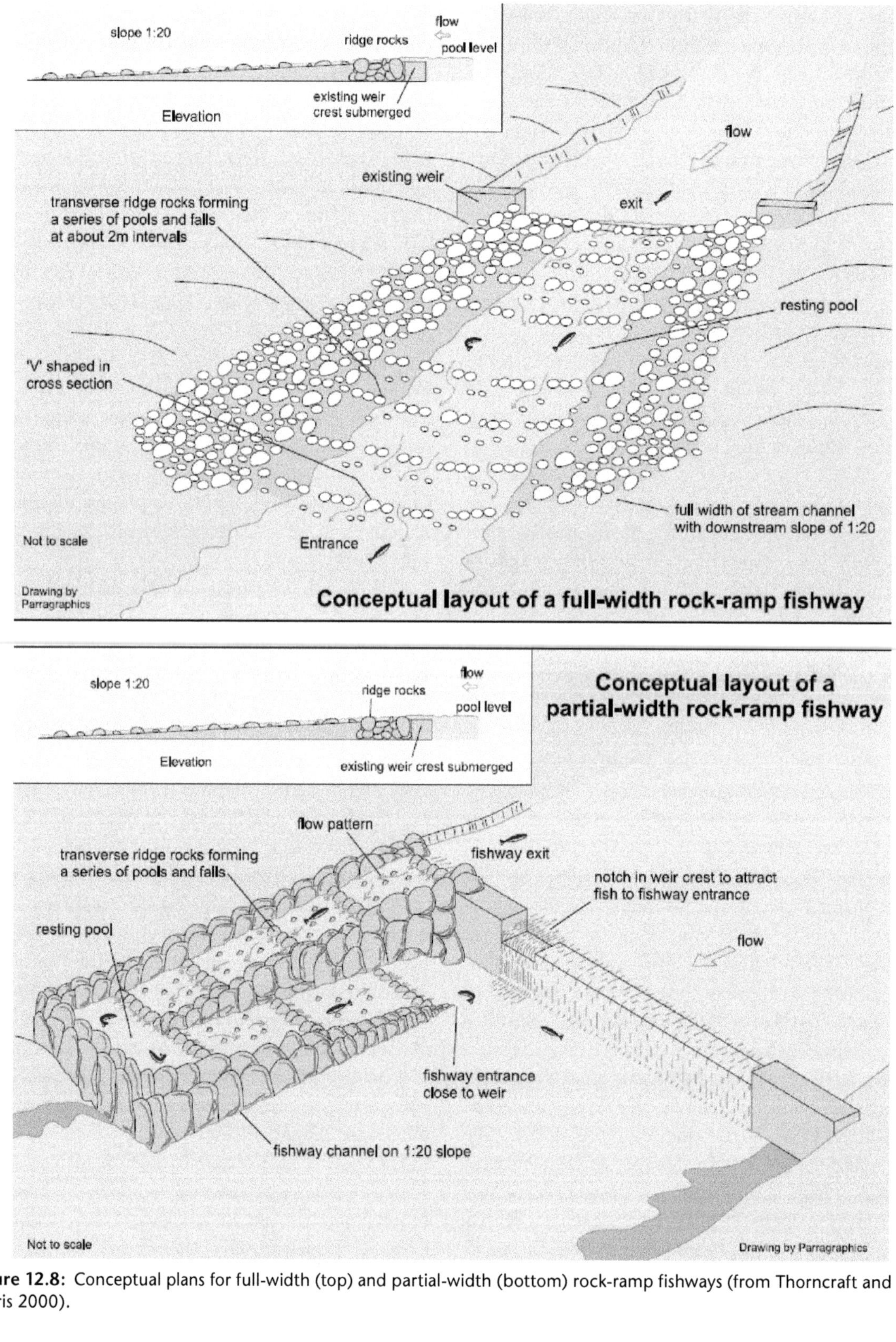

Figure 12.8: Conceptual plans for full-width (top) and partial-width (bottom) rock-ramp fishways (from Thorncraft and Harris 2000).

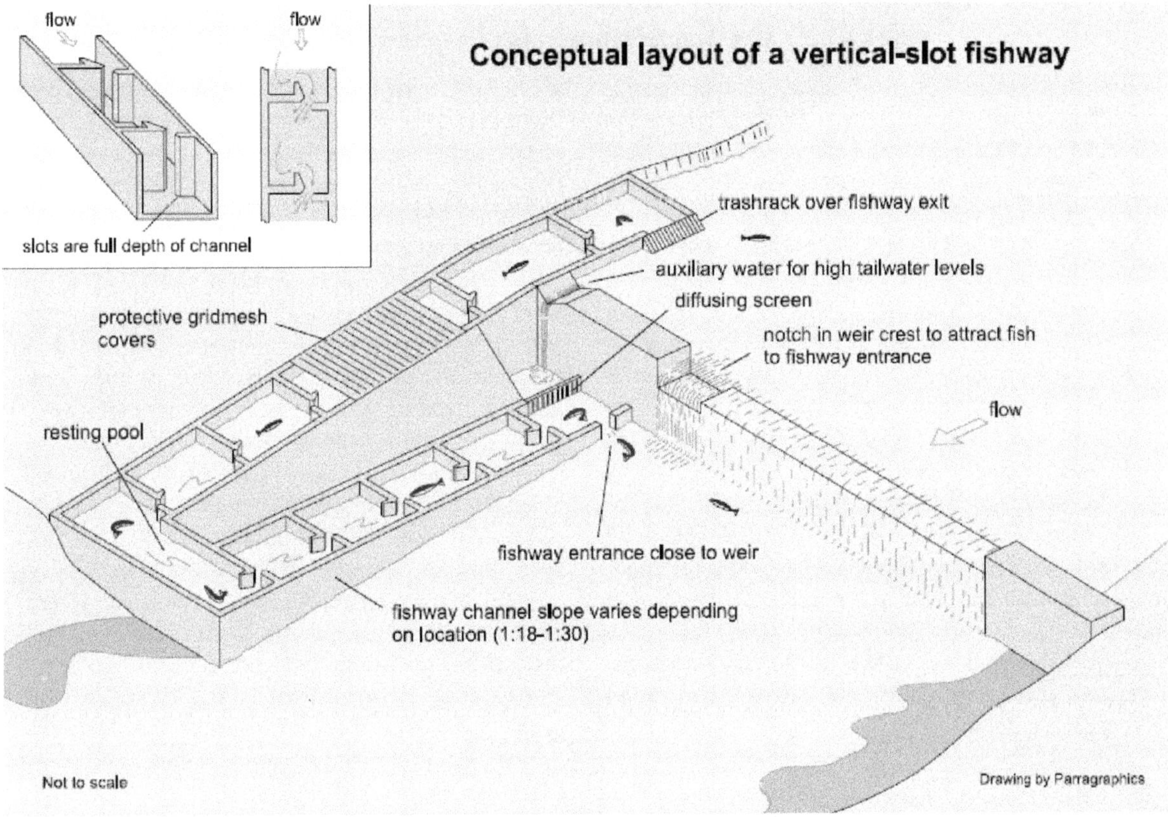

Figure 12.9: Conceptual plan of a vertical slot fishway (from Thorncraft and Harris 2000).

Dams, weirs and road crossings continue to be built, and most jurisdictions now consider fish passage issues as part of the approval process. Policy and engineering guidelines have been developed for fish passage at road crossings (Cotterell 1998; Witheridge 2002) and remedial measures include replacement of barriers with spanning structures (e.g. bridges), low-level box culverts or fords, drowning or reconstruction of perched outlets and the use of baffles or rock grouting to increase hydraulic complexity (Mallen-Cooper 2001; Dept of Industry and Development, New South Wales 2009).

Some barriers serve a useful purpose in excluding alien fishes. For example, Cotter Dam on the Cotter River, Australian Capital Territory, excludes common carp and redfin perch (Lintermans 2012). The Cotter River contains a number of threatened fishes and maintaining the barrier has become a priority for management (ACT Govt 2007). In other places, barriers have been built or augmented to protect native fishes from predatory trout (Lintermans 2000; Lintermans and Raadik 2003).

The Murray-Darling Basin is a showcase of problems and solutions in regard to fish passage. A range of fishway designs and techniques has been developed over the last 20 years, including vertical slot fishways (Mallen-Cooper and Brand 2007), Denil fishways (Mallen-Cooper and Stuart 2007), Deelder locks (Baumgartner and Harris 2007) and conventional fish locks (Stuart *et al.* 2010). For more information, see Box 12.3.

Alien species

An increasing focus on management of alien fishes over the last decade has resulted in several state and regional plans or strategies directed at alien fishes, particularly common carp (Carp Control Coordinating Group 2000; Dept of Primary Industries and Fisheries, Queensland 2000; Koehn *et al.* 2000; Dept of Industry and Development, New South Wales 2010a).

Box 12.3: The Sea to Hume Fish Passage Program

This is a system of world-class fishways along 2225 river km of the Murray River between the mouth and Hume Dam at Albury (Barrett and Mallen-Cooper 2006). Commenced in 2001 with a budget of $45 million, the program aimed to provide passage past 15 weirs and barrages, using purpose-built fishways. Construction was planned to be completed in 2011–12.

The program aimed to provide passage for large- and small-bodied fishes, a task made easier by the generally low height of instream barriers (< 8 m) and the low gradient of the river (150 m elevation change over 2225 km).

At Torrumbarry Weir near Echuca, Victoria, a fishway designed for native species (vertical slot, slope 1:18) was constructed in the 1990s and passed more than 20 000 fish over three years of monitoring (Mallen-Cooper and Brand 2007). The reconstruction of this weir in 1994–96 included decommissioning and replacement of the fishway with a modified vertical slot design.

Initially, the Sea to Hume Program employed vertical slot fishways of slope 1:32, designed to pass fishes of 40–1000 mm total length. Monitoring identified thousands of small fishes and crustaceans (12+ mm long) attempting to use the fishway, so the design was changed to combine features of a vertical slot fishway with slope 1:18, used by larger fishes (120+ mm), with an automated fish lock for smaller fishes (12+ mm). The fish lock is for low and medium river flows, when small fishes migrate, while the vertical slot operates also at high flows, when large fishes migrate. This combination provided capital cost savings and avoided the need for a very long vertical slot fishway.

The program incorporated modifications to the existing fish lift at Yarrawonga Weir, Denil fishways at Euston and Mildura and other prototype fishways at the barrages near the Murray Mouth. The construction and operation of the fishways were overseen by a multi-jurisdictional taskforce, supported by monitoring in a context of adaptive management (Murray-Darling Basin Commission 2008).

The program was one of the top 25 Australasian projects listed by the Global Restoration Network (<http://www.globalrestorationnetwork.org/countries/australianew–zealand/>).

The *Australian Pest Animal Strategy* (Natural Resource Management Ministerial Council 2007) includes alien freshwater fishes in its scope, and a separate national plan is being developed for alien freshwater fish management. There is also a national strategy for ornamental fishes (Dept of Agriculture, Fisheries and Forestry 2007), providing for alignment of state and federal lists of noxious species (Koehn and MacKenzie 2004).

Preventing the establishment of alien species costs less and environmentally is more desirable than having to control or remove them (Natural Resource Management Ministerial Council 2007; Fig. 12.10). Risk assessment models have been developed to support proactive management (Bomford and Glover 2004). These evaluate the risks of establishment, spread and environmental impacts (these may not be correlated: Ricciardi and Cohen 2007) and are used to screen both prospective and established alien species (Kolar and Lodge 2002; Bomford and Glover 2004; Rowe *et al.* 2008).

If prevention fails, early detection and rapid response can increase the likelihood of successful eradication or control, but there are few monitoring programs with the spatial or temporal capacity to detect new invasions quickly and, even when a new incursion is detected, the response may not be sufficient to prevent spread (Ansell and Jackson 2007). A framework for rapid response is being developed as part of national arrangements for biosecurity (Ayres and Clunie 2010b).

For established alien species, risk assessments are needed to assist with prioritisation, planning responses and selection of control methods. There are broadly four kinds of response, namely eradication, containment, control and mitigation (Wittenberg and Cock 2001). These are elaborated below.

Eradication is feasible only if the target individuals can be removed faster than they are replaced, if there is no immigration and the entire population is vulnerable (Bomford and O'Brien 1995). There have been many attempts at eradication of alien fishes but few

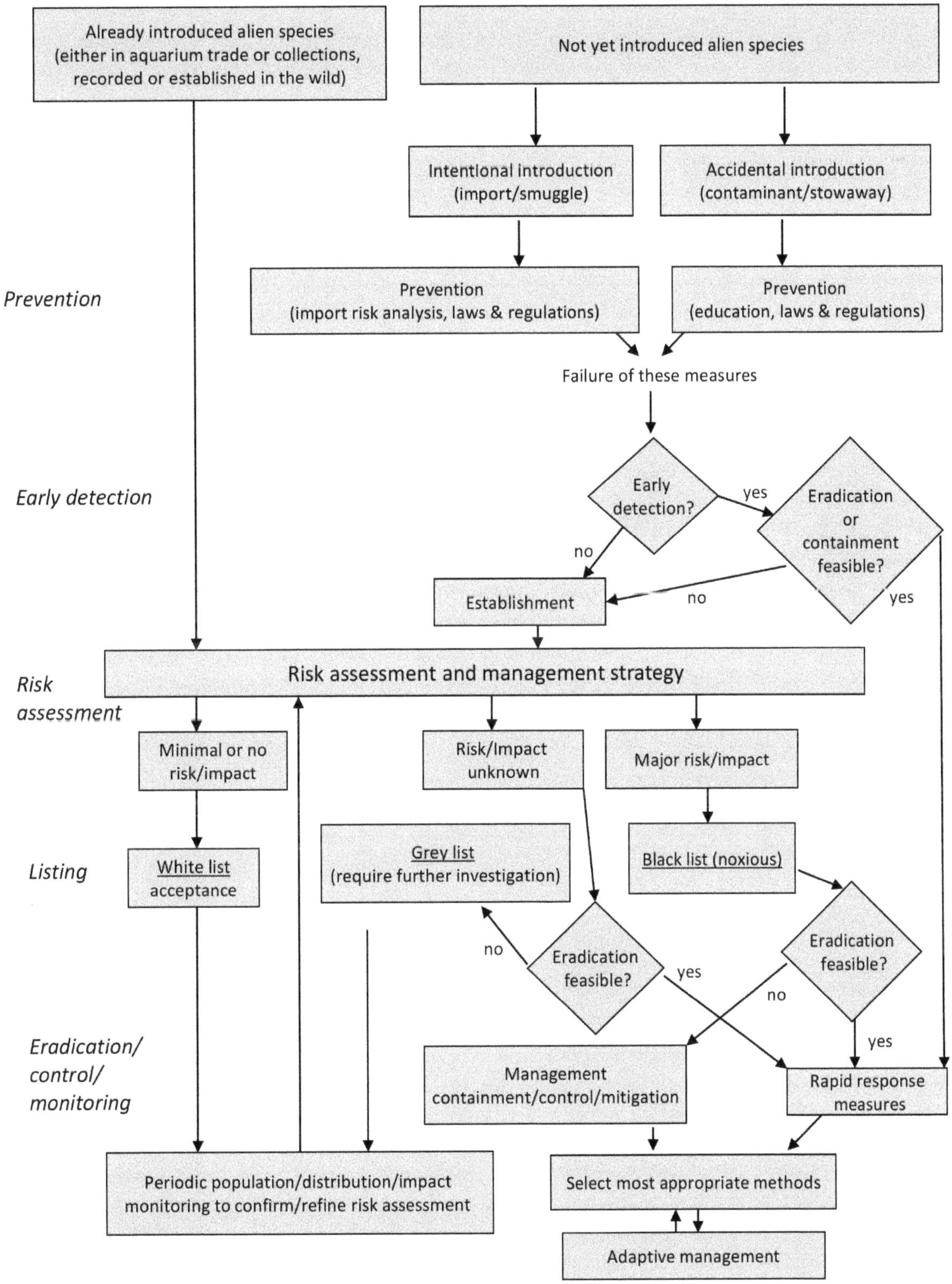

Figure 12.10: A 'decision support system' for managing alien fish species (modified from Wittenberg and Cock 2001; Nehring and Klingenstein 2008).

Table 12.5: Examples of containment or control measures for alien fishes

Species	Legislation	Management plans	Barriers	Physical removal	Chemicals	Education	References
Common carp	✓	✓	✓	✓	✓	✓	Carp Control Coordinating Group (2000); Dept of Agriculture, Fisheries and Forestry (2007); Koehn *et al.* (2000); Pinto *et al.* (2005); Stuart *et al.* (2006); Hillyard *et al.* (2010); Dept of Industry and Development, New South Wales (2010a)
Goldfish	✓			✓		✓	Pinto *et al.* (2005); Morgan and Beatty (2006b)
Redfin perch	✓		✓		✓	✓	ACT Govt (2007)
Eastern gambusia	✓	✓	✓	✓	✓	✓	Dept of Primary Industries and Fisheries, Queensland (2000); National Parks and Wildlife Service, New South Wales (2003); Kerezsy (2010b)
Tilapia	✓	✓	✓	✓	✓	✓	Dept of Agriculture, Fisheries and Forestry (2007); Greiner and Gregg (2008); Bradford *et al.* (2011); Thuesen *et al.* (2011b)
Rainbow and brown trout	✓		✓	✓	✓		Raadik (1995); Lintermans (2000); Lintermans and Raadik (2003); Dept of Primary Industries, Water and Energy, Tasmania (2006)
Oriental weatherloach	✓	✓			✓	✓	Burchmore *et al.* (1990); Koster *et al.* (2002)
One-spot livebearer	✓			✓	✓	✓	Morgan and Beatty (2006c); Rayner and Creese (2006); McNeil *et al.* (2010)

have been successful (Lintermans 2004) and documentation is missing or difficult to access. The most extensive efforts have been directed at common carp. In 1962, following the illegal importation and stocking of the invasive 'Boolarra strain' of common carp to Victoria, more than 1300 localities were poisoned (Koehn *et al.* 2000). The species had already spread beyond these sites, however, and widespread flooding in 1974–75 assisted its spread throughout the Murray-Darling Basin. Common carp is now the dominant fish species in the basin, and extensive sampling in 2004–07 showed that it accounted for nearly 60% of total fish biomass (Davies *et al.* 2010).

Eradication attempts are difficult, and have most chance of success in confined areas (Lintermans 2000, 2004; Rayner and Creese 2006; Dept of Primary Industries, New South Wales 2009). For example, common carp were discovered in Lakes Crescent and Sorell in Tasmania in early 1995. After attempts at eradication in 1995–2009, the population in Lake Sorell was reduced to fewer than 20 adult fish and none has been captured in Lake Crescent since 2007 (Inland Fisheries Service Tasmania 2010). Management actions included closing the lakes to public access, manipulation of water levels, exclosures to deny access to spawning habitats, containment of eggs, juveniles and adults using screens, localised poisoning, electrofishing and netting, pheromone attractants and the use of radio-tagged Judas fish to locate aggregations (Inland Fisheries Service Tasmania 2004, 2010). In late 2009, however, a spawning event occurred in Lake Sorell and carp numbers have rebounded (Inland Fisheries Service Tasmania 2010).

Containment and control measures are more likely to succeed than eradication, but they are difficult to sustain (Wittenberg and Cock 2001). Containment screens and barriers are prone to failure through floods, fire and vandalism or other human interference, but may provide a period of grace for development of other control methods. Control programs ideally involve reduction of pest numbers below a level where damage is considered acceptable, but thresholds have rarely been determined for Australian inland waters and control programs thereby lack clear goals. Multiple control methods may increase the chance of success, as different life stages can be targeted. Known spawning sites, refugia or patterns of behaviour may help to prioritise plans for control. For example, major recruitment of common carp in the Murray-Darling Basin may occur at relatively few 'hotspots' under certain conditions (e.g. Barmah-Millewa Forest, Macquarie Marshes, lower Murray wetlands: Smith and Walker 2004; Stuart and Jones 2006a).

Physical methods of control include harvesting (netting, trapping, electrofishing, commercial and recreational harvesting), exclusion and habitat manipulation (e.g. dewatering spawning sites) (Table 12.5). Commercial harvesting can remove large quantities of fishes rapidly but is limited to situations where there is a viable, marketable product (Jackson 2009). Recreational harvests (e.g. carp-athons) are a popular community activity with educational value but have little lasting effect on this species (Dept of Industry and Development, New South Wales 2010a). Chemical methods are usually non-selective, with adverse effects on non-target species (Table 12.5). Rotenone is registered nationally for this purpose and has been used in recent attempts at control or eradication (Rayner and Creese 2006; McNeil *et al.* 2010). In farm dams, ponds and other confined areas, lime has been used to raise the pH and control fish, and use of copper sulfate to control algae may also remove unwanted fish. Biological methods have not yet been used, but 'daughterless' technology and Koi herpes virus are being investigated as possible methods to control common carp (Thresher 2008; Saunders *et al.* 2010).

Mitigation (accept the pest but manage its impacts) is an option where eradication, containment and control are not feasible (Wittenberg and Cock 2001). The introduction of threatened galaxiids to salmonid-free waters is one example (Dept of Primary Industries, Water and Energy, Tasmania 2006). Many control activities directed at alien fishes are really a type of mitigation, in that impacts have not been quantified and there are no damage thresholds or known empirical relationships between pest abundance and damage. These programs aim to mitigate impacts by reducing numbers of the alien species, and their value is difficult to assess.

Most management activities depend on goodwill and cooperation from the general public (Mack *et al.* 2000), with public ignorance or misguided fervour about particular species often defeating management intentions. Consequently, involvement of all stakeholders and social viewpoints, through education, communication and collaboration, are vital in effective pest management programs (Stokes *et al.* 2006).

Translocation and stocking

Translocations of native fishes were commonplace prior to development of hatchery breeding programs, but many early transfers were *ad hoc* and ill-considered, and there have been comparatively few in the last 50 years. Development of a national translocation policy (Ministerial Council on Forestry Fisheries and Aquaculture 1999) has encouraged preparation of state policies and protocols (Sinclair Knight Merz 2008) that have greatly improved the practice. Translocations for conservation purposes have become more popular in the last decade or two, and have included freshwater catfish (*Tandanus tandanus*), Macquarie perch, Murray hardyhead, southern pygmy perch, Yarra pygmy perch, western pygmy perch (*Nannoperca vittata*), southern purple-spotted gudgeon, Arthurs galaxias (*Paragalaxias mesotes*), eastern dwarf galaxias (*Galaxiella pusilla*), golden galaxias, Pedder galaxias, Swan galaxias (*G. fontanus*) and western galaxias (*G. occidentalis*) (Dept of Primary Industries, Water and Energy, Tasmania 2006; Sinclair Knight Merz 2008).

Translocation is usually reliant on populations of wild fish and so avoids issues with domestication. In many recent translocations, the donor population was considered doomed and the translocation was a last-ditch rescue effort (e.g. Lintermans 2006). In other cases, care is needed to ensure that the removal of individuals for translocation does not adversely impact the donor population.

There are fish stocking programs in all jurisdictions, with native species stocked in all except South Australia, Western Australia and Tasmania. Commercial hatcheries have been producing native fishes since the early 1980s (Rowland and Tully 2004) and salmonid production has been underway since the 1860s (Clements 1988; Ch. 11). Most stocking is concerned with recreational fisheries but some, particularly in eastern Australia, is to enhance populations of threatened species. In rivers where recruitment fails as a result of human impacts (Humphries *et al.* 2008a, b), stocking may be the sole factor maintaining some species. Hatchery practices have been improved by quality assurance and accreditation schemes (Rowland and Tully 2004) leading to better management of genetic resources and diseases, with obvious consequences for wild populations that interact with stocked fish.

Since 1960, more than 137 million freshwater fish have been stocked into public waters in New South Wales and since 2000 more than 78 million fishes (15 species) have been stocked in New South Wales and Victoria (Table 12.6). More than 60 million native fishes have been stocked in public waters throughout the Murray-Darling Basin. The number stocked in private waters (farm dams) is unknown but certainly substantial, as many hatcheries produce fingerlings of native and alien species for this purpose. About 90% of stocking is for five species, including 44% brown trout and rainbow trout (*Oncorhynchus mykiss*) and 47% Murray cod, golden perch and silver perch. In 2000–10, the ratio of stocked native and alien fishes was near 50:50.

Concern about the ecological consequences of stocking is relatively new (Molony *et al.* 2003; State Fisheries, New South Wales 2003; Phillips 2003), but it has brought some changes to policy and practice. In New South Wales, an environmental impact statement on the potential impacts of fish stocking (State Fisheries, New South Wales 2003) was a milestone; the approach is yet to be embraced by other jurisdictions. Most jurisdictions now have stocking policies (Phillips 2003) but recreational interests still overshadow conservation concerns, particularly in regard to salmonid stocking (Jackson *et al.* 2004). Salmonid stocking policies in some jurisdictions invoke a perceived lower predation threat to native species from rainbow trout than brown trout (State Fisheries, New South Wales 2003), but it remains a questionable practice, particularly where there are threatened fauna. Rainbow trout are efficient predators on fishes and other species (Blinn *et al.* 1993; Marsh and Douglas 1997; Vigliano *et al.* 2009).

There has been little research into the effectiveness of stocking in maintaining recreational fisheries, particularly in rivers. Even in impoundments, where spawning habitats are limited and recreational fisheries depend on stocking, more research is needed. Some native species will breed in impoundments (e.g. golden perch breed in urban lakes and reservoirs near Canberra), but the contribution of natural recruitment is usually small (Lintermans, unpubl.).

The significance of stocking has increased with the utilisation of fishes as indicators of river health (Davies *et al.* 2010), because the stocked fish could confound those assessments. Developments in marking techniques now permit easy differentiation of wild and hatchery-produced fish (Munro *et al.* 2008; Crook *et al.* 2009). Recent assessments show that stocked golden perch may be 30–40% (in some cases, 80–100%) of populations in rivers of the southern Murray-Darling Basin (Crook *et al.* 2011).

Management of recreational fisheries

In south-eastern Australia, most commercial fishing of native freshwater fishes ceased in 2003, when the last licences for the Murray-Darling Basin were revoked. Some commercial harvest of common carp continues, and there is a small fishery in Lakes Alexandrina and Albert in South Australia, but commercial river-fishing generally has ceased. This was partly in response to declining catch rates and profitability, and partly in response to concerns about declining native fish stocks.

Recreational fishing continues, however, in virtually all Australian rivers, and employs thousands of people. Indeed, the recreational take of some species equals or exceeds the commercial take (Henry and Lyle 2003; West and Gordon 1994), and could contribute to the decline of species (see Box 12.4). In the 12 months commencing May 2000, $1.86 billion was spent by recreational anglers (Henry and Lyle 2003) and almost 20% of Australians aged five years or more fished at least once in 2000–01 (Henry and Lyle 2003). Some 20% of the fishing effort during this period was in fresh water and a further 35% was in estuaries

Table 12.6: Numbers of fishes stocked into public waters in New South Wales and Victoria in 2000–10

	2000	2001	2002	2003	2004	2005	2006	2007	2008	2009	2010	Total
Atlantic salmon	9700	460 000	424 370	271 050	424 361	400 360	226 000	155 290	252 410	252 480	202	2 876 223
Brook trout	7950	105 000	30 500	57 150	97 230	23 580	50 000	30 850	61 431	40 350	4160	508 201
Brown trout	336 915	1 204 321	997 447	1 009 790	955 977	948 019	990 630	659 921	801 363	782 841	347 370	9 034 594
Rainbow trout	749 610	2 537 977	3 135 312	3 458 068	3 024 944	3 163 357	2 238 769	2 168 157	2 517 833	2 210 568	786 292	25 990 887
Chinook salmon	2500	6839	0	9750	27 500	34 685	20 650	6150	0	0	0	108 074
Freshwater catfish	600	0	0	0	0	0	0	0	0	0	0	600
Australian bass	93 568	290 200	410 899	503 302	169 950	419 562	368 011	159 655	226 616	408 540	52 000	3 102 303
Golden perch	1 692 420	1 306 584	2 165 459	1 965 773	2 148 672	2 752 008	1 833 863	2 385 909	1 686 804	2 178 456	2 630 875	22 746 823
Silver perch	409 260	10 000	851 600	484 000	879 900	411 000	300 000	437 000	467 000	126 000	423 000	4 798 760
Murray cod	450 832	576 102	822 137	796 775	1 011 305	1 099 356	909 430	805 987	1 072 443	787 740	707 024	9 039 131
Trout cod	122 430	68 700	77 000	95 260	157 500	97 900	58 400	24 000	8300	75 300	17 000	801 790
Clarence River cod	2000	0	0	0	0	0	0	0	0	0	0	2000
Macquarie perch	15 000	0	0	0	0	0	0	0	0	0	3500	18 500
Southern purple-spotted gudgeon	0	0	0	0	0	0	0	101	2000	116	0	2217
Southern pygmy perch	0	0	0	0	0	0	0	397	75	37	106	615
Total	**3 894 785**	**6 567 724**	**8 916 726**	**8 652 921**	**8 899 343**	**9 351 832**	**6 997 759**	**6 835 424**	**7 098 283**	**6 864 437**	**4 973 539**	**79 030 718**

Sources: New South Wales Department of Industry and Investment, fish stocking database: <http://new.dpi.vic.gov.au/fisheries/recreational–fishing/fish–stocking>.

Box 12.4: Management of Murray cod

Murray cod (*Maccullochella peelii*) is the largest species of freshwater fish in Australia, attaining a record 1800 mm length and 113.6 kg weight. It is naturally restricted to the Murray-Darling Basin, where it was commercially harvested until the early 2000s. The species was listed as nationally vulnerable in 2003, in response to declining abundance, habitat losses and reduced recruitment. It remains a popular target for recreational anglers, with more than 108 000 fish captured in the 12 months to April 2001 (91% from rivers); 78% of these were released (Henry and Lyle 2003). In Victoria in 2008–09 it was estimated that 44% of anglers targeted Murray cod, with estimated expenditure of $166.7 million in 2009–10 (Dept of Primary Industries, Victoria 2010).

More than a million Murray cod are stocked annually in the Murray-Darling Basin (Lintermans *et al.* 2005). Until 2005, the minimum legal length in New South Wales, Victoria and the Australian Capital Territory was 500 mm total length (TL). As one in two Murray cod reaches maturity at 500–550 mm TL (Rowland 1998b), many fish were being harvested before they could breed. The minimum legal length has been raised to 600 mm and it may have to be increased further if catches increase (Allen *et al.* 2009). There is some evidence that numbers have increased in recent years, particularly in New South Wales (Rowland 2005), probably as a result of stronger natural recruitment, protection through regulations, stocking and lower numbers of alien common carp and redfin perch.

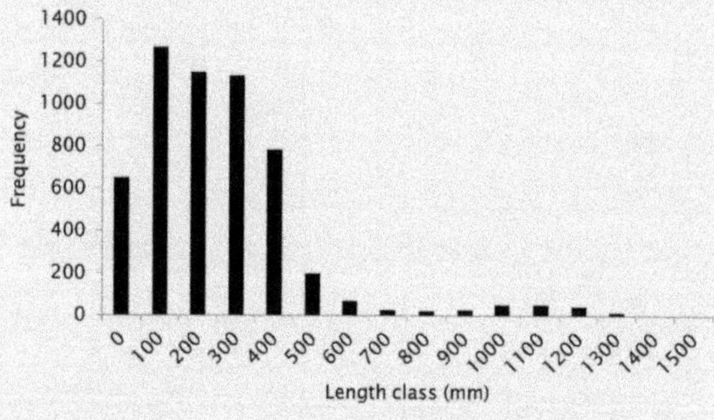

Box Figure 12.4: The length–frequency distribution of Murray cod from the Murray River between Yarrawonga and Tocumwal in 1995–2004, redrawn from Nicol *et al.* (2005). Note the rapid decline in abundance of individuals > 500 mm, the minimum legal length at that time.

(Henry and Lyle 2003). The freshwater (finfish) harvest was predominantly common carp, redfin perch, golden perch and salmonids, followed by Australian bass, barramundi and Murray cod.

Most agencies regulate recreational harvest by declaring size, bag and gear limits, closed seasons and closed waters. Some kinds of gear (e.g. set lines) have been phased out for recreational fishing, in keeping with current philosophies that favour active rather than passive gear. This may reduce the quantities of fishes taken, although some species are subject to catch-and-release angling, including accidental and deliberate catches (e.g. Murray cod: Douglas *et al.* 2010). Catch-and-release is now widely practised, and a national survey in 2000–01 suggested that more than 70% of captured Australian bass, barramundi and Murray cod were released alive (Henry and Lyle 2003). Although the impacts of catch-and-release are widely assumed to be minor, confirmation is needed and research should include investigations of sublethal effects (Arlinghaus *et al.* 2007; Hall *et al.* 2009b).

Aside from regulating harvest, active enhancement of recreational fisheries largely relies on stocking programs (Table 12.6). Stocking is seen by many as a panacea for declining fisheries (Phillips 2003; Chs 9, 11) although the drawbacks are well-documented (see 'Translocation and stocking'). Stocking programs enhance fish abundance in impoundments (Hunt *et al.* 2010) and can contribute significantly to riverine populations (Crook *et al.* 2011). Stocking in reservoirs can deflect pressure from stream fishes (Hunt *et al.* 2010). However, care is needed to ensure that stocking does not have deleterious effects on adjacent wild populations. Australia has not gone down the path of enhancing fish habitat in freshwater lakes and reservoirs to enhance recreational fisheries (see review by Bolding *et al.* 2004), with most freshwater habitat enhancement occurring in rivers and streams. The claimed recreational benefits of riverine enhancements (e.g. resnagging) are largely anecdotal, and more research is required.

There has been a shift in angling ethics since the 1980s, from the food harvesting approach of previous decades towards recreational and environmentally sympathetic practices (e.g. catch-and-release, cessation of set line use). There has also been a shift

Table 12.7: Regulatory measures in recreational fishery management

Measure	Explanation	Rationale	Examples
Fishing licence	Mechanism for recovering costs (e.g. education, stocking, enforcement, habitat enhancement), tracks angler participation	Applies user-pays principle, tracks recreational effort for species or fisheries	Common in fresh waters in NSW, Q, T, V
Daily bag limit	Number (species-specific) that can be taken by angler in a day	To control harvest per angler	Bag limit of five golden perch in rivers of NSW, V
Possession limit	Number (species-specific) in an angler's possession at any one time (often twice daily bag limit)	To control harvest per angler	NSW, ACT legislation for Murray cod, golden perch, silver perch
Minimum length limit	Minimum length of individuals that can be retained	Allow fishes to reach minimum size (cf. length at maturity)	Barramundi in NT, Q, WA
Slot length limit	Slot size (minimum, maximum) of individuals that can be retained	To restrict harvest of small, immature fishes and large, mature adults	Barramundi in Gulf of Carpentaria, Q (600–1200 mm), Murray cod in V (600–1000 mm)
Maximum length limit	Maximum length of individuals that can be retained	To protect large, mature adults	Jungle perch in Q (350 mm), Murray cod in Q (1100 mm)
Gear limits	Allowable number of rods, lines or bait types (artificial or natural) that can be used	To control harvest per angler	Two attended lines in NSW general inland waters
Gear specifications	Dimensions of gear types (e.g. crayfish traps) and definitions (e.g. a rod and line may contain no more than two hooks)	To control harvest per angler	Yabbie traps, landing nets in ACT, NSW
Water designations	To distinguish impoundment and riverine fisheries, to designate open, trout and trophy fisheries (salmonid fisheries)	To control harvest per angler	ACT, NSW trout streams, riverine and impoundment fisheries in NSW, V
Closed waters	Specified waters where fishing is not permitted	To protect spawning streams, habitats for threatened native species or where access is not desirable (e.g. reservoirs)	Section of Murrumbidgee River in ACT to protect trout cod, closed streams for salmonid breeding in ACT, NSW, T
Open seasons	Start and finish dates when angling is allowed in a waterway	To regulate activity in specified streams and to control harvest per angler	Trout streams in ACT, NSW
Closed seasons	Closures to prevent harvest of species or groups (e.g. salmonids)	To protect species or groups, often in breeding seasons	Murray cod closed seasons in ACT, NSW, V
Protected species	Species that cannot be retained if caught (e.g. threatened species)	To protect threatened species	Trout cod in ACT, NSW, sawfish in WA
Noxious species	Non-desirable species (usually alien), usually cannot be returned alive to the water	To protect spread of alien species and reduce impacts on desirable native species	Common carp in Q, V, redfin perch in ACT, NSW

ACT = Australian Capital Territory; NSW = New South Wales; NT = Northern Territory; Q = Queensland; SA = South Australia; T = Tasmania; V = Victoria; WA = Western Australia.

in management focus from salmonids to native fishes since the 1960s and 1970s. Fisheries legislation has been updated and regulatory controls are in place for individual species and public waters (Table 12.7). In addition, anglers themselves have developed national and state codes of practice. Engagement with anglers is an integral part of successful fisheries and conservation management (Granek *et al.* 2008). National recreational fishing organisations and advisory groups and public consultations are routinely part of recreational management, as agencies have become more sensitive to the needs and expectations of anglers. Population models have been constructed for some angling species to demonstrate the possible impacts of changed regulations on harvest and sustainability (Todd *et al.* 2004; Allen *et al.* 2009; Todd 2009).

MANAGEMENT, MONITORING AND RESEARCH

Learning by doing

'Adaptive management' (AM) (learning by doing) has been a popular concept for many years (Holling 1978; Walters 1986). There are three types:

1 *evolutionary* AM (trial and error) is where early management choices are essentially undirected or haphazard and later choices are based on the most successful results;
2 *passive* AM is where the learning from early choices is formulated as best practice, subject to review after implementation and a period of monitoring;
3 *active* AM is where implementation of one or more policies or actions is treated as an experiment, with monitoring, review and comparisons (Allan and Jacobson 2009).

AM acknowledges that there is usually imperfect understanding of systems being managed and uncertainty over how they will respond and interact. It concedes that it is often not possible to wait for knowledge to accumulate; rather, it is better to commit to learning by doing, to review the outcomes and incorporate what is learned into practice. Many natural resource management groups claim to have embraced AM but in many plans and reports it appears to be a hollow gesture. There is often little appreciation of what is needed and a lack of will to implement it fully (Stankey and Allan 2009). Perhaps what is lacking most is a commitment to a role for research, supported by monitoring. AM does not replace or bypass research; on the contrary, it helps to direct research in areas that match management needs and priorities.

There are six steps in an AM cycle (Fig. 12.11).

1 **Assess the problem and specify objectives:** a failing of many policy/management actions and programs is that the objectives are not specified, are vague and cannot be evaluated, or do not represent the values of stakeholders (Johnson 1999).
2 **Model existing knowledge:** models must be documented so that they can be reviewed in light of new information. Simple models are preferred, as they are easier to develop and understand, require fewer data and allow outputs to be easily compared (Schreiber *et al.* 2004; Argent 2009). Quantitative models are preferred, but non-quantitative, conceptual models can be used.
3 **Model potential management options:** alternative management options are an essential component, as AM has little value if an agency is committed to a fixed solution. A reluctance to accept more risk for higher gain is an impediment to applications of AM.
4 **Implement chosen option:** the preferred option is implemented.
5 **Monitoring:** information is gathered to evaluate performance and outcomes and to enable learning.
6 **Review and adjust actions:** new knowledge from monitoring is evaluated, the management or policy regime is modified if necessary and the cycle begins to repeat.

Many interventions in Australian rivers could be categorised loosely as passive AM, but these are merely a first step towards the active version, where management, monitoring and research are engaged fully as parts of a continuing process (see Box 12.5).

Role of monitoring

There is much confusion about the purpose and practice of monitoring, what information it can supply and what scale and type of monitoring are required in

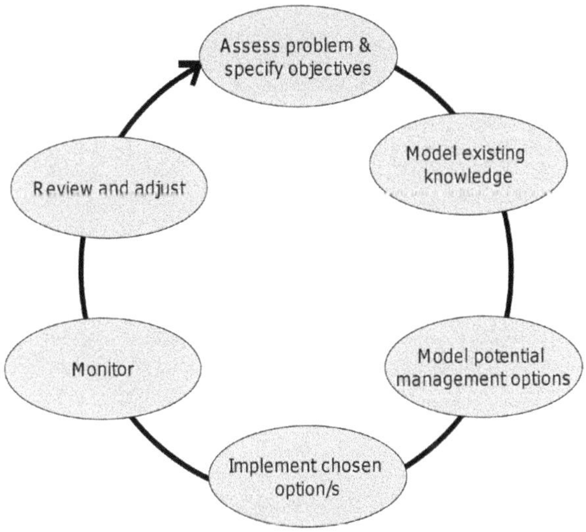

Figure 12.11: The adaptive management cycle.

particular situations. It is often inadequately resourced, superficial, poorly designed or reported and considered not to be 'real science' (Lindenmayer and Likens 2009; Stankey and Allan 2009). Agencies that profess a commitment to AM but do not support monitoring demonstrate a lack of basic understanding. Imagine if, as owner of a business, you were denied information about cash inflow, stock inventories, debts and market preferences – that basic information is akin to the data that should flow from monitoring the environment. Two kinds of monitoring are most relevant in environmental management (Lindenmayer and Likens 2010), namely 'surveillance monitoring' and 'intervention monitoring'.

Surveillance monitoring is often at inappropriate scales for the core questions in AM but it is important nevertheless. The Murray-Darling Basin Authority's Sustainable Rivers Audit, for example, is a surveillance monitoring program designed to assess the ecological health of rivers in the basin (Davies *et al.* 2010). It operates at the river valley scale, and synthesises and integrates data gathered with regard for a number of themes (hydrology, macroinvertebrates, fishes). The audit does not indicate the cause of an observation (e.g. poor fish condition in a particular area); rather, it identifies issues that may need investigation. Surveillance monitoring can therefore inform AM (Bunn *et al.* 2010), but it has a different purpose.

Box 12.5: Fishways as an example of Adaptive Management

The Sea to Hume Fish Passage Program (Box 12.3) exemplifies the AM cycle.

Step 1: Assess the problem and specify objectives: To counter restrictions on upstream passage caused by weirs, the program planned to restore passage for fishes 40–1000 mm length along the Murray River between the sea and Hume Dam.

Step 2: Model existing knowledge: Knowledge of fish aggregations below weirs, movement needs (spawning, dispersal) for various species (e.g. congolli, golden perch, Murray cod, silver perch) and the risks of facilitating passage of alien species (e.g. common carp) was reviewed.

Step 3: Potential management options: These included consideration of which fishway type and slope were most appropriate given the position of fishway on the weir (mid-channel, left/right bank), the hydraulic conditions at each site and existing weir management requirements.

Step 4: Implement chosen option: Low-slope (1:32) fishways were selected for most weirs, and construction was undertaken at Locks 1 and 7–10.

Step 5: Monitoring: Intervention monitoring was undertaken at Locks 7–10 to determine whether the fishways were effective passing fishes in the specified range, and whether fish aggregations were still occurring below weirs.

Step 6: Review and adjust actions: Monitoring of fishways at Locks 1 and 7–10 showed that many small fishes (12+ mm) were unable to complete passage. New measures were trialled, including rocks grouted to the base of fishway chambers to create a low-flow barrier layer, adding hydraulic modifiers to reduce turbulence and operating the fishway as a lock (Stuart *et al.* 2008b). Alternative designs were trialled at Lock 8, involving a high-gradient (1:18) vertical slot fishway for large fishes (120+ mm) at all river discharges and a fish lock for small fishes (12+ mm) at low–medium discharges. This design was planned for Locks 3–6 and 15. Monitoring is ongoing, completing the AM cycle and highlighting any need for adjustments.

Intervention monitoring (question-driven monitoring) is most informative for AM applications. It provides information about specific interventions (e.g. a resnagging program, mitigation of thermal pollution, alien fish control) and generally is conducted at smaller temporal and spatial scales than surveillance monitoring. There are different levels of intervention monitoring, from basic unreplicated, uncontrolled anecdotal observations to replicated, controlled systematic measurements, including 'before' and 'after' treatments (Downes *et al.* 2002). Far from being an incidental issue, monitoring is inherently part of AM and part of environmental science.

CHALLENGES FOR CONSERVATION AND MANAGEMENT

Spatial and temporal scales

Most initiatives in aquatic ecosystem management have been at site or regional scales rather than at landscape scale. The 'riverscape concept' promotes the need for holistic perspectives in fish management but this is still in its infancy in Australia (Fausch *et al.* 2002). Uncoordinated, small-scale projects have dominated in the past and continue to do so, and we have learned little from them to guide other efforts (Bond and Lake 2003a; Lake 2005; Palmer *et al.* 2005). This is true, for example, of alien fish management (Koehn and MacKenzie 2004). Other examples are the ways that fishery and water resource regulations change at jurisdictional boundaries, when it is clear that fauna and flora are insensitive to political borders. These are critical concerns, because on-ground management is the responsibility of regional catchment authorities or councils who must work within boundaries imposed by planning and political processes. Management needs to transcend these boundaries if worthwhile conservation outcomes are to be achieved.

The time scales for management of fishes and other aquatic resources also need careful consideration. Ecosystems and communities that have evolved over protracted time scales then degraded over a matter of years are unlikely to recover as a result of simple manipulations in short periods. For threatened species recovery, it must be acknowledged that some species may be in severe extinction debt and may not survive, let alone recover (Kuussaari *et al.* 2009). The need to manage aquatic resources as part of ecosystems is increasingly recognised (Likens *et al.* 2009). Complex systems involve patterns and processes that operate and interact, sometimes with substantial time lags, and individual components cannot be isolated and manipulated at scales that suit managers before fishes. Getting managers, policy-makers and governments to embrace and implement long-term, 'whole of valley' and 'riverscape' thinking will not be a trivial task.

Climate change

Climate change has been a reality for millennia but there is overwhelming evidence that the nature and pace of change are now responding to human industry. Changes in coming decades may cause some species and ecological communities to expand their range and abundance; others will contract and all, it is safe to say, will be affected in some way.

Sea levels are rising, river water temperatures are rising (Kaushal *et al.* 2010), rainfall patterns have changed dramatically in some areas (e.g. Perth: Bureau of Meteorology 2011), Australia experienced the Millennium Drought and the decade ending in 2010 was the hottest on record (since 1910: Bureau of Meteorology 2011). Lower rainfall, coupled with warmer air temperatures, indicate substantial declines in runoff and river flows for much of Australia. Increasing aridity is likely to increase fire risk, with the potential to affect fish populations (Lyon and O'Connor 2008). Fires alter vegetation, which may then alter runoff and stream flow, resulting in increased sediment delivery to streams (Lane *et al.* 2009; Wittenberg and Inbar 2009) followed by reduced runoff in subsequent decades. Some areas will experience increased rainfall or an increased frequency of severe climatic events such as storms and flooding. Rising sea levels will result in saltwater intrusions and the loss of coastal wetlands.

Coupled with these changing climatic factors, anthropogenic changes in land cover (e.g. urbanisation, land clearing) have altered water infiltration and runoff patterns. Dams are also stressing riverine ecosystems (Bond *et al.* 2008). Australia appears to be entering a new 'infrastructure solution' phase, with new dams and pipelines to facilitate intercatchment movement of water proposed or constructed in various regions. Australia already has the highest per capita level of water storage in the world, and the underlying philosophy will not change quickly. Increased populations and new dams will further change streamflow patterns, as water resource managers endeavour to capture and deliver water to meet increased demands.

Changed rainfall, runoff and temperature regimes will inevitably modify ecosystem processes and the dynamics of populations and test the physiological tolerances, parasite interactions and distributional limits of fishes (Collares-Pereira and Cowx 2004; Perry and Bond 2009; Balcombe *et al.* 2011b; Morrongiello *et al.* 2011a). These changes could have revolutionary implications for conservation of native fishes, rehabilitation of habitats and management of alien fishes.

Stemming the tide of aliens

The number of alien species in Australia has risen quickly since the 1980s, and will continue to rise unless management intervenes (Lintermans 2004; Ch. 11). Most recent invasions have been aquarium fishes, especially in the tropical or subtropical waterways of northern Australia. Some ornamental species with serious potential impacts are already at Australia's northern boundaries, with climbing perch (*Anabus testudineus*), snakehead (*Channa striata*) and walking catfish (*Clarias batrachus*) having reached Papua New Guinea. Climbing perch are also reported from islands in Torres Strait (Hitchcock 2002; Burrows and Perna 2009). Some 700 aquarium species known to have been in Australia in the past are no longer approved for importation (McNee 2002), but species like these pose a persistent risk. The continuing spread of alien fish species already in Australia is cause for alarm (Lintermans 2004; Rowley *et al.* 2005; McNeil *et al.* 2010) and a coordinated national strategy is needed to galvanise action.

Freshwater protected areas

The concept of 'freshwater protected areas' has been applied only patchily in Australia, but there are successful examples in other countries (Abell *et al.* 2007; Suski and Cooke 2007; Pittock *et al.* 2008). Australia does not have a comprehensive, representative network of freshwater reserves (Nevill 2007), and it is unlikely that existing terrestrial reserves adequately conserve native freshwater fishes (Keith 2000; Nel *et al.* 2007; Roux *et al.* 2008). National parks do not adequately protect rivers; for example, 12 of 15 rivers in Kosciuszko National Park, New South Wales, are affected by dams and diversions (Lake 2005). The same may be said of the degree of protection afforded by initiatives such as the *Directory of Important Wetlands in Australia* (Dept of Environment Water Heritage and the Arts 2011) and identification of 'high conservation value aquatic ecosystems' (Sinclair Knight Merz 2007), 'heritage' or 'wild' rivers and 'wetlands of international importance' listed under the 1971 'Ramsar Convention'. Such initiatives do not approach a comprehensive and representative system relevant to fishes. For example, of 64 Ramsar wetlands declared in Australia since 1974, none is declared primarily for its native fish values despite some strong candidates (e.g. the Murray River between Yarrawonga and Cobram, Victoria, with the only natural wild population of trout cod). In most national parks and reserves, freshwater fishes are largely unprotected, and they are not considered 'animals' under many nature conservation statutes.

Freshwater protected areas may not be a complete solution but they do have virtues that offset the inadequacies of existing reserves (Abellan *et al.* 2007), particularly in regard to fishes (Keith 2000; Cucherousset *et al.* 2007; Hermoso *et al.* 2009). There are several problems associated with freshwater protected areas, because upstream management may compromise the value of a downstream reserve (e.g. flow abstraction, thermal and chemical pollution: Saunders *et al.* 2002; Moilanen *et al.* 2008). There is resistance to declaring freshwater protected areas in some states, as the move is perceived by some as 'locking up' rivers and there is a fear that angling might be restricted or banned (Koehn 2010). Involvement of key stakeholders from the outset is essential if freshwater protected areas are to be effective (Barmuta *et al.* 2011), and well-informed and engaged anglers will recognise the benefits of having core populations of valued species that can disperse and maintain regional fishing opportunities. Freshwater protected areas do not necessarily mean 'no fishing'; they may encompass a mosaic of gear, spatial and temporal restrictions or closures, much as presently applies for alien and native fish species. A model for a system of freshwater protected areas in the Murray-Darling Basin has been proposed as a starting point (Phillips and Butcher 2005; Phillips 2008).

Community engagement

Engaging community interest in the future of native freshwater fishes is difficult. While it may be easy to awaken interest in recreationally important species such as Murray cod and barramundi, non-recreational species and the many small species do not attract the same empathy. Iconic taxa such as the Australian lungfish (*Neoceratodus forsteri*) receive national attention (Arthington 2009) and some species have local profiles (e.g. southern pygmy perch and southern purple-spotted gudgeon in South Australia: Hammer *et al.* 2009; Mary River cod in Queensland: Arthington 2009), but these are often responses to immediate threats and interest is not sustained. Indigenous consultation and engagement have been poor, and need to be addressed. Government resources are strained; without community interest and active involvement in fish-related activities, resources will be directed elsewhere. As water becomes increasingly scarce, the debate over consumptive versus non-consumptive aquatic values will intensify: if native fishes are to be considered, we need the community to be barracking for the fish. To paraphrase a political aphorism, 'Ask not what fishes can do for you, but what you can do for fishes'!

CONCLUSION

Many Australian freshwater fishes are in decline and that trend will continue without substantial new investments in conservation and management. Some threatened species have been rescued from the brink of extinction (e.g. barred galaxias, Pedder galaxias) and others are at the brink. There is a strong groundswell of opinion calling for action, and although management of fishes and fisheries has improved in the last three decades there is much room for improvement and innovation. It is a challenging, exciting time to be involved in fish conservation and management.

13 Looking ahead

Paul Humphries and Keith F. Walker

INTRODUCTION

For more than 150 years, our ecological and biological knowledge of freshwater fishes languished in the wake of alien introductions, environmental changes and unsympathetic management. In the last 50 years the gap has narrowed through new awareness, research and management, illustrated in the foregoing chapters. But fortunes have changed – now there are good prospects for ecologists, less so for many native fishes.

Conservation, management and ecology complement one another. They are overlapping endeavours separable only at their margins, where conservation and management are linked to cultural values and ecology is linked to other natural sciences. This interface is likely to change in the future as cultural and environmental challenges emerge, and preparedness surely will be a better strategy than prediction. Our science may need to become more adaptive than conservative, and more proactive than reactive.

In this final chapter, we touch lightly on these challenges. We begin by recounting briefly the emergence of scientific studies of native fishes and sketch some earlier visions of the future. We highlight some ideas that could have special portents for those who study fishes and manage their environment and pose some ideas for research, as 'teasers' for those contemplating more study. Finally, we venture a scenario of issues for ecology and conservation in the next 50 years.

BACK TO THE FUTURE

Awareness

Some defining features of the 19th century, from a piscatorial viewpoint, include early encounters between Europeans and *Terra Australis Incognita*, the plunder of what (by European standards) were virgin fisheries (Ch. 9) and the rise of acclimatisation societies dedicated to importing animals and plants from other parts of the world (Ch. 11). In the vanguard of exploration, natural historians were few and often amateurish; their records were fragmentary (albeit now priceless) and most of their collections were despatched to European taxonomists with little knowledge of Australia (Chs 1, 2). Few of the colonists were sensitive to the culture and traditions of aboriginal Australians and fewer still were able to bridge the cultural divide, so that much indigenous wisdom was lost (Humphries 2007).

As early as the 1850s, there was alarm at the decline of native fisheries (Humphries and Winemiller 2009). These concerns are reflected in many historical recollections, newspapers and reports (Chs 9, 11); as an aside, many of these articles are accessible in digitised form through the National Library of Australia (<http://trove.nla.gov.au>) and various state libraries. These kinds of resources are often overlooked but, used critically, they can provide invaluable scientific information (Robertson *et al.* 2000; Trueman 2007; Jackson *et al.* 2011). New fisheries generally provide bumper yields before catches are reduced to a level where, ideally, a balance is struck between yield and recruitment. In practice, the ideal generally is not realised: catches are not sustainable and the fishery fails.

The first half of the 20th century brought a new federal system of government to Australia and a rush to develop resources, twice arrested by war. In the inland fisheries, catches varied with the vagaries of climate but the scant records kept in those days were of little value for fishery management. The second post-war economic boom brought new irrigation and hydropower schemes, encouraged by a decade of exceptional rainfall and floods. It became clear that there must be limits to growth and that inland water resources, fisheries included, needed to be managed if they were to be sustainable. The downward trend in fisheries was matched by evidence of general declines in native fishes and their habitats. Amid much wider changes, terms such as 'environment', 'biodiversity', 'conservation', 'heritage' and 'sustainability' progressively became part of the argot of government (Ch. 12).

In the writings of early naturalists it is difficult to find evidence of a conservation ethic applied to native fishes. There were reflections on the fortunes of fisheries but issues of biodiversity and conservation had little gravitas until the 1950s, when scientific studies of native fishes began to gather pace, in company with ideas about resource management (Chs 1, 12). Among the first scientists to speak out was Gilbert Whitley (Australian Museum, Sydney), who in 1955 listed several species that he believed were in need of protection (Whitley 1955). These included members of the Galaxiidae, the Australian grayling (*Prototroctes maraena*), Balston's pygmy perch (*Nannatherina balstoni*), blackmast (*Craterocephalus stramineus*), cave gudgeon (*Milyeringa veritas*), congolli (*Pseudaphritis urvillii*), desert goby (*Chlamydogobius eremius*), freshwater sole (*Brachirus selheimi*), nightfish (*Bostockia porosa*), nurseryfish (*Kurtus gulliveri*), primitive archerfish (*Toxotes lorentzi*), river blackfish (*Gadopsis marmoratus*), saltpan sole (*Brachirus salinarum*), saratogas (*Scleropages* spp.) and western pygmy perch (*Nannoperca vittata*). The names here are updated as appropriate, but the list was a perceptive one in the *Zeitgeist* of the day. Tellingly, most of these fish feature in lists of threatened species today (Ch. 12).

Scenario for 2000

In 1986, Pollard and Burchmore (1986) ventured a scenario of the future of Australian freshwater fishes as it could have been in 2000. The scenario (Table 13.1) was intended to suggest what might have happened, given maximal human population growth and minimal environmental controls. It is interesting (perhaps encouraging) how few of these speculations were actually realised in 2000 or indeed are true now, more than a decade later.

The scenario for 2000 did not consider inland fisheries, although they clearly were in decline (in the Murray-Darling Basin, most commercial operations ceased well before 2001: Gilligan 2005). Nor did it mention salmonids, vigorously supported by anglers despite their effects on galaxiids and other native fishes (Cadwallader 1996). It dismissed common carp (the Boolarra strain of *Cyprinus carpio*), implying that this would prove to be a benign introduction. In fact, it was a game changer and the carp since have confirmed their credentials as ecosystem engineers, emphatically so in the Murray-Darling Basin (Ch. 11). Note that Nile perch (*Lates niloticus*) features in the scenario because around that time there were moves to introduce this species, but the plan was defeated (Williams 1982; Barlow and Lisle 1987).

An implicit message in the scenario for 2000 was its acceptance of a common perception that, for many native fishes, the future is one of decline and extinction, and that for alien species it means further spread and more new arrivals. In that context, the scenario referred to trends that are clearly evident today, and to events that still might occur. For example, it suggested:

- invasions of ornamental fishes (e.g. platys, swordtails, small cichlids) throughout northern Australia, threatening small species of Atherinidae, Melanotaeniidae and Pseudomugilidae (Ch. 12);

Table 13.1: An imagined scenario for Australian freshwater fishes in 2000, based on Pollard and Burchmore (1986). This was presented originally as a narrative, but is adapted here as a table. Several states (NSW, Q, SA, V) overlap Central Australia and the Murray-Darling Basin. Queensland and Western Australia include northern (NE, NW) and southern (SE, SW) regions.

Region	Native species Major decline or extinction	Alien species Common or dominant
Western Australia (SW)	*Galaxiella* spp. Nightfish Salamanderfish	Gambusia
Western Australia (NW)	Blind cave eel Atherinidae Eleotridae, incl. cave gudgeon Melanotaeniidae	Gambusia Nile perch Piranhas[1] Platys Swordtails Tilapia Walking catfish[2] Cichlidae (small cichlids)
Northern Territory	Barramundi Northern saratoga Nurseryfish Primitive archerfish Atherinidae Melanotaeniidae Pseudomugilidae Plotosidae Terapontidae	
Queensland (NE)	Southern saratoga[3,4] Gobiidae Eleotridae Melanotaeniidae Plotosidae Pseudomugilidae Terapontidae, incl. leathery grunter[3]	
Queensland (SE)	Australian lungfish[4,5] Mary River cod Oxleyan pygmy perch	
New South Wales	Australian bass[4] Australian grayling[6] Clarence River cod[4] Freshwater herring Non-parasitic lamprey Oxleyan pygmy perch	Gambusia Japanese sea bass[7] Tilapia Yellowfin goby[8]
Victoria	Macquarie perch[9] River blackfish Yarra pygmy perch	Goldfish[10] Oriental weatherloach Redfin perch[11]
South Australia	Many species[12]	
Tasmania	Australian grayling[13] Galaxiidae, incl. Tasmanian whitebait	
Central Australia (part NSW, NT, Q, SA)	Barcoo grunter Dalhousie hardyhead Desert goby Flinders Ranges purple-spotted gudgeon Plotosidae	Common carp Gambusia
Murray-Darling Basin (part NSW, Q, SA, V)	Freshwater catfish[14] Golden perch[14] Murray cod[14] Silver perch[14] Trout cod[9]	Common carp[15] Grass carp[16] Nile perch

1 *Serrasalmus* spp. (Queensland).
2 *Clarias batrachus*.
3 Lost from Fitzroy system.
4 Maintained in captivity (aquaria or hatcheries).
5 Extinct in Mary and Burnett rivers; may occur in upper Brisbane River.
6 Survives in Tasmania, but rare. Vulnerable to disease.
7 *Lateolabrax japonicus* (estuaries).
8 *Acanthogobius flavimanus*.
9 Protected Hawkesbury-Nepean storages (NSW) harbour Murray cod × trout cod hybrids and Macquarie perch.
10 Disease vector.
11 Range stable since 1960s.
12 Lower Murray River now too saline for many native species.
13 Rare, in few rivers
14 Displaced from northern basin by grass carp, Nile perch.
15 No major effects (southern basin).
16 *Ctenopharyngodon idella* (northern basin).

- a continuing diaspora of alien species, particularly common carp, eastern gambusia (*Gambusia holbrooki*), oriental weatherloach (*Misgurnus anguillicaudatus*) and tilapia (*Oreochromis mossambicus*). These are a formidable quartet, each with a record of destruction and potential to spread further afield (Ch. 11). In contrast, the salmonids have been given every encouragement and probably do occupy their potential ranges fully, or nearly so;
- members of the Plotosidae (eel-tailed catfish) and Terapontidae (grunters) are vulnerable. Some are small, little-known species with restricted distributions (e.g. freshwater cobbler, *Tandanus bostocki*); others, such as the freshwater catfish (*T. tandanus*), are more widespread and confronted by alien competitors and predators;
- native species with ancient lineages, few relatives, small ranges or other unusual characteristics are also vulnerable. Examples include the Australian lungfish (*Neoceratodus forsteri*), cave gudgeon, desert goby, nightfish, saratogas, non-parasitic lamprey (*Mordacia praecox*) and salamanderfish (*Lepidogalaxias salamandroides*). For some, the future may be not in natural habitats, but as captives in aquaria or in hatchery ponds.

Other scenarios exist for native fishes, but as risk assessments (Koehn and MacKenzie 2004; Pollino *et al.* 2007a, b; Whittington and Chong 2007; Ch. 12). This methodical approach is preferred where there are scant data and one of its virtues, ideally, is that the assessments are transparent and open to review.

We now revisit earlier chapters to select some points to illustrate something of the potential and the challenges for ecological research on Australian freshwater fishes. As an accompaniment, we note some questions for new researchers to consider (see Box 13.1) and conclude with another scenario, with regard for issues that are likely to shape our outlook some decades from now.

Box 13.1: Questions for research and discussion

These are some broad areas where there are opportunities for research. The list is neither ordered or comprehensive.

- How can we incorporate genetics in ecological studies? For example, is it feasible to develop genetic methods to identify species in the field?
- What is the role of fishes as apex predators in structuring assemblages and food webs?
- What drives recruitment and year–class strength in freshwater fishes? How can we most easily estimate/measure these parameters in wild populations?
- How do variations in growth, age at maturity and longevity affect the dynamics of populations?
- What are the relative influences of biotic and abiotic factors in assemblage structure?
- What should be done about alien fishes? Are some acceptable and some not? What are the prospects for control/eradication using genetic methods?
- Is it feasible to eradicate salmonids from a stream and re-establish and maintain an assemblage of native fishes?
- How should we monitor and manage the effects of stocking with hatchery-bred native fishes?
- How should we measure the resilience of fishes at the levels of individuals, populations, metapopulations and species?
- Which native species are predisposed to adapt to environments changed by humans? Which species have most invasive potential? Which are best adapted to endure climate change?
- How should environmental water allocations be deployed and monitored to benefit fishes and other aquatic organisms?
- What new technological methods might be applied or developed to enhance data collection through monitoring of fishes and their habitats?
- What prospects are there to re-establish commercial fisheries in inland waters?
- How can management of recreational fisheries in inland waters be improved?
- Are there fish species (or ecological communities) that are not listed as threatened by state or federal agencies, but ought to be? How effective are the various listing processes?
- What kinds of models could be developed as tools for research and management?

CHALLENGES

Biodiversity, biogeography and evolution

Studies of biogeography and systematics have ballooned in recent years with the advent of new technology, and perceptions are changing rapidly (Chs 2, 3). The revolution is both sobering and exhilarating. While our native freshwater fish fauna has been seen as depauperate compared to other continents, the number of known species has reached 256 (Ch. 2), more than claimed by McDowall (1996b: 180 species) and twice the number reported by Lake (1971: 130 species). Many of the new taxa are in well-explored regions, such as the Murray-Darling Basin and Tasmania, as well as in the more remote areas of inland and northern Australia. The cryptic nature of some taxa means that species may not always be distinguishable in the field, and the term 'species complex' is likely to become more popular in future. There is abundant evidence to show that genetic structure can be significant even within populations. The taxonomic revolution also adds a new dimension to issues of conservation and management. It is another game changer.

Habitat and movement

Australian ecologists were slow to realise that the propensities of our native fishes to move, migrate and negotiate barriers are unlike those of salmonids and that we cannot simply adopt research findings from the Northern Hemisphere (Chs 4, 5). Again, technological advances (radio-tracking, innovative marking techniques, otolith microchemistry) have shown the way. For example, it is now confirmed that congolli and Australian grayling are amphidromous (Ch. 8). There has also been progress in understanding habitat use and movements by smaller, less conspicuous fish such as gudgeons (Eleotridae) and smelt (Retropinnidae).

The precarious conservation status of some native species suggests that their remnant populations occupy suboptimal habitats and are denied opportunities for free movement, so that populations are not truly self-sustaining. While a riverscape-scale project requires research and costly infrastructure, as in the Hume to the Sea fish-passage program on the Murray River (Chs 5, 12), programs of that kind must be replicated. Restoring connectivity between habitats, promoting genetic mixing, recruitment and resilience, is surely one of the most potent ways to restore native fish populations.

Trophic ecology

The trophic ecology of Australian freshwater fishes has attracted many studies but most are concerned with the diets of individual species, especially predators, rather than concepts and ideas (Ch. 6). This is remarkable considering the emphasis on trophic studies in freshwater ecology (e.g. trophic cascades) and their significance in conservation and management. The challenge now is to integrate the information we have, considering the ways that morphology and feeding styles are related, how nutrient and other chemical requirements match the prey ingested and how fish influence, and are influenced by, food webs. In Australia, the main emphasis in river restoration and rehabilitation has been bottom-up, in the belief that providing appropriate habitats or flows will restore systems to functional integrity (field of dreams hypothesis: Palmer *et al.* 1997). The roles of fishes, especially apex predators, in structuring freshwater food webs are little understood (Chs 6, 10).

Reproduction and early life history

Research on reproduction and early life history is another field that has been constrained by a species focus, as biologists have tried to comprehend the life cycles of fishes that are rarely seen, hard to catch and prone to appear or disappear as if by chance. Technological advances in underwater filming, however, offer new ways to monitor the behaviour of fish in their natural habitats. Progress from hatchery to field studies has dispelled some misconceptions (Ch. 7), but hatchery studies and laboratory experimentation are perhaps the only ways to tease apart circadian rhythms, photoperiod, temperature and other factors that influence the timing of breeding. Attempts to generalise about life history strategies have been fruitful (Ch. 7) and show promise as a way to view Australian freshwater fishes in a broader conceptual framework.

Paramount among the challenges is to understand recruitment – the factors governing the progression from one generation to the next (Chs 7, 9). Field-based studies are vital, not least because they generate hypotheses, but they are limited because the complexities of natural systems and the vagaries of sampling taunt us with correlations rather than confirmations. Even extended studies over multiple years are likely to lack controls, as experimental manipulations in the field are easily confounded by other factors and chance events. Well-conceived laboratory experiments may reveal the basic mechanisms of recruitment, but combinations of experiments, field work and modelling are needed for conceptual understanding and applications in management.

Age and growth

Information about rates of growth, age at maturity, longevity and the interplay of related ecological and evolutionary variables is vital for fish ecology (Ch. 8). Although much remains to be done, Australia has passed through a validation phase, regarding otoliths as monitors of growth, and has entered an application phase. Otolith chemistry is complemented by the use of RNA/DNA ratios and methods of sclerochronology to monitor short- and long-term patterns of growth. With our enhanced ability to determine age and describe patterns of growth, the challenge is to understand how factors such as temperature, flow, food, competition, predation and pollutants influence population parameters and, ultimately, the dynamics of populations, metapopulations and species. Again, there is scope for laboratory experiments and synergistic field programs.

Population dynamics

If past and present attitudes to gathering data about the dynamics of fish and fisheries and the environment persist (Ch. 9), fish biologists in 2050 could know little more than we do today. Monitoring is a low priority for some environmental agencies. No doubt this is because it is costly in terms of money and other resources, because it is rarely clear what should be monitored and how the data should be used, and perhaps because the data could be used to hold governments to account. Scientists and managers need to respond; advances in data-gathering technology (Ch. 5) could provide the impetus for new approaches and adaptive management (Ch. 12) could provide the working environment.

Knowledge of the dynamics of populations is the basis of effective conservation and management of freshwater fishes (Ch. 9). Australia is still recovering from a legacy of poorly managed commercial native fisheries, coupled with declining health of its inland water bodies. Only alien species are now fished commercially. Recreational fishing, on the other hand, remains an important leisure pursuit for a large number of Australians, and the increasing conservation ethic makes fishers potentially important allies in the task of ensuring the sustainability of native fish stocks.

Targeted monitoring, management and research must be priorities for management of fisheries into the future. The sampling, statistical and modelling toolkits now available to fisheries scientists provide new and exciting opportunities to describe and understand the factors affecting fluctuations in populations (Ch. 9). Applying properly validated models will allow scientists to predict the effects of future water resource development, climate change or changes to fishing regulations or pressure, which surely will contribute to better conservation and management of our fishes. The formation of interdisciplinary teams comprising scientists, water managers and other key professionals is now recognised as the best way to effect these types of programs so that all aspects of fresh waters, their inhabitants and the pressures placed on them are considered.

Assemblages

Another limitation of the species focus mentioned above is that, not surprisingly, we have underestimated the importance of assemblages (Ch. 10). This is a fast-growing area for research, exemplified by surveys and other studies where geomorphic, hydrologic, climatic and faunal data have been collected to define assemblages and communities, in keeping with agency moves toward management of ecological communities as well as species. The next challenge could be to develop ideas about assemblage–environment relationships and metacommunity dynamics, capitalising on environmental water allocations as a means to test hypotheses in a context of adaptive management (Ch. 12). Another challenge is to tease apart the influences of abiotic and biotic factors in structuring assemblages. There may be opportunities for large-scale experiments to investigate the effects of changes in fishing regulations. Modelling approaches in particular may assist us to test scenarios and examine sensitivities in ways that cannot be achieved experimentally.

Alien species

Australia, like many other parts of the world, has a strong contingent of alien freshwater fishes (Ch. 11). Some have wrought havoc, some remain sleepers with the potential to do so, and others have failed to become established. In the wider community, some attitudes to alien species are little different from those of the 19th-century acclimatisation societies (Francis 1862); others favour particular alien groups, such as salmonids or aquarium fishes. And surprisingly, many people are unsure which species are native and which are alien. Some may think that the distinction is unimportant.

The legacies of common carp, eastern gambusia and redfin perch (*Perca fluviatilis*) are testaments to the effects that alien species may have. While there have been efforts to measure their impacts, managing them is a far more difficult problem. Some argue that species such as brown trout (*Salmo trutta*) and rainbow trout (*Oncorhynchus mykiss*) are socially and financially beneficial and should remain; others argue that they are alien species that cause unacceptable damage to natural ecosystems. Which alien species are acceptable? If we do accept trout, why not accept common carp? These are questions with cultural dimensions and without simple answers, and they are issues where the role of scientists is advisory rather than decisive (Sarewitz 2004).

If the community decides that alien species should not be promoted, how should we contain, manage or eradicate them? Management thus far has been spectacularly unsuccessful but this is not a reason to accept defeat, as major advances in the past were made as new technologies became available. Genetic approaches, such as 'daughterless carp' (Ch. 11), seem promising and one day might be another game changer.

Conservation and management

Conservation of freshwater fishes is an issue on the agenda of all Australian governments. For example, the Murray-Darling Basin Native Fish Strategy (<http://www.mdba.gov.au/programs/nativefishstrategy>) has fostered many initiatives to better understand the ecology of fishes, to enhance management and increase awareness. Consider also the many 'threatened species' nominations and recovery plans prepared by agencies and scientists (Ch. 12). Despite these activities, big challenges remain.

- *What is the best strategy to deploy funds for conservation?* Should we focus our efforts on sharply defined, solvable problems or spread them across many species, big areas and long periods? A short answer may be that neglecting some fronts in a battle may open the way for new incursions by the foe!
- *How should we set targets for restoration*? The idea of a shifting baseline (Jackson *et al.* 2011) is that we tend to accept our first experiences of the environment as an ideal, when the environment was much changed, of course, even 100–150 years ago. A pristine natural environment is not a realistic target for restoration; rather, our goal could be to manage our impacts so that we are part of a balanced system that provides sustainable resources and retains some natural character. Again, there are cultural issues here that transcend ecology.
- *How do we reconcile conservation of threatened fish species that sustain recreational fisheries*? Species such as Murray cod (*Maccullochella peelii*) are considered vulnerable nationally, yet we allow them to be fished. There is no essential conflict, however, between management for conservation and exploitation; there is even a case for the endangered trout cod (*M. macquariensis*) to be exposed to sustainable recreational fishing (Ch. 12). This would call upon anglers, as an underused source of information and support for conservation. Conservation ideally has more to with managing impacts than with managing the environment.

Climate change

Discussions of climate change are spread across several chapters because this is another game changer. If projections are realised, Australian environments will undergo profound changes (CSIRO/ABM 2007; Steffen 2009; Steffen *et al.* 2009). Of course, climates have changed throughout time and will continue to do so; the issue here is the effects of human industry, manifest as global warming. The threat is not change itself but change on short time scales, as unfamiliar to fishes as to humans. In addition, the capacity of fishes to endure changes (Ch. 4) or move to other areas (Ch. 5) may be constrained by human factors, including barriers to movement and intensified demands for water.

The vulnerability of species to change is a product of exposure, governed by extrinsic (climatic) factors, and sensitivity, governed by intrinsic ecological, physiological and genetic traits (Smith *et al.* 2000; Turner *et al.* 2003; Williams *et al.* 2008). Predictions of exposure generally rely on modelling of temperature, rainfall and other climatic variables, guided by forecast greenhouse gas emissions under different controls. Ecological scenarios are needed to complement these, but we know too little of the sensitivity of most freshwater fishes to predict their responses accurately and, in any case, predictions could be undermined by feedback effects or stochastic interventions. This is underscored by the prevalence of terms like 'uncertainty' and 'surprise' in the climate change literature (Steffen 2009). Strategic responses to climate change will therefore need to include ways to respond to rather than overcome uncertainty, and to apply theory and practice in adaptive management with unprecedented rigour. The implications of climate change for freshwater fishes and other species are reviewed in the papers of a symposium introduced by Koehn (2011).

CODA: A SCENARIO FOR 2050

The scenario for species discussed earlier was not realised, although most of its speculations remain possibilities for the future. Thus, the time scale could be in error, rather than the prescriptions for particular events. In regard to trends, however, the scenario was nearer the mark. In this concluding section, we venture another scenario to indicate some of the factors that could influence freshwater fishes and ecologists in 2050. As before, the aim is not to challenge Nostradamus but to provoke argument and discussion.

- The perception that native freshwater fishes generally will experience declines or extinctions and that alien species will dominate should be tempered, as some native species may adapt to change better than others. The climbing galaxias (*Galaxias brevipinnis*) and spangled perch (*Leiopotherapon unicolor*) may be better equipped; another possible example could be golden perch (*Macquaria ambigua*).
- The adaptive capacities of species depend upon the nature of the challenge. It could be abiotic (e.g. climate change) or biotic (e.g. interactions with an invasive species), or both. It could be another game changer that resets the ecological and evolutionary trajectories of a wide range of species. For example, invasion of Lake Eyre Basin rivers by common carp is a daunting prospect, although there could be some perverse consolation in knowing whether they could adapt to an environment so variable as Cooper Creek and real satisfaction in knowing that they could not.
- The continuing spread of tilapia in northern Australia, and the possibility of that species invading the Murray-Darling Basin, could be another game changer. Tilapia in Australia is potentially an ecosystem engineer (Ehrenfeld 2010) and one of a quartet of alien species that are well-established and yet to fully occupy their potential ranges (see 'Scenario for 2000').
- Climate change will surely have consequences for freshwater fishes but apprehension over its likely effects threatens to overwhelm broader perspectives (Strayer 2010). The impacts of flow regulation, in all its manifestations, and interactions with alien species are challenges that, even in the future, will be of no less consequence for the native fauna.
- In the past, the Murray-Darling Basin has been a melting pot for fishes, and one-quarter of the species in the basin now are of alien origin (Davies *et al.* 2010). It may be overshadowed, however, by tropical northern Australia, where there already are more than 20 alien species including tilapia and numerous aquarium escapees (Ch. 12).
- There is likely to be continuing tension over alien salmonids which sustain an important recreational industry but have major impacts on native fishes, especially in upland streams (Chs 11, 12). The arguments remain polarised, when a reasonable compromise might be to stock salmonids in some streams and not others, in deference to the need to protect galaxiids (McDowall 2006) and

other native fishes. This would be consistent with proposals to establish freshwater protected areas (Humphries and Winemiller 2009).

It is distressing how the threats to our native freshwater fishes have intensified even as we have learned more about them, but this reflects the role of science as informant rather than arbiter in government policy (Sarewitz 2004). Political and cultural values will change in coming decades, and we will need to move with the times. Perhaps too many scientists have adopted a reactive approach whereby priorities are determined by managers seeking to resolve existing problems. On the other hand, there are managers and scientists concerned with forward-looking, proactive approaches (Bond *et al.* 2008). In the face of uncertainty, we can learn only so much from hindsight; predictions may fail and long-term planning is a gamble. We must prepare for change, rather than resist it. If you will pardon a tenuous metaphor, it is an evolutionary strategy that has served our freshwater fishes well!

Appendix – Species list of Australian freshwater fishes

There have been many changes to the names of species, genera and families in the past and more are to come, especially following applications of molecular methods (Chs 2, 3). There are complexes of species awaiting formal description, species with genetic substructure that may warrant taxonomic recognition and species that have been reassigned to different genera or families. These changes are a normal part of progress in systematics, and they follow the conventions of zoological nomenclature. Common names are more problematic because some have changed erratically over time and, despite some standardisation, they often differ between regions. The best course is to be wary, and to avoid perpetuating confusion.

The following provisional list was prepared by Michael Hammer and Peter Unmack. It is adapted from Allen *et al.* (2002) but includes recent published records and a review of alien species (Ch. 11). It includes undescribed forms which are visually distinctive, and flags others where taxonomic issues are indicated by genetic data. In fact, most genera and species need to be examined genetically for the presence of cryptic species.

The taxonomic framework follows Eschmeyer's (2012) *Catalog of Fishes*. Most common names follow the recommendations of CSIRO Marine Research, the Fisheries Research and Development Corporation and Seafood Services Australia (<http://www.marine.csiro.au/caab/standard_names.htm>). Other record sources include Musyl and Keenan (1992), Larson and Hoese (1996), Wager and Unmack (2000), Hogan and Vallance (2005), Morgan and Beatty (2006a) and Ebner and Thuesen (2011).

The inclusion of marine/estuarine species here is somewhat arbitrary, including only some of the taxa commonly reported in fresh water. Many more species occur occasionally in the lowermost reaches of streams and rivers.

Family	Common name	Species	Authority or [Reference]
Native species			
Ambassidae	Agassiz's glassfish	*Ambassis agassizii*[1]	Steindachner, 1867
Ambassidae	Sailfin glassfish	*Ambassis agrammus*	Günther, 1867
Ambassidae	Elongate glassfish	*Ambassis elongatus*[1]	(Castelnau, 1878)
Ambassidae	Macleay's glassfish	*Ambassis macleayi*	(Castelnau, 1878)
Ambassidae	Flagtail glassfish	*Ambassis miops*	Günther, 1872
Ambassidae	Northwest glassfish	*Ambassis* sp.[1,5]	[Allen *et al.* (2002)]
Ambassidae	Pennyfish	*Denariusa australis*	(Steindachner, 1867)
Ambassidae	Giant glassfish	*Parambassis gulliveri*	(Castelnau, 1878)
Anguillidae	Southern shortfin eel	*Anguilla australis*[4]	Richardson, 1841
Anguillidae	Indonesian shortfin eel	*Anguilla bicolor*	McClelland, 1844
Anguillidae	Giant mottled eel	*Anguilla marmorata*	Quoy & Gaimard, 1824
Anguillidae	Pacific shortfin eel	*Anguilla obscura*	Günther, 1872
Anguillidae	Longfin eel	*Anguilla reinhardtii*	Steindachner, 1867
Apogonidae	Mouth almighty	*Glossamia aprion*[2]	(Richardson, 1842)
Ariidae	Smallmouth catfish	*Cinetodus froggatti*	(Ramsey & Ogilby, 1886)
Ariidae	Warrior catfish	*Hemiarius dioctes*	(Kailola, 2000)

Family	Common name	Species	Authority or [Reference]
Ariidae	Highfin catfish	*Neoarius berneyi*	(Whitley, 1941)
Ariidae	Blue catfish	*Neoarius graeffei*	(Kner & Steindachner, 1867)
Ariidae	Silver cobbler	*Neoarius midgleyi*	(Kailola & Pierce, 1988)
Ariidae	Boofhead catfish	*Sciades leptaspis*	(Bleeker, 1862)
Ariidae	Shovelnose catfish	*Sciades paucus*	(Kailola, 2000)
Atherinidae	Smallmouth hardyhead	*Atherinosoma microstoma*	(Günther, 1861)
Atherinidae	Darling hardyhead	*Craterocephalus amniculus*[1]	Crowley & Ivantsoff, 1990
Atherinidae	Finke hardyhead	*Craterocephalus centralis*[3]	Crowley & Ivantsoff, 1990
Atherinidae	Deep hardyhead	*Craterocephalus cuneiceps*	Whitley, 1944
Atherinidae	Dalhousie hardyhead	*Craterocephalus dalhousiensis*	Ivantsoff & Glover, 1974
Atherinidae	Lake Eyre hardyhead	*Craterocephalus eyresii*	(Steindachner, 1883)
Atherinidae	Murray hardyhead	*Craterocephalus fluviatilis*	McCulloch, 1912
Atherinidae	Glover's hardyhead	*Craterocephalus gloveri*[3]	Crowley & Ivantsoff, 1990
Atherinidae	Drysdale hardyhead	*Craterocephalus helenae*	Ivantsoff, Crowley & Allen, 1987
Atherinidae	Freckled hardyhead	*Craterocephalus lentiginosus*	Ivantsoff, Crowley & Allen, 1987
Atherinidae	Mariana's hardyhead	*Craterocephalus marianae*	Ivantsoff, Crowley & Allen, 1987
Atherinidae	Marjorie's hardyhead	*Craterocephalus marjoriae*	Whitley, 1948
Atherinidae	Fly-specked hardyhead	*Craterocephalus stercusmuscarum*[2,4]	(Günther, 1867)
Atherinidae	Blackmast	*Craterocephalus stramineus*[2]	(Whitley, 1950)
Atherinidae	Western hardyhead	*Leptatherina wallacei*	(Prince, Ivantsoff & Potter, 1982)
Belonidae	Freshwater longtom	*Strongylura krefftii*	(Günther, 1866)
Carcharhinidae	Bull shark	*Carcharhinus leucas*	(Müller & Henle, 1839)
Carcharhinidae	Speartooth shark	*Glyphis garricki*	Compagno, White & Last, 2008
Carcharhinidae	Northern river shark	*Glyphis glyphis*	(Müller & Henle, 1839)
Clupeidae	River herring	*Clupeoides* sp.[5]	[Hogan & Wallace (2005)]
Clupeidae	Bony herring	*Nematalosa erebi*[1]	(Günther, 1868)
Clupeidae	Freshwater herring	*Potamalosa richmondia*	(Macleay, 1879)
Cynoglossidae	Freshwater tongue sole	*Cynoglossus heterolepis*	Weber, 1910
Dasyatidae	Freshwater whipray	*Himantura dalyensis*	Last & Manjaji-Matsumoto, 2008
Eleotridae	Sunset gudgeon	*Bostrychus zonatus*	Weber, 1907
Eleotridae	Greenback gudgeon	*Bunaka gyrinoides*	(Bleeker, 1853)
Eleotridae	Snakehead gudgeon	*Giurus margaritacea*	(Valenciennes in Cuvier & Valenciennes, 1837)

Family	Common name	Species	Authority or [Reference]
Eleotridae	Striped gudgeon	*Gobiomorphus australis*	(Krefft, 1864)
Eleotridae	Cox's gudgeon	*Gobiomorphus coxii*	(Krefft, 1864)
Eleotridae	Golden carp gudgeon	*Hypseleotris aurea*	(Shipway, 1950)
Eleotridae	Barraway's gudgeon	*Hypseleotris barrawayi*	Larson, 2007
Eleotridae	Empire gudgeon	*Hypseleotris compressa*	(Krefft, 1864)
Eleotridae	Slender carp gudgeon	*Hypseleotris ejuncida*	Hoese & Allen in Allen, 1982
Eleotridae	Firetail gudgeon	*Hypseleotris galii*[1]	(Ogilby, 1898)
Eleotridae	Barnett River gudgeon	*Hypseleotris kimberleyensis*	Hoese & Allen in Allen, 1982
Eleotridae	Western carp gudgeon	*Hypseleotris klunzingeri*[2]	(Ogilby, 1898)
Eleotridae	Regent carp gudgeon	*Hypseleotris regalis*	Hoese & Allen in Allen, 1982
Eleotridae	Midgley's carp gudgeon	*Hypseleotris* sp.[2,5]	[Larson & Hoese (1996)]
Eleotridae	Lake's carp gudgeon	*Hypseleotris* sp.[3,5]	[Larson & Hoese (1996)]
Eleotridae	Murray-Darling carp gudgeon	*Hypseleotris* sp.[5]	[Allen *et al.* (2002)]
Eleotridae	Mitchell gudgeon	*Kimberleyeleotris hutchinsi*	Hoese & Allen, 1987
Eleotridae	Drysdale gudgeon	*Kimberleyeleotris notata*	Hoese & Allen, 1987
Eleotridae	Southern purple-spotted gudgeon	*Mogurnda adspersa*[2]	(Castelnau, 1878)
Eleotridae	Flinders Ranges purple-spotted gudgeon	*Mogurnda clivicola*	Allen & Jenkins, 1999
Eleotridae	Finke purple-spotted gudgeon	*Mogurnda larapintae*	(Zietz, 1896)
Eleotridae	Northern purple-spotted gudgeon	*Mogurnda mogurnda*[2]	(Richardson, 1844)
Eleotridae	Kimberley purple-spotted gudgeon	*Mogurnda oligolepis*[1]	Allen & Jenkins, 1999
Eleotridae	Dalhousie purple-spotted gudgeon	*Mogurnda thermophila*	Allen & Jenkins, 1999
Eleotridae	Aru gudgeon	*Oxyeleotris aruensis*	(Weber, 1911)
Eleotridae	Fimbriate gudgeon	*Oxyeleotris fimbriata*	Weber, 1907
Eleotridae	Sleepy cod	*Oxyeleotris lineolata*	(Steindachner, 1867)
Eleotridae	Poreless gudgeon	*Oxyeleotris nullipora*[3]	Roberts, 1978
Eleotridae	Blackbanded gudgeon	*Oxyeleotris selheimi*	(Macleay, 1884)
Eleotridae	Flathead gudgeon	*Philypnodon grandiceps*[1]	(Krefft, 1864)
Eleotridae	Dwarf flathead gudgeon	*Philypnodon macrostomus*[1]	Hoese & Reader, 2006
Engraulidae	Freshwater anchovy	*Thryssa scratchleyi*	(Ramsey & Ogilby, 1886)
Galaxiidae	Golden galaxias	*Galaxias auratus*	Johnston, 1883
Galaxiidae	Climbing galaxias	*Galaxias brevipinnis*[2]	Günther, 1866
Galaxiidae	Swan galaxias	*Galaxias fontanus*	Fulton, 1978
Galaxiidae	Barred galaxias	*Galaxias fuscus*	Mack, 1936
Galaxiidae	Clarence galaxias	*Galaxias johnstoni*	Scott, 1936
Galaxiidae	Common galaxias	*Galaxias maculatus*	(Jenyns, 1842)
Galaxiidae	Western galaxias	*Galaxias occidentalis*[1]	Ogilby, 1899
Galaxiidae	Mountain galaxias	*Galaxias olidus*[2]	Günther, 1866

Family	Common name	Species	Authority or [Reference]
Galaxiidae	Swamp galaxias	*Galaxias parvus*	Frankenberg, 1968
Galaxiidae	Pedder galaxias	*Galaxias pedderensis*	Frankenberg, 1968
Galaxiidae	Flathead galaxias	*Galaxias rostratus*	Klunzinger, 1872
Galaxiidae	Saddled galaxias	*Galaxias tanycephalus*	Fulton, 1978
Galaxiidae	Spotted galaxias	*Galaxias truttaceus*[4]	Valenciennes (ex Cuvier) in Cuvier & Valenciennes, 1846
Galaxiidae	Western dwarf galaxias	*Galaxiella munda*	McDowall, 1978
Galaxiidae	Blackstriped dwarf galaxias	*Galaxiella nigrostriata*	(Shipway, 1953)
Galaxiidae	Eastern dwarf galaxias	*Galaxiella pusilla*[1]	(Mack, 1936)
Galaxiidae	Tasmanian whitebait	*Lovettia sealii*	(Johnston, 1883)
Galaxiidae	Tasmanian mudfish	*Neochanna cleaveri*	(Scott, 1934)
Galaxiidae	Shannon galaxias	*Paragalaxias dissimilis*	(Regan, 1906)
Galaxiidae	Great Lake galaxias	*Paragalaxias eleotroides*	McDowall & Fulton, 1978
Galaxiidae	Julian galaxias	*Paragalaxias julianus*	McDowall & Fulton, 1978
Galaxiidae	Arthurs galaxias	*Paragalaxias mesotes*	McDowall & Fulton, 1978
Geotriidae	Pouch lamprey	*Geotria australis*	Gray, 1851
Gobiidae	Roman-nose goby	*Awaous acritosus*	Watson, 1994
Gobiidae	River goby	*Awaous ocellaris*	(Broussonet, 1782)
Gobiidae	Desert goby	*Chlamydogobius eremius*	(Zietz, 1896)
Gobiidae	Dalhousie goby	*Chlamydogobius gloveri*	Larson, 1995
Gobiidae	Finke goby	*Chlamydogobius japalpa*	Larson, 1995
Gobiidae	Elizabeth Springs goby	*Chlamydogobius micropterus*	Larson, 1995
Gobiidae	Tadpole goby	*Chlamydogobius ranunculus*	Larson, 1995
Gobiidae	Edgbaston goby	*Chlamydogobius squamigenus*	Larson, 1995
Gobiidae	Golden flathead goby	*Glossogobius aureus*[1]	Akihito & Meguro, 1975
Gobiidae	Mulgrave goby	*Glossogobius bellendenensis*	Hoese & Allen, 2009
Gobiidae	Concave flathead goby	*Glossogobius concavifrons*	(Ramsey & Ogilby, 1886)
Gobiidae	Tank goby	*Glossogobius giuris*[1]	(Hamilton, 1822)
Gobiidae	False Celebes goby	*Glossogobius illimis*	Hoese & Allen, 2011
Gobiidae	Square blotch goby	*Glossogobius munroi*	Hoese & Allen, 2011
Gobiidae	Dwarf goby	*Glossogobius* sp.[5]	[Allen *et al.* (2002)]
Gobiidae	Freshwater mangrove goby	*Mugilogobius notospilus*	(Günther, 1877)
Gobiidae	Drain mangrove goby	*Mugilogobius rivulus*	Larson, 2001
Gobiidae	Swan River goby	*Pseudogobius olorum*	(Sauvage, 1880)
Gobiidae	Blue-spot goby	*Pseudogobius* sp.[5]	[Larson & Hoese (1996)]
Gobiidae	Speckled goby	*Redigobius bikolanus*	(Herre, 1927)
Gobiidae	Scaleless goby	*Schismatogobius insignus*	(Herre, 1927)
Gobiidae	Blue stream goby	*Sicyopterus lagocephalus*	(Pallas, 1770)

Family	Common name	Species	Authority or [Reference]
Gobiidae	Unnamed cling goby	*Sicyopus discordipinnis*	Watson, 1995
Gobiidae	Allen's Stiphodon	*Stiphodon allen*	Watson, 1996
Gobiidae	Black Stiphodon	*Stiphodon atratus*	Watson, 1996
Gobiidae	Unnamed Stiphodon	*Stiphodon birdsong*	Watson, 1996
Gobiidae	Golden-red Stiphodon	*Stiphodon rutilaureus*	Watson, 1996
Gobiidae	Neon goby	*Stiphodon semoni*	Weber, 1895
Gobiidae	Scary's Tasman goby	*Tasmanogobius lasti*	Hoese, 1991
Gobiidae	Lord's Tasman goby	*Tasmanogobius lordi*	Scott, 1935
Hemiramphidae	Snub-nosed garfish	*Arrhamphus sclerolepis*[4]	Günther, 1866
Hemiramphidae	Long-jawed river garfish	*Zenarchopterus caudovittatus*	(Weber, 1907)
Hemiramphidae	Fly River garfish	*Zenarchopterus novaeguineae*	Weber, 1913
Kuhliidae	Spotted flagtail	*Kuhlia marginata*	(Cuvier, 1829)
Kuhliidae	Jungle perch	*Kuhlia rupestris*	(Lacépède, 1802)
Kurtidae	Nurseryfish	*Kurtus gulliveri*	Castelnau, 1878
Latidae	Barramundi	*Lates calcarifer*	(Bloch, 1790)
Lepidogalaxiidae	Salamanderfish	*Lepidogalaxias salamandroides*	Mees, 1961
Megalopidae	Oxeye herring	*Megalops cyprinoides*	(Broussonet, 1782)
Melanotaeniidae	Cairns rainbowfish	*Cairnsichthys rhombosomoides*	(Nichols & Raven, 1928)
Melanotaeniidae	Threadfin rainbowfish	*Iriatherina werneri*[1]	Meinken, 1974
Melanotaeniidae	Western rainbowfish	*Melanotaenia australis*[1]	(Castelnau, 1875)
Melanotaeniidae	Crimson-spotted rainbowfish	*Melanotaenia duboulayi*	(Castelnau, 1878)
Melanotaeniidae	Lake Eacham rainbowfish	*Melanotaenia eachamensis*	Allen & Cross, 1982
Melanotaeniidae	Exquisite rainbowfish	*Melanotaenia exquisita*	Allen, 1978
Melanotaeniidae	Murray River rainbowfish	*Melanotaenia fluviatilis*	(Castelnau, 1878)
Melanotaeniidae	Slender rainbowfish	*Melanotaenia gracilis*	Allen, 1978
Melanotaeniidae	McCulloch's rainbowfish	*Melanotaenia maccullochi*[1]	Ogilby, 1915
Melanotaeniidae	Blackbanded rainbowfish	*Melanotaenia nigrans*	(Richardson, 1843)
Melanotaeniidae	Pygmy rainbowfish	*Melanotaenia pygmaea*	Allen, 1978
Melanotaeniidae	Eastern rainbowfish	*Melanotaenia splendida*[4]	(Peters, 1866)
Melanotaeniidae	Banded rainbowfish	*Melanotaenia trifasciata*[1]	(Rendahl, 1922)
Melanotaeniidae	Utchee rainbowfish	*Melanotaenia utcheensis*	McGuigan, 2000
Melanotaeniidae	Ornate rainbowfish	*Rhadinocentrus ornatus*[2]	Regan, 1914
Mordaciidae	Shorthead lamprey	*Mordacia mordax*	(Richardson, 1846)
Mordaciidae	Non-parasitic lamprey	*Mordacia praecox*	Potter, 1968
Mugilidae	Diamond mullet	*Liza alata*[3]	(Steindachner, 1892)
Mugilidae	Sea mullet	*Mugil cephalus*	Linnaeus, 1758
Mugilidae	Freshwater mullet	*Trachystoma petardi*	(Castelnau, 1875)

Family	Common name	Species	Authority or [Reference]
Muraenidae	Freshwater moray	*Gymnothorax polyuranodon*	(Bleeker, 1853)
Neoceratodontidae	Australian lungfish	*Neoceratodus forsteri*	(Krefft, 1870)
Odontobutidae	Cave gudgeon	*Milyeringa veritas*	Whitley, 1945
Osteoglossidae	Northern saratoga	*Scleropages jardinii*	(Saville-Kent, 1892)
Osteoglossidae	Southern saratoga	*Scleropages leichhardti*	Günther, 1864
Percichthyidae	Nightfish	*Bostockia porosa*[1]	Castelnau, 1873
Percichthyidae	Two-spine blackfish	*Gadopsis bispinosus*[1]	Sanger, 1984
Percichthyidae	River blackfish	*Gadopsis marmoratus*[2]	Richardson, 1848
Percichthyidae	Bloomfield River cod	*Guyu wujalwujalensis*[3]	(Pusey & Kennard, 2001)
Percichthyidae	Clarence River cod	*Maccullochella ikei*	Rowland, 1986
Percichthyidae	Trout cod	*Maccullochella macquariensis*	(Cuvier in Cuvier & Valenciennes, 1829)
Percichthyidae	Mary River cod	*Maccullochella mariensis*	Rowland, 1993
Percichthyidae	Murray cod	*Maccullochella peelii*	(Mitchell, 1838)
Percichthyidae	Golden perch	*Macquaria ambigua*[1,4]	(Richardson, 1845)
Percichthyidae	Macquarie perch	*Macquaria australasica*[2]	Cuvier in Cuvier & Valenciennes, 1830
Percichthyidae	Lake Eyre golden perch	*Macquaria* sp.[5]	[Musyl & Keenan (1992)]
Percichthyidae	Balston's pygmy perch	*Nannatherina balstoni*	Regan, 1906
Percichthyidae	Southern pygmy perch	*Nannoperca australis*[1]	Günther, 1861
Percichthyidae	Yarra pygmy perch	*Nannoperca obscura*	(Klunzinger, 1872)
Percichthyidae	Oxleyan pygmy perch	*Nannoperca oxleyana*	Whitley, 1940
Percichthyidae	Little pygmy perch	*Nannoperca* sp.[5]	[Morgan *et al.* (2011)]
Percichthyidae	Variegated pygmy perch	*Nannoperca variegata*	Kuiter & Allen, 1986
Percichthyidae	Western pygmy perch	*Nannoperca vittata*[2]	(Castelnau, 1873)
Percichthyidae	Estuary perch	*Percalates colonorum*[3]	(Günther, 1863)
Percichthyidae	Australian bass	*Percalates novemaculeata*[3]	(Steindachner, 1866)
Plotosidae	Toothless catfish	*Anodontiglanis dahli*	Rendahl, 1922
Plotosidae	Cooper Creek catfish	*Neosiluroides cooperensis*	Allen & Feinberg, 1998
Plotosidae	Black catfish	*Neosilurus ater*[1]	(Perugia, 1894)
Plotosidae	Shortfin catfish	*Neosilurus brevidorsalis*	(Günther, 1867)
Plotosidae	Dalhousie catfish	*Neosilurus gloveri*	Allen & Feinberg, 1998
Plotosidae	Hyrtl's catfish	*Neosilurus hyrtlii*[2]	Steindachner, 1867
Plotosidae	Softspine catfish	*Neosilurus mollespiculum*	Allen & Feinberg, 1998
Plotosidae	Falsespine catfish	*Neosilurus pseudospinosus*[1]	Allen & Feinberg, 1998
Plotosidae	Bulloo falsespine catfish	*Neosilurus* sp.[5]	[Wager & Unmack (2000)]
Plotosidae	Silver catfish	*Porochilus argenteus*	(Zietz, 1896)
Plotosidae	Obbes' catfish	*Porochilus obbesi*	Weber, 1913
Plotosidae	Rendahl's catfish	*Porochilus rendahli*	(Whitley, 1928)
Plotosidae	Unnamed catfish	*Porochilus* sp.[5]	[Hogan & Wallace (2005)]

Family	Common name	Species	Authority or [Reference]
Plotosidae	Freshwater cobbler	*Tandanus bostocki*[3]	Whitley, 1944
Plotosidae	Freshwater catfish	*Tandanus tandanus*[2]	Mitchell, 1838
Pristidae	Freshwater sawfish	*Pristis microdon*	Latham, 1794
Pseudaphritidae	Congolli	*Pseudaphritis urvillii*	(Valenciennes in Cuvier & Valenciennes, 1832)
Pseudomugilidae	Spotted blue-eye	*Pseudomugil gertrudae*	Weber, 1911
Pseudomugilidae	Honey blue-eye	*Pseudomugil mellis*	Allen & Ivantsoff, 1982
Pseudomugilidae	Pacific blue-eye	*Pseudomugil signifer*[1]	Kner, 1866
Pseudomugilidae	Delicate blue-eye	*Pseudomugil tenellus*[1]	Taylor, 1964
Pseudomugilidae	Redfin blue-eye	*Scaturiginichthys vermeilipinnis*	(Ivantsoff, Unmack, Saeed & Crowley, 1991)
Retropinnidae	Australian grayling	*Prototroctes maraena*	Günther, 1864
Retropinnidae	Australian smelt	*Retropinna semoni*[2]	(Weber, 1895)
Retropinnidae	Tasmanian smelt	*Retropinna tasmanica*	McCulloch, 1920
Soleidae	Saltpan sole	*Brachirus salinarum*	Ogilby, 1910
Soleidae	Freshwater sole	*Brachirus selheimi*	(Macleay, 1882)
Soleidae	Darwin sole	*Leptachirus darwinensis*	Randall, 2007
Soleidae	Many-scale sole	*Leptachirus polylepis*	Randall, 2007
Soleidae	Three-line sole	*Leptachirus triramus*	Randall, 2007
Soleidae	Unnamed sole	*Synclidopus hogani*	Johnson & Randall, 2008
Synbranchidae	Belut	*Monopterus albus*	(Zuiew, 1793)
Synbranchidae	One-gilled eel	*Ophisternon bengalense*	McClelland, 1844
Synbranchidae	Blind cave eel	*Ophisternon candidum*	(Mees, 1962)
Synbranchidae	Swamp eel	*Ophisternon gutturale*	(Richardson, 1845)
Syngnathidae	Unnamed pipefish	*Dorichthys* sp.[5]	[Ebner & Theusen (2010)]
Syngnathidae	Reticulated pipefish	*Hippichthys heptagonus*	Bleeker, 1849
Syngnathidae	Steep-nosed pipefish	*Hippichthys penicillus*	(Cantor, 1849)
Syngnathidae	Banded pipefish	*Hippichthys spicifer*	(Rüppell, 1838)
Syngnathidae	Short-tailed pipefish	*Microphis brachyurus*	(Bleeker, 1854)
Terapontidae	Barred grunter	*Amniataba percoides*[1]	(Günther, 1864)
Terapontidae	Silver perch	*Bidyanus bidyanus*	(Mitchell, 1838)
Terapontidae	Welch's grunter	*Bidyanus welchi*	(McCulloch & Waite, 1917)
Terapontidae	Greenway's grunter	*Hannia greenwayi*	Vari, 1978
Terapontidae	Coal grunter	*Hephaestus carbo*	(Ogilby & McCulloch, 1916)
Terapontidae	Longnose sooty grunter	*Hephaestus epirrhinos*	Vari & Hutchins, 1978
Terapontidae	Sooty grunter	*Hephaestus fuliginosus*	(Macleay, 1883)
Terapontidae	Western sooty grunter	*Hephaestus jenkinsi*	(Whitley, 1945)
Terapontidae	Khaki grunter	*Hephaestus tulliensis*	De Vis, 1884

Family	Common name	Species	Authority or [Reference]
Terapontidae	Fortescue grunter	*Leiopotherapon aheneus*	(Mees, 1963)
Terapontidae	Kimberley spangled perch	*Leiopotherapon macrolepis*	Vari, 1978
Terapontidae	Spangled perch	*Leiopotherapon unicolor*	(Günther, 1859)
Terapontidae	Gilbert's grunter	*Pingalla gilberti*	Whitley, 1955
Terapontidae	Lorentz grunter	*Pingalla lorentzi*	(Weber, 1910)
Terapontidae	Midgley's grunter	*Pingalla midgleyi*	Allen & Merrick, 1984
Terapontidae	Barcoo grunter	*Scortum barcoo*	(McCulloch & Waite, 1917)
Terapontidae	Leathery grunter	*Scortum hillii*	(Castelnau, 1878)
Terapontidae	Angalarri grunter	*Scortum neili*	Allen, Larson & Midgley, 1993
Terapontidae	Gulf grunter	*Scortum ogilbyi*	Whitley, 1951
Terapontidae	Smallhead grunter	*Scortum parviceps*	(Macleay, 1883)
Terapontidae	Sharpnose grunter	*Syncomistes butleri*[1]	Vari, 1978
Terapontidae	Kimberley grunter	*Syncomistes kimberleyensis*	Vari, 1978
Terapontidae	Drysdale grunter	*Syncomistes rastellus*	Vari & Hutchins, 1978
Terapontidae	Longnose grunter	*Syncomistes trigonicus*	Vari, 1978
Terapontidae	Lake grunter	*Variichthys lacustris*	(Mees & Kailola, 1977)
Tetrarogidae	Bullrout	*Notesthes robusta*	(Günther, 1860)
Toxotidae	Sevenspot archerfish	*Toxotes chatareus*	(Hamilton, 1822)
Toxotidae	Kimberley archerfish	*Toxotes kimberleyensis*	Allen, 2004
Toxotidae	Primitive archerfish	*Toxotes lorentzi*	Weber, 1910
Alien species			
Cichlidae	Convict cichlid	*Amatitlania nigrofasciata*	(Günther, 1867)
Cichlidae	Midas cichlid	*Amphilophus citrinellus*	(Günther, 1864)
Cichlidae	Red devil	*Amphilophus labiatus*	(Günther, 1864)
Cichlidae	Blue acara	*Andinoacara pulcher*	(Gill, 1858)
Cichlidae	Green terror	*Andinoacara rivulatus*	(Günther, 1860)
Cichlidae	Oscar	*Astronotus ocellatus*	(Agassiz, 1831)
Cichlidae	Three-spot cichlid	*Cichlasoma trimaculatum*	(Günther, 1867)
Cichlidae	Pearl cichlid	*Geophagus brasiliensis*	(Quoy & Gaimard, 1824)
Cichlidae	Victoria Burton's Haplochromis	*Haplochromis burtoni*	(Günther, 1894)
Cichlidae	Jewel cichlid	*Hemichromis bimaculatus*	Gill, 1862
Cichlidae	Green severum	*Heros severus*	Heckel, 1840
Cichlidae	Tilapia	*Oreochromis mossambicus*	(Peters, 1852)
Cichlidae	Redhead cichlid	*Paraneetroplus synspilus*	(Hubbs, 1935)
Cichlidae	Jack Dempsey cichlid	*Rocio octofasciata*	(Regan, 1903)
Cichlidae	Firemouth	*Thorichthys meeki*	Brind, 1918
Cichlidae	Black mangrove cichlid	*Tilapia mariae*	Boulenger, 1899

Family	Common name	Species	Authority or [Reference]
Cichlidae	Redbelly tilapia	*Tilapia zillii*	(Gervais, 1848)
Cobitidae	Oriental weatherloach	*Misgurnus anguillicaudatus*[1]	(Cantor, 1842)
Cyprinidae	Goldfish	*Carassius auratus*	(Linnaeus, 1758)
Cyprinidae	Crucian carp	*Carassius carassius*	Linnaeus, 1758
Cyprinidae	Common carp	*Cyprinus carpio*	Linnaeus, 1758
Cyprinidae	Rosy barb	*Puntius conchonius*	(Hamilton, 1822)
Cyprinidae	Roach	*Rutilus rutilus*	(Linnaeus, 1758)
Cyprinidae	Sumatra barb	*Systomus tetrazona*	(Bleeker, 1855)
Cyprinidae	Whitecloud mountain minnow	*Tanichthys albonubes*	Lin, 1932
Cyprinidae	Tench	*Tinca tinca*	(Linnaeus, 1758)
Cyprinodontidae	American flagfish	*Jordanella floridae*	Goode & Bean, 1879
Gobiidae	Yellowfin goby	*Acanthogobius flavimanus*	(Temminck & Schlegel, 1845)
Gobiidae	Streaked goby	*Acentrogobius pflaumii*	(Bleeker, 1853)
Percidae	Redfin perch	*Perca fluviatilis*	Linnaeus, 1758
Poeciliidae	Eastern gambusia	*Gambusia holbrooki*	Girard, 1859
Poeciliidae	One-spot livebearer	*Phalloceros caudimaculatus*	(Hensel, 1868)
Poeciliidae	Sailfin molly	*Poecilia latipinna*	(Lesueur, 1821)
Poeciliidae	Guppy	*Poecilia reticulata*	Peters, 1859
Poeciliidae	Green swordtail	*Xiphophorus hellerii*	Heckel, 1848
Poeciliidae	Platy	*Xiphophorus maculatus*	(Günther, 1866)
Salmonidae	Rainbow trout	*Oncorhynchus mykiss*	(Walbaum, 1792)
Salmonidae	Chinook salmon	*Oncorhynchus tshawytscha*	(Walbaum, 1792)
Salmonidae	Atlantic salmon	*Salmo salar*	Linnaeus, 1758
Salmonidae	Brown trout	*Salmo trutta*	Linnaeus, 1758
Salmonidae	Brook trout	*Salvelinus fontinalis*	(Mitchell, 1814)

[1] Cryptic species may be present, based on preliminary genetic data.
[2] Two or more cryptic species present, based on genetic data.
[3] Need for taxonomic review indicated by genetic data.
[4] Includes recognised subspecies.
[5] Not formally described.

References

Aarn and Ivantsoff W (2001) Predation on native fishes by *Gambusia holbrooki* in the Orara River, New South Wales. *Fishes of Sahul* **15**, 726–732.

Abbe TB and Montgomery DR (1996) Large woody debris jams, channel hydraulics and habitat formation in large rivers. *Regulated Rivers: Research and Management* **12**, 201–221.

Abell R, Allan JD and Lehner B (2007) Unlocking the potential of protected areas for freshwaters. *Biological Conservation* **134**, 48–63.

Abell R, Thieme M *et al.* (2008) Freshwater ecoregions of the world: a new map of biogeographic units for freshwater biodiversity conservation. *BioScience* **58**, 403–414.

Abellan P, Sanchez-Fernandez D, Velasco D and Millan A (2007) Effectiveness of protected area networks in representing freshwater biodiversity: the case of a Mediterranean river basin (south-eastern Spain). *Aquatic Conservation: Marine and Freshwater Ecosystems* **17**, 361–374.

Abery NW, Gunasekera RM and De Silva SS (2002) Growth and nutrient utilization of Murray cod *Maccullochella peelii peelii* (Mitchell) fingerlings fed diets with varying levels of soybean meal and blood meal. *Aquaculture Research* **33**, 279–289.

ACT Government (2007) 'Ribbons of life: ACT aquatic species and riparian zone conservation strategy. Action plan 29'. Department of Arts, Heritage and Environment, Canberra.

Adams M and Humphreys WF (1993) Patterns of genetic diversity within selected subterranean fauna of the North West Cape peninsula, Western Australia: systematic and biogeographic implications. *Records of the Western Australian Museum*, Supplement **45**, 145–164.

Adams SM and Breck JE (1990) Bioenergetics. In *Methods for Fish Biology*. (Eds CB Schreck and PB Moyle) pp. 389–416. American Fisheries Society, Bethesda.

Adams M, Wedderburn SD, Unmack PJ, Hammer MP *et al.* (2011) Congeneric assessment demonstrates the linked genetic histories of two threatened fishes (Atherinidae: *Craterocephalus*) in Australia's Murray-Darling Basin. *Conservation Biology* **25**, 767–776.

Akçakaya H (2005) *RAMAS Metapop: Viability Analysis for Stage-Structured Metapopulations (version 5.0)*. Applied Biomathematics, Setauket, New York.

Akçakaya H, Burgman M and Ginzburg L (1997) *Applied Population Ecology*, 2nd edn. Sinauer Associates, New York.

Alerstam T, Hedenström A and Åkesson S (2003) Long-distance migration: evolution and determinants. *Oikos* **103**, 247–260.

Ali MA, Nicieza A and Wootton RJ (2003) Compensatory growth in fishes: a response to growth depression. *Fish and Fisheries* **4**, 147–190.

Allan C and Jacobson C (2009) Passive/active adaptive management. In *Adaptive Environmental Management: A Practitioner's Guide*. (Eds C Allan and GH Stankey). Springer, Dordrecht/CSIRO Publishing, Melbourne.

Allan JD (2004) Landscapes and riverscapes: the influence of land use on stream ecosystems. *Annual Review of Ecology, Evolution and Systematics* **35**, 257–284.

Allee W (1978) *Animal Aggregations: A Study in General Sociology*. University of Chicago Press, Chicago.

Allen GR (1982) *A Field Guide to the Inland Fishes of Western Australia*. Western Australian Museum, Perth.

Allen GR (1989) *Freshwater Fishes of Australia*. TFH Publications, New Jersey.

Allen GR (1991) *Field Guide to the Freshwater Fishes of New Guinea*. Christensen Research Institute, Madang.

Allen GR (1995) *Rainbowfishes in Nature and in the Aquarium*. Tetra-Verlag, Osnabruck.

Allen GR and Jenkins AP (1999) A review of the Australian freshwater gudgeon genus *Mogurnda* (Eleotridae) with descriptions of three new species. *Aqua, Journal of Ichthyology and Aquatic Biology* **3**, 141–156.

Allen GR, Midgley SH and Allen M (2002) *Field Guide to the Freshwater Fish of Australia*. CSIRO Publishing, Melbourne.

Allen H (1972) Where the crow flies backwards: man and land in the Darling Basin. PhD thesis. Australian National University, Canberra.

Allen KR (1951) 'The Horokiwi Stream: a study of a trout population'. Marine Department Fisheries Bulletin No. 10. Wellington, New Zealand.

Allen MG, Morgan DL and Gill HS (2005) Distribution, zoogeography and biology of the Murchison River hardyhead (*Craterocephalus cuneiceps* Whitley, 1944), an atherinid endemic to the Indian Ocean (Pilbara) Drainage Division of Western Australia. *Ecology of Freshwater Fish* **14**, 209–224.

Allen MS, Brown P, Douglas J, Fulton W *et al.* (2009) An assessment of recreational fishery harvest policies for Murray cod in southeast Australia. *Fisheries Research* **95**, 260–267.

Allen-Ankins S, Stoffels RJ, Pridmore PA and Vogel MT (in press) The effects of turbidity, prey density and environmental complexity on the feeding of juvenile Murray cod *Maccullochella peelii*. *Journal of Fish Biology*.

Alley NF and Benbow MC (1995) Torrens Basin. In *The Geology of South Australia*. Vol. 2, *The Phanerozoic*. (Eds JF Drexel and WV Preiss) Department of Manufacturing, Innovation, Trade, Resources and Energy, Adelaide: *Bulletin* **54**, 187–188.

Allibone RM, Crowl TA, Holmes JW, King TM *et al.* (1996) Isozyme analysis of *Galaxias* species from the Taieri River, South Island, New Zealand: a species complex revealed. *Biological Journal of the Linnean Society* **57**, 107–127.

Allison GB, Cook PG, Barnett SR, Walker GR *et al.* (1990) Land clearance and river salination in the western Murray Basin. *Journal of Hydrology* **119**, 1–20.

Amoros C, Roux AL, Reygrobellet JL, Bravard JP *et al.* (1987) A method for applied ecological studies of fluvial hydrosystems. *Regulated Rivers: Research and Management* **1**, 17–36.

Amundsen P-A, Knudsen R and Klemetsen A (2007) Intraspecific competition and density dependence of food consumption and growth in Arctic charr. *Journal of Animal Ecology* **76**, 149–158.

Anderson J, Lake J and Mackay N (1971) Notes on reproductive behaviour and ontogeny in two species of *Hypseleotris* (=*Carassiops*) (Gobiidae: Teleostei). *Marine and Freshwater Research* **22**, 139–146.

Anderson JR, Morison AK and Ray DJ (1992a) Age and growth of Murray cod, *Maccullochella peelii* (Perciformes: Percichthyidae), in the Lower Murray-Darling Basin, Australia, from thin-sectioned otoliths. *Marine and Freshwater Research* **43**, 983–1013.

Anderson JR, Morison AK and Ray DJ (1992b) Validation of the use of thin-sectioned otoliths for determining the age and growth of golden perch *Macquaria ambigua* (Perciformes: Percichthyidae), in the lower Murray-Darling Basin, Australia. *Australian Journal of Marine and Freshwater Research* **43**, 1103–1128.

Anderson TR, Boersma M and Raubenheimer D (2004) Stoichiometry: linking elements to biochemicals. *Ecology* **85**, 1193–1202.

Andrews AEJ (1981) *Hume and Hovell 1824*. Blubber Head Press, Hobart.

Andrews AP (1976) A revision of the family Galaxiidae (Pisces) in Tasmania. *Australian Journal of Marine and Freshwater Research* **27**, 297–349.

Angermeier PL (1992) Predation by rock bass on other stream fishes: experimental effects of depth and cover. *Environmental Biology of Fishes* **34**, 171–180.

Anonymous (1971) Murray cod takes the long distance title. *The Fisherman*, spring 1971.

Ansell D and Jackson P (2007) *Emerging Issues in Alien Fish Management in the Murray-Darling Basin: Statement, Recommendations and Supporting Papers.* Proceedings of a workshop held in Brisbane, May 2006. (Eds D Ansell and P Jackson). Murray-Darling Basin Commission, Canberra.

Appel TA (1987) *The Cuvier-Geoffroy Debate: French Biology in the Decades before Darwin.* Oxford University Press, New York.

Appleford P, Anderson TA and Gooley GJ (1998) Reproductive cycle and gonadal development of Macquarie perch, *Macquaria australasica* Cuvier (Percichthyidae), in Lake Dartmouth and tributaries of the Murray-Darling Basin, Victoria, Australia. *Marine and Freshwater Research* **49**, 163–169.

Arai T, Kotake A, Lokman PM, Miller MJ *et al.* (2004) Evidence of different habitat use by New Zealand freshwater eels *Anguilla australis* and *A. dieffenbachii*, as revealed by otolith microchemistry. *Marine Ecology Progress Series* **266**, 213–225.

Arai T, Otake T, Jellyman DJ and Tsukamoto K (1999) Differences in the early life history of the Australasian shortfinned eel *Anguilla australis* from Australia and New Zealand, as revealed by otolith microstructure and microchemistry. *Marine Biology* **135**, 381–389.

Argent RM (2009) Components of adaptive management. In *Adaptive Environmental Management: A Practitioner's Guide.* (Eds C Allan and GH Starkey) pp. 11–32. Springer, Dordrecht/CSIRO Publishing, Melbourne.

Arlinghaus R, Cooke SJ, Lyman J, Policansky D *et al.* (2007) Understanding the complexity of catch-and-release in recreational fishing: an integrative synthesis of global knowledge from historical, ethical, social, and biological perspectives. *Reviews in Fisheries Science* **15**, 75–167.

ARMCANZ and ANZECC (1996) 'National principles for the provision of water for ecosystems'. Occasional Paper SWR 3. Agricultural Resource Management Council of Australia and New Zealand/Australian and New Zealand Environment and Conservation Council, Canberra.

Armstrong JD and Herbert NA (1997) Homing movements of displaced stream-dwelling brown trout. *Journal of Fish Biology* **50**, 445–449.

Armstrong JD, Braithwaite VA and Huntingford FA (1997) Spatial strategies of wild Atlantic salmon parr: exploration and settlement in unfamiliar areas. *Journal of Animal Ecology* **66**, 203–211.

Arnott SA, Barber I and Huntingford FA (2000) Parasite-associated growth enhancement in a fish-cestode system. *Proceedings of the Royal Society of London, Series B: Biological Sciences* **267**, 657–663.

Arthington A (1991) Ecological and genetic impacts of introduced and translocated freshwater fishes in Australia. *Canadian Journal of Fisheries and Aquatic Sciences* **48** (Supplement 1), 33–43.

Arthington AH (2009) Australian lungfish, *Neoceratodus forsteri*, threatened by a new dam. *Environmental Biology of Fishes* **84**, 211–221.

Arthington AH and Balcombe SR (2011) Extreme flow variability and the 'boom and bust' ecology of fish in arid-zone floodplain rivers: a case history with implications for environmental flows, conservation and management. *Ecohydrology* **4**, 708–720.

Arthington AH and Marshall CJ (1999) Diet of the exotic mosquitofish, *Gambusia holbrooki*, in an Australian lake and potential for competition with indigenous fish species. *Asian Fisheries Science* **12**, 1–16.

Arthington AH and McKenzie F (1997) 'Review of impacts of displaced/introduced fauna associated with inland waters, Australia. State of the Environment'. Department of the Environment, Canberra.

Arthington AH and Pusey BJ (2003) Flow restoration and protection in Australian rivers. *River Research and Applications* **19**, 377–395.

Arthington AH, Balcombe SR, Wilson GW, Thoms MC *et al.* (2005) Spatial and temporal variation in fish-assemblage structure in isolated waterholes during the 2001 dry season of an arid-zone floodplain river, Cooper Creek, Australia. *Marine and Freshwater Research* **56**, 25–35.

Arthington AH, Bunn SE, Poff NL and Naiman RJ (2006) The challenge of providing environmental flow rules to sustain river ecosystems. *Ecological Applications* **16**, 1311–1318.

Arthington AH, Burton HB, Williams RW and Outridge PM (1986) Ecology of humic and non-humic dune lakes, Fraser Island, with emphasis on the effects of sandy infilling in Lake Wabby. *Australian Journal of Marine and Freshwater Research* **37**, 743–764.

Arthington AH, Kailola PJ, Woodland DJ and Zaluki JM (1999) 'Baseline environmental data relevant to an evaluation of quarantine risk potentially associated with the importation to Australia of ornamental finfish'. Report to the Australian Quarantine and Inspection Service. Department of Agriculture, Fisheries and Forestry, Canberra.

Arthington AH, Naiman RJ, McClain ME and Nilsson C (2010a) Preserving the biodiversity and ecological services of rivers: new challenges and research opportunities. *Freshwater Biology* **55**, 1–16.

Arthington AH, Olden JD, Balcombe SR and Thoms MC (2010b) Multi-scale environmental factors explain fish losses and refuge quality in drying waterholes of Cooper Creek, an Australian arid-zone river. *Marine and Freshwater Research* **61**, 842–856.

Astles KL, Winstanley RK, Harris JH and Gehrke PC (2003) 'Experimental study of the effects of cold water pollution on native fish'. A final report for the Regulated Rivers and Fisheries Restoration Project. NSW Fisheries Final Report Series 44, Sydney.

Auer NA (1996) Importance of habitat and migration to sturgeons with emphasis on lake sturgeon. *Canadian Journal of Fisheries and Aquatic Sciences* **53**, 152–160.

Ault TR and White RWG (1994) Effects of habitat structure and the presence of brown trout on the population density of *Galaxias truttaceus* in Tasmania, Australia. *Transactions of the American Fisheries Society* **123**, 939–949.

Australian and New Zealand Environment and Conservation Council (2000) 'Australian and New Zealand guidelines for fresh and marine water quality'. Australian and New Zealand Environment and Conservation Council/Agricultural Resource Management Council of Australia and New Zealand, Canberra.

Australian Bureau of Agricultural and Resource Economics/Bureau of Rural Sciences (2010) 'Australian Fisheries Statistics 2009'. Australian Bureau of Agricultural and Resource Economics/Bureau of Rural Sciences, Canberra.

Australian Quarantine and Inspection Service (1999) 'Exotic fish trade'. Report to the National Fisheries Compliance Committee. Australian Quarantine and Inspection Service, Canberra.

Australian Society for Fish Biology (2009) Report of the Threatened Fishes Committee. *Australian Society for Fish Biology Newsletter* **39**, 103–125.

Auty EH (1978) Reproductive behaviour and early development of the empire fish *Hypseleotris compressus* (Eleotridae). *Marine and Freshwater Research* **29**, 585–597.

Avise JC (2000) *Phylogeography: The History and Formation of Species*. Harvard University Press, Cambridge.

Avise JC (2002) *Genetics in the Wild*. Smithsonian Institution Press, Washington.

Avise JC (2004) *Molecular Markers, Natural History and Evolution*, 2nd edn. Sinauer Associates, New York.

Avise JC (2010) Perspective: conservation genetics enters the genomics era. *Conservation Genetics* **11**, 665–669.

Ayliffe LK, Marianelli PC, Moriarty KC, Wells RT *et al.* (1998) 500 ka precipitation record from southeastern Australia: evidence for interglacial relative aridity. *Geology* **26**, 147–150.

Ayres R and Clunie P (2010a) 'Management of freshwater fish incursions: a review'. Arthur Rylah Institute for Environmental Research, Department of Sustainability and Environment, Melbourne.

Ayres R and Clunie P (2010b) 'Towards a national emergency response system for freshwater fish incursions'. PestSmart Toolkit publication. Invasive Animals Cooperative Research Centre, Canberra.

Bacher GL and O'Brien TA (1989) Salinity tolerance of the eggs and larvae of the Australian grayling, *Prototroctes maraena* Günther (Salmoniformes: Prototroctidae). *Marine and Freshwater Research* **40**, 227–230.

Bagenal TB (1969) The relationship between food supply and fecundity in brown trout *Salmo trutta* L. *Journal of Fish Biology* **1**, 167–182.

Bagenal TB (1978) *Methods for Assessment of Fish Production in Fresh Waters*, 3rd edn. Blackwell Scientific, Oxford.

Bailey KM (1994) Predation on juvenile flatfish and recruitment variability. *Netherlands Journal of Sea Research* **32**, 175–189.

Bailey KM and Houde ED (1989) Predation on eggs and larvae of marine fishes and the recruitment problem. *Advances in Marine Biology* **25**, 251–281.

Bain MB and Stevenson NJ (Eds) (1999) *Aquatic Habitat Assessment: Common Methods.* American Fisheries Society, Bethesda.

Baker JA and Ross ST (1981) Spatial and temporal resource utilization by southeastern cyprinids. *Copeia* **1981**, 178–189.

Baker N, De Bruyn M and Mather PB (2008) Patterns of molecular diversity in wild stocks of the redclaw crayfish (*Cherax quadricarinatus*) from northern Australia and Papua New Guinea: impacts of Plio-Pleistocene landscape evolution. *Freshwater Biology* **53**, 1592–1605.

Balcombe SR and Arthington AH (2009) Temporal changes in fish abundance in response to hydrological variability in a dryland floodplain river. *Marine and Freshwater Research* **60**, 146–159.

Balcombe SR and Closs GP (2004) Spatial and temporal variability in a littoral macrophyte fish assemblage. *Marine and Freshwater Research* **55**, 603–617.

Balcombe SR and Humphries P (2006) Diet of the western carp gudgeon (*Hypseleotris klunzingeri* Ogilby) in an Australian floodplain lake: the role of water level stability. *Journal of Fish Biology* **68**, 1484–1493.

Balcombe SR, Arthington AH, Foster ND, Thoms MC *et al.* (2006) Fish assemblages of an Australian dryland river: abundance, assemblage structure and recruitment patterns in the Warrego River, Murray-Darling Basin. *Marine and Freshwater Research* **57**, 619–633.

Balcombe SR, Arthington AH, Thoms MC and Wilson G (2011a) Fish assemblage patterns across a gradient of flow regulation in an Australian dryland river system. *River Research and Applications* **27**, 168–183.

Balcombe SR, Bunn SE, Arthington AH, Fawcett JH *et al.* (2007) Fish larvae, growth and biomass relationships in an Australian arid zone river: links between floodplains and waterholes. *Freshwater Biology* **52**, 2385–2398.

Balcombe SR, Bunn SE, McKenzie-Smith FJ and Davies PM (2005) Variability of fish diets between dry and flood periods in an arid zone floodplain river. *Journal of Fish Biology* **67**, 1552–1567.

Balcombe SR, Sheldon F, Capon SJ, Bond NR *et al.* (2011b) Climate-change threats to native fish in degraded rivers and floodplains of the Murray-Darling Basin, Australia. *Marine and Freshwater Research* **62**, 1099–1114.

Baldwin D and Whitworth K (2009) 'Current conditions in the Wakool River and the potential for blackwater events resulting in fish deaths'. Murray-Darling Freshwater Research Centre, Albury.

Ballard JWO and Whitlock MC (2004) The incomplete natural history of mitochondria. *Molecular Ecology* **13**, 729–744.

Balme J (1995) 30,000 years of fishery in western New South Wales. *Archaeology in Oceania* **30**, 1–21.

Balon EK (1975) Reproductive guilds of fishes: a proposal and definition. *Journal of Fisheries Research Board of Canada* **32**, 821–64.

Balon EK (1981) Additions and amendments to the classification of reproductive styles in fishes. *Environmental Biology of Fishes* **6**, 377–389.

Balon EK (1986) Types of feeding in the ontogeny of fishes and the life-history model. *Environmental Biology of Fishes* **16**, 11–24.

Balon EK (1990) Epigenesis of an epigeneticist: the devlopment of some alternative concepts on the early ontogeny and evolution of fishes. *Guelph Ichthyology Reviews* **1**, 1–48.

Balon EK (1999) Alternative ways to become a juvenile or a definitive phenotype (and on some persisting linguistic offenses). *Environmental Biology of Fishes* **56**, 17–38.

Baltz DM (1990) Autecology. In *Methods for Fish Biology.* (Eds CB Schreck and PB Moyle) pp. 585–607. American Fisheries Society, Bethesda.

Banarescu P (1990) *Zoogeography of Fresh Waters*, AULA-Verlag, Wiesbaden.

Banavar JR, Maritan A and Rinaldo A (1999) Size and form in efficient transportation networks. *Nature* **399**, 130–132.

Bansemer CS and Bennett MB (2008) Multi-year validation of photographic identification of grey nurse sharks, *Carcharias taurus*, and applications for non-invasive conservation research. *Marine and Freshwater Research* **59**, 322–331.

Barbee NC and Swearer SE (2007) Characterizing natal source population signatures in the diadromous fish *Galaxias maculatus*, using embryonic otolith chemistry. *Marine Ecology Progress Series* **343**, 273–282.

Bari MA and Schofield NJ (1992) Lowering of a shallow, saline water table by extensive eucalypt reforestation. *Journal of Hydrology* **133**, 273–291.

Barlow CG and Lisle A (1987) Biology of the Nile perch *Lates niloticus* (Pisces, Centropomidae) with reference to its proposed role as a sport fish in Australia. *Biological Conservation* **39**, 269–289.

Barlow CG, Hogan AE and Rodgers LJ (1987) Implication of translocated fishes in the apparent extinction in the wild of the Lake Eacham rainbowfish, *Melanotaenia eachamensis. Australian Journal of Marine and Freshwater Research* **38**, 897–902.

Barluenga M, Stoelting KN, Salzburger W, Muschick M *et al.* (2006) Sympatric speciation in Nicaraguan crater lake cichlid fish. *Nature* **439**, 719–723.

Barmuta LA, Linke S and Turak E (2011) Bridging the gap between 'planning' and 'doing' for biodiversity conservation in freshwaters. *Freshwater Biology* **56**, 180–195.

Barrett J (Ed.) (2008) 'The Sea to Hume Dam: restoring fish passage in the Murray River'. Murray-Darling Basin Commission, Canberra.

Barrett J and Ansell D (2005) Demonstration reaches for native fish: moving from theory to practice. In *Proceed-*

ings of the 4th Australian Stream Management Conference: Linking Rivers to Landscapes.* (Eds ID Rutherfurd, I Wiszniewski, M Askey-Doran and R Glazik). Department of Primary Industries, Water and Environment, Hobart.

Barrett J and Mallen-Cooper M (2006) The Murray River's 'Sea to Hume Dam' fish passage program: progress to date and lessons learned. *Ecological Management and Restoration* **7**, 173–183.

Bartholomew A and Bohnsack JA (2005) A review of catch-and-release angling mortality with implications for no-take reserves. *Reviews in Fish Biology and Fisheries* **15**, 129–154.

Bartley R and Rutherfurd ID (2005) Measuring the reach-scale geomorphic diversity of streams: application to a stream disturbed by a sediment slug. *River Research and Applications* **21**, 39–59.

Bartolino V, Colloca F, Sartor P and Ardizzone G (2008) Modelling recruitment dynamics of hake, *Merluccius merluccius*, in the central Mediterranean in relation to key environmental variables. *Fisheries Research* **92**, 277–288.

Barton M (2007) *Bond's Biology of Fishes.* Thomson Brooks/Cole, Belmont.

Battaglene S (1985) A creel survey of the Lake Keepit recreational fishery. *Australian Zoologist* **21**, 565–578.

Battaglene SC (1991) The golden perch, *Macquaria ambigua* (Pisces: Percichthyidae) of Lake Keepit, NSW. Masters thesis. University of New South Wales, Sydney.

Batty J and Lim R (1999) Morphological and reproductive characteristics of male mosquitofish (*Gambusia affinis holbrooki*) inhabiting sewage-contaminated waters in New South Wales, Australia. *Archives of Environmental Contamination and Toxicology* **36**, 301–307.

Baumgartner L, McPherson B, Doyle J, Cory F *et al.* (2010a) 'Quantifying and mitigating the impacts of weirs on downstream passage of native fish in the Murray-Darling Basin'. Report to Murray-Darling Basin Authority. NSW Department of Industry and Investment, Sydney.

Baumgartner LJ (2006) Population estimation methods to quantify temporal variation in fish accumulations downstream of a weir. *Fisheries Management and Ecology* **13**, 355–364.

Baumgartner LJ (2007) Diet and feeding habits of predatory fishes upstream and downstream of a low-level weir. *Journal of Fish Biology* **70**, 879–894.

Baumgartner LJ and Harris JH (2007) Passage of non-salmonid fish through a Deelder lock on a lowland river. *River Research and Applications* **23**, 1058–1069.

Baumgartner LJ, Boys CA, Stuart IG and Zampatti BP (2010b) Evaluating migratory behaviour and fishway performance: testing a combined assessment methodology. *Australian Journal of Zoology* **58**, 154–164.

Baumgartner LJ, Reynoldson N and Gilligan DM (2006b) Mortality of larval Murray cod (*Maccullochella peelii peelii*) and golden perch (*Macquaria ambigua*) associated with passage through two types of low-head weirs. *Marine and Freshwater Research* **57**, 187–191.

Baumgartner LJ, Reynoldson N, Cameron L and Stanger J (2006a) 'Assessment of a Dual-frequency Identification SONar (DIDSON) for application in fish migration studies'. Department of Primary Industries, Sydney.

Baumgartner LJ, Reynoldson NK, Cameron L and Stanger J (2009) Effects of irrigation pumps on riverine fish. *Fisheries Management and Ecology* **16**, 429–437.

Baumgartner LJ, Stuart IG and Zampatti BP (2008) Determining diel variation in fish assemblages downstream of three weirs in a regulated lowland river. *Journal of Fish Biology* **72**, 218–232.

Baverstock PR and Adams M (1984) Comparative rates of molecular, chromosomal and morphological evolution in some Australian vertebrates. In *Rates of Evolution.* (Eds KSW Campbell and MF Day) pp. 175–188. Allen and Unwin, London.

Bax NJ and Thresher RE (2009) Ecological, behavioral, and genetic factors influencing the recombinant control of invasive pests. *Ecological Applications* **19**, 873–888.

Baxter AF, Vallis SL and Hume DJ (1985) 'Predation of recently released rainbow trout fingerlings, *Salmo gairdneri*, by redfin, *Perca fluviatilis*, in Lake Burrumbeet, October–December 1983'. Arthur Rylah Institute for Environmental Research, Department of Sustainability and Environment, Melbourne.

Bayley PB (1991) The flood pulse advantage and the restoration of river-floodplain systems. *Regulated Rivers: Research and Management* **6**, 75–86.

Bayley PB (1995) Understanding large river-floodplain ecosystems. *BioScience* **45**, 153–158.

Beamish FWH (1978) Swimming capacity. In *Fish Physiology*, Vol. 7. (Eds WS Hoar and DJ Randall) pp. 101–189. Academic Press, New York.

Beamish RJ and McFarlane GA (1983) The forgotten requirement for age validation in fisheries biology. *Transactions of the American Fisheries Society* **112**, 735–743.

Bearlin AR and Tikel D (2003) Conservation genetics of Murray-Darling Basin fish; silver perch (*Bidyanus bidyanus*), Murray cod (*Maccullochella peelii*), and trout cod (*M. macquariensis*). In *Managing Fish Translocation and Stocking in the Murray-Darling Basin: Statement, Recommendations and Supporting Papers.* Canberra, 25–26 September 2002. (Ed. B Phillips). WWF Australia, Sydney.

Bearlin AR, Schreiber ESG, Nicol SJ, Starfield AM *et al.* (2002) Identifying the weakest link: simulating adaptive management of the reintroduction of a threatened fish. *Canadian Journal of Fisheries and Aquatic Sciences* **59**, 1709–1716.

Beatty SJ, Morgan DL and Torre A (2007) Restoring ecological connectivity in the Margaret River: Western Australia's first rock-ramp fishways. *Ecological Management and Restoration* **8**, 224–228.

Beatty SJ, Morgan DL, McAleer FJ and Ramsay AR (2010) Groundwater contribution to baseflow maintains habitat connectivity for *Tandanus bostocki* (Teleostei: Plotosidae) in a south-western Australian river. *Ecology of Freshwater Fish* **19**, 595–608.

Beckman BR (2010) Perspectives on concordant and discordant relations between insulin-like growth factor 1 (IGF1) and growth in fishes. *General and Comparative Endocrinology* **170**, 233–252.

Beesley LS and Prince J (2010) Fish community structure in an intermittent river: the importance of environmental stability, landscape factors and within-pool habitat descriptors. *Marine and Freshwater Research* **61**, 605–614.

Begon M, Harper JL and Townsend C, R. (1996) *Ecology: Individuals, Populations and Communities*, 3rd edn. Blackwell Science, Oxford.

Beheregaray LB (2008) Twenty years of phylogeography: the state of the field and the challenges for the Southern Hemisphere. *Molecular Ecology* **17**, 3754–3774.

Beheregaray LB and Caccone A (2007) Cryptic biodiversity in a changing world. *Journal of Biology* **6**, 9.1–9.5.

Beissinger SR (1995) Modeling extinction in periodic environments: Everglades water levels and snail kite population viability. *Ecological Applications* **5**, 618–631.

Beitinger TL and Fitzpatrick LC (1979) Physiological and ecological correlates of preferred temperature in fish. *American Zoologist* **19**, 319–329.

Beitzel M (2002) Genetic population structure of the two-spined blackfish *Gadopsis bispinosus*. Hons thesis. University of Canberra, Canberra.

Bell JD, Berra TM, Jackson PD, Last PR *et al.* (1980) Recent records of the Australian grayling *Prototroctes maraena* Günther (Pisces: Prototroctidae) with notes on its distribution. *Australian Zoologist* **20**, 419–431.

Bellwood DR, Wainwright PC, Fulton CJ and Hoey AS (2006) Functional versatility supports coral reef biodiversity. *Proceedings of the Royal Society Series B – Biological Sciences* **273**, 101–107.

Benbow MC (1990) Tertiary coastal dunes of the Eucla Basin, Australia. *Geomorphology* **3**, 9–29.

Benke AC, Van Arsdall TC and Gillespie DM (1984) Invertebrate productivity in a subtropical blackwater river: the importance of habitat and life history. *Ecological Monographs* **54**, 25–63.

Bennett G (1834) *Wanderings in New South Wales, Batavia, Pedir Coast, Singapore, and China; being the Journal of a Naturalist in those Countries during 1832, 1833, and 1834*, Vols 1–2. Richard Bentley, London.

Bennett G (1982) *Gatherings of a Naturalist in Australasia, 1860*. Currawong Press, Sydney.

Bennett KD (1990) Milankovitch cycles and their effects on species in ecological and evolutionary time. *Paleobiology* **16**, 11–21.

Bensink AHA and Burton H (1975) North Stradbroke Island: a place for freshwater invertebrates. *Proceedings of the Royal Society of Queensland* **86**, 29–45.

Benson KR (2000) The emergence of ecology from natural history. *Endeavour* **24**, 59–62.

Benzie V (1968) The life history of *Galaxias vulgaris* Stokell, with a comparison with *G. maculatus attenuatus*. *New Zealand Journal of Marine and Freshwater Research* **2**, 628–653.

Berendzen PB, Olson WM and Barron SM (2009) The utility of molecular hypotheses for uncovering morphological diversity in the *Notropis rubellus* species complex (Cypriniformes: Cyprinidae). *Copeia* **2009**, 661–673.

Berg OK and Finstad AG (2008) Energentic trade-offs in reproduction: cost–benefit considerations and plasticity. In *Fish Reproduction*. (Eds MJ Rocha, A Arukwe and BG Kapoor) pp. 351–396. Science Publishers, New Hampshire.

Bergeron JP (1997) Nucleic acids in ichthyoplankton ecology: a review, with emphasis on recent advances for new perspectives. *Journal of Fish Biology* **51**, 284–302.

Berghuis A (2008) 'Performance of a single frequency split beam hydroacoustic system: an innovative fish counting technology'. Technical Report Series 177. Arthur Rylah Institute for Environmental Research, Department of Sustainability and Environment, Melbourne.

Berglund A (1997) Mating systems and sex allocation. In *Behavioural Ecology of Teleost Fishes*. (Ed. JGJ Godin) pp. 237–265. Oxford University Press, Oxford.

Bernhardt ES, Palmer MA *et al.* (2005) Synthesizing US river restoration efforts. *Science* **308**, 636–637.

Bernhardt ES, Sudduth EB *et al.* (2007) Restoring rivers one reach at a time: results from a survey of US river restoration practitioners. *Restoration Ecology* **15**, 482–493.

Berra TM (1973) A home range study of *Galaxias bongbong* in Australia. *Copeia* **1973**, 363–366.

Berra TM (1974) The trout cod, *Maccullochella macquariensis*, a rare freshwater fish of eastern Australia. *Biological Conservation* **6**, 53–56.

Berra TM (1982) Life history of the Australian grayling, *Prototroctes maraena* (Salmoniformes: Prototroctidae) in the Tambo River, Victoria. *Copeia* **1982**, 795–805.

Berra TM (1997) Some 20th century fish discoveries. *Environmental Biology of Fishes* **50**, 1–12.

Berra TM (2007) *Freshwater Fish Distribution*. University of Chicago Press, Chicago.

Berra TM and Aday DD (2004) Otolith description and age and growth of *Kurtus gulliveri* from northern Australia. *Journal of Fish Biology* **65**, 354–362.

Berra TM and Allen GR (1989) Burrowing, emergence, behavior, and functional morphology of the Australian salamanderfish, *Lepidogalaxias salamandroides*. *Fisheries* **14**, 2–10.

Berra TM and Cadwallader PL (1983) Age and growth of Australian grayling, *Prototroctes maraena* Günther (Salmoniformes: Prototroctidae), in the Tambo River, Victoria. *Australian Journal of Marine and Freshwater Research* **34**, 451–460.

Berra TM and Humphrey JD (2002) Gross anatomy and histology of the hook and skin of forehead brooding male nurseryfish, *Kurtus gulliveri*, from northern Australia. *Environmental Biology of Fishes* **65**, 263–270.

Berra TM and Weatherley AH (1972) A systematic study of the Australian freshwater serranid fish genus *Maccullochella*. *Copeia* **1**, 53–64.

Bertozzi T, Adams M and Walker KF (2000) Species boundaries in carp gudgeons (Eleotridae: *Hypseleotris*) from the River Murray, South Australia: evidence for multiple species and extensive hybridization. *Marine and Freshwater Research* **51**, 805–815.

Beumer JP (1979a) Feeding and movement of *Anguilla australis* and *A. reinhardtii* in Macleods Morass, Victoria, Australia. *Journal of Fish Biology* **14**, 573–592.

Beumer JP (1979b) Temperature and salinity tolerance of the spangled perch *Therapon unicolor* Günther, 1859 and the east Queensland rainbowfish *Nematocentris splendida* Peters, 1866. *Australian Journal of Marine and Freshwater Research* **90**, 85–91.

Beumer JP (1996) Family Anguillidae – freshwater eels. In *Freshwater Fishes of Southeastern Australia*. (Ed. R McDowall) pp. 39–43. Reed, Sydney.

Beumer JP and Harrington DJ (1980a) Pre-impoundment distribution of fishes in the Lerderderg River, Victoria. *Victorian Naturalist* **97**, 68–72.

Beumer JP and Harrington DJ (1980b) Techniques for collecting glass eels and brown elvers. *Australian Fisheries* **39**, 16–22.

Beverton RJH (1954) 'Notes on the use of theoretical models in the study of the dynamics of exploited fish populations'. US Fishery Laboratory, Beaufort.

Beverton RJH and Holt SJ (1957) *On the Dynamics of Exploited Fish Populations*. Chapman and Hall, London.

Bice CM and Zampatti BP (2011) Engineered water level management facilitates recruitment of non-native common carp, *Cyprinus carpio*, in a regulated lowland river. *Ecological Engineering* **37**, 1901–1904.

Bickford D, Lohman DJ, Sodhi NS, Ng PKL *et al.* (2007) Cryptic species as a window on diversity and conservation. *Trends in Ecology and Evolution* **22**, 148–155.

Bigarre L, Cabon J, Baud M, Heimann M *et al.* (2009) Outbreak of betanodavirus infection in tilapia, *Oreochromis niloticus* (L.), in fresh water. *Journal of Fish Diseases* **32**, 667–673.

Bilby RE and Likens GE (1980) Importance of organic debris dams in the structure and function of stream ecosystems. *Ecology* **61**, 1107–1113.

Bilby RE and Ward JW (1991) Characteristics and function of large woody debris in streams draining old-growth, clear-cut, and second-growth forests in southwestern Washington. *Canadian Journal of Fisheries and Aquatic Sciences* **48**, 2499–2507.

Birch WD (Ed.) (2003) *Geology of Victoria*. Geological Society of Australia (Victoria Division), Melbourne.

Bird DJ and Potter IC (1981) Proximate body composition of the larval, metamorphosing and downstream migrant stages in the life cycle of the Southern Hemisphere lamprey, *Geotria australis*. *Environmental Biology of Fishes* **6**, 285–297.

Birindelli JLO, Zanata AM, Sousa LM and Netto-Ferreira AL (2009) New species of *Jupiaba zanata* (Characiformes: Characidae) from Serra do Cachimbo, with comments on the endemism of upper Rio Curuá, Rio Xingu Basin, Brazil. *Neotropical Ichthyology* **7**, 11–18.

Birkeland C and Dayton PK (2005) The importance in fishery management of leaving the big ones. *Trends in Ecology and Evolution* **20**, 356–358.

Bishop KA (1980) Fish kills in relation to physical and chemical changes in Magela Creek (East Alligator River system, Northern Territory) at the beginning of the tropical wet season. *Australian Zoologist* **20**, 485–500.

Bishop KA and Bell JD (1978) Observations of the fish fauna below Tallowa Dam (Shoalhaven River, New South Wales) during flow stoppages. *Australian Journal of Marine and Freshwater Research* **29**, 543–549.

Bishop KA and Forbes MA (1991) The freshwater fishes of northern Australia. In *Monsoonal Australia. Landscape, Ecology and Man in the Northern Lowlands*. (Eds CD Haynes, MG Ridpath and MAJ Williams) pp. 79–107. AA Balkema, Rotterdam.

Bishop KA, Allen SA, Pollard DA and Cook MG (2001) 'Ecological studies on the freshwater fishes of the Alligator Rivers Region, Northern Territory: Autoecology'. Supervising Scientist Report 145. Supervising Scientist, Darwin.

Bishop KA, Pidgeon RW and Walden DJ (1995) Studies on fish movement dynamics in a tropical floodplain river: prerequisites for a procedure to monitor the impacts of mining. *Australian Journal of Ecology* **20**, 81–107.

Bishop P (1995) Drainage rearrangement by river capture, beheading and diversion. *Progress in Physical Geography* **19**, 449–473.

Black BA (2009) Climate-driven synchrony across tree, bivalve, and rockfish growth-increment chronologies of the northeast Pacific. *Marine Ecology Progress Series* **378**, 37–46.

Blackburn M (1950) The Tasmanian whitebait, *Lovettia seali* (Johnston) and the whitebait fishery. *Australian Journal of Marine and Freshwater Research* **1**, 155–198.

Blackley T (2003) Screening irrigation offtakes in the Murray-Darling Basin to reduce the loss of native fish. In

Downstream Movement of Fish in the Murray-Darling Basin. (Eds M Lintermans and B Phillips). Murray-Darling Basin Commission, Canberra.

Blake RW (1983) *Fish Locomotion.* Cambridge University Press, Cambridge.

Blandowski W (1857) Recent discoveries in natural history on the lower Murray. *Transactions of the Philosophical Institute of Victoria* **2**, 124–137.

Blinn DW, Runck C, Clark DA and Rinne JN (1993) Effects of rainbow trout predation on little Colorado spinedace. *Transactions of the American Fisheries Society* **122**, 139–143.

Blühdorn DR and Arthington AH (1990) Somatic characteristics of an Australian population of *Oreochromis mossambicus* (Pisces: Cichlidae). *Environmental Biology of Fishes* **29**, 277–291.

Boersma M and Elser JJ (2006) Too much of a good thing: on stoichiometrically balanced diets and maximal growth. *Ecology* **87**, 1325–1330.

Boeuf G and Le Bail P-Y (1999) Does light have an influence on fish growth? *Aquaculture* **177**, 129–152.

Bolding B, Bonar S and Divens M (2004) Use of artificial structure to enhance angler benefits in lakes, ponds, and reservoirs: a literature review. *Reviews in Fisheries Science* **12**, 75–96.

Bolger T and Connolly PL (1989) The selection of suitable indexes for the measurement and analysis of fish condition. *Journal of Fish Biology* **34**, 171–182.

Bomford M and Glover J (2004) 'Risk assessment model for the import and keeping of exotic freshwater and estuarine finfish'. Bureau of Rural Sciences, Canberra.

Bomford M and O'Brien P (1995) Eradication or control for vertebrate pests. *Wildlife Society Bulletin* **23**, 249–255.

Bonar SA, Hubert WA and Willis DW (2009) *Standard Methods for Sampling North American Freshwater Fishes.* American Fisheries Society, Bethesda.

Bond NR and Lake PS (2003a) Local habitat restoration in streams: constraints on the effectiveness of restoration for stream biota. *Ecological Management and Restoration* **4**, 193–198.

Bond NR and Lake PS (2003b) Characterizing fish–habitat associations in streams as the first step in ecological restoration. *Austral Ecology* **28**, 611–621.

Bond NR and Lake PS (2005) Ecological restoration and large-scale ecological disturbance: the effects of drought on the response by fish to a habitat restoration experiment. *Restoration Ecology* **13**, 39–48.

Bond NR, Lake PS and Arthington AH (2008) The impacts of drought on freshwater ecosystems: an Australian perspective. *Hydrobiologia* **600**, 3–16.

Bond NR, Thomson J, Reich P and Stein J (2011) Using species distribution models to infer environmental filters and climate induced range shifts. *Marine and Freshwater Research* **62**, 1043–1061.

Borg D, Rutherfurd ID and Stewardson M (2007) The geomorphic and ecological effectiveness of habitat rehabilitation works: continuous measurement of scour and fill around large logs in sand-bed streams. *Geomorphology* **89**, 205–216.

Bortolus A (2008) Error cascades in the biological sciences: the unwanted consequences of using bad taxonomy in ecology. *Ambio* **37**, 114–118.

Bostock BM, Adams M, Austin CM and Laurenson LJB (2006) The molecular systematics of *Leiopotherapon unicolor* (Günther, 1859): testing for cryptic speciation in Australia's most widespread freshwater fish. *Biological Journal of the Linnean Society* **87**, 537–552.

Bottrill MC, Joseph LN *et al.* (2008) Is conservation triage just smart decision making? *Trends in Ecology and Evolution* **23**, 649–654.

Boulton AJ (1993) Stream ecology and surface-hyporheic exchange: implications, techniques and limitations. *Australian Journal of Marine and Freshwater Research* **44**, 553–564.

Boulton AJ and Brock MA (1999) *Australian Freshwater Ecology: Processes and Management.* Gleneagles Publishing, Adelaide.

Bovee KD (1982) 'A guide to stream habitat analysis using the instream flow incremental methodology'. Instream Flow Information Paper 12. FWS/OBS–82/26. USDI Fish and Wildlife Services, Office of Biology Services, Washington DC.

Bowen SH, Lutz EV and Ahlgren MO (1995) Dietary protein and energy as determinants of food quality: trophic strategies compared. *Ecology* **76**, 899–907.

Bowler JM, Jones R, Allen H and Thorne AG (1970) Pleistocene human remains from Australia: a living site and human cremation from Lake Mungo, western New South Wales. *World Archaeology* **2**, 39–60.

Bowmaker JK, Govardovskii VI, Shukolyukov SA, Zueva LV *et al.* (1994) Visual pigments and the photic environment: the cottoid fish of Lake Baikal. *Vision Research* **34**, 591–605.

Boxall GD, Sandberg JJ and Kroon FJ (2002) Population structure, movement and habitat preferences of the purple-spotted gudgeon, *Mogurnda adspersa. Marine and Freshwater Research* **53**, 909–917.

Boys C, Miles N and Rayner T (2009) 'Scoping options for the ecological assessment of cold water pollution mitigation downstream of Keepit Dam, Namoi River'. Murray-Darling Basin Authority, Canberra.

Boys CA, Robinson W, Butcher A, Zampatti B and Lyon J (2008) Framework for developing and implementing ecological monitoring and evaluation of aquatic rehabilitation in demonstration reaches. MDBA Publication 43/08. Murray-Darling Basin Commission, Canberra.

Boys CA and Thoms MC (2006) A large-scale, hierarchical approach for assessing habitat associations of fish assemblages in large dryland rivers. *Hydrobiologia* **572**, 11–31.

Bradford M, Kroon FJ and Russell DJ (2011) The biology and management of *Tilapia mariae* (Pisces: Cichlidae) as a native and invasive species: a review. *Marine and Freshwater Research* **62**, 902–917.

Bradshaw A (1996) Underlying principles of restoration. *Canadian Journal of Fisheries and Aquatic Sciences* **53**, 3–9.

Braun J, Burbidge DR, Gesto FN, Sandiford M *et al.* (2009) Constraints on the current rate of deformation and surface uplift of the Australian continent from a new seismic database and low-T thermochronological data. *Australian Journal of Earth Sciences* **56**, 99–110.

Breckwoldt R, Boden R and Andrew J (2004) *The Darling.* Murray-Darling Basin Commission, Canberra.

Brett JR and Blackburn JM (1981) Oxygen requirements for growth of young coho (*Oncorhynchus kisutch*) and sockeye (*O. nerka*) salmon at 15°C. *Canadian Journal of Fisheries and Aquatic Sciences* **38**, 399–404.

Brickhill MJ, Lee SY and Connolly RM (2005) Fishes associated with artificial reefs: attributing changes to attraction or production using novel approaches. *Journal of Fish Biology* **67**, 53–71.

Briggs IC and McDowall RM (1996) Family Clupeidae. Herrings. In *Freshwater Fishes of Southeastern Australia.* (Ed. RM McDowall) pp. 44–47. Reed, Sydney.

Briggs SV (2009) Priorities and paradigms: directions in threatened species recovery. *Conservation Letters* **2**, 101–108.

Brinsley W (2010) *The Speckled Success: The Story of Trout in New South Wales.* Paddington, New South Wales.

Brisbane Declaration (2007) 'The Brisbane Declaration. Environmental flows are essential for freshwater ecosystem health and human well-being'. Declaration of the 10th International Rivers Symposium and International Environmental Flows Conference, Brisbane, Australia.

Brito P and Edwards SV (2009) Multilocus phylogeography and phylogenetics using sequence-based markers. *Genetica* **135**, 439–455.

Bromham L (2009) Why do species vary in their rate of molecular evolution? *Biology Letters* **5**, 401–404.

Brookes A (1995) River channel restoration: theory and practice. In *Changing River Channels.* (Eds A Gurnell and G Petts) pp. 369–388. Wiley, Chichester.

Brooks AP, Gehrke PC, Jansen JD and Abbe TB (2004) Experimental reintroduction of woody debris on the Williams River, NSW: geomorphic and ecological responses. *River Research and Applications* **20**, 513–536.

Brooks JL and Dodson SI (1965) Predation, body size and the composition of plankton. *Science* **150**, 28–35.

Brooks SG and Kind PK (2002) 'Ecology and demography of the Queensland lungfish (*Neoceratodus forsteri*) in the Burnett River, Queensland with reference to the impacts of Walla Weir and future water infrastructure development'. Report QO02004. Department of Primary Industries, Queensland, Agency for Food and Fibre Sciences, Brisbane.

Brooks SS and Lake PS (2007) River restoration in Victoria, Australia: change is in the wind, and none too soon. *Restoration Ecology* **15**, 584–591.

Browman HI (1995) Commentaries on current research trends in recruitment studies. *Marine Ecology Progress Series* **128**, 305–310.

Brown AV and Armstrong ML (1985) Propensity to drift downstream among various species of fish. *Journal of Freshwater Ecology* **3**, 3–17.

Brown CM and Radke BM (1989) Stratigraphy and sedimentology of mid-Tertiary permeability barriers in the subsurface of the Murray Basin, southeastern Australia. *BMR Journal of Australian Geology and Geophysics* **11**, 367–385.

Brown JA (1985) The adaptive significance of behavioural ontogeny in some centrachid fishes. *Environmental Biology of Fishes* **13**, 25–34.

Brown JH, Gillooly JF, Allen AP, Savage VM *et al.* (2004a) Toward a metabolic theory of ecology. *Ecology* **85**, 1771–1789.

Brown MC (2008) Discussion and reply – no mountains to snow on: major post-Eocene uplift of the East Victoria Highlands; evidence from Cenozoic deposits. *Australian Journal of Earth Sciences* **55**, 1181–1187.

Brown P (2004) Predicting growth and mortality of brown trout (*Salmo trutta*) in the Goulburn River after mitigation of cold-water discharge from Lake Eildon, Australia. *New Zealand Journal of Marine and Freshwater Research* **38**, 279–287.

Brown P and Walker TI (2004) CARPSIM: stochastic simulation modelling of wild carp (*Cyprinus carpio* L.) population dynamics, with applications to pest control. *Ecological Modelling* **176**, 83–97.

Brown P and Wooden I (2007) Age at first increment formation and validation of daily growth increments in golden perch (*Macquaria ambigua*: Percichthyidae) otoliths. *New Zealand Journal of Marine and Freshwater Research* **41**, 157–161.

Brown P, Green C, Sivakumaran KP, Stoessel D *et al.* (2004b) Validating otolith annuli for annual age determination of common carp. *Transactions of the American Fisheries Society* **133**, 190–196.

Brown P, Sivakumaran KP, Stoessel D and Giles A (2005) Population biology of carp (*Cyprinus carpio* L.) in the mid-Murray River and Barmah Forest Wetlands, Australia. *Marine and Freshwater Research* **56**, 1151–1164.

Brown P, Sivakumaran KP, Stoessel D, Giles A *et al.* (2003) 'Carp population biology in Victoria'. Department of Primary Industries, Melbourne.

Brown R and Severin KP (1999) Elemental distribution within polymorphic inconnu (*Stenodus leucichthys*) otoliths is affected by crystal structure. *Canadian Journal of Fisheries and Aquatic Sciences* **56**, 1898–1903.

Brumley AR (1987) Past and present distributions of golden perch in Victoria,with reference to releases of hatchery-produced fry. *Proceedings of the Royal Society of Victoria* **99**, 111–116.

Bruton MN (1990) The conservation of the fishes of Lake Victoria, Africa: an ecological perspective. *Environmental Biology of Fishes* **27**, 161–175.

Bryant SL and Harris S (1996) Down the track: a critique of recovery programs in Tasmania. In *Back from the Brink: Refining the Threatened Species Recovery Process.* (Eds S Stephens and S Maxwell) pp. 180–184. Surrey Beatty and Sons, Sydney.

Bucater L, Cheshire K and Ye Q (2009) 'Fish spawning in the lower River Murray icon sites, South Australia: with reference to drought intervention'. SARDI Research Report Series 377. SARDI Aquatic Sciences, Adelaide.

Buckley L, Caldarone E and Ong TL (1999) RNA-DNA ratio and other nucleic acid-based indicators for growth and condition of marine fishes. *Hydrobiologia* **401**, 265–277.

Buckley LJ, Caldarone EM and Clemmesen C (2008) Multi-species larval fish growth model based on temperature and fluorometrically derived RNA/DNA ratios: results from a meta-analysis. *Marine Ecology Progress Series* **371**, 221–232.

Buckworth R, Ovenden JR, Broderick D, McPherson GR *et al.* (2007) Genetag: monitoring fishing mortality rates and catchability using remote biopsy and genetic mark–recapture. In *Cutting-edge Technologies in Fish and Fisheries Science.* August 2006, Hobart. (Eds JM Lyle, DM Furlani and CD Buxton), pp. 44–49.

Builth H (2004) Mt Eccles and the Gunditjmara connection: a landform for all seasons. *Proceedings of the Royal Society of Victoria* **11**, 163–181.

Bunn SE and Arthington AH (2002) Basic principles and ecological consequences of altered flow regimes for aquatic biodiversity. *Environmental Management* **30**, 492–507.

Bunn SE, Abal EG, Smith MJ, Choy SC *et al.* (2010) Integration of science and monitoring of river ecosystem health to guide investments in catchment protection and rehabilitation. *Freshwater Biology* **55**, 223–240.

Bunn SE, Balcombe SR, Davies PM, Fellows CS *et al.* (2006a) Aquatic productivity and food webs of desert river ecosystems. In *Ecology of Desert Rivers.* (Ed. RT Kingsford) pp. 76–99. Cambridge University Press, Melbourne.

Bunn SE, Thoms MC, Hamilton SK and Capon SJ (2006b) Flow variability in dryland rivers: boom, bust and the bits in between. *River Research and Applications* **22**, 179–186.

Burbidge AA and Jenkins RWG (1984) *Endangered Vertebrates of Australia and its Island Territories.* Australian National Parks and Wildlife Service, Canberra.

Burchmore JJ, Farragher R and Thorncraft G (1990) 'Occurrence of the introduced oriental weatherloach (*Misgurnus anguillicaudatus*) in the Wingecarribee River New South Wales'. Bureau of Rural Resources Proceedings 8. Australian Government Publishing Service, Canberra.

Bureau of Agricultural Resource Economics and Sciences (2010) 'Fisheries status reports'. Department of Agriculture, Fisheries and Forestry, Canberra.

Bureau of Meteorology (2011) <http://www.bom.gov.au/climate/change/>. 2/2/2011.

Burggren WW and Blank T (2009) Physiological study of larval fishes: challenges and opportunities. *Scientia Marina* **73**, 99–110.

Burgman M and Possingham H (2000) Population viability analysis for conservation: the good, the bad and the undescribed. In *Genetics, Demography and Viability of Fragmented Populations.* (Eds AG Young and GM Clarke) pp. 97–112. Cambridge University Press, London.

Burnet AMR (1965) Observations on the spawning migrations of *Galaxias attenuatus* (Jenyns). *New Zealand Journal of Science* **8**, 79–87.

Burridge CP, Craw D and Waters JM (2006) River capture, range expansion, and cladogenesis: the genetic signature of freshwater vicariance. *Evolution* **60**, 1038–1049.

Burridge CP, McDowall RM, Craw D, Wilson MVH *et al.* (2012) Marine dispersal as a pre-requisite for Gondwanan vicariance among elements of the galaxiid fish fauna. *Journal of Biogeography* **39**, 306–321.

Burrows DW (2004) 'Translocated fishes in streams of the Wet Tropics Region, North Queensland: distribution and potential impact'. Cooperative Research Centre for Tropical Rainforest Ecology and Management, Cairns.

Burrows DW and Perna C (2009) 'A survey for exotic freshwater fish on Saibai Island and Thursday Island, Torres Strait'. ACTFR Report 09/01. Australian Centre for Tropical Freshwater Research, James Cook University, Townsville.

Burt WH (1943) Territoriality and home range concepts as applied to mammals. *Journal of Mammalogy* **24**, 346–352.

Burton LD (2010) *Fish and Wildlife: Principles of Zoology and Ecology.* Delmar Cengage Learning, New York.

Busacker GP, Adelman IR and Goolish EM (1990) Growth. In *Methods for Fish Biology.* (Eds CB Schrek and PB Moyle) pp. 363–387. American Fisheries Society, Bethesda.

Butchardt SHM (2010) Global biodiversity: indicators of recent declines. *Science* **328**, 1164–1168.

Butcher AD (1945) The food of indigenous and non-indigenous freshwater fish in Victoria, with special reference to trout. *Fisheries and Game Department, Victoria, Fisheries Pamphlet* **2**, 1–48.

Butcher AD (1946) 'The freshwater fish of Victoria and their food.' Fisheries and Game Department, Victoria, Melbourne.

Butler GL (2009a) Biology and conservation of the endangered eastern freshwater cod, *Maccullochella ikei* Rowland. PhD thesis. Southern Cross University, Lismore.

Butler GL (2009b) 'Review of the biology and conservation of the endangered eastern freshwater cod *Maccullochella ikei*, with particular reference to defining environmental water requirements. Unregulated river water sharing plan ecological monitoring and reporting'. NSW Department of Water and Energy, Sydney.

Butler GL and Rowland SJ (2009) Using underwater cameras to describe the reproductive behaviour of the endangered eastern freshwater cod *Maccullochella ikei*. *Ecology of Freshwater Fish* **18**, 337–349.

Butler VL (2000) Resource depression on the northwest coast of North America. *Antiquity* **74**, 649–661.

Byrne M, Yeates DK and Joseph L (2008) Birth of a biome: insights into the assembly and maintenance of the Australian arid zone biota. *Molecular Ecology* **17**, 4398–4417.

Cadwallader P (1978) Some causes of decline in range and abundance of native fish in the Murray-Darling River system. *Proceedings of the Royal Society of Victoria* **90**, 211–224.

Cadwallader P and Backhouse G (1983) *A Guide to the Freshwater Fish of Victoria*. Fisheries and Wildlife Division, Ministry for Conservation, Melbourne.

Cadwallader PL (1976a) Breeding biology of a non-diadromous galaxiid, *Galaxias vulgaris* Stokell, in a New Zealand river. *Journal of Fish Biology* **8**, 157–177.

Cadwallader PL (1976b) Home range and movements of the common river galaxias, *Galaxias vulgaris*, in the Glentui River, New Zealand. *Australian Journal of Marine and Freshwater Research* **27**, 33.

Cadwallader PL (1977) 'J.O. Langtry's 1949–50 Murray River investigations'. Fisheries and Wildlife Paper 13. Fisheries and Wildlife Division, Ministry for Conservation, Melbourne.

Cadwallader PL (1978) Some causes of the decline in range and abundance of native fish in the Murray-Darling River system. *Proceedings of the Royal Society of Victoria* **90**, 211–224.

Cadwallader PL (1979) Distribution of native and introduced fish in the Seven Creeks River system, Victoria. *Australian Journal of Ecology* **4**, 361–385.

Cadwallader PL (1981) Past and present distributions and translocations of Macquarie perch *Macquaria australasica* (Pisces: Percichthyidae) in Victoria. *Proceedings of the Royal Society of Victoria* **93**, 23–30.

Cadwallader PL (1996) 'Overview of the impacts of introduced salmonids on Australian native fauna'. Australian Nature Conservation Agency, Canberra.

Cadwallader PL and Douglas J (1986) Changing food habits of Macquarie perch, *Macquaria australasica* Cuvier (Pisces: Percichthyidae), during the initial filling phase of Lake Dartmouth, Victoria. *Australian Journal of Marine and Freshwater Research* **37**, 647–657.

Cadwallader PL and Gooley GJ (1984) Past and present distributions and translocations of Murray cod *Maccullochella peeli* and trout cod *M. macquariensis* (Pisces: Percichthyidae). *Proceedings of the Royal Society of Victoria* **96**, 33–43.

Cadwallader PL and Rogan PL (1977) The Macquarie perch, *Macquaria australasica* (Pisces: Percicthyidae), of Lake Eildon, Victoria. *Australian Journal of Ecology* **2**, 409–418.

Cadwallader PL, Eden AK and Hook RA (1980) Role of streamside vegetation as a food source for *Galaxias olidus* Günther (Pisces: Galaxiidae). *Marine and Freshwater Research* **31**, 257–262.

Caldarone EM, Clemmesen CM, Berdalet E, Miller TJ *et al.* (2006) Intercalibration of four spectrofluorometric protocols for measuring RNA/DNA ratios in larval and juvenile fish. *Limnology and Oceanography* **4**, 153–163.

Campana SE (1990) How reliable are growth back-calculations based on otoliths. *Canadian Journal of Fisheries and Aquatic Sciences* **47**, 2219–2227.

Campana SE (1997) Use of radiocarbon from nuclear fallout as a dated marker in the otoliths of haddock, *Melanogrammus aeglefinus*. *Marine Ecology Progress Series* **150**, 49–56.

Campana SE (1999) Chemistry and composition of fish otoliths: pathways, mechanisms and applications. *Marine Ecology Progress Series* **188**, 263–297.

Campana SE (2001) Accuracy, precision and quality control in age determination, including a review of the use and abuse of age validation methods. *Journal of Fish Biology* **59**, 197–242.

Campana SE and Jones CM (1992) Analysis of otolith microstructure data. In *Otolith Microstructure Examination and Analysis*. (Eds DK Stevenson and SE Campana) pp. 73–100. Canada Communications Group, Ottawa.

Campana SE and Neilson JD (1985) Microstructure of fish otoliths. *Canadian Journal of Fisheries and Aquatic Sciences* **42**, 1014–1032.

Campana SE and Thorrold SR (2001) Otoliths, increments, and elements: keys to a comprehensive understanding of fish populations? *Canadian Journal of Fisheries and Aquatic Sciences* **58**, 30–38.

Canonico GC, Arthington AH, McCrary JK and Thieme ML (2005) The effects of introduced tilapias on native biodiversity. *Aquatic Conservation: Marine and Freshwater Ecosystems* **15**, 463–483.

Cappo M, De'ath G, Boyle S, Aumend J *et al.* (2005) Development of a robust classifier of freshwater residence in barramundi (*Lates calcarifer*) life histories using elemental ratios in scales and boosted regression trees. *Marine and Freshwater Research* **56**, 713–723.

Caraco T and Gillespie RG (1986) Risk-sensitivity: foraging mode in an ambush predator. *Ecology* **67**, 1180–1185.

Cardinale M and Arrhenius F (2000) The relationship between stock and recruitment: are the assumptions valid? *Marine Ecology Progress Series* **196**, 305–309.

Carlander KD (1977) *Handbook of Freshwater Fishery Biology*, Vol. 2. Iowa State University Press, Ames.

Carlander KD (1987) A history of scale age and growth studies of North American freshwater fish. In *Age and Growth of Fish*. (Eds RC Summerfelt and GE Hall) pp. 3–14. Iowa State University Press, Ames.

Carleton KL, Parry JWL, Bowmaker JK, Hunt DM *et al.* (2005) Colour vision and speciation in Lake Victoria cichlids of the genus *Pundamilia*. *Molecular Ecology* **14**, 4341–4353.

Carp Control Coordinating Group (2000) 'National management strategy for carp control 2000–2005'. Murray-Darling Basin Commission, Canberra.

Carpenter SR and Kitchell JF (Eds) (1993) *The Trophic Cascade in Lakes*. Cambridge University Press, New York.

Carpenter SR, Cole JJ *et al.* (2001) Trophic cascades, nutrients, and lake productivity: whole-lake experiments. *Ecological Monographs* **71**, 163–186.

Carroll AM, Wainwright PC, Huskey SH, Collar DC *et al.* (2004) Morphology predicts suction feeding performance in centrarchid fishes. *Journal of Experimental Biology* **207**, 3873–3881.

Carvalho DC, Sasaki M, Hammer MP and Beheregaray LB (2012) Development of 18 microsatellite markers for the southern purple-spotted gudgeon (*Mogurnda adspersa*) from the lower Murray-Darling Basin through 454 pyrosequencing. *Conservation Genetics Resources* **4**, 339–341.

Castelnau FL (1872) Contribution to the ichthyology of Australia. 1. The Melbourne Fish Market. *Proceedings of the Zoological and Acclimatisation Society of Victoria* **1**, 29–242.

Castelnau FL (1878) On some new Australian (chiefly) freshwater fishes. *Proceedings of the Linnean Society of New South Wales* **2**, 140–144.

Castro-Santos T, Haro A and Walk S (1996) A Passive Integrated Transponder (PIT) tag system for monitoring fishways. *Fisheries Research* **28**, 253–261.

Caswell H (2000) *Matrix Population Models: Construction, Analysis and Interpretation*, 2nd edn. Sinauer Associates, New York.

Chakrabarty P (2010) Status and phylogeny of Milyeringidae (Teleostei: Gobiiformes), with the description of a new blind cave-fish from Australia, *Milyeringa brooksi*, n. sp. *Zootaxa* **2557**, 19–28.

Chambers RC and Leggett WC (1989) Event analysis applied to timing in marine fish ontogeny. *Canadian Journal of Fisheries and Aquatic Sciences* **46**, 1633–1641.

Chan T, Hart B, Kennard MJ, Pusey BJ *et al.* (2012) Bayesian network models for environmental flow decision-making in the Daly River, Northern Territory, Australia. *River Research and Applications* **28**, 283–301.

Chang R (1994) *Chemistry*, 5th edn. McGraw-Hill, New York.

Chapman A, Hobbs RP, Morgan DL and Gill HS (2006a) Helminth parasitism of *Galaxias maculatus* (Jenyns 1842) in southwestern Australia. *Ecology of Freshwater Fish* **15**, 559–564.

Chapman A, Morgan DL, Beatty SJ and Gill HS (2006b) Variation in life history of land-locked lacustrine and riverine populations of *Galaxias maculatus* (Jenyns 1842) in Western Australia. *Environmental Biology of Fishes* **77**, 21–37.

Chappaz R, Brun G and Olivari G (1989) Contribution à l'étude du régime alimentaire, de la croissance et de la fécondité de la perche (*Perca fluviatilis* L.) dans un lac oligotrophe de region tempérée. *Revue des Sciences de l'eau* **2**, 95–107.

Chenoweth SF and Hughes JM (1997) Genetic population structure of the catadromous perciform: *Macquaria novemaculeata* (Percichthyidae). *Journal of Fish Biology* **50**, 721–733.

Chenoweth SF and Hughes JM (2003) Speciation and phylogeography in *Caridina indistincta*, a complex of freshwater shrimps from Australian heathland streams. *Marine and Freshwater Research* **54**, 807–812.

Chenoweth SF, Hughes JM, Keenan CP and Lavery S (1998) Concordance between dispersal and mitochondrial gene flow: isolation by distance in a tropical teleost, *Lates calcarifer* (Australian barramundi). *Heredity* **80**, 187–197.

Cheshire KJ (2005) Using otolith chemistry to track the movements of a diadromous fish: the congolli *Pseudaphritis urvillii* (Teleostei: Pseudaphritidae). Hons thesis. University of Adelaide, Adelaide.

Cheshire KJ (2010) Larval fish assemblages in the lower River Murray, Australia: examining the influence of hydrology, habitat and food. PhD thesis. University of Adelaide, Adelaide.

Chessman BC (1971) Studies of Victorian salt lake fishes, with species reference to *Galaxias maculatus* (Jenyns) 1842 (Pisces: Salmoniformes: Galaxiidae). Hons thesis. Monash University, Melbourne.

Chessman BC and Williams WD (1974) Distribution of fish in inland saline waters in Victoria, Australia. *Australian Journal of Marine and Freshwater Research* **25**, 167–172.

Chessman BC and Williams WD (1975) Salinity tolerance and osmoregulatory ability of *Galaxias maculatus* (Jenyns) (Pisces, Salmoniformes, Galaxiidae). *Freshwater Biology* **5**, 167–172.

Chesson P (2000) Mechanisms of maintenance of species diversity. *Annual Review of Ecology and Systematics* **31**, 343–366.

Clack J (2009) The fish–tetrapod transition: new fossils and interpretations. *Evolution: Education and Outreach* **2**, 213–223.

Clark M (2001) Are deepwater fisheries sustainable? The example of orange roughy (*Hoplostethus atlanticus*) in New Zealand. *Fisheries Research* **51**, 123–135.

Clark PU and Mix AC (2002) Ice sheets and sea level of the Last Glacial Maximum. *Quaternary Science Reviews* **21**, 1–7.

Clark TD, Ryan T, Ingram BA, Woakes AJ *et al.* (2005) Factorial aerobic scope is independent of temperature and primarily modulated by heart rate in exercising Murray cod (*Maccullochella peelii peelii*). *Physiological and Biochemical Zoology* **78**, 347–355.

Clarke A and Johnston NM (1999) Scaling of metabolic rate with body mass and temperature in teleost fish. *Journal of Animal Ecology* **68**, 893–905.

Clarke KR and Warwick RM (2001) *Change in Marine Communities: An Approach to Statistical Analysis and Interpretation*, 2nd edn. PRIMER-E Ltd, Plymouth.

Clements J (1988) *Salmon at the Antipodes: A History and Review of Trout, Salmon and Char and Introduced Coarse Fish in Australasia.* Published by the author.

Clements KD and Raubenheimer D (2006) Feeding and nutrition. In *The Physiology of Fishes*, 3rd edn. (Eds DH Evans and JB Claiborne) pp. 47–82. Taylor and Francis, Boca Raton.

Closs G, Downes B and Boulton A (2004) *Freshwater Ecology: A Scientific Introduction.* Blackwell Science, Malden.

Closs GP (1996) Effects of a predatory fish (*Galaxias olidus*) on the structure of intermittent stream pool communities in southeast Australia. *Australian Journal of Ecology* **21**, 217–223.

Closs GP and Lake PS (1996) Drought, differential mortality and the coexistance of a native and an introduced fish species in a south east Australian intermittent stream. *Environmental Biology of Fishes* **47**, 17–26.

Closs GP, Smith M, Barry B and Markwitz A (2003) Non-diadromous recruitment in coastal populations of common bully (*Gobiomorphus cotidianus*). *New Zealand Journal of Marine and Freshwater Research* **37**, 301–313.

Clunie P and Koehn JD (2001a) 'Freshwater catfish: a resource document'. Arthur Rylah Institute for Environmental Research, Department of Natural Resources and Environment, Melbourne.

Clunie P and Koehn JD (2001b) 'Silver perch: a resource document'. Arthur Rylah Institute for Environmental Research, Department of Natural Resources and Environment, Melbourne.

Clunie P, Ryan T, James K and Cant B (2002a) 'Implications for rivers from salinity hazards: scoping study'. Report produced for Murray-Darling Basin Commission, Strategic Investigations and Riverine Program, Project R2003. Arthur Rylah Institute for Environmental Research, Department of Natural Resources and Environment, Melbourne.

Clunie P, Stuart I, Jones M, Crowther D *et al.* (2002b) 'A risk assessment of the impacts of pest species in the riverine environment in the Murray-Darling Basin'. Arthur Rylah Institute for Environmental Research, Department of Natural Resources and Environment, Melbourne.

Cole LC (1954) The population consequences of life history phenomena. *Quarterly Review of Biology* **29**, 103–137.

Coleman RA, Pettigrove V, Raadik TA, Hoffmann AA *et al.* (2010) Microsatellite markers and mtDNA data indicate two distinct groups in dwarf galaxias, *Galaxiella pusilla* (Mack) (Pisces: Galaxiidae), a threatened freshwater fish from south-eastern Australia. *Conservation Genetics* **11**, 1911–1928.

Collares-Pereira MJ and Cowx IG (2004) The role of catchment scale environmental management in freshwater fish conservation. *Fisheries Management and Ecology* **11**, 303–312.

Collins AL and Anderson TA (1995) The regulation of endogeneous energy stores during starvation and refeeding in the somatic tissues of the golden perch. *Journal of Fish Biology* **47**, 1004–1015.

Collins AL and Anderson TA (1997) The influence of changes in food availability on the activities of key degradative and metabolic enzymes in the liver and epaxial muscle of the golden perch. *Journal of Fish Biology* **50**, 1158–1165.

Collins AL and Anderson TA (1999) The role of food availability in regulating reproductive development in female golden perch. *Journal of Fish Biology* **55**, 94–104.

Conallin AJ, Hillyard KA, Walker KF, Gillanders BM and Smith BB (2010) Offstream movements of fish during drought in a regulated lowland river. *River Research and Applications* **26**, 1–16.

Conallin AJ, Smith BB, Thwaites LA, Walker KF *et al.* (2012) Environmental water allocations in regulated lowland rivers may encourage offstream movements and spawning by common carp, *Cyprinus carpio*: implications for wetland rehabilitation. *Marine and Freshwater Research* **63**, 865–877.

Condrey RE (1982) Ingestion-limited growth of aquatic animals: the case for Blackman kinetics. *Canadian Journal of Fisheries and Aquatic Sciences* **39**, 1585–1595.

Connell JH (1978) Diversity in tropical rain forests and coral reefs. High diversity of trees and corals is maintained only in a nonequilibrium state. *Science* **199**, 1302–1310.

Cook BD and Hughes JM (2010) Historical population connectivity and fragmentation in a tropical freshwater fish with a disjunct distribution (pennyfish, *Denariusa bandata*). *Journal of the North American Benthological Society* **29**, 1119–1131.

Cook BD, Bunn SE and Hughes JM (2007) Molecular genetic and stable isotope signatures reveal complementary patterns of population connectivity in the regionally vulnerable southern pygmy perch (*Nannoperca australis*). *Biological Conservation* **138**, 60–72.

Cook BD, Kennard MJ, Real K, Pusey BJ *et al.* (2010) Landscape genetic analysis of the tropical freshwater fish *Mogurnda mogurnda* (Eleotridae) in a monsoonal river basin: importance of hydrographic factors and population history. *Freshwater Biology* **56**, 812–827.

Cooke SJ and Cowx IG (2004) The role of recreational fishing in global fish crises. *BioScience* **54**, 857–859.

Cooke SJ and Suski CD (2005) Do we need species-specific guidelines for catch-and-release recreational angling to effectively conserve diverse fishery resources? *Biodiversity and Conservation* **14**, 1195–1209.

Cooke SJ, Hinch SG, Wikelski M, Andrews RD *et al.* (2004) Biotelemetry: a mechanistic approach to ecology. *Trends in Ecology and Evolution* **19**, 334–343.

Coombs S and Braun CB (2003) Information processing by the lateral line system. In *Sensory Processing in Aquatic Environments.* (Eds SP Collin and NJ Marshall) pp. 123–138. Springer-Verlag, New York.

Cooper WE (2005) The foraging mode controversy: both continuous variation and clustering of foraging movements occur. *Journal of Zoology* **267**, 179–190.

Copp GH, Kováč V and Hensel K (1999) Prelude: looking at early development in fishes. *Environmental Biology of Fishes* **56**, 7–14.

Corfield J, Diggles B, Jubb C, McDowall RM *et al.* (2008) 'Review of the impacts of introduced ornamental fish species that have established wild populations in Australia'. Department of the Environment, Water, Heritage and the Arts, Canberra.

Costelloe JF, Reid JRW, Pritchard JC, Puckridge JT *et al.* (2010) Are alien fish disadvantaged by extremely variable flow regimes in arid-zone rivers? *Marine and Freshwater Research* **61**, 857–863.

Cotterell E (1998) 'Fish passage in streams: fisheries guidelines for design of stream crossings'. Queensland Department of Primary Industries, Brisbane.

Cowan JH, Houde ED and Rose KA (1996) Size-dependent vulnerability of marine fish larvae to predation: an individual-based numerical experiment. *ICES Journal of Marine Science* **53**, 23–37.

Cowx I (1994) *Rehabilitation of Freshwater Fisheries.* Fishing News Books, Oxford.

Cowx IG and Welcomme RL (Eds) (1998) *Rehabilitation of Rivers for Fish.* FAO and Fishing News Books, Blackwell Science, Oxford.

Coyne JA (2007) Sympatric speciation. *Current Biology* **17**, 787–788.

Coyne JA and Orr HA (2004) *Speciation.* Sinauer Associates, New York.

Crandall KA, Bininda-Emonds ORP, Mace GM and Wayne RK (2000) Considering evolutionary processes in conservation biology. *Trends in Ecology and Evolution* **15**, 290–295.

Croke JC and Hairsine PB (2006) Sediment delivery in managed forests: a review. *Environmental Reviews* **14**, 59–87.

Croke JC, Hairsine PB and Fogarty P (1999) Sediment transport, redistribution and storage on logged forest hillslopes in southeastern Australia. *Hydrological Processes* **13**, 2705–2720.

Crook D and Sanger A (1997) 'Recovery plan for the Pedder, Swan, Clarence, swamp and saddled galaxias'. Inland Fisheries Commission, Hobart.

Crook DA (2004a) Is the home range concept compatible with fish movements of two species of lowland fiver fish? *Journal of Animal Ecology* **73**, 353–399.

Crook DA (2004b) A method for externally attaching radio transmitters to minimize dermal irritation. *Journal of Fish Biology* **64**, 258–261.

Crook DA (2004c) Movements associated with home-range establishment by two species of lowland river fish. *Canadian Journal of Fisheries and Aquatic Sciences* **61**, 2183–2193.

Crook DA and Gillanders BM (2006) Use of otolith chemical signatures to estimate carp recruitment sources in the mid-Murray River, Australia. *River Research and Applications* **22**, 871–879.

Crook DA and Koster WM (2006) Temporal change in fish assemblages in the lower Goulburn River, south-eastern Australia: comment on Pollino *et al.* (2004). *Marine and Freshwater Research* **57**, 303–308.

Crook DA and Robertson AI (1999) Relationships between riverine fish and woody debris: implications for lowland rivers. *Marine and Freshwater Research* **50**, 941–953.

Crook DA and Sanger AC (1998) Threatened fishes of the world: *Galaxias Johnstoni* Scott,1936 (Galaxiidae). *Environmental Biology of Fishes* **53**, 154.

Crook DA and White RWG (1995) Evaluation of subcutaneously implanted visual implant tags and coded wire tags for marking and benign recovery in a small scaleless fish, *Galaxias truttaceus. Marine and Freshwater Research* **46**, 943–946.

Crook DA, Gillanders BM, Sanger AC, Munro AR *et al.* (2011) 'Methods for discriminating hatchery fish and outcomes of stocking in the Murray-Darling Basin'. Final report to the Murray-Darling Basin Authority, Project MD741. Arthur Rylah Institute for Environmental Research, Department of Sustainability and Environment, Melbourne.

Crook DA, Koster WM *et al.* (2010a) Catadromous migrations by female tupong (*Pseudaphritis urvillii*) in coastal streams in Victoria, Australia. *Marine and Freshwater Research* **61**, 474–483.

Crook DA, Macdonald JI and Raadik TA (2008) Evidence of diadromous movements in a coastal population of southern smelts (Retropinninae: *Retropinna*) from Victoria, Australia. *Marine and Freshwater Research* **59**, 638–646.

Crook DA, Macdonald JI, O'Connor JP and Barry B (2006) Use of otolith chemistry to examine patterns of diadromy in the threatened Australian grayling *Prototroctes maraena*. *Journal of Fish Biology* **69**, 1330–1344.

Crook DA, O'Mahony D, Sanger AC, Munro AR *et al.* (2009) Development and evaluation of methods for osmotic induction marking of golden perch (*Macquaria ambigua*) with calcein and alizarin red S. *North American Journal of Fisheries Management* **29**, 279–287.

Crook DA, Reich P, Bond NR, McMaster D *et al.* (2010b) Using biological information to support proactive strategies for managing freshwater fish during drought. *Marine and Freshwater Research* **61**, 379–387.

Crook DA, Robertson AI, King AJ and Humphries P (2001) The influence of spatial scale and habitat arrangement on diel patterns of habitat use by two lowland river fishes. *Oecologia* **129**, 525–533.

Crooks JA (2002) Characterizing ecosystem-level consequences of biological invasions: the role of ecosystem engineers. *Oikos* **97**, 153–166.

Crosby AW (2000) *Ecological Imperialism: The Biological Expansion of Europe, 900–1900*. Cambridge University Press, Cambridge.

Crossin GT and Hinch SG (2005) A nonlethal, rapid method for assessing the somatic energy content of migrating adult Pacific salmon. *Transactions of the American Fisheries Society* **134**, 184–191.

Crowder LB and Cooper WE (1982) Habitat structural complexity and the interaction between bluegills and their prey. *Ecology* **63**, 1802–1813.

Crowl TA, Townsend CR and McIntosh AR (1992) The impact of introduced brown and rainbow trout on native fish: the case of Australasia. *Reviews in Fish Biology and Fisheries* **2**, 217–241.

Crowley LELM and Ivantsoff W (1990) A review of species previously identified as *Craterocephalus eyresii* (Pisces: Atherinidae). *Proceedings of the Linnean Society of New South Wales* **112**, 87–103.

CSIRO (2008) 'Water availability in the Murray-Darling Basin'. Report to the Australian Government from the CSIRO Murray-Darling Basin Sustainable Yields Project. CSIRO, Canberra.

CSIRO/ABM (2007) 'Climate change in Australia. Projections developed by CSIRO and the Australian Bureau of Meteorology for the Australian Climate Change Science Programme. Observed changes and predictions'. Department of Environment and Water Resources, Canberra.

Cucherousset J, Paillisson JM, Carpentier A, Thoby V *et al.* (2007) Freshwater protected areas: an effective measure to reconcile conservation and exploitation of the threatened European eels (*Anguilla anguilla*)? *Ecology of Freshwater Fish* **16**, 528–538.

Cushing DH (1968) *Fisheries Biology: A Study in Population Dynamics*. University of Wisconsin Press, Madison.

Cushing DH (1990) Plankton production and year-class strength in fish populations: an update of the match/mismatch hypothesis. *Advances in Marine Biology* **26**, 250–293.

Cushing DH (1996) *Towards a Science of Recruitment in Fish Populations*. Ecology Institute, Germany.

Cussac V, Ortubay S, Iglesias G, Milano D *et al.* (2004) The distribution of South American galaxiid fishes: the role of biological traits and post glacial history. *Journal of Biogeography* **31**, 103–122.

Dakin WJ and Kesteven GL (1938) The Murray cod (*Maccullochella macquariensis* (Cuv. et Val.)). *New South Wales State Fisheries Research Bulletin* **1**, 1–18.

Dannevig HC (1903) 'Murray cod fisheries. Extracts from evidence collected by the South Australian authorities during October 1900'. Department of Fisheries, Sydney.

Dargin P (1976) *Aboriginal Fisheries of the Darling-Barwon Rivers*. Brewarina Historical Society, Dubbo.

Darwin C (1871) *Sexual Selection and the Descent of Man*. Murray, London.

Davey JL and Blaxter MW (2010) RADSeq: next-generation population genetics. *Briefings in Functional Genomics* **9**, 416–423.

David BO and Closs GP (2003) Seasonal variation in diel activity and microhabitat use of an endemic New Zealand stream-dwelling galaxiid fish. *Freshwater Biology* **48**, 1765–1781.

Davies P and Nelson M (1994) Relationships between riparian buffer widths and the effects of logging on stream habitat, invertebrate community composition and fish abundance. *Australian Journal of Marine and Freshwater Research* **45**, 1289–1305.

Davies P, Cook B, Rutherfurd K and Walshe T (2004) 'Managing high in-stream temperature using riparian vegetation. River and Riparian Land Management Technical Guideline 5'. Land and Water Australia, Canberra.

Davies PE (1995) Mail surveys of Tasmanian inland water recreational fisheries: preliminary results and sources of error. In *Recreational Fishing: What's the Catch?* (Ed. DA Hancock). Australian Society for Fish Biology Workshop Proceedings, Canberra.

Davies PE and Sloane RD (1987) Characteristics of the spawning migrations of brown trout, *Salmo trutta* L., and rainbow trout, *S. gairdneri* Richardson, in Great Lake, Tasmania. *Journal of Fish Biology* **31**, 353–373.

Davies PE, Harris JH, Hillman TJ and Walker KF (2008) 'Sustainable Rivers Audit. SRA Report-1. A report on the ecological health of rivers in the Murray-Darling Basin, 2004–2007'. Murray-Darling Basin Commission, Canberra.

Davies PE, Harris JH, Hillman TJ and Walker KF (2010) The Sustainable Rivers Audit: assessing river ecosystem health in the Murray-Darling Basin, Australia. *Marine and Freshwater Research* **61**, 764–777.

Davies PE, Sloane RD and Andrew J (1988) Effects of hydrological change and the cessation of stocking on a stream population of *Salmo trutta* L. *Australian Journal of Marine and Freshwater Research* **39**, 337–354.

Davis AM, Pearson RG, Pusey BJ, Perna C *et al.* (2011) Trophic ecology of northern Australia's terapontids: ontogenetic dietary shifts and feeding classification. *Journal of Fish Biology* **78**, 265–286.

Davis AM, Pusey BJ, Thorburn DC, Dowe JL *et al.* (2010a) Riparian contributions to the diet of terapontid grunters (Pisces: Terapontidae) in wet–dry tropical rivers. *Journal of Fish Biology* **76**, 862–879.

Davis J and Finlayson B (2000) 'Sand slugs and stream degradation: the case of the Granite Creeks, north-east Victoria'. Technical Report 7/2000. Cooperative Research Centre for Freshwater Ecology, Canberra.

Davis J, Sim L and Chambers J (2010b) Multiple stressors and regime shifts in shallow aquatic ecosystems in antipodean landscapes. *Freshwater Biology* **55**, 5–18.

Davis KM, Dixon PI and Harris JH (1999) Allozyme and mitochondrial DNA analysis of carp, *Cyprinus carpio* L., from south-eastern Australia. *Marine and Freshwater Research* **50**, 253–260.

Davis MA (2009) *Invasion Biology*. Oxford University Press, New York.

Davis TLO (1977a) Age determination and growth of the freshwater catfish, *Tandanus tandanus* Mitchell, in the Gwydir River, Australia. *Marine and Freshwater Research* **28**, 119–137.

Davis TLO (1977b) Food habits of the freshwater catfish, *Tandanus tandanus* Mitchell, in the Gwydir River, Australia, and effects associated with impoundment of this river by the Copeton Dam. *Australian Journal of Marine and Freshwater Research* **28**, 455–465.

Davis TLO (1977c) Reproductive biology of the freshwater catfish, *Tandanus tandanus* Mitchell, in the Gwydir River, Australia. I. Structure of the gonads. *Australian Journal of Marine and Freshwater Research* **28**, 139–158.

Davis TLO (1982) Maturity and sexuality in barramundi, *Lates calcarifer* (Bloch), in the Northern Territory and south-eastern Gulf of Carpentaria. *Australian Journal of Marine and Freshwater Research* **33**, 529–545.

Davis TLO (1984) Estimation of fecundity in barramundi, *Lates calcarifer* (Bloch), using an automatic particle counter. *Australian Journal of Marine and Freshwater Research* **35**, 111–118.

Davis TLO and Kirkwood GP (1984) Age and growth studies on barramundi, *Lates calcarifer* (Bloch), in Northern Australia. *Australian Journal of Marine and Freshwater Research* **35**, 673–689.

Day T and Taylor PD (1997) von Bertalanffy's growth equation should not be used to model age and size at maturity. *American Naturalist* **149**, 381–393.

de Bruyn M, Parenti LR and Carvalho GR (2011) Successful extraction of DNA from archived alcohol-fixed white-eye fish specimens using an ancient DNA protocol. *Journal of Fish Biology* **78**, 2074–2079.

de Bruyn M, Wilson JC and Mather PB (2004) Reconciling geography and genealogy: phylogeography of giant freshwater prawns from the Lake Carpentaria region. *Molecular Ecology* **13**, 3515–3526.

De Deckker P and Williams WD (Eds) (1986) *Limnology in Australia*. Dr W Junk, Dordrecht/CSIRO Publishing, Melbourne.

de Queiroz K (2007) Species concepts and species delimitation. *Systematic Biology* **56**, 879–886.

De Silva SS, Gunasekera RM, Collins RA and Ingram BA (2002) Performance of juvenile Murray cod, *Maccullochella peelii peelii* (Mitchell), fed with diets of different protein to energy ratio. *Aquaculture Nutrition* **8**, 79–85.

DeAngelis DL, Loftus WF, Trexler JC and Ulanowicz RE (1997) Modeling fish dynamics and effects of stress in a hydrologically pulsed ecosystem. *Journal of Aquatic Ecosystem Stress and Recovery* **6**, 1–13.

DeAngelis DL, Trexler JC and Loftus WF (2005) Life history trade-offs and community dynamics of small fishes in a seasonally pulsed wetland. *Canadian Journal of Fisheries and Aquatic Sciences* **62**, 781–790.

Deap L, Degen P and van Zalinge N (2003) 'Fishing gears of the Cambodian Mekong', Vol. 4. Cambodia Fisheries Technical Paper Series, Phnom Penh.

Décamps H and Naiman RJ (1989) L'écologie des fleuves. *La Recherche* **20**, 310–318.

Dendy JS (1963) Farm ponds. In *Limnology in North America*. (Ed. DG Frey) pp. 595–620. University of Wisconsin Press, Madison.

Department of Agriculture, Fisheries and Forestry (2007) 'A strategic approach to the management of ornamental fish in Australia'. Department of Agriculture, Fisheries and Forestry, Canberra.

Department of Environment and Resource Management, Queensland (2011) Edgbaston hardyhead. <http://www.derm.qld.gov.au/wildlife-ecosystems/wildlife/az_of_animals/edgbaston_hardyhead.html>. December 2011.

Department of Environment, Climate Change and Water, New South Wales (2010) Northern Rivers Regional Biodiversity Management Plan, National Recovery Plan for the Northern Rivers Region. <http://www.environment.gov.au/biodiversity/threatened/publications/recovery/pubs/ northern-rivers.pdf>. March 2011.

Department of Environment, Water, Heritage and the Arts (2011) Australian Wetlands Database <http://www.environment.gov.au/water/topics/wetlands/database/index.html>. November 2010.

Department of Fisheries, Western Australia (2003) 'The translocation of golden perch, Murray cod and Australian bass, into and within Western Australia, for the purposes of recreational stocking, domestic stocking and commercial and non-commercial aquaculture'. Fisheries Management Paper 174. Department of Fisheries Western Australia, Perth.

Department of Industry and Development, New South Wales (2009) 'Bringing back the fish: improving fish passage and aquatic habitat in coastal NSW'. Final report to the Southern Rivers Catchment Management Authority. Department of Industry and Development, Sydney.

Department of Industry and Development, New South Wales (2010a) 'NSW control plan for the noxious fish carp *Cyprinus carpio*'. Aquatic Biosecurity and Risk Management Unit, Department of Industry and Development, Orange.

Department of Industry and Development, New South Wales (2010b) 'Preliminary identification of critical habitat for the Oxleyan pygmy perch (*Nannoperca oxleyana*): public consultation document'. Fisheries Conservation and Aquaculture Branch, Port Stephens Fisheries Institute, Department of Industry and Investment, Port Stephens.

Department of Industry and Development, New South Wales (2011a) What is currently listed (threatened fish)? <http://www.dpi.nsw.gov.au/fisheries/species-protection/conservation/what-current>. February 2011.

Department of Industry and Investment, New South Wales (2011b) Black water events causing fish kills in the Murray and Murrumbidgee. <http://www.dpi.nsw.gov.au/fisheries/habitat/threats/fish-kills/black-water-events-causing-fish-kills-in-the-murray-and-murrumbidgee-river-catchments>. February 2011.

Department of Primary Industries and Fisheries, Queensland (2000) 'Control of exotic pest fishes: an operational strategy for Queensland waters 2000–2005'. Queensland Department of Primary Industries, Brisbane.

Department of Primary Industries and Water, Tasmania (2008) 'Water level manipulation in lakes Crescent and Sorell, Tasmania to assist spawning of the threatened golden galaxias (*Galaxias auratus*): a strategy to mitigate impacts of drought-induced habitat alterations'. Report WA 08/49. Water Assessment Aquatic Ecology Report Series. Water Resources Division, Department of Primary Industries and Water, Hobart.

Department of Primary Industries, New South Wales (2006) 'The assessment and modification of barriers to fish passage in the Namoi Catchment'. Department of Primary Industries, Tamworth.

Department of Primary Industries, New South Wales (2009) 'Eradication of one-spot livebearer, *Phalloceros caudimaculatus*, at Long Reef Golf Course. A post-eradication report'. <http://www.dpi.nsw.gov.au/fisheries/pests-diseases/freshwater-pests/species/speckled-mosquitofish#Eradication-of-one-spot-livebearer>. January 2011.

Department of Primary Industries, Parks, Water and Environment, Tasmania (2009) 'An evaluation of the environmental water release from Parangana Dam to the lower Mersey River (1999 to 2009)'. Report WA 09/09. Water Assessment Aquatic Ecology Report Series. Water and Marine Resources Division. Department of Primary Industries, Parks, Water and Environment, Hobart.

Department of Primary Industries, Victoria (2010) Economic study of recreational fishing in Victoria: Murray cod assessment. <http://www.new.dpi.vic.gov.au/fisheries/recreational-fishing/fishing-grants-program/your-licence-fees-at-work-research-reports/economic-study-of-recreational-fishing-in-victoria-murray-cod-assessment>. March 2011.

Department of Primary Industries, Water and Energy, Tasmania (2006) 'Recovery plan: Tasmanian Galaxiidae 2006–2010'. Threatened Species Section, Department of Primary Industries, Water and Energy, Hobart.

Department of Sustainability and Environment, Victoria (2007) 'Advisory list of threatened vertebrate fauna in Victoria – 2007'. Department of Sustainability and Environment, Melbourne.

Department of Sustainability and Environment, Victoria (2010) 'Securing our rivers for future generations. Victorian River Health Program report card 2002–2009'. Department of Sustainability and Environment, Melbourne.

Department of Sustainability, Environment, Water, Population and Communities (2011a) 'State of the Environment 2011'. Department of Sustainability, Environment, Water, Population and Communities, Canberra.

Department of Sustainability, Environment, Water, Population and Communities (2011b) *Galaxias pedderensis* (Pedder galaxias). Advice to the Minister for the Environment and Heritage from the Threatened Species Scientific Committee on Amendments to the list of Threatened Species under the Environment Protection and Biodiversity Conservation Act 1999. <http://www.environment.gov.au/index.htm>. December 2011.

Derby CD and Sorensen PW (2008) Neural processing, perception, and behavioral responses to natural chemical stimuli by fish and crustaceans. *Journal of Chemical Ecology* **34**, 898–914.

Desvilettes C, Bourdier G and Breton JC (1997) Changes in lipid class and fatty acid composition during development in pike (*Esox lucius* L) eggs and larvae. *Fish Physiology and Biochemistry* **16**, 381–393.

Diehl S (1992) Fish predation and benthic community structure: the role of omnivory and habitat complexity. *Ecology* **73**, 1646–1661.

Dill LM (1977) Refraction and the spitting behavior of the archerfish (*Toxotes chatareus*). *Behavioral Ecology and Sociobiology* **2**, 169–184.

Dill LM, Yvenberg RC and Fraser AHG (1981) Food abundance and territory size in juvenile coho salmon (*Oncorhynchus kisutch*). *Canadian Journal of Zoology* **59**, 1801–1809.

Dobson SH and Holmes RM (1984) Compensatory growth in the rainbow trout, *Salmo gairdneri* Richardson. *Journal of Fish Biology* **25**, 649–656.

Doeg TJ and Koehn JD (1994) Effects of draining and desilting a small weir on downstream fish and macroinvertebrates. *Regulated Rivers: Research and Management* **9**, 263–277.

Dolloff A, Kershner J and Thurow R (1996) Underwater observation. In *Fisheries Techniques.* (Eds BR Murphy and DW Willis) pp. 533–554. American Fisheries Society, Bethesda.

Dolloff CA and Warren ML (2003) Fish relationships with large wood in small streams. *American Fisheries Society Symposium* **37**, 179–193.

Domenici P and Blake R (1997) The kinematics and performance of fish fast-start swimming. *Journal of Experimental Biology* **200**, 1165–1178.

Domergue F and Bietrix E (1897) La periode critique post-larvaire des poissons marins. *Bulletin Museum National d'Histoire Naturelle, Paris* **3**, 57–58.

Doremus H and Pagel JE (2001) Why listing may be forever: perspectives on delisting under the U.S. Endangered Species Act. *Conservation Biology* **15**, 1258–1268.

Dorsey SE, Houde ED and Gamble JC (1996) Cohort abundances and daily variability in mortality of eggs and yolk-sac larvae of bay anchovy, *Anchoa mitchilli*, in Chesapeake Bay. *Fishery Bulletin* **94**, 257–267.

Douglas J, Brown P, Hunt T, Rogers M *et al.* (2010) Evaluating relative impacts of recreational fishing harvest and discard mortality on Murray cod (*Maccullochella peelii peelii*). *Fisheries Research* **106**, 18–21.

Douglas JW and Brown P (2000) Notes on successful spawning and recruitment of a stocked population of the endangered Australian freshwater fish, trout cod, *Maccullochella macquariensis* (Cuvier) (Percichthyidae). *Proceedings of the Linnean Society of New South Wales* **122**, 143–147.

Douglas JW, Gooley GJ and Ingram BA (1994) 'Trout cod, *Maccullochella macquariensis* (Cuvier) (Pisces: Percichthyidae). Resource handbook and research and recovery plan'. Department of Conservation and Natural Resources, Melbourne.

Douglas JW, Gooley GJ, Ingram BA, Murray ND *et al.* (1995) Natural hybridisation between Murray cod, *Maccullochella peelii peelii* (Mitchell), and trout cod, *Maccullochella macquariensis* (Cuvier) (Percichthyidae), in the Murray River, Australia. *Marine and Freshwater Research* **46**, 729–734.

Douglas MM, Bunn SE and Davies PM (2005) River and wetland food webs in Australia's wet–dry tropics: general principles and implications for management. *Marine and Freshwater Research* **56**, 329–342.

Doupé RG and Burrows DW (2008) Thirty years later, should we be more concerned for the ongoing invasion of Mozambique tilapia in Australia? *Pacific Conservation Biology* **14**, 236–239.

Doupé RG and Knott MJ (2010) Rapid digestion of fish prey by the highly invasive 'detritivore' *Oreochromis mossambicus*. *Journal of Fish Biology* **76**, 1019–1024.

Doupé RG, Knott MJ, Schaffer J and Burrows DW (2009) Investigational piscivory of some juvenile Australian freshwater fishes by the introduced Mozambique tilapia *Oreochromis mossambicus*. *Journal of Fish Biology* **74**, 2386–2400.

Dove ADM and Ernst I (1998) Concurrent invaders: four exotic species of Monogenea now established on exotic freshwater fishes in Australia. *International Journal of Parasitology* **28**, 1755–1764.

Dove ADM, Cribb TH, Mockler SP and Lintermans M (1997) The Asian fish tapeworm *Bothriocephalus acheilognathi* in Australian freshwater fishes. *Marine and Freshwater Research* **48**, 181–183.

Downes BJ, Barmuta LA, Fairweather PG, Faith DP *et al.* (2002) *Monitoring Ecological Impacts: Concepts and Practice in Flowing Waters.* Cambridge University Press, Cambridge.

Downs CJ, Hayes JP and Tracy CR (2008) Scaling metabolic rate with body mass and inverse body temperature: a test of the Arrhenius fractal supply model. *Functional Ecology* **22**, 239–244.

Drexel JF and Preiss WV (Eds) (1995) *The Geology of South Australia.* Geological Survey of South Australia, Adelaide.

Driver PD, Closs GP and Koen T (2005a) The effects of size and density of carp (*Cyprinus carpio* L.) on water quality in an experimental pond. *Archiv für Hydrobiologie* **163**, 117–131.

Driver PD, Harris JH, Closs GP and Koen TB (2005b) Effects of flow regulation on carp (*Cyprinus carpio* L.) recruitment in the Murray-Darling Basin, Australia. *River Research and Applications* **21**, 237–335.

Duan C (1997) The insulin-like growth factor system and its biological actions in fish. *American Zoologist* **37**, 491–503.

Dudgeon D, Arthington AH *et al.* (2006) Freshwater biodiversity: importance, threats, status and conservation challenges. *Biological Reviews of the Cambridge Philosophical Society* **81**, 163–182.

Duffy GA, Pike TW and Laland KN (2009) Size-dependent directed social learning in nine-spined sticklebacks. *Animal Behaviour* **78**, 371–375.

Dufour S, Prevost E, Rochard E and Williot P (Eds) (2008) *Fish and Diadromy in Europe (Ecology, Management, Conservation).* Springer, Dordrecht.

Dufty S (1986) Genetic and morphological divergence between populations of Macquarie perch (*Macquaria australasica*) east and west of the Great Dividing Range. Hons thesis. University of New South Wales, Sydney.

Dugatkin LA and FitzGerald GJ (1997) Sexual selection. In *Behavioural Ecology of Teleost Fishes.* (Ed. JGJ Godin) pp. 266–291. Oxford University Press, Oxford.

Duguid A, Barnetson J, Clifford B, Pavey C *et al.* (2005) 'Wetlands in the arid Northern Territory'. Department of the Environment and Heritage, Alice Springs.

Dunstan DJ (1959) The barramundi *Lates calcarifer* (Bloch) in Queensland waters. *Technical Paper of the Division of Fisheries, Australia* **5**, 1–22.

Durance I, Lepichon C and Ormerod SJ (2006) Recognizing the importance of scale in the ecology and management of riverine fish. *River Research and Applications* **22**, 1143–1152.

Duyker E (2006) *François Péron: An Impetuous Life: Naturalist and Voyager.* Miegunyah Press, Melbourne.

Eadie JM and Keast A (1984) Resource heterogeneity and fish species-diversity in lakes. *Canadian Journal of Zoology* **62**, 1689–1695.

Ebner B (2006) Murray cod an apex predator in the Murray River, Australia. *Ecology of Freshwater Fish* **15**, 510–520.

Ebner B and Thiem J (2006) Movement of hatchery-reared and wild-reared *Maccullochella macquariensis* stocked in a large lowland river. In 'An ecological approach to re-establishing Australian freshwater cod populations: an application to trout cod in the Murrumbidgee catchment'. Final report to Fisheries Research and Development Corporation, Project 2003/034. (Eds B Ebner, J Thiem, M Lintermans and D Gilligan). Parks, Conservation and Lands, Canberra.

Ebner BC and Thuesen P (2011) Discovery of stream cling-goby assemblages (*Stiphodon* species) in the Australian Wet Tropics. *Australian Journal of Zoology* **58**, 331–340.

Ebner BC, Broadhurst B, Lintermans M and Jekabsons M (2007) A possible false negative: lack of evidence for trout predation on a remnant population of the endangered Macquarie perch, *Macquaria australasica*, in Cotter Reservoir, Australia. *New Zealand Journal of Marine and Freshwater Research* **41**, 231–237.

Ebner BC, Johnston L and Lintermans M (2009a) Radio-tagging and tracking of translocated trout cod (*Maccullochella macquariensis*: Percichthyidae) in an upland river. *Marine and Freshwater Research* **60**, 346–355.

Ebner BC, Lintermans M and Dunford M (2011) A reservoir serves as refuge for adults of the endangered Macquarie perch. *Lakes and Reservoirs: Research and Management* **16**, 23–33.

Ebner BC, Lintermans M, Jekabsons M and Dunford M (2010) Convoluted shorelines confound diel-range estimates of radio-tracked fish. *Marine and Freshwater Research* **61**, 1360–1365.

Ebner BC, Scholz O and Gawne B (2009b) Golden perch *Macquaria ambigua* are flexible spawners in the Darling River, Australia. *New Zealand Journal of Marine and Freshwater Research* **43**, 571–578.

Egerton FN (2001) A history of the ecological sciences: early Greek origins. *Bulletin of the Ecological Society of America*, 93–97.

Egge JJD and Simons AM (2006) The challenge of truly cryptic diversity: diagnosis and description of a new madtom catfish (Ictaluridae: *Noturus*). *Zoologica Scripta* **35**, 581–595.

Ehlers J and Gibbard PL (2007) The extent and chronology of Cenozoic global glaciation. *Quaternary International* **164–165**, 6–20.

Ehrenfeld JG (2010) Ecosystem consequences of biological invasions. *Annual Review of Ecology, Evolution and Systematics* **41**, 59–80.

Eiler JH, Alcorn DJ and Neuman MR (Eds) (2000) *Biotelemetry 15: Proceedings of the 15th International Symposium on Biotelemetry.* Juneau, 10–15 May 1999. International Society on Biotelemetry, Wageningen, Netherlands.

Elder JF Jr and Schlosser IJ (1995) Extreme clonal uniformity of *Phoxinus eos/Neogaeus gynogens* (Pisces: Cyprinidae) among variable habitats in northern Minnesota beaver ponds. *Proceedings of the National Academy of Sciences of the United States of America* **92**, 5001–5005.

Elliot JM (1976) Body composition of brown trout (*Salmo trutta* L.) in relation to temperature and ration size. *Journal of Animal Ecology* **44**, 805–821.

Elliott JM (1966) Downstream movements of trout fry (*Salmo trutta*) in a Dartmoor stream. *Journal of the Fisheries Research Board of Canada* **23**, 157–159.

Ellis I and Pyke L (2010) 'Conservation of Murray hardyhead *Craterocephalus fluviatilis* in Victoria: status of population monitoring, translocation and captive breeding programs'. Report prepared for the Mallee Catchment Management Authority. Murray-Darling Freshwater Research Centre, Albury.

Elser JJ, Sterner RW *et al.* (2000) Biological stoichiometry from genes to ecosystems. *Ecology Letters* **3**, 540–550.

Endler JA (1986) *Natural Selection in the Wild.* Princeton University Press, Princeton.

Engelmann J, Hanke W and Bleckmann H (2002) Lateral line reception in still and running-water. *Journal of Comparative Physiology A – Neuroethology Sensory Neural and Behavioral Physiology* **188**, 513–526.

Engström-Öst J, Karjalainen M and Viitasalo M (2006) Feeding and refuge use by small fish in the presence of cyanobacteria blooms. *Environmental Biology of Fishes* **76**, 109–117.

Eros T (2005) Life-history diversification in the Middle Danubian fish fauna: a conservation perspective. *Archiv für Hydrobiologie, Supplement on Large Rivers* **16**, 289–304.

Erskine WD and Harris JH (2004) Australia. In *Fishes and Forestry: Worldwide Watershed Interactions and Management.* (Eds TG Northcote and GF Hartman) pp. 678–703. Blackwell Science, Oxford.

Erskine WD and Webb AA (2003) Desnagging to resnagging: new directions in river rehabilitation in southeastern Australia. *River Research and Applications* **19**, 233–249.

Eschmeyer WN (2012) Catalog of fishes, online version (updated 12 January 2012). California Academy of Sciences. <http://research.calacademy.org/research/ichthyology/catalog>. January 2012.

Eschmeyer WN and Fong JD (2012) Species of fishes by family/subfamily. <http://research.calacademy.org/research/ichthyology/catalog/SpeciesByFamily.asp>. October 2010.

Eschmeyer WN, Fricke R, Fong JD and Polack DA (2010) Marine fish diversity: history of knowledge and discovery (Pisces). *Zootaxa* **2525**, 19–50.

ESD Steering Committee (1992) *National Strategy for Ecologically Sustainable Development.* Australian Government Publishing Service, Canberra.

Estes JA, Terborgh J *et al.* (2011) Trophic downgrading of planet Earth. *Science* **333**, 301–306.

Evans JP, Box TM, Brooshooft P, Tatler JR *et al.* (2010) Females increase egg deposition in favor of large males in the rainbowfish, *Melanotaenia australis. Behavioral Ecology* **21**, 465–469.

Everett RA and Ruiz GM (1993) Coarse woody debris as a refuge from predation in aquatic communities: an experimental test. *Oecologia* **93**, 475–486.

Everhart WH, Eipper AW and Youngs WD (1975) *Principles of Fishery Science.* Cornell University Press, London.

Excoffier L and Heckel G (2006) Computer programs for population genetics data analysis: a survival guide. *Nature Reviews Genetics* 7, 745–758.

Facey DE and Grossman GD (1991) The relationship between water velocity, energetic costs, and microhabitat use in four North American stream fishes. *Hydrobiologia* **200**, 1–6.

Fagan MB (2007) Wallace, Darwin, and the practice of natural history. *Journal of the History of Biology* **40**, 601–635.

Fagan WF (2002) Connectivity, fragmentation, and extinction risk in dendritic metapopulations. *Ecology* **83**, 3243–3249.

Falcón J, Migaud H, Muñoz-Cueto JA and Carrillo M (2010) Current knowledge on the melatonin system in teleost fish. *General and Comparative Endocrinology* **165**, 469–482.

Falke J and Fausch KD (2010) From metapopulations to metacommunities in stream fish: linking theory with empirical observations of the spatial dynamics of stream fishes. In *Community Ecology of Stream Fishes: Concepts, Approaches, and Techniques, Symposium 73.* (Eds KB Gido and DA Jackson) pp. 207–233. American Fisheries Society, Bethesda.

Faragher R and Gordon G (1992) Comparative exploitation exploitation by recreational anglers of brown trout, *Salmo trutta* L., and rainbow trout, *Oncorhynchus mykiss* (Walbaum), in Lake Eucumbene, New South Wales. *Marine and Freshwater Research* **43**, 835–845.

Faragher RA, Pogonoski JJ, Cameron L, Baumgartner L *et al.* (2007) 'Assessment of a stocking program: findings and recommendations for the Snowy Lakes trout strategy'. Department of Primary Industries, Sydney.

Farrington LW, Austin CM, Burridge CP, Gooley GJ *et al.* (2004) Allozyme diversity in Australian rainbow trout, *Oncorhynchus mykiss. Fisheries Management and Ecology* **11**, 97–106.

Faulks LK, Gilligan DM and Beheregaray LB (2008) Phylogeography of a threatened freshwater fish (*Mogurnda adspersa*) in eastern Australia: conservation implications. *Marine and Freshwater Research* **59**, 89–96.

Faulks LK, Gilligan DM and Beheregaray LB (2010a) Evolution and maintenance of divergent lineages in an endangered freshwater fish, *Macquaria australasica. Conservation Genetics* **11**, 921–934.

Faulks LK, Gilligan DM and Beheregaray LB (2010b) Clarifying an ambiguous evolutionary history: range-wide phylogeography of an Australian freshwater fish, the golden perch (*Macquaria ambigua*). *Journal of Biogeography* **37**, 1329–1340.

Faulks LK, Gilligan DM and Beheregaray LB (2010c) Islands of water in a sea of dry land: hydrological regime predicts genetic diversity and dispersal in a widespread fish from Australia's arid zone, the golden perch (*Macquaria ambigua*). *Molecular Ecology* **19**, 4723–4737.

Faulks LK, Gilligan DM and Beheregaray LB (2011) The role of anthropogenic vs. natural in-stream structures in determining connectivity and genetic diversity in an endangered freshwater fish, Macquarie perch (*Macquaria australasica*). *Evolutionary Applications* **19**, 4723–4737.

Fausch KD (1984) Profitable stream positions for salmonids: relating specific growth rate to net energy gain. *Canadian Journal of Zoology* **62**, 441–451.

Fausch KD (1993) Experimental analysis of microhabitat selection by juvenile steelhead (*Oncorhynchus mykiss*) and coho salmon (*O. kisutch*) in a British Columbia stream. *Canadian Journal of Fisheries and Aquatic Sciences* **50**, 1198–1207.

Fausch KD, Taniguchi Y, Nakano S, Grossman GD *et al.* (2001) Flood disturbance regimes influence rainbow trout invasion success among five holarctic regions. *Ecological Applications* **11**, 1438–1455.

Fausch KD, Torgersen CE, Baxter CE and Li HW (2002) Landscapes to riverscapes: bridging the gap between research and conservation of stream fishes. *BioScience* **52**, 483–498.

Fellows CS, Bunn SE, Sheldon F and Beard NJ (2009) Benthic metabolism in two turbid dryland rivers. *Freshwater Biology* **54**, 236–253.

Fensham RJ, Ponder WF and Fairfax RJ (2010) 'Recovery plan for the community of native species dependent on natural discharge of groundwater from the Great Artesian Basin'. Report to the Department of Environment, Water, Heritage and the Arts, Canberra. Queensland Department of Environment and Resources Management, Brisbane.

Ferguson A (2004) The importance of identifying conservation units: brown trout and pollan biodiversity in Ireland. *Biology and Environment: Proceedings of the Royal Irish Academy* **104B**, 33–41.

Ferron A and Leggett WC (1994) An appraisal of condition measures for marine fish larvae. *Advances in Marine Biology* **30**, 217–303.

Field IC, Charters R, Buckworth RC, Meekan MG *et al.* (2008) 'Distribution and abundance of *Glyphis* and sawfishes in northern Australia and their potential interactions with commercial fisheries'. Department of the Environment, Water, Heritage and the Arts, Canberra.

Figuerola J, Green AJ and Michot TC (2005) Invertebrate eggs can fly: evidence of waterfowl mediated gene-flow in aquatic invertebrates. *American Naturalist* **165**, 274–280.

Figuerola J, Green AJ and Santamaria L (2003) Passive internal transport of aquatic organisms by waterfowl in Doñana, south-west Spain. *Global Ecology and Biogeography* **12**, 427–436.

Fiksen Ŏ, Aksnes DL, Flyum MH and Giske J (2002) The influence of turbidity on growth and survival of fish larvae: a numerical analysis. *Hydrobiologia* **48**, 449–59.

Finney CM (1984) *To Sail Beyond the Sunset: Natural History in Australia 1699–1829.* Rigby Publishers, Adelaide.

Finney CM (1993) *Paradise Revealed: Natural History in Nineteenth Century Australia.* Museum of Victoria, Melbourne.

Finstad AG, Ugedal O and Berg OK (2006) Growing large in a low grade environment: size dependent foraging gain and niche shifts to cannibalism in Arctic char. *Oikos* **112**, 73–82.

Firestone KB, Elphinstone MS, Sherwin WB and Houlden BA (1999) Phylogeographical population structure of tiger quolls *Dasyurus maculatus* (Dasyuridae: Marsupialia), an endangered carnivorous marsupial. *Molecular Ecology* **8**, 1613–1625.

Fishelson L, Delarea Y and Zverdling A (2004) Taste bud form and distribution on lips and in the oropharyngeal cavity of cardinal fish species (Apogonidae, Teleostei), with remarks on their dentition. *Journal of Morphology* **259**, 316–327.

Fisher HS, Wong BBM and Rosenthal GG (2006) Alteration of the chemical environment disrupts communication in a freshwater fish. *Proceedings of the Royal Society of London B* **273**, 1187–1193.

Fleming IA (1996) Reproductive strategies of Atlantic salmon: ecology and evolution. *Reviews in Fish Biology and Fisheries* **6**, 379–416.

Fletcher A and Whittington ID (1998) A parasite–host checklist for Monogenea from freshwater fishes in Australia, with comments on biodiversity. *Systematic Parasitology* **41**, 159–168.

Fletcher AR (1979) Effects of *Salmo trutta* on *Galaxias olidus* and macroinvertebrates in stream communities. Masters thesis. Monash University, Melbourne.

Fletcher AR, Morison AK and Hume DJ (1985) Effects of carp, *Cyprinus carpio* L., on communities of aquatic vegetation and turbidity of waterbodies in the lower Goulburn River Basin. *Australian Journal of Marine and Freshwater Research* **36**, 311–327.

Floyd KB, Hoyt RD and Timbrook S (1984) Chronology of appearance and habitat partitioning by stream larval fishes. *Transactions of the American Fisheries Society* **113**, 217–223.

Forbes SA (1878) The food of Illinois fishes. *Bulletin of the Illinois State Laboratory of Natural History* **1**, 71–89.

Forbes SA (1887) The lake as a microcosm. *Bulletin of the Scientific Association* (Peoria, Illinois), **1887**, 77–87.

Ford MJ (2002) Selection in captivity during supportive breeding may reduce fitness in the wild. *Conservation Biology* **16**, 815–825.

Fortier L and Leggett WC (1985) A drift of larval fish survival. *Marine Ecology Progress Series* **25**, 245–257.

Fournier DA, Sibert JR, Majkowski J and Hampton J (1990) MULTIFAN: a likelihood-based method for estimating growth parameters and age composition from multiple length frequency data sets illustrated using data for southern bluefin tuna (*Thunnus maccoyii*). *Canadian Journal of Fisheries and Aquatic Sciences* **47**, 301–317.

Frakes LA, Burger D *et al.* (1987a) Australian Cretaceous shorelines, stage by stage. *Palaeogeography, Palaeoclimatology, Palaeoecology* **59**, 31–48.

Frakes LA, McGowran B and Bowler JM (1987b) Evolution of Australian environments. In *Fauna of Australia*, Vol. 1a. (Eds GR Dyne and DW Walton) pp. 1–16. Australian Government Publishing Service, Canberra.

Frances J, Allan GL and Nowak BF (1998) The effects of nitrite on the short-term growth of silver perch (*Bidyanus bidyanus*). *Aquaculture* **163**, 63–72.

Frances J, Nowak BF and Allan GL (2000) Effects of ammonia on juvenile silver perch (*Bidyanus bidyanus*). *Aquaculture* **183**, 95–103.

Frances J, Ogburn D, Fielder S and Kirkland P (2005) Investigations of viral nervous necrosis: the Australian bass outbreak. In *FRDC Aquatic Animal Health Subprogram Scientific Conference.* Fisheries Research and Development Corporation, Cairns.

Francis DS, Turchini GM, Jones PL and De Silva SS (2006) Effects of dietary oil source on growth and fillet fatty acid composition of Murray cod, *Maccullochella peelii peelii*. *Aquaculture* **253**, 547–556.

Francis GF (1862) *The Acclimatisation of Harmless, Useful, Interesting, and Ornamental, Animals and Plants, being Read before The Philosophical Society, Adelaide, South Australia, on May 13th, 1862.* Philosophical Society, Adelaide.

Francis RICC (1988a) Are growth-parameters estimated from tagging and age length data comparable? *Canadian Journal of Fisheries and Aquatic Sciences* **45**, 936–942.

Francis RICC (1988b) Maximum likelihood estimation of growth and growth variability from tagging data. *New*

Zealand Journal of Marine and Freshwater Research* **22**, 43–51.

Francis RICC (1990) Back-calculation of fish length: a critical review. *Journal of Fish Biology* **36**, 883–902.

Francis RICC (1995) An alternative mark–recapture analog of Schnute's growth model. *Fisheries Research* **23**, 95–111.

Frankham R, Ballou JD and Briscoe DA (2010) *Introduction to Conservation Genetics.* 2nd edn. Cambridge University Press, London.

Frentiu FD, Ovenden JR and Street R (2001) Australian lungfish (*Neoceratodus forsteri*: Dipnoi) have low genetic variation at allozyme and mitochondrial DNA loci: a conservation alert? *Conservation Genetics* **2**, 63–67.

Fretwell SD and Lucas HL (1970) On territorial behaviour and other factors influencing habitat distribution in birds. I. Theoretical development. *Acta Biotheoretica* **19**, 16–36.

Friedland KD, Haas RE and Sheehan TF (1996) Post-smolt growth, maturation, and survival of two stocks of Atlantic salmon. *Fishery Bulletin* **94**, 654–663.

Frissell CA, Liss WJ, Warren CE and Hurley MD (1986) A hierarchical framework for stream habitat classification: viewing streams in a watershed context. *Environmental Management* **10**, 199–214.

Froese R and Pauly D (2010) FishBase. <http://www.fishbase.org>. February 2012.

Fromm P (1980) A review of some physiological and toxicological responses of freshwater fish to acid stress. *Environmental Biology of Fishes* **5**, 79–93.

Fryirs K, Hillman M and Spink A (2007) Challenges faced in the integration of science in river management in Australia. In *Proceedings of the 5th Australian Stream Management Conference. Australian Rivers: Making a Difference.* (Eds AL Wilson, RL Dehaan, RJ Watts, KJ Page, KH Bowmer and A Curtis). Charles Sturt University, Thurgoona, New South Wales.

Fu SJ and Xie XJ (2004) Nutritional homeostasis in carnivorous southern catfish (*Silurus meridionalis*): is there a mechanism for increased energy expenditure during carbohydrate overfeeding? *Comparative Biochemistry and Physiology A: Molecular and Integrative Physiology* **139**, 359–363.

Fu SJ, Zeng LQ, Li XM, Pang X *et al.* (2009) The behavioural, digestive and metabolic characteristics of fishes with different foraging strategies. *Journal of Experimental Biology* **212**, 2296–2302.

Fuiman LA and Werner RG (2002) *Fishery Science: The Unique Contributions of Early Life Stages.* Blackwell Science, Oxford.

Fulton W (1982) Observations on the ecology of four species of the genus *Paragalaxias* (Pisces: Galaxiidae) from Tasmania. *Australian Journal of Marine and Freshwater Research* **33**, 999–1016.

Fulton W (1990) *Tasmanian Freshwater Fishes.* University of Tasmania, Hobart.

Fulton W and Pavuk N (1988) 'The Tasmanian whitebait fishery: summary of present knowledge and outline of future management plans'. Inland Fisheries Commission, Hobart.

Funk DJ and Omland KE (2003) Species-level paraphyly and polyphyly: frequency, causes and consequences, with insights from animal mitochondrial DNA. *Annual Review of Ecology, Evolution and Systematics* **34**, 397–423.

Funk JL (1957) Movement of stream fishes in Missouri. *Transactions of the American Fisheries Society* **85**, 39–57.

Futuyama DJ (2005) *Evolution.* Sinauer Associates, Massachusetts.

Gadomski D and Parsley M (2005) Effects of turbidity, light level, and cover on predation of white sturgeon larvae by prickly sculpins. *Transactions of the American Fisheries Society* **134**, 369–374.

Gale SJ (1992) Long-term landscape evolution in Australia. *Earth Surface Processes and Landforms* **17**, 323–343.

Gallagher T and Hutchison M (2004) Effects of tag location on detection rates of Visible Implanted Elastomer (VIE) tags in juvenile silver perch, *Bidyanus bidyanus* (Mitchell) and Australian bass, *Macquaria novemaculeata* (Steindachner). *Proceedings of the Royal Society of Queensland* **111**, 13–18.

Gamito S (1998) Growth models and their use in ecological modelling: an application to a fish population. *Ecological Modelling* **113**, 83–94.

Gangl SR, Periera DL and Walsh RJ (2000) Seasonal movements, habitat use, and spawning areas of walleye *Stizostedon vitreum* and sauger *S. canadense* in Pool 2 of the upper Mississippi River. In 'Section of Fisheries Investigation Report 482'. Department of Natural Resources, Division of Fish and Wildlife, Minnesota.

Garrett RN, MacKinnon MR and Russell DJ (1987) Wild barramundi breeding and its implications for culture. *Australian Fisheries Magazine*, 4–6.

Gaston KJ and Blackburn TM (1996) Conservation implications of geographic range size–body size relationship. *Conservation Biology* **10**, 638–646.

Gaughan DJ and Potter IC (1997) Analysis of diet and feeding strategies within an assemblage of estuarine larval fish and an objective assessment of dietary niche overlap. *Fishery Bulletin* **95**, 722–731.

Gauldie RW (1996) Effects of temperature and vaterite replacement on the chemistry of metal ions in the otoliths of *Oncorhynchus tshawytscha. Canadian Journal of Fisheries and Aquatic Sciences* **53**, 2015–2026.

Gehrke P (1991) Avoidance of inundated floodplain habitat by larvae of golden perch (*Macquaria ambigua* Richardson): influence of water quality or food distribution? *Australian Journal of Marine and Freshwater Research* **42**, 707–719.

Gehrke PC (1988) Feeding energetics and angling catches of spangled perch, *Leiopotherapon unicolor* (Günther 1859) (Percoidei, Teraponidae). *Australian Journal of Marine and Freshwater Research* **39**, 569–577.

Gehrke PC (1992) Diel abundance, migration and feeding of fish larvae (Eleotridae) in a floodplain billabong. *Journal of Fish Biology* **40**, 695–707.

Gehrke PC (1994) Influence of light intensity and wavelength on phototactic behaviour of larval silver perch *Bidyanus bidyanus* and golden perch *Macquaria ambigua* and the effectiveness of light traps. *Journal of Fish Biology* **44**, 741–751.

Gehrke PC and Fielder DR (1988) Effects of temperature and dissolved oxygen on heart rate, ventilation rate and oxygen consumption of spangled perch, *Leiopotherapon unicolor* (Günther 1859) (Percoidei, Teraponidae). *Journal of Comparative Physiology B: Biochemical, Systemic, and Environmental Physiology* **157**, 771–782.

Gehrke PC and Harris JH (1994) The role of fish in cyanobacterial blooms in Australia. *Australian Journal of Marine and Freshwater Research* **45**, 175–186.

Gehrke PC and Harris JH (2000) Large-scale patterns in species richness and composition of temperate riverine fish communities, south-eastern Australia. *Marine and Freshwater Research* **51**, 165–182.

Gehrke PC and Harris JH (2001) Regional-scale effects of flow regulation on lowland riverine fish communities in New South Wales, Australia. *River Research and Applications* **17**, 369–391.

Gehrke PC, Astles KL and Harris JH (1999) Within-catchment effects of flow alteration on fish assemblages in the Hawkesbury-Nepean river system, Australia. *Regulated Rivers: Research and Management* **15**, 181–198.

Gehrke PC, Brown P, Schiller CB, Moffatt DB *et al.* (1995) River regulation and fish communities in the Murray-Darling river system, Australia. *Regulated Rivers: Research and Management* **11**, 363–375.

Gehrke PC, Clarke M, Matveev V, St Pierre S *et al.* (2011) Carp control improves the health of aquatic ecosystems. *Water (Journal of the Australian Water Association)* **35**, 91–95.

Gehrke PC, Gilligan DM and Barwick M (2002) Changes in fish communities of the Shoalhaven River 20 years after construction of Tallowa Dam, Australia. *River Research and Applications* **18**, 265–286.

Gehrke PC, St Pierre S, Matveev V and Clarke M (2010) 'Ecosystem responses to carp population reduction in the Murray-Darling Basin'. SMEC Australia, Brisbane.

Georges A and Cottingham P (Eds) (2002) 'Biodiversity in inland waters: priorities for its protection and management'. Technical Report 1/2002. Cooperative Research Centre for Freshwater Ecology, Canberra.

Gerking SD (1949) Characteristics of stream fish populations. *Investigations of Indiana Lakes and Streams* **3**, 283–309.

Gerking SD (1950) Stability of stream fish populations. *Journal of Wildlife Management* **14**, 193–202.

Gerking SD (1953) Evidence for the concepts of home range and territory in stream fishes. *Ecology* **34**, 347–364.

Gerking SD (1959) The restricted movements of fish populations. *Biological Reviews* **34**, 221–242.

Gerking SD (1978) *Ecology of Fresh Water Fish Production*. Wiley, New York.

Gido KB and Jackson DA (2010) *Community Ecology of Stream Fishes: Concepts, Approaches, and Techniques*, Symposium 73. American Fisheries Society, Bethesda.

Gilk SE, Wang IA, Hoover CL, Smoker WW *et al.* (2004) Outbreeding depression in hybrids between spatially separated pink salmon, *Oncorhynchus gorbuscha*, populations: marine survival, homing ability, and variability in family size. *Environmental Biology of Fishes* **69**, 287–297.

Gill HS and Morgan DL (1998) Larval development of *Nannatherina balstoni* Regan (Nannopercidae), with a description of ontogenetic changes in diet. *Ecology of Freshwater Fish* **7**, 132–139.

Gill HS and Morgan DL (1999) Larval development of the salamanderfish, *Lepidogalaxias salamandroides* Mees (Lepidogalaxiidae). *Copeia* **1999**, 219–224.

Gill HS and Morgan DL (2003) Ontogenetic changes in the diet of the black-stripe minnow *Galaxiella nigrostriata* (Shipway, 1953) (Galaxiidae) and the salamanderfish *Lepidogalaxias salamandroides* (Mees, 1961) (Lepidogalaxiidae). *Ecology of Freshwater Fish* **12**, 151–158.

Gillanders BM (2005) Otolith chemistry to determine movements of diadromous and freshwater fish. *Aquatic Living Resources* **18**, 291–300.

Gillanders BM, Elsdon TS and Munro AR (2006) 'Impacts of native fish stocking on fish within the Murray-Darling Basin'. University of Adelaide, Adelaide.

Gilligan D (2005) 'Fish communities of the Murrumbidgee Catchment: status and trends'. Department of Primary Industries, Sydney.

Gilligan D and Rayner T (2007) 'The distribution, spread, ecological impacts and potential control of carp in the upper Murray River'. Department of Primary Industry, Sydney.

Gilligan D and Schiller C (2003) 'Downstream transport of larval and juvenile fish in the Murray River'. NSW Fisheries Final Report 50. NSW Fisheries, Narrandera, New South Wales.

Gillooly JF, Brown JH, West GB, Savage VM *et al.* (2001) Effects of size and temperature on metabolic rate. *Science* **293**, 2248–2251.

Gilmore M (1934) *Old Days, Old Ways*. Angus and Robertson, Sydney.

Gippel CJ and Stewardson MJ (1998) Use of wetted perimeter in defining minimum environmental flows. *Regulated Rivers: Research and Management* **14**, 53–67.

Gisbert E, Conklin DB and Piedrahita RH (2004) Effects of delayed first feeding on the nutritional condition and mortality of California halibut larvae. *Journal of Fish Biology* **64**, 116–132.

Glenn TC (2011) Field guide to next-generation DNA sequencers. *Molecular Ecology Resources* **11**, 759–769.

Global Invasive Species Database (2004) 100 of the world's worst invasive alien species. Global invasive species database, International Union for the Conservation of Nature.

Glover CJM (1979) Studies on central Australian fishes: further observations and records, Part 1. *South Australian Naturalist* **53**, 58–62.

Glover CJM (1982) Adaptations of fishes in arid Australia. In *Evolution of the Flora and Fauna of Arid Australia.* (Eds WR Barker and PJM Greenslade) pp. 241–246. Peacock Publications, Adelaide.

Glover CJM (1990) Fishes. In *Natural History of the North-East Deserts.* (Eds MJ Tyler, CR Twidale, M Davies and CB Wells) pp. 189–198. Royal Society of South Australia, Adelaide.

Glover CJM and Sim TC (1978a) Studies on central Australian fish: a progress report. *South Australian Naturalist* **52**, 35–44.

Glover CJM and Sim TC (1978b) A survey of central Australian ichthyology. *Australian Zoologist* **19**, 245–256.

Glozier NE, Culp JM and Scrimgeour GJ (1997) Transferability of habitat suitability curves for a benthic minnow, *Rhinichthys cataractae. Journal of Freshwater Ecology* **12**, 379–393.

Go J and Whittington R (2006) Experimental transmission and virulence of a megalocytivirus (family Iridoviridae) of dwarf gourami (*Colisa lalia*) from Asia in Murray cod (*Maccullochella peelii peelii*) in Australia. *Aquaculture* **258**, 140–149.

Go J, Lancaster M, Deece K, Dhungyel O *et al.* (2006) The molecular epidemiology of iridovirus in Murray cod (*Maccullochella peelii peelii*) and dwarf gourami (*Colisa lalia*) from distant biogeographical regions suggests a link between trade in ornamental fish and emerging iridoviral diseases. *Molecular and Cellular Probes* **20**, 212–222.

Gomahr A, Palzenberger M and Kotraschal K (1992) Density and distribution of external taste-buds in cyprinids. *Environmental Biology of Fishes* **33**, 125–134.

Gooley GJ (1992) Validation of the use of otoliths to determine the age and growth of Murray cod, *Maccullochella peelii* (Mitchell) (Percichthyidae), in Lake Charlegrark, Western Victoria. *Australian Journal of Marine and Freshwater Research* **43**, 1091–1102.

Gooley GJ and McDonald GL (1988) 'Preliminary study on the hormone-induced spawning of Macquarie perch, *Macquaria australasica* (Cuvier) (Percichthyidae), from Lake Dartmouth, Victoria'. Arthur Rylah Institute for Environmental Research, Department of Sustainability and Environment, Melbourne.

Gordon ND, McMahon TA, Finlayson BL, Gippel CJ *et al.* (2004) *Stream Hydrology. An Introduction for Ecologists*, 2nd edn. Wiley, Chichester.

Gore JA and Shields FD Jr (1995) Can large rivers be restored? *BioScience* **45**, 142–152.

Gorodilov YN (1996) Description of the early ontogeny of the Atlantic salmon, *Salmo salar*, with a novel system of interval (state) identification. *Environmental Biology of Fishes* **47**, 109–127.

Gotceitas V, Puvanendran V, Leader LL and Brown JA (1996) An experimental investigation of the 'match/mismatch' hypothesis using larval Atlantic cod. *Marine Ecology Progress Series* **130**, 29–37.

Goudswaard KPC, Witte F and Katunzi EFB (2008) The invasion of an introduced predator, Nile perch (*Lates niloticus* L.) in Lake Victoria (East Africa): chronology and causes. *Environmental Biology of Fishes* **81**, 127–139.

Goulding M (1981a) *Fishes and the Forest: Explorations in Amazonian Natural History.* University of California Press, California.

Goulding M (1981b) *Man and Fisheries on an Amazon Frontier.* Dr W Junk, The Hague.

Gowan C and Fausch KD (2002) Why do foraging stream salmonids move during summer? *Environmental Biology of Fishes* **64**, 153.

Gowan C, Young MK, Fausch KD and Riley SC (1994) Restricted movement in resident stream salmonids: a paradigm lost? *Canadian Journal of Fisheries and Aquatic Sciences* **51**, 2626–2637.

Gozlan RE, Britton JR, Cowx I and Copp GH (2010) Current knowledge on non-native freshwater fish introductions. *Journal of Fish Biology* **76**, 751–786.

Graham R and Harris JH (2005) 'Floodplain inundation and fish dynamics in the Murray-Darling Basin: current concepts and future research'. Scoping study. Cooperative Research Centre for Freshwater Ecology, Canberra.

Granek EF, Madin EMP *et al.* (2008) Engaging recreational fishers in management and conservation: global case studies. *Conservation Biology* **22**, 1125–1134.

Grant JWA (1997) Territoriality. In *Behavioural Ecology of Teleost Fishes.* (Ed. J-GJ Godin) pp. 81–103. Oxford University Press, Oxford.

Graynoth E and Taylor MJ (2000) Influence of different rations and water temperatures on the growth rates of shortfinned eels and longfinned eels. *Journal of Fish Biology* **57**, 681–699.

Grayson DK (2001) The archaeological record of human impacts on animal populations. *Journal of World Prehistory* **15**, 1–68.

Greek-Walker M and Pull GA (1975) A survey of red and white muscle in marine fish. *Journal of Fish Biology* **7**, 295–300.

Green K (2008) Fragmented distribution of a rock climbing fish, the mountain galaxias (*Galaxias olidus*), in the Snowy Mountains. *Proceedings of the Linnaean Society of New South Wales* **129**, 175–182.

Greenslade PJM (1983) Adversity selection and the habitat templet. *American Naturalist* **122**, 352–365.

Greiner R and Gregg D (2008) 'Tilapia in north Queensland waterways: risks and potential economic impacts'. Report prepared for the Australian Centre for Tropical Freshwater Research. River Consulting, Townsville.

Gresswell RE and Hendricks SR (2007) Population-scale movement of coastal cutthroat trout in a naturally isolated stream network. *Transactions of the American Fisheries Society* **136**, 238–253.

Griffin RK (1987) Life history, distribution and seasonal migration of barramundi in the Daly River, Northern Territory, Australia. *American Fisheries Society Symposium* **1**, 358–363.

Griffin RK (1993) 'The recreational fishery for barramundi (*Lates calcarifer*) in the Mary River, Northern Territory 1986–1992'. Fishery Report 30. Fisheries Division, Northern Territory Department of Primary Industry and Fisheries, Darwin.

Griffiths SP (1999) Consequences of artificially opening coastal lagoons on their fish assemblages. *International Journal of Salt Lake Research* **8**, 307–327.

Griffiths T (1997) *Ecology and Empire: Environmental History of Settler Societies*. Keele University Press, Edinburgh.

Grigg GC (1965) Studies on the Queensland lungfish, *Neoceratodus forsteri* (Krefft). 3. Aerial respiration in relation to habits. *Australian Journal of Zoology* **13**, 413–422.

Grime JP (1977) Evidence for the existence of three primary strategies in plants and its relevance to ecological and evolutionary theory. *American Naturalist* **111**, 1169–1194.

Grimes CB and Isely JJ (1996) Influence of size-selective mortality on growth of gulf menhaden and king mackerel larvae. *Transactions of the American Fisheries Society* **125**, 741–752.

Grimes CB and Kingsford MJ (1996) How do riverine plumes of different sizes influence fish larvae: do they enhance recruitment? *Marine and Freshwater Research* **47**, 191–208.

Groshens TP and Orth DJ (1994) Transferability of habitat suitability criteria for smallmouth bass, *Micropterus dolomieu*. *Rivers* **4**, 194–212.

Gross MR, Coleman RM and McDowall RM (1988) Aquatic productivity and the evolution of diadromous fish migration. *Science* **239**, 1291–1293.

Grossman GD and Freeman MC (1987) Microhabitat use in a stream fish assemblage. *Journal of Zoology* **212**, 151–176.

Grossman GD and Sabo JL (2010) Preface: structure and dynamics of stream fish assemblages. In *Community Ecology of Stream Fishes: Concepts, Approaches, and Techniques, Symposium 73*. (Eds KB Gido and DA Jackson) pp. 401–405. American Fisheries Society, Bethesda.

Grossman GD, Ratajczak RE Jr, Farr MD, Wagner CM *et al.* (2010) Why there are fewer fish upstream? In *Community Ecology of Stream Fishes: Concepts, Approaches, and Techniques, Symposium 73*. (Eds KB Gido and DA Jackson) pp. 63–81. American Fisheries Society, Bethesda.

Grossman GD, Ratajczak RE, Crawford M and Freeman MC (1998) Assemblage organization in stream fishes: effects of environmental variation and interspecific interactions. *Ecological Monographs* **68**, 395–420.

Growns I (2004) A numerical classification of reproductive guilds of the freshwater fishes of south-eastern Australia and their application to river management. *Fisheries Management and Ecology* **11**, 369–377.

Growns I and West G (2008) Classification of aquatic bioregions through the use of distributional modelling of freshwater fish. *Ecological Modelling* **217**, 79–86.

Growns IO, Pollard DA and Gehrke PC (1998) Changes in river fish assemblages associated with vegetated and degraded banks, upstream of and within nutrient-enriched zones. *Fisheries Management and Ecology* **5**, 55–69.

Gulland JA (1977) *Fish Population Dynamics*. Wiley, New York.

Gunasekera RM, De Silva SS and Ingram BA (1999) Early ontogeny-related changes of the fatty acid composition in the percichthyid fishes trout cod, *Maccullochella macquariensis* and Murray cod, *M. peelii peelii*. *Aquatic Living Resources* **12**, 219–227.

Gunasekera RM, De Silva SS, Collins RA, Gooley G *et al.* (2000) Effect of dietary protein level on growth and food utilization in juvenile Murray cod *Maccullochella peelii peelii* (Mitchell). *Aquaculture Research* **31**, 181–187.

Günther ACLG (1880) *An Introduction to the Study of Fishes*. Adam and Charles Black, Edinburgh.

Guppy M and Withers PC (1999) Metabolic depression in animals: physiological perspectives and biochemical generalizations. *Biological Reviews of the Cambridge Philosophical Society* **74**, 1–40.

Gustafson RG, Waples RS, Myers JM, Weitkamp LA *et al.* (2006) Pacific salmon extinctions: quantifying lost and remaining diversity. *Conservation Biology* **21**, 1009–1020.

Guy CS, Blankenship HL and Nielsen LA (1996) Tagging and marking. In *Fisheries Techniques*. (Eds BR Murphy and DW Willis) pp. 353–383. American Fisheries Society, Bethesda.

Habermehl MA (1982) 'Springs in the Great Artesian Basin, Australia: their origin and nature'. Report 235. Bureau of Mineral Resources, Geology and Geophysics, Canberra.

Haddon M (2001) *Modelling and Quantitative Methods in Fisheries*. Chapman Hall/CRC, New York.

Hadwen WL and Arthington AH (2007) Food webs of two intermittently open estuaries receiving N-15-enriched sewage effluent. *Estuarine Coastal and Shelf Science* **71**, 347–358.

Hadwen WL, Russell GL and Arthington AH (2007) Gut content- and stable isotope-derived diets of four commercially and recreationally important fish species in two intermittently open estuaries. *Marine and Freshwater Research* **58**, 363–375.

Haeckel E (1876) *History of Creation*. Appleton, London.

Hairsine PB, Croke JC, Mathews H, Fogarty P *et al.* (2002) Modelling plumes of overland flow from roads and logging tracks. *Hydrological Processes* **16**, 2311–2327.

Hale R and Swearer SE (2008) Otolith microstructural and microchemical changes associated with settlement in the diadromous fish *Galaxias maculatus*. *Marine Ecology Progress Series* **354**, 229–234.

Hale R, Downes BJ and Swearer SE (2008) Habitat selection as a source of inter-specific differences in recruitment of two diadromous fish species. *Freshwater Biology* **53**, 2145–2157.

Hale R, Swearer SE and Downes BJ (2009) Is settlement at small spatial scales by diadromous fishes from the family Galaxiidae passive or active in a small coastal river? *Marine and Freshwater Research* **60**, 971–975.

Hall A, Higham J, Hammer M, Bice C *et al.* (2009a) 'Draft drought action plan for South Australian Murray-Darling Basin threatened freshwater fish populations'. South Australian Department for Environment and Heritage, Adelaide.

Hall DN and Tunbridge BR (1988) Distribution of native and introduced freshwater fishes in the Barwon River and its upper tributaries, Victoria. *Proceedings of the Royal Society of Victoria* **100**, 61–65.

Hall DN, Harrington DJ and Fairbrother PS (1990) 'The commercial eels fishery in Victoria'. Arthur Rylah Institute for Environmental Research Technical Report Series 100. Fisheries Division, Ministry for Conservation, Melbourne.

Hall KC, Broadhurst MK, Butcher PA and Rowland SJ (2009b) Effects of angling on post-release mortality, gonadal development and somatic condition of Australian bass *Macquaria novemaculeata*. *Journal of Fish Biology* **75**, 2737–2755.

Halse SA, Ruprecht JK and Pinder AM (2003) Salinisation and prospects for biodiversity in rivers and wetlands of south-west Western Australia. *Australian Journal of Botany* **51**, 673–688.

Halverson A (2010) *An Entirely Synthetic Fish: How Rainbow Trout beguiled America and Overran the World*. Yale University Press, New York.

Hamilton SK, Bunn SE, Thoms MC and Marshall JC (2005) Persistence of aquatic refugia between flow pulses in a dryland river system (Cooper Creek, Australia). *Limnology and Oceanography* **50**, 743–754.

Hammer M (2001) Molecular systematics and conservation biology of the southern pygmy perch *Nannoperca australis* (Günther, 1861) (Teleostei: Percichthyidae) in south-eastern Australia. Hons thesis. University of Adelaide, Adelaide.

Hammer M (2008) A molecular genetic appraisal of biodiversity and conservation units in freshwater fishes from southern Australia. PhD thesis. University of Adelaide, Adelaide.

Hammer M, Wedderburn S and Van Weenen J (2009) 'Action plan for South Australian freshwater fishes'. Native Fish Australia (SA), Adelaide.

Hammer MP, Adams M, Unmack PJ and Walker KF (2007) A rethink on *Retropinna*: conservation implications of new taxa and significant genetic substructure in Australian smelts (Pisces: Retropinnidae). *Marine and Freshwater Research* **58**, 327–341.

Hammer MP, Unmack PJ, Adams M, Johnson JB *et al.* (2010) Phylogeographic structure in the threatened Yarra pygmy perch *Nannoperca obscura* (Teleostei: Percichthyidae) has major implications for declining populations. *Conservation Genetics* **11**, 213–223.

Hancock DA (Ed.) (1989) Tagging: solution or problem? *Proceedings No. 5*, Australian Society for Fish Biology tagging workshop. Australian Society for Fish Biology, Sydney.

Hancock DA (Ed.) (1994) *Population Dynamics for Fisheries Management: Workshop Proceedings*. Perth, 24–25 August 1993. Australian Society for Fish Biology, Perth.

Hancock DA, Smith DC and Koehn JD (Eds) (2000) *Fish Movement and Migration: Workshop Proceedings*. Bendigo. Australian Society for Fish Biology, Sydney.

Hansen MJ, Boisclair D, Brandt SB, Hewett SW *et al.* (1993) Applications of bioenergetics models to fish ecology and management: where do we go from here? *Transactions of the American Fisheries Society* **122**, 1019–1030.

Hanski I and Gaggiotti O (Eds) (2004) *Ecology, Genetics, and Evolution of Metapopulations*. Elsevier, Burlington.

Hanski I, Alho J and Moilanen A (2000) Estimating the parameters of survival and migration of individuals in metapopulations. *Ecology* **81**, 239–251.

Hara TJ (2007) Gustation. In *Sensory Systems Neuroscience*. Vol. 25, *Fish Physiology*. (Eds TJ Hara and BS Zielinski) pp. 45–96. Elsevier, New York.

Harcup MF, Williams R and Ellis DM (1984) Movements of brown trout in the River Gwyddon,South Wales. *Journal of Fish Biology* **24**, 415–426.

Hardie SA, Barmuta LA and White RWG (2005) Spawning related seasonal variation in fyke net catches of golden galaxias (*Galaxias auratus*): implications for monitoring lacustrine galaxiid populations. *Fisheries Management and Ecology* **12**, 407–409.

Hardie SA, Barmuta LA and White RWG (2006) Comparison of day and night fyke netting, electrofishing and snorkelling for monitoring a population of the threatened golden galaxias (*Galaxias auratus*). *Hydrobiologia* **560**, 145–158.

Hardie SA, White RWG and Barmuta LA (2007) Reproductive biology of the threatened golden galaxias *Galaxias auratus* Johnston and the influence of lake hydrology. *Journal of Fish Biology* **71**, 1820–1840.

Hardman M, Page LM, Sabaj MH, Armbruster JW *et al.* (2002) A comparison of fish surveys made in 1908 and 1998 of the Potaro, Essequibo, Demerara, and coastal river drainages of Guyana. *Ichthyological Exploration of Freshwaters* **13**, 225–238.

Hardy CM, Adams M, Jerry DR, Court LN *et al.* (2011) DNA barcoding to support conservation: species identification, genetic structure and biogeography of fishes in the Murray-Darling River Basin, Australia. *Marine and Freshwater Research* **62**, 1–15.

Harris JH (1983) Impoundment of coastal drainages of south-eastern Australia, and a review of its relevance to fish migrations. *Australian Zoologist* **21**, 235–250.

Harris JH (1984) A survey of fishways in streams of coastal south-eastern Australia. *Australian Zoologist* **21**, 219–233.

Harris JH (1985) Age of Australian bass, *Macquaria novemaculeata* (Perciformes: Percichthyidae), in the Sydney Basin. *Australian Journal of Marine and Freshwater Research* **36**, 235–246.

Harris JH (1986) Reproduction of the Australian bass, *Macquaria novemaculeata* (Perciformes: Percichthyidae) in the Sydney Basin. *Australian Journal of Marine and Freshwater Research* **37**, 209–235.

Harris JH (1987a) Growth of Australian bass *Maquaria novemaculeata* (Perciformes: Percichthyidae) in the Sydney Basin. *Australian Journal of Marine and Freshwater Research* **38**, 351–361.

Harris JH (1987b) *Proceedings of the Conference on Australian Threatened Fishes.* Australian Society for Fish Biology/Department of Agriculture, Sydney.

Harris JH (1988) Demography of Australian bass, *Macquaria novemaculeata* (Perciformes, Percichthyidae), in the Sydney Basin. *Australian Journal of Marine and Freshwater Research* **39**, 355–369.

Harris JH (1995) Zoogeography of the Australian freshwater fish fauna. In *Vertebrate Zoogeography and Evolution in Australasia.* (Eds M Archer and G Clayton) pp. 211–218. Hesperian Press, Perth.

Harris JH (2001) Fish passage in Australia: experience, challenges and projections. In *3rd Australian Technical Workshop on Fishways.* (Eds RJ Keller and P Peterken) pp. 1–17. Monash University, Melbourne.

Harris JH and Dixon PI (1988) Hybridization between trout cod and Murray cod. *Isozyme Bulletin* **19**, 39.

Harris JH and Gehrke PC (1994) 'Development of predictive models linking fish population recruitment with streamflow'. *Proceedings of Australian Society for Fish Biology Workshop: Population Dynamics for Fisheries Management.* (Ed. DA Hancock) pp. 195–199. Bureau of Rural Resources Proceedings, Canberra.

Harris JH and Gehrke PC (Eds) (1997) 'Fish and rivers in stress: the NSW Rivers Survey'. NSW Fisheries Office of Conservation/Cooperative Research Centre for Freshwater Ecology, Sydney.

Harris JH and Rowland SJ (1996) Family Percichthyidae: Australian freshwater cods and basses. In *Freshwater Fishes of South-Eastern Australia.* (Ed. RM McDowall) pp. 150–163. Reed, Sydney.

Harris JH and Silveira R (1999) Large-scale assessments of river health using an index of biotic integrity with low-diversity fish communities. *Freshwater Biology* **41**, 235–252.

Harris JH, Bowling L, Keller R, Kress J *et al.* (2006) 'The Tooma River Project: interdisciplinary probes into ill-defined and unpredictable contamination'. Technical Report. Cooperative Research Centre for Freshwater Ecology, Canberra.

Harris JH, Thorncraft G and Wem P (1998) Evaluation of rock-ramp fishways in Australia. In *Fish Migration and Fish Bypasses.* (Eds M Jungwirth, S Schmutz and S Weiss) pp. 331–347. Fishing News Books, University Press, Cambridge.

Hart BT, Bailey P, Edwards R, Hortle K *et al.* (1990) Effects of salinity on river, stream and wetland ecosystems in Victoria, Australia. *Water Research* **24**, 1103–1117.

Hart BT, Bailey P, Edwards R, Hortle K *et al.* (1991) A review of the salt sensitivity of the Australian freshwater biota. *Hydrobiologia* **210**, 105–144.

Hart PBJ and Reynolds JD (Eds) (2002) *Handbook of Fish Biology and Fisheries*, Vol. 2. Blackwell Science, Oxford.

Hart PR, Hutchinson WG and Purser GJ (1996) Effects of photoperiod, temperature and salinity on hatchery-reared larvae of the greenback flounder (*Rhombosolea tapirina* Günther, 1862). *Aquaculture* **144**, 303–311.

Hartman KJ and Kitchell JF (2008) Bioenergetics modeling progress since the 1992 symposium. *Transactions of the American Fisheries Society* **137**, 216–223.

Hasler AD and Scholz AT (1983) *Olfactory Imprinting and Homing in Salmon: Investigations into the Mechanisms of Imprinting Processes.* Springer-Verlag, Berlin.

Hawke A (2009) The Australian Environment Act: Report of the Independent Review of the Environment Protection and Biodiversity Conservation Act 1999. <http://www.environment.gov.au/epbc/review/publications/final-report.html>. March 2010.

Hayes DB, Ferreri CP and Taylor WW (1996) Linking fish habitat to their population dynamics. *Canadian Journal of Fisheries and Aquatic Sciences* **53**, 383–390.

Haynes GD, Gilligan DM, Grewe P and Nicholas FW (2009) Population genetics and management units of invasive common carp *Cyprinus carpio* in the Murray-Darling Basin, Australia. *Journal of Fish Biology* **75**, 295–320.

Heath M and Richardson K (1989) Comparative study of early-life survival variaiblity of herring, *Clupea haren-*

gus, in the north-eastern Atlantic. *Journal of Fish Biology* **35** (Supplement A), 49–57.

Heggenes J (1996) Habitat selection by brown trout (*Salmo trutta*) and young Atlantic salmon (*S. salar*) in streams: static and dynamic hydraulic modelling. *Regulated Rivers: Research and Management* **12**, 155–169.

Heibo E and Vøllestad LA (2002) Life-history variation in perch (*Perca fluviatilis* L.) in five neighbouring Norwegian lakes. *Ecology of Freshwater Fish* **11**, 270–280.

Heino M and Kaitala V (1999) Evolution of resource allocation between growth and reproduction in animals with indeterminate growth. *Journal of Evolutionary Biology* **12**, 423–429.

Helfman G, Collette BB, Facey DE and Bowen BW (2009) *The Diversity of Fishes: Biology, Evolution and Ecology*, 2nd edn. Wiley-Blackwell, Chichester.

Helfman S (2007) *Fish Conservation: A Guide to Understanding and Restoring Global Aquatic Biodiversity and Fishery Resources*. Island Press, Washington DC.

Helle JH and Hoffman MS (1995) Size decline and older age at maturity of two chum salmon (*Oncorhynchus keta*) stocks in western North America, 1972–1992. In *Climate Change and Northern Fish Populations*. (Ed. RJ Beamish). *Canadian Special Publication of Fisheries and Aquatic Sciences* **121**, 245–260.

Henderson PA and Corps M (1997) The role of temperature and cannibalism in interannual recruitment variation of bass in British waters. *Journal of Fish Biology* **50**, 280–295.

Henderson WR (2009) 'Pathogens in vertebrate pests in Australia'. Invasive Animals Cooperative Research Centre, Canberra.

Hendrixson HA, Sterner RW and Kay AD (2007) Elemental stoichiometry of freshwater fishes in relation to phylogeny, allometry and ecology. *Journal of Fish Biology* **70**, 121–140.

Henley WF, Patterson MA, Neves RJ and Lemly AD (2000) Effects of sedimentation and turbidity on lotic food webs: a concise review for natural resource managers. *Reviews in Fisheries Science* **8**, 125–139.

Henry GW and Lyle JM (2003) 'The National Recreational and Indigenous Fishing Survey'. FRDC Project No. 99/158. NSW Fisheries Final Report Series 48, Sydney.

Herbert B and Graham P (2004) Breeding and fecundity of the endemic Australian gudgeon, sleepy cod *Oxyeleotris lineolatus* (Steindachner 1867) (Eleotridae). *Aquaculture* **236**, 241–252.

Herbert B and Peeters J (1995) *Freshwater Fishes of Far North Queensland*. Queensland Department of Primary Industries, Brisbane.

Herbert BW, Graham PA and Foster SD (2003) Effects of added shelter and stocking density on growth of sleepy cod *Oxyeleotris lineolatus* in ponds. *Journal of the World Aquaculture Society* **34**, 433–440.

Herder F, Nolte AW, Pfaender J, Schwarzer J *et al.* (2006) Adaptive radiation and hybridization in Wallace's Dreamponds: evidence from sailfin silversides in the Malili Lakes of Sulawesi. *Proceedings of the Royal Society – Biological Sciences, Series B* **273**, 2209–2217.

Herder F, Pfaender J and Schliewen UK (2008) Adaptive sympatric speciation of polychromatic 'roundfin' sailfin silverside fish in Lake Matano (Sulawesi). *Evolution* **62**, 2178–2195.

Hermoso V, Linke S and Prenda J (2009) Identifying priority sites for the conservation of freshwater fish biodiversity in a Mediterranean basin with a high degree of threatened endemics. *Hydrobiologia* **623**, 127–140.

Hesp SA and Potter IC (2003) Reproductive biology of *Rhabdosargus sarba* (Sparidae) in Western Australian waters, in which it is a rudimentary hermaphrodite. *Journal of the Marine Biological Association of the United Kingdom* **83**, 1333–1346.

Hesp SA, Potter IC and Hall NG (2004) Reproductive biology and protandrous hermaphroditism in *Acanthopagrus latus*. *Environmental Biology of Fishes* **70**, 257–272.

Hesthagen T (1988) Movements of brown trout, *Salmo trutta*, and juvenile Atlantic salmon, *Salmo salar*, in a coastal stream in northern Norway. *Journal of Fish Biology* **32**, 639–653.

Hicks AS, Closs GP and Swearer SE (2010) Otolith microchemistry of two amphidromous galaxiids across an experimental salinity gradient: a multi-element approach for tracking diadromous migrations. *Journal of Experimental Marine Biology and Ecology* **394**, 86–97.

Higgs ES (1997) What is good ecological restoration? *Conservation Biology* **11**, 338–348.

Higham TE, Day SW and Wainwright PC (2005) Sucking while swimming: evaluating the effects of ram speed on suction generation in bluegill sunfish *Lepomis macrochirus* using digital particle image velocimetry. *Journal of Experimental Biology* **208**, 2653–2660.

Higham TE, Day SW and Wainwright PC (2006) Multidimensional analysis of suction feeding performance in fishes: fluid speed, acceleration, strike accuracy and the ingested volume of water. *Journal of Experimental Biology* **209**, 2713–2725.

Hilborn R and Walters CJ (1992) *Quantitative Fisheries Stock Asessment: Choice, Dynamics and Uncertainty*. Chapman Hall, New York.

Hilderbrand RH, Watts AC and Randle AM (2005) The myths of restoration ecology. *Ecology and Society* **10**, 1–11.

Hill J and Grossman GD (1993) An energetic model of microhabitat use for rainbow trout and rosyside dace. *Ecology* **74**, 685–698.

Hillis DM, Moritz C and Mabel BK (Eds) (1996) *Molecular Systematics*, 2nd edn. Sinauer Associates, Massachusetts.

Hills S (1934) Tertiary fresh water fishes from southern Queensland. *Memoirs of the Queensland Museum* **10**, 157–173.

Hills S (1943) Tertiary fresh-water fishes and crocodilian remains from Gladstone and Duaringa, Queensland. *Memoirs of the Queensland Museum* **12**, 96–100.

Hills S (1956) A brief review of Australian fossil vertebrates. In *Studies on Fossil Vertebrates.* (Ed. TS Westoll) pp. 86–107. Athlone Press, London.

Hillyard KA (2011) Carp exclusion screens on wetland inlets: their value for control of common carp (*Cyprinus carpio* L.) and effects on offstream movements by other fish species in the River Murray, Australia. PhD thesis. University of Adelaide, South Australia.

Hillyard KA, Smith BB, Conallin AJ and Gillanders BM (2010) Optimising exclusion screens to control exotic carp in an Australian lowland river. *Marine and Freshwater Research* **61**, 418–429.

Hindell JS (2007) Determining patterns of use by black bream *Acanthopagrus butcheri* (Munro, 1949) of re-established habitat in a south-eastern Australian estuary. *Journal of Fish Biology* **71**, 1331–1346.

Hirshfield MF (1980) An experimental analysis of reproductive effort and cost in the Japanese medaka, *Oryzias latipes. Ecology* **61**, 282–292.

Hitchcock G (2002) Fish fauna of the Bensbach River, southwest Papua New Guinea. *Memoirs of the Queensland Museum* **48**, 119–122.

Hjort J (1914) Fluctuations in the great fisheries of Northern Europe viewed in the light of biological research. *Rapports et Procés-Verbaux des Reunions, Conseil International pour l'Exploration de la Mer* **20**, 1–228.

Hoare ME (Ed.) (1982) *The 'Resolution' Journal of Johann Reinhold Forster, 1772–1775*, 4 vols. Hakluyt Society, London.

Hochachka PW and Somero GN (2002) *Biochemical Adaptation: Mechanism and Process in Physiological Evolution.* Oxford University Press, New York.

Hoeinghaus DJ, Winemiller KO and Birnbaum JS (2007) Local and regional determinants of stream fish assemblage structure: inferences based on taxonomic vs. functional groups. *Journal of Biogeography* **34**, 324–338.

Hoffmann RC (1996) Economic development and aquatic ecosystems in medieval Europe. *American Historical Review* **101**, 631–669.

Hogan AE and Vallance TD (2005) 'Rapid assessment of fish biodiversity in southern Gulf of Carpentaria catchments'. Project Report QI04074. Queensland Department of Primary Industries and Fisheries, Brisbane.

Holderegger R and Wagner HH (2008) Landscape genetics. *BioScience* **58**, 199–207.

Holdgate GR, Wallace MW, Gallagher SJ, Wagstaff BE *et al.* (2008) No mountains to snow on: major post-Eocene uplift of the East Victoria Highlands: evidence from Cenozoic deposits. *Australian Journal of Earth Sciences* **55**, 211–234.

Holdgate GR, Wallace MW, Gallagher SJ, Wagstaff BE *et al.* (2011) Paleogene basalts prove early uplift of Victoria's Eastern Uplands. *Australian Journal of Earth Sciences* **58**, 95–99.

Holdway DA, Hefferman J and Smith A (2008) Multigeneration assessment of nonylphenol and endosulfan using a model Australian freshwater fish, *Melanotaenia fluviatilis. Environmental Toxicology* **23**, 253–262.

Holling CS (1978) *Adaptive Environmental Assessment and Management.* Wiley, Chichester.

Horn MH (1998) Feeding and digestion. In *The Physiology of Fishes.* (Ed. DH Evans) pp. 43–63. CRC Press, New York.

Hortle KG and Lake PS (1982) Macroinvertebrate assemblages in channelized and unchannelized section of the Bunyip River, Victoria. *Australian Journal of Marine and Freshwater Research* **33**, 1071–1082.

Hortle KG and Pearson RG (1990) Fauna of the Annan River system, far North Queensland, with reference to the impact of tin mining. 1. Fishes. *Australian Journal of Marine and Freshwater Research* **41**, 677–694.

Hortle ME (1979) The ecology of the sandy, *Pseudaphritis urvillii*, in south-east Tasmania. Hons thesis. University of Tasmania, Hobart.

Hortle ME and White RWG (1980) Diet of *Pseudaphritis urvillii* (Cuvier and Valenciennes) (Pisces, Bovichthyidae) from southeastern Tasmania. *Australian Journal of Marine and Freshwater Research* **31**, 533–539.

Horwitz P, Lindsay M and O'Connor M (2001) Biodiversity, endemism, sense of place and public health: inter-relationships for Australian inland aquatic systems. *Ecosystem Health* **7**, 253–265.

Horwitz RJ (1978) Temporal variability patterns and the distributional patterns of stream fishes. *Ecological Monographs* **48**, 307–321.

Hose GC, Lim RP and Hyne RV (2003) The transport, fate and effects of endosulfan in the Australian freshwater environment. *Australian Journal of Ecotoxicology* **9**, 101–111.

Houde ED (1987) Fish early life history dynamics and recruitment variability. *American Fisheries Society Symposium* **2**, 17–29.

Houde ED (1989) Subtleties and episodes in the early life of fishes. *Journal of Fish Biology* **35** (Supplement A), 29–38.

Houde ED (1997) Patterns and consequences of selective processes in teleost early life histories. In *Early Life History and Recruitment in Fish Populations.* (Eds RC Chambers and EA Trippel) pp. 173–196. Chapman and Hall, London.

Houde ED (2002) Mortality. In *Fishery Science: The Unique Contributions of Early Life Stages.* (Eds LA Fuiman and RG Werner) pp. 64–87. Blackwell Science, Oxford.

Howe E, Howe C, Lim R and Burchett M (1997) Impact of the introduced poeciliid *Gambusia holbrooki* (Girard, 1859) on the growth and reproduction of *Pseudomugil signifer* (Kner, 1865) in Australia. *Marine and Freshwater Research* **48**, 425–434.

Howell TD, Arthington AH, Pusey BJ, Brooks AP *et al.* (2011) Responses of fish to experimental introduction of structural woody habitat in riffles and pools. *Restoration Ecology* **20**, 43–55.

Howitt JA, Baldwin DS, Rees GN and Williams JL (2007) Modelling blackwater: predicting water quality during flooding of lowland river forests. *Ecological Modelling* **203**, 229–242.

Howson TJ, Robson BJ and Mitchell BD (2009) Fish assemblage response to rehabilitation of a sand-slugged lowland river. *River Research and Applications* **25**, 1251–1267.

Hoyle SD and Jellyman DJ (2002) Longfin eels need reserves: modelling the effects of commercial harvest on stocks of New Zealand eels. *Marine and Freshwater Research* **53**, 887–895.

Hubbs CL (1955) Hybridization between fish species in nature. *Systematic Zoology* **4**, 1–20.

Hubbs CL and Lagler KF (2004) *Fishes of the Great Lakes Region*, revised edn. University of Michigan Press, Ann Arbor.

Huber R and Rylander MK (1992) Brain morphology and turbidity preference in *Notropis* and related genera (Cyprinidae, Teleostei). *Environmental Biology of Fishes* **33**, 153–165.

Huber R, van Staaden MJ, Kaufman LS and Liem KF (1997) Microhabitat use, trophic patterns, and the evolution of brain structure in African cichlids. *Brain Behavior and Evolution* **50**, 167–182.

Hubert N, Hanner R *et al.* (2008) Identifying Canadian freshwater fishes through DNA barcodes. *PLoS ONE* **3** e2940, 8 pp.

Huet M (1959) Profiles and biology of Western European streams as related to fish management. *Transactions of the American Fisheries Society* **88**, 155–163.

Huey JA, Baker AM and Hughes JM (2008) The effect of landscape processes upon gene flow and genetic diversity in an Australian freshwater fish, *Neosilurus hyrtlii*. *Freshwater Biology* **53**, 1393–1408.

Huey JA, Baker AM and Hughes JM (2010) High levels of genetic structure in the Australian freshwater fish, *Ambassis macleayi*. *Journal of the North American Benthological Society* **29**, 1148–1160.

Huey JA, Baker AM and Hughes JM (2011a) Evidence for multiple historical colonizations of an endoreic drainage basin by an Australian freshwater fish. *Journal of Fish Biology* **79**, 1047–1067.

Huey JA, Hughes JM and Baker AM (2006) Patterns of gene flow in two species of eel-tailed catfish, *Neosilurus hyrtlii* and *Porochilus argenteus* (Siluriformes: Plotosidae) in western Queensland's dry inland rivers. *Biological Journal of the Linnean Society* **87**, 457–467.

Huey JA, Schmidt DJ, Balcombe SR, Marshall JC *et al.* (2011b) High gene flow and metapopulation dynamics detected for three species in a dryland river system. *Freshwater Biology* **56**, 2378–2390.

Huey RB (1991) Physiological consequences of habitat selection. *American Naturalist* **137**, 95–115.

Huey RB and Pianka ER (1981) Ecological consequences of foraging mode. *Ecology* **62**, 991–999.

Hughes J, Ponniah M, Hurwood D, Chenoweth S *et al.* (1999) Strong genetic structuring in a habitat specialist, the Oxleyan pygmy perch *Nannoperca oxleyana*. *Heredity* **83**, 5–14.

Hughes JM (2007) Constraints on recovery: using molecular methods to study connectivity of aquatic biota in rivers and streams. *Freshwater Biology* **52**, 616–631.

Hughes JM and Hillyer MJ (2006) Mitochondrial DNA and allozymes reveal high dispersal abilities and historical movement across drainage boundaries in two species of freshwater fishes from inland rivers in Queensland, Australia. *Journal of Fish Biology* **68**, 270–291.

Hughes JM and Stewart J (2006) Reproductive biology of three commercially important hemiramphid species in south-eastern Australia. *Environmental Biology of Fishes* **75**, 237–256.

Hughes JM, Goudkamp K, Hurwood DA, Hancock M *et al.* (2003) Translocation causes extinction of a local population of the glass shrimp, *Paratya australiensis*. *Conservation Biology* **17**, 1007–1012.

Hughes JM, Schmidt DJ and Finn DS (2009) Genes in streams: a molecular approach to understanding movement of freshwater fauna and their riverine habitat. *BioScience* **59**, 573–583.

Hughes RL and Potter IC (1969) Studies on gametogenesis and fecundity in the lampreys *Mordacia praecox* and *M. mordax* (Petromyzonidae) *Australian Journal of Zoology* **17**, 447–464.

Hughes RN and Croy MI (1993) An experimental analysis of frequency-dependent predation (switching) in the 15-spined stickleback, *Spinachia spinachia*. *Journal of Animal Ecology* **62**, 341–352.

Hughes V, Thoms M, Nicol SJ and Koehn JD (2007) Physical–ecological interactions in a lowland river sytem: large wood, hydraulic complexity and native fish associations in the River Murray, Australia. In *Hydroecology and Ecohydrology: Past, Present and Future*. (Eds PJ Wood, DM Hannah and JP Sadler) pp. 388–403. Wiley, London.

Hume DJ, Fletcher AR and Morison AK (1983) 'Final Report. Carp Program Report 10'. Arthur Rylah Institute, Department of Natural Resources and Environment, Melbourne.

Humphrey C, Klumpp DW and Pearson RG (2003) Early development and growth of the eastern rainbowfish,

Melanotaenia splendida splendida (Peters). II. Otolith development, increment validation and larval growth. *Marine and Freshwater Research* **54**, 105–111.

Humphrey JD and Ashburner LD (1993) Spread of the bacterial fish pathogen *Aeromonas salmonicida* after importation of infected goldfish, *Carassius auratus*, into Australia. *Australian Veterinary Journal* **70**, 453–454.

Humphrey JD and Langdon JS (Eds) (1986) 'Diseases of Australian fish and shellfish', Australian Fish Health Reference Laboratory, Benalla.

Humphreys W, Shiao JC, Iizuka Y and Tzeng WN (2006) Can otolith microchemistry reveal whether the blind cave gudgeon, *Milyeringa veritas* (Eleotridae), is diadromous within a subterranean estuary? *Environmental Biology of Fishes* **75**, 439–453.

Humphreys WF (2001) *Milyeringa veritas* (Eleotridae), a remarkably versatile cave fish from the arid tropics of northwestern Australia. *Environmental Biology of Fishes* **62**, 297–313.

Humphries P (1989) Variation in the life history of diadromous and landlocked populations of the spotted galaxias, *Galaxias truttaceus* Valenciennes, in Tasmania. *Australian Journal of Marine and Freshwater Research* **40**, 501–518.

Humphries P (1990) Morphological variation in diadromous and landlocked populations of the spotted galaxias, *Galaxias truttaceus*, in Tasmania, south-eastern Australia. *Environmental Biology of Fishes* **27**, 97–105.

Humphries P (1995) Life history, food and habitat of southern pygmy perch, *Nannoperca australis*, in the Macquarie River, Tasmania. *Australian Journal of Marine and Freshwater Research* **46**, 1159–1169.

Humphries P (2003a) Blandowski misses out: ichthyological etiquette in 19th-century Australia. *Endeavour* **27**, 160–165.

Humphries P (2003b) From Van Diemen's Land to Paris: the scientific discovery of a Tasmanian fish. *Forty Degrees South* **30**, 44–47.

Humphries P (2005) Spawning time and early life history of Murray cod, *Maccullochella peelii peelii* (Mitchell) in an Australian river. *Environmental Biology of Fishes* **72**, 393–407.

Humphries P (2007) Historical indigenous use of aquatic resources in Australia's Murray-Darling Basin, and its implications for river management. *Ecological Management and Restoration* **8**, 106–113.

Humphries P (2009) Wilhelm Blandowski's contributions to the ichthyology of the Murray-Darling Basin. *Proceedings of the Royal Society of Victoria* **121**, 90–108.

Humphries P and Baldwin DS (2003) Drought and aquatic ecosystems: an introduction. *Freshwater Biology* **48**, 1141–1146.

Humphries P and King AJ (2004) Drifting fish larvae in Murray-Darling Basin rivers: composition, spatial and temporal patterns and distance drifted. In *Downstream Movement of Fish in the Murray-Darling Basin: Statement, Recommendations and Supporting Papers*. Canberra, 3–4 June 2003. (Eds M Lintermans and B Phillips) pp. 51–58. Murray-Darling Basin Commission, Canberra.

Humphries P and Lake PS (2000) Fish larvae and the management of regulated rivers. *Regulated Rivers: Research and Management* **16**, 421–432.

Humphries P and Winemiller KO (2009) Historical impacts on river fauna, shifting baselines and challenges for restoration. *BioScience* **59**, 673–684.

Humphries P, Brown P, Douglas J, Pickworth A *et al.* (2008a) Flow-related patterns in abundance and composition of the fish fauna of a degraded Australian lowland river. *Freshwater Biology* **53**, 789–813.

Humphries P, Cook RA, Richardson AJ and Serafini LG (2006) Creating a disturbance: manipulating slackwaters in a lowland river. *River Research and Applications* **22**, 525–542.

Humphries P, King AJ and Koehn JD (1999) Fish, flows and flood plains: links between freshwater fishes and their environment in the Murray-Darling River system, Australia. *Environmental Biology of Fishes* **56**, 129–151.

Humphries P, Serafini LG and King AJ (2002) River regulation and fish larvae: variation through space and time. *Freshwater Biology* **47**, 1307–1331.

Humphries P, Wilson G, Richardson AJ and Ellison T (2008b) 'Impacts of managed flows on fish spawning and recruitment'. Report prepared for the Murray-Darling Basin Commission. Charles Sturt University, Albury.

Hunt TL, Allen MS, Douglas J and Gason A (2010) Evaluation of a sport fish stocking program in lakes of the southern Murray-Darling Basin, Australia. *North American Journal of Fisheries Management* **30**, 805–811.

Hunt TL, Allen MS, Douglas J and Gason A (2011a) 'Evaluation of past and new stocking practices for recreational fisheries in the Kerang Lakes'. Recreational Fishing Grants Program Research Report. Department of Primary Industries, Melbourne.

Hunt TL, Douglas JW, Allen MS, Gwinn DC *et al.* (2011b) Evaluation of population decline and fishing sustainability of the endangered Australian freshwater fish *Macquaria australasica*. *Fisheries Management and Ecology* **18**, 513–520.

Hurwood DA and Hughes JM (1998) Phylogeography of the freshwater fish, *Mogurnda adspersa*, in streams of north-eastern Queensland, Australia: evidence for altered drainage patterns. *Molecular Ecology* **7**, 1507–1517.

Hutchison M, Gallagher T, Chilcott K, Simpson R *et al.* (2006) 'Impoundment stocking strategies for Australian native fishes in eastern and northern Australia: with an assessment of the value of scales as tags for stocked barramundi'. Final Report FRDC Project 98/221. Department of Pimary Industries and Fisheries, Brisbane.

Hutchison MJ (1991) Distribution patterns of red fin perch *Perca fluviatilis* Linnaeus and western pygmy perch *Edelia vittata* Castelnau in the Murray River. *Records of the Western Australian Museum* **15**, 295–301.

Huusko A, Mäki-Petäys A, Stickler M and Mykrä H (2011) Fish can shrink under harsh living conditions. *Functional Ecology* **25**, 628–633.

Huxley JS (1932) *Problems of Relative Growth.* Methuen, London.

Huynh H (2010) Evaluating the use of alternative proteins in feed on growth and nutrient utilization of Australian catfish, *Tandanus tandanus*, with emphasis on environmental contamination. Hons thesis. RMIT University, Melbourne.

Hynes HBN (1970) *The Ecology of Running Waters.* University of Toronto Press, Toronto.

Hynes HBN (1975) The stream and its valley. *Verhandlungen internationale für theoretische und angewandte Limnologie* **19**, 1–15.

Ilerbaig J (1999) Allied sciences and fundamental problems: C.C. Adams and the search for method in early American ecology. *Journal of the History of Biology* **32**, 439–463.

Ingram BA (1993) Evaluation of coded wire tags for marking fingerling golden perch, *Macquaria ambigua*, and silver perch, *Bidyanus bidyanus. Australian Journal of Marine and Freshwater Research* **44**, 817–824.

Ingram BA (2009) Culture of juvenile Murray cod, trout cod and Macquarie perch (Percichthyidae) in fertilised earthen ponds. *Aquaculture* **287**, 98–106.

Ingram BA, Hayes B and Rourke ML (2011) Impacts of stock enhancement strategies on the effective population size of Murray cod, *Maccullochella peelii*, a threatened Australian fish. *Fisheries Management and Ecology* **18**, 467–481.

Ingram BA, McKinnon LJ and Gooley GJ (2002) Growth and survival of selected aquatic animals in two saline groundwater evaporation basins: an Australian case study. *Aquaculture Research* **33**, 425–436.

Inland Fisheries Service Tasmania (2004) 'Carp Management Program Report, Lakes Crescent and Sorell: 1995 – June 2004'. Inland Fisheries Service, Hobart.

Inland Fisheries Service Tasmania (2010) 'Carp Management Program Annual Report 2009–2010'. Inland Fisheries Service, Hobart.

Inoue M and Nakano S (1998) Effects of woody debris on the habitat of juvenile masou salmon (*Oncorhynchus masou*) in northern Japanese streams. *Freshwater Biology* **40**, 1–16.

International Union for the Conservation of Nature (1987) Position statement on the translocation of living organisms. <http://www.iucn.org/about/work/programmes/species/publicationstechnicaldocuments/publications/iucnguidelinesandpolicystatements/>. February 2011.

Iredale T and Whitley GP (1938) The fluvifaunulae of Australia. *South Australian Naturalist* **18**, 64–68.

Isley JJ and Grabowski TB (2007) Age and growth. In *Analysis and Interpretation of Freshwater Fisheries Data.* (Eds CS Guy and ML Brown) pp. 187–228. American Fisheries Society, Bethesda.

Ivantsoff W and Aarn (1999) Detection of predation on Australian native fishes by *Gambusia holbrooki. Marine and Freshwater Research* **50**, 467–468.

Ivantsoff W, Unmack PJ, Saeed B and Crowley LELM (1991) A redfinned blue-eye, a new species and genus of the family Pseudomugilidae from central Western Queensland. *Fishes of Sahul* **6**, 277–282.

Iwama GK, Vijayan MM, Forsyth RB and Ackerman PA (1999) Heat shock proteins and physiological stress in fish. *American Zoologist* **39**, 901–909.

Jachowski DS and Kesler DC (2009) Allowing extinction: should we let species go? *Trends in Ecology and Evolution* **24**, 180.

Jackson DA and Harvey HH (1989) Biogeographic associations in fish assemblages: local vs regional processes. *Ecology* **70**, 1472–1484.

Jackson DA, Peres-Neto PR and Olden JD (2001a) What controls who is where in freshwater fish communities: the roles of biotic, abiotic, and spatial factors. *Canadian Journal of Fisheries and Aquatic Sciences* **58**, 157–170.

Jackson JBC, Alexander K and Sala E (Eds) (2011) *Shifting Baselines: The Past and the Future of Ocean Fisheries.* Island Press, Washington DC.

Jackson JBC, Kirby MX *et al.* (2001b) Historical overfishing and the recent collapse of coastal ecosystems. *Science* **293**, 629–637.

Jackson JE, Raadik TA, Lintermans M and Hammer M (2004) Alien salmonids in Australia: impediments to effective impact management, and future directions. *New Zealand Journal of Marine and Freshwater Research* **38**, 447–455.

Jackson PD (1976) A note on the food of the Australian grayling, *Prototroctes maraena* Günther (Galaxioidei: Prototroctidae). *Marine and Freshwater Research* **27**, 525–528.

Jackson PD (1978) Spawning and early development of the river blackfish, *Gadopsis marmoratus* Richardson (Gadopsiformes: Gadopsidae), in the McKenzie River, Victoria. *Australian Journal of Marine and Freshwater Research* **29**, 293–298.

Jackson PD (1980) Movement and home range of brown trout, *Salmo trutta* Linnaeus, in the Aberfeldy River, Victoria. *Australian Journal of Marine and Freshwater Research* **31**, 837–845.

Jackson PD (2009) Carp harvest and disposal at Murray River fishways. In *Proceedings of the Murray-Darling Basin Authority Native Fish Forum 2009.* Albury, 1–2 September. (Ed. J Pritchard) pp. 118–121. Murray-Darling Basin Authority, Canberra.

Jackson PD and Davies JN (1983) Survey of the fish fauna in the Grampians region, south-western Victoria. *Proceedings of the Royal Society of Victoria* **95**, 39–51.

Jackson PD and Williams WD (1980) Effects of brown trout, *Salmo trutta* L. on the distribution of some native fishes in three areas of southern Victoria. *Australian Journal of Marine and Freshwater Research* **31**, 61–67.

Jacobsen L and Berg S (1998) Diel variation in habitat use by planktivores in field enclosure experiments: the effect of submerged macrophytes and predation. *Journal of Fish Biology* **53**, 1207–1219.

Jager H, Cardwell H, Sale M and Bevelhimer M (1997) Modelling the linkages between flow management and salmon recruitment in rivers. *Ecological Modelling* **103**, 171–191.

James K, Fallon S, McDougall A, Espinoza T *et al.* (2010) Assessing the potential for radiocarbon dating the scales of Australian lungfish (*Neoceratodus forsteri*). *Radiocarbon* **52**, 1084–1089.

Janssen J (1997) Comparison of response distance to prey via the lateral line in the ruffe and yellow perch. *Journal of Fish Biology* **51**, 921–930.

Janssen J, Sideleva V and Biga H (1999) Use of the lateral line for feeding in two Lake Baikal sculpins. *Journal of Fish Biology* **54**, 404–416.

Jansson R, Nilsson C and Malmqvist B (2007) Restoring freshwater ecosystems in riverine landscapes: the roles of connectivity and recovery processes. *Freshwater Biology* **52**, 589–596.

Jardine TD, Pusey BJ, Hamilton SK, Pettit NE *et al.* (2011) Fish mediate high food web connectivity in the lower reaches of a tropical floodplain river. *Oecologia* in press.

Jax K (1998) Holocoen and ecosystem: on the origin and historical consequences of two concepts. *Journal of the History of Biology* **31**, 113–142.

Jelks HL, Walsh SJ *et al.* (2008) Conservation status of imperiled North American freshwater and diadromous fishes. *Fisheries* **33**, 372–407.

Jellyman DJ and Bowen MM (2009) Modelling larval migration routes and spawning areas of anguillid eels of New Zealand and Australia. *American Fisheries Society Symposium* **69**, 255–274.

Jellyman DJ and Tsukamoto K (2002) First use of archival transmitters to track migrating freshwater eels *Anguilla dieffenbachii* at sea. *Marine Ecology Progress Series* **233**, 207–215.

Jellyman DJ and Sykes JRE (2003) Diel and seasonal movements of radio-tagged freshwater eels, *Anguilla* spp., in two New Zealand streams. *Environmental Biology of Fishes* **66**, 143–154.

Jerry DR (1997) Population genetic structure of the catadromous Australian bass from throughout its range. *Journal of Fish Biology* **51**, 909–920.

Jerry DR (2008) Phylogeography and evidence for cryptic speciation in the Australian freshwater catfish *Tandanus tandanus*: a model species to understand biogeographical processes and evolution of the eastern Australian freshwater fish fauna (Plotosidae). *Marine and Freshwater Research* **59**, 351–360.

Jerry DR and Baverstock PR (1998) Consequences of a catadromous life-strategy for levels of mitochondrial DNA differentiation among populations of the Australian bass, *Macquaria novemaculeata*. *Molecular Ecology* **7**, 1003–1013.

Jerry DR and Cairns SC (1998) Morphological variation in the catadromous Australian bass, from seven geographically distinct riverine drainages. *Journal of Fish Biology* **52**, 829–843.

Jerry DR, Elphinstone MS and Baverstock PR (2001) Phylogenetic relationships of Australian members of the family Percichthyidae inferred from mitochondrial 12S rRNA sequence data. *Molecular Phylogenetics and Evolution* **18**, 335–347.

Jerry DR, Raadik TA, Cairns SC and Baverstock PR (1999) Evidence for natural interspecific hybridization between the Australian bass (*Macquaria novemaculeata*) and estuary perch (*M. colonorum*). *Marine and Freshwater Research* **50**, 661–666.

Jobling M (1995) *Environmental Biology of Fishes*. Chapman and Hall, London.

Johnson BL (1999) Introduction to the special feature: adaptive management – scientifically sound, socially challenged? *Conservation Ecology* **3**, 10.

Johnson BL, Richardson WB and Naimo TJ (1995a) Past, present, and future concepts in large river ecology. *BioScience* **45**, 134–141.

Johnson DH (1980) The comparison of usage and availability measurements for evaluating resource preference. *Ecology* **61**, 65–71.

Johnson DL (2004) *The Geology of Australia*. Cambridge University Press, Cambridge.

Johnson DL, Beaumier RA and Lynch WE (1988) Selection of habitat structure interstice size by bluegills and largemouth bass in ponds. *Transactions of the American Fisheries Society* **117**, 171–179.

Johnson GD, Paxton JR, Sutton TT, Satoh TP *et al.* (2009) Deep-sea mystery solved: astonishing larval transformations and extreme sexual dimorphism unite three fish families. *Biology Letters* **5**, 235–239.

Johnson IW, Elliot CRN and Gustard A (1995b) Modelling the effect of groundwater abstraction on salmonid habitat availability in the River Allen, Dorset, England. *Regulated Rivers: Research and Management* **10**, 229–238.

Johnson JB (2002) Evolution after the flood: phylogeography of the desert fish Utah chub. *Evolution* **56**, 948–960.

Johnson JH and Dropkin DJ (1995) Effects of prey density and short term food deprivation on the growth and survival of American shad larvae. *Journal of Fish Biology* **46**, 872–879.

Johnson JW and Randall JE (2008) *Synclidopus hogani*, a new species of soleid fish from northeastern Queensland, Australia. *Memoirs of the Queensland Museum* **52**, 245–254.

Johnson LB and Host GE (2010) Recent developments in landscape approaches for the study of aquatic ecosystems. *Journal of the North American Benthological Society* **29**, 41–66.

Johnson RW (1989) *Intraplate Volcanism in Eastern Australia and New Zealand*. Cambridge University Press, Cambridge.

Johnson TC, Scholz CA, Talbot MR, Kelts K *et al.* (1996) Late Pleistocene desiccation of Lake Victoria and rapid evolution of cichlid fishes. *Science* **273**, 1091–1093.

Jonas JL and Wahl DH (1998) Relative importance of direct and indirect effects of starvation for young walleyes. *Transactions of the American Fisheries Society* **127**, 192–205.

Jones CG, Lawton JH and Shachak M (1994) Organisms as ecosystem engineers. *Oikos* **69**, 373–386.

Jones CM (2000) Fitting growth curves to retrospective size-at-age data. *Fisheries Research* **46**, 123–129.

Jones JC and Reynolds JD (1997) Effects of pollution on reproductive behaviour of fishes. *Reviews in Fish Biology and Fisheries* **7**, 463–491.

Jones MJ and Stuart IG (2007) Movements and habitat use of common carp (*Cyprinus carpio*) and Murray cod (*Maccullochella peelii peelii*) juveniles in a large lowland Australian river. *Ecology of Freshwater Fish* **16**, 210–220.

Jones MJ and Stuart IG (2008) Regulated floodplains: a trap for unwary fish. *Fisheries Management and Ecology* **15**, 71–79.

Jones MJ and Stuart IG (2009) Lateral movement of common carp (*Cyprinus carpio* L.) in a large lowland river and floodplain. *Ecology of Freshwater Fish* **18**, 72–82.

Jonsson B (2006) Life-history effects of migratory costs in anadromous brown trout. *Journal of Fish Biology* **69**, 860–869.

Jonsson B and Jonsson N (1993) Partial migration: niche shift versus sexual maturation in fishes. *Reviews in Fish Biology and Fisheries* **3**, 348–365.

Jonsson B and Jonsson N (2009) A review of the likely effects of climate change on anadromous Atlantic salmon *Salmo salar* and brown trout *Salmo trutta*, with particular reference to water temperature and flow. *Journal of Fish Biology* **75**, 2381–2447.

Joseph LN, Maloney RF and Possingham HP (2009) Optimal allocation of resources among threatened species: a project prioritization protocol. *Conservation Biology* **23**, 328–338.

Jowett IG and Richardson J (1995) Habitat preferences of common riverine New Zealand native fishes and implication for flow management. *New Zealand Journal of Marine and Freshwater Research* **29**, 13–23.

Joyce EB, Webb JA *et al.* (2003) Geomorphology, the evolution of Victorian landscapes. Geological Society of Australia Special Publication 23. In *Geology of Victoria*. (Ed. WD Birch) pp. 533–561. Geological Society of Australia (Victoria Division), Melbourne.

Junk WJ and Bayley PB (2008) The scope of the flood pulse concept regarding riverine fish and fisheries, given geographic and man-made differences among systems. In *Reconciling Fisheries with Conservation:Proceedings of the 4th World Fisheries Congress*. (Eds J Nielsen, JJ Dodson, K Friedland, TR Hamon, J Musick and E Verspoor) pp. 1907–1923. Symposium 49(2). American Fisheries Society, Bethesda.

Junk WJ, Bayley PB and Sparks RE (1989) The flood pulse concept in river-floodplain systems. *Canadian Special Publication of Fisheries and Aquatic Sciences* **106**, 110–127.

Kacperczyk A and Daczewska M (2008) The Australian lungfish (*Neoceratodus forsteri*): fish or amphibian pattern of muscle development? *International Journal of Developmental Biology* **52**, 279–286.

Kailola PJ (2000) 'Development of an alert list for non-native freshwater fishes'. Report to Environment Australia. Patricia J. Kailola, Consultant, Newnham, Tasmania.

Kailola PJ, Williams MJ, Stewart PC, Reichelt RE *et al.* (1993) 'Australian fishery resources'. Bureau of Resource Sciences/Fisheries Research and Development Corporation, Canberra.

Kalish JM (1993) Pre- and post-bomb radiocarbon in fish otoliths. *Earth and Planetary Science Letters* **114**, 549–554.

Kaminskas S and Humphries P (2009) Diet of Murray cod (*Maccullochella peelii peelii*) (Mitchell) larvae in an Australian lowland river in low flow and high flow years. *Hydrobiologia* **636**, 449–461.

Kamler E (2002) Ontogeny of yolk-feeding fish: an ecological perspective. *Reviews in Fish Biology and Fisheries* **12**, 79–103.

Karolak S (2006) 'Alien fish in the Murray-Darling Basin'. Murray-Darling Basin Commission, Canberra.

Karr JR (1981) Assessment of biotic integrity using fish communities. *Fisheries* **6**, 21–27.

Katersky RS and Carter CG (2005) Growth efficiency of juvenile barramundi, *Lates calcarifer*, at high temperatures. *Aquaculture* **250**, 775–780.

Katersky RS and Carter CG (2007a) High growth efficiency occurs over a wide temperature range for juvenile barramundi *Lates calcarifer* fed a balanced diet. *Aquaculture* **272**, 444–450.

Katersky RS and Carter CG (2007b) A preliminary study on growth and protein synthesis of juvenile barramundi, *Lates calcarifer* at different temperatures. *Aquaculture* **267**, 157–164.

Katharios P, Papadaki M, Papandroulakis N and Divanach P (2008) Severe mortality in mesocosm-reared sharp-snout sea bream *Dioplodus puntazzo* larvae due to epith-leliocytis infection. *Diseases of Aquatic Organisms* **82**, 55–60.

Kaushal SS, Likens GE, Jaworski NA, Pace ML *et al.* (2010) Rising stream and river temperatures in the United States. *Frontiers in Ecology and the Environment* **8**, 461–466.

Kawasaki T (1980) Fundamental relations among the selections of life history in marine teleosts. *Bulletin of the Japanese Society of Scientific Fisheries* **46**, 289–293.

Keane JP and Neira FJ (2004) First record of mosquitofish, *Gambusia holbrooki*, in Tasmania, Australia: strock structure and reproductive biology. *New Zealand Journal of Marine and Freshwater Research* **38**, 857–867.

Kearney R (1995) Recreational fishing: what's the catch? In *Recreational Fishing: What's the Catch?* (Ed. DA Hancock). Australian Society for Fish Biology Workshop Proceedings, Canberra.

Kearney R, Buxton C and Goodsell P (2012) Questionable interpretation of the precautionary principle in Australia's implementation of 'no-take' marine protected areas. *Marine Policy* in press.

Keast A (1965) Resource subdivision amongst cohabitating fish species in a bag, Lake Opincon, Ontario. *University of Michigan, Great Lakes Research Division* **13**, 106–132.

Keenan CP, Watts RJ and Serafini LG (1996) Population genetics of golden perch, silver perch and eel-tailed catfish within the Murray-Darling Basin. In *1995 Riverine Environment Research Forum.* (Eds RJ Banens and R Lehane) pp. 17–26. Murray-Darling Basin Commission, Canberra.

Keenleyside MHA and Mackereth RW (1992) Effects of loss of male parent on brood survival in a biparental cichlid fish. *Environmental Biology of Fishes* **34**, 207–212.

Keith P (2000) The part played by protected areas in the conservation of threatened French freshwater fish. *Biological Conservation* **92**, 265–273.

Kellaway P (1946) The part played by electric fish in the early history of bioelectricity and electrotherapy. *Bulletin of Historical Medicine* **20**, 112–137.

Keller K and Brown C (2008) Behavioural interactions between the introduced plague minnow *Gambusia holbrooki* and the vulnerable native Australian ornate rainbowfish *Rhadinocentrus ornatus*, under experimental conditions. *Journal of Fish Biology* **73**, 1714–1729.

Keller RJ and Peterken C (2001) '3rd Australian technical workshop on fishways'. Monash University, Conference Management Office, Melbourne.

Keller RP, Lodge DM, Lewis MA and Shogren JF (2009) *Bioeconomics of Invasive Species: Integrating Ecology, Economics Policy and Management.* Oxford University Press, Oxford.

Kemp A (1984) Spawning of the Australian lungfish, *Neoceratodus forsteri* (Krefft) in the Brisbane River and in Enoggera Reservior, Queensland. *Memoirs of the Queensland Museum* **21**, 391–399.

Kemp A (1996) The biology of the Australian lungfish, *Neoceratodus forsteri. Journal of Morphology* **1**, 181–198.

Kennard MJ (1995) Factors influencing freshwater fish assemblages in floodplain lagoons of the Normanby River, Cape York Peninsula, a large tropical Australian river. MPhil thesis. Griffith University, Brisbane.

Kennard MJ, Arthington AH, Pusey BJ and Harch BD (2005) Are alien fish a reliable indicator of river health? *Freshwater Biology* **50**, 174–193.

Kennard MJ, Olden JD, Arthington AH, Pusey BJ *et al.* (2007) Multiscale effects of flow regime and habitat and their interaction on fish assemblage structure in eastern Australia. *Canadian Journal of Fisheries and Aquatic Sciences* **64**, 1346–1359.

Kennard MJ, Pusey BJ, Arthington AH, Harch BD *et al.* (2006a) Development and application of a predictive model of freshwater fish assemblage composition to evaluate river health in eastern Australia. *Hydrobiologia* **572**, 33–57.

Kennard MJ, Pusey BJ, Harch BD, Dore E *et al.* (2006b) Estimating local stream fish assemblage attributes: sampling effort and efficiency at two spatial scales. *Marine and Freshwater Research* **57**, 635–653.

Kennard MJ, Pusey BJ, Olden JD, Mackay SJ *et al.* (2010) Classification of natural flow regimes in Australia to support environmental flow management. *Freshwater Biology* **55**, 171–193.

Kennedy BP, Blum JD, Folt Cl and Nislow KH (2000) Using natural strontium isotopic signatures as fish markers: methodology and application. *Canadian Journal of Fisheries and Aquatic Sciences* **57**, 2292.

Kerezsy A (2010a) The distribution, recruitment and movement of fish in far western Queensland. PhD thesis. Griffith University, Brisbane

Kerezsy A (2010b) Everything's endangered: ecological restoration work in the springs at Edgbaston. *Australian Society for Fish Biology Newsletter* **40**, 5–7.

Kerezsy A, Balcombe SR, Arthington AH and Bunn SE (2011) Continuous recruitment underpins fish persistence in the arid rivers of far-western Queensland, Australia. *Marine and Freshwater Research* **62**, 1178–1190.

Kerkvliet J and Langpap C (2007) Learning from endangered and threatened species recovery programs: a case study using US Endangered Species Act recovery scores. *Ecological Economics* **63**, 499–510.

Kerr LA, Secor DH and Piccoli PM (2009) Partial migration of fishes as exemplified by the estuarine-dependent white perch. *Fisheries* **34**, 114–123.

Kershaw AP, Martin HA and McEwen Mason JRC (1994) The Neogene: a period of transition. In *History of the*

Australian Vegetation: Cretaceous to Recent. (Ed. RS Hill) pp. 299–327. University of Cambridge Press, Cambridge.

Khan MT, Khan TA and Wilson ME (2004) Habitat use and movement of river blackfish (*Gadopsis marmoratus* R.) in a highly modified Victorian stream, Australia. *Ecology of Freshwater Fish* **13**, 285–293.

Khan TA, Wilson ME and Khan MT (2003) Evidence for invasive carp mediated trophic cascade in shallow lakes of western Victoria, Australia. *Hydrobiologia* **506**, 465–472.

Kilsby NN and Walker KF (2012) Behaviour of two small pelagic and demersal fish species in diverse hydraulic environments. *River Research and Applications.* DOI: 10.1002/rra.1466.

King AJ (2004a) Density and distribution of potential prey for larval fish in the main channel of a floodplain river: pelagic versus epibenthic meiofauna. *River Research and Applications* **20**, 883–897.

King AJ (2004b) Ontogenetic patterns of habitat use by fishes within the main channel of an Australian floodplain river. *Journal of Fish Biology* **65**, 1582–1603.

King AJ (2005) Ontogenetic dietary shifts of fishes in an Australian floodplain river. *Marine and Freshwater Research* **56**, 215–225.

King AJ and O'Connor JP (2007) Native fish entrapment in irrigation systems: a step towards understanding the significance of the problem. *Ecological Management and Restoration* **8**, 32–37.

King AJ, Crook DA, Koster WM, Mahoney J *et al.* (2005) Comparison of larval fish drift in the Lower Goulburn and mid-Murray Rivers. *Ecological Management and Restoration* **6**, 136–138.

King AJ, Humphries P and Lake PS (2003a) Fish recruitment on floodplains: the roles of patterns of flooding and life history characteristics. *Canadian Journal of Fisheries and Aquatic Sciences* **60**, 773–786.

King AJ, Robertson AI and Healey MR (1997) Experimental manipulations of the biomass of introduced carp (*Cyprinus carpio*) in billabongs. I. Impacts on water-column properties. *Marine and Freshwater Research* **48**, 435–443.

King AJ, Tonkin Z and Mahoney J (2009) Environmental flow enhances native fish spawning and recruitment in the Murray River. Australia. *River Research and Applications* **25**, 1205–1218.

King AJ, Tonkin Z and Mahoney JM (2007) 'Assessing the effectiveness of environmental flows on fish recruitment in Barmah-Millewa Forest'. Report to Murray-Darling Basin Commission. Arthur Rylah Institute for Environmental Research, Department of Sustainability and Environment, Melbourne.

King AJ, Ward KA, O'Connor P, Green D *et al.* (2010) Adaptive management of an environmental watering event to enhance native fish spawning and recruitment. *Freshwater Biology* **55**, 17–31.

King J, Brown C and Sabet H (2003b) A scenario-based holistic approach to environmental flow assessments for rivers. *River Research and Applications* **19**, 619–639.

King JR and McFarlane GA (2003) Marine fish life history strategies: applications to fishery management. *Fisheries Management and Ecology* **10**, 249–264.

King JR, Shuter BJ and Zimmerman AP (1999) Empirical links between thermal habitat, fish growth, and climate change. *Transactions of the American Fisheries Society* **128**, 656–665.

King M (1995) *Fisheries Biology, Assessment and Management.* Blackwell Science, Oxford.

King S and Warburton K (2007) The environmental preferences of three species of Australian freshwater fish in relation to the effects of riparian degradation. *Environmental Biology of Fishes* **78**, 307–316.

Kingsford RT (2000) Ecological impacts of dams, water diversions and river management on floodplain wetlands in Australia. *Austral Ecology* **25**, 109–127.

Kingsford RT, Curtin AL and Porter J (1999) Water flows on Cooper Creek in arid Australia determine 'boom' and 'bust' periods for waterbirds. *Biological Conservation* **88**, 231–248.

Kirk JTO (1985) Effects of suspensoids (turbidity) on penetration of solar radiation in aquatic ecosystems. *Hydrobiologia* **125**, 195–208.

Knapp S (1999) *Footsteps in the Forest: Alfred Russel Wallace in the Amazon*. Natural History Museum, London.

Knight JT and Arthington AH (2008) Distribution and habitat associations of the endangered Oxleyan pygmy perch, *Nannoperca oxleyana* Whitley, in eastern Australia. *Aquatic Conservation: Marine and Freshwater Ecosystems* **18**, 1240–1254.

Knight JT, Butler GL, Smith PS and Wager RNE (2007) Reproductive biology of the endangered Oxleyan pygmy perch *Nannoperca oxleyana* Whitley. *Journal of Fish Biology* **71**, 1494–1511.

Knight JT, Nock CJ, Elphinstone MS and Baverstock PR (2009) Conservation implications of distinct genetic structuring in the endangered freshwater fish *Nannoperca oxleyana* (Percichthyidae). *Marine and Freshwater Research* **60**, 34–44.

Knight W (1968) Asymptotic growth: an example of nonsense disguised as mathematics. *Journal of the Fisheries Research Board of Canada* **25**, 1303–1307.

Knights BC, Vallazza JM and Dewey MR (2002) Habitat and movement of lake sturgeon in the upper Mississippi system, USA. *Transactions of the American Fisheries Society* **131**, 507–522.

Knowles LL (2004) The burgeoning field of statistical phylogeography. *Journal of Evolutionary Biology* **17**, 1–10.

Koed A, Mejlhede P, Balleby K and Aarestrup K (2000) Annual movement and migration of adult pikeperch in a lowland river. *Journal of Fish Biology* **57**, 1266–1279.

Koehn J (1986) Approaches to determining flow and habitat requirements for native freshwater fish in Victoria. In *Stream Protection: The Management of Rivers for Instream Uses*. (Ed. IC Campbell) pp. 95–115. Water Studies Centre, Chisholm Institute of Technology, Melbourne.

Koehn J (2005) The loss of valuable Murray cod in fish kills: a science and management perspective. In *Management of Murray Cod in the Murray-Darling Basin: Statement, Recommendations and Supporting Papers*. Canberra, 3–4 June 2004. (Eds M Lintermans and B Phillips) pp. 73–82. Murray-Darling Basin Commission/Cooperative Research Centre for Freshwater Ecology, Canberra.

Koehn JD (1987) 'Artificial habitat increases abundance of two-spined blackfish *Gadopsis bispinosus* in Ovens River, Victoria'. Technical Report Series 56. Arthur Rylah Institute for Environmental Research, Department of Conservation, Forests and Lands, Melbourne.

Koehn JD (1996) The key criteria to sustaining the wild stock Murray cod fishery in Lake Mulwala. In *Proceedings of the 2nd World Fisheries Congress. Developing and Sustaining World Fish Resources: The State of Science and Management*. (Eds DA Hancock and JP Beumer), p. 119. Brisbane.

Koehn JD (2003) Riverine aquatic protected areas: protecting species, communities or ecosystem processes? In *Proceedings of the World Congress on Aquatic Protected Areas. Aquatic Protected Areas: What Works Best and How do we Know?* (Eds JP Beumer, A Grant and DC Smith) pp. 614–624. University of Queensland, Brisbane.

Koehn JD (2004) Carp (*Cyprinus carpio*) as a powerful invader in Australian waterways. *Freshwater Biology* **49**, 882–894.

Koehn JD (2006) The ecology and conservation management of Murray cod *Maccullochella peelii peelii*. PhD thesis. University of Melbourne, Melbourne.

Koehn JD (2009a) Multi-scale habitat selection by Murray cod *Maccullochella peelii peelii* in two lowland rivers. *Journal of Fish Biology* **75**, 113–129.

Koehn JD (2009b) Using radio telemetry to evaluate the depths inhabited by Murray cod (*Maccullochella peelii peelii*). *Marine and Freshwater Research* **60**, 317–320.

Koehn JD (2010) Conservation and utilisation: harnessing forces for better outcomes for native fishes. *Ecological Management and Restoration* **11**, 86–87.

Koehn JD (2011) Spatial management of freshwater fish: a case study for Murray cod. In *Spatial Management in Fisheries. Proceedings of the Australian Society for Fish Biology Workshop*. Canberra, 11–12 September. (Eds MA Treloar and RD Tilzey) pp. 14–30. Australian Society for Fish Biology, Canberra.

Koehn JD and Harrington DJ (2005) Collection and distribution of the early life stages of the Murray cod (*Maccullochella peelii peelii*) in a regulated river. *Australian Journal of Zoology* **53**, 137–144.

Koehn JD and Harrington DJ (2006) Environmental conditions and timing for the spawning of Murray cod (*Maccullochella peelii peelii*) and the endangered trout cod (*M. macquariensis*) in southeastern Australian rivers. *River Research and Applications* **22**, 327–342.

Koehn JD and Lintermans M (2012) A strategy to rehabilitate fishes of the Murray-Darling Basin, south-eastern Australia. *Endangered Species Research* **16**, 165–181.

Koehn JD and MacKenzie RF (2004) Priority management actions for alien freshwater fish species in Australia. *New Zealand Journal of Marine and Freshwater Research* **38**, 457–472.

Koehn JD and McDowall RM (2004) Invasive species: fish and fisheries workshop overview, then and now. *New Zealand Journal of Marine and Freshwater Research* **38**, 383–389.

Koehn JD and Nicol S (1998) *Habitat and Movement Requirements of Fish*. Murray-Darling Basin Commission, Canberra.

Koehn JD and O'Connor WG (1990a) *Biological Information for Management of Native Freshwater Fish in Victoria*. Department of Conservation and Environment, Melbourne.

Koehn JD and O'Connor WG (1990b) Distribution of freshwater fish in the Otway region, south-western Victoria. *Proceedings of the Royal Society of Victoria* **102**, 29–39.

Koehn JD and O'Connor WG (1990c) Threats to Victorian native freshwater fish. *Victorian Naturalist* **107**, 5–12.

Koehn JD and Raadik TA (1991) The Tasmanian mudfish *Galaxias cleaveri* Scott, 1934, in Victoria. *Proceedings of the Royal Society of Victoria* **103**, 77–86.

Koehn JD, Brumley AR and Gehrke PC (2000) *Managing the Impacts of Carp*. Bureau of Resource Sciences, Canberra.

Koehn JD, Doeg T, Harrington D and Milledge G (1997) Dartmouth Dam: effects on the downstream aquatic fauna. In *1995 Riverine Environment Research Forum*. (Eds RJ Banens and R Lehane) pp. 49–56. Murray-Darling Basin Commission, Canberra.

Koehn JD, Eiler JG, McKenzie JA and O'Connor WG (2012) An improved method for obtaining fine-scale location of radio tags when tracking by boat. In *Proceedings of the 2nd International Symposium on Advances in Fish Tagging and Marking Technology*. (Eds JR McKenzie, B Parsons, AC Seitz, RK Kopf, M Mesa and Q Phelps) pp. 379–384. American Fisheries Society Symposium 76, Bethesda, Maryland.

Koehn JD, Hobday AJ, Pratchett MS and Gillanders BM (2011) Climate change and Australian marine and freshwater environments, fishes and fisheries: synthesis and options for adaptation. *Marine and Freshwater Research* **62**, 1148–1164.

Koehn JD, McKenzie JA, O'Mahony DJ, Nicol SJ *et al.* (2009) Movements of Murray cod (*Maccullochella peelii*

peelii) in a large Australian lowland river. *Ecology of Freshwater Fish* **18**, 594–602.

Koehn JD, Nicol SJ and Fairbrother PS (2004a) Spatial arrangement and physical characteristics of structural woody habitat in a lowland river in south-eastern Australia. *Aquatic Conservation: Marine and Freshwater Ecosystems* **14**, 457–464.

Koehn JD, Nicol SJ, McKenzie JA, Lieschke JA *et al.* (2008) Spatial ecology of an endangered native Australian percichthyid fish, the trout cod *Maccullochella macquariensis*. *Endangered Species Research* **4**, 219–225.

Koehn JD, O'Connor NA and Jackson PD (1994) Seasonal and size related variation in microhabitat use by a southern Victorian stream fish assemblage. *Australian Journal of Marine and Freshwater Research* **45**, 1353–1366.

Koehn JD, Stuart I and Crook DA (2004b) Linking the ecological importance of downstream fish movements to management of Murray-Darling Basin fish populations. In *Downstream Movement of Fish in the Murray-Darling Basin: Statement, Recommendations and Supporting Papers.* Canberra, 3–4 June 2003. (Eds M Lintermans and B Phillips) pp. 67–78. Murray-Darling Basin Commission, Canberra.

Kolar CS and Lodge DM (2001) Progress in invasion biology: predicting invaders. *Trends in Ecology and Evolution* **16**, 199–204.

Kolar CS and Lodge DM (2002) Ecological predictions and risk assessment for alien fishes in North America. *Science* **298**, 1233.

Koster W and Dawson D (2011) 'Investigation of the movement and spawning of Australian grayling'. Arthur Rylah Institute for Environmental Research, Department of Sustainability and Environment, Melbourne.

Koster WM (2003) Threatened fishes of the world: *Galaxiella pusilla* (Mack 1936) (Galaxiidae). *Environmental Biology of Fishes* **68**, 268.

Koster WM and Crook DA (2008) Diurnal and nocturnal movements of river blackfish (*Gadopsis marmoratus*) in a south-eastern Australian upland stream. *Ecology of Freshwater Fish* **17**, 146–154.

Koster WM, Raadik TA and Clunie P (2002) 'Scoping study of the potential spread and impact of the exotic fish oriental weatherloach in the Murray-Darling Basin, Australia: a resource document'. Arthur Rylah Institute for Environmental Research, Department of Sustainability and Environment, Melbourne.

Kotrschal K and Palzenberger M (1992) Neuroecology of cyprinids: comparative, quantitative histology reveals diverse brain patterns. *Environmental Biology of Fishes* **33**, 135–152.

Kowarsky J and Ross A (1981) Fish movement upstream through a central Queensland (Fitzroy River) coastal fishway. *Marine and Freshwater Research* **32**, 93–109.

Kramer DL and Chapman MR (1999) Implications of fish home range size and relocation for marine reserve function. *Environmental Biology of Fishes* **55**, 65–79.

Krebs CJ (2009) *Ecology: The Experimental Analysis of Distribution and Abundance.* Pearson Benjamin Cummings, New York.

Krebs JR and Kacelnik A (1991) Decision-making. In *Behavioural Ecology: An Evolutionary Approach*, 3rd edn. (Eds JD Krebs and NB Davies) pp. 105–136. Blackwell Scientific, Oxford.

Krefft G (1866) On the vertebrated animals of the Lower Murray and Darling, their habits, economy, and geographical distribution, by Gerard Krefft. *Transactions of the Philosophical Society of New South Wales* **1862–1865**, 1–33.

Krieg GW (1989) Geology. In *Natural History of Dalhousie Springs.* (Eds W Zeidler and WF Ponder) pp. 19–26. South Australian Museum, Adelaide.

Kuiter RH and Allen GR (1986) A synopsis of the Australian pygmy perches (Percichthyidae) with the description of a new species. *Revue Française d'Aquariologie* **12**, 109–116.

Kuiter RH, Humphries P and Arthington AH (1996) Pygmy perches: Family Nannopercidae. In *Freshwater Fishes of South-eastern Australia*, 2nd edn. (Ed. RM McDowall) pp. 168–175. Reed, Sydney.

Kumar A and Chapman JC (1998) Profenofos toxicity to the eastern rainbow fish (*Melanotaenia duboulayi*). *Environmental Toxicology and Chemistry* **17**, 1799–1806.

Kumazawa Y and Nishida M (2000) Molecular phylogeny of osteoglossoids: a new model for Gondwanian origin and plate tectonic transportation of the Asian arowana. *Molecular Biology and Evolution* **17**, 1869–1878.

Kuussaari M, Bommarco R *et al.* (2009) Extinction debt: a challenge for biodiversity conservation. *Trends in Ecology and Evolution* **24**, 564–571.

Lake JS (1967a) *Freshwater Fish of the Murray-Darling River System.* Chief Secretary's Department, Sydney.

Lake JS (1967b) Principal fishes of the Murray-Darling. In *Australian Inland Waters and their Fauna: Eleven Studies.* (Ed. AH Weatherley) pp. 192–216. Australian National University Press, Canberra.

Lake JS (1967c) Rearing experiments with five species of Australian freshwater fishes. II. Morphogenesis and ontogeny. *Australian Journal of Marine and Freshwater Research* **18**, 155–176.

Lake JS (1971) *Freshwater Fishes and Rivers of Australia.* Thomas Nelson, Melbourne.

Lake JS (1978) *Freshwater Fishes of Australia: An Illustrated Field Guide.* Thomas Nelson, Melbourne.

Lake JS and Midgley SH (1970) Reproduction of freshwater Ariidae in Australia. *Australian Journal of Science* **32**, 441.

Lake PS (1998) After the inundation: long-term changes in the fauna of Lake Pedder, Tasmania, Australia. In *Wetlands for the Future*. (Ed. AJ McComb). Gleneagles Publishing, South Australia.

Lake PS (2000) Disturbance, patchiness, and diversity in streams. *Journal of the North American Benthological Society* **19**, 573–592.

Lake PS (2001) The fauna of Lake Pedder: changes after the flooding and thoughts on restoration. In *Lake Pedder: Values and Restoration*. Occasional Paper 72, pp. 87–98. Centre for Environmental Studies, University of Tasmania, Hobart.

Lake PS (2003) Ecological effects of perturbation by drought in flowing waters. *Freshwater Biology* **48**, 1161–1172.

Lake PS (2005) Perturbation, restoration and seeking ecological sustainability in Australian flowing waters. *Hydrobiologia* **552**, 109–120.

Lake PS and Fulton W (1981) Observations on the freshwater fish of a small Tasmanian coastal stream. *Papers and Proceedings of the Royal Society of Tasmania* **115**, 163–172.

Lake PS, Bond N and Reich P (2007) Linking ecological theory with stream restoration. *Freshwater Biology* **52**, 597–615.

Lammens EHRR and Hoogenboezem W (1991) Diets and feeding behaviour. In *Cyprinid Fishes: Systematics, Biology and Exploitation*. (Eds IJ Winfield and JS Nelson) pp. 353–376. Chapman and Hall, London.

Lamouroux N and Cattaneo F (2006) Fish assemblages and stream hydraulics: consistent relations across spatial scales and regions. *River Research and Applications* **22**, 727–737.

Lamouroux N, Mérigoux S, Capra H, Dolédec S *et al.* (2010) The generality of abundance-environment relationships in microhabitats: a comment on Lancaster and Downes (2009). *River Research and Applications* **26**, 915–920.

Lancaster J and Downes BJ (2010a) Ecohydraulics needs to embrace ecology and sound science, and to avoid mathematical artefacts. *River Research and Applications* **26**, 921–929.

Lancaster J and Downes BJ (2010b) Linking the hydraulic world of individual organisms to ecological processes: putting ecology into ecohydraulics. *River Research and Applications* **26**, 385–403.

Lane P, Sheridan G, Noske P, Costenaro J *et al.* (2009) 'Dynamics of sediment and nutrient fluxes from burnt forest catchments'. Land and Water Australia, Canberra.

Lane PNJ, Hairsine PB, Croke JC and Takken I (2006) Quantifying diffuse pathways for overland flow between the roads and streams of the mountain ash forests of central Victoria, Australia. *Hydrological Processes* **20**, 1875–1884.

Langdon JS (1987) Active osmoregulation in the Australian bass, *Macquaria novemaculeata* (Steindachner), and the golden perch, *Macquaria ambigua* (Richardson) (Percichthyidae). *Marine and Freshwater Research* **38**, 771–776.

Langdon JS (1989) Experimental transmission and pathogenicity of Epizootic Haematopoietic Necrosis Virus (EHNV) in redfin perch, *Perca fluviatilis* L., and 11 other teleosts. *Journal of Fish Diseases* **12**, 295–310.

Langdon JS (1990) Disease risks of fish introductions and translocations. In *Introduced and Translocated Fishes and their Ecological Effects*. Proceedings of Australian Society for Fish Biology Workshop. (Ed. DA Pollard) pp. 98–107. Australian Society for Fish Biology, Sydney.

Langdon JS, Gudkovs N, Humphrey JD and Saxon EC (1985) Deaths in Australian freshwater fishes associated with *Chilodonella hexasticha* infection. *Australian Veterinary Journal* **62**, 409–413.

Larned ST, Datry T, Arscott DB and Tockner K (2010) Emerging concepts in temporary-river ecology. *Freshwater Biology* **55**, 717–738.

Larson HK (1995) A review of the Australian endemic gobiid fish genus *Chlamydogobius*, with description of five new species. *The Beagle, Records of the Museums and Art Galleries of the Northern Territory* **12**, 19–51.

Larson HK and Hoese DF (1996) Family Gobiidae, subfamilies Eleotridinae and Butinae: gudgeons. In *Freshwater Fishes of South-eastern Australia*. (Ed. RM McDowall) pp. 200–219. Reed, Sydney.

Larson HK and Martin KC (1990) *Freshwater Fishes of the Northern Territory*. Northern Territory Museum of Arts and Sciences, Darwin.

Lasker R (1978) Factors contributing to the variable recruitment of the northern anchovy (*Engraulis mordax*) in the Californian current: contrasting years 1975–1978. *Rapports et Porcčs-Verbaux des Réunions, Conseil International pour l'Exploration de la Mer* **178**, 375–388.

Lathrop RC, Johnson BM *et al.* (2002) Stocking piscivores to improve fishing and water clarity: a synthesis of the Lake Mendota biomanipulation project. *Freshwater Biology* **47**, 2410–2424.

Lauder GV (1985) Aquatic feeding in lower vertebrates. In *Functional Vertebrate Morphology*. (Eds M Hildebrand, DM Bramble, KF Liem and DB Wake) pp. 210–229. Harvard University Press, Cambridge.

Layman CA and Winemiller KO (2004) Size-based responses of prey to piscivore exclusion in a species-rich Neotropical river. *Ecology* **85**, 1311–1320.

Le Cren ED (1965) Some factors regulating the size of populations of freshwater fish. *Mitteilungen der Internationalen Vereinigung für theoretische und angewandte Limnologiae* **13**, 88–105.

Le Quesne T, Kendy E and Weston D (2010) 'The implementation challenge: taking stock of government policies to protect and restore environmental flows' Nature Conservancy/WWF, Godalming.

Le Vin AL, Adam A, Tedder A, Arnold KE *et al.* (2011) Validation of swabs as a non-destructive and relatively non-invasive DNA sampling method in fish. *Molecular Ecology Resources* **11**, 107–109.

Leggett WC and Deblois E (1994) Recruitment in marine fishes: is it regulated by starvation and predation in the egg and larval stages? *Netherlands Journal of Sea Research* **32**, 119–134.

Leibold MA, Holyoak M *et al.* (2004) The metacommunity concept: a framework for multi-scale community ecology. *Ecology Letters* **7**, 601–613.

Leichhardt L (1847) *Journal of an Overland Expedition in Australia from Moreton Bay to Port Essington.* T and W Boone, London.

Leis J, Gomon M and Larson HK (2007) Australian fish taxonomists: an endangered species. *Australian Society for Fish Biology Newsletter* **37**, 8–11.

Leprieur F, Brosse S, Garcia-Bertou E, Oberdorff T *et al.* (2009) Scientific uncertainty and the assessment of risks posed by non-native freshwater fishes. *Fish and Fisheries* **10**, 88–97.

Lester NP, Shuter BJ and Abrams PA (2004) Interpreting the von Bertalanffy model of somatic growth in fishes: the cost of reproduction. *Proceedings of the Royal Society of London. Series B: Biological Sciences* **271**, 1625–1631.

Lester RE and Boulton AJ (2008) Rehabilitating agricultural streams in Australia with wood: a review. *Environmental Management* **42**, 310–326.

Lester RE and Wright W (2009) Reintroducing wood to streams in agricultural landscapes: changes in velocity profile, stage and erosion rates. *River Research and Applications* **25**, 376–392.

Lévêque C, Oberdorff T, Paugy D, Stiassny MLJ *et al.* (2008) Global diversity of fish (Pisces) in freshwater. *Hydrobiologia* **595**, 545–567.

Lewin W, Arlinghaus R and Mehner T (2006) Documented and potential biological impacts of recreational fishing: insights for management and conservation. *Reviews in Fisheries Science* **14**, 305–367.

Lewis WM and Flickinger S (1967) Home range tendency of the largemouth bass (*Micropterus salmoides*). *Ecology* **48**, 1020–1023.

Li J, Rong Xia R, McDowall RM, López JA *et al.* (2010) Phylogenetic position of the enigmatic *Lepidogalaxias salamandroides* with comment on the orders of lower Euteleostean fishes. *Molecular Phylogenetics and Evolution* **57**, 932–936.

Li S and Mathias JA (1987) The critical period of high mortality of larvae fish: a discussion based on current research. *Chinese Journal of Oceanology and Limnology* **5**, 80–97.

Li Z, Gao Y and Feng Q (2009) Hierarchical structure of the otolith of adult wild carp. *Materials Science and Engineering: C* **29**, 919–924.

Liem KF (1980) Adaptive significance of intra- and interspecific differences in the feeding repertoires of cichlid fishes. *American Zoologist* **20**, 295–314.

Liem KF (1990) Aquatic versus terrestrial feeding modes: possible impacts on the trophic ecology of vertebrates. *American Zoologist* **30**, 209–221.

Lieschke JA and Closs GP (1999) Regulation of zooplankton composition and distribution by a zooplanktivorous fish in a shallow, eutrophic floodplain lake in south east Australia. *Archiv für Hydrobiologie* **146**, 397–412.

Likens GE, Walker KF *et al.* (2009) Ecosystem science: toward a new paradigm for managing Australia's inland aquatic ecosystems. *Marine and Freshwater Research* **60**, 271–279.

Lind PR, Robson BJ and Mitchell BD (2007) Multiple lines of evidence for the beneficial effects of environmental flows in two lowland rivers in Victoria, Australia. *River Research and Applications* **23**, 933–946.

Lindeman RL (1942) The trophic-dynamic aspect of ecology. *Ecology* **23**, 399–418.

Lindenmayer DB and Likens GE (2009) Adaptive monitoring: a new paradigm for long-term research and monitoring. *Trends in Ecology and Evolution* **24**, 482–486.

Lindenmayer DB and Likens GE (2010) The science and application of ecological monitoring. *Biological Conservation* **143**, 1317–1328.

Linfield RSJ (1974) The errors likely in ageing roach *Rutilus rutilus* L., with special reference to stunted populations. In *The Ageing of Fish: Symposium Proceedings, Reading 1973*. (Ed. TB Bagenal) pp. 167–172. Unwin, Old Woking.

Linfield RSJ (1985) An alternative concept to home range theory with respect to populations of cyprinids in major river systems. *Journal of Fish Biology* **27**, 187–196.

Lintermans M (1998) 'The ecology of the two-spined blackfish *Gadopsis bispinosus* (Pisces: Gadopsidae)'. Masters thesis. Australian National University, Canberra.

Lintermans M (2000) Recolonization by the mountain galaxias *Galaxias olidus* of a montane stream after the eradication of rainbow trout *Oncorhynchus mykiss*. *Marine and Freshwater Research* **51**, 799–804.

Lintermans M (2004) Human-assisted dispersal of alien freshwater fish in Australia. *New Zealand Journal of Marine and Freshwater Research* **38**, 481–501.

Lintermans M (2005) Sand slugs and fish movement in the Murrumbidgee River: can you restore and maintain deepwater habitats in sediment-impacted large rivers? In *Native Fish Habitat Rehabilitation and Management in the Murray-Darling Basin: Statement, Recommendations and Supporting Papers*. Albury, NSW, 10–11 February 2004. (Eds M Lintermans, P Cottingham and R O'Connor) pp. 77–87. Murray-Darling Basin Commission, and Cooperative Research Centre for Freshwater Ecology, Canberra.

Lintermans M (2006) 'The re-establishment of endangered Macquarie perch *Macquaria australasica* in the Queanbeyan River, New South Wales, with an examination of dietary overlap with alien trout'. Technical Report. Cooperative Research Centre for Freshwater Ecology, Canberra.

Lintermans M (2007) *Fishes of the Murray-Darling Basin: An Introductory Guide*. Murray-Darling Basin Commission, Canberra.

Lintermans M (2012) Managing potential impacts of reservoir enlargement on threatened *Macquaria australasica* and *Gadopsis bispinosus* in southeastern Australia. *Endangered Species Research* **16**, 1–16.

Lintermans M and Osborne WS (2002) *Wet and Wild: A Field Guide to the Freshwater Animals of the Southern Tablelands and High Country of the ACT and NSW.* Environment ACT, Canberra.

Lintermans M and Phillips B (2004) *Downstream Movement of Fish in the Murray-Darling Basin: Statement, Recommendations and Supporting Papers.* Canberra, 3–4 June 2003. Murray-Darling Basin Commission, Canberra.

Lintermans M and Raadik T (2003) Local eradication of trout from streams using rotenone: the Australian experience. In *Managing Invasive Freshwater Fish in New Zealand*. pp. 95–101. Department of Conservation, Wellington.

Lintermans M, Broadhurst B, Thiem JD, Ebner BC *et al.* (2010) 'Constructed homes for threatened fishes in the Cotter River catchment: Phase 2'. Final Report. Institute for Applied Ecology, University of Canberra, Canberra.

Lintermans M, Rowland SJ, Koehn J, Butler G *et al.* (2005) The status, threats and management of freshwater cod species in Australia. In *Management of Murray Cod in the Murray-Darling Basin: Statement, Recommendations and Supporting Papers.* Canberra, 3–4 June 2004. (Eds M Lintermans and B Phillips) pp. 15–29. Murray-Darling Basin Commission, Canberra.

Llewellyn LC (1966) Age determination of native inland fish of NSW. *The Fisherman(New South Wales)* **2**, 14–19.

Llewellyn LC (1968) Tagging gives answers to fish growth queries. *The Fisherman (New South Wales)* **3**, 1–5.

Llewellyn LC (1973) Spawning, development, and temperature tolerance of the spangled perch, *Madigania unicolor* (Günther), from inland waters in Australia. *Australian Journal of Marine and Freshwater Research* **24**, 73–94.

Llewellyn LC (1974) Spawning, development and distribution of the southern pigmy perch *Nannoperca australis australis* Günther from inland waters in eastern Australia. *Australian Journal of Marine and Freshwater Research* **25**, 121–149.

Llewellyn LC (1979) Some observations on the spawning and development of the Mitchellian freshwater hardyhead *Craterocephalus fluviatilis* McCulloch from inland waters in New South Wales. *Australian Zoologist* **20**, 269–288.

Llewellyn LC and MacDonald MC (1980) Family Percichthyidae: Australian freshwater basses and cods. In *Freshwater Fishes of South-eastern Australia*. (Ed. RM McDowall) pp. 142–149. Reed, Sydney.

Lloyd LN (1990) Ecological interactions of *Gambusia holbrooki* with Australian native fishes. In *Introduced and Translocated Fshes and their Ecological Effects*. (Ed. DA Pollard) pp. 94–97. Bureau of Rural Resources, Canberra.

Lloyd LN and Walker KF (1986) Distribution and conservation status of small freshwater fish in the River Murray, South Australia. *Transactions of the Royal Society of South Australia* **110**, 49–57.

Lockett MM (1988) The effect of rotenone on fishes and its use as a sampling technique: a survey. *Zeitschrift für Fischkunde* **5**, 13–45.

Lodge DM (1993) Biological invasions: lessons for ecology. *Trends in Ecology and Evolution* **8**, 133–137.

Lomnicki A (1988) *Population Ecology of Individuals.* Princeton University Press, New York.

Lomolino MV, Riddle BR, Whittaker RJ and Brown JH (2009) *Biogeography*, 4th edn. Sinauer Associates, Massachusetts.

Long JA (2010) *The Rise of Fishes: 500 million Years of Evolution*, 2nd edn. Johns Hopkins University Press, Baltimore.

Long JA, Young GC, Holland T, Senden TJ *et al.* (2006) An exceptional Devonian fish from Australia sheds light on tetrapod origins. *Nature* **444**, 199–202.

Loot G, Poulin R, Lek S and Guegan J-F (2002) The differential effects of *Ligula intestinalis* (L.) plerocercoids on host growth in three natural population of roach, *Rutilus rutilus* (L.). *Ecology of Freshwater Fish* **11**, 1–10.

Lorenzen K (2012) Allometry of natural mortality as a basis for assessing optimal release size in fish-stocking programmes. *Canadian Journal of Fisheries and Aquatic Sciences* **57**, 2374–2381.

Lorenzen K, Amarasinghe US, Bartley DM, Bell JD *et al.* (2000) Strategic review of enhancements and culture-based fisheries. In *Conference on Aquaculture in the Third Millennium*. Bangkok, 20–25 February, pp. 221–237.

Low T (2001) *Feral Future*. Penguin, Melbourne.

Lowe S, Browne M, Boudjelas S and De Poorter M (2000) '100 of the world's worst invasive alien species: a selection from the Global Invasive Species Database'. Invasive Species Specialist Group, International Union for Conservation of Nature, Auckland.

Lowe WH, Likens GE and Power ME (2006) Linking scales in stream ecology. *BioScience* **56**, 591–597.

Lowe-McConnell R (1975) *Fish Communities in Tropical Freshwaters.* Longman, London.

Lowe-McConnell RH (1964) The fishes of the Rupununi Savanna district of British Guiana, South America. Part 1. Ecological grouping of fish species and effects of the seasonal cycle on the fish. *Journal of the Linnean Society (Zoology)* **45**, 103–144.

Lowe-McConnell RH (1987) *Ecological Studies in Tropical Fish Communities.* Cambridge University Press, Cambridge.

Lucas MC and Baras E (2000) Methods for studying spatial behaviour of freshwater fishes in the natural environment. *Fish and Fisheries* **1**, 283–316.

Lucas MC, Baras E, Thom TJ, Duncan A *et al.* (2001) *Migration of Freshwater Fishes.* Blackwell Science, Oxford.

Lucas MC and Batley E (1996) Seasonal movements and behaviour of adult barbel *Barbus barbus*, a riverine cyprinid fish: implications for river management. *Journal of Applied Ecology* **33**, 1345–1358.

Ludgate BG and Closs GP (2003) 'Responses of fish communities to sustained removals of perch (*Perca fluviatilis*)'. Department of Conservation, Wellington, New Zealand.

Lundberg JG, Kottelat M, Smith GR, Stiassny MLJ *et al.* (2000) So many fishes, so little time: an overview of recent ichthyological discovery in continental waters. *Annals of the Missouri Botanical Garden* **87**, 26–62.

Lundquist CJ, Diehl JM, Harvey E and Botsford LW (2002) Factors affecting implementation of recovery plans. *Ecological Applications* **12**, 713–718.

Lüthi D, Le Floch M *et al.* (2008) High-resolution carbon dioxide concentration record 650,000–800,000 years before present. *Nature* **453**, 379–382.

Lymbery AJ, Hassan M, Morgan DL, Beatty SJD *et al.* (2010) Parasites of native and exotic freshwater fishes in southwestern Australia. *Journal of Fish Biology* **76**, 1770–1785.

Lyon J, McDonald A, Stoessel D, Ingram B *et al.* (2008a) 'Population, growth and genetic status of trout cod *Maccullochella macquariensis* in the Ovens River, Victoria'. Arthur Rylah Institute for Environmental Research, Department of Sustainability and Environment, Melbourne.

Lyon J, Stuart I, Ramsey D and O'Mahony J (2010) The effect of water level on lateral movements of fish between river and off-channel habitats and implications for management. *Marine and Freshwater Research* **61**, 271–278.

Lyon JP and O'Connor JP (2008) Smoke on the water: can riverine fish populations recover following a catastrophic fire-related sediment slug? *Austral Ecology* **33**, 794–806.

Lyon JP, Ryan TJ and Scroggie MP (2008b) Effects of temperature on the fast-start swimming performance of an Australian freshwater fish. *Ecology of Freshwater Fish* **17**, 184–188.

Lythgoe JN (1979) *The Ecology of Vision.* University of California Press, San Diego.

Lythgoe JN (1980) Vision in fishes: ecological adaptations. In *Environmental Physiology of Fishes.* (Ed. MA Ali) pp. 431–445. Plenum Press, New York.

Lythgoe JN and Partridge JC (1989) Visual pigments and the acquisition of visual information. *Journal of Experimental Biology* **146**, 1–20.

Lytle DA and Poff NL (2004) Adaptation to natural flow regimes. *Trends in Ecology and Evolution* **19**, 94–100.

Macarthur RH and Pianka ER (1966) On optimal use of a patchy environment. *American Naturalist* **100**, 603–609.

MacArthur RH and Wilson EO (1967) *The Theory of Island Biogeography.* Princeton University Press, New York.

Macdonald J and Tonkin Z (2008) 'A review of the impact of eastern gambusia on native fishes of the Murray-Darling Basin'. Arthur Rylah Institute for Environmental Research, Department of Sustainability and Environment, Melbourne.

Macdonald JA, Montgomery JC and Wells RM (1987) Comparative physiology of Antarctic fishes. *Advances in Marine Biology* **24**, 321–388.

Macdonald JI and Crook DA (2010) Variability in Sr:Ca and Ba:Ca ratios in water and fish otoliths across an estuarine salinity gradient. *Marine Ecology Progress Series* **413**, 147–161.

Macdonald JI and Davies PE (2007) Improving the upstream passage of two galaxiid fish species through a pipe culvert. *Fisheries Management and Ecology* **14**, 221–230.

Mack RN, Simberloff D, Lonsdale MW, Evans H *et al.* (2000) Biotic invasions: causes, epidemiology, global consequences, and control. *Ecological Applications* **10**, 689–710.

Mackaness G (1965) *Fourteen Journeys over the Blue Mountains of New South Wales 1813–1841.* Horwitz-Grahame, Sydney.

Mackay N and Eastburn D (Eds) (1990) *The Murray.* Murray-Darling Basin Commission, Canberra.

Mackay NJ (1973a) The effects of methallibure (I.C.I. 33,828) and thiourea on gametogenesis in the firetail gudgeon, *Hypseleotris galii. General and Comparative Endocrinology* **20**, 221–235.

Mackay NJ (1973b) Histological changes in the ovaries of the golden perch, *Plectroplites ambiguus*, associated with the reproductive cycle. *Marine and Freshwater Research* **24**, 95–102.

Mackay NJ (1973c) The reproductive cycle of the firetail gudgeon, *Hypseleotris galii*. II. Seasonal histological changes in the testes. *Australian Journal of Zoology* **21**, 67–74.

Mackay NJ (1974) The reproductive cycle in the firetail gudgeon, *Hypseleotris galii*. III. Environmental regulation of vitellogenesis. *Australian Journal of Zoology* **22**, 449–456.

Maddern MG, Morgan DL and Gill HS (2007) Distribution, diet and potential ecological impacts of the introduced Mozambique mouthbrooder *Oreochromis mossambicus* Peters (Pisces: Cichlidae) in Western Australia. *Journal of the Royal Society of Western Australia* **90**, 203–214.

Maddock I (1999) The importance of physical habitat assessment for evaluating river health. *Freshwater Biology* **41**, 373–391.

Magellan K, Pettersson LB and Magurran AE (2005) Quantifying male attractiveness and mating behaviour through phenotypic size manipulation in the Trinidadian guppy, *Poecilia reticulata*. *Behavioral Ecology and Sociobiology* **58**, 366–374.

Magnhagen C (2008) Parental care and predation risk in fish: defend, desert or devour. In *Fish Reproduction*. (Eds MJ Rocha, A Arukwe and BG Kapoor) pp. 397–420. Science Publishers, New Hampshire.

Magnuson JJ (1991) Fish and fisheries ecology. *Ecological Applications* **1**, 13–26.

Malinin LK (1969) Home range and homing instinct of fish. *Journal of Zoology* **48**, 381–391.

Mallen-Cooper M (1992) Swimming ability of juvenile Australian bass, *Macquaria novemaculeata* (Steindachner), and juvenile barramundi, *Lates calcarifer* (Bloch), in an experimental vertical-slot fishway. *Australian Journal of Marine and Freshwater Research* **43**, 823–834.

Mallen-Cooper M (1994) Swimming ability of adult golden perch *Macquaria ambigua* (Percichthyidae), and adult silver perch, *Bidyanus bidyanus* (Teraponidae), in an experimental vertical-slot fishway. *Australian Journal of Marine and Freshwater Research* **45**, 191–198.

Mallen-Cooper M (1996) Fishways and freshwater fish migration in south-eastern Australia. PhD thesis. University of Technology, Sydney.

Mallen-Cooper M (1999) Developing fishways for non-salmonid fishes: a case study from the Murray River. In *Innovations in Fish Passage Technology*. (Ed. M Odeh) pp. 173–195. American Fisheries Society, Bethesda.

Mallen-Cooper M (2000) Taking the mystery out of migration. In *Proceedings of Workshop on Fish Movement and Migration*. Bendigo, September 1999. (Eds DA Hancock, DC Smith and JD Koehn) pp. 101–112. Australian Society for Fish Biology, Sydney.

Mallen-Cooper M (2001) 'Fish passage in off-channel habitats of the lower Murray river'. Report to Wetland Care Australia. Fishway Consulting Services, Sydney.

Mallen-Cooper M and Brand DA (1992), 'Assessment of two fishways on the River Murray and historical changes in fish movement'. Report to the Murray-Darling Basin Commission. Fisheries Research Institute, Sydney.

Mallen-Cooper M and Brand DA (2007) Non-salmonids in a salmonid fishway: what do 50 years of data tell us about past and future fish passage? *Fisheries Management and Ecology* **14**, 319–332.

Mallen-Cooper M and Harris J (1990) Fishways in mainland south-eastern Australia. In *Proceedings of the International Symposium on Fishways '90*. Gifu, 8–10 October. (Ed. S Kimura) pp. 211–230. Organizing Committee, Gifu.

Mallen-Cooper M and Stuart IG (2003) Age, growth and non-flood recruitment of two potamodromous fishes in a large semi-arid/temperate river system. *River Research and Applications* **19**, 697–719.

Mallen-Cooper M and Stuart IG (2007) Optimising Denil fishways for passage of small and large fishes. *Fisheries Management and Ecology* **14**, 61–71.

Mallen-Cooper M, Koehn J, King A, Stuart I *et al.* (2008) 'Risk assessment of the proposed Chowilla regulator and managed floodplain inundations on fish'. Department of Water, Land and Biodiversity Conservation, Berri.

Mallen-Cooper M, Stuart IG, Hides-Pearson F and Harris JH (Eds) (1996) 'Fish migration in the Murray River and assessment of the Torrumbarry fishway'. In *Proceedings of the Inaugural Riverine Environment Research Forum of MDBC Natural Resource Management Strategy Funded Projects*. 4–6 October 1995, Atwood. Murray-Darling Basin Commission, Canberra.

Malmqvist B and Rundle S (2002) Threats to the running water ecosystems of the world. *Environmental Conservation* **29**, 134–153.

Malzahn AM, Aberle N, Clemmesen C and Boersma M (2007) Nutrient limitation of primary producers affects planktivorous fish condition. *Limnology and Oceanography* **52**, 2062–2071.

Mancinelli G, Costantini ML and Rossi L (2002) Cascading effects of predatory fish exclusion on the detritus-based food web of a lake littoral zone (Lake Vico, central Italy). *Oecologia* **133**, 402–411.

Manly BFJ, McDonald LL and Thomas DL (2002) *Resource Selection by Animals: Statistical Design and Analysis for Field Studies*, 2nd edn. Chapman and Hall, London.

Marchetti MP, Moyle PB and Levine R (2004) Invasive species profiling? Exploring the characteristics of non-native fishes across invasion stages in California. *Freshwater Biology* **49**, 646–661.

Margalef R (1960) Ideas for a synthetic approach to the ecology of running waters. *International Revue gesamten Hydrobiologie* **45**, 133–153.

Marina H, Beatty SJ, Morgan DL, Doupé RG *et al.* (2008) An introduced parasite, *Lernaea cyprinacea*, found on native freshwater fish in the south west of Western Australia. *Journal of the Royal Society of Western Australia* **91**, 149–153.

Mark W, Hofer R and Wieser W (1987) Diet spectra and resource partitioning in the larvae and juveniles of three species and six cohorts of cyprinids from a subalpine lake. *Oecologia* **71**, 388–396.

Marr SM, Marchetti MP, Olden JD, García-Berthou E *et al.* (2010) Freshwater fish introductions in Mediterranean-climate regions: are there commonalities in the conservation problem? *Diversity and Distributions* **16**, 606–619.

Marsh PC and Douglas ME (1997) Predation by introduced fishes on endangered humpback chub and other native species in the Little Colorado River, Arizona. *Transactions of the American Fisheries Society* **126**, 343–346.

Marshall DJ, Monro K, Bode M, Keough MJ *et al.* (2010) Phenotype–environment mismatches reduce connectivity in the sea. *Ecology Letters* **13**, 128–140.

Martin HA (1994) Australian Tertiary phytogeography: evidence from palynology. In *History of the Australian*

Vegetation: Cretaceous to Recent. (Ed. RS Hill) pp. 104–142. Cambridge University Press, Cambridge.

Masci KD, Ponniah M and Hughes JM (2008) Patterns of connectivity between the Lake Eyre and Gulf drainages, Australia: a phylogeographic approach. *Marine and Freshwater Research* **59**, 760–761.

Mathews RH (1903) The aboriginal fisheries at Brewarrina. *Journal and Proceedings of the Royal Society of New South Wales* **37**, 146–157.

Matsuzaki SS, Usio N, Takamura N and Washitani I (2009) Contrasting impacts of invasive engineers on freshwater ecosystems: an experiment and meta-analysis. *Oecologia* **158**, 673–686.

Matthews WJ (1998) *Patterns in Freshwater Fish Ecology.* Chapman and Hall, New York.

Matthews WJ and Heins DC (Eds) (1987) *Community and Evolutionary Ecology of North American Stream Fishes.* University of Oklahoma Press, Norman.

Matveev VF (2007) Assessing the biomass of small fish with a split-beam sonar in the Murray River, Australia. *Fisheries Research* **88**, 139–145.

May RC (1974) Larval mortality in fishes. In *The Early Life History of Fish.* (Ed. JHS Blaxter) pp. 3–19. Springer-Verlag, Berlin.

Mayden RL (1999) Consilience and a hierarchy of species concepts: advances toward closure on the species puzzle. *Journal of Nematology* **31**, 95–116.

McCann KS and Rooney N (2009) The more food webs change, the more they stay the same. *Philosophical Transactions of the Royal Society B – Biological Sciences* **364**, 1789–1801.

McCann KS, Rasmussen JB and Umbanhowar J (2005) The dynamics of spatially coupled food webs. *Ecology Letters* **8**, 513–523.

McCarthy B, Conallin A, D'Santos P and Baldwin D (2006) Acidification, salinization and fish kills at an inland wetland in south-eastern Australia following partial drying. *Ecological Management and Restoration* **7**, 218–223.

McCasker NG (2009) Of life and death in a lowland river: investigating mortality during the early life stages of Murray-Darling fishes. PhD thesis. Charles Sturt University, Albury.

McCoy F (1885) *Prodromus of the Zoology of Victoria or Figures and Descriptions of the Living Species of all Classes of the Victorian Indigenous Animals,* Vol. 1. John Ferres, Government Printer, Melbourne.

McCulloch M, Cappo M, Aumend J and Muller W (2005) Tracing the life history of individual barramundi using laser ablation MC-ICP-MS Sr-isotopic and Sr/Ba ratios in otoliths. *Marine and Freshwater Research* **56**, 637–644.

McDougall A (2004) Assessing the use of sectioned otoliths and other methods to determine the age of the centropomid fish, barramundi (*Lates calcarifer*) (Bloch), using known-age fish. *Fisheries Research* **67**, 129–141.

McDowall RM (1981) The relationships of Australian freshwater fishes. *Monographiae Biologicae* **41**, 1253–1273. Dr W Junk, The Hague.

McDowall RM (1988) *Diadromy in Fishes.* Croom Helm, London.

McDowall RM (1990) *New Zealand Freshwater Fishes*, 2nd edn. Heinemann Reed, Auckland.

McDowall RM (1994) On size and growth in freshwater fish. *Ecology of Freshwater Fish* **3**, 67–79.

McDowall RM (1996a) Family Poeciliidae: Livebearers. In *Freshwater Fishes of South-eastern Australia.* (Ed. RM McDowall) pp. 116–122. Reed, Sydney.

McDowall RM (1996b) *Freshwater Fishes of South-eastern Australia.* Reed, Sydney.

McDowall RM (1997) An accessory lateral line in some New Zealand and Australian galaxiids (Teleostei: Galaxiidae). *Ecology of Freshwater Fish* **6**, 217–224.

McDowall RM (2001) Diadromy, diversity and divergence: implications for speciation processes in fishes. *Reviews in Fish Biology and Fisheries* **7**, 443–462.

McDowall RM (2002) Accumulating evidence for a dispersal biogeography of southern cool temperate freshwater fishes. *Journal of Biogeography* **29**, 207–219.

McDowall RM (2006) Crying wolf, crying foul, or crying shame: alien salmonids and a biodiversity crisis in the southern cool-temperate galaxioid fishes? *Reviews in Fish Biology and Fisheries* **16**, 233–422.

McDowall RM (2007) On amphidromy, a distinct form of diadromy in aquatic organisms. *Fish and Fisheries* **8**, 1–13.

McDowall RM (2010) *New Zealand Freshwater Fishes: An Historical and Ecological Biogeography.* Fish and Fisheries Series, no. 32. Springer, New York.

McDowall RM and Charteris SC (2006) The possible adaptive advantages of terrestrial egg deposition in some fluvial diadromous galaxiid fishes (Teleostei: Galaxiidae). *Fish and Fisheries* **7**, 153–164.

McDowall RM and Waters JM (2004) Phylogenetic relationships in a small group of diminutive galaxiid fishes and the evolution of sexual dimorphism. *Journal of the Royal Society of New Zealand* **34**, 23–57.

McGarvey DJ and Hughes RM (2008) Longitudinal zonation of Pacific Northwest (USA) fish assemblages and the species–discharge relationship. *Copeia* **2008**, 311–321.

McGlashan DJ and Hughes JM (2000) Reconciling patterns of genetic variation with stream structure, Earth history and biology in the Australian freshwater fish *Craterocephalus stercusmuscarum* (Atherinidae). *Molecular Ecology* **9**, 1737–1751.

McGlashan DJ and Hughes JM (2001) Low levels of genetic differentiation among populations of the freshwater fish *Hypseleotris compressa* (Gobiidae: Eleotridinae): implications for its biology, population connectivity and history. *Heredity* **86**, 222–233.

McGlashan DJ and Hughes JM (2002) Extensive genetic divergence among populations of the Australian freshwater fish, *Pseudomugil signifer* (Pseudomugilidae), at different hierarchical scales. *Marine and Freshwater Research* **53**, 897–907.

McGuigan K, Franklin CE, Moritz C and Blows MW (2003) Adaptation of rainbowfish to lake and stream habitats. *Evolution* **57**, 104–118.

McGuigan K, Zhu D, Allen GR and Moritz C (2000) Phylogenetic relationships and historical biogeography of melanotaeniid fishes in Australia and New Guinea. *Marine and Freshwater Research* **51**, 713–723.

McGuigan KL (2001) An addition to the rainbowfish (Melanotaeniidae) fauna of north Queensland. **46**, 647–655.

McIntosh AR and Towsend CR (1996) Interactions between fish, grazing invertebrates and algae in a New Zealand stream: a trophic cascade mediated by fish-induced changes to grazer behaviour? *Oecologia* **108**, 174–181.

McKay R (1977) 'The Australian aquarium fish industry and the possibility of the introduction of exotic fish species and diseases'. Fisheries Paper 25. Fisheries Division, Department of Primary Industry, Canberra.

McKinnon LJ (1997) 'Monitoring of fish aspects of the flooding of Barmah forest'. Final report to the Murray-Darling Basin Commission for Natural Resources Management Strategy Project V014. Marine and Freshwater Resources Institute, Queenscliff.

McLaren S, Wallace MW, Gallagher SJ, Miranda JA *et al.* (2011) Palaeogeographic, climatic and tectonic change in southeastern Australia: the Late Neogene evolution of the Murray Basin. *Quaternary Science Reviews* **30**, 1086–1111.

McLeod R (2004) 'Counting the cost: impact of invasive animals in Australia'. Cooperative Research Centre for Pest Animal Control, Canberra.

McMahon TA and Finlayson BL (2003) Droughts and anti-droughts: the low flow hydrology of Australian rivers. *Freshwater Biology* **48**, 1147–1160.

McMaster D and Bond N (2008) A field and experimental study on the tolerances of fish to *Eucalyptus camaldulensis* leachate and low dissolved oxygen concentrations. *Marine and Freshwater Research* **59**, 177–185.

McNee A (2002) 'A national approach to the management of exotic fish species in the aquarium trade: an inventory of exotic freshwater fish species'. Report for Fisheries Resources Research Fund. Department of Agriculture, Fisheries and Forestry, Canberra.

McNeil DG and Closs GP (2007) Behavioural responses of a south-east Australian floodplain fish community to gradual hypoxia. *Freshwater Biology* **52**, 412–420.

McNeil DG, Westergaard S and Hartwell D (2010) 'Preliminary investigations into the control of speckled livebearers (*Phalloceros caudimaculatus*)'. Report to Primary Industries and Resources South Australia. SARDI Research Report Series 452. South Australian Research and Development Institute, Adelaide.

McPeek MA and Peckarsky BL (1998) Life histories and the strengths of species interactions: combining mortality, growth, and fecundity effects. *Ecology* **79**, 867–879.

Mees GF (1961) Description of a new fish of the family Galaxiidae from Western Australia. *Journal of the Royal Society of Western Australia* **44**, 33–38.

Meffe GK and Vrijenhoek RC (1988) Conservation genetics in the management of desert fishes. *Conservation Biology* **2**, 157–169.

Menge BA and Sutherland JP (1976) Species diversity gradients: synthesis of the roles of predation, competition and temporal variability. *American Naturalist* **110**, 351–359.

Meredith SN (2005) Ecology and bioenergetics of the gudgeon (*Hypseleotris* spp.) in Maroon Dam: a zooplanktivorous fish in a whole-lake biomanipulation. PhD thesis. Queensland University of Technology, Brisbane.

Meredith SN, Matveev VF and Mayes P (2003) Spatial and temporal variability in the distribution and diet of the gudgeon (Eleotridae: *Hypseleotris* spp.) in a subtropical Australian reservoir. *Marine and Freshwater Research* **54**, 1009–1017.

Merrick JR (2006) Australasian freshwater fish faunas: diversity, interrelationships, radiations and conservation. In *Evolution and Biogeography of Australasian Vertebrates.* (Eds JR Merrick, M Archer, G Hickey and MSY Lee) pp. 195–224. Auscipubs, Sydney.

Merrick JR and Midgley SH (1976) Reproduction and development in the freshwater grunters, *Therapon fuliginosus* and *T. welchi* (Theraponidae; Teleostei). *Newsletter of the Australian Society for Limnology* **13**, 19–20.

Merrick JR and Midgley SH (1981) Spawning behaviour of the freshwater catfish *Tandanus tandanus* (Plotosidae). *Australian Journal of Marine and Freshwater Research* **32**, 1003–1006.

Merrick JR and Schmida GE (1984) *Australian Freshwater Fishes: Biology and Management.* Griffith Press, Netley.

Merrick JR, Midgley SH and Rimmer MA (1983) Preliminary studies on growth of two spotted barramundi species, *Scleropages leichardti* and *S. jardini* (Pisces: Osteoglossidae). *Aquaculture* **32**, 195–199.

Metcalfe NB and Thorpe JE (1992) Anorexia and defended energy levels in over-wintering juvenile salmon. *Journal of Animal Ecology* **61**, 175–181.

Michaelis FB (1985) 'Threatened fish: a report on the threatened fish on inland waters in Australia'. Australian National Parks and Wildlife Service Report 3. Australian National Parks and Wildlife Service, Canberra.

Midgley SH (1977) 'Burdekin River'. Queensland Fisheries Service Special Report 10. Queensland Fisheries Service, Brisbane.

Milano D, Lozada M and Zagarese H (2010) Predator-induced reaction patterns of landlocked *Galaxias maculatus* to visual and chemical cues. *Aquatic Ecology* **44**, 741–748.

Miles NG, West RJ and Norman MD (2009) Does otolith chemistry indicate diadromous life cycles for five Australian riverine fishes? *Marine and Freshwater Research* **60**, 904–911.

Milinski M (1988) Games fish play: making decisions as a social forager. *Trends in Ecology and Evolution* **3**, 325–330.

Milla S, Depiereux S and Kestemont P (2011) The effects of estrogenic and androgenic endocrine disruptors on the immune system of fish: a review. *Ecotoxicology* **20**, 305–319.

Miller AD, Waggy G, Ryan SG and Austin CM (2004) Mitochondrial 12S rRNA sequences support the existence of a third species of freshwater blackfish (Percichthyidae: *Gadopsis*) from south-eastern Australia. *Memoirs of Museum Victoria* **61**, 121–127.

Miller KG, Kominz MA *et al.* (2005) The Phanerozoic record of global sea-level change. *Science* **310**, 1293–1298.

Miller TJ (2002) Assemblages, communities and species interactions. In *Fishery Science: The Unique Contribution of Early Life Stages.* (Eds LA Fuiman and RG Werner) pp. 183–205. Blackwell Science, Oxford.

Miller TJ, Crowder LB, Rice JA and Marschell EA (1988) Larval size and recruitment mechanisms in fishes: toward a conceptual framework. *Canadian Journal of Fisheries and Aquatic Sciences* **45**, 1657–1670.

Mills CE, Hadwen WL and Hughes JM (2008) Looking through glassfish: marine genetic structure in estuarine species. *Marine and Freshwater Research* **59**, 627–637.

Mills LS, Soulé ME and Doak DF (1993) The keystone-species concept in ecology and conservation. *BioScience* **43**, 219–224.

Milton DA and Arthington AH (1984) Reproductive strategy and growth of the crimson spotted rainbowfish, *Melanotaenia splendida fluviatilis* (Castelnau) (Pisces, Melanotaeniidae) in southeastern Queensland. *Australian Journal of Marine and Freshwater Research* **35**, 75–83.

Milton DA and Arthington AH (1985) Reproductive strategy and growth of the Australian smelt, *Retropinna semoni* (Weber) (Pisces: Retropinnidae), and the olive perchlet, *Ambassis nigripinnis* (De Vis) (Pisces: Ambassidae), in Brisbane, south-eastern Queensland. *Australian Journal of Marine and Freshwater Research* **36**, 329–341.

Milton DA and Chenery SR (2005) Movement patterns of barramundi *Lates calcarifer*, inferred from 87Sr/86Sr and Sr/Ca ratios in otoliths, indicate non-participation in spawning. *Marine Ecology Progress Series* **301**, 279–291.

Mims MC, Olden JD, Shattuck ZR and Poff NL (2010) Life history trait diversity of native freshwater fishes in North America. *Ecology of Freshwater Fish* **19**, 390–400.

Ministerial Council on Forestry Fisheries and Aquaculture (1999) 'National policy for the translocation of live organisms: issues, principles and guidelines for implementation'. Ministerial Council on Forestry, Fisheries and Aquaculture, Canberra.

Mion JB, Stein RA and Marschall EA (1998) River discharge drives survival of larval walleye. *Ecological Applications* **8**, 88–103.

Miranda LE and Bettoli PW (2007) Mortality. In *Analysis and Interpretation of Freshwater Fisheries Data.* (Eds CS Guy and ML Brown) pp. 229–277. American Fisheries Society, Bethesda.

Mitchell TL (1839) *Three Expeditions into the Interior of Eastern Australia, with Descriptions of the Recently Explored Region of Australia Felix, and of the Present Colony of New South Wales*, 2 vols. T and W Boone, London.

Mittelbach G (1986) Predator-mediated habitat use: some consequences for species interactions. *Environmental Biology of Fishes* **16**, 159–169.

Mittelbach GG (1988) Competition among refuging sunfishes and effects of fish density on littoral zone invertebrates. *Ecology* **69**, 614–623.

Moffatt D and Voller J (2002) 'Fish and fish habitat of the Queensland Murray-Darling Basin'. Queensland Department of Primary Industries, Brisbane.

Moilanen A, Leathwick J and Elith J (2008) A method for spatial freshwater conservation prioritization. *Freshwater Biology* **53**, 577–592.

Molony BW, Lenanton R, Jackson G and Norriss J (2003) Stock enhancement as a fisheries management tool. *Reviews in Fish Biology and Fisheries* **13**, 409–432.

Mommsen TP (1998) Growth and metabolism. In *The Physiology of Fishes*, 2nd edn. (Ed. DH Evans) pp. 65–97. CRC Press, New York.

Mommsen TP (2001) Paradigms of growth in fish. *Comparative Biochemistry and Physiology B: Biochemistry and Molecular Biology* **129**, 207–219.

Mommsen TP and Moon TW (2001) Hormonal regulation of muscle growth. *Fish Physiology* **18**, 251–308.

Mommsen TP, French CJ and Hochachka PW (1980) Sites and pattern of protein and amino acid utlilization during the spawning migration of salmon. *Canadian Journal of Zoology* **58**, 1758–1799.

Montgomery J, Coombs S and Halstead M (1995) Biology of the mechanosensory lateral line in fishes. *Reviews in Fish Biology and Fisheries* **5**, 399–416.

Mooi RD and Gill AC (2010) Phylogenies without synapomorphies: a crisis in fish systematics: time to show some character. *Zootaxa* **2450**, 26–40.

Mooney HA and Cleland EE (2001) The evolutionary impact of invasive species. *Proceedings of the National Academy of Sciences of the United States of America* **98**, 5446–5451.

Moore A (2007) 'Freshwater stocking in Queensland: a position paper for use in the development of future ecologically sustainable management practices'. Department of Primary Industries and Fisheries, Brisbane.

Moore A, Ingram B, Friend S, Hui KH *et al.* (2010) 'Management of genetic resources for fish and crustaceans in the Murray-Darling Basin'. Bureau of Rural Sciences, Canberra.

Moore R (1979) Natural sex inversion in the giant perch (*Lates calcarifer*). *Australian Journal of Marine and Freshwater Research* **30**, 803–13.

Morales-Nin B (2000) Review of the growth regulation processes of otolith daily increment formation. *Fisheries Research* **46**, 53–67.

Morgan D and Beatty S (2006a) 'Fish and freshwater crayfish communities of the Brunswick and Preston rivers'. Report to the Leschenault Catchment Council. Freshwater Fish Research, Murdoch University, Perth.

Morgan DL (2003) Distribution and biology of *Galaxias truttaceus* (Galaxiidae) in south-western Australia, including first evidence of parasitism of fishes in Western Australia by *Ligula intestinalis* (Cestoda). *Environmental Biology of Fishes* **66**, 155–167.

Morgan DL and Beatty SJ (2006b) 'Overview of the feral goldfish control programme in the Vasse River, Western Australia: 2004–2006'. Report to Geocatch. Centre for Fish and Fisheries Research, Murdoch University, Perth.

Morgan DL and Beatty SJ (2006c) 'Re-establishment of native freshwater fishes in Bull Creek'. Report to Melville City Council, Perth.

Morgan DL and Beatty SJ (2006d) Use of a vertical-slot fishway by galaxiids in Western Australia. *Ecology of Freshwater Fish* **15**, 500–509.

Morgan DL and Gill HS (2001) The green swordtail *Xiphophorus helleri* Heckel (Poeciliidae): another aquarium fish established in the wild in Western Australia. *Records of the Western Australian Museum* **20**, 349–352.

Morgan DL and Gill HS (2004) Fish fauna in inland waters of the Pilbara (Indian Ocean) drainage division of Western Australia: evidence for three subprovinces. *Zootaxa* **636**, 1–43.

Morgan DL, Allen GR, Pusey BJ and Burrows DW (2011a) A review of the freshwater fishes of the Kimberley region of Western Australia. *Zootaxa* **2816**, 1–64.

Morgan DL, Allen M, Bedford P and Horstman M (2004a) Fish fauna of the Fitzroy River in the Kimberley region of Western Australia – including the Bunuba, Gooniyandi, Ngarinyin, Nyikina and Walmajarri aboriginal names. *Records of the Western Australian Museum* **22**, 147–161.

Morgan DL, Beatty SJ, Klunzinger MW, Allen MG *et al.* (2011b) *A Field Guide to Freshwater Fishes, Crayfishes and Mussels of South-western Australia*. SERCUL/Centre for Fish and Fisheries Research, Murdoch University, Perth.

Morgan DL, Gill HS and Potter IC (1995) Life cycle, growth and diet of Balston's pygmy perch in its natural habitat of acidic pools in south-western Australia. *Journal of Fish Biology* **47**, 808–825.

Morgan DL, Gill HS and Potter IC (1998) Distribution, identification and biology of freshwater fishes in south-western Australia. *Records of the Western Australian Museum* Supplement 56.

Morgan DL, Gill HS and Potter IC (2000) Age composition, growth and reproductive biology of the salamanderfish *Lepidogalaxias salamandroides*: a re-examination. *Environmental Biology of Fishes* **57**, 191–204.

Morgan DL, Gill HS, Maddern MG and Beatty SJ (2004b) Distribution and impacts of introduced freshwater fishes in Western Australia. *New Zealand Journal of Marine and Freshwater Research* **38**, 511–523.

Morgan DL, Hambleton SJ, Gill HS and Beatty SJ (2002) Distribution, biology and likely impacts of the introduced redfin perch (*Perca fluviatilis*) (Percidae) in Western Australia. *Marine and Freshwater Research* **53**, 1211–1221.

Morgan DL, Rowland AJ, Gill HS and Doupé RG (2004c) The implications of introducing a large piscivore (*Lates calcarifer*) into a regulated northern Australian river (Lake Kununurra, Western Australia). *Lakes and Reservoirs: Research and Management* **9**, 181–193.

Morgan DL, Thorburn DC and Gill HS (2003) Salinization of southwestern Western Australian rivers and the implications for the inland fish fauna: the Blackwood River, a case study. *Pacific Conservation Biology* **9**, 161–171.

Moritz C (1994) Defining 'Evolutionarily Significant Units' for conservation. *Trends in Ecology and Evolution* **9**, 373–375.

Moritz C (2002) Strategies to protect biological diversity and the evolutionary processes that sustain it. *Systematic Biology* **51**, 238–254.

Moritz C and Cicero C (2004) DNA barcoding: promise and pitfalls. *PLoS Biology* **2**, 1529–1531.

Moritz C, Lavery S and Slade R (1995) Using allele frequency and phylogeny to define units for conservation and management. In *Evolution and the Aquatic Ecosystem: Defining Unique Units in Population Conservation*. (Ed. JL Nielsen) pp. 249–262. American Fisheries Society, Bethesda.

Morris DW (1987) Spatial scale and the cost of density-dependent habitat selection. *Evolutionary Ecology* **1**, 379–88.

Morris DW and Brown JS (1992) The role of habitat selection in landscape ecology. *Evolutionary Ecology* **6**, 357–359.

Morrongiello JR, Beatty SJ *et al.* (2011a) Climate change and its implications for Australia's freshwater fish. *Marine and Freshwater Research* **62**, 1082–1098.

Morrongiello JR, Bond NR, Crook DA and Wong BBM (2011b) *Eucalyptus* leachate inhibits reproduction in a freshwater fish. *Freshwater Biology* **56**, 1736–1745.

Morrongiello JR, Crook DA, King AJ, Ramsey DSL *et al.* (2011c) Impacts of drought and predicted effects of climate change on fish growth in temperate Australian lakes. *Global Change Biology* **17**, 745–755.

Moyes CD and Schulte PM (2008) *Principles of Animal Physiology*, 2nd edn. Pearson Education, San Francisco.

Moyle PB and Cech JJ (2004) *Fishes: An Introduction to Ichthyology*, 5th edn. Prentice Hall, New Jersey.

Moyle PB and Herbold B (1987) Life-history patterns and community structure in stream fishes of western North America: comparisons with eastern North America and Europe. In *Community and Evolutionary Ecology of North American Stream Fishes.* (Eds WJ Matthews and DC Heins) pp. 25–32. University of Oklahoma Press, Norman.

Moyle PB and Light T (1996) Fish invasions in California: do abiotic factors determine success? *Ecology* **77**, 1666–1670.

Mugiya Y (1987) Effects of photoperiods on the formation of otolith increments in the embryonic and laval rainbow trout *Salmo gairdneri. Nippon Suisan Gakkaishi* **53**, 1979–1984.

Mukhina NV, Marshall CT and Yaragina NA (2003) Tracking the signal in year-class strength of Northeast Arctic cod through multiple survey estimates of egg, larval and juvenile abundance. *Journal of Sea Research* **50**, 57–75.

Munro AR, Gillanders BM, Elsdon TS, Crook DA *et al.* (2008) Enriched stable isotope marking of juvenile golden perch (*Macquaria ambigua*) otoliths. *Canadian Journal of Fisheries and Aquatic Sciences* **65**, 276–285.

Munro AR, Gillanders BM, Thurstan S, Crook DA *et al.* (2009) Transgenerational marking of freshwater fishes with enriched stable isotopes: a tool for fisheries management and research. *Journal of Fish Biology* **75**, 668–684.

Munz II (1979) Morphology and innervation of the lateral line system in *Sarotherodon niloticus* (L) (Cichlidae, Teleostei). *Zoomorphologie* **93**, 73–86.

Murphy BF and Timbal B (2008) A review of recent climate variability and climate change in southeastern Australia. *International Journal of Climatology* **28**, 859–879.

Murray M and Roach J (2002) Whitley, Gilbert Percy (1903–1975). *Australian Dictionary of Biography.* Vol. 16, pp. 542–543. Melbourne University Press, Melbourne. <http://adb.anu.edu.au/biography/whitley-gilbert-percy-12022/text21563>. 7/5/2012.

Murray-Darling Basin Authority (2011) Update on mid-Murray and Edward-Wakool system blackwater event (09/12/2010). <http://www.mdba.gov.au/media_centre/media_releases/ update-on-mid-murray-and-edward-wakool-system-blackwater-event>. February 2011.

Murray-Darling Basin Commission (2004) 'Native fish strategy for the Murray-Darling Basin 2003–2013'. Murray-Darling Basin Commission, Canberra.

Murray-Darling Basin Commission (2005) 'The Living Murray Foundation report on the significant ecological assets targeted in the First Step Decision'. Murray-Darling Basin Commission, Canberra.

Murray-Darling Basin Commission (2008) 'The Sea to Lake Hume: restoring fish passage in the Murray River'. MDBC Publication 32/08. Murray-Darling Basin Commission, Canberra.

Murray-Darling Basin Ministerial Council (2001) 'Basin salinity management strategy 2001–2015'. Murray-Darling Basin Commission, Canberra.

Musyl MK and Keenan CP (1992) Population genetics and zoogeography of Australian freshwater golden perch, *Macquaria ambigua* (Richardson 1845) (Teleostei: Percichthyidae), and electrophoretic identification of a new species from the Lake Eyre Basin. *Australian Journal of Marine and Freshwater Research* **43**, 1585–1601.

Musyl MK and Keenan CP (1996) Evidence for cryptic speciation in Australian freshwater eel-tailed catfish, *Tandanus tandanus* (Teleostei: Plotosidae). *Copeia* **1996**, 526–534.

Myers GS (1949) Usage of anadromous, catadromous and allied terms for migratory fishes. *Copeia* **1949**, 89–97.

Myers N, Mittermeier RA, Mittermeier CG, da Fonseca GAB *et al.* (2000) Biodiversity hotspots for conservation priorities. *Nature* **403**, 853–858.

Myers R (1995) Recruitment of marine fish: the relative roles of density-dependent and density-independent mortality in the egg, larval, and juveniles stages. *Marine Ecology Progress Series* **128**, 308–309.

Myers RA and Worm B (2003) Rapid worldwide depletion of predatory fish communities. *Nature* **423**, 280–283.

Naiman RJ and Décamps H (1997) The ecology of interfaces: riparian zones. *Annual Review of Ecology and Systematics* **28**, 621–658.

Naiman RJ, Latterell JJ, Pettit NE and Olden JD (2008) Flow variability and the biophysical vitality of river systems. *Comptes Rendus Geoscience* **340**, 629–643.

Nakano S (1995a) Competitive interactions for foraging microhabitats in a size-structured interspecific dominance hierarchy of two sympatric stream salmonids in a natural habitat. *Canadian Journal of Fisheries and Aquatic Sciences* **73**, 1845–1854.

Nakano S (1995b) Individual differences in resource use, growth and emigration under the influence of a dominance hierarchy in fluvial red-spotted masu salmon in a natural habitat. *Journal of Animal Ecology* **64**, 75–84.

Nakano S, Fausch KD and Kitano S (1999) Flexible niche partitioning via a foraging mode shift: a proposed mechanism for coexistence in stream-dwelling charrs. *Journal of Animal Ecology* **68**, 1079–1092.

Nancarrow J (2009) Gerard Krefft: a singular man. *Proceedings of the Royal Society of Victoria* **121**, 146–154.

Nankervis L and Southgate PC (2006) An integrated assessment of gross marine protein sources used in formulated microbound diets for barramundi (*Lates calcarifer*) larvae. *Aquaculture* **257**, 453–464.

Nankervis L, Matthews SJ and Appleford P (2000) Effect of dietary non-protein energy source on growth, nutrient retention and circulating insulin-like growth factor I and triiodothyronine levels in juvenile barramundi, *Lates calcarifer. Aquaculture* **191**, 323–335.

Nathan R (2008) An emerging movement ecology paradigm. *Proceedings of the National Academy of Science of the United States of America* **105**, 19050–19051.

National Land and Water Resources Audit (2001) 'Australian water resources assessment 2000'. National Water Commission, Canberra.

National Land and Water Resources Audit (2002) 'Australian catchment, river and estuary assessment 2002'. National Water Commission, Canberra.

National Parks and Wildlife Service, New South Wales (2003) 'NSW threat abatement plan. Predation by *Gambusia holbrooki* – the plague minnow'. National Parks and Wildlife Service, Sydney.

Native Fish Australia (2006) Proposal for the establishment of recreational fisheries in Victoria for the bluenose or trout cod *Maccullochella macquariensis*. <http://www.nativefish.asn.au/files/NFA_VIC_trout_cod_recreational_plan.pdf>. January 2011.

Natsumeda T (1998) Home range of the Japanese fluvial sculpin, *Cottus pollux*, in relation to nocturnal activity patterns. *Environmental Biology of Fishes* **53**, 295–301.

Natural Resource Management Ministerial Council (2007) 'Australian pest animal strategy: a national strategy for the management of vertebrate pest animals in Australia'. Natural Resource Management Ministerial Council, Canberra.

Near TJ, Sandel M, Kuhn KL, Unmack PJ *et al.* (2012) Nuclear gene inferred phylogenies resolve the relationships of the enigmatic pygmy sunfishes, *Elassoma* (Teleostei: Percomorpha). *Molecular Phylogenetics and Evolution* **63**, 388–395.

Nehring S and Klingenstein F (2008) Aquatic alien species in Germany – listing system and options for action. *Neobiota* **7**, 19–33.

Neilson JL, Arrizabalaga H, Fragoso N, Hobday A *et al.* (Eds) (1999) *Tagging and Tracking Marine Mammals with Electronic Devices*. Springer, New York.

Nel JL, Roux DJ, Maree G, Kleynhans CJ *et al.* (2007) Rivers in peril inside and outside protected areas: a systematic approach to conservation assessment of river ecosystems. *Diversity and Distributions* **13**, 341–352.

Nevill J (2007) Policy failure: Australian freshwater protected area networks. *Australasian Journal of Environmental Management* **14**, 35–47.

Neville H, Dunham J and Peacock M (2006) Landscape attributes and life history variability shape genetic structure of trout populations in a stream network. *Landscape Ecology* **21**, 901–916.

Nevitt GA (1991) Do fish sniff? A new mechanism of olfactory sampling in pleuronectid flounders. *Journal of Experimental Biology* **157**, 1–18.

Nevo E, Beiles A and Ben-Shlomo R (1984) The evolutionary significance of genetic diversity: ecological, demographic and life history correlates. In *Lecture Notes in Biomathematics 53: Evolutionary Dynamics of Genetic Diversity*. (Ed. GS Mani) pp. 13–213. Springer-Verlag, Berlin.

New JG (2002) Multimodal integration in the feeding behaviors of predatory teleost fishes. *Brain Behavior and Evolution* **59**, 177–189.

Newbury RW and Gaboury MC (1993) *Stream Analysis and Fish Habitat Design: A Field Manual*. Newbury Hydraulics, British Columbia.

Newman DM, Jones PL and Ingram BA (2008) Age-related changes in ovarian characteristics, plasma sex steroids and fertility during pubertal development in captive female Murray cod *Maccullochella peelii peelii*. *Comparative Biochemistry and Physiology A – Molecular and Integrative Physiology* **150**, 444–451.

Ngarrindjeri-Tendi (2006) 'Ngarrindjeri Nation Yarluwar-Ruwe plan: caring for Ngarrindjeri sea country and culture'. Ngarrindjeri Heritage Committee/Ngarrindjeri Native Title Management Committee, Meningie.

Nicholls AG (1958) The population of a trout stream and the survival of released fish. *Australian Journal of Marine and Freshwater Research* **9**, 526–536.

Nicol S, Bearlin A, Robley A, Koehn J *et al.* (2001) Distribution of large woody debris in the mid-reaches of the Murray River: a template for reconstruction. *Ecological Management and Restoration* **2**, 64–67.

Nicol S, Todd C, Koehn J and Lieschke J (2005) How can recreational angling regulations help meet the multiple objectives for the management of Murray cod populations? In *Management of Murray Cod in the Murray-Darling Basin: Statement, Recommendations and Supporting Papers*. Canberra, 3–4 June 2004. (Eds M Lintermans and B Phillips) pp. 98–106. Murray-Darling Basin Commission, Canberra.

Nicol SJ, Barker RJ, Koehn JD and Burgman MA (2007) Structural habitat selection by the critically endangered trout cod, *Maccullochella macquariensis*, Cuvier. *Biological Conservation* **138**, 30–37.

Nicol SJ, Lieschke J, Lyon J and Hughes V (2002) 'Resnagging revolution: river habitat rehabilitation through resnagging'. Report to MD 2001 Fish Rehab Program, Department of Agriculture Fisheries Forestry Australia. Department of Natural Resources and Environment, Melbourne.

Nicol SJ, Lieschke JA, Lyon JP and Koehn JD (2004) Observations on the distribution and abundance of carp and native fish, and their responses to a habitat restoration trial in the Murray River, Australia. *New Zealand Journal of Marine and Freshwater Research* **38**, 541–551.

Nicols A (1882) *The Acclimatisation of the Salmonidae at the Antipodes: Its History and Results*. Sampson Low, Marston Searle and Rivington, London.

Nielsen DL, Brock MA, Rees GN and Baldwin DS (2003) Effects of increasing salinity on freshwater ecosystems in Australia. *Australian Journal of Botany* **51**, 655–665.

Nielsen DL, Hillman TJ, Smith FJ and Shiel RJ (2000a) The influence of a planktivorous fish on zooplankton assemblages in experimental billabongs. *Hydrobiologia* **434**, 1–9.

Nielsen DL, Smith FJ, Hillman TJ and Shiel RJ (2000b) Impact of water regime and fish predation on zooplankton resting egg production and emergence. *Journal of Plankton Research* **22**, 433–446.

Nielson JL, Arrizabalaga H, Fragoso N, Hobday A *et al.* (Eds) (2009) *Tagging and Tracking Marine Mammals with Electronic Devices.* Springer, New York.

Nikolskii GV (1969) *Fish Population Dynamics.* Oliver and Boyd, Edinburgh.

Nikolsky GV (1963) *The Ecology of Fishes.* Academic Press, New York.

Nikora VI, Aberle J, Biggs BJF, Jowett IG *et al.* (2003) Effects of fish size, time-to-fatigue and turbulence on swimming performance: a case study of *Galaxias maculatus. Journal of Fish Biology* **63**, 1365–1382.

Nilsson C, Reidy CA, Dynesius M and Revenga C (2005) Fragmentation and flow regulation of the world's large river systems. *Science* **308**, 405–408.

Ning NSP, Nielsen DL, Hillman TJ and Suter PJ (2010) The influence of planktivorous fish on zooplankton communities in riverine slackwaters. *Freshwater Biology* **55**, 360–374.

Nock CJ, Elphinstone MS, Rowland SJ and Baverstock PR (2010) Phylogenetics and revised taxonomy of the Australian freshwater cod genus, *Maccullochella* (Percichthyidae). *Marine and Freshwater Research* **61**, 980–991.

Nock CJ, Ovenden JR, Butler GL, Wooden I *et al.* (2011) Population structure, effective population size and adverse effects of stocking in the endangered Australian eastern freshwater cod *Maccullochella ikei. Journal of Fish Biology* **78**, 303–321.

Norris A and Burgin S (2011) Apparent rapid loss of endocrine disruptors from wetlands used to store either tertiary treated sewage effluent or stormwater runoff. *Water, Air and Soil Pollution* **219**, 285–295.

Norris R, Lake P and Swain R (1982) Ecological effects of mine effluents on the South Esk River, north-eastern Tasmania. III. Benthic macroinvertebrates. *Marine and Freshwater Research* **33**, 789–809.

Northcote TG (1978) *Migratory Strategies and Production in Freshwater Fishes.* Blackwell Scientific, Oxford.

Northcote TG (1992) Migration and residency in stream salmonids: some ecological considerations and evolutionary consequences. *Nordic Journal of Freshwater Research* **62**, 5–17.

Northcote TG (2010) Controls for trout and char migratory/resident behaviour mainly in stream systems above and below waterfalls/barriers: a multidecadal and broad geographical review. *Ecology of Freshwater Fish* **19**, 487–509.

Northcote TG and Hartman GF (Eds) (2004) *Fishes and Forestry: Worldwide Watershed Interactions and Management.* Blackwell Science, Oxford.

Norton SF and Brainerd EL (1993) Convergence in the feeding mechanics of ecomorphologically similar species in the Centrarchidae and Cichlidae. *Journal of Experimental Biology* **176**, 11–29.

Norvick MS (2011) Paleogene basalts prove early uplift of Victoria's Eastern Uplands. *Australian Journal of Earth Sciences* **58**, 93–94.

O'Connor NA (1991) The effect of habitat complexity on the macroinvertebrates colonising wood substrates in a lowland stream. *Oecologia* **85**, 504–512.

Oberdorff T, Hugueny B and Guégan JF (1997) Is there an influence of historical events on contemporary fish species richness in rivers? Comparisons between Western Europe and North America. *Journal of Biogeography* **24**, 461–467.

O'Connor JP and Mahoney JC (2004) Observations of ovarian involution in the Australian grayling (*Prototroctes maraena*). *Ecology of Freshwater Fish* **13**, 70–73.

O'Connor JP and Zampatti BP (2006) Spawning season and site location of *Gadopsis bispinosus* Sanger (Pisces: Gadopsidae) in a montane stream of southeastern Australia. *Transactions of the Royal Society of South Australia* **130**, 227–232.

O'Connor JP, Koehn JD, Nicol SJ, O'Mahony DJ *et al.* (2009) Retention of radio tags in golden perch (*Macquaria ambigua*), silver perch (*Bidyanus bidyanus*) and carp (*Cyprinus carpio*). *Marine and Freshwater Research* **60**, 334–340.

O'Connor JP, O'Mahony DJ and O'Mahony JM (2005) Movements of *Macquaria ambigua*, in the Murray River, south-eastern Australia. *Journal of Fish Biology* **66**, 392–403.

O'Connor JP, O'Mahony DJ, O'Mahony JM and Glenane TJ (2006) Some impacts of low and medium head weirs on downstream fish movement in the Murray-Darling Basin in southeastern Australia. *Ecology of Freshwater Fish* **15**, 419–427.

O'Connor WG and Koehn JD (1998) Spawning of the broad-finned galaxias, *Galaxias brevipinnis* Günther (Pisces: Galaxiidae) in coastal streams of southeastern Australia. *Ecology of Freshwater Fish* **7**, 95–100.

Odeh M (Ed.) (1999) *Innovations in Fish Passage Technology.* American Fisheries Society, Bethesda.

Odum HT (1957) Trophic structure and productivity of Silver Springs, Florida. *Ecological Monographs* **27**, 55–112.

Okuda N, Tayasu I and Yanagisawa Y (1998) Determinate growth in a paternal mouthbrooding fish whose reproductive success is limited by buccal capacity. *Evolutionary Ecology* **12**, 681–699.

Olden JD (2006) Biotic homogenization: a new research agenda for conservation biogeography. *Journal of Biogeography* **33**, 2027–2039.

Olden JD and Kennard MJ (2010) Intercontinental convergence of fish life history strategies along a gradient of hydrologic variability. In *Community Ecology of Stream Fishes: Concepts, Approaches, and Techniques. Symposium 73.* (Eds KB Gido and DA Jackson) pp. 83–107. American Fisheries Society, Bethesda.

Olden JD and Naiman RJ (2010) Incorporating thermal regimes into environmental flows assessments: modifying dam operations to restore freshwater ecosystem integrity. *Freshwater Biology* **55**, 86–107.

Olden JD, Jackson DA and Peres-Neto PR (2001) Spatial isolation and fish communities in drainage lakes. *Oecologia* **127**, 572–585.

Olden JD, Kennard MJ and Pusey BJ (2008) Species invasions and the changing biogeography of Australian freshwater fishes. *Global Ecology and Biogeography* **17**, 25–37.

Olden JD, Kennard MJ, Leprieur F, Tedesco PA *et al.* (2010) Conservation biogeography of freshwater fishes: recent progress and future challenges. *Diversity and Distributions* **16**, 496–513.

Oliveira EJ, Padua JG, Zucchi MI, Vencovsky R *et al.* (2006) Origin, evolution and genome distribution of microsatellites. *Genetics and Molecular Biology* **29**, 294–307.

Olley JM and Wasson RJ (2003) Changes in the flux of sediment in the upper Murrumbidgee catchment, southeastern Australia, since European settlement. *Hydrological Processes* **17**, 3307–3320.

Omland KE (1997) Correlated rates of molecular and morphological evolution. *Evolution* **51**, 1381–1393.

Ormerod SJ, Dobson M, Hildrew AG and Townsend CR (2010) Multiple stressors in freshwater ecosystems. *Freshwater Biology* **55** (Supplement 1), 1–4.

Orr TM and Milward NE (1984) Reproduction and development of *Neosilurus ater* (Perugia) and *Neosilurus hyrtlii* Steindachner (Teleostei: Plotosidae) in a tropical Queensland stream. *Australian Journal of Marine and Freshwater Research* **35**, 187–195.

Ovenden JR, White RWG and Adams M (1993) Mitochondrial and allozyme genetics of two Tasmanian galaxiids (*Galaxias auratus* and *G. tanycephalus*, Pisces: Galaxiidae) with restricted lacustrine distributions. *Heredity* **70**, 223–230.

Oxley J (1820) *Journals of Two Expeditions into the Interior of New South Wales undertaken by order of the British Government in the years 1817–18.* John Murray, London.

Page RDM and Holmes EC (1988) *Molecular Evolution: A Phylogenetic Approach.* Blackwell, Oxford.

Page TJ and Hughes JM (2010) Comparing the performance of multiple mitochondrial genes in the analysis of Australian freshwater fishes. *Journal of Fish Biology* **77**, 2093–2122.

Page TJ, Sharma S and Hughes JM (2004) Deep phylogenetic structure has conservation implications for ornate rainbowfish (Melanotaeniidae: *Rhadinocentrus ornatus*) in Queensland, eastern Australia. *Marine and Freshwater Research* **55**, 165–172.

Palmer M, Allan JD, Meyer J and Bernhardt ES (2007) River restoration in the twenty-first century: data and experiential future efforts. *Restoration Ecology* **15**, 472–481.

Palmer MA, Ambrose RF and Poff NL (1997) Ecological theory and community restoration ecology. *Restoration Ecology* **5**, 291–300.

Palmer MA, Bernhardt ES, Allan JD, Lake P *et al.* (2005) Standards for ecologically successful river restoration. *Journal of Applied Ecology* **42**, 208–217.

Palmer MA, Liermann CAR, Nilsson C, Florke M *et al.* (2008) Climate change and the world's river basins: anticipating management options. *Frontiers in Ecology and the Environment* **6**, 81–89.

Palsbøll PJ, Berube M and Allendorf FW (2007) Identification of management units using population genetic data. *Trends in Ecology and Evolution* **22**, 12–16.

Palstra AP, de Graaf M and Sibbing FA (2004) Riverine spawning and reproductive segregation in a lacustrine cyprinid species flock, facilitated by homing? *Animal Biology* **54**, 393–415.

Pankhurst NW and King HR (2010) Temperature and salmonid reproduction: implications for aquaculture. *Journal of Fish Biology* **76**, 69–85.

Pankhurst NW and Munday PL (2011) Effects of climate change on fish reproduction and early life stages. *Marine and Freshwater Research* **62**, 1015–1026.

Pannella G (1971) Fish otoliths: daily growth layers and periodical patterns. *Science* **173**, 1124–1127.

Paragamian VL and Kruse G (2001) Kootenai River white sturgeon spawning migration behavior and a predictive model. *North American Journal of Fisheries Management* **21**, 10–21.

Parker GA (1992) The evolution of sexual size dimorphism in fish. *Journal of Fish Biology* **41**, 1–20.

Parkinson D, Philippart J and Baras E (1999) A preliminary investigation of spawning migrations of grayling in a small stream as determined by radio-tracking. *Journal of Fish Biology* **55**, 182.

Partridge TC, Bond GC, Hartnady CJH, de Menocal PB *et al.* (1995) Climatic effects of Late Neogene tectonism and volcanism. In *Paleoclimate and Evolution, with emphasis on Human Origins.* (Eds ES Vrba, GH Denton, TC Partridge and LH Burckle) pp. 8–23. Yale University Press, New Haven.

Pasonson IM (2000) *The Australian Ark: A History of Domesticated Animals in Australia.* CSIRO Publishing, Melbourne.

Paterson JA and Chapman LJ (2009) Fishing down and fishing hard: ecological change in the Nile perch of Lake Nabugabo, Uganda. *Ecology of Freshwater Fish* **18**, 380–394.

Patrick J (1998) Aquarium fish culture. In *Proceedings of the Queensland Warmwater Aquaculture Conference (Status and Potential).* Aquaculture Information Technologies, Tarome.

Patterson TA, Thomas L, Wilcox C, Ovaskainen O *et al.* (2007) State-space models of individual animal movement. *Trends in Ecology and Evolution* **23,** 87–94.

Patzner RA (2008) Reproductive strategies of fish. In *Fish Reproduction.* (Eds MJ Rocha, A Arukwe and BG Kapoor) pp. 311–350. Science Publishers, New Hampshire.

Pauly D, Christensen V, Dalsgaard J, Froese R *et al.* (1998) Fishing down marine food webs. *Science* **279,** 860–863.

Pavlov DA, Emil'yanova NG and Novikov GG (2009) Reproductive dynamics. In *Fish Reproductive Biology: Implications for Assessment and Management.* (Ed. T Jakobsen) pp. 48–90. Wiley-Blackwell, Chichester.

Pavuk NC (1994) Population genetics of selected whitebait species: *Galaxias maculatus* (Jenyns) and *Lovettia sealii* (Johnston). Masters thesis. University of Tasmania, Hobart.

Paxton JR and McGrouther MA (1997) A history of the fish collection at the Australian Museum (1860–1968), with a summary of current Australian fish collections. In *Collection Building in Ichthyology and Herpetology,* Vol. 3. (Eds TW Pietsch and WDJ Anderson) pp. 183–205. American Society of Ichthyologists and Herpetologists, Lawrence.

Pease BC, Reynolds DP and Walsh CT (2003) Validation of otolith age determination in Australian longfinned river eels, *Anguilla reinhardtii. Marine and Freshwater Research* **54,** 995–1004.

Peig J and Green AJ (2009) New perspectives for estimating body condition from mass/length data: the scaled mass index as an alternative method. *Oikos* **118,** 1883–1891.

Peig J and Green AJ (2010) The paradigm of body condition: a critical reappraisal of current methods based on mass and length. *Functional Ecology* **24,** 1323–1332.

Pen LJ and Potter IC (1990) Biology of the nightfish, *Bostockia porosa* Castelnau, in a south-western Australian river. *Australian Journal of Marine and Freshwater Research* **41,** 627–645.

Pen LJ and Potter IC (1991a) Biology of the western minnow, *Galaxias occidentalis* Ogilby (Teleostei: Galaxiidae), in a south-western Australian river. *Hydrobiologia* **211,** 77–88.

Pen LJ and Potter IC (1991b) Reproduction, growth and diet of *Gambusia holbrooki* (Girard) in a temperate Australian river. *Aquatic Conservation* **1,** 159–172.

Peňáz M (2001) A general framework of fish ontogeny: a review of the ongoing debate. *Folia Zoologica* **50,** 241–256.

Pender PJ and Griffin RK (1996) Habitat history of barramundi *Lates calcarifer* in a north Australian river system based on barium and strontium levels in scales. *Transactions of the American Fisheries Society* **125,** 679–689.

Pepin P, Shears TH and de Lafontaine Y (1992) Significance of body size to the interaction between a larval fish (*Mallotus villosus*) and a vertebrate predator (*Gasterosteus aculeatus*). *Marine Ecology Progress Series* **81,** 1–12.

Perry GLW and Bond NR (2009) Spatially explicit modeling of habitat dynamics and fish population persistence in an intermittent lowland stream. *Ecological Applications* **19,** 731–746.

Persson L and De Roos AM (2006) Food-dependent individual growth and population dynamics in fishes. *Journal of Fish Biology* **69,** 1–20.

Persson L and Greenberg LA (1990) Juvenile competitive bottlenecks: the perch (*Perca fluviatilis*)–roach (*Rutilus rutilus*) interaction. *Ecology* **71,** 41–56.

Peterson DL, Vecsei P and Jennings CA (2007) Ecology and biology of the lake sturgeon: a synthesis of current knowledge of threatened North American Acipenseridae. *Reviews in Fish Biology and Fisheries* **17,** 59–76.

Peterson RH, Daye PG, Lacroix GL and Garside ET (1982) Reproduction in fish experiencing acid and metal stress. In *Acid Rain/Fisheries.* (Eds TA Haines and RE Johnson) pp. 177–196. American Fisheries Society, Bethesda.

Petit RJ and Excoffier L (2009) Gene flow and species delimitation. *Trends in Ecology and Evolution* **24,** 386–393.

Peverell SC (2005) Distribution of sawfishes (Pristidae) in the Queensland Gulf of Carpentaria, Australia, with notes on sawfish ecology. *Environmental Biology of Fishes* **73,** 391–402.

Pfenninger M and Schwenk K (2007) Cryptic animal species are homogeneously distributed among taxa and biogeographical regions. *Evolutionary Biology* **7,** 121–127.

Philippart JC (1995) Is captive breeding an effective solution for the preservation of endemic species? *Biological Conservation* **72,** 281–295.

Phillips B (2001) 'Thermal pollution of the Murray-Darling Basin waterways'. Workshop at Lake Hume, 18–19 June. (Ed. B. Phillips). Inland Rivers Network/World Wide Fund for Nature Australia, Sydney.

Phillips B (2008) 'Audit of freshwater protected areas of the Murray-Darling Basin and development of a Habitat Management Areas "toolkit"'. Murray-Darling Basin Commission, Canberra.

Phillips B (Ed.) (2003) *Managing Fish Translocation and Stocking in the Murray-Darling Basin: Statement, Recommendations and Supporting Papers.* Canberra, 25–26 September 2002. WWF Australia, Sydney.

Phillips B and Butcher R (2005) 'River parks: building a system of "Habitat Management Areas" across the Murray-Darling. An international and national review of freshwater "protected areas" for conserving aquatic biodiversity and river health'. Murray-Darling Basin Commission, Canberra.

Phillips NM, Whitty JM, Morgan DL, Chaplin JA *et al.* (2009b) 'Freshwater sawfish (*Pristis microdon*) movements and demographics in the Fitzroy River, Western Australia and genetic analysis of *P. microdon* and *Pristis zijsron*'. Centre for Fish and Fisheries Research, Murdoch University, Perth.

Phillips RD, Storey AW and Johnson MS (2009a) Genetic structure of *Melanotaenia australis* at local and regional scales in the east Kimberley, Western Australia. *Journal of Fish Biology* **74**, 437–451.

Phillipson MJ, Byrne J and Rohan G (1986) 'Fisheries Research Paper 16'. South Australian Department of Fisheries, Adelaide.

Pianka ER (1970) On r- and K-selection. *American Naturalist* **104**, 592–597.

Piasecki W, Goodwin AE, Eiras JC and Nowak BF (2004) Importance of Copepoda in freshwater aquaculture. *Zoological Studies* **43**, 193–205.

Picha ME, Turano MJ, Beckman BR and Borski RJ (2008) Endocrine biomarkers of growth and applications to aquaculture: a minireview of growth hormone, insulin-like growth factor (IGF)-I, and IGF-binding proteins as potential growth indicators in fish. *North American Journal of Aquaculture* **70**, 196–211.

Pidgeon RWJ (1989) Age and growth of freshwater herring, *Potamalosa richmondia* Macleay (Clupeidae: Hyperlophinae), in the Bellinger River, NSW. *Australian Journal of Marine and Freshwater Research* **40**, 679–692.

Pimentel D, Zuniga R and Morrison D (2005) Update on the environmental and economic costs associated with alien-invasive species in the United States. *Ecological Economics* **52**, 273–288.

Pine WEI, Ludsin SA and DeVries DR (2000) First-summer survival of largemouth bass cohorts: is early spawning really best? *Transactions of the American Fisheries Society* **129**, 504–513.

Pinto L, Chandrasena N, Pera J, Hawkins P *et al.* (2005) Managing invasive carp (*Cyprinus carpio* L.) for habitat enhancement at Botany Wetlands, Australia. *Aquatic Conservation: Marine and Freshwater Ecosystems* **15**, 447–462.

Pires DF, Pires AM, Collares-Pereira MJ and Magalhães MF (2010) Variation in fish assemblages across dry-season pools in a Mediterranean stream: effects of pool morphology, physicochemical factors and spatial context. *Ecology of Freshwater Fish* **19**, 74–86.

Pitcher TJ, Green DA and Magurran AE (1986) Dicing with death: predator inspection strategy in minnow shoals. *Journal of Fish Biology* **28**, 439–448.

Pittock J, Hansen LJ and Abell R (2008) Running dry: freshwater biodiversity, protected areas and climate change. *Biodiversity* **9**, 30–38.

Poff NL (1997) Landscape filters and species traits: towards mechanistic understanding and prediction in stream ecology. *Journal of the North American Benthological Society* **16**, 391–409.

Poff NL and Allan JD (1995) Functional organization of stream fish assemblages in relation to hydrological variability. *Ecology* **76**, 606–627.

Poff NL and Zimmerman JKH (2010) Ecological responses to altered flow regimes: a literature review to inform the science and management of environmental flows. *Freshwater Biology* **55**, 194–205.

Poff NL, Allan JD, Bain MB, Karr JR *et al.* (1997) The natural flow regime: a paradigm for river conservation and restoration. *BioScience* **47**, 769–784.

Poff NL, Allan JD, Palmer MA, Hart DD *et al.* (2003) River flows and water wars: emerging science for environmental decision making. *Frontiers in Ecology and the Environment* **1**, 298–306.

Poff NL, Olden JD, Vieira NKM, Finn DS *et al.* (2006) Functional trait niches of North American lotic insects: traits-based ecological applications in light of phylogenetic relationships. *Journal of the North American Benthological Society* **25**, 730–755.

Poff NL, Richter BD, Arthington AH, Bunn SE *et al.* (2010) The ecological limits of hydrologic alteration (ELOHA): a new framework for developing regional environmental flow standards. *Freshwater Biology* **55**, 147–170.

Pohlmann K, Atema J and Breithaupt T (2004) The importance of the lateral line in nocturnal predation of piscivorous catfish. *Journal of Experimental Biology* **207**, 2971–2978.

Polis GA, Sears ALW, Huxel GR, Strong DR *et al.* (2000) When is a trophic cascade a trophic cascade? *Trends in Ecology and Evolution* **15**, 473–475.

Pollard DA (1971a) The biology of a landlocked form of the normally catadromous salmoniform fish *Galaxias maculatus* (Jenyns). II. Morphology and systematic relationships. *Australian Journal of Marine and Freshwater Research* **22**, 125–137.

Pollard DA (1971b) The biology of a landlocked form of the normally catadromous salmoniform fish *Galaxias maculatus* (Jenyns). I. Life cycle and origin. *Australian Journal of Marine and Freshwater Research* **22**, 91–124.

Pollard DA (1972) The biology of a landlocked form of the normally catadromous salmoniform fish *Galaxias maculatus* (Jenyns). V. Composition of the diet. *Australian Journal of Marine and Freshwater Research* **23**, 39–48.

Pollard DA (1974) The biology of a landlocked form of the normally catadromous salmoniform fish *Galaxias maculatus* (Jenyns). IV. Effects of cestode and nematode parasites. *Australian Journal of Marine and Freshwater Research* **25**, 105–120.

Pollard DA (1990) 'Introduced and translocated fishes and their ecological effects'. In *Proceedings of Australian Society for Fish Biology Workshop*. Magnetic Island. (Ed. DA Pollard). Bureau of Rural Resources, Canberra.

Pollard DA (1994) A comparison of fish assemblages and fisheries in intermittently open and permanently open coastal lagoons on the south coast of New South Wales, south-eastern Australia. *Estuaries* **17**, 631–646.

Pollard DA and Burchmore JJ (1986) A possible scenario for the future of Australia's freshwater fish fauna. In *Limnology in Australia.* (Eds P De Deckker and WD Williams) pp. 615–636. CSIRO, Australia/Dr W Junk, Dordrecht.

Pollard DA and Parker P (1996) Family Scorpionidae. Scorpionfishes. In *Freshwater Fishes of South-eastern Australia.* (Ed. RM McDowall) pp. 144–145. Reed, Sydney.

Pollino CA, Georgiades E and Holdway DA (2007a) Use of the Australian crimson-spotted rainbowfish (*Melanotaenia fluviatilis*) as a model test species for investigating the effects of endocrine disruptors. *Environmental Toxicology and Chemistry* **26**, 2171–2178.

Pollino CA, Georgiades E and Holdway DA (2009) Physiological changes in reproductively active rainbowfish (*Melanotaenia fluviatilis*) following exposure to naphthalene. *Ecotoxicology and Environmental Safety* **72**, 1265–1270.

Pollino CA and Holdway DA (2002) Toxicity testing of crude oil and related compounds using early life stages of the crimson-spotted rainbowfish (*Melanotaenia fluviatilis*). *Ecotoxicology and Environmental Safety* **52**, 180–189.

Pollino CA, Woodberry O, Nicholson A, Korb K *et al.* (2007b) Parameterisation and evaluation of a Bayesian network for use in an ecological risk assessment. *Environmental Modelling and Software* **22**, 1140–1152.

Pollock KH, Jones CM and Brown TL (1994) *Angler Survey Methods and their Applications in Fisheries Management.* American Fisheries Society, Bethesda.

Poole D (1984) Inland fishery on the River Murray: some historical facts. *SAFIC* **8**, 20–22.

Pope KL and Willis DW (1998) Early life history and recruitment of black crappie (*Pomoxis nigromaculatus*) in two South Dakota waters. *Ecology of Freshwater Fish* **7**, 56–68.

Popper AN and Lu Z (2000) Structure and function relationships in fish otolith organs. *Fisheries Research* **46**, 15–25.

Possingham HP, Andelman SJ, Burgman MA, Medellin RA *et al.* (2002) Limits to the use of threatened species lists. *Trends in Ecology and Evolution* **17**, 503–507.

Possingham HP, Lindenmayer DB and Norton TW (1993) A framework for the improved management of threatened species based on population viability analysis (PVA). *Pacific Conservation Biology* **1**, 39–45.

Post JR, Mushens C, Paul A and Sullivan M (2003) Assessment of alternative harvest regulations for sustaining recreational fisheries: model development and application to bull trout. *North American Journal of Fisheries Management* **23**, 22–34.

Potter IC (1970) The life cycle and ecology of Australian lampreys of the genus *Mordacia*. *Journal of Zoology, London* **161**, 487–511.

Potter IC (1996) Shortheaded lampreys. In *Freshwater Fishes of South-eastern Australia.* (Ed. RM McDowall) pp. 32–35. Reed, Sydney.

Potter IC and Strahan R (1968) The taxonomy of the lampreys *Geotria* and *Mordacia* and their distribution in Australia. *Proceedings of the Linnean Society, London* **179**, 229–240.

Power ME (1984) Habitat quality and the distribution of algae-grazing catfish in a Panamanian stream. *Journal of Animal Ecology* **53**, 357–374.

Power ME, Matthews WJ and Stewart AJ (1985) Grazing minnows, piscivorous bass, and stream algae: dynamics of a strong interaction. *Ecology* **66**, 1448–1456.

Power ME, Parker MS and Dietrich WE (2008) Seasonal reassembly of a river food web: floods, droughts, and impacts of fish. *Ecological Monographs* **7**, 263–282.

Power ME, Stout RJ, Cushing CE, Harper PP *et al.* (1988) Biotic and abiotic controls in river and stream communities. *Journal of the North American Benthological Society* **7**, 456–479.

Preece R (2004) 'Cold water pollution below dams in New South Wales: a desktop assessment'. NSW Department of Infrastructure, Planning and Natural Resources, Sydney.

Preece RM and Jones HA (2002) The effect of Keepit Dam on the temperature regime of the Namoi River, Australia. *River Research and Applications* **18**, 397–414.

Prentice EF, Flagg TA and McCutcheon NC (1990) Feasibility of using implantable passive integrated transponder (PIT) tags in salmonids. In *Fish Marking Techniques.* Symposium. (Eds NC Parker, AE Giorgi, RC Heidinger, DB Jester Jr, ED Prince and GA Winans) pp. 317–322. American Fisheries Society, Bethesda.

Price A (2007) Utilisation of still-water patches by fish and shrimp in a lowland river, with particular emphasis on early-life stages. PhD thesis. University of Canberra, Canberra.

Price P, Lovett S and Davies P (2009) 'A national synthesis of river restoration projects'. Waterlines Report Series 23. National Water Commission, Canberra.

Priede IG (1980) An analysis of objectives in telemetry studies of fish in the natural environment. In *A Handbook on Biotelemetry and Radio Tracking: Proceedings of an International Conference on Telemetry and Radio Tracking in Biology and Medicine.* Oxford, 20–22 March 1979. (Eds CJ Amlaner and DW MacDonald) pp. 105–118. Pergamon Press, Oxford.

Priede IG (1985) Metabolic scope in fishes. In *Fish Energetics: New Perspectives.* (Eds P Tytler and P Calow) pp. 33–64. Croom-Helm, London.

Pringle CM (2001) Hydrological connectivity and the management of biological reserves: a global perspective. *Ecological Applications* **11**, 981–998.

Pringle CM, Naiman RJ, Bretschko G, Karr JR *et al.* (1988) Patch dynamics in lotic systems: the stream as a mosaic.

Journal of the North American Benthological Society **7**, 503–524.

Pritchard J, Hammer M, Hall A, Lugg A *et al.* (2009) Drought and threatened species. In *Proceedings of the Murray-Darling Basin Authority Native Fish Forum 2009.* (Ed. J Pritchard) pp. 63–73. Murray-Darling Basin Authority, Canberra.

Pritchard JC (2004) Linking fish growth and climate across modern space and through evolutionary time. PhD thesis. Australian National University, Canberra.

Prosek J (2010) *Eels: An Exploration, from New Zealand to the Sargasso, of the World's Most Mysterious Fish.* Harper Collins, New York.

Prosser IP, Rutherfurd ID, Olley JM, Young WJ *et al.* (2001) Large-scale patterns of erosion and sediment transport in river networks, with examples from Australia. *Marine and Freshwater Research* **52**, 81–99.

Proudlove GS (2006) 'Subterranean fishes of the world. An account of the subterranean (hypogean) fishes described to 2003 with a bibliography 1541–2004'. International Society for Subterranean Biology, Moulis.

Puckridge JT and Drewien M (1988) The aquatic fauna of the north-west branch of Cooper Creek. In *The Coongie Lakes Study.* (Eds JR Reid and J Gillen) pp. 69–108. South Australian Department of Environment and Planning, Adelaide.

Puckridge JT and Walker KF (1990) Reproductive biology and larval development of a gizzard shad, *Nematalosa erebi* (Günther) (Dorosomatinae: Teleostei), in the River Murray, South Australia. *Australian Journal of Marine and Freshwater Research* **41**, 695–712.

Puckridge JT, Sheldon F, Walker KF and Boulton AJ (1998) Flow variability and the ecology of large rivers. *Marine and Freshwater Research* **49**, 55–72.

Puckridge JT, Walker KF and Costelloe JF (2000) Hydrological persistence and the ecology of dryland rivers. *River Research and Applications* **16**, 385–402.

Puckridge JT, Walker KF, Langdon JS, Daley C *et al.* (1989) Mycotic dermatitis in a fresh-water gizzard shad, the bony bream, *Nematalosa erebi* (Günther), in the River Murray, South Australia. *Journal of Fish Diseases* **12**, 205–221.

Pusey B, Kennard M, Burrows D, Perna C *et al.* (2011) Freshwater fish. In *Aquatic Biodiversity in Northern Australia: Patterns, Threats and Future.* (Ed. BJ Pusey) pp. 71–92. Charles Darwin University Press, Darwin.

Pusey BJ (1986) The effect of starvation on oxygen consumption and nitrogen excretion in *Lepidogalaxias salamandroides. Journal of Comparative Physiology B* **156**, 701–705.

Pusey BJ (1989) Aestivation in the teleost fish *Lepidogalaxias salamandroides* (Mees). *Comparative Biochemistry and Physiology A: Physiology* **92**, 137–138.

Pusey BJ (1990) Seasonality, aestivation and the life history of the salamanderfish *Lepidogalaxias salamandroides* (Pisces: Lepidogalaxiidae). *Environmental Biology of Fishes* **29**, 15–26.

Pusey BJ (1998) Methods addressing the flow requirements of fish. In 'Comparative evaluation of environmental flow assessment techniques: review of methods'. Occasional Paper 27/98. (Eds AH Arthington and JM Zalucki) pp. 64–103. Land and Water Resources Research and Development Corporation, Canberra.

Pusey BJ and Arthington AH (2003) Importance of the riparian zone to the conservation and management of freshwater fish: a review. *Marine and Freshwater Research* **54**, 1–16.

Pusey BJ and Bradshaw SD (1996) Diet and dietary overlap in fishes of temporary waters of southwestern Australia. *Ecology of Freshwater Fish* **5**, 183–194.

Pusey BJ and Kennard MJ (1996) Species richness and geographical variation in assemblage structure of the freshwater fish fauna of the Wet Tropics region of northern Queensland. *Marine and Freshwater Research* **47**, 563–573.

Pusey BJ and Kennard MJ (2001) *Guyu wujalwujalensis*, a new genus and species (Pisces: Percichthyidae) from north-eastern Queensland. *Ichthyological Exploration of Freshwaters* **12**, 17–28.

Pusey BJ and Kennard MJ (2009) Aquatic ecosystems of northern Australia. In *Northern Australia Land and Water Science Review.* (Ed. P Stone). Final report to the Northern Australia Land and Water Taskforce. CSIRO Publishing, Melbourne.

Pusey BJ and Stewart T (1989) Internal fertilization in *Lepidogalaxias salamandroides* Mees (Pisces: Lepidogalaxiidae). *Zoological Journal of the Linnean Society* **97**, 69–79.

Pusey BJ, Arthington AH and Flanders TJ (1994) An out door replicated artificial stream system: design, operating conditions, and initial invertebrate colonization. *Ecotoxicology and Environmental Safety* **27**, 177–191.

Pusey BJ, Arthington AH and Read MG (1993) Spatial and temporal variation in fish assemblage structure in the Mary River, south-eastern Queensland: the influence of habitat structure. *Environmental Biology of Fishes* **37**, 355–380.

Pusey BJ, Arthington AH and Read MG (1995) Species richness and spatial variation in fish assemblage structure in two rivers of the Wet Tropics of northern Queensland, Australia. *Environmental Biology of Fishes* **42**, 181–199.

Pusey BJ, Arthington AH and Read MG (1998a) Freshwater fishes of the Burdekin River, Australia: biogeography, history and spatial variation in community structure. *Environmental Biology of Fishes* **53**, 303–318.

Pusey BJ, Arthington AH and Read MG (2000a) The dry-season diet of freshwater fishes in monsoonal tropical rivers of Cape York Peninsula, Australia. *Ecology of Freshwater Fish* **9**, 177–190.

Pusey BJ, Arthington AH, Bird JR and Close PG (2001) Reproduction in three species of rainbowfish (Melanotaeniidae) from rainforest streams in northern Queensland, Australia. *Ecology of Freshwater Fish* **10**, 75–87.

Pusey BJ, Arthington AH, Close PG and Bird J (2002) Larval fishes in rainforest streams: recruitment and microhabitat use. *Proceedings of the Royal Society of Queensland* **110**, 27–46.

Pusey BJ, Arthington AH, Stewart-Koster B, Kennard MJ *et al.* (2010) Widespread omnivory and low temporal and spatial variation in the diet of fishes in a hydrologically variable northern Australian river. *Journal of Fish Biology* **77**, 731–753.

Pusey BJ, Bird J, Kennard MJ and Arthington AH (1997) Distribution of the Lake Eacham rainbowfish in the Wet Tropics region, North Queensland. *Australian Journal of Zoology* **45**, 75–84.

Pusey BJ, Kennard MJ and Arthington AH (2000b) Discharge variability and the development of predictive models relating stream fish assemblage structure to habitat in northeastern Australia. *Ecology of Freshwater Fish* **9**, 30–50.

Pusey BJ, Kennard MJ and Arthington AH (2004) *Freshwater Fishes of North-eastern Australia.* CSIRO Publishing, Melbourne.

Pusey BJ, Kennard MJ and Arthington AH (2008) Origins and maintenance of freshwater fish biodiversity in the Wet Tropics region. In *Living in a Dynamic Tropical Forest Landscape.* (Eds N Stork and S Turton). Wiley-Blackwell, Oxford.

Pusey BJ, Kennard MJ and Bird J (2000c) Fishes of the dune fields of Cape Flattery, northern Queensland and other dune systems in north-eastern Australia. *Icthyological Exploration of Freshwaters* **11**, 65–74.

Pusey BJ, Kennard MJ, Arthur JM and Arthington AH (1998b) Quantitative sampling of stream fish assemblages: single- vs. multiple-pass electrofishing. *Austral Ecology* **23**, 365–374.

Pusey BJ, Storey AW, Davies PM and Edward DHD (1989) Spatial variation in fish communities in two south-western Western Australian river systems. *Journal of Royal Society of Western Australia* **71**, 69–75.

Pyke GH (2008) Plague minnow or mosquito fish? A review of the biology and impacts of introduced *Gambusia* species. *Annual Review of Ecology, Evolution and Systematics* **39**, 171–191.

Pyron M (1999) Relationships between geographic range size, body size, local abundance, and habitat breadth in North American suckers and sunfish. *Journal of Biogeography* **26**, 549–558.

Quinn TJ and Deriso RB (1999) *Quantitative Fish Dynamics.* Oxford University Press, New York.

Quist MC and Guy CS (2001) Growth and mortality of prairie stream fishes: relations with fish community and instream habitat characteristics. *Ecology of Freshwater Fish* **10**, 88–96.

Raadik TA (1995) 'A research recovery plan for the barred galaxias in south-eastern Australia'. Department of Conservation and Natural Resources, Victoria.

Raadik TA (2011). Systematic revision of the mountain galaxias (*Galaxias olidus* Günther) species complex (Teleostei: Galaxiidae) in eastern Australia. PhD thesis. University of Canberra, Canberra.

Rahel FJ (2002) Homogenisation of freshwater faunas. *Annual Review of Ecology and Systematics* **33**, 291–315.

Rahel FJ (2007) Biogeographic barriers, connectivity and homogenization of freshwater faunas: it's a small world after all. *Freshwater Biology* **52**, 696–710.

Raine A, Hardwick L *et al.* (2009) Managing impacts from thermal pollution on NSW rivers: policy and practice. In *Proceedings of the Australian Water Association Specialty Conference on Rivers and Reservoirs.* Canberra. Australian Water Association, Sydney.

Rakes PL, Shute JR and Shute PW (1999) Reproductive behavior, captive breeding, and restoration ecology of endangered fishes. *Environmental Biology of Fishes* **55**, 31–42.

Raso S and Anderson TA (2003) Effects of dietary fish oil replacement on growth and carcass proximate composition of juvenile barramundi (*Lates calcarifer*). *Aquaculture Research* **34**, 813–819.

Ratnasingham S and Hebert PDN (2007) BOLD: the barcode of life data system (<http://www.barcodinglife.org>). *Molecular Ecology Notes* **7**, 355–364.

Raubenheimer D and Simpson SJ (1994) The analysis of nutrient budgets. *Functional Ecology* **8**, 783–791.

Raubenheimer D and Simpson SJ (2004) Organismal stoichiometry: quantifying non-independence among food components. *Ecology* **85**, 1203–1216.

Raubenheimer D, Simpson SJ and Mayntz D (2009) Nutrition, ecology and nutritional ecology: toward an integrated framework. *Functional Ecology* **23**, 4–16.

Raubenheimer D, Zemke-White WL, Phillips RJ and Clements KD (2005) Algal macronutrients and food selection by the omnivorous marine fish *Girella tricuspidata. Ecology* **86**, 2601–2610.

Rawson CA, Lim RP and Warne MSJ (2008) Skeletal morphology and maturation of male *Gambusia holbrooki* exposed to sewage treatment plant effluent. *Ecotoxicology and Environmental Safety* **70**, 453–461.

Rayner TS and Creese RG (2006) A review of rotenone use for the control of non-indigenous fish in Australian fresh waters, and an attempted eradication of the noxious fish, *Phalloceros caudimaculatus. New Zealand Journal of Marine and Freshwater Research* **40**, 477–486.

Rayner TS, Jenkins KM and Kingsford RT (2009a) Small environmental flows, drought and the role of refugia for freshwater fish in the Macquarie Marshes, arid Australia. *Ecohydrology* **2**, 440–453.

Rayner TS, Pusey BJ and Pearson RG (2008) Seasonal flooding, instream habitat structure and fish assemblages in the Mulgrave River, north-east Queensland: towards a new conceptual framework for understanding fish-habitat dynamics in small tropical rivers. *Marine and Freshwater Research* **59**, 97–116.

Rayner TS, Pusey BJ and Pearson RG (2009b) Spatio-temporal dynamics of fish feeding in the lower Mulgrave River, north-eastern Queensland: the influence of seasonal flooding, instream productivity and invertebrate abundance. *Marine and Freshwater Research* **60**, 97–111.

Reckel F, Melzer RR, Parry JWL and Bowmaker JK (2002) The retina of five atherinomorph teleosts: photoreceptors, patterns and spectral sensitivities. *Brain Behavior and Evolution* **60**, 249–264.

Reeves GH, Everest FH and Sedell JR (1993) Diversity of juvenile anadromous salmonid assemblages in coastal Oregon basins with different levels of timber harvest. *Transactions of the American Fisheries Society* **122**, 309–317.

Reeves JM, Chivas AR, García A, Holt S *et al.* (2008) The sedimentary record of palaeoenvironments and sea-level change in the Gulf of Carpentaria, Australia, through the last glacial cycle. *Quaternary International* **183**, 3–22.

Reice SR, Wissmar RC and Naiman RJ (1990) Disturbance regimes, resilience and recovery of animal communities and habitats in lotic ecosystems. *Environmental Management* **14**, 647–659.

Reich P, McMaster D, Bond N, Metzeling L *et al.* (2010) Examining the ecological consequences of restoring flow intermittency to artificially perennial lowland streams: patterns and predictions from the Broken-Boosey Creek system in northern Victoria, Australia. *River Research and Applications* **26**, 529–545.

Reid DD, Harris JH and Chapman DJ (1997) 'NSW inland commercial fishery data analysis'. Fisheries Research and Development Corporation, Sydney.

Reis RE, Kullander SO and Ferraris CJ (2003) *Checklist of Freshwater Fishes of South and Central America.* EDIPUCRS, Porto Alegre.

Resh VW, Brown AV *et al.* (1988) The role of disturbance in stream ecology. *Journal of the North American Benthological Society* **7**, 433–455.

Reynolds LF (1976) Inland fisheries of South Australia: description, production, fishing methods. *SAFIC* **15**, 3–6.

Reynolds LF (1983) Migration patterns of five fish species in the Murray-Darling River system. *Australian Journal of Marine and Freshwater Research* **34**, 857–871.

Reynolds SJ (2009) Impact of the introduced poeciliid *Gambusia holbrooki* on amphibians in southwestern Australia. *Copeia* **2009**, 296–302.

Reynolds WW and Casterlin ME (1979) Behavioral thermoregulation and the 'final preferendum' paradigm. *American Zoologist* **19**, 211–224.

Reznick D (1983) The structure of guppy life histories: the tradeoff between growth and reproduction. *Ecology* **64**, 862–873.

Reznick D and Endler JA (1982) The impact of predation on life-history evolution in Trinidadian guppies (*Poecilia reticulata*). *Evolution* **36**, 160–177.

Reznick DA, Bryga H and Endler JA (1990) Experimentally-induced life-history evolution in a natural population. *Nature* **346**, 357–359.

Ricciardi A and Cohen J (2007) The invasiveness of an introduced species does not predict its impact. *Biological Invasions* **9**, 309–315.

Ricciardi A and Rasmussen JB (1999) Extinction rates of North American freshwater fauna. *Conservation Biology* **13**, 1220–1222.

Rice WR (1989) Analyzing tables of statistical tests. *Evolution* **43**, 223–225.

Richardson BJ, Baverstock PR and Adams M (1986) *Allozyme Electrophoresis: A Handbook for Animal Systematics and Population Studies.* Academic Press, Sydney.

Richardson NA, Anderson AJ, Rimmer MA and Sara VR (1995) Localization of insulin-like growth factor-I immunoreactivity in larval and juvenile barramundi (*Lates calcarifer*). *General and Comparative Endocrinology* **100**, 282–292.

Richmond AD and Fausch KD (1995) Characteristics and function of large woody debris in subalpine Rocky Mountain streams in northern Colorado. *Canadian Journal of Fisheries and Aquatic Sciences* **52**, 1789–1802.

Richter BD, Baumgartner JV, Wigington R and Braun DP (1997) How much water does a river need? *Freshwater Biology* **37**, 231–249.

Ricker WE (1954) Stock and recruitment. *Journal of the Fisheries Research Board of Canada* **11**, 559–623.

Ricker WE (1975) *Computation and Interpretation of Biological Statistics of Fish Populations.* Fisheries Research Board of Canada, Ottawa.

Rimmer MA (1985a) Early development and buccal incubation in the fork-tailed catfish *Arius graeffei* Kner and Steindachner (Pisces: Ariidae) from the Clarence River, New South Wales. *Australian Journal of Marine and Freshwater Research* **36**, 405–411.

Rimmer MA (1985b) Growth, feeding and condition of the fork-tailed catfish *Arius graeffei* Kner and Steindachner (Pisces, Ariidae) from the Clarence River, New South Wales. *Australian Journal of Marine and Freshwater Research* **36**, 33–39.

Rimmer MA (1985c) Reproductive cycle of the fork-tailed catfish *Arius graeffei* Kner and Steindachner (Pisces: Ariidae) from the Clarence River, New South Wales. *Australian Journal of Marine and Freshwater Research* **36**, 23–32.

Rimmer MA and Merrick JR (1983) A review of reproduction and development in the fork-tailed catfishes (Ariidae). *Proceedings of the Linnean Society of New South Wales* **107**, 41–50.

Roberts DT, Duivenvoorden LJ and Stuart IG (2008) Factors influencing recruitment patterns of golden perch (*Macquaria ambigua oriens*) within a hydrologically variable and regulated Australian tropical river system. *Ecology of Freshwater Fish* **17**, 577–589.

Roberts J and Ebner B (1997) 'An overview of carp *Cyprinus carpio* L. in Australia. Part 1'. CSIRO Land and Water, Griffith.

Roberts J and Sainty G (1996) *Listening to the Lachlan.* Murray-Darling Basin Commission, Canberra.

Roberts J and Tilzey R (Eds) (1996) *Controlling Carp: Exploring the Options for Australia.* CSIRO Land and Water, Griffith.

Roberts J, Chick A, Oswald L and Thompson P (1995) Effect of carp, *Cyprinus carpio* L., an exotic benthivorous fish, on aquatic plants and water quality in experimental ponds. *Marine and Freshwater Research* **46**, 1171–1180.

Roberts JH and Angermeier PL (2007) Spatiotemporal variability of stream habitat and movement of three species of fish. *Oecologia* **151**, 417–430.

Robertson AI and Rowling RW (2000) Effects of livestock on riparian zone vegetation in an Australian dryland river. *Regulated Rivers: Research and Management* **16**, 527–541.

Robertson AI, Healey MR and King AJ (1997) Experimental manipulations of the biomass of introduced carp (*Cyprinus carpio*) in billabongs. II. Impacts on benthic properties and processes. *Marine and Freshwater Research* **48**, 445–454.

Robertson M, Nichols P, Horwitz P, Bradby K *et al.* (2000) Environmental narratives and the need for multiple perspectives to restore degraded landscapes in Australia. *Ecosystem Health* **6**, 119–133.

Robins JB, Halliday IA, Staunton-Smith J, Mayer DG *et al.* (2005) Freshwater-flow requirements of estuarine fisheries in tropical Australia: a review of the state of knowledge and application of a suggested approach. *Marine and Freshwater Research* **56**, 343–360.

Robinson AT, Clarkson RW and Forrest ER (1998) Dispersal of larval fishes in a regulated river tributary. *Transactions of the American Fisheries Society* **127**, 772–786.

Robinson BW and Wilson DS (1998) Optimal foraging, specialization, and a solution to Liem's Paradox. *American Naturalist* **151**, 223–235.

Robson BJ and Mitchell BD (2010) Metastability in a river subject to multiple disturbances may constrain restoration options. *Marine and Freshwater Research* **61**, 778–785.

Rodd FH, Reznick DN and Sokolowski MB (1997) Phenotypic plasticity in the life history traits of guppies: responses to social environment. *Ecology* **78**, 419–433.

Rodehutscord M, Gregus Z and Pfeffer E (2000) Effect of phosphorus intake on faecal and non-faecal phosphorus excretion in rainbow trout (*Oncorhynchus mykiss*) and the consequences for comparative phosphorus availability studies. *Aquaculture* **188**, 383–398.

Rodríguez MA (2002) Restricted movement in stream fish: the paradigm is incomplete, not lost. *Ecology* **83**, 1–13.

Roff DA (1980) A motion for the retirement of the von Bertalanffy function. *Canadian Journal of Fisheries and Aquatic Sciences* **37**, 127–129.

Roff DA (1983) An allocation model of growth and reproduction in fish. *Canadian Journal of Fisheries and Aquatic Sciences* **40**, 1395–1404.

Rohan G (1989) River (Murray) fishery review. *SAFISH* **13**, 4–7.

Rolls RJ and Wilson GG (2010) Spatial and temporal patterns in fish assemblages following an artificially extended floodplain inundation event, Northern Murray-Darling Basin, Australia. *Environmental Management* **45**, 822–833.

Rolls RJ, Boulton AJ, Growns IO and Maxwell SE (2011) Response by fish assemblages to an environmental flow release in a temperate coastal Australian river: a paired catchment analysis. *River Research and Applications* **27**, 867–880.

Romanuk TN, Jackson LJ, Post JR, McCauley E *et al.* (2006) The structure of food webs along river networks. *Ecography* **29**, 3–10.

Romero A and Vanselow PBS (2000) Threatened fishes of the world: *Milyeringa veritas* Whitley, 1945 (Eleotridae). *Environmental Biology of Fishes* **57**, 36.

Roni P, Hanson K, Beechie T, Pess G *et al.* (2005) 'Habitat rehabilitation for inland fisheries. Global review of effectiveness and guidance for rehabilitation of freshwater ecosystems'. FAO Fisheries Technical Paper 484. Food and Agriculture Organization, Rome.

Rooney N, McCann K, Gellner G and Moore JC (2006) Structural asymmetry and the stability of diverse food webs. *Nature* **442**, 265–269.

Roscoe DW and Hinch SG (2010) Effectiveness monitoring of fish passage facilities: historical trends, geographic patterns and future directions. *Fish and Fisheries* **11**, 12–33.

Rosenfeld J (2003) Assessing the habitat requirements of stream fishes: an overview and evaluation of different approaches. *Transactions of the American Fisheries Society* **132**, 953–968.

Rosenfeld JS and Hatfield T (2006) Information needs for assessing critical habitat of freshwater fish. *Canadian Journal of Fisheries and Aquatic Sciences* **63**, 683–698.

Ross ST (1986) Resource partitioning in fish assemblages: a review of field studies. *Copeia* **1986**, 352–388.

Rossel S, Corlija J and Schuster S (2002) Predicting three-dimensional target motion: how archer fish determine where to catch their dislodged prey. *Journal of Experimental Biology* **205**, 3321–3326.

Rothschild BJ and Osborn TR (1988) Small-scale turbulence and plankton contact rates. *Journal of Plankton Research* **10**, 465–474.

Roughley TC (1916) *Fishes of Australia and their Technology*. William Applegate Gullick, Sydney.

Roughley TC (1951) *Fish and Fisheries of Australia*. Angus and Robertson, Sydney.

Rounsefell GA and Everhart WH (1953) *Fishery Science: Its Methods and Applications*. Wiley, New York.

Rourke ML (2007) Population genetic structure of Murray cod (*Maccullochella peelii peelii*) and impacts of stocking in the Murray-Darling Basin. PhD thesis. Monash University, Melbourne.

Rourke ML, McPartlan HC, Ingram BA and Taylor AC (2009) Polygamy and low effective population size in a captive Murray cod (*Maccullochella peelii peelii*) population: genetic implications for wild restocking programs. *Marine and Freshwater Research* **60**, 873–883.

Rourke ML, McPartlan HC, Ingram BA and Taylor AC (2010) Biogeography and life history ameliorate the potentially negative genetic effects of stocking on Murray cod (*Maccullochella peelii peelii*). *Marine and Freshwater Research* **61**, 918–927.

Roux DJ, Nel JL *et al.* (2008) Designing protected areas to conserve riverine biodiversity: lessons from a hypothetical redesign of the Kruger National Park. *Biological Conservation* **141**, 100–117.

Rowe DK, Moore A, Giogetti A, Maclean C *et al.* (2008) 'Review of the impacts of gambusia, redfin perch, tench, roach, yellowfin, goby and streaked goby in Australia'. Department of the Environment, Water, Heritage and the Arts, Canberra.

Rowland SJ (1983) Spawning of the Australian freshwater fish Murray cod, *Maccullochella peeli* (Mitchell), in earthen ponds. *Journal of Fish Biology* **23**, 525–534.

Rowland SJ (1984) Hybridization between the estuarine fishes yellowfin bream, *Acanthopagrus australis* (Günther), and black bream *A. butcheri* (Munro) (Pisces: Sparidae). *Australian Journal of Marine and Freshwater Research* **35**, 427–440.

Rowland SJ (1985) Aspects of the biology and artificial breeding of the Murray cod *Maccullochella peeli*, and the eastern freshwater cod *M. ikei* sp. *nov.* (Pisces: Percichthyidae). PhD thesis. Macquarie University, Sydney.

Rowland SJ (1989) Aspects of the history and fishery of the Murray cod *Maccullochella peelii* (Mitchell) (Percichthyidae). *Proceedings of the Linnaean Society of New South Wales* **111**, 201–213.

Rowland SJ (1993) *Maccullochella ikei*, an endangered species of freshwater cod (Pisces: Percichthyidae) from the Clarence River system, NSW, and *M. peelii mariensis*, a new subspecies from the Mary River System, Qld. *Records of the Australian Museum* **45**, 121–145.

Rowland SJ (1996) Development of techniques for the large-scale rearing of the larvae of the Australian freshwater fish golden perch, *Macquaria ambigua* (Richardson, 1845). *Marine and Freshwater Research* **47**, 233–242.

Rowland SJ (1998a) Age and growth of the Australian freshwater fish Murray cod, *Maccullochella peelii peelii*. *Proceedings of the Linnean Society of New South Wales* **120**, 163–180.

Rowland SJ (1998b) Aspects of the reproductive biology of Murray cod, *Maccullochella peelii peelii*. *Proceedings of the Linnean Society of New South Wales* **120**, 147–162.

Rowland SJ (2001) Record of the banded grunter *Amniataba percoides* (Teraponidae) from the Clarence River, New South Wales. *Australian Zoologist* **31**, 603–607.

Rowland SJ (2004) Domestication of silver perch, *Bidyanus bidyanus*, broodfish. *Journal of Applied Aquaculture* **16**, 75–83.

Rowland SJ (2005) Overview of the history, fishery, biology and aquaculture of Murray cod (*Maccullochella peelii peelii*). In *Management of Murray Cod in the Murray-Darling Basin: Statement, Recommendations and Supporting Papers*. Canberra, 3–4 June 2004. (Eds M Lintermans and B Phillips). Murray-Darling Basin Commission, Canberra.

Rowland SJ (2009) Review of aquaculture research and development of the Australian freshwater fish silver perch, *Bidyanus bidyanus*. *Journal of the World Aquaculture Society* **40**, 291–324.

Rowland SJ and Ingram BA (1991) 'Diseases of Australian native freshwater fishes, with particular emphasis on the ectoparasitic and fungal diseases of Murray cod (*Maccullochella peeli*), golden perch (*Macquaria ambigua*) and silver perch (*Bidyanus bidyanus*)'. New South Wales Agriculture and Fisheries, Sydney.

Rowland SJ and Tully P (2004) 'Hatchery quality assurance program for Murray cod (*Maccullochella peelii peelii*), golden perch (*Macquaria ambigua*) and silver perch (*Bidyanus bidyanus*)'. Department of Primary Industries, Sydney.

Rowland SJ, Allan GL, Clark K, Hollis M *et al.* (1994) Production of fingerling *Bidyanus bidyanus* (Teraponidae) at two densities in earthen ponds in Australia. *Progressive Fish-Culturist* **56**, 296–298.

Rowland SJ, Mifsud C, Nixon M and Boyd P (2006) Effects of stocking density on the performance of the Australian freshwater silver perch (*Bidyanus bidyanus*) in cages. *Aquaculture* **253**, 301–308.

Rowlerson A, Mascarello F, Radaelli G and Veggetti A (1995) Differentiation and growth of muscle in the fish *Sparus aurata* (L). II. Hyperplastic and hypertrophic growth of lateral muscle from hatching to adult. *Journal of Muscle Research and Cell Motility* **16**, 223–236.

Rowley JJL, Rayner TS and Pyke GH (2005) New records and invasive potential of the poeciliid fish *Phalloceros caudimaculatus*. *New Zealand Journal of Marine and Freshwater Research* **39**, 1013–1022.

Rubenstein DI (1981) Individual variation and competition in the Everglades pygmy sunfish. *Journal of Animal Ecology* **50**, 337–350.

Ruello NV (1976) Observations on some massive fish kills in Lake Eyre. *Australian Journal of Marine and Freshwater Research* **27**, 667–672.

Rundle HD (2002) A test of ecologically dependent postmating isolation between sympatric sticklebacks. *Evolution* **56**, 322–329.

Russell DJ and McDougall AJ (2005) Movement and juvenile recruitment of mangrove jack, *Lutjanus argentimaculatus* (Forsskål, 1775), in northern Australia. *Marine and Freshwater Research* **56**, 465–475.

Rutherford JC, Lintermans M, Groves J, Liston P *et al.* (2009) 'The effect of cold-water releases in upland streams'. Technical Report. eWater Cooperative Research Centre, Canberra.

Rutherfurd ID, Jerie K and Marsh N (2000) 'A rehabilitation manual for Australian streams'. Land and Water Resources Research and Development Corporation/Cooperative Research Centre for Catchment Hydrology, Canberra.

Rutherfurd ID, Ladson TL and Stewardson MJ (2004) Evaluating stream rehabilitation projects: reasons not to and approaches if you have to. *Australian Journal of Water Resources* **8**, 1–12.

Ruxton GD, Armstrong JD and Humphries S (1999) Modelling territorial behaviour of animals in variable environments. *Animal Behaviour* **58**, 120.

Ryan T, Lennie R and Lyon J (2003) 'Thermal rehabilitation of the Southern Murray-Darling Basin'. Final Report to Agriculture, Forestry, Fisheries Australia. MD 2001 Fish Rehab Program. Arthur Rylah Institute for Environmental Research, Department of Sustainability and Environment, Melbourne.

Ryder D, Brierley GJ, Hobbs R, Kyle G and Leishman M (2008) Vision generation: what do we seek to achieve in river rehabilitation? In *River Futures: An Integrative Scientific Approach to River Repair.* (Eds GJ Brierley and KA Fryirs). Island Press, Washington DC.

Ryder OA (1986) Species conservation and systematics: the dilemma of subspecies. *Trends in Ecology and Evolution* **1**, 9–10.

Sabo MJ and Orth DJ (1994) Temporal variation in microhabitat use by age-0 smallmouth bass in the North Anna River. *Transactions of the American Fisheries Society* **291**, 157–178.

Saddlier S, O'Mahony J and Ramsey D (2008) 'Protection and enhancement of Murray cod populations'. Technical Report Series 172. Arthur Rylah Institute for Environmental Research, Department of Sustainability and Environment, Melbourne.

Saddlier SR and Doeg TJ (1997) 'The fish and aquatic macroinvertebrate fauna of the Yarra River: the impact of 4 years of an environmental water release from the Upper Yarra Reservoir'. Marine and Freshwater Resources Institute/Arthur Rylah Institute for Environmental Research, Department of Sustainability and Environment, Melbourne.

Sainsbury KJ (1980) Effect of individual variability on the von Bertalanffy growth equation. *Canadian Journal of Fisheries and Aquatic Sciences* **37**, 241–247.

Sakaris PC and Irwin ER (2010) Tuning stochastic matrix models with hydrologic data to predict the population dynamics of a riverine fish. *Ecological Applications* **20**, 483–496.

Salzburger W and Meyer A (2004) The species flocks of East African cichlid fishes: recent advances in molecular phylogenetics and population genetics. *Naturwissenschaften* **91**, 277–290.

Sammut J, Melville MD, Callinan RB and Fraser GC (1995) Estuarine acidification: impacts on aquatic biota of draining acid sulphate soils. *Australian Geographical Studies* **33**, 89–100.

Sarewitz D (2004) How science makes environmental controversies worse. *Environmental Science and Policy* **7**, 385–403.

Sargent RC (1997) Parental care. In *Behavioural Ecology of Teleost Fishes.* (Ed. JGJ Godin) pp. 292–315. Oxford University Press, Oxford.

Sargent RC and Gross MR (1993) William's Principle, an explanation of parental care in teleost fishes. In *Behaviour of Teleost Fishes.* (Ed. TJ Pitcher) pp. 333–361. Chapman and Hall, London.

Saunders DL, Meeuwig JJ and Vincent CJ (2002) Freshwater protected areas: strategies for conservation. *Conservation Biology* **16**, 30–41.

Saunders G, Cooke B, McColl K, Shine R *et al.* (2010) Modern approaches for the biological control of vertebrate pests: an Australian perspective. *Biological Control* **52**, 288–295.

Savage VM, Gillooly JF, Woodruff WH, West GB *et al.* (2004) The predominance of quarter-power scaling in biology. *Functional Ecology* **18**, 257–282.

Savino JF and Stein RA (1982) Predator–prey interaction between largemouth bass and bluegills as influenced by simulated, submersed vegetation. *Transactions of the American Fisheries Society* **111**, 255–266.

Scheffer M, Baveco JM, DeAngelis DL, Rose KA *et al.* (1995) Super-individuals, a simple solution for modelling large populations on an individual basis. *Ecological Modelling* **80**, 161–170.

Schick RS and Lindley ST (2007) Directed connectivity among fish populations in a riverine network. *Journal of Applied Ecology* **44**, 1116–1126.

Schiemer F, Keckeis H, Reckendorfer W and Winkler G (2001) The 'inshore retention concept' and its significance for large rivers. *Archiv fur Hydrobiologie*, Supplement **135**, 509–516.

Schiller CB and Harris JH (2001) Native and alien fish. In *Rivers as Ecological Systems: The Murray-Darling Basin.* (Ed. WJ Young) pp. 229–258. Murray-Darling Basin Commission, Canberra.

Schiller CB, Bruce AM and Gehrke PC (1997) Distribution and abundance of native fish in New South Wales rivers. In *Fish and Rivers in Stress: The NSW Rivers Survey.* (Eds JH Harris and PC Gehrke) pp. 71–102. New South Wales Fisheries Office of Conservation/Cooperative Research Centre for Freshwater Ecology, Canberra.

Schlosser IJ (1987a) A conceptual framework for fish communities in warmwater streams. In *Community and Evolutionary Ecology of North American Stream Fishes.* (Eds WJ Matthews and DC Heins) pp. 17–24. University of Oklahoma Press, Norman.

Schlosser IJ (1987b) The role of predation in age- and size-related habitat use by stream fishes. *Ecology* **68**, 651–659.

Schlosser IJ (1991) Stream fish ecology: a landscape perspective. *BioScience* **41**, 704–712.

Schlosser IJ (1995) Dispersal, boundary processes, and trophic-level interactions in streams adjacent to beaver ponds. *Ecology* **76**, 908–925.

Schlosser IJ and Angermeier PL (1995) Spatial variation in demographic processes of lotic fishes: conceptual models, empirical evidence, and implications for conservation. *American Fisheries Society Symposium* **17**, 392–401.

Schlosser IJ and Kallemeyn LW (2000) Spatial variation in fish assemblages across a beaver-influenced successional landscape. *Ecology* **81**, 1371–1382.

Schmidt DJ, Bond NR, Adams M and Hughes JM (2011a) Cytonuclear evidence for hybridogenetic reproduction in natural populations of the Australian carp gudgeon (*Hypseleotris*: Eleotridae). *Molecular Ecology* **20**, 3367–3380.

Schmidt DJ, Crook DA, O'Connor JP and Hughes JM (2011b) Genetic analysis of threatened Australian grayling *Prototroctes maraena* suggests recruitment to coastal rivers from an unstructured marine larval source population. *Journal of Fish Biology* **78**, 98–111.

Schmidt-Nielsen K (1997) *Animal Physiology: Adaptation and Environment*, 5th edn. Cambridge University Press, Cambridge.

Schmitz OJ, Krivan V and Ovadia O (2004) Trophic cascades: the primacy of trait-mediated indirect interactions. *Ecology Letters* **7**, 153–163.

Schnute J (1981) A versatile growth model with statistically stable parameters. *Canadian Journal of Fisheries and Aquatic Sciences* **38**, 1128–1140.

Scholz O and Boon PI (1993) Biofilm development and extracellular enzyme activities on wood in billabongs of south-eastern Australia. *Freshwater Biology* **30**, 359–368.

Schreiber ESG, Bearlin AR, Nicol SJ and Todd CR (2004) Adaptive management: a synthesis of current understanding and effective application. *Ecological Management and Restoration* **5**, 177–182.

Schwartz MW (2008) The performance of the Endangered Species Act. *Annual Review of Ecology, Evolution and Systematics* **39**, 279–299.

Schwartz TS and Beheregaray LB (2008) Using genotype simulations and Bayesian analyses to identify individuals of hybrid origin in Australian bass: lessons for fisheries management. *Journal of Fish Biology* **72**, 435–450.

Scott A (2005) 'Historical evidence of native fish in the Murray-Darling Basin at the time of European settlement'. Cooperative Research Centre for Freshwater Ecology, Canberra.

Scott JM, Goble DD, Wiens JA, Wilcove DS *et al.* (2005) Recovery of imperiled species under the Endangered Species Act: the need for a new approach. *Frontiers in Ecology and the Environment* **3**, 383–389.

Scott WB and Crossman EJ (1973) *Freshwater Fishes of Canada.* Bulletin 184. Fisheries Research Board of Canada, Ottawa.

Seager PS (1888) 'Concise history of the acclimatisation of the Salmonidae by P.S. Seager, Secretary of the Fisheries Board of Tasmania'. Fisheries Board of Tasmania, Hobart.

Sebens KP (1987) The ecology of indeterminate growth in animals. *Annual Review of Ecology and Systematics* **18**, 371–407.

Secor DH, Dean JM and Lban EH (1992) Otolith removal and preparation for microstructural examination. *Special Publication, Canadian Journal of Fisheries Aquatic Sciences* **117**, 19–57.

Secor SM (2009) Specific dynamic action: a review of the postprandial metabolic response. *Journal of Comparative Physiology B – Biochemical Systemic and Environmental Physiology* **179**, 1–56.

Sergio F, Caro T, Brown D, Clucas B *et al.* (2008) Top predators as conservation tools: ecological rationale, assumptions, and efficacy. *Annual Review of Ecology, Evolution, and Systematics* **39**, 1–19.

Shaddick K, Burridge CP, Jerry DR, Gilligan DM *et al.* (2011a) Historical divergence with recent connectivity in a catadromous fish, the estuary perch (*Macquaria colonorum*). *Canadian Journal of Fisheries and Aquatic Sciences* **68**, 304–318.

Shaddick K, Burridge CP, Jerry DR, Schwartz TS *et al.* (2011b) A hybrid zone and bidirectional introgression between two catadromous species: Australian bass *Macquaria novemaculeata* and estuary perch *Macquaria colonorum*. *Journal of Fish Biology* **79**, 1214–1235.

Sheard MJ and Smith PC (1995) Karst and mound spring deposits. In *The Geology of South Australia*, Vol. 2. (Eds

JF Drexel and WV Preiss) pp. 257–263. Bulletin 54. Geological Survey of South Australia.

Shearer KD and Mulley JC (1978) The introduction and distribution of the carp, *Cyprinus carpio* Linnaeus, in Australia. *Australian Journal of Marine and Freshwater Research* **29**, 551–63.

Sheldon F and Walker KF (1997) Changes in biofilms induced by flow regulation could explain extinctions of aquatic snails in the Lower River Murray, Australia. *Hydrobiologia* **347**, 97–108.

Sheldon F, Bunn SE, Hughes JM, Arthington AH *et al.* (2010) Ecological roles and threats to aquatic refugia in arid landscapes: dryland river waterholes. *Marine and Freshwater Research* **61**, 885–895.

Shelford VE (1911a) Ecological succession. I. Stream fishes and the method of physiographic analysis. *Biological Bulletin* **21**, 9–35.

Shelford VE (1911b) Ecological succession. II. Pond fishes. *Biological Bulletin* **21**, 127–151.

Shelford VE (1911c) Ecological succession. III. A reconnaissance of its causes in ponds with particular reference to fish. *Biological Bulletin* **22**, 1–38.

Shepherd JG and Cushing DH (1980) A mechanism for density-dependent survival of larval fish as the basis of a stock-recruitment relationship. *Journal du Conseil Internationale pour l'exploration de la mer* **39**, 160–167.

Sheridan MA (1988) Lipid dynamics in fish: aspects of absorption, transportation, deposition and mobilization. *Comparative Biochemistry and Physiology B: Comparative Biochemistry* **90**, 679–690.

Sherman B, Todd CR, Koehn JD and Ryan T (2007) Modelling the impact and potential mitigation of cold water pollution on Murray cod populations downstream of Hume Dam, Australia. *River Research and Applications* **23**, 377–389.

Sherman BS (2000) 'Scoping options for mitigating cold water discharges from dams'. Consultancy Report 00/21. CSIRO Land and Water, Canberra.

Shipway B (1947) Rains of fishes? *Western Australian Naturalist* **1**, 47–48.

Shirley MJ and Raadik TA (1997) Apects of the ecology and breeding biology of *Galaxias fuscus* Mack, in the Goulburn River system, Victoria. *Proceedings of the Royal Society of Victoria* **109**, 157–166.

Silva EIL and Davies RW (1986) Movement of some indigenous riverine fish in Sri Lanka. *Hydrobiologia* **137**, 265–270.

Simon KS and Townsend CR (2003) Impacts of freshwater invaders at different levels of ecological organisation, with emphasis on salmonids and ecosystem consequences. *Freshwater Biology* **48**, 982–994.

Simpson B, Hutchison M, Gallagher T and Chilcott K (2002) 'Fish stocking in impoundments: a best practice fish stocking manual for impoundments in eastern and northern Australia'. Agency for Food and Fibre Science, Department of Primary Industries, Brisbane.

Simpson R and Jackson P (1996) 'The Mary River cod research and recovery plan'. Fisheries Group, Department of Primary Industries, Brisbane.

Simpson RR and Mapleston AJ (2002) Movements and habitat use by the endangered Australian freshwater Mary River cod, *Maccullochella peelii mariensis*. *Environmental Biology of Fishes* **65**, 401–410.

Simpson SJ, Raubenheimer D, Charleston MA, Clissold FJ *et al.* (2010) Modelling nutritional interactions: from individuals to communities. *Trends in Ecology and Evolution* **25**, 53–60.

Sinclair Knight Merz (2007) High conservation value aquatic ecosystems project: identifying, categorising and managing HCVAE. Final report. <http://www.environment.gov.au/water/publications/environmental/ecosystems/pubs/hcvae.pdf>. July 2010.

Sinclair Knight Merz (2008) 'An overview of the impacts of translocated native fish species in Australia'. Report to the Department of the Environment and Water Resources, Canberra. Sinclair Knight Merz, Melbourne.

Sinclair P (2005) The loss of valuable Murray cod in fish kills: a community and conservation perspective. In *Management of Murray Cod in the Murray-Darling Basin: Statement, Recommendations and Supporting Papers*. Canberra, 3–4 June 2004. (Eds M Lintermans and B Phillips) pp. 83–87. Murray-Darling Basin Commission, Canberra.

Sirois P and Dodson JJ (2000) Critical periods and growth-dependent survival of larvae of an estuarine fish, the rainbow smelt *Osmerus mordax*. *Marine Ecology Progress Series* **203**, 233–245.

Slatkin M (1985) Gene flow in natural populations. *Annual Review of Ecology and Systematics* **16**, 393–430.

Sloane RD (1984a) Preliminary observations of migrating adult freshwater eels (*Anguilla australis australis* Richardson) in Tasmania. *Australian Journal of Marine and Freshwater Research* **35**, 471–476.

Sloane RD (1984b) Upstream migration by young pigmented freshwater eels (*Anguilla australis australis* Richardson) in Tasmania. *Australian Journal of Marine and Freshwater Research* **35**, 61–73.

Sloane RD (1984c) The upstream movements of fish in the Plenty River, Tasmania. *Papers and Proceedings of the Royal Society of Tasmania* **118**, 163–171.

Sloman KS, Wilson RW and Balshine S (2006) *Behaviour and Physiology of Fish*. Elsevier Academic Press, San Diego.

Smith B (2005) 'The state of the art: a synopsis of information on common carp (*Cyprinus carpio*) in Australia'. South Australian Research and Development Institute, Adelaide.

Smith B, Burton I, Klein RJT and Wandel J (2000) An anatomy of adaptation to climate change and variability. *Climatic Change* **45**, 223–251.

Smith BB and Walker KF (2003) Validation of the ageing of 0+ carp (*Cyprinus carpio* L.). *Marine and Freshwater Research* **54**, 1005–1008.

Smith BB and Walker KF (2004) Spawning dynamics of carp (*Cyprinus carpio* L.) in the River Murray, South Australia, shown by macroscopic and histological staging of gonads. *Journal of Fish Biology* **64**, 1–19.

Smith CL and Powell CR (1971) The summer fish communities of Brier Creek, Marshall County, Oklahoma. *American Museum Novitates* **2458**, 1–30.

Smith GR (1978) Biogeography of intermountain fishes. *Great Basin Naturalist Memoirs* **2**, 17–42.

Smith JA, Baumgartner LJ, Suthers IM and Taylor MD (2011) Generalist niche, specialist strategy: the diet of an Australian percichthyid. *Journal of Fish Biology* **78**, 1183–1199.

Smith WR (1970) *Myths and Legends of the Australian Aborigines*. George G Harrap, London.

Smithson EB and Johnston CE (1999) Movement patterns of stream fishes in a Ouachita Highlands stream: an examination of the restricted movement paradigm. *Transactions of the American Fisheries Society* **128**, 847–853.

Smokorowski KE, Withers KJ and Kelso JRM (1998) Does habitat creation contribute to management goals? An evaluation of literature documenting freshwater habitat rehabilitation on enhancement projects. *Canadian Technical Report of Fisheries and Aquatic Sciences* **2249**, 72 pp.

Sola C (1995) Chemoattraction of upstream migrating glass eels *Anguilla anguilla* to earthy and green odorants. *Environmental Biology of Fishes* **43**, 179–185.

Southwood R (2000) *Ecological Methods*, 3rd edn. Blackwell Science, Oxford.

Southwood TRE (1977) Habitat: the templet for ecological strategies? *Journal of Animal Ecology* **46**, 337–365.

Sparks JS and Smith WL (2004) Phylogeny and biogeography of the Malagasy and Australasian rainbowfishes (Teleostei: Melanotaenioidei): Gondwanan vicariance and evolution in freshwater. *Molecular Phylogenetics and Evolution* **33**, 719–734.

Sparks JS and Stiassny MLJ (2003) The freshwater fishes of Madagascar: introduction to the freshwater fishes. In *The Natural History of Madagascar*. (Eds SM Goodman and JP Benstead) pp. 849–863. University of Chicago Press, Chicago.

Spencer SR, Cameron GN and Swihart RK (1990) Operationally defining home range: temporal dependence exhibited by hispid cotton rats. *Ecology* **71**, 1817–1822.

Stamps JA (1993) Sexual size dimorphism in species with asymptotic growth after maturity. *Biological Journal of the Linnean Society* **50**, 123–145.

Stankey GH and Allan C (2009) Introduction. In *Adaptive Environmental Management: A Practitioner's Guide*. (Eds C Allan and GH Stankey) pp. 3–10. Springer, Dordrecht/CSIRO Publishing, Melbourne.

State Fisheries, New South Wales (2003) *Environmental Impact Statement on Freshwater Fish Stocking in NSW*. Public consultation document, 3 vols. State Fisheries, Sydney.

Staunton-Smith J, Robins JB, Mayer DG, Sellin MJ *et al.* (2004) Does the quantity and timing of fresh water flowing into a dry tropical estuary affect year-class strength of barramundi (*Lates calcarifer*)? *Marine and Freshwater Research* **55**, 787–797.

Stearns SC (1984) Plasticity for age and size at maturity: a life history response to unavoidable stress. In *Fish Reproduction*. (Eds G Potts and RJ Wootton) pp. 13–33. Academic Press, London.

Stearns SC (1992) *The Evolution of Life Histories*. Oxford University Press.

Stearns SC and Koella JC (1986) The evolution of phenotypic plasticity in life-history traits: predictions of reaction norms for age and size at maturity. *Evolution* **40**, 893–913.

Stearns WC and Etnier DA (1986) Drainage evolution and fish biogeography of the Tennessee and Cumberland rivers drainage realm. In *The Zoogeography of North American Freshwater Fishes*. (Eds CH Hocutt and EO Wiley) pp. 325–361. Wiley, New York.

Steffen W (2009) *Climatic Change 2009: Faster Change and More Serious Risks*. Department of Climate Change, Canberra.

Steffen W, Burbidge AA, Hughes L, Kitching R *et al.* (2009) 'Australia's biodiversity and climate change: a strategic assessment of the vulnerability of Australia's biodiversity to climate change'. Report to the NRM Ministerial Council by Biodiversity and Climate Change Expert Advisory Group. CSIRO Publishing, Melbourne.

Stephenson AE (1986) Lake Bungunnia: a Plio-Pleistocene megalake in southern Australia. *Palaeogeography, Palaeoclimatology, Palaeoecology* **57**, 137–156.

Stephenson AE and Brown CM (1989) The ancient Murray River system. *BMR Journal of Australian Geology and Geophysics* **11**, 387–95.

Sternberg D, Balcombe SR, Marshall J and Lobegeiger J (2008) Food resource variability in an Australian dryland river: evidence from the diet of two generalist native fish species. *Marine and Freshwater Research* **59**, 137–144.

Sterner RW and Elser JJ (2002) *Ecological Stoichiometry: The Biology of Elements from Molecules to the Biosphere*. Princeton University Press, New York.

Stevenson RD and Woods WA (2006) Condition indices for conservation: new uses for evolving tools. *Integrative and Comparative Biology* **46**, 1169–1190.

Stewart-Koster B, Kennard MJ, Harch BD, Sheldon F *et al.* (2007) Partitioning the variation in stream fish assemblages within a spatio-temporal hierarchy. *Marine and Freshwater Research* **58**, 675–686.

Stewart-Koster B, Olden JD, Kennard MJ, Pusey BJ *et al.* (2011) Fish response to the temporal hierarchy of the natural flow regime in the Daly River, northern Australia. *Journal of Fish Biology* **79**, 1525–1544.

Stoddard JL, Larsen DP, Hawkins CP, Johnson RK *et al.* (2006) Setting expectations for the ecological condition of streams: the concept of reference condition. *Ecological Applications* **16**, 1267–1276.

Stoffels RJ and Humphries P (2003) Ontogenetic variation in the diurnal food and habitat associations of an endemic and an exotic fish in floodplain ponds: consequences for niche partitioning. *Environmental Biology of Fishes* **66**, 293–305.

Stokes KE, O'Neill KP, Montgomery WI, Dick JTA *et al.* (2006) The importance of stakeholder engagement in invasive species management: a cross-jurisdictional perspective in Ireland. *Biodiversity and Conservation* **8**, 2829–2852.

Stone T (2006) The late-Holocene origin of the modern Murray River course, southeastern Australia. *The Holocene* **16**, 771–778.

Strayer DL (2010) Alien species in fresh waters: ecological effects, interactions with other stressors, and prospects for the future. *Freshwater Biology* **55**, 152–174.

Strayer DL, Eviner DT, Jeschke JM and Pace ML (2006) Understanding the long-term effects of species invasions. *Trends in Ecology and Evolution* **21**, 645–651.

Stuart IG (2006) Validation of otoliths for determining age of golden perch, a long-lived freshwater fish of Australia. *North American Journal of Fisheries Management* **26**, 52–55.

Stuart IG (2008) 'The Mark V Williams' cage for coordinated trapping of Murray fishways'. Final report to the Murray-Darling Basin Commission, MDBC Project MD1003. Murray-Darling Basin Commission, Canberra.

Stuart IG and Berghuis AP (2002) Upstream passage of fish through a vertical-slot fishway in an Australian subtropical river. *Fisheries Management and Ecology* **9**, 111–122.

Stuart IG and Jones M (2002) 'Ecology and management of common carp in the Barmah-Millewa Forest'. Final report to Agriculture, Fisheries and Forestry Australia. Arthur Rylah Institute, Department of Sustainability and Environment, Melbourne.

Stuart IG and Jones M (2006a) Large, regulated forest floodplain is an ideal recruitment zone for non-native common carp (*Cyprinus carpio* L.). *Marine and Freshwater Research* **57**, 333–347.

Stuart IG and Jones MJ (2006b) Movement of common carp, *Cyprinus carpio*, in a regulated lowland Australian river: implications for management. *Fisheries Management and Ecology* **13**, 213–219.

Stuart IG and Koehn JD (2007) Big cod do breed. *Freshwater Fishing* **85**, 62–64.

Stuart IG and Mallen-Cooper M (1999) An assessment of the effectiveness of a vertical-slot fishway for non-salmonid fish at a tidal barrier on a large tropical/subtropical river. *Regulated Rivers: Research and Management* **15**, 575–590.

Stuart IG and McKillup SC (2002) The use of sectioned otoliths to age barramundi (*Lates calcarifer*) (Bloch, 1790) (Centropomidae). *Hydrobiologia* **479**, 231–236.

Stuart IG, Baumgartner LJ and Zampatti BP (2008a) Lock gates improve passage of small-bodied fish and crustaceans in a low gradient vertical-slot fishway. *Fisheries Management and Ecology* **15**, 241–248.

Stuart IG, Berghuis AP, Long PE and Mallen-Cooper M (2007) Do fish locks have potential in tropical rivers? *River Research and Applications* **23**, 269–286.

Stuart IG, Koehn JD, O'Brien TA, McKenzie JA *et al.* (2010) Too close for comfort: a fishway exit and a hydro-power station inlet. *Marine and Freshwater Research* **61**, 23–33.

Stuart IG, Williams A, McKenzie J and Holt T (2006) Managing a migratory pest species: a selective trap for common carp. *North American Journal of Fisheries Management* **26**, 888–893.

Stuart IG, Zampatti BP and Baumgartner LJ (2008b) Can a low-gradient vertical-slot fishway provide passage for a lowland river fish community? *Marine and Freshwater Research* **59**, 332–346.

Stuart-Smith RD, Barmuta LA and White RWG (2006) Nocturnal and diurnal feeding by *Galaxias auratus*, a lentic galaxiid fish. *Ecology of Freshwater Fish* **15**, 521–531.

Stuart-Smith RD, White RWG and Barmuta LA (2008) A shift in the habitat use pattern of a lentic galaxiid fish: an acute behavioural response to an introduced predator. *Environmental Biology of Fishes* **82**, 93–100.

Sturt C (1982) *Two Expeditions into the Interior of Southern Australia during the Years 1828, 1829, 1830, and 1831 with Observations on the Soil, Climate, and General Resources of the Colony of New South Wales.* Doubleday, Sydney.

Summerfelt RC and Hall GE (1987) *Age and Growth of Fish.* Iowa State University Press, Ames.

Sunderam RIM, Thompson GB and Cheng DMH (1992) Toxicity of endosulfan to native and introduced fish in Australia. *Environmental Toxicology and Chemistry* **11**, 1469–1476.

Suski CD and Cooke SJ (2007) Conservation of aquatic resources through the use of freshwater protected areas: opportunities and challenges. *Biodiversity and Conservation* **16**, 2015–2029.

Suthers IM (1998) Bigger? Fatter? Or is faster growth better? Considerations on condition in larval and juvenile coral-reef fish. *Australian Journal of Ecology* **23**, 265–273.

Sweka JA and Hartman KJ (2003) Reduction of reactive distance and foraging success in smallmouth bass, *Micropterus dolomieu*, exposed to elevated turbidity levels. *Environmental Biology of Fishes* **67**, 341–347.

Switzer PV (1993) Site fidelity in predictable and unpredictable habitats. *Evolutionary Ecology* **7**, 533–555.

Switzer PV (1997) Factors affecting site fidelity in a territorial animal, *Perithemis tenera*. *Animal Behaviour* **53**, 865–877.

Tappin AR (1997) *Mogurnda adspersa* – the purple-spotted gudgeon. *Australia New Guinea Fishes Association Bulletin* **32**, 1–3.

Tatarenkov A, Healey CIM and Avise JC (2010) Microgeographic population structure of green swordtail fish: genetic differentiation despite abundant migration. *Molecular Ecology* **19**, 257–268.

Taylor BW, Flecker AS and Hall RO (2006) Loss of a harvested fish species disrupts carbon flow in a diverse tropical river. *Science* **313**, 833–836.

Taylor MFJ, Suckling KF and Rachlinski JJ (2005) The effectiveness of the Endangered Species Act: a quantitative analysis. *BioScience* **55**, 360–367.

Taylor MI and Knight ME (2008) Mating systems in fishes. In *Fish Reproduction*. (Eds MJ Rocha, A Arukwe and BG Kapoor) pp. 277–309. Science Publishers, New Hampshire.

Taylor WW and Freeburg MH (1984) Effects of food abundance on larval lake whitefish, *Coregonus clupeaformis* Mitchell, growth and survival. *Journal of Fish Biology* **24**, 733–741.

Teal JM (1957) Community metabolism in a temperate cold spring. *Ecological Monographs* **27**, 283–302.

Tenison-Woods JE (1883) *Fish and Fisheries of New South Wales*. Thomas Richards, Government Printer, Sydney.

Thacker C and Unmack PJ (2005) Phylogeny and biogeography of the eleotrid genus *Hypseleotris* (Teleostei: Gobioidei: Eleotridae). *Records of the Australian Museum* **57**, 1–13.

Thacker C, Unmack PJ, Matsui L, Duong P *et al.* (2008) Phylogeography of *Philypnodon* species (Teleostei: Perciformes: Gobioidei: Eleotridae) across eastern Australia: comparative evolution of a sister pair. *Biological Journal of the Linnean Society* **95**, 175–192.

Thacker CE, Unmack PJ, Matsui L and Rifenbark N (2007) Comparative phylogeography of five sympatric *Hypseleotris* species (Teleostei: Eleotridae) in south-eastern Australia reveals a complex pattern of drainage basin exchanges with little congruence across species. *Journal of Biogeography* **34**, 1518–1533.

Tharme RE (2003) A global perspective on environmental flow assessment: emerging trends in the development and application of environmental flow methodologies for rivers. *River Research and Applications* **19**, 397–441.

Theilacker GH, Bailey KM, Canino MF and Porter SM (1996) Variations in larval walleye pollock feeding and condition: a synthesis. *Fisheries Oceanography* **5**, 112–123.

Thiem JD, Ebner BC and Broadhurst BT (2008) Diel activity of the endangered trout cod (*Maccullochella macquariensis*) in the Murrumbidgee River. *Proceedings of the Linnean Society of New South Wales* **129**, 167–173.

Thomas JA and Bovee KD (1993) Application and testing of a procedure to evaluate transferability of habitat suitability criteria. *Regulated Rivers: Research and Management* **8**, 285–294.

Thomas JA, Welch JJ, Lanfear R and Bromham L (2010) A generation time effect on the rate of molecular evolution in invertebrates. *Molecular Biology and Evolution* **27**, 1173–1180.

Thompson GG and Withers PC (1999) The metabolic response to hypoxia and emersion of aestivating fishes (*Lepidogalaxias salamandroides* and *Galaxiella nigrostriata*) and a non-aestivating fish (*Bostockia porosa*) from south-western Australia. *Australian Journal of Zoology* **47**, 295–305.

Thompson LG (1893) *History of the Fisheries of New South Wales, with a Sketch of the Laws by Which They Have Been Regulated*. Charles Potter, Government Printer, Sydney.

Thomson JM (1955) The movements and migrations of mullet (*Mugil cephalus* L.). *Australian Journal of Marine and Freshwater Research* **6**, 328–347.

Thorncraft G and Harris JH (2000) 'Fish passage and fishways in New South Wales: a status report'. Office of Conservation, NSW Fisheries, Sydney.

Thorp JH and Delong MD (1994) The riverine productivity model: an heuristic view of carbon sources and organic processing in large river ecosystems. *Oikos* **70**, 305–308.

Thorp JH, Delong MD, Greenwood KS and Casper AF (1998) Isotopic analysis of three food web theories in constricted and floodplain regions of a large river. *Oecologia* **117**, 551–563.

Thorp JH, Thoms MC and Delong MD (2006) The riverine ecosystem synthesis: biocomplexity in river networks across space and time. *River Research and Applications* **22**, 123–147.

Thorp JH, Thoms MC and Delong MD (2008) *The Riverine Ecosystem Synthesis*. Elsevier, San Diego.

Thresher RE (2008) Autocidal technology for the control of invasive fish. *Fisheries* **33**, 114–121.

Thresher RE (2007) Genetic options for the control of invasive vertebrate pests: prospects and constraints. In *Managing Vertebrate Invasive Species: Proceedings of an International Symposium*. (Eds GW Witmer, WC Pitt and K Fagerstone) pp. 318–331. US Department of Agriculture, APHIS Wildlife Services, Fort Collins.

Thuesen PA, Ebner BC, Larson H, Keith P *et al.* (2011a) Amphidromy links a newly documented fish community of continental Australian streams to oceanic islands of the West Pacific. *PLoS ONE* **6**(10), 10–11.

Thuesen PA, Pusey BJ, Peck DR, Pearson R *et al.* (2008) Genetic differentiation over small spatial scales in the absence of physical barriers in an Australian rainforest stream fish. *Journal of Fish Biology* **72**, 1174–1187.

Thuesen PA, Russell DJ, Thomson FE, Pearce MG *et al.* (2011b) An evaluation of electrofishing as a control measure for an invasive tilapia (*Oreochromis mossambicus*) population in northern Australia. *Marine and Freshwater Research* **62**, 110–118.

Tian X and Qin JG (2003) A single phase of food deprivation provoked compensatory growth in barramundi *Lates calcarifer*. *Aquaculture* **224**, 169–179.

Tibbets CA and Dowling TE (1996) Effects of intrinsic and extrinsic factors on population fragmentation in three species of North American minnows (Teleostei: Cyprinidae). *Evolution* **50**, 1280–1292.

Tilzey RDJ (1976) Observations on interactions between indigenous Galaxiidae and introduced Salmonidae in the Lake Eucumbene catchment, New South Wales. *Australian Journal of Marine and Freshwater Research* **27**, 551–564.

Tilzey RDJ (1977a) Key factors in the establishment and success of trout in Australia. *Proceedings of the Ecological Society of Australia* **10**, 97–105.

Tilzey RDJ (1977b) Repeat homing of brown trout (*Salmo trutta*) in Lake Eucumbene, New South Wales, Australia. *Journal of the Fisheries Research Board of Canada* **34**, 1085–1094.

Timms BV (1986) The coastal dune lakes of eastern Australia. In *Limnology in Australia*. (Eds P De Dekker and WD Williams) pp. 421–432. CSIRO Publishing, Melbourne.

Timms BV (1999) Local runoff, Paroo floods and water extraction impacts on wetlands of Currawinya National Park. In *A Free-flowing River: The Ecology of the Paroo River*. (Ed. RT Kingsford) pp. 51–66. National Parks and Wildlife Service, Sydney.

Todd CR (2009) 'Murray cod management model: an application of essential modelling software'. Arthur Rylah Institute for Environmental Research, Department of Sustainability and Environment, Melbourne.

Todd CR, Nicol SJ and Koehn JD (2004) Density-dependence uncertainty in population models for the conservation management of trout cod, *Maccullochella macquariensis*. *Ecological Modelling* **171**, 359–380.

Todd CR, Ryan T, Nicol SJ and Bearlin AR (2005) The impact of cold water releases on the critical period of post-spawning survival and its implications for Murray cod (*Maccullochella peelii peelii*): a case study of the Mitta Mitta River, southeastern Australia. *River Research and Applications* **21**, 1035–1052.

Tomás J, Geffen AJ, Allen IS and Berges J (2004) Analysis of the soluble matrix of vaterite otoliths of juvenile herring (*Clupea harengus*): do crystalline otoliths have less protein? *Comparative Biochemistry and Physiology A: Molecular and Integrative Physiology* **139**, 301–308.

Tonkin Z, King A, Mahoney J and Morrongiello J (2007) Diel and spatial drifting patterns of silver perch *Bidyanus bidyanus* eggs in an Australian lowland river. *Journal of Fish Biology* **70**, 313–317.

Tonkin Z, King AJ and Mahoney J (2008a) Effects of flooding on recruitment and dispersal of the southern pygmy perch (*Nannoperca australis*) at a Murray River floodplain wetland. *Ecological Management and Restoration* **9**, 196–201.

Tonkin Z, King AJ and Robertson A (2008b) Validation of daily increment formation and the effects of different temperatures and feeding regimes on short-term otolith growth in Australian smelt *Retropinna semoni*. *Ecology of Freshwater Fish* **17**, 312–317.

Tonkin ZD, Humphries P and Pridmore PA (2006) Ontogeny of feeding in two native and one alien fish species from the Murray-Darling Basin, Australia. *Environmental Biology of Fishes* **76**, 303–315.

Tonkin ZD, Lyon J and Pickworth A (2010) Spawning behaviour of the endangered Macquarie perch *Macquaria australasica* in an upland Australian river. *Ecological Management and Restoration* **11**, 223–226.

Tonkin ZD, Ramsey DSL and King AJ (2008c) Larval and juvenile Australian smelt *Retropinna semoni* somatic and otolith growth parameters: implications for growth interpretation of wild fish. *Ecology of Freshwater Fish* **17**, 489–494.

Tonn WM (1990) Climate change and fish communities: a conceptual framework. *Transactions of the American Fisheries Society* **119**, 337–352.

Torrente-Vilara G, Zuanon J, Leprieur F, Oberdorff T *et al.* (2011) Effects of natural rapids and waterfalls on fish assemblage structure in the Madeira River (Amazon Basin). *Ecology of Freshwater Fish* **20**, 588–597.

Townsend CR (1989) The patch dynamics concept of stream community ecology. *Journal of the North American Benthological Society* **8**, 36–50.

Townsend CR (1996) Invasion biology and ecological impacts of brown trout *Salmo trutta* in New Zealand. *Biological Conservation* **78**, 13–22.

Townsend CR and Hildrew AG (1994) Species traits in relation to a habitat templet for river systems. *Freshwater Biology* **31**, 265–275.

Townsend SA and Edwards CA (2003) A fish kill event, hypoxia and other limnological impacts associated with early wet season flow into a lake on the Mary River floodplain, tropical northern Australia. *Lakes and Reservoirs: Research and Management* **8**, 169–176.

Tracey DM and Horn PL (1999) Background and review of ageing orange roughy (*Hoplostethus atlanticus*, Trachichthyidae) from New Zealand and elsewhere. *New Zealand Journal of Marine and Freshwater Research* **33**, 67–86.

Treadwell SA, Koehn JD and Bunn SE (1999) Large woody debris and other aquatic habitat. In *Riparian Land Management Technical Guidelines*, Vol. 1. (Eds S Lovett and P

Prince) pp. 79–96. Land and Water Resources Research and Development Corporation, Canberra.

Trnski T, Hay AC and Fielder S (2005) Larval development of estuary perch (*Macquaria colonorum*) and Australian bass (*M. novemaculeata*) (Perciformes: Percichthyidae), and comments on their life history. *Fishery Bulletin* **103**, 183–194.

Trout Cod Recovery Team (2008) 'National recovery plan for the trout cod *Maccullochella macquariensis*'. Department of Sustainability and Environment, Melbourne.

Trueman WT (2007) 'Some recollections of native fish in the Murray-Darling system with special reference to the trout cod *Maccullochella macquariensis*. Summary and source material from the draft publication "Bluenose: the lost world of the trout cod"'. Interim report for Native Fish Australia (Victoria). <http://www.native-fish.asn.au>. February 2012.

Tucker JW, Mackinnon MR, Russell DJ, O'Brien JJ *et al.* (1988) Growth of juvenile barramundi (*Lates calcarifer*) on dry feeds. *Progressive Fish-Culturist* **50**, 81–85.

Turchini GM, Francis DS and De Silva SS (2007) Finishing diets stimulate compensatory growth: results of a study on Murray cod, *Maccullochella peelii peelii*. *Aquaculture Nutrition* **13**, 351–360.

Turner BL, Kasperson RE *et al.* (2003) A framework for vulnerability analysis in sustainability science. *Proceedings of the National Academy of Sciences of the United States of America* **100**, 8074.

Turner MG, O'Neill RV and Gardner RH (2001) *Landscape Ecology in Theory and Practice: Pattern and Process*. Springer, New York.

Twidale CR, Lindsay JM and Bourne JA (1978) Age and origin of the Murray River and gorge in South Australia. *Proceedings of the Royal Society of Victoria* **90**, 27–42.

Tyler JA and Gilliam JF (1995) Ideal free distributions of stream fish: a model and test with minnows, *Rhinicthys atratulus*. *Ecology* **76**, 580–592

Tyler PA and Buckney RT (1973) Pollution of a Tasmanian river by mine effluents. I. Chemical evidence. *Internationale Revue der gesamten Hydrobiologie und Hydrographie* **58**, 873–883.

Tzeng WN, Chang CW, Wang CH, Shiao JC *et al.* (2007) Misidentification of the migratory history of anguillid eels by Sr/Ca ratios of vaterite otoliths. *Marine Ecology Progress Series* **348**, 285–295.

Unmack PJ (1996) The unique Cooper Creek catfish from central Australia. *Fishes of Sahul* **10**, 460–464.

Unmack PJ (2000) The genus *Hypseleotris* of southeastern Australia: its identification and breeding biology. *Fishes of Sahul* **14**, 645–656.

Unmack PJ (2001a) Biogeography of Australian freshwater fishes. *Journal of Biogeography* **28**, 1053–1089.

Unmack PJ (2001b) Fish persistence and fluvial geomorphology in central Australia. *Journal of Arid Environments* **49**, 653–659.

Unmack PJ (2005) Historical biogeography and *a priori* hypotheses based on freshwater fishes. PhD thesis. Arizona State University, Phoenix.

Unmack PJ and Dowling TE (2010) Biogeography of the genus *Craterocephalus* (Teleostei: Atherinidae) in Australia. *Molecular Phylogenetics and Evolution* **55**, 968–984.

Unmack PJ, Hammer MP, Adams M and Dowling TE (2011) A phylogenetic analysis of pygmy perches (Teleostei: Percichthyidae) with an assessment of the major historical influences on aquatic biogeography in southern Australia. *Systematic Biology* **60**, 797–812.

Upchurch P (2008) Gondwanan break-up: legacies of a lost world? *Trends in Ecology and Evolution* **23**, 229–236.

Utne-Palm AC (2002) Visual feeding of fish in a turbid environment: physical and behavioural aspects. *Marine and Freshwater Behaviour and Physiology* **35**, 111–128.

van de Graaff WJE, Crowe RWA, Bunting JA, Jackson P *et al.* (1977) Relict Early Cainozoic drainages in arid Western Australia. *Zeitschrift für Geomorphologie* **21**, 379–400.

van der Veer JW, Berghahn R, Miller JM and Rijnsdorp AD (2000) Recruitment in flatfish, with special emphasis on North Atlantic species: progress made by the Flatfish Symposia. *ICES Journal of Marine Science* **57**, 202–215.

van der Walt B and Faragher RA (2003) Otolith marking of rainbow trout fry by immersion in low concentrations of alizarin complexone. *North American Journal of Fisheries Management* **23**, 141–148.

van der Walt B, Faragher R and Harris J (2005a) Comparative angler catches of Australian bass (*Macquaria novemaculeata*) in three major river systems in New South Wales, Australia. *Asian Fisheries Science* **18**, 175–193.

van der Walt B, Faragher RA and Lowry MB (2005b) Hooking mortality of released silver perch (*Bidyanus bidyanus*) after capture by hook-and-line fishing in New South Wales, Australia. *Asian Fisheries Science* **18**, 205–216.

Van Oosterzee P (1997) *Where Worlds Collide: The Wallace Line*. Cornell University Press, New York.

Van Winkle W, Rose KA, Winemiller KO, DeAngelis DL *et al.* (2011) Linking life history theory, environmental setting, and individual-based modeling to compare responses of different fish species to environmental change. *Transactions of the American Fisheries Society* **122**, 459–466.

Vandenberg AHM (2010) Paleogene basalts prove early uplift of Victoria's Eastern Uplands. *Australian Journal of Earth Sciences* **57**, 291–315.

Vandenberg AHM (2011) Paleogene basalts prove early uplift of Victoria's Eastern Uplands. *Australian Journal of Earth Sciences* **58**, 101–108.

Vanni MJ (2002) Nutrient cycling by animals in freshwater ecosystems. *Annual Review of Ecology and Systematics* **33**, 341–370.

Vannote RL, Minshall GW, Cummins KW, Sedell JR *et al.* (1980) The river continuum concept. *Canadian Journal of Fisheries and Aquatic Sciences* **37**, 130–137.

Vari RP (1978) The terapon perches (Percoidei, Teraponidae): a cladistic analysis and taxonomic revision. *Bulletin of the American Museum of Natural History* **159**, 175–340.

Vari RP and Hutchins JB (1978) New species of terapon perches (Percoidei, Teraponidae) from Australia. *American Museum Novitates* **2654**, 1–8.

Veevers JJ (1984) *Phanerozoic Earth History of Australia.* Clarendon Press, Oxford.

Veevers JJ (1991) Mid-Cretaceous tectonic climax, Late Cretaceous recovery, and Cainozoic relaxation in the Australian region. In *The Cainozoic in Australia: A Reappraisal of the Evidence.* (Eds MAJ Williams, P De Deckker and AP Kershaw) pp. 1–14. Geological Society of Australia, Sydney.

Veinott GI, Porter TR and Nasdala L (2009) Using Mg as a proxy for crystal structure and Sr as an indicator of marine growth in vaterite and aragonite otoliths of aquaculture rainbow trout. *Transactions of the American Fisheries Society* **138**, 1157–1165.

Vermeij GJ (1996) An agenda for invasion biology. *Biological Conservation* **78**, 3–9.

Verspoor E and Hammar J (1991) Introgressive hybridization in fishes: the biochemical evidence. *Journal of Fish Biology* **39** (Supplement A), 309–334.

Videler JJ (1993) *Fish Swimming.* Chapman and Hall, London.

Vielma J, Koskela J and Ruohonen K (2002) Growth, bone mineralization, and heat and low oxygen tolerance in European whitefish (*Coregonus lavaretus* L.) fed with graded levels of phosphorus. *Aquaculture* **212**, 321–333.

Vigliano PH, Beauchamp DA *et al.* (2009) Quantifying predation on galaxiids and other native organisms by introduced rainbow trout in an ultraoligotrophic lake in Northern Patagonia, Argentina: a bioenergetics modeling approach. *Transactions of the American Fisheries Society* **138**, 1405–1419.

Vila-Gispert A and Moreno-Amich R (2002) Life history patterns of 25 species from European freshwater fish communities. *Environmental Biology of Fishes* **65**, 387–400.

Vilizzi L (1998a) Age, growth and cohort composition of 0+ carp in the River Murray, Australia. *Journal of Fish Biology* **52**, 997–1013.

Vilizzi L (1998b) Observations on ontogenetic shifts in the diet of 0+ carp, *Cyprinus carpio*, from the River Murray, Australia. *Folia Zoologica* **47**, 225–229.

Vilizzi L and Walker KF (1995) Otoliths as a potential indicator of age in common carp, *Cyprinus carpio* L. (Cyprinidae, Teleostei). *Transactions of the Royal Society of South Australia* **119**, 97–98.

Vilizzi L and Walker KF (1999a) Age and growth of the common carp, *Cyprinus carpio*, in the River Murray, Australia: validation, consistency of age interpretation, and growth models. *Environmental Biology of Fishes* **54**, 77–106.

Vilizzi L and Walker KF (1999b) The onset of the juvenile period in carp, *Cyprinus carpio*: a literature survey. *Environmental Biology of Fishes* **56**, 93–102.

Vilizzi L, Wallace T, Fraser P, Ellis I *et al.* (2007) 'Native fish recruitment and flood pulse water quality monitoring on Lindsay Island'. Technical Report 07/07. Murray-Darling Freshwater Research Centre, Mildura.

Vischer HA (1990) Morphology of the lateral line system in three species of Pacific cottoid fishes occupying disparate habitats. *Experientia* **46**, 244–250.

Vladimirov VI (1975) Critical periods in the development of fishes. *Journal of Ichthyology* **15**, 851–868.

Vogel M (2003) The effects of varying temperature and feeding levels on somatic and otolith growth in Murray cod larvae. Hons thesis. La Trobe University, Wodonga.

Vogler AP and Desalle R (1994) Diagnosing units of conservation management. *Conservation Biology* **8**, 354–363.

Vokoun JC and Rabeni CF (2005) Variation in an annual movement cycle of flathead catfish within and between two Missouri watersheds. *North American Journal of Fisheries Management* **25**, 563–572.

von Bertalanffy L (1938) A quantitative theory of organic growth. *Human Biology* **10**, 181–213.

von Humboldt A (1995) *Personal Narrative of a Journey to the Equinoctial Regions of the New Continent.* Penguin, London.

von Humboldt A and Bonpland A (1805) Observations sur l'anguille electrique (*Gymnotus electricus* L.) du nouveau continent. In *Recueil d'Observations de Zoologie et d'Anatomie Comparée.* (Eds A von Humboldt and A Bonpland). Levrault, Schoell et Cie Librairies, Paris.

Vucic-Pestic O, Ehnes RB, Rall BC and Brose U (2011) Warming up the system: higher predator feeding rates but lower energetic efficiencies. *Global Change Biology* **17**, 1301–1310.

Wager R and Jackson PD (1993) 'Action plan for Australian freshwater fishes'. Australian Nature Conservation Agency, Canberra.

Wager R and Unmack PJ (2000) 'Fishes of the Lake Eyre Catchment of Central Australia'. Department of Primary Industries/Queensland Fisheries Service, Brisbane.

Wager R and Unmack PJ (2004) Threatened fishes of the world, *Scaturiginichthys vermeilipinnis. Environmental Biology of Fishes* **70**, 300.

Wainwright PC (1996) Ecological explanation through functional morphology: the feeding biology of sunfishes. *Ecology* **77**, 1336–1343.

Wainwright PC, Alfaro ME, Bolnick DI and Hulsey CD (2005) Many-to-one mapping of form to function: a general principle in organismal design? *Integrative and Comparative Biology* **45**, 256–262.

Wainwright PC, Bellwood DR, Westneat MW, Grubich JR *et al.* (2004) A functional morphospace for the skull of labrid fishes: patterns of diversity in a complex biomechanical system. *Biological Journal of the Linnean Society* **82**, 1–25.

Wajon S (1983) Hybridisation between Murray cod and trout cod in Cataract Dam, NSW. Hons thesis. University of New South Wales, Sydney.

Walford LA (1946) A new graphic method of describing the growth of animals. *Biological Bulletin* **90**, 141–147.

Walker KF (1981) 'Ecology of freshwater mussels in the River Murray'. Technical Paper 63. Australian Water Resources Council.

Walker KF (2006) Serial weirs, cumulative effects: the lower River Murray, Australia. In *The Ecology of Desert Rivers.* (Ed. R Kingsford) pp. 248–279. Cambridge University Press, Cambridge.

Walker KF, Sheldon F and Puckridge JT (1995) A perspective on dryland river ecosystems. *Regulated Rivers: Research and Management* **11**, 85–104.

Walker WC (1937) Animal electricity before Galvani. *Annals of Science* **2**, 84–113.

Wallace JB and Benke AC (1984) Quantification of wood habitat in subtropical coastal plain streams. *Canadian Journal of Fisheries and Aquatic Sciences* **41**, 1643–1652.

Wallace JB, Webster JR and Meyer JL (1995) Influence of log additions on physical and biotic characteristics of a mountain stream. *Canadian Journal of Fisheries and Aquatic Sciences* **52**, 2120–2137.

Walsh CT, Gray CA, West RJ, van der Meulen DE *et al.* (2010) Growth, episodic recruitment and age truncation in populations of a catadromous percichthyid, *Macquaria colonorum. Marine and Freshwater Research* **61**, 397–407.

Walsh CT, Pease BC and Booth DJ (2004) Variation in the sex ratio, size and age of longfinned eels within and among coastal catchments of south-eastern Australia. *Journal of Fish Biology* **64**, 1297–1312.

Walsh CT, Pease BC, Hoyle SD and Booth DJ (2006) Variability in growth of longfinned eels among coastal catchments of south-eastern Australia. *Journal of Fish Biology* **68**, 1693–1706.

Walsh PJ (1998) Nitrogen excretion and metabolism. In *The Physiology of Fishes*, 2nd edn. (Ed. DH Evans) pp. 201–214. CRC Press, New York.

Walters CJ (1986) *Adaptive Management of Renewable Resources.* McGraw-Hill, New York.

Walters CJ and Martell SJD (2004) *Fisheries Ecology and Management.* Princeton University Press, New York.

Wang T, Lefevre J, Huong DT, Cong NV *et al.* (2009) The effects of hypoxia on growth and digestion. In *Fish Physiology.* Vol. 27, *Hypoxia.* (Eds JG Richards, AP Farrell and CJ Brauner) pp. 361–396. Elsevier, New York.

Waples RS (1995) Evolutionary significant units and the conservation of biological diversity under the Endangered Species Act. *American Fisheries Society Symposium* **17**, 8–27.

Ward RD, Hanner R and Hebert PDN (2009) The campaign to DNA barcode all fishes, FISH-BOL. *Journal of Fish Biology* **74**, 329–356.

Ward RD, Woodwark M and Skibinski DOF (1994) A comparison of genetic diversity levels in marine, freshwater, and anadromous fishes. *Journal of Fish Biology* **44**, 213–232.

Warfe DM, Pettit NE, Davies PM, Pusey BJ *et al.* (2011) The 'wet–dry' in the Wet–Dry Tropics drives ecosystem structure and function in northern Australian rivers. *Freshwater Biology* **56**, 2169–2195.

Waters JM (2011) Competitive exclusion: phylogeography's 'elephant in the room'? *Molecular Ecology* **20**, 4388–4394.

Waters JM and Burridge CP (1999) Extreme intraspecific mitochondrial DNA sequence divergence in *Galaxias maculatus* (Osteichthys: Galaxiidae), one of the world's most widespread freshwater fish. *Molecular Phylogenetics and Evolution* **11**, 1–12.

Waters JM and McDowall RA (2005) Phylogenetics of the Australasian mudfishes: evolution of an eel-like body plan. *Molecular Phylogenetics and Evolution* **37**, 417–425.

Waters JM and Wallis GP (2001) Cladogenesis and loss of the marine life-history phase in freshwater galaxiid fishes (Osmeriformes: Galaxiidae). *Evolution* **55**, 587–597.

Waters JM, Dijkstra LH and Wallis GP (2000a) Biogeography of a southern hemisphere freshwater fish: how important is marine dispersal? *Molecular Ecology* **9**, 1815–1821.

Waters JM, Lintermans M and White RWG (1994) Mitochondrial DNA variation suggests river capture as a source of vicariance in *Gadopsis bispinosus* (Pisces: Gadopsidae). *Journal of Fish Biology* **44**, 549–551.

Waters JM, Lopez JA and Wallis GP (2000b) Molecular phylogenetics and biogeography of galaxiid fishes (Osteichthyes: Galaxiidae): dispersal, vicariance, and the position of *Lepidogalaxias salamandroides. Systematic Biology* **49**, 777–795.

Waters JM, Rowe DL, Burridge CP and Wallis GP (2010) Gene trees versus species trees: reassessing life-history evolution in a freshwater fish radiation. *Systematic Biology* **59**, 504–517.

Waters JM, Shirley M and Closs GP (2002) Hydroelectric development and translocation of *Galaxias brevipinnis*: a cloud at the end of the tunnel? *Canadian Journal of Fisheries and Aquatic Sciences* **59**, 49–56.

Watkins G, Saul W, Holm E, Watson C *et al.* (2004) The fish fauna of the Iwokrama Forest. *Proceedings of the Academy of Natural Sciences of the United States of America* **15**, 39–53.

Watkins S, Quinn G and Gawne B (2010) Changes in organic matter dynamics and physicochemistry associated with riparian vegetation loss and river regulation in floodplain wetlands of the Murray River, Australia. *Marine and Freshwater Research* **61**, 1207–1217.

Watts RJ, Storey AW, Hebbert DR and Edward DHD (1995) Genetic and morphological differences between populations of the western minnow, *Galaxias occidentalis*, from two river systems in south-western Australia. *Marine and Freshwater Research* **46**, 769–777.

Weatherley AH (1958) 'Tasmanian farm dams in relation to fish culture'. CSIRO Fisheries and Oceanography, Canberra.

Weatherley AH (1959) Some features of the biology of the tench *Tinca tinca* (Linnaeus) in Tasmania. *Journal of Animal Ecology* **28**, 73–87.

Weatherley AH (1961) Notes on distribution, taonomy and behaviour of tench *Tinca tinca* (L.) in Tasmania. *Annals and Magazine of Natural History* **4**, 713–719.

Weatherley AH (1963) A note on the head kidney and kidney of the perch *Perca fluviatilis* (Linnaeus) with special reference to the blood vascular system. *Proceedings of the Zoological Society of London* **140**, 161–167.

Weatherley AH (1972) *Growth and Ecology of Fish Populations*. Academic Press, London.

Weatherley AH and Gill HS (1983) Relative growth of tissues at somatic growth rates in rainbow trout *Salmo gairdneri*. *Journal of Fish Biology* **22**, 43–60.

Weatherley AH and Gill HS (1987) *The Biology of Fish Growth*. Academic Press, London.

Weatherley AH and Lake JS (1967) Introduced fish species in Australian inland waters. In *Australian Inland Waters and their Fauna: Eleven Studies*. (Ed. AH Weatherley) pp. 217–239. Australian National University Press, Canberra.

Weatherley AH, Beevers JR and Lake PS (1967) The ecology of a zinc-polluted river. In *Australian Inland Waters and their Fauna: Eleven Studies*. (Ed. AH Weatherley) pp. 252–278. Australian National University Press, Canberra.

Webb AC (2007) Status of non-native freshwater fishes in tropical northern Queensland, including establishment success, rates of spread, range and introduction pathways. *Journal and Proceedings of the Royal Society of New South Wales* **140**, 63–78.

Webb JF, Montgomery JC and Mogdans J (2008) Bioacoustics and the lateral line system of fishes. In *Fish Bioacoustics*. (Eds JF Webb, RR Fay and AN Popper) pp. 145–182. Springer, New York.

Webb PW (1975) Hydrodynamics and energetics of fish propulsion. *Bulletin of the Fisheries Research Board of Canada* **190**, 1–156.

Webb PW (1984a) Body form, locomotion and foraging in aquatic vertebrates. *American Zoologist* **24**, 107–120.

Webb PW (1984b) Form and function in fish swimming. *Scientific American* **251**, 58–68.

Weber MJ and Brown ML (2009) Effects of common carp on aquatic ecosystems 80 years after 'carp as a dominant': ecological insights for fisheries management. *Reviews in Fisheries Science* **19**, 524–537.

Webster PJ and Streten NA (1972) Aspects of Late Quaternary climate in tropical Australasia. In *Bridge and Barrier: The Natural and Cultural History of Torres Strait*. (Ed. D Walker) pp. 39–60. Australian National University Press, Canberra.

Wedderburn SD, Walker KF and Zampatti BP (2007) Habitat separation of *Craterocephalus* (Atherinidae) species and populations in off-channel areas of the lower River Murray, Australia. *Ecology of Freshwater Fish* **16**, 442–449.

Wedderburn SD, Walker KF and Zampatti BP (2008) Salinity may cause fragmentation of hardyhead (Teleostei: Atherinidae) populations in the River Murray, Australia. *Marine and Freshwater Research* **59**, 1–5.

Weeks AR, Sgro CM, Young AG, Frankham R *et al.* (2011) Assessing the benefits and risks of translocations in changing environments: a genetic perspective. *Evolutionary Applications* **4**, 709–725.

Welcomme R (1979) *Fisheries Ecology of Floodplain Rivers*. Longman, London.

Welcomme RL (1985) *River Fisheries*. FAO, Rome.

Welcomme RL and Bartley DM (1998) Current approaches to the enhancement of fisheries. *Fisheries Management and Ecology* **5**, 351–382.

Welcomme RL, Bene C, Brown CA, Arthington AH *et al.* (2006) Predicting the water requirements of river fisheries. In *Wetlands and Natural Resource Management: Ecological Studies*, Vol. 90. (Eds JTA Verhoeven, B Beltman, R Bobbink and DF Whigham) pp. 123–154. Springer-Verlag, Berlin.

Wellborn GA (1994) Size-biased predation and prey life-histories: a comparative study of freshwater amphipod populations. *Ecology* **75**, 2104–2117.

Wells RMG, Baldwin J, Seymour RS and Weber RE (1997) Blood oxygen transport and hemoglobin function in three tropical fish species from northern Australian freshwater billabongs. *Fish Physiology and Biochemistry* **16**, 247–258.

Werner EE (1977) Species packing and niche complementarity in three sunfishes. *American Naturalist* **111**, 553–578.

Werner EE and Gilliam JF (1984) The ontogenetic niche and species interactions in size-structured populations. *Annual Review of Ecology and Systematics* **15**, 393–425.

Werner EE and Hall DJ (1977) Competition and habitat shift in two sunfishes (Centrarchidae). *Ecology* **58**, 869–876.

Werner EE, Mittelbach GG, Hall DJ and Gilliam JF (1983) Experimental tests of optimal habitat use in fish: the role of relative habitat profitability. *Ecology* **64**, 1525–1539.

Werner RG (2002) Habitat requirements. In *Fishery Science: The Unique Contributions of Early Life Stages*. (Eds LA Fuiman and RG Werner). Blackwell Science, Oxford.

West GB and Brown JH (2005) The origin of allometric scaling laws in biology from genomes to ecosystems: towards a quantitative unifying theory of biological structure and organization. *Journal of Experimental Biology* **208**, 1575–1592.

West GB, Brown JH and Enquist BJ (1997) A general model for the origin of allometric scaling laws in biology. *Science* **276**, 122–126.

West GB, Brown JH and Enquist BJ (1999) The fourth dimension of life: fractal geometry and allometric scaling of organisms. *Science* **284**, 1677–1679.

West GB and Gordon GNG (1994) Commercial and recreational harvest of fish from two Australian coastal rivers. *Australian Journal of Marine and Freshwater Research* **45**, 1259–1279.

White GC and Burnham KP (1999) Program MARK: survival estimation from populations of marked animals. *Bird Study* **46** (Supplement), 120–138.

White GC and Garrott RA (1990) *Analysis of Wildlife Radio-Tracking Data*. Academic Press, New York.

White LJ, Harris JH and Keller RJ (2011) Movement of three non-salmonid fish species through a low-gradient vertical-slot fishway. *River Research and Applications* **27**, 499–510.

White ME (1994) *After the Greening: The Browning of Australia*. Kangaroo Press, Sydney.

White ME (1998) *The Greening of Gondwana: The 400 million Year Story of Australian Plants*, 3rd edn. Kangaroo Press, Sydney.

White ME (2000) *Running Down: Water in a Changing Land*. Kangaroo Press, Sydney.

White PS and Pickett STA (1985) Natural disturbance and patch dynamics: an introduction. In *The Ecology of Natural Disturbances and Patch Dynamics*. (Eds STA Pickett and PS White). Academic Press, Orlando.

White RJ, Karr JR and Nehlsen W (1995) Better roles for fish stocking in aquatic resources management. *American Fisheries Society Symposium* **15**, 527–547.

Whiterod NR (2010a) The bioenergetics implications for Murray cod (*Maccullochella peelii*) below a large hypolimnetic-releasing impoundment. PhD thesis. Charles Sturt University, Albury.

Whiterod NS (2010b) Calibration of a rapid non-lethal method to measure energetic status of a freshwater fish (Murray cod, *Maccullochella peelii peelii*). *Marine and Freshwater Research* **61**, 527–531.

Whitley GP (1927) The Queensland lungfish. *Australian Museum Magazine* **3**, 50.

Whitley GP (1929) The discovery of the Queensland lungfish. *Australian Museum Magazine* **3**, 363.

Whitley GP (1955) Conservation of freshwater fishes and shoreline fauna. *Australian Museum Magazine* **11**, 359–364.

Whitley GP (1964) A survey of Australian ichthyology. *Proceedings of the Linnean Society of New South Wales* **89**, 11–127.

Whitley GP (1970) *Early History of Australian Zoology*. Royal Zoological Society of New South Wales, Sydney.

Whitley GP (1972) Rains of fishes in Australia. *Australian Natural History* **17**, 154–159.

Whittington RJ and Chong R (2007) Global trade in ornamental fish from an Australian perspective: the case for revised import risk analysis and management strategies. *Preventive Veterinary Medicine* **81**, 92–116.

Whittington RJ, Becker JA and Dennis MM (2010) Iridovirus infections in finfish: critical review with emphasis on ranaviruses. *Journal of Fish Diseases* **33**, 95–122.

Whittington RJ, Kearns C, Hyatt AD, Hengstberger S *et al.* (1996) Spread of Epizootic Haematopoietic Necrosis Virus (EHNV) in redfin perch (*Perca fluviatilis*) in southern Australia. *Australian Veterinary Journal* **73**, 112–114.

Whitty JM, Phillips NM, Morgan DL, Chaplin JA *et al.* (2008) 'Habitat associations of freshwater sawfish (*Pristis microdon*) and northern river shark (*Glyphis* sp. C): including genetic analysis of *P. microdon* across northern Australia'. Report to Department of the Environment, Water, Heritage and the Arts, Canberra. Centre for Fish and Fisheries Research, Murdoch University, Perth.

Wiens JA (2002a) Central concepts and issues of landscape ecology. In *Applying Landscape Ecology in Biological Conservation*. (Ed. K Gutzwiller) pp. 3-21. Springer, New York.

Wiens JA (2002b) Riverine landscapes: taking landscape ecology into the water. *Freshwater Biology* **47**, 501–515.

Wieser W (1995) Energetics of fish larvae, the smallest vertebrates. *Acta Physiologica Scandinavica* **154**, 279–290.

Wieser W, Krumschnabel G and Ojwang-Okwor JP (1992) The energetics of starvation and growth after re-feeding in juveniles of three cyprinid species. *Environmental Biology of Fishes* **33**, 63–71.

Wikelski M and Thom C (2000) Marine iguanas shrink to survive El Niño. *Nature* **403**, 37–38.

Wilde GR and Sawynok W (2005) Growth rate and mortality of Australian bass, *Macquaria novemaculeata*, in four freshwater impoundments in south-eastern Queensland, Australia. *Fisheries Management and Ecology* **12**, 1–7.

Wiley MJ, Kohler SL and Seelbach PW (1997) Reconciling landscape and local views of aquatic communities: lessons from Michigan trout streams. *Freshwater Biology* **37**, 133–148.

Will KW, Mishler BD and Wheeler QD (2005) The perils of DNA barcoding and the need for integrative taxonomy. *Systematic Biology* **54**, 844–851.

Willet DJ (1996) Use of scale patterns to evaluate stocking success of siver perch, *Bidyanus bidyanus*, released at two different sizes. *Marine and Freshwater Research* **47**, 761.

Williams JG, Armstrong G, Katopodis C, Larinier M *et al.* (2011) Thinking like a fish: a key ingredient for development of effective fish passage facilities at river obstructions. *River Research and Applications* **28**, 407–417.

Williams MAJ (1984) Cenozoic evolution of arid Australia. In *Arid Australia*. (Eds HC Cogger and EE Cameron) pp. 59–77. Australian Museum, Sydney.

Williams NJ (1970) A comparison of the two species of the genus *Percalates* Ramsey and Ogilby (Percomorphi: Macquariidae) and their taxonomy. *New South Wales State Fisheries Bulletin* **11**, 1–59.

Williams SE, Shoo LP, Isaac JL, Hoffmann AA *et al.* (2008) Towards an integrated framework for assessing the vulnerability of species to climate change. *PLoS Biology* **6** e325, 2621–2626.

Williams WD (1982) The argument against the introduction of the Nile perch to northern Australian fresh waters. *Search* **13**, 67–70.

Willis SC, Nunes M, Montaña CG, Farias IP *et al.* (2010) The Casiquiare River acts as a corridor between the Amazonas and Orinoco river basins: biogeographic analysis of the genus *Cichla*. *Molecular Ecology* **19**, 1014–1030.

Wilson AB, Noack-Kunnmann K and Meyer A (2000) Incipient speciation in sympatric Nicaraguan crater lake cichlid fishes: sexual selection versus ecological diversification. *Proceedings of the Royal Society Biological Sciences B* **267**, 2133–2141.

Wilson E (1859) *Rambles at the Antipodes.* WH Smith and Son, London.

Wilson GG (2005) Impact of invasive exotic fishes on wetland ecosystems in the Murray-Darling Basin. In *Proceedings of Workshop on Native Fish and Wetlands in the Murray-Darling Basin.* Murray-Darling Basin Commission, Canberra.

Wilson PJ, McNeil DG and Gillanders BM (2008) 'Impacts of introduced redfin perch on native flathead gudgeons in the South Para River'. South Australian Research and Development Institute, Adelaide.

Winemiller KO (2007) Interplay between scale, resolution, life history and food web properties. In *Energetics to Ecosystems: The Dynamics and Structure of Ecological Systems.* (Eds K McCann, D Noakes and N Rooney) pp. 101–126. Springer, New York.

Winemiller KO and Rose KA (1992) Patterns of life-history diversification in North American fishes: implications for population regulation. *Canadian Journal of Fisheries and Aquatic Sciences* **49**, 2196–2218.

Winemiller KO, Flecker AS and Hoeinghaus DJ (2010) Patch dynamics and environmental heterogeneity in lotic ecosystems. *Journal of the North American Benthological Society* **29**, 84–99.

Winemiller KO, López-Fernández H, Taphorn DC, Nico L *et al.* (2008) Fish assemblages of the Casiquiare River, a corridor and zoogeographical filter for dispersal between the Orinoco and Amazon basins. *Journal of Biogeography* **35**, 1551–1563.

Winter JD (1983) Underwater biotelemetry. In *Fisheries Techniques.* (Eds LA Nielsen and DL Johnson) pp. 371–395. American Fisheries Society, Bethesda.

Winter JD (1996) Advances in underwater biotelemetry. In *Fisheries Techniques.* (Eds BR Murphy and DW Willis) pp. 555–590. American Fisheries Society, Bethesda.

Wisenden BD (1999) Alloparental care in fishes. *Reviews in Fish Biology and Fisheries* **9**, 45–70.

Witheridge G (2002) 'Fish passage requirements for waterways crossings: engineering guidelines'. Catchment and Creeks, Brisbane.

Wittenberg L and Inbar M (2009) The role of fire disturbance on runoff and erosion processes: a long-term approach. Mt Carmel case study, Israel. *Geographical Research* **47**, 46–56.

Wittenberg R and Cock MJW (Eds) (2001) 'Invasive alien species: a toolkit of best prevention and management practices'. CAB International, Wallingford.

Woinarski JCZ and Fisher A (1999) The Australian Endangered Species Protection Act 1992. *Conservation Biology* **13**, 959–962.

Wolanski E and Jones M (1981) Physical properties of Great Barrier Reef lagoon waters near Townsville. I. Effects of Burdekin River floods. *Australian Journal of Marine and Freshwater Research* **32**, 305–319.

Wong BBM (2004) Superior fighters make mediocre fathers in the Pacific blue-eye fish. *Animal Behaviour* **67**, 583–590.

Wong BBM and Svensson PA (2009) Strategic male signalling effort in a desert-dwelling fish. *Behavioral Ecology and Sociobiology* **63**, 543–549.

Wong BBM, Jennions MD and Keogh JS (2004a) Sequential male mate choice in a fish, the Pacific blue-eye *Pseudomugil signifer*. *Behavioral Ecology and Sociobiology* **56**, 253–256.

Wong BBM, Keogh JS and Jennions MD (2004b) Mate recognition in a freshwater fish: geographic distance, genetic differentiation, and variation in female preference for local over foreign males. *Journal of Evolutionary Biology* **17**, 701–708.

Wong BBM, Keogh JS and McGlashan DJ (2004c) Current and historical patterns of drainage connectivity in eastern Australia inferred from population genetic structuring in a widespread freshwater fish *Pseudomugil signifer* (Pseudomugilidae). *Molecular Ecology* **13**, 391–401.

Wood CC and Gross MR (2008) Elemental conservation units: communicating extinction risk without dictating targets for protection. *Conservation Biology* **22**, 36–47.

Woodcock SH, Gillanders BM, Munro AR, Crook DA *et al.* (2011) Determining mark success of 15 combinations of enriched stable isotopes for the batch marking of larval otoliths. *North American Journal of Fisheries Management* **31**, 843–851.

Woodcock SH, Gillanders BM, Munro AR, McGovern F *et al.* (2010) Using enriched stable isotopes of barium and magnesium to batch mark otoliths of larval golden perch (*Macquaria ambigua* Richardson). *Ecology of Freshwater Fish* **20**, 157–165.

Woodford DJ and McIntosh AR (2010) Evidence of source-sink metapopulations in a vulnerable native galaxiid fish driven by introduced trout. *Ecological Applications* **20**, 967–977.

Woods RJ, Macdonald JI, Crook DA, Schmidt DJ *et al.* (2010) Contemporary and historical patterns of connectivity among populations of an inland river fish species inferred from genetics and otolith chemistry. *Canadian Journal of Fisheries and Aquatic Sciences* **67**, 1098–1115.

Woodward G and Warren P (2007) Body size and predatory interactions in freshwaters: scaling from individuals to communities. In *Body Size: The Structure and Function of Aquatic Ecosystems*. (Eds AG Hildrew, DG Raffaelli and R Edmonds-Brown) pp. 98–117. Cambridge University Press, Cambridge.

Woodward G, Papantoniou G, Edwards F and Lauridsen RB (2008) Trophic trickles and cascades in a complex food web: impacts of a keystone predator on stream community structure and ecosystem processes. *Oikos* **117**, 683–692.

Wootton RJ (1998) *Ecology of Teleost Fishes*. Kluwer Academic, Dordrecht.

Worrall K, Carter C, Wilkinson R and Porter M (2011) The effects of continuous photoperiod (24L:0D) on growth of juvenile barramundi (*Lates calcarifer*). *Aquaculture International* **19**, 1–8.

Wright S (1969) *Evolution and the Genetics of Populations*. Vol. 2, *The Theory of Gene Frequencies*. University of Chicago Press, Chicago.

Wu CH (1984) Electric fish and the discovery of animal electricity. *American Scientist* **72**, 598–607.

Wu JG and Loucks OL (1995a) From balance of nature to hierarchical patch dynamics: a paradigm shift in ecology. *Quarterly Review of Biology* **70**, 439–466.

Wu JG and Loucks OL (1995b) A hierarchical patch dynamics paradigm in ecology: elements and implications. *Bulletin of the Ecological Society of America* **76**, 405.

Yamamoto T, Shima T, Furuita H, Shiraishi M *et al.* (2000) Self-selection of diets with different amino acid profiles by rainbow trout (*Oncorhynchus mykiss*). *Aquaculture* **187**, 375–386.

Yang SD, Lin TS, Liu FG and Liou CH (2006) Influence of dietary phosphorus levels on growth, metabolic response and body composition of juvenile silver perch (*Bidyanus bidyanus*). *Aquaculture* **253**, 592–601.

Ye Q and Hammer M (2009) Fishes. In *Natural History of the Riverland and Murraylands*. (Ed. J Jennings) pp. 334–352. Royal Society of South Australia, Adelaide.

Ye Q, Cheshire KJ and McNeil DG (2010) 'Influences of salinity, water quality and hydrology on early life stages of fishes of the lower River Murray, South Australia'. Publication F2009/00470-1-6, Report Series 446. South Australian Research and Development Institute, Adelaide.

Young GC (2006) Biostratigraphic and biogeographic context for tetrapod origins during the Devonian: Australian evidence. *Alcheringa* **30**, 409–428.

Young MJ, Simmons LW and Evans JP (2010) Pre- and post-mating sexual selection both favor large males in a rainbowfish. *Behavioral Ecology and Sociobiology* **64**, 915–925.

Young MK (1999) Summer diel activity and movement of adult brown trout in high-elevation streams in Wyoming, USA. *Journal of Fish Biology* **54**, 181–189.

Young WJ (Ed.) (2001) *Rivers as Ecological Systems: The Murray-Darling Basin*. Murray-Darling Basin Commission, Canberra.

Zafar M, Mussaddeq Y, Akhter S and Sultan A (2003) Weight–length and condition factor relationship of thaila, *Catla catla*, from Rawal Dam, Islamabad, Pakistan. *Pakistan Journal of Biological Sciences* **6**, 1532–1534.

Zalewski M and Naiman RJ (1985) The regulation of riverine fish communities by a continuum of abiotic–biotic factors. In *Habitat Modification and Freshwater Fisheries*. (Ed. JS Alabaster) pp. 3–9. Butterworth, London.

Zampatti BP, Bice CM and Jennings PR (2010) Temporal variation in fish assemblage structure and recruitment in a freshwater-deprived estuary: the Coorong, Australia. *Marine and Freshwater Research* **61**, 1298–1312.

Zeller D and Pauly D (2005) Good news, bad news: global fisheries discards are declining, but so are total catches. *Fish and Fisheries* **6**, 156–159.

Zhu DQ, Degnan S and Moritz C (1998) Evolutionary distinctiveness and status of the endangered Lake Eacham rainbowfish (*Melanotaenia eachamensis*). *Conservation Biology* **12**, 80–93.

Zielinski BS and Hara TJ (2007) Olfaction. In *Sensory Systems Neuroscience*. Vol. 25, *Fish Physiology*. (Eds TJ Hara and BS Zielinski) pp. 1–43. Elsevier, New York.

Index to species

Index to subjects